Theory of Third-Order Differential Equations

Seshadev Padhi · Smita Pati

Theory of Third-Order Differential Equations

Springer

Seshadev Padhi
Department of Applied Mathematics
Birla Institute of Technology, Mesra
Ranchi, Jharkhand, India

Smita Pati
Department of Applied Mathematics
Birla Institute of Technology, Mesra
Ranchi, Jharkhand, India

ISBN 978-81-322-1613-1 ISBN 978-81-322-1614-8 (eBook)
DOI 10.1007/978-81-322-1614-8
Springer New Delhi Heidelberg New York Dordrecht London

Library of Congress Control Number: 2013951151

Printed on acid-free paper

Springer is part of Springer Science+Business Media (www.springer.com)

Dedicated to Suchitra and Aryadev

Preface

Over the last four decades, intensive work has been carried out in the field of theory of nonautonomous third-order ordinary and delay differential equations. However, it is only recently that the global attractivity of third-order equations has been given a serious study. During these years, new investigations were developed and results of principal importance were obtained. In particular, suitable oscillation criteria for third-order linear and nonlinear differential equations were established, with emphasis on the oscillation of third-order nonhomogeneous differential equations with respect to oscillation and nonoscillation of third-order homogeneous differential equations. Existence and nonexistence conditions for oscillatory and nonoscillatory solutions of various types were found and asymptotic formulae for solutions of a sufficiently wide class of linear and nonlinear equations were derived. These studies notwithstanding, one would observe that oscillation and asymptotic behaviour of nonoscillatory solutions of third-order delay differential equations has rarely been studied.

In this book, an attempt is made to sum up these results. The necessity of such an attempt is felt especially as the well-known monograph of M. Greguš, Third Order Linear Differential Equations, Reidel, Dordrecht, The Netherlands, 1982, devoted to related topics, reflects results which are 30 years old.

This book consists of seven chapters. In the first one, we consider the third-order linear differential equation, with constant coefficients, of the form

$$x''' + ax'' + bx' + cx = 0, \tag{1}$$

where a, b and c are constants. Eight different cases of a, b and c were considered while dealing with (1). Further, it is observed that different structures of solution spaces of (1) appears for different cases on a, b and c. In this chapter, an introduction is given to the oscillation theory of the nonhomogeneous equation

$$x''' + ax'' + bx' + cx = f, \tag{2}$$

where a, b, c and f are constants. An apparatus of comparison theorems is developed, which allows us to establish criteria for (2) to have oscillatory or nonoscillatory solutions with the help of oscillation and nonoscillation of (1). An overall idea on oscillation of third-order delay differential equations is also included in this chapter. Additionally, some basic results are incorporated, which are needed in the forthcoming chapters.

Chapter 2 deals with the linear equation

$$x''' + a(t)x'' + b(t)x' + c(t)x = 0, \tag{3}$$

where $a, b \in C^1([\sigma, \infty), R)$ and $c \in C([\sigma, \infty), R), \sigma \in R$. This chapter contains seven sections, where most of the results obtained for (1) have been generalised to (3). Many problems have been proposed in the sections or at the end of the chapter. Apart from these, some new nontrivial sufficient conditions for the oscillation of (3) are given. Further, several sufficient conditions, different from earlier ones, are given which can be applied to the third-order Euler equation

$$x''' + \frac{a_0}{t}x'' + \frac{b_0}{t^2}x' + \frac{c_0}{t^3}x = 0,$$

where a_0, b_0 and c_0 are real constants. Several comparison principles have been used to establish the nonoscillation of (3). We have established criteria for (3) to have a family of solutions asymptotically, equivalent to solutions of (3), and we studied the properties of this family. Finally, existence criteria for solutions vanishing at infinity have been explored in this chapter.

Chapter 3 is concerned with the oscillation of the equation

$$x''' + a(t)x'' + b(t)x' + c(t)x = f(t), \tag{4}$$

where $a(t)$, $b(t)$ and $c(t)$ are as defined earlier and $f \in C([\sigma, \infty), R)$. Some of the eight different cases of $a(t)$, $b(t)$ and $c(t)$ have been considered while the remaining have been retained as open problems. We have applied the results of Chap. 2 to obtain sufficient conditions for the oscillation and nonoscillation of (4).

In Chap. 4, emphasis is given to the oscillation and nonoscillation of solutions and the asymptotic behaviour of the third-order nonlinear differential equations

$$x''' + b(t)x' + c(t)x^\alpha = 0,$$
$$x''' + b(t)x' + c(t)f(x) = 0,$$

and results of more general equations

$$x''' + a(t)x'' + b(t)x' + c(t)x^\alpha = 0,$$

and

$$x''' + a(t)x'' + b(t)x' + c(t)f(x) = 0,$$

where $a(t)$, $b(t)$ and $c(t)$ were defined earlier in (4) and $f : R \to R$ and some other restrictions on f. In this chapter, we solve the Kneser problem on monotone solutions and prove, under certain restrictions on f and α, that monotone solutions tend to zero eventually.

Chapter 5 is quite interesting. We have studied oscillation of solutions of the third-order nonlinear and nonhomogeneous equations of the form

$$\big(r(t)x''\big)' + q(t)\big(x'\big)^\beta + p(t)x^\alpha = f(t), \tag{5}$$

where r, p, q and $f \in C([\sigma,\infty), R)$ and $r(t) > 0$, α and β are quotient of odd integers. Along the way, we have given some interesting results on oscillation of the nonlinear equation

$$x''' + a(t)x'' + b(t)x' + c(t)x = f\big(t, x, x', x''\big), \tag{6}$$

where a, b and $c \in C([\sigma,\infty), R)$ and $f : [\sigma,\infty) \times R^3 \to R$.

In Chap. 6, we apply the results of Chap. 2 to obtain oscillation and asymptotic behaviour of nonoscillation of the third-order delay differential equation

$$x'''(t) + a(t)x''(t) + b(t)x'(t) + c(t)x(g(t)) = 0, \tag{7}$$

where $a(t)$, $b(t)$ and $c(t)$ are as defined earlier and $g : R \to R$ with the property that $0 \le g(t) \le t$ and $\lim_{t\to\infty} g(t) = \infty$. It has been observed that the presence of the delay term $g(t)$ makes it difficult to make a direct study of the asymptotic behaviour of solutions of (7). However, the canonical transformations makes it possible to study the properties of solutions of (7). Equations (3) and (7) have been transformed to equivalent canonical equations and then the asymptotic behaviour of solutions of (7) have been studied by knowing the oscillation and nonoscillation of (3). We have classified the nonoscillatory solutions of (7) as Property A and Property B, according to their asymptotic behaviour.

In the last chapter, we study the global attractivity of solutions of equations

$$x''' + \psi\big(x, x'\big)x'' + f\big(x, x'\big) = 0 \tag{8}$$

and

$$x''' + \psi\big(x, x'\big)x'' + f\big(x, x'\big) = p(t), \tag{9}$$

where ψ, f, ψ_x, $f_x \in C(R \times R, R)$ and $p \in C([0,\infty), R)$. Lyapunov function has been used to obtain the results. A direct step by step method has been used to find the asymptotic stability of solutions of the third-order equation

$$\begin{aligned} x'''(t) = {} & p_1x''(t) + p_2x''(t-\tau) + q_1x'(t) + q_2x'(t-\tau) + r_1x(t) \\ & + r_2x(t-\tau), \quad t \ge 0, \end{aligned} \tag{10}$$

with the initial condition

$$x(t) = \phi(t), \quad t \in [-\tau, 0], \tag{11}$$

where p_1, p_2, q_1, q_2, r_1 and r_2 are real constants, $\tau > 0$ is a real number and $\phi \in C([-\tau,0), R)$ is an initial function. The results are then applied to obtain stability and asymptotic stability of solutions of (11).

Some results are given in the form of open problems to be solved by the readers. We also state some problems whose solutions we do not know.

Seshadev Padhi
Smita Pati

Acknowledgements

It is almost impossible to thank everyone who helped us over the years to understand the theory of third-order differential equations. However, three people must be singled out: Prof. N. Parhi who taught us oscillation theory of third-order differential equations, Prof. John R. Graef for his constant inspiration and valuable suggestions, and Prof. Chuanxi Qian who inspired us to learn stability theory of third-order differential equations. We are also thankful to Prof. Julio G. Dix, Prof. E. Thandapani, Prof. P.D.N. Srinivasu and Dr. B.S.R.V. Prasad for their constant inspiration and encouragement in preparation of this book. We should also mention here that it was a great pleasure to work with Mr. Shamim Ahmad and Ms. Nupoor Singh, the editorial team of Springer and the production team, especially Mr. Donatas Akmanavičius, who took utmost care during the preparation of the book. We are thankful to our family members, especially to Suchitra Panigrahi, Aryadev Padhi, Mr. Bansidhar Padhi, Mr. Sitanath Panigrahi and Mr. Ashok Kumar Pati, for their constant inspirations in preparing this book.

Contents

Chapter 1
Introduction

1.1 Preliminaries

In recent years, a great deal of work has been done on various aspects of differential equations of third order. There are different problems concerning third-order differential equations which have drawn the attention of researchers throughout the world. Third-order differential equations describe many mathematical models of great interest in engineering, biology and physics. Equations of the form

$$x''' + a(t)x'' + b(t)x' + c(t)x = f(t)$$

arise in the study of entry-flow phenomenon, a problem of hydrodynamics which is of considerable importance in many branches of engineering (see [29]). Reynolds [51] has studied a model equation of the type

$$x'''(t) - \lambda x'(t) - 2x(t)x'(t) = \mu_1\big(Dx'\big)(t) + \mu_2 \sin t, \tag{1.1a}$$

$$x(t+2\pi) = x(t), \qquad \int_{-\pi}^{\pi} x(t)\,dt = 0 \tag{1.1b}$$

which describes the steady flow of water in a long rectangular tank, oscillating horizontally near a resonant frequency. The integral damping operator D in (1.1a) is formally defined by

$$(Du)(t) = \left(\frac{2}{\pi}\right)^{1/2} \int_0^{\infty} u(t-s)s^{-1/2}\,ds$$

for all functions u with period 2π and mean zero. Cox and Mortell [8] have considered (1.1a) and (1.1b) with $(Dx')(t) = x(t)$. In [28], Jackiewicz et al. have investigated the asymptotic behaviour of the solutions of Volterra integro-differential equations of the form

$$x'(t) = \gamma x(t) + \int_0^1 (\lambda + \mu t + \upsilon s)x(s)\,ds, \quad t \ge 0,$$

$$x(0) = 1$$

S. Padhi, S. Pati, *Theory of Third-Order Differential Equations*,
DOI 10.1007/978-81-322-1614-8_1, © Springer India 2014

with the help of third-order differential equations of the type

$$x''' = \gamma x'' + \big(\lambda + (\mu + \upsilon)t\big)x' + (2\mu + \upsilon)x,$$

where λ, γ, μ and υ are real parameters and $\upsilon + \mu \neq 0$. In the development of a mathematical theory of thyroid-pituitary interaction, Danziger and Elemergreen (see [18], p. 133) have obtained the following third-order linear differential equations:

$$\alpha_3\theta''' + \alpha_2\theta'' + \alpha_1\theta' + (1+k)\theta = kc, \quad \theta < c$$

and

$$\alpha_3\theta''' + \alpha_2\theta'' + \alpha_1\theta' + \theta = 0, \quad \theta > c.$$

These equations describe the variation of thyroid hormone with time. Here $\theta = \theta(t)$ is the concentration of thyroid hormone at time t and $\alpha_3, \alpha_2, \alpha_1, k$ and c are constants. For other application of third-order differential equations, one may refer to Chap. 4 in [21].

In early 1950s, Alan Llyod Hodgkin and Andrew Huxley developed a mathematical model for the propagation of electrical pulses in the nerve of a squid. The original model describes the ionic mechanisms underlying the initiation and propagation of action potentials in the squid giant axom. The Hodgkin–Huxley model is a set of nonlinear differential equations that approximates the electrical characteristics of excitable cells such as neurons and cardiac myocytes. The model has played a vital role in biophysics and neuronal modelling. Alan Llyod Hodgkin and Andrew Huxley were awarded a Nobel prize in 1963 for this work. A reduced version of the Hodgkin–Huxley model was proposed by Nagumo. He suggested a third-order differential equation of the form

$$x''' - cx'' + f'(x)x' - \frac{b}{c}x = 0$$

as a model exhibiting many of the features of the Hodgkin–Huxley equations, where the function f is an entire function. For more details of Nagumo's equations, we refer the reader to the paper by McKeen [44] who gave some of the background of these equations and summarised some of the numerical results of this model. For application in physics, Vreeke and Sandquist [62] proposed the following system of differential equations (which is equivalent to a system of third-order differential equations)

$$\begin{aligned}
\frac{dx_1}{dt} &= x_1\big(\gamma_1(1-x_2) + \gamma_2(1-x_3)\big),\\
\frac{dx_2}{dt} &= \gamma_3(x_1 - x_2),\\
\frac{dx_3}{dt} &= \gamma_4(x_1 - x_3)
\end{aligned}$$

to describe the two temperature feedback nuclear reactor problem, where x_1 is normalised neutron density, x_2 and x_3 are normalised temperatures, x_2 being associated with fuel and x_3 with the moderator or coolant, γ_3 and γ_4 are positive heat transfer coefficients, γ_1 and γ_2 are normalised effective neutron lifetime parameters associated with the temperature feedbacks, and the expression $\gamma_1(1-x_2)+\gamma_2(1-x_3)$ in the first equation is called the *reactivity* and is a measure of multiplication factor of the neutrons in the fusion reactor.

The Kuramoto–Sivashinsky equation

$$u_t + u_{xxxx} + u_{xx} + \frac{1}{2}u^2 = 0$$

arises in a wide variety fascinating physical phenomena. For instance, we recall that the Kuramoto–Sivashinsky equation is introduced to describe pattern formulation in reaction diffusion systems, and to model the instability of flame front propagation (see Y. Kuramoto and T. Yamada [37] and D. Michelson [45]). To find the travelling wave solutions of this partial differential equation, we may use the substitution of the form $u(x, ct) = u(x - ct)$ with period c and one has to solve the nonlinear third-order differential equation of the form

$$\lambda u'''(x) + u'(x) + f(u) = 0,$$

where λ is a parameter and f is an even function. On the other hand, by the qualitative behaviour of the above third-order differential equation, one can predict the behaviour of the above-mentioned partial differential equation. Hence, it would be interesting to study the qualitative behaviour of solutions of third-order linear and nonlinear differential equations. One may observe that the oscillation theory of second-order differential equations is well developed. It is only recently that the oscillation theory of third-order equations has been given a serious study.

In 1908, Birkhoff [6] applied a method of projective geometry and started the study of separation and comparison theorems for third-order equations. While Birkhoff's paper is a necessary reference to many papers on differential equations, his results or methods are seldom quoted. In 1961, Hanan [25] studied the oscillation and nonoscillation of two different types of third-order differential equation. The works due to Hanan [25] and Lazer [42] are the starting point of many investigations on the oscillation and asymptotic behaviour of solutions of third-order differential equations. In 1969, Barrett [5] made a self contained study on development of third-order equations. We may note that the third order is the lowest order where truly non-self-adjoint equations do occur. For example, $x''' + tx' + t^2x = 0$ is a non-self-adjoint third-order differential equation. The self-adjoint form of the third-order equation is

$$x''' + q(t)x' + \frac{1}{2}q'(t)x = 0, \tag{1.2}$$

where $q \in C^1([\sigma, \infty), R)$ and $\sigma \in R$. In the study of second- and fourth-order equations, the self-adjoint form of an equation plays a significant role. However, the self-adjoint form of the third-order equation does not play any special role in the theory. The general solution of (1.2) is given by

$$x(t) = c_1 u^2(t) + c_2 u(t)v(t) + c_3 v^2(t),$$

where $u(t)$ and $v(t)$ are linearly independent solutions of the second-order equation

$$z'' + \frac{1}{4}q(t)z = 0,$$

and c_1, c_2 and c_3 are constants. Obviously $u^2(t), u(t)v(t)$ and $v^2(t)$ are linearly independent solutions of (1.2).

During last three decades, a considerable amount of work has been done on oscillation theory of linear homogeneous third-order differential equations with variable coefficients of the form

$$x''' + a(t)x'' + b(t)x' + c(t)x = 0, \tag{LH1}$$

under various sign restrictions on the coefficient functions $a(t), b(t)$ and $c(t)$, where a, b and $c \in C([\sigma, \infty), R)$. If a is twice continuously differentiable, then the transformation

$$x = y \exp\left(-\frac{1}{3}\int_\sigma^t a(s)\,ds\right)$$

transforms (LH1) to an equation of the form

$$y''' + Q(t)y' + P(t)y = 0, \tag{1.3}$$

where

$$P(t) = \frac{1}{27}\left[2a^3(t) - 9a(t)b(t) + 27c(t) - 9a''(t)\right]$$

and

$$Q(t) = \frac{1}{3}\left[3b(t) - 3a'(t) - a^2(t)\right].$$

Third-order differential equations of the form

$$\left(r(t)x''\right)' + q(t)x' + p(t)x = 0, \tag{LH2}$$

where $p, q, r \in C([\sigma, \infty), R)$ such that $r(t) > 0$, are more general than (LH1). In fact, (LH1) may be written as (LH2), where $r(t) = \exp(\int_\sigma^t a(s)\,ds)$, $q(t) = r(t)b(t)$ and $p(t) = r(t)c(t)$.

In [21], Greguš discussed oscillatory and asymptotic behaviour of solutions of equations of the type

$$x''' + 2A(t)x' + \left(A'(t) + B(t)\right)x = 0. \tag{1.4}$$

Equation (1.4) is obtained from (1.3) by putting $A(t) = \frac{1}{2}Q(t)$ and $B(t) = P(t) - A'(t)$. Greguš has assumed that $a \in C^2([\sigma,\infty), R)$ and $b \in C^1([\sigma,\infty), R)$. The form (1.4) of a differential equation is called the *normal form* and the function $B(t)$ is referred to as the *Laguerre invariant*. Although many authors (see [4, 11, 13, 14, 17, 19, 20, 25–27, 30–33, 36, 42, 55–57, 59]) have considered Eq. (1.3), it is always interesting to obtain sufficient conditions for oscillation/nonoscillation of solutions of (LH1) explicitly in terms of the coefficient functions $a(t)$, $b(t)$ and $c(t)$. For this reason, Greguš has discussed Eq. (LH1) in Chap. 2 in his monograph (see [21]). However, very little work has been done (see [9, 10, 47–49]) on the linear nonhomogeneous differential equations of third order of the type

$$x''' + a(t)x'' + b(t)x' + c(t)x = f(t), \tag{NH1}$$

or

$$\big(r(t)x''\big)' + q(t)x' + p(t)x = f(t), \tag{NH2}$$

where $a, b, c, p, q, r \in C([\sigma,\infty), R)$ and $r(t) > 0$. Unlike second-order equations, (LH2) cannot be transformed to an equation of the form (1.3), and (NH1) cannot be transformed to an equation of the type

$$x''' + Q(t)x' + P(t)x = f(t).$$

Thus, the oscillation theory for Eqs. (LH2) and (NH2) is to be developed systematically.

A function $x \in C([\sigma,\infty), R)$ is said to be nonoscillatory, if there exists a $t_1 \geq \sigma$ such that $x(t) > 0$ or < 0 for $t \geq t_1$; otherwise, $x(t)$ is called oscillatory. By a solution of any of the above-mentioned differential equations, we mean a real-valued solution of that equation which exists on $[T,\infty)$ and is nontrivial in any neighbourhood of infinity, where $T \geq \sigma$ depends on the solution. Equation (LH1) is said to be nonoscillatory if all its solutions are nonoscillatory. It is said to be oscillatory if it admits an oscillatory solution. Equation (LH1) is said to be strongly oscillatory if all of its solutions are oscillatory. These definitions also hold good for other differential equations without delay. Such a classification of definitions is necessary because we have third-order differential equations which admit both oscillatory and nonoscillatory solutions. For example, $x''' - x = 0$ admits solutions $x_1(t) = e^t$ and $x_2(t) = e^{-t/2}\sin\frac{\sqrt{3}}{2}t$, $t \geq 0$. It is easy to see that all solutions of $x''' - x' = 0$ are nonoscillatory. The nonhomogeneous equation

$$x''' - 2x' + e^t x = 1 + e^{-t}, \quad t \geq 0$$

has a nonoscillatory solution $x(t) = e^{-t}$. From Lazer's work (Theorem 1.3 in [42]), it follows that $x''' - 2x' + e^t x = 0$ is oscillatory. Consequently, the above nonhomogeneous equation is oscillatory (see Corollary 4.2.4 in [9]). The equation

$$x''' - x = -\frac{9}{16}e^{-\frac{t}{2}}, \quad t \geq 0$$

admits a nonoscillatory solution, $x_1(t) = \frac{1}{2}e^{-t/2}$. Clearly, $x_2(t) = e^{-t/2}(\frac{1}{2} + \sin\frac{\sqrt{3}}{2}t)$ is an oscillatory solution of the equation. Further, the nonhomogeneous delay differential equation

$$x''' - e^{\pi/2}x\left(t - \frac{\pi}{2}\right) = \left(e^{\pi/2} - 1\right)\cos t, \quad t \geq 0$$

admits an oscillatory solution $x_1(t) = \sin t$ and a nonoscillatory solution $x_2(t) = \sin t + e^t$. The homogeneous delay differential equation

$$x'''(t) + 2x'(t) - x\left(t - \frac{3\pi}{2}\right) = 0$$

has an oscillatory solution $x_1(t) = \sin t$. If $\phi(m) = m^3 + 2m - e^{-(3\pi/2)m} = 0$, then $\phi(0) = -1 < 0$ and $\phi(m) \to \infty$ as $m \to \infty$. Hence there exists $m_1 > 0$ such that $\phi(m_1) = 0$. Consequently, $x_2(t) = e^{m_1 t}$ is a nonoscillatory solution of the equation.

It would be interesting to obtain an example of third-order differential equation which has only oscillatory solutions. Equation (LH1) is said to be disconjugate in $[T, \infty)$, $T \geq \sigma$ if no solution of (LH1) has more than two zeros in $[T, \infty)$, counting multiplicities. However, it carries no meaning for nonhomogeneous and nonlinear equations.

Now, we state oscillation of a solution of (LH1) in two different ways and show that these definitions are equivalent. Both definitions are used freely for linear homogeneous equations of the form (LH1) or (LH2). However, these definitions differ in case of linear nonhomogeneous and nonlinear equations.

Definition 1.1.1 A solution $x(t)$ of (LH1) is said to be oscillatory on $[T_x, \infty)$, $T_x \geq \sigma$, if it has arbitrarily large zeros in $[T_x, \infty)$, that is, there exists a sequence $\{t_n\} \subset [T_x, \infty)$ such that $t_n \to \infty$ as $n \to \infty$ and $x(t_n) = 0$ for $n = 1, 2, 3, \ldots$.

Definition 1.1.2 A solution $x(t)$ of (LH1) is said to be oscillatory on $[T_x, \infty)$, $T_x \geq \sigma$ if it has an infinite number of zeros in $[T_x, \infty)$.

These two definitions are equivalent. Suppose that a solution $x(t)$ of (LH1) is oscillatory on $[T_x, \infty)$ in the sense of Definition 1.1.2, then $x(t)$ has infinite number of zeros in $[T_x, \infty)$. We claim that the infinite set $A \subset [T_x, \infty)$ of zeros of $x(t)$ is unbounded. If not, then A is bounded and hence there exist real numbers a and b $(T_x \leq a < b < \infty)$ such that $A \subset [a, b] \subset [T_x, \infty)$. From the Weierstrass theorem, it follows that A has a limit point, say t_0. Hence there exists a sequence $\{t_n\} \subset A$ such that $t_n \to t_0$ as $n \to \infty$. Thus $t_0 \in [a, b]$ and $x(t_n) = 0$ for $n = 1, 2, 3, \ldots$. Since x is continuous, $x(t_0) = 0$. If $x'(t_0) \neq 0$, then $x'(t_0) > 0$ or < 0. We may assume, without any loss of generality, that $x'(t_0) > 0$. From the continuity of $x'(t)$, it follows that there exists a $\delta > 0$ such that $x'(t) > 0$ for $t \in (t_0 - \delta, t_0 + \delta)$. Hence $x(t)$ is increasing in $(t_0 - \delta, t_0 + \delta)$. Since $x(t_0) = 0$, we have $x(t) < 0$ for $t \in (t_0 - \delta, t_0)$ and $x(t) > 0$ for $t \in (t_0, t_0 + \delta)$. On the other hand, there exists a $t^* \in A$, $t^* \neq t_0$ such that $t^* \in (t_0 - \delta, t_0 + \delta)$ as t_0 is a limit point of A. Hence $x(t^*) = 0$, a contradiction. Thus $x'(t_0) = 0$. Further, if $x''(t_0) \neq 0$, then we may assume, without any loss of generality, that $x''(t_0) > 0$. Since $x''(t)$ is continuous, there exists an $\eta > 0$ such that $x''(t) > 0$ for $t \in (t_0 - \eta, t_0 + \eta)$. Hence $x'(t)$ is continuous and increasing in $(t_0 - \eta, t_0 + \eta)$. As $x'(t_0) = 0$, then $x'(t) < 0$ for $t \in (t_0 - \eta, t_0)$ and $x'(t) > 0$ for $t \in (t_0, t_0 + \eta)$. Hence $x(t)$ is decreasing in $(t_0 - \eta, t_0)$ and increasing in $(t_0, t_0 + \eta)$. Since $x(t_0) = 0$, we have $x(t) > 0$ for $t \in (t_0 - \eta, t_0) \cup (t_0, t_0 + \eta)$. However, there exists a $t^{**} \neq t_0$, $t^{**} \in A$ such that $t^{**} \in (t_0 - \eta, t_0 + \eta)$ because t_0 is a limit point of A. Consequently, $0 < x(t^{**}) = 0$, a contradiction. Hence $x''(t_0) = 0$. Thus from uniqueness it follows that $x(t) \equiv 0$ on $[T_x, \infty)$, a contradiction. Then our claim holds, that is, A is unbounded. Hence there exists a sequence $\{t_n\} \subset A$ such that $t_n \to \infty$ as $n \to \infty$. As $x(t_n) = 0$ for $n = 1, 2, 3, \ldots$, then $x(t)$ is oscillatory in the sense of Definition 1.1.1.

Conversely, if $x(t)$ is oscillatory in the sense of Definition 1.1.1, then obviously, it is oscillatory in the sense of Definition 1.1.2.

These two definitions are equivalent for solutions of linear homogeneous second-order differential equations of the form

$$\bigl(r(t)x'\bigr)' + p(t)x = 0,$$

but not equivalent for a function $x \in C([\sigma, \infty), R)$. We may note that the equation

$$x'' = t^2\bigl(30t^2 - 1\bigr)\sin\frac{1}{t} - 10t^3\cos\frac{1}{t} + 21t^5, \quad t > 0$$

admits a solution $x(t) = t^6(\sin\frac{1}{t} + \frac{t}{2})$, which has an infinite number of zeros in $(0, 1]$, a finite number of zeros in $[1, 2]$ and it is positive on $(2, \infty)$. Thus $x(t)$ is oscillatory on $(0, \infty)$ in the sense of Definition 1.1.2, but is nonoscillatory by Definition 1.1.1. This is because the equation is nonhomogeneous one. Further, one may observe that the zeros of an oscillatory solution of (LH1) are isolated. Indeed, if $x(t)$ is an oscillatory solution of (LH1) on $[T_x, \infty)$, then there exists a sequence $\{t_n\} \subset [T_x, \infty)$ such that $t_n \to \infty$ as $n \to \infty$ and $x(t_n) = 0, n = 1, 2, 3, \ldots$. We claim that t_n is isolated for all n. If not, then t_m is not isolated for some $m \geq 1$. Thus t_m is a limit point of $\{t_n; n \geq 1\}$. Hence there exists a subsequence $\{t_{n_j}\}$ of $\{t_n\}$ such that $t_{n_j} \to t_m$ as $j \to \infty$. But $t_n \to \infty$ as $n \to \infty$ implies that $t_{n_j} \to \infty$ as $j \to \infty$, a contradiction. Hence our assertion stands.

The adjoint of (LH1) is given by

$$\bigl(\bigl(z' - a(t)z\bigr)' + b(t)z\bigr)' - c(t)z = 0. \tag{LH1*}$$

Definitions 1.1.1 and 1.1.2 are equivalent for (LH1) $\{$(LH1*)$\}$. However, if we consider (LH1) $\{$(LH1*)$\}$ on $(0, d)$, where $d < \infty$, then Definition 1.1.1 has no meaning. Here we repeat that a nontrivial solution $x(t)$ of (LH1) $\{$(LH1*)$\}$ is said to be nonoscillatory if it is not oscillatory. If Eq. (LH1) $\{$(LH1*)$\}$ has a nontrivial oscillatory solution, then it is said to be oscillatory; otherwise, Eq. (LH1) $\{$(LH1*)$\}$ is said to be nonoscillatory.

Let S and S^* denote the solution space of (LH1) and (LH1*), respectively. Thus each of them is a three-dimensional vector space over the field of real numbers. Let S_1 $\{S_1^*\}$ denote a nontrivial subspace of S $\{S^*\}$. Then S_1 $\{S_1^*\}$ is said to be nonoscillatory, if every nonzero member of S_1 $\{S_1^*\}$ is nonoscillatory. S_1 $\{S_1^*\}$ is said to be weakly oscillatory, if it contains a nontrivial oscillatory and nonoscillatory solution. S_1 $\{S_1^*\}$ is said to be strongly oscillatory, if every nonzero member of S_1 $\{S_1^*\}$ oscillates, and S_1 $\{S_1^*\}$ is said to be oscillatory if S_1 $\{S_1^*\}$ is either weakly oscillatory or strongly oscillatory. It may be noted that weakly oscillatory definition applies only to subspaces of dimension greater than or equal to two. If S $\{S^*\}$ is nonoscillatory, weakly oscillatory or strongly oscillatory, then (LH1) {(LH1*)} is said to be nonoscillatory, weakly oscillatory or strongly oscillatory, respectively.

1.2 Third-Order Linear Differential Equations with Constant Coefficients

Consider the linear third-order differential equation with constant coefficients of the form

$$x''' + ax'' + bx' + cx = 0, \quad t \geq \sigma, \tag{1.5}$$

where a, b and c are real constants. The auxiliary or the characteristic equation of (1.5) is given by

$$m^3 + am^2 + bm + c = 0. \tag{1.6}$$

The transformation $n = m + \frac{a}{3}$ transforms (1.6) to

$$n^3 + 3Hn + G = 0, \tag{1.7}$$

where $H = \frac{1}{3}(b - \frac{a^2}{3})$ and $G = c - \frac{ab}{3} + \frac{2a^3}{27}$.

From "Theory of Equations" (see [7]), it is well known that $G^2 + 4H^3 > 0$ implies that (1.7) has two imaginary roots and a real root. If $G^2 + 4H^3 \leq 0$, then all the three roots of (1.7) are real. In particular, $G^2 + 4H^3 = 0$ implies that two of these three roots are equal. If $G = 0$ and $H = 0$, then all the three roots are equal. Two imaginary roots $\alpha + i\beta$ and $\alpha - i\beta$ of (1.7) give rise to two oscillatory solutions

$$x_1(t) = e^{(\alpha - \frac{a}{3})t} \cos \beta t \quad \text{and} \quad x_2(t) = e^{(\alpha - \frac{a}{3})t} \sin \beta t$$

of (1.5). It is well known that an algebraic equation of an odd degree has at least one real root of sign opposite to that of its last term. Consequently, Eq. (1.5) always admits a nonoscillatory solution of the type $x_3(t) = e^{mt}$, where $m < 0$ if $c > 0$ and $m > 0$ if $c < 0$. If $c = 0$, then $m = 0$. Thus there is no linear homogeneous third-order differential equation with constant coefficients, which is strongly oscillatory.

According to the signs of a, b and c, one may consider the following eight different cases, viz., (i) $a \geq 0, b \leq 0, c > 0$, (ii) $a \leq 0, b \leq 0, c > 0$, (iii) $a \leq 0, b \leq 0, c < 0$, (iv) $a \geq 0, b \leq 0, c < 0$, (v) $a \geq 0, b \geq 0, c > 0$, (vi) $a \leq 0, b \geq 0, c > 0$, (vii) $a \geq 0, b \geq 0, c < 0$ and (viii) $a \leq 0, b \geq 0, c < 0$.

Proposition 1.2.1 *Suppose that $a \geq 0$, $b \leq 0$ and $c > 0$. Then the following holds:*

(i) *Equation* (1.5) *admits oscillatory solutions if and only if*

$$\frac{2a^3}{27} - \frac{ab}{3} + c - \frac{2}{3\sqrt{3}}\left(\frac{a^2}{3} - b\right)^{3/2} > 0 \tag{1.8}$$

holds, and

(ii) *When* (1.8) *is satisfied, all solutions of* (1.5) *are oscillatory except constant multiples of one solution which does not vanish on* $[\sigma, \infty)$ *and which together with all of its derivatives is monotonic on* $[\sigma, \infty)$ *and approaches zero as t tends to ∞. In fact, when* (1.8) *is satisfied, nonoscillatory solutions of* (1.5) *form a one-dimensional subspace of the solution space of* (1.5).

Proof (i) Clearly $G > 0$ and $H \leq 0$. Further, $G^2 + 4H^3 > 0$ if and only if $(G - 2(-H)^{3/2})(G + 2(-H)^{3/2}) > 0$ if and only if $G - 2(-H)^{3/2} > 0$, that is (1.8) holds. Hence (1.8) holds if and only if (1.5) admits oscillatory solutions.

(ii) Since $G > 0$ and (1.8) holds, (1.7) admits a negative root γ and two imaginary roots $\alpha + i\beta$ and $\alpha - i\beta$. From "Theory of equations", it is clear that $\gamma + (\alpha + i\beta) + (\alpha - i\beta) = 0$ holds and hence $\gamma = -2\alpha$. This in turn implies that $\alpha > 0$. Thus,

$$\left\{e^{(\alpha - \frac{a}{3})t}\cos\beta t, e^{(\alpha - \frac{a}{3})t}\sin\beta t, e^{(\gamma - \frac{a}{3})t}\right\}$$

is a basis of the solution space of (1.5). Clearly, constant multiples of $e^{(\gamma - \frac{a}{3})t}$ are nonoscillatory solutions of (1.5). These solutions and their derivatives are monotonic and tend to zero as t tends to ∞. To complete the proof of the proposition, it is enough to show that

$$x(t) = e^{(\alpha - \frac{a}{3})t}(\lambda_1 \cos\beta t + \lambda_2 \sin\beta t) + \lambda_3 e^{(\gamma - \frac{a}{3})t}$$

is oscillatory, where λ_1, λ_2 and λ_3 are constants such that λ_1 and λ_2 are not simultaneously equal to zero. If $\lambda_3 = 0$, then $x(t)$ is oscillatory. Without any loss of generality, we assume that $\lambda_3 > 0$. If possible, let $x(t) < 0$ for $t > t_0$. Since $u(t) = \lambda_1 \cos\beta t + \lambda_2 \sin\beta t$ is oscillatory, there exists a $t_1 > t_0$ such that $u(t_1) = 0$. So $x(t_1) > 0$, a contradiction. Suppose that $x(t) > 0$ for $t > t_0$. So $e^{(\alpha-\gamma)t}u(t) > -\lambda_3$ for $t > t_0$, a contradiction, because $\alpha - \gamma > 0$ implies that $e^{(\alpha-\gamma)t}u(t)$ oscillates between $-\infty$ and ∞. This completes the proof. □

Proposition 1.2.2 *Suppose that $a \leq 0$, $b \leq 0$ and $c > 0$. Then*

(i) *Equation* (1.5) *admits oscillatory solutions if and only if* (1.8) *is true, and*

(ii) *Further, if* (1.8) *is satisfied, then the conclusion of Proposition* 1.2.1 *remains true.*

Proof (i) Clearly, $H \leq 0$. (1.8) implies that $G > 0$, $G - 2(-H)^{3/2} > 0$ and hence $G^2 + 4H^3 > 0$. Consequently, (1.5) admits oscillatory solutions. On the other hand, if (1.5) admits oscillatory solutions, then (1.6) has two imaginary roots and a real root. So (1.7) has two imaginary roots and a real root. This in turn implies that $G^2 + 4H^3 > 0$. Since $c > 0$, the real root of (1.6) is negative. Thus $\gamma - \frac{a}{3} < 0$. Since $a < 0$, the real root γ of (1.7) is negative. Consequently, $G > 0$ and (1.8) holds.

(ii) If $\alpha + i\beta$ and $\alpha - i\beta$ are imaginary roots of (1.7), then $\gamma + (\alpha + i\beta) + (\alpha - i\beta) = 0$ implies that $\alpha > 0$ and

$$\left\{e^{(\alpha-\frac{a}{3})t}\cos\beta t,\, e^{(\alpha-\frac{a}{3})t}\sin\beta t,\, e^{(\gamma-\frac{a}{3})t}\right\}$$

defines a basis of the solution space of (1.5). The rest of the proof is same as that of Proposition 1.2.1. □

Proposition 1.2.3 *Suppose that $a \le 0$, $b \le 0$ and $c < 0$. Then*

(i) *Equation* (1.5) *admits oscillatory solutions if and only if*

$$-\frac{2a^3}{27} + \frac{ab}{3} - c - \frac{2}{3\sqrt{3}}\left(\frac{a^2}{3} - b\right)^{3/2} > 0 \tag{1.9}$$

holds,

(ii) *When* (1.9) *holds, oscillatory solutions of* (1.5) *form a two-dimensional subspace of the solutions space of* (1.5), *the zeros of any two oscillatory solutions of* (1.5) *separate on* $[\sigma, \infty)$ *and the absolute values of the successive maxima and minima form a decreasing sequence, these oscillatory solutions tend to zero as* $t \to \infty$, *and*

(iii) *Equation* (1.5) *admits a positive solution which tends to infinity as t tends to ∞ and whose successive derivatives are positive.*

Proof (i) In this case $G < 0$ and $H \le 0$. Clearly, (1.9) holds if and only if (1.5) admits oscillatory solutions.

(ii) If (1.9) holds, then (1.7) has two imaginary roots, say $\alpha + i\beta$ and $\alpha - i\beta$ and a positive root γ, say. So $\gamma + (\alpha + i\beta) + (\alpha - i\beta) = 0$ implies that $\alpha < 0$ and

$$\left\{e^{(\alpha-\frac{a}{3})t}\cos\beta t,\, e^{(\alpha-\frac{a}{3})t}\sin\beta t,\, e^{(\gamma-\frac{a}{3})t}\right\}$$

defines a basis of the solution space of (1.5). The general solution of (1.5) is given by

$$x(t) = \lambda_1 e^{(\alpha-\frac{a}{3})t}\cos\beta t + \lambda_2 e^{(\alpha-\frac{a}{3})t}\sin\beta t + \lambda_3 e^{(\gamma-\frac{a}{3})t},$$

where λ_1, λ_2 and λ_3 are reals. We may note that $x(t)$ is oscillatory, if and only if $\lambda_3 = 0$. Indeed, $\lambda_3 = 0$ implies that $x(t)$ is oscillatory. If $\lambda_3 \ne 0$, then

$$\left|x(t)\right| \ge |\lambda_3| e^{(\gamma-\frac{a}{3})t} - e^{(\alpha-\frac{a}{3})t}\left(|\lambda_1| + |\lambda_2|\right)$$

implies that $x(t)$ is nonoscillatory. So oscillatory solutions of (1.5) form a two-dimensional subspace of the solution space of (1.5) with the basis

$$\left\{e^{(\alpha-\frac{a}{3})t}\cos\beta t,\, e^{(\alpha-\frac{a}{3})t}\sin\beta t\right\}.$$

Since the elements of the above basis are solutions of the second-order linear homogeneous equation

$$z'' - 2\left(\alpha - \frac{a}{3}\right)z' + \left[\left(\alpha - \frac{a}{3}\right)^2 + \beta^2\right] = 0, \tag{1.10}$$

from Sturm's separation theorem, it follows that the zeros of any two oscillatory solutions of (1.5) separate on $[\sigma, \infty)$. Further,

$$2\left(\alpha - \frac{a}{3}\right)\left(\gamma - \frac{a}{3}\right) + \left(\alpha - \frac{a}{3}\right)^2 + \beta^2 = b$$

and $b \le 0$ implies that $(\alpha - \frac{a}{3}) < 0$. This, in turn, implies that the absolute values of the successive maxima and minima of an oscillatory solution of (1.5) form a decreasing sequence, and these oscillatory solutions tend to zero as $t \to \infty$.

(iii) Clearly, $e^{(\gamma - \frac{a}{3})t}$ is the required nonoscillatory solution of (1.5) which together with all its derivatives tends to ∞ as $t \to \infty$. This completes the proof of the proposition. □

Proposition 1.2.4 *Suppose that $a \ge 0$, $b \le 0$ and $c < 0$. Then the following hold:*

(i) *Equation* (1.5) *admits oscillatory solutions, if and only if* (1.9) *holds,*
(ii) *if* (1.9) *holds, then oscillatory solutions of* (1.5) *form a two-dimensional subspace of the solution space of* (1.5). *The zeros of any two linearly independent oscillatory solutions of* (1.5) *separate each other on* $[\sigma, \infty)$, *and these oscillatory solutions tend to zero as $t \to \infty$,*
(iii) *Equation* (1.5) *admits an oscillatory solution, if and only if all nonoscillatory solutions of* (1.5) *tend to $\pm\infty$ as $t \to \infty$, and*
(iv) *Equation* (1.5) *admits a positive solution, which tends to ∞ as $t \to \infty$ and whose successive derivative are positive and tend to ∞ as $t \to \infty$.*

Proof (i) We may notice that $H \le 0$ and $-G - 2(-H)^{3/2} > 0$ if and only if (1.9) holds. Clearly, (1.9) implies that $G < 0$. Since $G^2 + 4H^3 = (-G - 2(-H)^{3/2})(-G + 2(-H)^{3/2})$, (1.9) implies that $G^2 + 4H^3 > 0$. Thus, (1.7) has two imaginary roots and a real root. Consequently, (1.6) has two imaginary roots, say $(\alpha - \frac{a}{3}) + i\beta$ and $(\alpha - \frac{a}{3}) - i\beta$, and a real root $(\gamma - \frac{a}{3})$, where $\alpha + i\beta$, $\alpha - i\beta$ and γ are the roots of (1.7). Thus (1.5) admits oscillatory solutions. On the other hand, if (1.5) admits an oscillatory solution, then (1.6) has two imaginary roots and a real root. This real root is positive because $c < 0$. Consequently, $\gamma - \frac{a}{3} > 0$ because $c < 0$. Thus (1.7) has two imaginary roots and a positive root. Hence $G^2 + 4H^3 > 0$ and $G < 0$ because $\gamma - \frac{a}{3} > 0$. This in turn implies that (1.9) holds. Thus Eq. (1.5) admits an oscillatory solution, if and only if (1.9) holds.

(ii) If (1.9) holds, then a basis of the solution space of (1.5) is given by

$$\left\{e^{(\alpha - \frac{a}{3})t}\cos\beta t, e^{(\alpha - \frac{a}{3})t}\sin\beta t, e^{(\gamma - \frac{a}{3})t}\right\}.$$

If we set

$$x(t) = \lambda_1 e^{(\alpha - \frac{a}{3})t} \cos \beta t + \lambda_2 e^{(\alpha - \frac{a}{3})t} \sin \beta t + \lambda_3 e^{(\gamma - \frac{a}{3})t}, \tag{1.11}$$

where λ_1, λ_2 and λ_3 are reals such that $\lambda_3 \neq 0$, then $x(t)$ is nonoscillatory because $\gamma - \frac{a}{3} > 0$. Again

$$\left(\alpha - \frac{a}{3}\right) + i\beta + \left(\alpha - \frac{a}{3}\right) - i\beta + \left(\gamma - \frac{a}{3}\right) = -a$$

implies that $\alpha < 0$ and, we may write $x(t)$ as

$$x(t) = e^{(\gamma - \frac{a}{3})t}\left[(\lambda_1 \cos \beta t + \lambda_2 \sin \beta t)e^{(\alpha-\gamma)t} + \lambda_3\right].$$

Hence, if (1.9) holds, then the oscillatory solutions of (1.5) form a two-dimensional subspace of the solution space of (1.5). Further, these oscillatory solutions of (1.5) tend to zero as $t \to \infty$, because $\alpha - \frac{a}{3} < 0$. Since $e^{(\alpha - \frac{a}{3})t} \cos \beta t$ and $e^{(\alpha - \frac{a}{3})t} \sin \beta t$ are solutions of (1.10), from Sturm's separation theorem, it follows that the zeros of any two linearly independent oscillatory solutions of (1.5) separate each other on $[\sigma, \infty)$.

(iii) If (1.5) admits an oscillatory solution and if $x(t)$ is a nonoscillatory solution of (1.5), then we may write

$$x(t) = e^{(\alpha - \frac{a}{3})t}(\mu_1 \cos \beta t + \mu_2 \sin \beta t) + \mu_3 e^{(\gamma - \frac{a}{3})t},$$

where μ_1, μ_2 and μ_3 are reals such that $\mu_3 \neq 0$. As

$$\left|x(t)\right| \geq |\mu_3| e^{(\gamma - \frac{a}{3})t} - \left(|\mu_1| + |\mu_2|\right)e^{(\alpha - \frac{a}{3})t},$$

then $|x(t)| \to \infty$ as $t \to \infty$. Conversely, if every nonoscillatory solution of (1.5) tend to $\pm\infty$ as $t \to \infty$, then (1.5) admits an oscillatory solution. Indeed, if all solutions of (1.5) are nonoscillatory, then all roots of (1.6) are positive. Hence the sum of the product of these roots taken two at a time is positive. But from Eq. (1.6), it follows that this sum is equal to $b \leq 0$, a contradiction.

(iv) Clearly, $x(t) = e^{m_3 t}$, $m_3 > 0$ is a nonoscillatory solution of (1.5) which together with all its derivatives tends to ∞ as $t \to \infty$. We note that $m_3 > 0$, because $c < 0$. Hence the proposition is proved. □

Proposition 1.2.5 *Suppose that $a \geq 0$, $b \geq 0$ and $c > 0$. Then*

(i) *Equation* (1.5) *admits a nonoscillatory solution which together with all of its derivatives tend to zero as $t \to \infty$,*
(ii) *if $3b > a^2$, then* (1.5) *is oscillatory,*
(iii) *if $3b \leq a^2$, then* (1.5) *is oscillatory if and only if*

$$\left|\frac{2a^3}{27} - \frac{ab}{3} + c\right| - \frac{2}{3\sqrt{3}}\left(\frac{a^2}{3} - b\right)^{3/2} > 0 \tag{1.12}$$

holds,

(iv) *if* $2c \geq ab$, *then* (1.5) *is oscillatory,*

(v) *if* (1.5) *admits an oscillatory solution and*

$$\frac{2a^3}{27} - \frac{ab}{3} + c > 0,$$

then nonoscillatory solutions of (1.5) *form a one-dimensional subspace of the solution space of* (1.5); *moreover, nonoscillatory solutions of* (1.5) *together with their derivatives tend to zero as* $t \to \infty$,

(vi) *if* (1.5) *admits an oscillatory solution and*

$$\frac{2a^3}{27} - \frac{ab}{3} + c < 0,$$

then oscillatory solutions of (1.5) *form a two-dimensional subspace of the solution space of* (1.5), *the zeros of any two linearly independent oscillatory solutions of* (1.5) *separate each other and every oscillatory solution of* (1.5) *tends to zero as* $t \to \infty$, *and*

(vii) *if*

$$\frac{2a^3}{27} - \frac{ab}{3} + c = 0 \quad \textit{and} \quad 3b \leq a^2,$$

then Eq. (1.5) *is nonoscillatory.*

Proof (i) Since $c > 0$, Eq. (1.5) admits a solution of the form $x(t) = e^{mt}$, where $m < 0$ is a root of Eq. (1.6). Thus (1.5) admits a nonoscillatory solution which together with all of its derivatives tends to zero as $t \to \infty$.

(ii) If $3b > a^2$, then $H > 0$ and hence $G^2 + 4H^3 > 0$. Consequently, (1.5) admits an oscillatory solution.

(iii) Let $3b \leq a^2$. Clearly (1.5) admits an oscillatory solution, if and only if $G^2 + 4H^3 > 0$, if and only if $(|G| + 2(-H)^{3/2})(|G| - 2(-H)^{3/2}) > 0$, if and only if $|G| - 2(-H)^{3/2} > 0$. That is, (1.12) holds.

(iv) If $2c \geq ab$, then

$$\frac{2a^3}{27} - \frac{ab}{3} + c \geq \frac{ab}{6} + \frac{2a^3}{27} \geq 0.$$

Hence, for $a > 0$ and $b > 0$,

$$\begin{aligned} G^2 + 4H^3 &= \left(\frac{2a^3}{27} - \frac{ab}{3} + c\right)^2 + \frac{4}{27}\left(b - \frac{a^2}{3}\right)^3 \\ &\geq \frac{b}{108}\left(8a^4 + 16b^2 - 13a^2b\right) \\ &\geq \frac{343b^3}{108 \times 32} > 0, \end{aligned}$$

because $8a^4 + 16b^2 - 13a^2b$ attains the minimum $\frac{343}{32}b^2$ at $\frac{13}{16}b$. In each of the other three cases, viz., $a = 0, b = 0$; $a > 0, b = 0$ and $a = 0, b > 0$, we observe that $G^2 + 4H^3 > 0$. Thus $2c \geq ab$ implies that (1.5) is oscillatory.

(v) If (1.5) has an oscillatory solution and $G > 0$, then (1.7) admits two imaginary roots $\alpha + i\beta$ and $\alpha - i\beta$ and a negative root γ. As $\alpha + i\beta + \alpha - i\beta + \gamma = 0$, then $\alpha > 0$. Clearly,

$$\left\{e^{(\alpha-\frac{a}{3})t}\cos\beta t, e^{(\alpha-\frac{a}{3})t}\sin\beta t, e^{(\gamma-\frac{a}{3})t}\right\}$$

is a basis of the solution space of (1.5). We claim that nonoscillatory solutions of (1.5), along with the trivial solution, form a one-dimensional subspace of the solution space of (1.5). For this, it is enough to show that the function $x(t)$ given in (1.11) is oscillatory, where λ_1, λ_2 and λ_3 are constants such that $\lambda_1^2 + \lambda_2^2 \neq 0$ and $\lambda_3 \neq 0$. Without any loss of generality, we may assume that $\lambda_3 > 0$. If possible, let $x(t) < 0$ for $t > t_0 > \sigma$. Since $u(t) = \lambda_1 \cos\beta t + \lambda_2 \sin\beta t$ is oscillatory, there exists a $t_1 > t_0$ such that $u(t_1) = 0$. Hence $x(t_1) > 0$, a contradiction. Next, suppose that $x(t) > 0$ for $t > t_0 > \sigma$. Then $e^{(\alpha-\gamma)t}u(t) > -\lambda_3$ for $t > t_0$. This is impossible, since $u(t)$ changes sign and $\alpha - \gamma > 0$. Hence our claim holds.

(vi) If (1.5) has an oscillatory solution and $G < 0$, then (1.7) has two imaginary roots $\alpha + i\beta$ and $\alpha - i\beta$ and a positive root γ. Since the sum of these roots is equal to zero, $\alpha < 0$. We wish to show that oscillatory solutions of (1.5), along with the trivial solution, form a two-dimensional subspace of the solution space of (1.5). Indeed, for this it is enough to prove that $x(t)$ given by (1.11) is nonoscillatory, if $\lambda_3 \neq 0$. Since

$$\left|x(t)\right| \geq e^{(\alpha-\frac{a}{3})t}\left[|\lambda_3|e^{(\gamma-\alpha)t} - \left(|\lambda_1| + |\lambda_2|\right)\right],$$

we have $|x(t)| > 0$ for large t. Thus $x(t)$ is nonoscillatory. We may observe that, in this case, the zeros of any two linearly independent oscillatory solutions of (1.5) separate each other, and every oscillatory solution of (1.5) tends to zero as $t \to \infty$.

(vii) Let $G = 0$. Then $G^2 + 4H^3 > 0$ or ≤ 0 if and only if $3b > a^2$ or $3b \leq a^2$. Thus $G = 0$ and $3b \leq a^2$ imply that (1.5) is nonoscillatory. This completes the proof of the proposition. □

Proposition 1.2.6 *Suppose that* $a \leq 0$, $b \geq 0$ *and* $c > 0$. *Then*

(i) *Equation* (1.5) *admits a nonoscillatory solution which together with all of its derivative tends to zero as* $t \to \infty$,

(ii) *if*

$$\frac{2a^3}{27} - \frac{ab}{3} + c < 0,$$

then all the solutions of (1.5) *are nonoscillatory,*

(iii) *if* $3b > a^2$, *then* (1.5) *is oscillatory,*

(iv) *if* $3b \leq a^2$, *then* (1.5) *is oscillatory if and only if* (1.8) *holds, and*

(v) *if* (1.5) *admits an oscillatory solution and*

$$\frac{2a^3}{27} - \frac{ab}{3} + c > 0,$$

then nonoscillatory solutions of (1.5) *form a one-dimensional subspace of the solution space of* (1.5). *Moreover, nonoscillatory solution space of* (1.5), *together with their derivatives, tend to zero as* $t \to \infty$.

Proof (i) The proof of the first observation is similar to that of Proposition 1.2.5.

(ii) Suppose that $G < 0$. If possible, let (1.5) admits an oscillatory solution. Then (1.7) admits two imaginary roots and a positive real root γ. Consequently, (1.6) admits two imaginary roots and a positive real root $\gamma - \frac{a}{3}$. On the other hand, $c > 0$ implies that $\gamma - \frac{a}{3} < 0$, a contradiction. Hence all solutions of (1.5) are nonoscillatory.

(iii) If $3b > a^2$, then $G^2 + 4H^3 > 0$ and hence (1.5) admits an oscillatory solution.

(iv) Suppose that $3b \le a^2$. If (1.5) admits oscillatory solutions, then $G^2 + 4H^3 > 0$. $G = 0$ implies that $G^2 + 4H^3 \le 0$, a contradiction. If $G < 0$, then from the above discussion, it follows that all solutions of (1.5) are nonoscillatory, a contradiction again. So, $G > 0$. Writing $G^2 + 4H^3 > 0$ as $(G - (-H)^{3/2})(G + (-H)^{3/2}) > 0$, we conclude that (1.8) holds. On the other hand, (1.8) implies that $G > 0$ and $G^2 + 4H^3 > 0$. Thus (1.5) admits oscillatory solutions.

One may proceed as in Proposition 1.2.5 to prove the last observation. Hence the proof is complete. □

Proposition 1.2.7 *Suppose that* $a \ge 0$, $b \ge 0$ *and* $c < 0$. *Then*

(i) *Equation* (1.5) *admits a positive solution which tends to* ∞ *as* $t \to \infty$ *and whose successive derivatives are positive,*

(ii) *if*

$$\frac{2a^3}{27} - \frac{ab}{3} + c > 0,$$

then all solutions of (1.5) *are nonoscillatory,*

(iii) *if* $3b > a^2$, *then* (1.5) *admits an oscillatory solution, and*

(iv) *if* $3b \le a^2$, *then* (1.5) *is oscillatory if and only if* (1.9) *holds. In this case, oscillatory solutions of* (1.5) *form a two-dimensional subspace of the solution space of* (1.5).

Proof (i) $c < 0$ implies that (1.6) has a positive root. This, in turn, implies that (1.5) admits a positive solution which tend to ∞ as $t \to \infty$ and whose successive derivatives are positive.

(ii) Given that $G > 0$. If possible, suppose that (1.5) has an oscillatory solution. So (1.7) has two imaginary roots and a negative root γ. Consequently, (1.6) has two imaginary roots and a negative root $\gamma - \frac{a}{3}$. But $c < 0$ implies that $\gamma - \frac{a}{3} > 0$, a contradiction. Hence all solutions of (1.5) are nonoscillatory.

(iii) Clearly, $3b > a^2$ implies that $G^2 + 4H^3 > 0$ and hence (1.5) admits oscillatory solutions.

(iv) Suppose that $3b \le a^2$. If (1.5) admits oscillatory solutions, then $G^2 + 4H^3 > 0$. From the above discussion, it follows that $G \not> 0$. Further, $G = 0$ implies that $G^2 + 4H^3 \le 0$, a contradiction. Thus $G < 0$. Writing $G^2 + 4H^3 > 0$ as $(-G - (-H)^{3/2})(-G + (-H)^{3/2}) > 0$, we obtain (1.9). Again (1.9) implies that $G < 0$, $-G + 2(-H)^{3/2} > 0$ and hence $G^2 + 4H^3 > 0$. Consequently, (1.5) admits oscillatory solutions. Now, one may proceed as in Proposition 1.2.5 to prove that oscillatory solutions of (1.5) form a two-dimensional subspace of the solution space of (1.5). The proposition is proved. □

Proposition 1.2.8 *Suppose that $a \le 0$, $b \ge 0$ and $c < 0$. Then*

(i) *Equation* (1.5) *admits a positive solution which tends to ∞ as $t \to \infty$ and whose successive derivatives are positive and tend to ∞ as $t \to \infty$,*
(ii) *if $3b > a^2$, then* (1.5) *is oscillatory,*
(iii) *if $3b \le a^2$, then* (1.5) *is oscillatory if and only if* (1.12) *holds,*
(iv) *if $2c \le ab$, then* (1.5) *is oscillatory,*
(v) *if* (1.5) *admits an oscillatory solution and*

$$\frac{2a^3}{27} - \frac{ab}{3} + c > 0,$$

then nonoscillatory solutions of (1.5) *form a one-dimensional subspace of the solution space of* (1.5); *moreover, every nonoscillatory solution of* (1.5) *together with their derivatives tends to $\pm\infty$ as $t \to \infty$,*
(vi) *if* (1.5) *admits an oscillatory solution and*

$$\frac{2a^3}{27} - \frac{ab}{3} + c < 0,$$

then oscillatory solutions of (1.5) *form a two-dimensional subspace of the solution space of* (1.5) *and the zeros of any two linearly independent oscillatory solutions of* (1.5) *separate each other, and*
(vii) *if*

$$\frac{2a^3}{27} - \frac{ab}{3} + c = 0 \quad \text{and} \quad 3b \le a^2,$$

then Eq. (1.5) *is nonoscillatory.*

Proof (i) As $c < 0$, Eq. (1.5) admits a solution of the form $x(t) = e^{mt}$, where $m > 0$ is a root of (1.6). Thus (1.5) admits a positive solution which together with all its derivatives tends to ∞ as $t \to \infty$.

(ii) If $3b > a^2$, then $H > 0$ and hence $G^2 + 4H^3 > 0$. Consequently, (1.5) is oscillatory.

(iii) Suppose that $3b \le a^2$. Then $H \le 0$. Hence, Eq. (1.5) is oscillatory, if and only if $G^2 + 4H^3 > 0$, if and only if $(|G| + 2(-H)^{3/2})(|G| - 2(-H)^{3/2}) > 0$, if and only if $|G| - 2(-H)^{3/2} > 0$, that is, (1.12) holds.

(iv) Let $2c \le ab$. Then

$$\frac{2a^3}{27} - \frac{ab}{3} + c \le \frac{ab}{6} + \frac{2a^3}{27} \le 0.$$

Hence, for $a < 0$ and $b > 0$,

$$\begin{aligned} G^2 + 4H^3 &= \left(\frac{2a^3}{27} - \frac{ab}{3} + c\right)^2 + \frac{4}{27}\left(b - \frac{a^2}{3}\right)^3 \\ &= \frac{b}{108}\left(8a^4 + 16b^2 - 13a^2b\right) \\ &= \frac{343b^3}{108 \times 32} > 0, \end{aligned}$$

because $8a^4 + 16b^2 - 13a^2b$ attains the minimum $\frac{343}{32}b^2$ at $\frac{13}{16}b$. In each of the other three cases, viz., $a = 0, b = 0$; $a < 0, b = 0$ and $a = 0, b < 0$, we have $G^2 + 4H^3 > 0$. Thus $2c \le ab$ implies that (1.5) is oscillatory.

(v) If (1.5) admits an oscillatory solution and $G > 0$, then (1.7) admits two imaginary roots $\alpha + i\beta$ and $\alpha - i\beta$ and a real negative root γ. As $\alpha + i\beta + \alpha - i\beta + \gamma = 0$, then $\alpha > 0$. Clearly

$$\left\{e^{(\alpha - \frac{a}{3})t} \cos \beta t, e^{(\alpha - \frac{a}{3})t} \sin \beta t, e^{(\gamma - \frac{a}{3})t}\right\}$$

is a basis of the solution space of (1.5). We show that nonoscillatory solutions of (1.5), along with the trivial solution, form a one-dimensional subspace of the solution space of (1.5). For this, it is enough to show that $x(t)$ given in (1.11) is oscillatory, where λ_1, λ_2 and λ_3 are constants such that $\lambda_1^2 + \lambda_2^2 \ne 0$ and $\lambda_3 \ne 0$. Without any loss of generality, we may assume that $\lambda_3 > 0$. If possible, let $x(t) < 0$ for $t > t_0 > \sigma$. Since $u(t) = \lambda_1 \cos \beta t + \lambda_2 \sin \beta t$ is oscillatory, there exists a $t_1 > t_0$ such that $u(t_1) = 0$. Hence $x(t_1) > 0$, a contradiction. Suppose that $x(t) > 0$ for $t > t_0 > \sigma$. Then $e^{(\alpha - \gamma)t}u(t) > -\lambda_3$ for $t > t_0$. This is impossible, because $u(t)$ is oscillatory and $\alpha - \gamma > 0$. Hence nonoscillatory solutions of (1.5) form a one-dimensional subspace of the solution space of (1.5). Since $\gamma - \frac{a}{3}$ is the only root of (1.6) and $c < 0$, we have $\gamma - \frac{a}{3} > 0$. As every nonoscillatory solution of (1.5) is a constant multiple of $e^{(\gamma - \frac{a}{3})t}$, the conclusion holds.

(vi) If Eq. (1.5) admits an oscillatory solution and $G < 0$, then (1.7) has two imaginary roots $\alpha + i\beta$ and $\alpha - i\beta$ and a positive root γ. Since $\alpha + i\beta + \alpha - i\beta + \gamma = 0$, we have $\alpha < 0$. We claim that oscillatory solutions of (1.5) form a two-dimensional subspace of the solution space of (1.5). For this, it is enough to show that $x(t)$ given by (1.11) is nonoscillatory, if $\lambda_3 \ne 0$. Since

$$|x(t)| \ge e^{(\alpha - \frac{a}{3})t}\left[|\lambda_3| e^{(\gamma - \alpha)t} - \left(|\lambda_1| + |\lambda_2|\right)\right],$$

we have $|x(t)| > 0$ for large t. Thus $x(t)$ is nonoscillatory. We may observe that in this case, the zeros of any two linearly independent oscillatory solutions of (1.5) separate each other.

(vii) Let $G = 0$. Then $G^2 + 4H^3 > 0$ or ≤ 0 if and only if $3b > a^2$ or $3b \le a^2$. Thus, $G = 0$ and $3b \le a^2$ imply that Eq. (1.5) is nonoscillatory. This completes the proof of the proposition. □

Now, we consider the third-order nonhomogeneous linear differential equation

$$y''' + ay'' + by' + cy = f, \quad t \ge \sigma, \tag{1.13}$$

where a, b, c and f are real constants and $f \neq 0$. We shall consider the following two cases:

(i) $a \ge 0$ (≤ 0), $b \le 0$, $c > 0$ and
(ii) $a \ge 0$, $b \ge 0$, $c > 0$.

First suppose that $a \ge 0$ (≤ 0), $b \le 0$ and $c > 0$. Clearly, all solutions of (1.5) are nonoscillatory, if and only if the characteristic equation (1.6) has only real roots, say $\gamma_i, i = 1, 2, 3$. Consequently, the general solution of (1.13) is given by

$$y(t) = \frac{f}{c} + \sum_{i=1}^{3} \lambda_i e^{\gamma_i t}, \quad \lambda_i \in R, \; c \neq 0,$$

which is nonoscillatory. Hence nonoscillation of (1.5) implies the nonoscillation of (1.13), that is, the oscillation of (1.13) implies the oscillation of (1.5). On the other hand, oscillation of (1.5) need not imply the oscillation of (1.13). Indeed, the solutions of

$$y''' - 2y' - 4y = 0$$

is given by $e^{-t}\cos t$, $e^{-t}\sin t$ and e^{2t}. So, the general solution of the corresponding nonhomogeneous equation

$$y''' - 2y' - 4y = f,$$

where $f \in R$ and $f \neq 0$, is of the form

$$y(t) = -\frac{f}{4} + \lambda_1 e^{-t}\cos t + \lambda_2 e^{-t}\sin t + \lambda_3 e^{2t},$$

which is nonoscillatory for all real λ_i, $i = 1, 2, 3$. However, for the above considered two cases (i) and (ii), the situation is different. The oscillation of (1.5) implies the oscillation of (1.13). Indeed, oscillation of (1.5) implies that (1.6) admits two imaginary roots $\alpha + i\beta$ and $\alpha - i\beta$ and a negative root γ. Using Descartes's rule of signs, the addition given by $b = \gamma(\alpha + i\beta) + \gamma(\alpha - i\beta) + (\alpha + i\beta)(\alpha - i\beta)$ implies that $\alpha > 0$. Consequently, $y(t) = \frac{f}{c} + \lambda e^{\alpha t}\cos\beta t$ is an oscillatory solution of (1.13). This yields the following proposition:

Proposition 1.2.9 *Suppose that $a \ge 0$ (≤ 0), $b \le 0$ and $c > 0$. Then* (1.5) *is oscillatory, if and only if* (1.13) *is oscillatory.*

Remark 1.2.1 From Propositions 1.2.9 and 1.2.1(i) (or Proposition 1.2.2(i)), it follows that (1.13) is oscillatory, if and only if (1.8) holds, when $a \geq 0$ (≤ 0), $b \leq 0$ and $c > 0$.

If $a \geq 0$, $b \geq 0$ and $c > 0$, then $\alpha + i\beta + \alpha - i\beta + \gamma = a$ implies that $\alpha > 0$, where $\alpha + i\beta$, $\alpha - i\beta$ and γ are the roots of (1.6). This leads to the following proposition:

Proposition 1.2.10 *Suppose that $a \geq 0$, $b \geq 0$ and $c > 0$. Then* (1.5) *is oscillatory, if and only if* (1.13) *is oscillatory.*

Remark 1.2.2 From Propositions 1.2.10 and 1.2.5(iii), it follows that (1.13) is oscillatory, if and only if (1.12) holds, provided that the coefficients a and b satisfies the property $3b \leq a^2$.

1.3 Introduction to Third-Order Delay Differential Equations

Let us consider the ordinary differential equation

$$x'(t) = f(t, x) \tag{1.14}$$

together with the initial condition

$$x(t_0) = x_0. \tag{1.15}$$

It is well known that under certain assumptions on f, the initial-value problem (1.14) and (1.15) has a unique solution and is equivalent to the integral equation

$$x(t) = x(t_0) + \int_{t_0}^{t} f\big(s, x(s)\big)\, ds \quad \text{for } t \geq t_0.$$

Now, we consider a differential equation of the form

$$x'(t) = f\big(t, x(t), x(t-\tau)\big) \quad \text{with } \tau > 0 \text{ and } t \geq t_0, \tag{1.16}$$

in which the right-hand side depends not only on the instantaneous position $x(t)$, but also on $x(t-\tau)$, the position at τ units back, that is to say, the equation has past memory. Such an equation is called a *delay differential equation*.

The integral equation equivalent to (1.16) is

$$x(t) = x(t_0) + \int_{t_0}^{t} f\big(s, x(s), x(s-\tau)\big)\, ds.$$

In order to define a solution of (1.16), we need to have a known function ϕ on $[t_0 - \tau, t_0]$, instead of just the initial condition (1.15).

Hence we define a basic initial-value problem of a delay differential equation as follows: On the interval $[t_0, T], T \le \infty$, we seek a continuous function $x(t)$ that satisfies (1.16) and an initial condition

$$x(t) = \phi(t) \quad \text{for all } t \in E_{t_0}, \tag{1.17}$$

where t_0 is an initial point, $E_{t_0} = [t_0 - \tau, t_0]$ is the initial set, the known function ϕ on E_{t_0} is called the *initial function*.

Under general assumptions, the existence and uniqueness of solutions to the initial-valued problem (1.16) and (1.17) can be established (see, for example, Gyori and Ladas [24] and Driver [12]). In the case of a variable delay $\tau = \tau(t) > 0$ in (1.16), it is also required to find a solution of this equation for $t \ge t_0$ such that the initial set

$$E_{t_0} = t_0 \cup \{t - \tau(t) : t - \tau(t) < t_0,\ t \ge t_0\},$$

$x(t)$ coincides with the given initial function ϕ. If it is required to determine the solution as the interval $[t_0, T]$, then the initial set is

$$E_{t_0,T} = \{t_0\} \cup \{t - \tau(t) : t - \tau(t) < t_0,\ t_0 \le t \le T\}.$$

Example 1.3.1 For the equation

$$x'(t) = f\big(t, x(t), x\big(t - \sin^2 t\big)\big),$$

$t_0 = 0$, $E_0 = [-1, 0]$, and the initial function ϕ must be given on the interval $[-1, 0]$.

Sometimes the initial set E_{t_0} depends on the initial point t_0, as can be seen from the following example.

Example 1.3.2 For the equation

$$x' = ax\left(\frac{t}{2}\right),$$

we have $\tau(t) = \frac{t}{2}$ so that $E_0 = \{0\}$ and $E_1 = [\frac{1}{2}, 1]$.

Now, we consider a differential equation of the nth order with l deviating arguments, of the form

$$\begin{aligned} x^{(m_0)}(t) = f\big(&t, x(t), \dots, x^{(m_0-1)}(t), x\big(t - \tau_1(t)\big), \dots, \\ &x^{(m_1-1)}\big(t - \tau_1(t)\big), \dots, x\big(t - \tau_l(t)\big), \dots, x^{(m_l-1)}\big(t - \tau_l(t)\big)\big), \end{aligned} \tag{1.18}$$

where the deviations $\tau_i(t) > 0$, and $\max_{0 \le i \le l} m_i = n$.

To formulate the initial-value problem for (1.18), we shall use the following notation. Let t_0 be a given initial point. Each deviation $\tau_i(t)$ defines the initial set $E_{t_0}^{(i)}$

given by

$$E_{t_0}^{(i)} = \{t_0\} \cup \{t - \tau_i(t) : t - \tau_i(t) < t_0,\ t \geq t_0\}.$$

We denote $E_{t_0} = \bigcup_{i=1}^{l} E_{t_0}^{(i)}$, and on E_{t_0} continuous functions ϕ_k, $k = 0, 1, \ldots, \mu$, must be given, with $\mu = \max_{0 \leq i \leq l} m_i$. In applications, it is most natural to consider the case where on E_{t_0},

$$\phi_k(t) = \phi_0^{(k)}(t) \quad \text{for } k = 0, 1, \ldots, \mu,$$

but it is not generally necessary.

For the nth order differential equation, there should be given initial values $x_0^{(k)}$, $k = 0, 1, 2, \ldots, n-1$. Let $x_0^{(k)} = \phi_k(t_0)$, $k = 0, 1, 2, \ldots, \mu$. If $\mu < n-1$, then, in addition, the numbers $x_0^{(\mu+1)}, x_0^{(\mu+2)}, \ldots, x_0^{(n-1)}$ are given. If the point t_0 is an isolated point of E_{t_0}, then $x_0^{(0)}, x_0^{(1)}, \ldots, x_0^{(n)}$ are also given.

For (1.18), the basic initial-value problem consists of the determination of an $(n-1)$ times continuously differentiable function x that satisfies (1.18) for $t > t_0$ and the conditions

$$x^{(k)}(t_0 + 0) = x_0^{(k)}, \quad k = 0, 1, 2, \ldots, n-1,$$

and

$$x^{(k)}\big(t - \tau_i(t)\big) = \phi_k\big(t - \tau_i(t)\big) \quad \text{if } t - \tau_i(t) < t_0$$

for $k = 0, 1, \ldots, \mu$ and $i = 1, 2, \ldots, l$.

Example 1.3.3 Consider

$$x''(t) = f\left(t, x(t), x'(t), x\big(t - \cos^2 t\big), x\left(\frac{t}{2}\right)\right).$$

For $t_0 = 0$, we have $n = 2$, $l = 2$, $\mu = 0$, the initial sets $E_0^{(1)} = [-1, 0]$, $E_0^{(2)} = \{0\}$ and $E_0 = [-1, 0]$, on which is given the initial function $\phi_0(t)$, $x_0^{(0)} = \phi_0(0)$ and $x_0^{(1)}$ is any given number.

Now, we consider the third-order linear delay differential equation of the form

$$x'''(t) + a(t)x''(t) + b(t)x'(t) + c(t)x\big(g(t)\big) = 0, \tag{DLH1}$$

where $a, b, c, g \in C([\sigma, \infty), R)$, $g(t) \leq t$ and $g(t) \to \infty$ as $t \to \infty$, or for more general delay differential equations of the form

$$\big(r(t)x''(t)\big)' + q(t)x'(t) + p(t)x\big(g(t)\big) = 0, \tag{DLH2}$$

where p, q and r are the same as in (LH2).

Definition 1.3.1 A solution $x(t)$ of (DLH1) or (DLH2) is said to be oscillatory, if there exists a sequence $\{t_n\}$ such that $t_n \to \infty$ as $t \to \infty$ and $x(t_n) = 0$ for $n = 1, 2, \ldots$, that is, $x(t)$ has arbitrarily large zeros; otherwise, $x(t)$ is called nonoscillatory.

Definition 1.3.2 Equation (LH1) or (LH2) or (DLH1) or (DLH2) is said to be disconjugate on an interval $[\sigma, \infty)$, in case no nontrivial solution of the equation has more than two zeros on $[\sigma, \infty)$.

Clearly, if (DLH1) or (DLH2) is disconjugate on some interval $[t_0, \infty)$, $t_0 \geq \sigma$, then all solutions of (DLH1) or (DLH2) are nonoscillatory. It is interesting to note that there are third-order delay differential equations which have only oscillatory solutions or have both oscillatory and nonoscillatory solutions. For example,

$$x'''(t) + x'(t) + x\left(t - \frac{\pi}{2}\right) = 0$$

admits an oscillatory solution $x_1(t) = \sin t$ and a nonoscillatory solution $x_2(t) = e^{\lambda t}$, where λ is a nonzero root of the characteristic equation of the above example. On the other hand, all solutions of

$$x'''(t) + x(t - \tau) = 0$$

are oscillatory if and only if $\tau > \frac{3}{e}$ (see [40]). But the corresponding ordinary differential equation

$$x'''(t) + x(t) = 0$$

admits a nonoscillatory solution $x_1(t) = e^{-t}$ and oscillatory solutions $x_2(t) = e^{t/2} \cos \frac{\sqrt{3}}{2} t$ and $x_3(t) = e^{t/2} \sin \frac{\sqrt{3}}{2} t$. This observation motivates the researchers to study the oscillation and asymptotic behaviour of nonoscillatory solutions of (DLH1) and (DLH2).

1.4 Introduction to Third-Order Canonical Differential Equations

Consider the third-order differential equation (LH1) and (DLH1). These equations can, respectively, be written as (LH2) and (DLH2), where $r(t) = \exp(\int_\sigma^t a(s)\,ds)$, $q(t) = r(t)b(t)$ and $p(t) = r(t)c(t)$.

If $g(t) \equiv t$, then (DLH1) and (DLH2) are reduced to (LH1) and (LH2), respectively. The method applied to study the oscillatory and asymptotic behaviour of solutions of (LH1) / (LH2) does not work for (DLH1) / (DLH2) due to the presence of the delay term. However, the asymptotic behaviour of solutions of (DLH2) is studied with the help of the solutions of (LH2). This has been possible because of:

- the canonical transformation due to Trench [61] and some comparison results by Kusano and Naito [38]. This is possible only when $b(t) \le 0$.
- the Riccati transformation and a comparison result, namely Lemma 4 due to Greguš and Gera [22] and some known inequalities by Erbe [15]. This method is possible when $b(t) \ge 0$.

By a canonical third-order linear homogeneous differential equation, we mean an equation of the type

$$Lx + p(t)x = 0, \tag{1.19}$$

where

$$Lx = \frac{1}{r_3(t)}\left(\frac{1}{r_2(t)}\left(\frac{1}{r_1(t)}\left(\frac{x}{r_0(t)}\right)'\right)'\right)', \tag{1.20}$$

$r_i \in C([\sigma,\infty), R)$ with $r_i(t) > 0$, $0 \le i \le 3$ and

$$\int_\sigma^\infty r_i(t)\,dt = \infty, \quad i = 1, 2. \tag{1.21}$$

This applies to all ordinary and delay differential equations of third order.

Suppose that $a \in C^1([\sigma,\infty), R)$, $b, c \in C([\sigma,\infty), R)$ with $b(t) \le 0$. Setting

$$Lx = \left(r(t)x''\right)' + q(t)x' \tag{1.22}$$

one can write (LH2) as (1.19).

It is easy to see that the operator (1.22) may be written in the form

$$Lx = \frac{1}{v(t)}\left(\frac{r(t)}{v^{-2}(t)}\left(\frac{x'}{v(t)}\right)'\right)', \tag{1.23}$$

where $v(t)$ is a positive solution of the second-order linear differential equation

$$\left(r(t)v'\right)' + q(t)v = 0. \tag{1.24}$$

Lemma 1.4.1 *Let $q(t) \le 0$. Equation* (1.24) *admits a positive increasing solution $v(t)$ with*

$$\int_\sigma^\infty v(t)\,dt = \infty \quad \text{and} \quad \int_\sigma^\infty \frac{dt}{v^2(t)r(t)} < \infty. \tag{1.25}$$

Proof Let $v(t)$ be a solution of (1.24) with $v(\sigma) > 0$ and $v'(\sigma) > 0$. From the continuity of $v'(t)$, it follows that there exists a $\delta > 0$ such that $v'(t) > 0$ for $t \in [\sigma, \sigma+\delta)$. We claim that $v'(t) > 0$ for $t \ge \sigma$. If not, then there exists a $t_1 > \sigma$ such that $v'(t_1) = 0$ and $v'(t) > 0$ for $t \in [\sigma, t_1)$. Integrating (1.24) from σ to t_1, we get

$$0 > r(t_1)v'(t_1) - r(\sigma)v'(\sigma) = -\int_\sigma^{t_1} q(t)v'(t)\,dt > 0,$$

a contradiction. Thus, $v(t) > 0$ and $v'(t) > 0$ for $t \geq \sigma$. Clearly, $\int_\sigma^\infty v(t)\,dt = \infty$. Further, $(r(t)v'(t))' \geq 0$ implies that $r(t)v'(t) \geq r(t_0)v'(t_0)$ for $t \geq t_0 \geq \sigma$. Hence

$$v^2(t) > \big(r(t_0)v'(t_0)\big)^2\left(\int_{t_0}^t \frac{ds}{r(s)}\right)^2.$$

Thus, for $t > t_2 > t_0$,

$$\begin{aligned}\int_{t_2}^t \frac{ds}{r(s)v^2(s)} &< \frac{1}{(r(t_0)v'(t_0))^2}\int_{t_2}^t \frac{ds}{r(s)(\int_{t_0}^s \frac{d\theta}{r(\theta)})^2}\\ &< \frac{1}{(r(t_0)v'(t_0))^2}\frac{t_2}{\int_{t_0}^{t_2}\frac{ds}{r(s)}} < \infty.\end{aligned}$$

Hence $\int_\sigma^\infty \frac{ds}{r(s)v^2(s)} < \infty$. This completes the proof of the lemma. □

Theorem 1.4.1 *Equation* (1.19) *with* Lx *defined in* (1.22) *can be represented essentially in the unique way in a canonical form* (1.19), *where* Lx *is defined by* (1.20) *and* $r_i(t)$, $i = 1, 2$ *satisfy the property* (1.21).

Proof In view of Lemma 1.4.1, the operator Lx given by (1.22) may be written in the form (1.23). Since (1.25) holds, proceeding as in Lemma 2 due to Trench [61], one may write (1.23) in the form

$$Lx = \frac{1}{\tilde{r}_3(t)}\left(\frac{1}{\tilde{r}_2(t)}\left(\frac{1}{\tilde{r}_1(t)}\left(\frac{x}{\tilde{r}_0(t)}\right)'\right)'\right)', \tag{1.26}$$

where

$$\begin{aligned}\tilde{r}_0(t) &= 1,\\ \tilde{r}_1(t) &= v(t)\int_t^\infty \frac{ds}{r(s)v^2(s)},\\ \tilde{r}_2(t) &= \frac{1}{r(t)v^2(t)}\left(\int_t^\infty \frac{ds}{r(s)v^2(s)}\right)^{-2}\end{aligned}$$

and

$$\tilde{r}_3(t) = v(t)\int_t^\infty \frac{ds}{r(s)v^2(s)}.$$

Clearly, $\int_\sigma^\infty \tilde{r}_2(t)\,dt = \infty$. If $\int_\sigma^\infty \tilde{r}_1(t)\,dt = \infty$, then we set $r_0(t) = \tilde{r}_0(t)$, $r_1(t) = \tilde{r}_1(t)$, $r_2(t) = \tilde{r}_2(t)$ and $r_3(t) = \tilde{r}_3(t)$. If $\int_\sigma^\infty \tilde{r}_1(t)\,dt < \infty$, then (1.26) may be expressed in the form

$$Lx = \frac{1}{r_3(t)}\left(\frac{1}{r_2(t)}\left(\frac{1}{r_1(t)}\left(\frac{x}{r_0(t)}\right)'\right)'\right)',$$

where

$$r_0(t) = \tilde{r}_0(t) \int_t^\infty \tilde{r}_1(s)\,ds,$$

$$r_1(t) = \tilde{r}_1(t)\left(\int_t^\infty \tilde{r}_1(s)\,ds\right)^{-2},$$

$$r_2(t) = \tilde{r}_2(t) \int_t^\infty \tilde{r}_1(s)\,ds$$

and

$$r_3(t) = \tilde{r}_3(t).$$

Clearly, $\int_\sigma^\infty r_1(t)\,dt = \infty$ and $\int_\sigma^\infty r_2(t)\,dt = \infty$. Indeed,

$$\int_\sigma^\infty r_1(t)\,dt = \int_\sigma^\infty \tilde{r}_1(t)\left(\int_t^\infty \tilde{r}_1(s)\,ds\right)^{-2} dt = \frac{1}{\int_t^\infty \tilde{r}_1(s)\,ds|_\sigma^\infty} = \infty.$$

Further, integration by parts gives, for $\sigma < T \le \infty$

$$\begin{aligned}\int_\sigma^T r_2(t)\,dt &= \int_\sigma^T \tilde{r}_2(t)\left(\int_t^\infty \tilde{r}_1(s)\,ds\right) dt \\ &= \left(\int_T^\infty \tilde{r}_1(s)\,ds\right)\left(\int_\sigma^T \tilde{r}_2(s)\,ds\right) + \int_\sigma^T \tilde{r}_1(t)\left(\int_\sigma^t \tilde{r}_2(s)\,ds\right) dt.\end{aligned}$$

However,

$$\int_\sigma^t \tilde{r}_2(s)\,ds = \left(\int_t^\infty \frac{v^{-2}(s)}{r(s)}\,ds\right)^{-1} - \left(\int_\sigma^\infty \frac{v^{-2}(s)}{r(s)}\,ds\right)^{-1}$$

implies that

$$\begin{aligned}\int_\sigma^T r_2(t)\,dt > &-\left(\int_T^\infty \tilde{r}_1(s)\,ds\right)\left(\int_\sigma^\infty \frac{v^{-2}(s)}{r(s)}\,ds\right)^{-1} \\ &+ \int_\sigma^T v(t)\,dt - \left(\int_\sigma^\infty \frac{v^{-2}(s)}{r(s)}\,ds\right)^{-1}\left(\int_\sigma^T \tilde{r}_1(t)\,dt\right).\end{aligned}$$

Taking limit $T \to \infty$ both sides, we obtain $\int_\sigma^\infty r_2(t)\,dt = \infty$ because $\int_\sigma^\infty v(t)\,dt = \infty$. Thus, the theorem is proved. □

Definition 1.4.1 Equation (LH1)/(LH2)/(DLH1)/(DLH2) is said to have property A, if every nonoscillatory solution $x(t)$ of the equation satisfies

$$(-1)^j x(t) x^{(j)}(t) > 0, \quad 0 \le j \le 3$$

for large t, and

$$\lim_{t\to\infty} x^{(i)}(t) = 0, \quad i = 0, 1, 2.$$

Definition 1.4.2 Equation (LH1)/(LH2)/(DLH1)/(DLH2) is said to have property B, if every nonoscillatory solution $x(t)$ of the equation satisfies

$$x(t)x^{(i)}(t) > 0, \quad 0 \le i \le 3$$

for large t.

If we denote $L_0 x = \frac{x}{r_0(t)}$, $L_i x = \frac{(L_{i-1}x)'}{r_i(t)}$, $1 \le i \le 3$, then (1.19) takes the form

$$L_3 x + p(t)x = 0. \tag{1.27}$$

Definition 1.4.3 Equation (1.27) is said to have the property A, if every nonoscillatory solution $x(t)$ of the equation satisfies

$$(-1)^i x(t)L_i x(t) > 0, \quad 0 \le i \le 3$$

for large t, and

$$\lim_{t\to\infty} L_i x(t) = 0, \quad i = 0, 1, 2.$$

Definition 1.4.4 Equation (1.27) is said to have the property B, if every nonoscillatory solution $x(t)$ of the equation satisfies

$$x(t)L_i x(t) > 0, \quad 0 \le i \le 3$$

for large t.

Remark 1.4.1 Definitions 1.4.1–1.4.4 also hold for the third-order canonical delay differential equation

$$L_3 x(t) + p(t)x\big(g(t)\big) = 0, \tag{1.28}$$

where $g \in C([\sigma, \infty), R)$ such that $0 < g(t) \le t$ and $g(t) \to \infty$ as $t \to \infty$.

Lemma 1.4.2 *Consider the canonical equation* (1.19) *with* (1.21). *Let* $p(t) > 0$ *hold. If* $x(t)$ *is a nonoscillatory solution of this equation, then either*

$$\operatorname{sgn} L_0 x(t) = \operatorname{sgn} L_1 x(t) = \operatorname{sgn} L_2 x(t) \ne \operatorname{sgn} L_3 x(t)$$

holds or

$$\operatorname{sgn} L_0 x(t) = \operatorname{sgn} L_2 x(t) \ne \operatorname{sgn} L_1 x(t) = \operatorname{sgn} L_3 x(t)$$

holds for large t.

Proof We may assume, without any loss of generality, that $x(t) > 0$ for $t \geq t_0 > \sigma$. Thus $L_0x(t) > 0$ and $L_3x(t) < 0$ for $t \geq t_0$. To complete the proof of the lemma it is enough, to show that $L_1x(t) > 0$ or < 0 and $L_2y(t) > 0$ for large t. Since $L_3x(t) < 0$ for $t \geq t_0$, $L_2x(t) > 0$ or < 0 for $t \geq t_1 \geq t_0$. If $L_2x(t) < 0$, then $L_3x(t) < 0$ and $\int_\sigma^\infty r_2(t)\,dt = \infty$ imply that $L_1x(t) < 0$ for large t, say for $t \geq t_2 \geq t_1$. Thus $L_0x(t) < 0$ for large t, because $\int_\sigma^\infty r_1(t)\,dt = \infty$, a contradiction. Hence $L_2x(t) > 0$ for $t \geq t_1$. Consequently, $L_1x(t) > 0$ or < 0 for large t. Thus the lemma is proved. □

Remark 1.4.2 We may note that Lemma 1.4.2 is true for Eq. (1.28) with (1.21).

We note that $x(t)$ is a solution of (LH2), if and only if $x(t)$ is a solution of (1.19). Indeed, it is clear that if $x(t)$ is a solution of (LH2), then it is a solution of (1.19). Next, we show that if $x(t)$ is a solution of (1.19), then $x(t)$ is a solution of (LH2). For this, it is enough to prove that

$$\frac{1}{r_3(t)}\left(\frac{1}{r_2(t)}\left(\frac{1}{r_1(t)}\left(\frac{x(t)}{r_0(t)}\right)'\right)'\right)' = \frac{1}{\tilde{r}_3(t)}\left(\frac{1}{\tilde{r}_2(t)}\left(\frac{1}{\tilde{r}_1(t)}\left(\frac{x(t)}{\tilde{r}_0(t)}\right)'\right)'\right)'$$

and

$$\frac{1}{\tilde{r}_3(t)}\left(\frac{1}{\tilde{r}_2(t)}\left(\frac{1}{\tilde{r}_1(t)}\left(\frac{x(t)}{\tilde{r}_0(t)}\right)'\right)'\right)' = \frac{1}{v(t)}\left(\frac{r(t)}{v^{-2}(t)}\left(\frac{x'(t)}{v(t)}\right)'\right)'$$

because from (1.22) and (1.23) it follows that

$$\frac{1}{v(t)}\left(\frac{r(t)}{v^{-2}(t)}\left(\frac{x'(t)}{v(t)}\right)'\right)' = \left(r(t)x''(t)\right)' + q(t)x(t).$$

Now,

$$\frac{x(t)}{r_0(t)} = x(t)\left(\int_t^\infty \tilde{r}_1(s)\,ds\right)^{-1}.$$

Differentiating both sides, we get

$$\frac{1}{r_1(t)}\left(\frac{x(t)}{r_0(t)}\right)' = \frac{x'(t)}{\tilde{r}_1(t)}\left(\int_t^\infty \tilde{r}_1(s)\,ds\right) + x(t).$$

Again differentiation of the above equation with respect to t yields

$$\frac{1}{r_2(t)}\left(\frac{1}{r_1(t)}\left(\frac{x(t)}{r_0(t)}\right)'\right)' = \frac{1}{\tilde{r}_2(t)}\left(\frac{x'(t)}{\tilde{r}_1(t)}\right)'.$$

Further differentiation gives

$$\frac{1}{r_3(t)}\left(\frac{1}{r_2(t)}\left(\frac{1}{r_1(t)}\left(\frac{x(t)}{r_0(t)}\right)'\right)'\right)' = \frac{1}{\tilde{r}_3(t)}\left(\frac{1}{\tilde{r}_2(t)}\left(\frac{1}{\tilde{r}_1(t)}\left(\frac{x(t)}{\tilde{r}_0(t)}\right)'\right)'\right)'$$

because $r_3(t) = \tilde{r}_3(t)$ and $\tilde{r}_0(t) = 1$. On the other hand, $\frac{x(t)}{\tilde{r}_0(t)} = x(t)$ implies that

$$\frac{1}{\tilde{r}_1(t)}\left(\frac{x(t)}{\tilde{r}_0(t)}\right)' = \frac{x'(t)}{v(t)}\left(\int_t^\infty \frac{v^{-2}(s)}{r(s)}\,ds\right)^{-1}.$$

Differentiation of this gives

$$\frac{1}{\tilde{r}_2(t)}\left(\frac{1}{\tilde{r}_1(t)}\left(\frac{x(t)}{\tilde{r}_0(t)}\right)'\right)' = \frac{r(t)}{v^{-2}(t)}\left(\frac{x'(t)}{v(t)}\right)'\left(\int_t^\infty \frac{v^{-2}(s)}{r(s)}\,ds\right) + \frac{x'(t)}{v(t)}.$$

Further differentiation yields

$$\begin{aligned}\left(\frac{1}{\tilde{r}_2(t)}\left(\frac{1}{\tilde{r}_1(t)}\left(\frac{x(t)}{\tilde{r}_0(t)}\right)'\right)'\right)' &= \left(\frac{r(t)}{v^{-2}(t)}\left(\frac{x'(t)}{v(t)}\right)'\right)'\left(\int_t^\infty \frac{v^{-2}(s)}{r(s)}\,ds\right)\\ &\quad - \frac{r(t)}{v^{-2}(t)}\left(\frac{x'(t)}{v(t)}\right)'\frac{v^{-2}(t)}{r(t)} + \left(\frac{x'(t)}{v(t)}\right)'\\ &= \left(\frac{r(t)}{v^{-2}(t)}\left(\frac{x'(t)}{v(t)}\right)'\right)'\left(\int_t^\infty \frac{v^{-2}(s)}{r(s)}\,ds\right).\end{aligned}$$

Hence

$$\frac{1}{\tilde{r}_3(t)}\left(\frac{1}{\tilde{r}_2(t)}\left(\frac{1}{\tilde{r}_1(t)}\left(\frac{x(t)}{\tilde{r}_0(t)}\right)'\right)'\right)' = \frac{1}{v(t)}\left(\frac{r(t)}{v^{-2}(t)}\left(\frac{x'(t)}{v(t)}\right)'\right)'.$$

Lemma 1.4.3 *Let $p(t) \le 0$. If $x(t)$ is a nonoscillatory solution of the canonical equation* (1.19) *with* (1.21), *then either*

$$\operatorname{sgn} L_0x(t) = \operatorname{sgn} L_1x(t) = \operatorname{sgn} L_2x(t) = \operatorname{sgn} L_3x(t)$$

or

$$\operatorname{sgn} L_0x(t) = \operatorname{sgn} L_1x(t) = \operatorname{sgn} L_3x(t) \ne \operatorname{sgn} L_2x(t)$$

for $t \ge t_0 \ge \sigma$.

Proof Without any loss of generality, we may assume that $x(t) > 0$ for $t \ge t_0 \ge \sigma$. Hence $L_0x(t) > 0$ and $L_3x(t) > 0$ for $t \ge t_0$. Since $L_3x(t) > 0$ for $t \ge t_0$, either $L_2x(t) > 0$ or < 0 for $t \ge t_1 \ge t_0$. Clearly, $L_2x(t) > 0$ for $t \ge t_1$ implies that $L_1x(t) > 0$ for large t, because of the assumption (1.21). Further, $L_2x(t) < 0$ for $t \ge t_1$ implies that $L_1x(t) > 0$ or < 0 for large t. However, $L_1x(t) < 0$ for large t and $\int_\sigma^\infty r_1(t)\,dt = \infty$ imply that $L_0x(t) < 0$ for large t. This contradiction completes the proof of the lemma. □

Remark 1.4.3 Lemma 1.4.3 is valid for the third-order canonical delay differential equation (1.28) with (1.21).

1.5 Some Basic Results

Definition 1.5.1 Equation (LH1) is of Class I or C_I, if for any of its solution $x(t)$ for which $x(t_0)=0$, $x'(t_0)=0$ and $x''(t_0)>0$ ($\sigma<t_0<\infty$) satisfies $x(t)>0$ for $t\in[\sigma,t_0)$. Equation (LH1) is of Class II or C_{II}, if for any of its solution $x(t)$ for which $x(t_0)=0$, $x'(t_0)=0$ and $x''(t_0)>0$, $t_0\geq\sigma$ satisfies $x(t)>0$ for $t>t_0$.

In the following, we list some results for ready reference. In what follows, we assume that a, b and $c\in C([\sigma,\infty),R)$.

Lemma 1.5.1 *Suppose that $a(t)\leq 0$ and $b(t)\geq 0$, and the second-order linear differential equation*

$$u''+a(t)u+b(t)u=0 \tag{1.29}$$

is nonoscillatory. If $z(t)$ is a nonoscillatory solution of (1.29), then $z(t)z'(t)>0$ for $t\geq t_0\geq\sigma$. If $v\in C^1([\alpha,\beta],R)$ such that $v(\alpha)=0$ and $v(t)\neq 0$ on (α,β), $\alpha>t_0$, then

$$\int_\alpha^\beta r(t)\big(v'(t)\big)^2\,dt\geq\int_\alpha^\beta q(t)v^2(t)\,dt,$$

where $r(t)=e^{\int_\sigma^t a(s)\,ds}$ and $q(t)=r(t)b(t)$.

Proof Clearly (1.29) may be rewritten as

$$\big(r(t)u'\big)'+q(t)u=0. \tag{1.30}$$

Let $z(t)$ be a nonoscillatory solution of (1.30). Without any loss of generality, we may assume that $z(t)>0$ for $t\geq T\geq\sigma$. From (1.30), it follows that $(r(t)z'(t))'<0$ for $t\geq T$. Hence $z'(t)$ is nonoscillatory. Thus there exists a real $t_0\geq T$ such that $z'(t)>0$ or <0 for large $t\geq t_0$. Suppose that $z'(t)<0$ for $t\geq t_0$. Since $a(t)\leq 0$, $r(t)$ is a decreasing function of t. Hence, for $t\geq t_0$, we have $r(t)\leq r(t_0)$. Integrating the inequality $(r(t)z'(t))'<0$ from t_0 to t, we obtain

$$z(t)<r(t_0)z'(t_0)\int_{t_0}^t\frac{1}{r(s)}\,ds,$$

which yields that $z(t)<0$ for large t, a contradiction. Hence $z(t)z'(t)>0$ for $t\geq t_0$. For $t_0<\alpha<t<\beta$,

$$\begin{aligned}
0&<\int_\alpha^\beta\left[r^{\frac12}(t)v'(t)-r^{\frac12}(t)\frac{z'(t)}{z(t)}v(t)\right]^2dt\\
&=\int_\alpha^\beta r(t)\big(v'(t)\big)^2\,dt+\int_\alpha^\beta r(t)\frac{(z'(t))^2}{z^2(t)}v^2(t)\,dt-2\int_\alpha^\beta r(t)\frac{z'(t)}{z(t)}v'(t)v(t)\,dt\\
&=\int_\alpha^\beta r(t)\big(v'(t)\big)^2\,dt+\int_\alpha^\beta r(t)\frac{(z'(t))^2}{z^2(t)}v^2(t)\,dt-\int_\alpha^\beta r(t)\frac{z'(t)}{z(t)}\big(v^2(t)\big)'\,dt
\end{aligned}$$

$$
\begin{aligned}
&= \int_\alpha^\beta r(t)\big(v'(t)\big)^2\,dt + \int_\alpha^\beta r(t)\frac{(z'(t))^2}{z^2(t)}v^2(t)\,dt \\
&\quad - \left[\frac{r(t)z'(t)}{z(t)}v^2(t)\right]_\alpha^\beta + \int_\alpha^\beta \left(\frac{r(t)z'(t)}{z(t)}\right)' v^2(t)\,dt \\
&= \int_\alpha^\beta r(t)\big(v'(t)\big)^2\,dt + \int_\alpha^\beta r(t)\frac{(z'(t))^2}{z^2(t)}v^2(t)\,dt - r(\beta)\frac{z'(\beta)}{z(\beta)}v^2(\beta) \\
&\quad + \int_\alpha^\beta \frac{(r(t)z'(t))'}{z(t)}v^2(t)\,dt - \int_\alpha^\beta r(t)\frac{(z'(t))^2}{z^2(t)}v^2(t)\,dt \\
&< \int_\alpha^\beta r(t)\big(v'(t)\big)^2\,dt + \int_\alpha^\beta \frac{(r(t)z'(t))'}{z(t)}v^2(t)\,dt \\
&= \int_\alpha^\beta r(t)\big(v'(t)\big)^2\,dt - \int_\alpha^\beta q(t)v^2(t)\,dt.
\end{aligned}
$$

Thus the lemma is proved. □

Remark 1.5.1 One important phenomenon as regards the nonoscillation of the linear second-order differential equation (1.29) is that if there exists a function $w \in C^2([\sigma,\infty))$ such that $w(t) > 0$ and $w''(t) + a(t)w'(t) + b(t)w(t) \le 0$ on $[\sigma,\infty)$, then the second-order linear differential equation (1.29) is nonoscillatory on $[\sigma,\infty)$.

Lemma 1.5.2 *Let $a(t) \le 0, b(t) \ge 0, c(t) < 0$ and $2c(t) - a(t)b(t) - b'(t) \le 0$. If* (LH1) *is nonoscillatory, then it admits a nonoscillatory solution $x(t)$ on $[\sigma,\infty)$ such that $x(t)x'(t) > 0$ for $t \ge t_0 > \sigma$ and $F[x(t)] > 0$ for $t \ge \sigma$, where*

$$F\big[x(t)\big] = r(t)\big(x'(t)\big)^2 - 2r(t)x(t)x''(t) - q(t)x^2(t).$$

Proof Clearly, $F[x(t)]$ may be written as

$$
\begin{aligned}
F\big[x(t)\big] &= F\big[x(\sigma)\big] + \int_\sigma^t r'(s)\big(x'(s)\big)^2\,ds + \int_\sigma^t \big(2p(s) - q'(s)\big)x^2(s)\,ds \\
&= F\big[x(\sigma)\big] + \int_\sigma^t a(s)r(s)\big(x'(s)\big)^2\,ds \\
&\quad + \int_\sigma^t r(s)\big(2c(s) - a(s)b(s) - b'(s)\big)x^2(s)\,ds. \qquad (1.31)
\end{aligned}
$$

Hence $F[x(t)]$ is a decreasing function of t for any solution $x(t)$ of (LH2).

Let $\{u_1(t), u_2(t), u_3(t)\}$ be a solution basis for (LH2), $t \in [\sigma,\infty)$. Define a sequence $\{x_n(t)\}$ of solutions of (LH2) such that

$$x_n(n) = 0 = x_n'(n) \quad \text{and} \quad x_n''(n) > 0$$

for $n \geq \sigma$. Then, one may write

$$x_n(t) = c_{1_n} u_1(t) + c_{2_n} u_2(t) + c_{3_n} u_3(t),$$

where $c_{1_n}^2 + c_{2_n}^2 + c_{3_n}^2 = 1$. Clearly, the bounded sequence $\{c_{i_n}\}, i = 1, 2, 3$, has a convergent subsequence, which we may call $\{c_{i_{n_j}}\}$. Let $c_{i_{n_j}} \to c_i, i = 1, 2, 3$ as $n_j \to \infty$. Then $c_1^2 + c_2^2 + c_3^2 = 1$. Setting $x(t) = c_1 u_1(t) + c_2 u_2(t) + c_3 u_3(t)$, we see that $x(t)$ is a nontrivial solution of (LH2) on $[\sigma, \infty)$. Then, $x(t)$ is nonoscillatory. Without any loss of generality, we may assume that $x(t) > 0$ for $t \geq T \geq \sigma$. Since $F[x_n(n)] = 0$ and $F[x_n(t)]$ is a decreasing function of t, for $t < n$, we have $0 = F[x_n(n)] < F[x_n(t)]$. Since the sequence $\{x_n(t)\}$ converges uniformly to $x(t)$ on any compact subinterval of $[\sigma, \infty)$, $F[x(t)] \geq 0$ for $t \in [\sigma, \infty)$. If $F[x(t_1)] = 0$ for some $t_1 \geq \sigma$, then $F[x(t)] < F[x(t_1)] = 0$ for $t \geq t_1$, a contradiction. Hence $F[x(t)] > 0$ for $t \in [\sigma, \infty)$. Let t_2 and t_3, $(T \leq t_2 < t_3)$ be two consecutive zeros of $x'(t)$. Then $x''(t_2) < 0$ and $x''(t_3) < 0$, which is impossible for any two consecutive zeros of $x'(t)$. Hence $x'(t)$ has at most one zero in $[T, \infty)$. Thus, there exists a $t_0 \geq T$ such that $x'(t) > 0$ or $x'(t) < 0$ for $t \geq t_0$. If $x'(t) < 0$ for $t \geq t_0$, then from (LH2), we have $(r(t)x''(t))' > 0$ for $t \geq t_0$ and hence $x''(t)$ is nonoscillatory. Clearly, $x''(t) < 0$ for large t implies that $x(t) < 0$ for large t, a contradiction. If $x''(t) > 0$ for large t, then from (LH1), we see that $x'''(t) > 0$ for large t. This, in turn, implies that $x'(t) > 0$ for large t, a contradiction. Hence $x'(t) > 0$ for $t \geq t_0$. The lemma is proved. □

Lemma 1.5.3 *Let $b(t) \leq 0$ and $c(t) > 0$. If $x(t)$ is a nonoscillatory solution of* (LH1), *then there exists a $t_0 \in [\sigma, \infty)$ such that either*

(i) $x(t)x'(t) < 0$ *or*
(ii) $x(t)x'(t) > 0$

holds for $t \geq t_0$. Further, if (i) *holds, then*

$$x(t)x'(t)x''(t) \neq 0,$$
$$\operatorname{sgn} x(t) = \operatorname{sgn} x''(t) \neq \operatorname{sgn} x'(t), \quad t \geq \sigma$$

and

$$\lim_{t\to\infty} x'(t) = \lim_{t\to\infty} x''(t) = 0, \qquad \lim_{t\to\infty} x(t) = \lambda \neq \pm\infty.$$

Lemma 1.5.4 *Let $b(t) \leq 0$ and $c(t) > 0$. A necessary and sufficient condition for* (LH1) *to have an oscillatory solution is that for any nonoscillatory solution $u(t)$ of* (LH1), *the following properties hold*:

$$u(t)u'(t)u''(t) \neq 0,$$
$$\operatorname{sgn} u(t) = \operatorname{sgn} u''(t) \neq \operatorname{sgn} u'(t), \quad t \geq \sigma$$

and

$$\lim_{t\to\infty} u'(t) = \lim_{t\to\infty} u''(t) = 0, \qquad \lim_{t\to\infty} u(t) = \lambda \neq \pm\infty.$$

Lemma 1.5.5 *Let $2c(t) - a(t)b(t) - b'(t) \geq 0$ and $a(t) \geq 0$. If* (LH1) *is nonoscillatory, then it admits a nonoscillatory solution $x(t)$ such that $x(t)x'(t) > 0$ for $t \geq t_0 > \sigma$.*

Remark 1.5.2 It can easily be proved that under the conditions $a(t) \geq 0$ and $2c(t) - a(t)b(t) - b'(t) \geq 0$, any nontrivial solution of (LH1) with a double zero at $t_0 \in (\sigma, \infty)$ does not have any simple zero to the left of t_0. In other words, under the above given condition, if a solution of (LH1) has a simple zero at $t_0 \in [\sigma, \infty)$, then any zero to the right of t_0 is simple. Further, if we proceed as in the lines of Theorems 1 and 2 in [52], it can be proved that Eq. (LH1) has at least one solution with no zeros in $[\sigma, \infty)$, and if (LH1) is oscillatory, then any nontrivial solution with one zero is oscillatory. This may be treated as a simple generalisation of a result due to Hanan [25], that is, Lemma 1.5.10 given below.

Lemma 1.5.6 *Let $a(t) \geq 0, 2c(t) - a(t)b(t) - b'(t) \geq 0$ and* (LH1) *be oscillatory. Then a nontrivial solution $x(t)$ of* (LH1) *is nonoscillatory, if and only if $F[x(t)] < 0$ for $t \geq \sigma$, where $F[x(t)]$ is same as in Lemma* 1.5.2.

Proof Let (LH1) be oscillatory and let x be a nontrivial solution of (LH1) such that $F[x(t_0)] \geq 0$ for some $t_0 \in [\sigma, \infty)$. We shall show that $x(t)$ is oscillatory. Suppose that $x(t)$ is nonoscillatory. Then by Remark 1.5.2, $x(t) \neq 0$ for $t \geq \sigma$. Let $y(t)$ be a solution of (LH1), satisfying the initial condition

$$y(t_0) = 0, \qquad y'(t_0) = x(t_0), \qquad y''(t_0) = x'(t_0).$$

Then again by Remark 1.5.2, $y(t)$ is a nontrivial oscillatory solution of (LH1), and all of its zeros to the right of t_0 are simple. Therefore, the function $W(x, y) = (\frac{y}{x})'x^2 = xy' - yx'$ is oscillatory on (t_0, ∞). Let $W(x, y)(t_1) = 0$, $t_1 > t_0$. Then there are constants c_1 and c_2 with $c_1^2 + c_2^2 \neq 0$ such that

$$c_1 x(t_1) + c_2 y(t_1) = 0,$$
$$c_1 x'(t_1) + c_2 y'(t_1) = 0.$$

Clearly, $c_2 \neq 0$, and the solution $z(t) = c_1 x(t) + c_2 y(t)$ of (LH1) satisfies the property $F[z(t_1)] > 0$. On the other hand, the integral identity (1.31) implies that $0 = F[z(t_1)] \geq F[z(t_0)] > 0$, a contradiction. This proves that $x(t)$ has at least one zero in $[\sigma, \infty)$ and hence again by Remark 1.5.2, $x(t)$ is oscillatory.

Now, suppose that a solution $x(t)$ of (LH1) satisfies $F[x(t)] < 0$ for $t \geq \sigma$. Then clearly, $x(t) \neq 0$ for all $t \geq \sigma$, and hence $x(t)$ is nonoscillatory. The lemma is proved. □

Lemma 1.5.7 *Let $a(t) \geq 0, 2c(t) - a(t)b(t) - b'(t) \geq 0$ and suppose that the second-order linear differential equation*

$$v'' + \frac{1}{4}b(t)v = 0 \tag{1.32}$$

is oscillatory. Then (LH1) *is oscillatory.*

Proof Suppose that (LH1) is nonoscillatory and let $x(t)$ be a solution of (LH1) such that $F[x(t_0)] > 0$ at $t_0 \in [\sigma, \infty)$. From the integral identity (1.31), it follows that $F[x(t)] > 0$ for $t \geq t_0$. Since $x(t)$ is nonoscillatory, there exists a $t_1 > t_0$ such that $x(t) \neq 0$ for $t > t_1$. Since $F[x] = F[-x]$, we can assume that $x(t) > 0$ for $t > t_1$. Let $X(t) = \sqrt{x(t)}$ for $t > t_1$. Then

$$x(t) = X^2(t), \qquad x'(t) = 2X(t)X'(t), \qquad x''(t) = 2X(t)X''(t) + 2X'^2(t)$$

and

$$0 < F\big[x(t)\big] = F\big[X^2(t)\big] = -2X^3(t)\Big[X''(t) + \frac{1}{4}b(t)X(t)\Big]e^{\int_{t_1}^t a(s)\,ds}$$

which implies that

$$X''(t) + \frac{1}{4}b(t)X(t) < 0$$

for $t > t_1$. Then by Remark 1.5.1, (1.32) is nonoscillatory. This contradiction proves the lemma. □

Remark 1.5.3 Let the conditions of Lemma 1.5.7 hold and $b(t) \geq 0$. If (LH1) is nonoscillatory and $x(t)$ is a nonoscillatory solution of (LH1) with $F[x(t_0)] > 0$ for some $t_0 \geq \sigma$, then there exists a $t_1 > t_0$ such that $x(t)x'(t) > 0$ for $t > t_1$.

Lemma 1.5.8 *Let $a(t) \geq 0, 2c(t) - a(t)b(t) - b'(t) \geq 0$ and suppose that the second-order linear differential equation* (1.29) *is oscillatory. Then* (LH1) *is oscillatory.*

Proof On the contrary, suppose that (LH1) is nonoscillatory. Then by Lemma 1.5.5, there exists a solution $x(t) > 0$ of (LH1) such that $x(t)x'(t) > 0$ for $t \geq t_1 \geq \sigma$. Let $x'(t) = w(t)$. Then it follows from (LH1) that

$$w''(t) + a(t)w'(t) + b(t)w(t) = -c(t)x(t) \leq 0$$

for $t \geq t_1$. Then by Remark 1.5.1, (1.29) is nonoscillatory. This contradiction proves that (LH1) is oscillatory. □

Lemma 1.5.9 *Let $u(t)$ be a solution of* (LH1) *with $u(c) = u(d) = 0, c < d, c \in [\sigma, \infty)$ with $u(t) \neq 0$ for $t \in (c, d)$. Then any nontrivial solution of* (LH1) *with a double zero at $s \in [c, d)$ has no zero in (s, d).*

Lemma 1.5.10 *If* (LH1) *is of type C_I and oscillatory, then any solution of this equation which vanishes at least once is oscillatory.*

Lemma 1.5.11 *A third-order differential equation which is of type C_I (or C_{II}) is oscillatory, if and only if its adjoint is oscillatory.*

Lemma 1.5.12 *A third-order differential equation is of type C_I, if and only if its adjoint is of C_{II}.*

Lemma 1.5.13 *Let $u(t)$ and $v(t)$ be of C^1 in (a, b), and let $v(t)$ be of constant sign in this interval. If $t = \alpha$ and $t = \beta$ $(a < \alpha < \beta < b)$ are the consecutive zeros of $u(t)$, then there exists a constant μ such that the function $u(t) - \mu v(t)$ has a double zero in (α, β).*

Lemma 1.5.14 *If $r(t) > 0$, $f(t) \geq 0$ and $q(t) \geq 0$ for $t \in [\sigma, \infty)$, then all solutions of the second-order linear differential equation*

$$\big(r(t)z'(t)\big)' - q(t)z(t) = f(t)$$

are nonoscillatory.

Lemma 1.5.15 *Let $z \in C^1([\sigma, \infty), R)$, and $\sigma \in R$. If $z \in L^2[\sigma, \infty)$ and $z'(t)$ is bounded, then $z(t) \to 0$ as $t \to \infty$.*

Proof Clearly the statement holds for any nonoscillatory function on $[\sigma, \infty)$. So, let $z(t)$ be an oscillatory function on $[\sigma, \infty)$. Since $z \in L^2[\sigma, \infty)$, $\int_\sigma^\infty z^2(t)\,dt < \infty$. This, in turn, implies that $\liminf_{t\to\infty} z(t) = 0$. To complete the proof of the theorem, it remains to show that $\limsup_{t\to\infty} z(t) = 0$. Suppose that $\limsup_{t\to\infty} z(t) \neq 0$. Then there exist an $\epsilon > 0$ and a sequence $\{t_n\}_{n=1}^\infty$ such that $t_n \to \infty$ as $n \to \infty$ and $z(t_n) > 2\epsilon$, for large n. Since $\liminf_{t\to\infty} z(t) = 0$, there exists a sequence $\{t_n^*\}_{n=1}^\infty$ such that $t_n^* \to \infty$ as $n \to \infty$ and $z(t_n^*) \to 0$ as $n \to \infty$. Thus $z(t_n^*) < \epsilon$ for $n \geq N_\epsilon$ for some positive integer N_ϵ. It is possible to extract sequences $\{s_n\}_{n=1}^\infty$ and $\{\sigma_n\}_{n=1}^\infty$ such that $\sigma_n < s_n < \sigma_{n+1}$, $\sigma_n \to \infty$ as $n \to \infty$, $z(s_n) > 2\epsilon$ and $z(\sigma_n) < \epsilon$ for $n \geq N_\epsilon$. Since $z(t)$ is continuous, there exist sequences $\{\tau_n\}_{n=1}^\infty$ and $\{\tau_n^*\}_{n=1}^\infty$ such that $\sigma_n < \tau_n^* < \tau_n < s_n$ with $z(\tau_n^*) = \epsilon$ and $z(\tau_n) = 2\epsilon$. It is clear that the intervals (τ_n^*, τ_n) are disjoint. Then

$$\sum_{n=1}^\infty (\tau_n - \tau_n^*)\epsilon^2 \leq \sum_{n=1}^\infty \int_{\tau_n^*}^{\tau_n} z^2(t)\,dt \leq \int_\sigma^\infty z^2(t)\,dt < \infty,$$

which implies that $\lim_{n\to\infty}(\tau_n - \tau_n^*) = 0$. By the mean-value theorem, the calculation

$$z'(\xi_n) = \frac{z(\tau_n) - z(\tau_n^*)}{\tau_n - \tau_n^*}, \qquad \tau_n^* < \xi_n < \tau_n,$$

implies that $z'(\xi_n) \to \infty$ as $n \to \infty$, which contradicts the hypothesis of the lemma. Hence our claim holds, that is, $\limsup_{t\to\infty} z(t) = 0$. Consequently, $z(t) \to 0$ as $t \to \infty$, and the lemma is proved. □

Lemma 1.5.16 *Let t_0 and T be such that $T \geq t_0 \geq \sigma$ and $g(t) \geq t_0$ for $t \geq T$. Assume that $u : [t_0, \infty) \to (0, \infty)$, $\omega^* : [T, \infty) \times (0, \infty) \to (0, \infty)$, $H : [T, \infty) \times$*

$[0,\infty)\to[0,\infty)$ *and* $\phi,\psi:\Delta\to[0,\infty)$ *are continuous, where* $\Delta=\{(t,s);t\ge s\ge T\}$, *and each of* ω^* *and* H *is nondecreasing in the second variable. Suppose that*

$$\int_T^\infty \psi^*(t)H\big(t,u\big(g(t)\big)\big)\,dt<\infty$$

and

$$u(t)\ge\omega^*\big(t,u(t)\big)+\int_T^t\phi(t,s)\left(\int_s^\infty\psi(\alpha,s)H\big(\alpha,u\big(g(\alpha)\big)\big)\,d\alpha\right)ds \tag{1.33}$$

for $t\ge T$, *where* $\psi^*(t)=\max\{\psi(s,t);s\in[T,t]\}$. *Then the integral equation*

$$v(t)=\omega^*\big(t,v(t)\big)+\int_T^t\phi(t,s)\left(\int_s^\infty\psi(\alpha,s)H\big(\alpha,v\big(g(\alpha)\big)\big)\,d\alpha\right)ds,$$

$t\ge T$, *has a solution* $v\in C([t_0,\infty),(0,\infty))$, *satisfying*

$$\omega^*\big(t,v(t)\big)\le v(t)\le u(t),\quad t\ge T.$$

Lemma 1.5.17 *If in Lemma* 1.5.16, *the inequality* (1.33) *is replaced by*

$$u(t)\ge w(t)+\int_t^\infty\psi(t,s)H\big(s,u\big(g(s)\big)\big)\,ds$$

for $t\ge T$, *where* $w:[T,\infty)\to(0,\infty)$ *is continuous, then the integral equation*

$$v(t)=w(t)+\int_t^\infty\psi(t,s)H\big(s,v\big(g(s)\big)\big)\,ds,\quad t\ge t_0$$

has a solution $v\in C([t_0,\infty),(0,\infty))$ *satisfying*

$$w(t)\le v(t)\le u(t),\quad t\ge T.$$

Lemma 1.5.18 *Let* t_0 *and* T *be such that* $T\ge t_0$ *and* $\tau(t)\ge t_0$ *for* $t\ge T$, $u:[t_0,\infty)\to(0,\infty)$, $\omega:[T,\infty)\to(0,\infty)$, $H:[T,\infty)\times[0,\infty)\to[0,\infty)$ *and* $\phi:\Delta\to[0,\infty)$ *are continuous, where* $\Delta=\{(t,s);t\ge s\ge t_0\}$ *and* H *is nondecreasing in the second variable. If* $u(t)$ *satisfies*

$$u(t)\ge\omega(t)+\int_T^t\phi(t,s)H\big(s,u\big(\tau(s)\big)\big)\,ds,\quad t\ge T,$$

then the integral equation

$$v(t)=\omega(t)+\int_T^t\phi(t,s)H\big(s,v\big(\tau(s)\big)\big)\,ds,\quad t\ge T,$$

has a solution $v\in C([t_0,\infty),(0,\infty))$, *satisfying*

$$\omega(t)\le v(t)\le u(t).$$

The above lemma holds for $\tau(t) \equiv t$.

Lemma 1.5.19 *Suppose that $a(t) \leq 0$, $b(t) \leq 0$ and $c(t) < 0$. The following statements are equivalent*:

(A) *There exists an oscillatory solution of* (LH1).
(B) *If $u(t)$ is a nonoscillatory solution of* (LH1), *then there exists a $t_0 \geq \sigma$ such that*

$$u(t)u'(t)u''(t) \neq 0 \quad \textit{for } t \geq t_0$$

and

$$\operatorname{sgn} u(t) = \operatorname{sgn} u'(t) = \operatorname{sgn} u''(t) \quad \textit{for } t \geq t_0.$$

Lemma 1.5.20 *If $g(t) < t$, $\lim_{t\to\infty} g(t) = \infty$, and either*

$$\liminf_{t\to\infty} \int_{g(t)}^{t} p(s)\, ds > \frac{1}{e}$$

or

$$\limsup_{t\to\infty} \int_{g(t)}^{t} p(s)\, ds > 1$$

holds, then

(i) *the delay differential inequality*

$$x'(t) + p(t)x(t-\tau) \leq 0$$

has no eventually positive solutions;
(ii) *the delay differential inequality*

$$x'(t) + p(t)x(t-\tau) \geq 0$$

has no eventually negative solutions, and
(iii) *all solutions of the delay differential equation*

$$x'(t) + p(t)x(t-\tau) = 0$$

are oscillatory.

Consider the nth order differential equation

$$x^{(n)}(t) + p(t)x\big(g(t)\big) = 0 \tag{1.34}$$

where $p \in C([\sigma,\infty),[0,\infty))$, $\sigma \in R$ and $g \in C([\sigma,\infty),R)$ be such that $0 < g(t) \leq t$ and $g(t) \to \infty$ as $t \to \infty$. Let $x(t)$ be a positive solution of (1.34) for $t \geq t_0 \geq \sigma$. Then there exists a $t_1 > t_0$ such that $x(g(t)) > 0$ for $t \geq t_1$. So $x^{(n)}(t) < 0$ for $t \geq t_1$.

Lemma 1.5.21 *If $x(t)$ is a positive solution of (1.34), then there exists an integer $l, 0 \leq l \leq n-1$ such that $n+l$ is odd and*

$$x^{(i)}(t) > 0, \quad i = 0, 1, 2, \dots, l,$$
$$(-1)^{i+l} x^{(i)}(t) > 0, \quad i = l+1, l+2, \dots, n$$

for large t. Again, for $l \in \{1, 2, \dots, n-1\}$, $n+l$ odd, the following inequality holds for large t, say for $t \geq t_0 > \sigma$:

$$\left|x(t)\right| \geq \frac{(t-t_0)^{n-1}}{(n-1)(n-2)\cdots(n-l)} \left|x^{(n-1)}(2^{n-l-1}t)\right|, \quad t \geq t_0.$$

Lemma 1.5.22 (Sturm's Comparison Theorem) *Suppose that $r(t) = R(t)$ and $q(t) < Q(t)$ in the bounded interval $\alpha < t < \beta$. If there exists a nontrivial real solution u of*

$$lu = \frac{d}{dt}\left(r(t)\frac{du}{dt}\right) + q(t)u = 0 \tag{1.35}$$

such that $u(\alpha) = u(\beta) = 0$, then every real solution of

$$Lv = \frac{d}{dt}\left(R(t)\frac{dv}{dt}\right) + Q(t)v = 0 \tag{1.36}$$

has at least one zero in (α, β).

Sturm obtained Lemma 1.5.22 in 1836 [58], but it was not until 1909 that Picone [60] disposed the case $r(t) \neq R(t)$. The modification due to Picone is as follows:

Lemma 1.5.23 (Sturm-Picone Comparison Theorem) *Suppose that $r(t) > R(t)$ and $q(t) < Q(t)$ in the bounded interval $\alpha < t < \beta$. Then the conclusion of Lemma 1.5.22 is valid.*

Lemma 1.5.24 (Sturm Separation Theorem) *The zeros of linearly independent oscillatory solutions of (1.35) separate each other.*

Lemma 1.5.25 (Leighton's Oscillation Criteria) *Let $r(t), r'(t)$ and $q(t)$ be continuous functions and $r(t) > 0$ for $t \geq \sigma$. Then (1.35) is oscillatory if*

$$\int_\sigma^\infty \frac{dt}{r(t)} = \infty \quad \textit{and} \quad \int_\sigma^\infty q(t)\,dt = \infty.$$

Let $r(t) \equiv 1$. An immediate consequence of Sturm comparison theorem is that (1.35) is nonoscillatory if $q(t) \leq 0$, but it may have this property also if $q(t) > 0$; in fact, the Euler equation

$$u''(t) + \gamma(t+1)^{-2}u(t) = 0, \quad \gamma = \text{constant}$$

with the general solution

$$u(t) = \begin{cases} k_1(t+1)^p + k_2(t+1)^{1-p}, & p = \frac{1}{2} + \frac{1}{2}(1-4\gamma)^{1/2},\ \gamma \neq \frac{1}{4}; \\ (t+1)^{1/2}[k_1 + k_2 \ln(t+1)], & \gamma = \frac{1}{4} \end{cases}$$

is nonoscillatory if $\gamma \leq \frac{1}{4}$ (and oscillatory if $\gamma > \frac{1}{4}$).

Lemma 1.5.26 (Kneser's Criterion) *Let* $r(t) \equiv 1$ *in* (1.35) *and* $q(t) > 0$. *Then* (1.35) *is oscillatory if*

$$\lim_{t\to\infty} t^2 q(t) > \frac{1}{4}$$

and nonoscillatory if

$$\lim_{t\to\infty} t^2 q(t) < \frac{1}{4}.$$

Later, Hille made a slight extension of the above Kneser's criterion which is given as

Lemma 1.5.27 (Hille–Kneser Theorem) *Let* $r(t) \equiv 1$. *Then* (1.35) *is oscillatory if* $w^* = \limsup_{t\to\infty} t^2 q(t) > \frac{1}{4}$ *and nonoscillatory if* $w_* = \liminf_{t\to\infty} t^2 q(t) < \frac{1}{4}$, *but it can be either oscillatory or nonoscillatory if either* w^* *or* w_* *equals* $\frac{1}{4}$.

Lemma 1.5.28 *If* c *is a nonnegative continuous function in* $[\sigma, \infty)$, *then the differential equation*

$$u''(t) - c(t)u(t) = 0 \tag{1.37}$$

has a positive non-increasing solution in $[\sigma, \infty)$.

Proof If u is a solution of (1.37) satisfying $u(\sigma) = u'(\sigma) = 1$, then $u'(t)$ is never decreasing, $u'(t) \geq 1$ and $u(t) \geq t$ for all $t \geq \sigma$. Hence the function $u_1(t)$ defined by

$$u_1(t) = u(t) \int_t^\infty \frac{ds}{u^2(s)}, \quad \sigma \leq t < \infty$$

exists, and satisfies the differential equation (1.37), and is positive and non-increasing in $[\sigma, \infty)$. In fact, since $u'(t)$ is nondecreasing, then

$$\begin{aligned} u_1'(t) &= -\frac{1}{u(t)} + u'(t) \int_t^\infty \frac{ds}{u^2(s)} \\ &\leq -\frac{1}{u(t)} + \int_t^\infty \frac{u'(s)ds}{u^2(s)} = 0 \end{aligned}$$

and $u_1''(t) - c(t)u_1(t) = [u''(t) - c(t)u(t)] \int_t^\infty \frac{ds}{u^2(s)} = 0$. The lemma is proved. □

Lemma 1.5.29 *If* (1.35) *is nonoscillatory, then every solution of*

$$\frac{d}{dt}\left(r(t)\frac{du}{dt}\right)+q(t)u=f(t)$$

is nonoscillatory, where f *is a real-valued continuous function on* $[\sigma,\infty)$ *such that* $f(t)\geq 0$ *for* $t\geq\sigma$.

1.6 Some Useful Results from Analysis

In this section, we state some results from analysis which are needed in deriving further results in this book. Among them are the Banach contraction principle, Schauder–Tychonoff fixed-point theorem and Lebesgue dominated convergence theorem. These results are employed, when we want to establish that a certain ordinary differential equation or a delay differential equation has a nonoscillatory solution.

First, we introduce some terminology. A vector space X on which we define a norm $\|\cdot\|$ is called a *normed vector space*. A sequence $\{x_n\}$ of vectors in a normed vector space X is said to *converge to a vector* $x\in X$, if and only if the sequence of nonnegative numbers $\|x_n-x\|$ converges to zero. In such a case, we write

$$\lim_{n\to\infty} x_n=x$$

and we call x is the *limit of* x_n.

We say that $\{x_n\}$ is a *Cauchy sequence* of vectors in X, if for every $\epsilon>0$ there exists an $N=N_\epsilon$ such that for all $n,m\geq N_\epsilon$, we have $\|x_n-x_m\|<\epsilon$. Clearly, a convergent sequence is a Cauchy sequence, but the converse may not be true. A space X, where every Cauchy sequence of elements of X converges to an element of X is called a *complete space*. A Banach space is a *complete normed vector space*, that is, a Banach space is a vector space X with a norm in which every Cauchy sequence of vectors converges to a vector in X.

An example of a Banach space is the space X of all real, bounded, and continuous functions f on some interval $[T,\infty)$ under the sup norm $\|f\|=\sup_{T\leq x<\infty}|f(x)|$.

Let M be a subset of a Banach space X. We say that M is compact, if every infinite sequence of elements of M has a subsequence which converges to an element of M. We say that M is *relatively compact*, if every sequence in M contains a subsequence which converges to an element in X.

The *Arzela–Ascoli theorem* states that a sequence of functions $\{f_n\}$ from $[a,b]$ into R^m, whose terms are equicontinuous and uniformly bounded, contains a subsequence $\{f_{n_k}\}$ which converges uniformly on $[a,b]$. Moreover, a set of functions in $C[a,b]$ with $\|f\|=\sup_{x\in[a,b]}|f(x)|$ is relatively compact, if and only if it is uniformly bounded and equicontinuous on $[a,b]$. By the Arzela–Ascoli theorem, a subset M of the space $C([a,b],R)$ with the sup norm is relatively compact provided that M is equicontinuous and uniformly bounded. We say that a family M of continuous functions from $[a, b]$ to R is equicontinuous, if for every $\epsilon > 0$ there exists a $\delta = \delta_\epsilon > 0$ such that for all $t_1, t_2 \in [a, b]$ with $|t_1 - t_2| < \delta$ and for all $f \in M$ we have $|f(t_1) - f(t_2)| < \epsilon$. The family M is called *uniformly bounded*, if there exists a positive number λ such that $|f(t)| \leq \lambda$ for all $t \in [a, b]$ and for all $f \in M$.

For example, a family M of differentiable functions from $[a, b]$ into R with uniformly bounded derivatives is equicontinuous. Indeed, if $|f'(t)| \leq k$ for every $f \in M$, then by the mean-value theorem, we have

$$\big|f(t_1) - f(t_2)\big| \leq k|t_1 - t_2| \tag{1.38}$$

and equicontinuity follows with $\delta = \frac{\epsilon}{k}$. If the interval $[a, b]$ is finite then (1.38) implies that the family M is also uniformly bounded. Indeed, by taking $t_1 = t$ and $t_2 = b$, (1.38) implies that

$$\big|f(t)\big| - \big|f(b)\big| \leq \big|f(t) - f(b)\big| \leq k|t - b|$$

and so $|f(t)| \leq |f(b)| + k(b - a)$.

We now define Schauder fixed-point theorem.

Theorem 1.6.1 (Schauder's Fixed-Point Theorem) *Let M be a closed, convex and nonempty subset of a Banach space X. Let $T : M \to M$ be a continuous function such that TM is a relatively compact subset of X. Then T has at least one fixed point in M.*

A topology T on a linear space X is called *locally convex*, if every neighbourhood of the element 0 includes a convex neighbourhood of 0. A complete metrizable locally convex space is called a *Frechet space*.

Theorem 1.6.2 (Schauder–Tychonoff fixed-point theorem) *Let X be a locally convex topological linear space, C be a compact convex subset of X, and $f : C \to C$ be a continuous mapping with $f(C)$ compact. Then f has a fixed point in C.*

Let X be any set. A metric in X is a function $d : X \times X \to R$, having the following properties for all $x, y, z \in X$ such that

(i) $d(x, y) \geq 0$, and $d(x, y) = 0$, if and only if $x = y$;
(ii) $d(x, y) = d(y, x)$;
(iii) $d(x, z) \leq d(x, y) + d(y, z)$.

A metric space is a set X together with a given metric in X. A complete metric space is a metric space X in which every Cauchy sequence converges to a point in X. A Banach space is a normed linear space that is complete with respect to the metric $d(x, y) = \|x - y\|$.

Let (X, d) be a metric space and let $T : X \to X$. If there exists a number $L \in [0, 1)$ such that $d(Tx, Ty) \leq L\, d(x, y)$ for all $x, y \in X$. Then we say that T is a *contraction mapping* on X.

Theorem 1.6.3 (Banach Contraction Principle) *A contraction mapping on a complete metric space has exactly one fixed point.*

Theorem 1.6.4 (Lebesgue's Dominated Convergence Theorem) *Let $\{f_n\}$ be a sequence of functions such that $\lim_{n\to\infty} f_n(x) = f(x)$ almost everywhere in A and such that for every $n = 1, 2, \ldots, |f_n(x)| \le g(x)$ almost everywhere in A, where g is integrable on A. Then*

$$\lim_{n\to\infty} \int_A f_n(x)\, d\mu = \int_A f(x)\, d\mu.$$

1.7 Notes

The results in Sect. 1.2 are taken from Das [9, 10] and Padhi [46, 47]. Lazer [42] proved some results on the oscillation and nonoscillation of (1.5) with $a(t) \equiv 0$, where the following three cases were considered: (I) $b \ge 0$, $c < 0$, (II) $b \le 0$, $c < 0$ and (III) $b \ge 0$, $c > 0$. Propositions 1.2.1–1.2.8 improve the Lazer results. Elementary discussions on delay differential equations in Sect. 1.3 are taken from Ladde, Lakshmikantham and Zhang [41] and Padhi [46]. Section 1.4 is taken from Padhi [46]. Lemma 1.5.1 and Lemma 1.5.2 are the improved versions of some results due to Padhi [46]. Lemmas 1.5.3 and 1.5.4 are taken from Erbe [14] while Lemma 1.5.5 can be found in [22]. Lemmas 1.5.6–1.5.9 are taken from [23]. The proofs of Lemmas 1.5.10–1.5.12 are given in [25]. Lemma 1.5.13 is proved by Leighton and Nehari [43]. Parhi and Nayak [50] proved Lemma 1.5.14. Lemma 1.5.15 is taken from Singh [55] and its proof is taken from Padhi [46]. The proof of Lemma 1.5.16 is similar to the proof of Lemma 3 in [38], and the proof of Lemma 1.5.17 is similar to the proof of Lemma 4 in [38]. Lemma 1.5.18 has been proved in [39]. Lemma 1.5.19 is due to Ahmad and Lazer [3], whereas Lemma 1.5.20 is taken from the monograph due to Erbe, Kong and Zhang [16]. Lemma 1.5.21 is due to Kiguradze [35], whereas Lemmas 1.5.22–1.5.28 are taken from the monograph due to Swanson [60]. Lemma 1.5.29 is taken from Keener [34]. Elementary results on analysis given in Sect. 1.6 are taken from Gyori and Ladas [24], Royden [53] and Rudin [54].

References

1. M. R. Abdollahpour and A. Najati; *Hyers-Ulam stability of a differential equation of third order*, International Journal of Mathematical Analysis, 6(59) (2012), 2943–2948.
2. R. P. Agarwal, M. F. Aktas and A. Tiryaki; *On oscillation criteria for third order nonlinear delay differential equations*, Archivum Mathematicum, 45(1) (2009), 1–18.
3. S. Ahmad and A. C. Lazer; *On the oscillatory behavior of a class of linear third order differential equations*, Journal of Mathematical Analysis and Applications, 28(3) (1970), 681–689.
4. J. H. Barrett; *Third order differential equations with nonnegative coefficients*, Journal of Mathematical Analysis and Applications, 24(1) (1968), 212–224.

5. J. H. Barrett; *Oscillation theory of ordinary linear differential equations*, Advances in Mathematics, 3(4) (1969), 415–509. (Reprinted in *Lectures in Ordinary Differential Equations*, Edited by R. McKelvev, Academic Press, New York, 1970).
6. G. D. Birkhoff; *On the solutions of ordinary linear homogeneous differential equations of the third order*, Annals of Mathematics, 12(3) (1911), 103–127.
7. W. S. Burnside and A. W. Panton; *The Theory of Equations*, Third Edition, S. Chand and Company Ltd., New Delhi, 1979.
8. E. A. Cox and M. P. Mortell; *The evolution of resonant water-wave oscillations*, Journal of Fluid Mechanics, 162 (1986), 99–116.
9. P. Das; *Contributions to the Study of Qualitative Behaviour of Solutions of Third-Order Differential Equations*, Ph.D. Thesis, Berhampur University, India, 1991.
10. P. Das; *On oscillation of third order forced equations*, Journal of Mathematical Analysis and Applications, 196(2) (1995), 502–513.
11. Z. Došlá; *On oscillatory solutions of third-order linear differential equations*, Časopis pro Pěstování Matematiky, 114(1) (1989), 28–34.
12. R. D. Driver; *Ordinary and Delay Differential Equations*, Springer-Verlag, New York, 1977.
13. L. Erbe; *Disconjugacy conditions for the third order linear differential equations*, Canadian Mathematical Bulletin, 12 (1969), 603–613.
14. L. Erbe; *Existence of oscillatory solutions and asymptotic behaviour for a class of third order linear differential equations*, Pacific Journal of Mathematics, 64(2) (1976), 369–385.
15. L. Erbe; *Oscillation and asymptotic behaviour of solutions of third order differential delay equations*, SIAM Journal of Mathematical Analysis, 7(4) (1976), 491–500.
16. L. Erbe, Q. Kong and B. G. Zhang; *Oscillation Theory for Functional Differential Equations*, Marcel Dekker Inc., New York, 1995.
17. G. J. Etgen and C. D. Shih; *Conditions for the nonoscillation of third order differential equations with nonnegative coefficients*, SIAM Journal of Mathematical Analysis, 6 (1975), 1–8.
18. N. Finizio and G. Ladas; *Ordinary Differential Equations with Modern Applications*, Third Edition, Wadsworth Pub. Co., Belmont, 1988.
19. M. Greguš; *On certain new properties of solutions of the differential equation* $y''' + Qy' + Q'y = 0$, Publ. Fac. Sci. Univ. Masaryk Brno Czech., 362 (1955), 237–251 (in Czech).
20. M. Greguš; *Oscillatory properties of solutions of a third order differential equation of the type* $y''' + 2A(x)y' + [A'(x) + b(x)]y = 0$, Acta Facultatis Rerum Naturalium Universitatis Comenianae, Mathematica, 6 (1961), 275–300.
21. M. Greguš; *Third Order Linear Differential Equations*, D. Reidel Publishing Company, Boston, 1987.
22. M. Greguš and M. Gera; *Some results in the theory of a third order linear differential equation*, Annales Polonici Mathematici, 42 (1983), 93–101.
23. M. Greguš, J. R. Graef and M. Gera; *Oscillating nonlinear third order differential equations*, Nonlinear Analysis; Theory Methods and Applications, 28(10) (1997), 1611–1622.
24. I. Gyori and G. Ladas; *Oscillation Theory of Delay Differential Equations with Applications*, Clarendon Press, Oxford, 1991.
25. M. Hanan; *Oscillation criteria for third order linear differential equations*, Pacific Journal of Mathematics, 11 (1961), 919–944.
26. P. Hartman; *Ordinary Differential Equations*, Wiley, New York, 1964, and Birkhäuser, Boston, 1982.
27. E. Hille; *Non-oscillation theorems*, Transactions of the American Mathematical Society, 64 (1948), 234–252.
28. Z. Jackiewicz, M. Klaus and C. O'Cinneide; *Asymptotic behaviour of solutions to Volterra integro-differential equations*, Journal of Integral Equations and Applications, 1(4) (1988), 501–516.
29. G. Jayaraman, N. Padmanabhan and R. Mehrotra; *Entry flow into a circular tube of slowly varying cross section*, Fluid Dynamics Research, 1(2) (1986), 131–144.
30. G. D. Jones; *A property of* $y''' + p(x)y' + \frac{1}{2}p'(x)y = 0$, Proceedings of American Mathematical Society, 33 (1972), 420–422.

31. G. D. Jones; *An asymptotic property of solutions of* $y''' + py' + qy = 0$, Pacific Journal of Mathematics, 48(1) (1973), 135–138.
32. G. D. Jones; *Oscillation properties of third order differential equations*, Rocky Mountain Journal of Mathematics, 3 (1973), 507–513.
33. G. D. Jones; *Properties of solutions of a class of third order differential equations*, Journal of Mathematical Analysis and Applications, 48(1) (1974), 165–169.
34. M. S. Keener; *On the solutions of certain linear nonhomogeneous second order differential equations*, Applicable Analysis, 1 (1971), 57–63.
35. I. T. Kiguradze and T. A. Chanturia; *Asymptotic Properties of Solutions of Nonautonomous Ordinary Differential Equations*, Kluwer Academic Publishers, Dordrecht, 1993.
36. W. J. Kim; *Oscillatory properties of linear third order differential equations*, Proceedings of American Mathematical Society, 26 (1970), 286–293.
37. Y. Kuramoto and T. Yamada; *Turbulent state in chemical reaction*, Progress of Theoretical Physics, 56 (1976), 679.
38. T. Kusano and M. Naito; *Comparison theorems for functional differential equations with deviating arguments*, Journal of the Mathematical Society of Japan, 3 (1981), 509–532.
39. T. Kusano, M. Naito and K. Tanaka; *Oscillatory and asymptotic behaviour of solutions of a class of linear ordinary differential equations*, Proceedings of the Royal Society of Edinburgh, 90A (1981), 25–40.
40. G. Ladas, Y. G. Sficas and I. P. Stavroulakis; *Necessary and sufficient conditions for oscillations of higher order delay differential equations*, Transactions of the American Mathematical Society, 285(1) (1984), 81–90.
41. G. S. Ladde, V. Laxmikantham and B. G. Zhang; *Oscillation Theory of Differential Equations with Deviating Arguments*, Marcel Dekker Inc., New York, 1987.
42. A. C. Lazer; *The behaviour of solutions of the differential equation* $y''' + p(x)y' + q(x)y = 0$, Pacific Journal of Mathematics, 17 (1966), 435–466.
43. W. Leighton and Z. Nehari; *On the oscillation of solutions of self adjoint linear differential equations of the fourth order*, Transactions of the American Mathematical Society, 89 (1958), 325–377.
44. H.P. McKean; *Nagumo's equation*, Advances in Mathematics, 4 (1970), 209–223.
45. D. Michelson; *Steady solutions of the Kuramoto-Sivashinsky equation*, Physica D, 19 (1986), 89–111.
46. S. Padhi; *Contributions to the Oscillation Theory of Ordinary and Delay Differential Equations of Third Order*, Ph.D. Thesis, Berhampur University, India, 1998.
47. S. Padhi; *On oscillatory linear third order forced differential equations*, Differential Equations and Dynamical Systems, 13 (2005), 343–358.
48. N. Parhi and P. Das; *On the zeros of solutions of nonhomogeneous third order differential equations*, Czechoslovak Mathematical Journal, 41(116) (1991), 641–652.
49. N. Parhi and P. Das; *Oscillation and nonoscillation of nonhomogeneous third order differential equations*, Czechoslovak Mathematical Journal, 44(119) (1994), 443–459.
50. N. Parhi and S. K. Nayak; *Nonoscillation of second order nonhomogeneous differential equations*, Journal of Mathematical Analysis and Applications, 102(1) (1984), 62–74.
51. D. W. Reynolds; *Bifurcation of harmonic solutions of an integro differential equation modelling resonant sloshing*, SIAM Journal of Applied Mathematics, 49(2) (1989), 362–372.
52. J. Rovder; *Oscillation criteria for third-order linear differential equations*, Matematický Časopis, 25(3) (1975), 231–244.
53. H. L. Royden; *Real Analysis*, Second Edition, McMillan Publishing Co. Inc., New York, 1968.
54. W. Rudin; *Principles of Mathematical Analysis*, Third Edition, McGraw-Hill Kogakusha Ltd., Tokyo, 1976.
55. Y. P. Singh; *The asymptotic behaviour of solutions of linear third order differential equations*, Proceedings of American Mathematical Society, 20(2) (1969), 309–314.
56. A. Škerlik; *Integral criteria of oscillation for a third order linear differential equation*, Mathematica Slovaca, 45(4) (1995), 403–412.

57. A. Škerlik; *An integral condition of oscillation for equation* $y''' + p(t)y' + q(t)y = 0$ *with nonnegative coefficients*, Archivum Mathematicum, 31(2) (1995), 155–161.
58. J. C. F. Sturm; *Memoire Sur les equations differentielles linearies du second order*, Journal de Mathématiques Pures et Appliquées, 11 (1836), 106–186.
59. M. Svec; *Some remarks on a third order linear differential equation*, Czechoslovak Mathematical Journal, 15 (1965), 42–49.
60. C. A. Swanson; *Comparison and Oscillation Theory of Linear Differential equations*, Academic Press, New York, 1968.
61. W. F. Trench; *Canonical forms and principal systems for general disconjugate equations*, Transactions of the American Mathematical Society, 189 (1974), 319–327.
62. D. D. Vreeke and G. M. Sandquist; *Phase plane analysis of reactor kinetics*, Nuclear Science and Engineering 42 (1970), 295–305.

Chapter 2
Behaviour of Solutions of Linear Homogeneous Differential Equations of Third Order

This chapter is concerned with the oscillation and nonoscillation of solutions of a class of third-order linear homogeneous differential equations of the form

$$x''' + a(t)x'' + b(t)x' + c(t)x = 0, \tag{2.1}$$

where $a \in C^2([\sigma, \infty), R)$, $b \in C^1([\sigma, \infty), R)$, $c \in C([\sigma, \infty), R)$ and $\sigma \in R$.

As dealt with in Sect. 1.2, according to the signs of $a(t)$, $b(t)$ and $c(t)$, one may consider the following eight different cases: (i) $a(t) \geq 0$, $b(t) \leq 0$, $c(t) > 0$, (ii) $a(t) \leq 0$, $b(t) \leq 0$, $c(t) > 0$, (iii) $a(t) \leq 0$, $b(t) \leq 0$, $c(t) < 0$, (iv) $a(t) \geq 0$, $b(t) \leq 0$, $c(t) < 0$, (v) $a(t) \geq 0$, $b(t) \geq 0$, $c(t) > 0$, (vi) $a(t) \leq 0$, $b(t) \geq 0$, $c(t) > 0$, (vii) $a(t) \geq 0$, $b(t) \geq 0$, $c(t) < 0$ and (viii) $a(t) \leq 0$, $b(t) \geq 0$, $c(t) < 0$.

Equation (2.1) may be written as

$$\big(r(t)x''\big)' + q(t)x' + p(t)x = 0, \tag{2.2}$$

where $r(t) = e^{\int_0^t a(s)\,ds}$, $q(t) = r(t)b(t)$ and $p(t) = r(t)c(t)$. Since Eqs. (2.1) and (2.2) are equivalent, we refer to either of them according to our convenience. The adjoint of (2.1) is given by

$$\big(\big(x' - a(t)x\big)' + b(t)x\big)' - c(t)x = 0,$$

that is,

$$x''' - a(t)x'' + \big(b(t) - 2a'(t)\big)x' - \big(c(t) - b'(t) + a''(t)\big)x = 0. \tag{2.3}$$

The transformation

$$x(t) = z(t)e^{-\frac{1}{3}\int_\sigma^t a(s)\,ds}$$

transforms (2.1) into

$$z''' + Q(t)z' + P(t)z = 0, \tag{2.4}$$

where $P(t) = \frac{1}{27}(2a^3(t) - 9a(t)b(t) + 27c(t) - 9a''(t))$ and $Q(t) = \frac{1}{3}(3b(t) - 3a'(t) - a^2(t))$. Clearly (2.1) is oscillatory, if and only if (2.4) is oscillatory.

S. Padhi, S. Pati, *Theory of Third-Order Differential Equations*,
DOI 10.1007/978-81-322-1614-8_2, © Springer India 2014

This chapter has been divided into nine sections. First-six sections deal with the oscillation, nonoscillation of solutions of (2.1) and their asymptotic behaviour for the six among the above eight different cases. Sufficient conditions have been obtained for oscillation of solutions of Eq. (2.1) of which the Euler equation

$$x'''(t) + \frac{a_0}{t}x''(t) + \frac{b_0}{t^2}x'(t) + \frac{c_0}{t^3}x(t) = 0, \tag{2.5}$$

where a_0, b_0 and c_0 are reals, is a particular case. Section 2.7 deals with the oscillation, nonoscillation, property A and B of a third-order disconjugate equation. Some remarks and open problems are given in Sect. 2.8. Section 2.9 contains a note for the readers for easy reference.

The results presented in this chapter on the oscillation and nonoscillation of (2.1) and (2.2) are new and were not presented in the monograph [14].

2.1 Behaviour of Solutions of $x''' + a(t)x'' + b(t)x' + c(t)x = 0$ with $a(t) \geq 0$, $b(t) \leq 0$ and $c(t) > 0$

This section deals with the oscillation and asymptotic behaviour of solutions of (2.1) with $a(t) \geq 0$, $b(t) \leq 0$ and $c(t) > 0$. The objective here is to present results generalising the observations in Proposition 1.2.1 to Eq. (2.1). Further, sufficient conditions are given for the oscillation of (2.1), which, in particular, is applicable to the Euler equation (2.5). Necessary and sufficient condition is given in Theorems 2.1.1 and 2.1.7 for (2.1) to be oscillatory. Theorems 2.1.9 and 2.1.10 give a sufficient condition for the nonoscillation of (2.1) when $a(t) \equiv 0$. Theorems 2.1.11 and 2.1.12 show that (2.1) admits at most one nonoscillatory solution $x(t)$ (neglecting linear dependence) with the property $x(t)x'(t) < 0$, $x(t)x''(t) > 0$ for $t \in [\sigma, \infty)$ and $\lim_{t\to\infty} x(t) = 0$, provided that (2.1) has an oscillatory solution.

In the particular case, when $a(t) \equiv 0$, $b(t) \equiv 0$ and $c(t) > 0$, $t \geq \sigma$, (2.1) reduces to

$$x''' + c(t)x = 0. \tag{2.6}$$

There is a well-known oscillation criterion for (2.6) of the form

$$\int^{\infty} t^{2-\epsilon} c(t)\, dt = \infty, \qquad \text{for some } \epsilon > 0. \tag{2.7}$$

This condition has been improved several times. Chanturiya [7] proved that if

$$\liminf_{t\to\infty} t^2 \int_t^{\infty} c(s)\, ds > \frac{2}{3\sqrt{3}}, \tag{2.8}$$

then (2.6) is oscillatory. Lazer [23] proved that if

$$\int^{\infty} \left[c(t) - \frac{2}{3\sqrt{3}}\left(-b(t)\right)^{3/2}\right] dt = \infty, \tag{2.9}$$

then

$$x''' + b(t)x' + c(t)x = 0 \tag{2.10}$$

is oscillatory. We note that (2.10) can be obtained from (2.1) with $a(t) \equiv 0$. We start this section with the following lemma.

Lemma 2.1.1 *Equation* (2.1) *is of type* C_I.

Proof Suppose that $x(t)$ is a solution of (2.1) with $x(t_0) = x'(t_0) = 0$ and $x''(t_0) > 0$ for some $t_0 > \sigma$. We show that $x(t) > 0$ for $t \in [\sigma, t_0)$. If possible, suppose that $x(t_1) = 0$ for some $t_1 \in [\sigma, t_0)$. There exists a $t_2 \in (t_1, t_0)$ such that $x'(t_2) = 0$. Consequently, there exists a $t_3 \in (t_2, t_0)$ such that $x''(t_3) = 0$ and $x(t) > 0$, $x'(t) < 0$ and $x''(t) > 0$ for $t \in (t_3, t_0)$. Since $x(t)$ is a solution of (2.1), it satisfies (2.2). Integrating (2.2) from t_3 to t_0, we obtain

$$0 < r(t_0)x''(t_0) = \int_{t_3}^{t_0} \left[-q(t)x'(t) - p(t)x(t)\right] dt < 0,$$

a contradiction. Thus $x(t) > 0$ for $t \in [\sigma, t_0)$. The lemma is proved. □

Remark 2.1.1 If $x(t)$ is a solution of (2.1) with $x(t_0) \geq 0$, $x'(t_0) \leq 0$ and $x''(t_0) > 0$, $t_0 \in (\sigma, \infty)$, then $x(t) > 0$, $x'(t) < 0$, $x''(t) > 0$ and $x'''(t) < 0$ for $t \in [\sigma, t_0)$.

Lemma 2.1.2 *If* $2c(t) - a(t)b(t) - b'(t) \geq 0$ *but* $\not\equiv 0$ *on any subinterval of* $[\sigma, \infty)$, *then* (2.1) *is of type* C_I.

Proof Suppose that $x(t)$ is a solution of (2.1) with $x(t_0) = x'(t_0) = 0$ and $x''(t_0) > 0$. If possible, suppose that $x(t_1) = 0$ for some $t_1 \in [\sigma, t_0)$. Since $x(t)$ is a solution of (2.1), it satisfies (2.2). Multiplying (2.2) through by $x(t)$ and integrating the resulting identity from t_1 to t_0, we obtain

$$-r(t_1)\big(x'(t_1)\big)^2 \geq \int_{t_1}^{t_0} \big(2p(t) - q'(t)\big)x^2(t)\, dt,$$

that is,

$$0 \geq -r(t_1)\big(x'(t_1)\big)^2 \geq \int_{t_1}^{t_0} \big(2c(t) - a(t)b(t) - b'(t)\big)r(t)x^2(t)\, dt > 0,$$

a contradiction. Hence $x(t) > 0$ for $t \in [\sigma, t_0)$. This completes the proof of the lemma. □

One may observe that the condition $b(t) \leq 0$ and $c(t) > 0$ has not been assumed explicitly in Lemma 2.1.2.

Theorem 2.1.1 *Equation* (2.1) *is oscillatory, if and only if all nonoscillatory solutions of the second-order differential equation*

$$z'' + 3zz' + a(t)z' + z^3 + a(t)z^2 + b(t)z + c(t) = 0 \tag{2.11}$$

are eventually negative.

Proof Suppose that all nonoscillatory solutions of (2.11) are eventually negative. It is required to show that (2.1) admits an oscillatory solution. If possible, let all solution of (2.1) be nonoscillatory. By Lemma 1.5.4, there exists at least one nonoscillatory solution $u(t)$ of (2.1) which does not satisfy the condition $u(t)u'(t) < 0$. Without any loss of generality, we may take $u(t) > 0$ for $t \geq t_0 \geq \sigma$. From Lemma 1.5.3, it follows that $u'(t) > 0$ for $t \geq t_1 > t_0$. Setting $z(t) = \frac{u'(t)}{u(t)}$, $t > t_1$, it is easy to verify that $z(t)$ is a nonnegative nonoscillatory solution of (2.11), a contradiction. Hence (2.1) admits an oscillatory solution.

Conversely, suppose that (2.1) has an oscillatory solution. If possible, let $z(t)$ be a positive nonoscillatory solution of (2.11). It may be verified that $v(t) = \exp(\int_{t_0}^t z(s)\,ds)$ is a positive increasing solution of (2.1), which contradicts Lemma 1.5.4. Hence the proof of the theorem is completed. □

Theorem 2.1.2 *Suppose that* $a'(t) \leq 0$. *If*

$$\int_\sigma^\infty \left[\frac{2a^3(t)}{27} - \frac{a(t)b(t)}{3} + c(t) - \frac{2}{3\sqrt{3}} \left(\frac{a^2(t)}{3} - b(t) \right)^{3/2} \right] dt = \infty, \tag{2.12}$$

then (2.1) *is oscillatory.*

Proof Let $x(t)$ be a nonoscillatory solution of (2.1). From Lemma 1.5.3, it follows that there exists a $t_0 \in [\sigma, \infty)$ such that $x'(t) \leq 0$ or ≥ 0 for $t \in [t_0, \infty)$. In view of Lemma 1.5.4 and the second part of Lemma 1.5.3, it is sufficient to prove that $x(t)x'(t) \geq 0$ for $t > t_0$ does not hold.

Suppose that $x(t)x'(t) \geq 0$ for $t > t_0$. Setting $u(t) = \frac{x'(t)}{x(t)}$, $t \geq t_0$, we see that $u(t)$ is a solution of the second-order Riccati equation

$$z'' + 3zz' + a(t)z' = -F\big(u(t), t\big), \tag{2.13}$$

where

$$F\big(u(t), t\big) = u^3(t) + a(t)u^2(t) + b(t)u(t) + c(t).$$

Clearly, $F(u(t), t)$ attains its minimum value for $u(t) \geq 0$ at

$$u(t) = \frac{1}{3}\left[-a(t) + \left(a^2(t) - 3b(t)\right)^{1/2}\right]$$

and the minimum of $F(u(t), t)$ is given by

$$\frac{2a^3(t)}{27} - \frac{a(t)b(t)}{3} + c(t) - \frac{2}{3\sqrt{3}} \left(\frac{a^2(t)}{3} - b(t) \right)^{3/2}.$$

So,

$$\begin{aligned} &u''(t) + 3u(t)u'(t) + a(t)u'(t) \\ &\quad \le -\left[\frac{2a^3(t)}{27} - \frac{a(t)b(t)}{3} + c(t) - \frac{2}{3\sqrt{3}}\left(\frac{a^2(t)}{3} - b(t)\right)^{3/2}\right]. \end{aligned}$$

Integrating the above inequality from t_0 to t, we obtain

$$\begin{aligned} u'(t) &\le u'(t_0) + \frac{3}{2}u^2(t_0) + a(t_0)u(t_0) - \frac{3}{2}u^2(t) - a(t)u(t) + \int_{t_0}^{t} a'(s)u(s)\,ds \\ &\quad - \int_{t_0}^{t}\left[\frac{2a^3(s)}{27} - \frac{a(s)b(s)}{3} + c(s) - \frac{2}{3\sqrt{3}}\left(\frac{a^2(s)}{3} - b(s)\right)^{3/2}\right]ds \\ &\le u'(t_0) + \frac{3}{2}u^2(t_0) + a(t_0)u(t_0) \\ &\quad - \int_{t_0}^{t}\left[\frac{2a^3(s)}{27} - \frac{a(s)b(s)}{3} + c(s) - \frac{2}{3\sqrt{3}}\left(\frac{a^2(s)}{3} - b(s)\right)^{3/2}\right]ds. \end{aligned}$$

From (2.12), it follows that $\lim_{t\to\infty} u'(t) = -\infty$. Hence $u(t) < 0$ for large t, a contradiction. This completes the proof of the theorem. □

Theorem 2.1.3 *Let* $b(t) - a'(t) \le 0$ *and*

$$\begin{aligned} &\int_{\sigma}^{\infty}\left[\frac{2a^3(t)}{27} - \frac{a(t)b(t)}{3} + \frac{a(t)a'(t)}{3} + c(t) - \frac{2}{3\sqrt{3}}\left(\frac{a^2(t)}{3} - b(t) + a'(t)\right)^{3/2}\right]dt \\ &\quad = \infty, \end{aligned} \tag{2.14}$$

then (2.1) *is oscillatory.*

Remark 2.1.2 Theorem 2.1.2 generalises the observation (i) in Proposition 1.2.1. However, it fails to hold for Eq. (2.5). Instead of setting $u(t) = \frac{x'(t)}{x(t)}$ in proof of Theorem 2.1.2, if we set $u(t) = \frac{t^2x'(t)}{x(t)}$ for $t \ge t_0$, we see that $u(t)$ is a positive solution of the second-order Riccati equation

$$\left[z' - \frac{4}{t}z + \frac{3}{2t^2}z^2 + a(t)z\right]' = -G\big(u(t), t\big), \tag{2.15}$$

where

$$\begin{aligned} G\big(u(t), t\big) = &\frac{u^3(t)}{t^4} - \frac{3u^2(t)}{t^3} + \frac{a(t)u^2(t)}{t^2} + \frac{2u(t)}{t^2} - a'(t)u(t) \\ &- \frac{2a(t)u(t)}{t} + b(t)u(t) + t^2c(t). \end{aligned} \tag{2.16}$$

If $ta(t) \le 3$ and $1 - t^2b(t) + \frac{1}{3}t^2a^2(t) + t^2a'(t) \ge 0$, then a simple calculation shows that $G(u(t),t)$ attains the minimum

$$\begin{aligned}\min G\big(u(t),t\big) = {} & \frac{2t^2a^3(t)}{27} - \frac{t^2a(t)b(t)}{3} + t^2c(t) - \frac{2}{3}a(t) + tb(t) + \frac{t^2a(t)a'(t)}{3} \\ & - ta'(t) - \frac{2}{3\sqrt{3}t}\left(1 - t^2b(t) + \frac{1}{3}t^2a^2(t) + t^2a'(t)\right)^{3/2}\end{aligned}$$

for $u(t) > 0$ at

$$u(t) = t\left[\left(1 - \frac{ta(t)}{3}\right) + \frac{1}{\sqrt{3}}\left(1 - t^2b(t) + \frac{1}{3}t^2a^2(t) + t^2a'(t)\right)^{1/2}\right].$$

Hence (2.15) yields

$$\left[u'(t) - \frac{4u(t)}{t} + \frac{3}{2}\frac{u^2(t)}{t^2} + a(t)u(t)\right]' \le -\min G\big(u(t),t\big).$$

Integrating the above inequality from t_0 to t, we obtain

$$u'(t) - \frac{4u(t)}{t} + \frac{3}{2}\frac{u^2(t)}{t^2} \le K - \int_{t_0}^{t} \min G\big(u(s),s\big)\,ds.$$

Since $\frac{3}{2}\frac{u^2(t)}{t^2} - \frac{4u(t)}{t}$ attains the minimum $-\frac{8}{3}$ for $u(t) > 0$ at $u(t) = \frac{4t}{3}$, the above integral inequality yields

$$u'(t) \le K + \frac{8}{3} - \int_{t_0}^{t} \min G\big(u(s),s\big)\,ds.$$

The above inequality leads us to the following theorem.

Theorem 2.1.4 *Let* $ta(t) \le 3$ *and* $1 - t^2b(t) + \frac{t^2a^2(t)}{3} + t^2a'(t) \ge 0$. *If*

$$\begin{aligned}\int_{\sigma}^{\infty} \Bigg[& \frac{2t^2a^3(t)}{27} - \frac{t^2a(t)b(t)}{3} + t^2c(t) - \frac{2}{3}a(t) + tb(t) + \frac{t^2a(t)a'(t)}{3} \\ & - ta'(t) - \frac{2}{3\sqrt{3}t}\left(1 - t^2b(t) + \frac{1}{3}t^2a^2(t) + t^2a'(t)\right)^{3/2}\Bigg]dt = \infty, \end{aligned} \qquad (2.17)$$

then (2.1) *is oscillatory.*

On the other hand, if $ta(t) \ge 3$ and $2ta(t) - t^2b(t) + t^2a'(t) > 2$, then

$$\left(1 - \frac{ta(t)}{3}\right) + \frac{1}{\sqrt{3}}\left(1 - t^2b(t) + \frac{1}{3}t^2a^2(t) + t^2a'(t)\right)^{1/2} > 0.$$

In this case, one can prove the following theorem:

Theorem 2.1.5 *Suppose that $ta(t) \geq 3$ and $2ta(t) - t^2b(t) + t^2a'(t) > 2$ hold. If* (2.17) *is satisfied, then* (2.1) *is oscillatory.*

Remark 2.1.3 Theorem 2.1.4 can be applied to find sufficient conditions for the oscillation of the Euler equation (2.5). If $0 \leq a_0 \leq 3$, $b_0 \leq 0$, $c_0 > 0$, $1 - b_0 + \frac{a_0^2}{3} - a_0 > 0$ and

$$\frac{2a_0^3}{27} + \frac{a_0}{3} + b_0 + c_0 - \frac{a_0^2}{3} - \frac{a_0 b_0}{3} - \frac{2}{3\sqrt{3}}\left(1 + \frac{1}{3}a_0^2 - a_0 - b_0\right)^{3/2} > 0, \quad (2.18)$$

then (2.5) is oscillatory.

Applying Theorem 2.1.5 to the Euler equation (2.5), we see that (2.5) is oscillatory if $a_0 \geq 3$, $b_0 \leq 0$, $c_0 > 0$, $a_0 - b_0 > 2$ and (2.18) hold.

Example 2.1.1 Consider

$$x''' + \frac{1}{t}x'' - \frac{1}{t^2}x' + \frac{3}{t^3}x = 0, \quad t \geq 1.$$

Since all the conditions of Theorem 2.1.4 are satisfied, this equation has an oscillatory solution. In particular, $x_1(t) = t^{3/2}\cos(\frac{\sqrt{3}}{2}\log t)$ and $x_2(t) = t^{3/2}\sin(\frac{\sqrt{3}}{2}\log t)$ are the oscillatory solutions of this equation.

Example 2.1.2 The equation

$$x''' + \frac{6}{t}x'' - \frac{1}{t^2}x' + \frac{25}{t^3}x = 0, \quad t \geq 1$$

satisfies the hypothesis of Theorem 2.1.5. Hence this equation has an oscillatory solution. In particular, $x_1(t) = t\cos(2\log t)$ and $x_2(t) = t\sin(2\log t)$ are the oscillatory solutions of this equation.

In the particular case, when $a(t) \equiv 0$ and $b(t) \equiv 0$, there is a well-known Kneser-type condition for oscillation of Eq. (2.6). Hanan [16] proved that (2.6) is oscillatory, if

$$\liminf_{t\to\infty} t^3 c(t) > \frac{2}{3\sqrt{3}}. \quad (2.19)$$

From (2.19), it follows that there exist $\epsilon > 0$ and $T \geq \sigma$ such that

$$c(t) - \frac{2}{3\sqrt{3}t^3} \geq \epsilon t^{-3}, \quad (2.20)$$

and

$$t^2\left[c(t) - \frac{2}{3\sqrt{3}t^3}\right] \geq \frac{\epsilon}{t} \quad (2.21)$$

for all $t \geq T$.

For Eq. (2.6), the conditions (2.8) and (2.17) may be rewritten as

$$\liminf_{t\to\infty} t^2 \int_\sigma^\infty \left(c(s) - \frac{2}{3\sqrt{3}s^3}\right) ds > 0 \tag{2.22}$$

and

$$\int^\infty s^2 \left(c(s) - \frac{2}{3\sqrt{3}s^3}\right) ds = \infty \tag{2.23}$$

respectively. Observe that (2.19) implies (2.22) and (2.21) implies (2.23). Now, we compare (2.22) and (2.23). For this, we suppose that

$$c(t) - \frac{2}{3\sqrt{3}t^3} \geq 0 \tag{2.24}$$

for sufficiently large t.

Remark 2.1.4 Assume that (2.24) is satisfied and (2.22) holds. So there exist $\delta > 0$ and $T_1 \geq \sigma$ such that

$$t^2 \int_t^\infty \left(c(s) - \frac{2}{3\sqrt{3}s^3}\right) ds \geq \delta \quad \text{for all } t \geq T_1.$$

If $\int^\infty s^2(c(s) - \frac{2}{3\sqrt{3}s^3})\, ds < \infty$, then there exists $T_2 \geq T_1$ such that

$$\int_{T_2}^\infty s^2 \left(c(s) - \frac{2}{3\sqrt{3}s^3}\right) ds \leq \frac{\delta}{2}.$$

So, we have

$$\begin{aligned}
\frac{\delta}{2} &\geq \int_{T_2}^\infty s^2 \left(c(s) - \frac{2}{3\sqrt{3}s^3}\right) ds \\
&= \liminf_{t\to\infty} \left[\int_{T_2}^t s^2 \left(c(s) - \frac{2}{3\sqrt{3}s^3}\right) ds + \int_t^\infty s^2 \left(c(s) - \frac{2}{3\sqrt{3}s^3}\right) ds\right] \\
&\geq \liminf_{t\to\infty} \int_t^\infty s^2 \left(c(s) - \frac{2}{3\sqrt{3}s^3}\right) ds \\
&\geq \liminf_{t\to\infty} t^2 \int_t^\infty \left(c(s) - \frac{2}{3\sqrt{3}s^3}\right) ds \\
&\geq \delta,
\end{aligned}$$

a contradiction. Hence (2.23) is satisfied.

Now, we consider the equation

$$x''' + b_0 t^\beta x' + c_0 t^\delta x = 0, \tag{2.25}$$

where $b_0 < 0$ and $c_0 > 0$ are some constants, and $\delta \geq -3$, $2\delta \geq 3\beta$.

For $\beta = -2$ and $\delta = -3$, Eq. (2.25) becomes the Euler equation. Lazer's condition (2.9) is not applicable to the Euler equation. The necessary and sufficient condition for oscillation of Euler's equation (2.25) is

$$b_0 + c_0 - \frac{2}{3\sqrt{3}}(1 - b_0)^{3/2} > 0, \tag{2.26}$$

which follows from (2.18) with $a_0 \equiv 0$.

Remark 2.1.5 Let $\delta = -3, \beta < -2$ and $c_0 > \frac{2}{3\sqrt{3}}$. Then (2.25) is oscillatory. Indeed, since

$$(1 + x)^{3/2} = 1 + \frac{3}{2}x + \frac{3}{8}x^2 + \frac{3x^2}{4}\sum_{k=1}^{\infty}\frac{(-1)^k(2k-1)!!}{2^k(k+2)!}x^k, \quad |x| < 1, \tag{2.27}$$

where $(2k-1)!! = (1)(3)(5)\cdots(2k-1)$, substituting the coefficient of Eq. (2.25) to the left-hand side of (2.17) for $t > \sigma_0 \geq (-b_0)^{-1/(\beta+2)}$ with $a(t) \equiv 0$, we obtain

$$\begin{aligned}
&\int_{\sigma_0}^{\infty}\frac{1}{t}\left[c_0 + b_0 t^{\beta+2} - \frac{2}{3\sqrt{3}}\left(1 - b_0 t^{\beta+2}\right)^{3/2}\right]dt \\
&= \int_{\sigma_0}^{\infty}\frac{1}{t}\left[c_0 + b_0 t^{\beta+2} - \frac{2}{3\sqrt{3}}\left(1 - \frac{3}{2}b_0 t^{\beta+2} + \frac{3}{8}b_0^2 t^{2\beta+4} + \cdots\right)\right]dt \\
&= \int_{\sigma_0}^{\infty}\frac{1}{t}\left[c_0 - \frac{2}{3\sqrt{3}} + b_0 t^{\beta+2} - \frac{2}{3\sqrt{3}}\left(-\frac{3}{2}b_0 t^{\beta+2} + \frac{3}{8}b_0^2 t^{2\beta+4} + \cdots\right)\right]dt.
\end{aligned}$$

Since $c_0 > \frac{2}{3\sqrt{3}}$ and $\beta + 2 < 0$, it is easy to see that condition (2.17) is satisfied. Thus, the proof follows immediately from Theorem 2.1.4.

In case $\delta > -3, 2\delta = 3\beta$, Lazer's condition (2.9) may be applied only when $\delta \geq -1$ and $c_0 > \frac{2(-b_0)^{3/2}}{3\sqrt{3}}$.

Remark 2.1.6 Let $\delta > -3, 2\delta = 3\beta$ and $c_0 > \frac{2(-b_0)^{3/2}}{3\sqrt{3}}$. Then (2.25) is oscillatory. Indeed, since $\delta > -3$ and $2\delta = 3\beta$, there exists $\epsilon > 0$ such that $\delta = -3 + \epsilon$ and $\beta = -2 + \frac{2}{3}\epsilon$. Let $t > \sigma_0 \geq (-b_0)^{\frac{-3}{2\epsilon}}$. Substituting the coefficients of Eq. (2.25) in the left-hand side of (2.17) with $a(t) \equiv 0$ and using (2.27), we get

$$\begin{aligned}
&\int_{\sigma_0}^{\infty}\left[c_0 t^{-1+\epsilon} + b_0 t^{-1+2\epsilon/3} - \frac{2}{3\sqrt{3}t}\left(-b_0 t^{\frac{2\epsilon}{3}}\right)\left(1 - \frac{t^{\frac{-2\epsilon}{3}}}{b_0}\right)^{3/2}\right]dt \\
&= \int_{\sigma_0}^{\infty} t^{-1+\epsilon}\left[c_0 + b_0 t^{\frac{-\epsilon}{3}} - \frac{2}{3\sqrt{3}}(-b_0)^{3/2}\left(1 - \frac{3t^{-\frac{2\epsilon}{3}}}{2b_0} + \frac{3t^{-\frac{4\epsilon}{3}}}{8b_0^2} + \cdots\right)\right]dt \\
&= \int_{\sigma_0}^{\infty} t^{-1+\epsilon}\left[c_0 - \frac{2}{3\sqrt{3}}(-b_0)^{3/2} + b_0 t^{\frac{-\epsilon}{3}} - \frac{2}{3\sqrt{3}}(-b_0)^{3/2}\right.
\end{aligned}$$

$$\times\left(-\frac{3t^{-\frac{2\epsilon}{3}}}{2b_0}+\frac{3t^{-\frac{4\epsilon}{3}}}{8b_0^2}+\cdots\right)\Bigg]dt$$

$$=\infty$$

since $c_0-\frac{2(-b_0)^{3/2}}{3\sqrt{3}}>0$. By Theorem 2.1.4, Eq. (2.25) is oscillatory.

Remark 2.1.7 Let $\delta>-3$ and $2\delta>3\beta$. Then Eq. (2.25) is oscillatory. In fact, similar to the above, after substituting the coefficients of Eq. (2.25) with $a(t)=0$ and using (2.27) for sufficiently large t, we obtain

$$\int_{\sigma_0}^{\infty}\left[c_0t^{\delta+2}+b_0t^{\beta+1}-\frac{2}{3\sqrt{3}t}(-b_0)^{3/2}t^{3(\beta+2)/2}\left(1-\frac{3t^{-\beta-2}}{2b_0}-\frac{3t^{-2\beta-4}}{8b_0^2}-\cdots\right)\right]dt$$

for $\beta+2>0$;

$$\int_{\sigma_0}^{\infty}\left[c_0t^{\delta+2}+b_0t^{-1}-\frac{2}{3\sqrt{3}t}(1-b_0)^{3/2}\right]dt,$$

for $\beta+2=0$; and

$$\int_{\sigma_0}^{\infty}\left[c_0t^{\delta+2}+b_0t^{\beta+1}-\frac{2}{3\sqrt{3}t}\left(1-\frac{3}{2}b_0t^{\beta+2}+\frac{3}{8}b_0^2t^{2\beta+4}+\cdots\right)\right]dt$$

for $\beta+2<0$. It is easy to check that $\delta+2>\beta-1$, and hence $\delta+2>-1$, and all the integrals above satisfy (2.17) with $a(t)\equiv 0$. Then Theorem 2.1.4 implies that (2.25) is oscillatory.

Theorem 2.1.6 *Suppose that* $a(t)+b(t)+1\le 0$. *If*

$$\begin{aligned}&\frac{2(a(t)+3)^3}{27}-\frac{(a(t)+3)(a(t)+b(t)+1)}{3}+c(t)\\&\quad-\frac{2}{3\sqrt{3}}\left(\frac{(a(t)+3)^2}{3}-\big(a(t)+b(t)+1\big)\right)^{3/2}>0\end{aligned}\tag{2.28}$$

for $t\ge t_0>\sigma$, *then* (2.1) *is oscillatory.*

Proof For the sake of contradiction, suppose that all solutions of (2.1) are nonoscillatory. Then by Lemma 1.5.4, there exists a solution $x(t)$ of (2.1) satisfying the property $x(t)x'(t)>0$ for $t\ge t_0\ge\sigma$. Let $x(t)>0$ and $x'(t)>0$ for $t\ge t_0$. Set

$$e^z=\frac{x'(t)}{x(t)}.\tag{2.29}$$

Differentiating successively, we obtain

$$\frac{x''(t)}{x(t)}=e^z+e^{2z}\tag{2.30}$$

and

$$\frac{x'''(t)}{x(t)} = e^{3z} + 3e^{2z} + e^z. \tag{2.31}$$

Dividing (2.1) throughout by $x(t)$ and using (2.29), (2.30) and (2.31) in the resulting equation, we obtain

$$F(e^z, t) = e^{3z} + \big(3 + a(t)\big)e^{2z} + \big(a(t) + b(t) + 1\big)e^z + c(t) = 0. \tag{2.32}$$

The minimum of $F(e^z, t)$ attains at

$$e^z = \frac{1}{3}\Big[-a(t) - 3 + \sqrt{\big(a(t)+3\big)^2 - 3\big(a(t) + b(t) + 1\big)}\Big]$$

and the minimum value of $F(e^z, t)$ is given by

$$\min_z F\big(e^z, t\big) = \frac{2A^3(t)}{27} - \frac{A(t)B(t)}{3} + c(t) - \frac{2}{3\sqrt{3}}\left(\frac{A^2(t)}{3} - B(t)\right)^{3/2}, \tag{2.33}$$

where $A(t) = a(t) + 3$ and $B(t) = a(t) + b(t) + 1$. Combining (2.32) and (2.33), we have the inequality

$$\frac{2A^3(t)}{27} - \frac{A(t)B(t)}{3} + c(t) - \frac{2}{3\sqrt{3}}\left(\frac{A^2(t)}{3} - B(t)\right)^{3/2} \le 0,$$

which contradicts (2.28). This completes the proof. □

Theorem 2.1.7 *Let $M(t)$ be a solution of* (2.1) *with $M(t) > 0$ and $M'(t) < 0$, $t \ge \sigma$. Equation* (2.1) *is oscillatory if and only if the second-order differential equation*

$$z'' + \left(\frac{3M'(t) + a(t)M(t)}{M(t)}\right)z' + \left(\frac{3M''(t) + 2a(t)M'(t) + b(t)M(t)}{M(t)}\right)z = 0 \tag{2.34}$$

is oscillatory.

Proof Let (2.1) be oscillatory. Let $x(t)$ be an oscillatory solution of the equation. Since $M(t) > 0$ for $t \ge \sigma$, it is easy to verify that $(\frac{x(t)}{M(t)})'$ is an oscillatory solution of (2.34), and hence (2.34) is oscillatory.

Conversely, suppose that (2.34) is oscillatory. If possible, let (2.1) be nonoscillatory. By Lemma 1.5.4, there exists at least one nonoscillatory solution $u(t)$ of (2.1) not satisfying the condition $u(t)u'(t) < 0$ for large t. Without any loss of generality we may assume that $u(t) > 0$ for $t \ge \sigma$. From Lemma 1.5.3, it follows that $u'(t) \ge 0$, $t \ge t_0 \ge \sigma$. Clearly, $x'(t)$ is an oscillatory solution of (2.34), where $x(t) = \frac{u(t)}{M(t)}$, $t > t_0 \ge \sigma$. But $x'(t) = (M(t)u'(t) - M'(t)u(t))M^{-2}(t) > 0$, $t \ge t_0$, a contradiction. Thus (2.1) is oscillatory. Hence the theorem is proved. □

Theorem 2.1.8 *Equation* (2.1) *has a solution* $x(t)$ *with the following properties*:

$$x'''(t)x''(t)x'(t)x(t) \neq 0, \quad t \in [\sigma, \infty),$$
$$\operatorname{sgn} x(t) = \operatorname{sgn} x''(t) \neq \operatorname{sgn} x'(t) = \operatorname{sgn} x'''(t),$$
$$\lim_{t\to\infty} x''(t) = \lim_{t\to\infty} x'(t) = 0,$$

and $x(t)$ *is asymptotic to a finite constant.*

Proof For every positive integer $n > \sigma$, let $x_n(t)$ be a solution of (2.1) satisfying the initial conditions

$$x_n(n) = 0, \qquad x_n'(n) = 0, \qquad x_n''(n) > 0.$$

By Remark 2.1.1, we have

$$x_n(t) > 0, \qquad x_n'(t) < 0, \qquad x_n''(t) > 0 \tag{2.35}$$

for $t \in [\sigma, n)$. Let $\{u_1(t), u_2(t), u_3(t)\}$ be a set of linearly independent solutions of (2.1). It is possible to write

$$x_n(t) = c_{1n}u_1(t) + c_{2n}u_2(t) + c_{3n}u_3(t), \tag{2.36}$$

where

$$c_{1n}^2 + c_{2n}^2 + c_{3n}^2 = 1. \tag{2.37}$$

Since the sequence $\langle c_{in}\rangle$, $i = 1, 2, 3$, is bounded, there exists a subsequence $\langle c_{in_j}\rangle$ which converges to c_i, $i = 1, 2, 3$. From (2.37), we obtain

$$c_1^2 + c_2^2 + c_3^2 = 1. \tag{2.38}$$

Now, consider the equation

$$x(t) = c_1u_1(t) + c_2u_2(t) + c_3u_3(t) \tag{2.39}$$

of (2.1). Since the sequences $\langle x_{nj}(t)\rangle$, $\langle x_{nj}'(t)\rangle$ and $\langle x_{nj}''(t)\rangle$ converge uniformly to $x(t)$, $x'(t)$ and $x''(t)$, respectively, on any finite subinterval of $[\sigma, \infty)$, it follows from (2.35) that

$$x(t) \geq 0, \qquad x'(t) \leq 0, \qquad x''(t) \geq 0$$

and

$$x'''(t) = -a(t)x''(t) - b(t)x'(t) - c(t)x(t) \leq 0 \tag{2.40}$$

for $t \in [\sigma, \infty)$. If $x(t_0) = 0$ for some $t_0 \geq \sigma$, then $x(t) \equiv 0$, $t \in [t_0, \infty)$, which contradicts (2.38) and (2.39). Thus, $x(t) > 0$ for $t \in [\sigma, \infty)$. If possible, suppose that $x'(t_0) = 0$ for some $t_0 \geq \sigma$. Since $x''(t) \geq 0$, it follows that $x'(t_0) \equiv 0$. Consequently, $x(t)$ reduces to a nonzero constant. From (2.1), we obtain $c(t)x = 0$, a contradiction.

Hence $x'(t) < 0$ for $t \in [\sigma, \infty)$. $x''(t_1) = 0$ for some $t_1 \geq \sigma$ implies that $x''(t) \equiv 0$, and hence a contradiction from (2.1). Hence $x''(t) > 0$ for $t \in [\sigma, \infty)$. From (2.40), it follows that $x'''(t) < 0$ for $t \in [\sigma, \infty)$. Thus

$$\lim_{t\to\infty} x'(t) = \lim_{t\to\infty} x''(t) = 0,$$

and $x(t)$ is asymptotic to a finite constant. This completes the proof of theorem. □

Lemma 2.1.3 *Consider* (2.1) *with* $a(t) \equiv 0$. *If* $u(t)$ *is a nonoscillatory solution of* (2.1) *and* $v(t)$ *is any solution of* (2.1), *then* $(u(t)v'(t) - u'(t)v(t))$ *is a solution of*

$$\left(\frac{x'}{u(t)}\right)' + \left(\frac{u''(t) + b(t)u(t)}{u^2(t)}\right)x = 0.$$

Theorem 2.1.9 *Consider* (2.1) *with* $a(t) \equiv 0$. *If* $\int_0^\infty c(t)\,dt < \infty$, $2c(t) - b'(t) \geq 0$, $\lim_{t\to\infty} b(t) = 0$ *and the second-order differential equation*

$$z'' + \left[b(t) + \frac{3}{2}\int_t^\infty c(s)\,ds\right]z = 0 \tag{2.41}$$

is nonoscillatory, then (2.1) *is nonoscillatory.*

Proof If possible, let $v(t)$ be an oscillatory solution of (2.1). From Theorem 1.1 of Lazer [23] and Theorem 6 due to Jones [19], it follows that (2.1) admits a nonoscillatory solution $u(t)$ such that $u(t) > 0$, $u'(t) < 0$, $u''(t) > 0$, $u'''(t) < 0$ for $t \in [\sigma, \infty)$ and

$$\lim_{t\to\infty} u(t) = \lim_{t\to\infty} u'(t) = \lim_{t\to\infty} u''(t) = 0.$$

If

$$F\big[u(t)\big] = \big(u'(t)\big)^2 - 2u(t)u''(t) - b(t)u^2(t), \tag{2.42}$$

then $\lim_{t\to\infty} F(u(t)) = 0$. Clearly, for $T > t$, we get

$$F\big[u(T)\big] = F\big[u(t)\big] + \int_t^T \big(2c(s) - b'(s)\big)u^2(s)\,ds.$$

Hence, taking limit as $T \to \infty$, we obtain

$$2u(t)u''(t) - \big(u'(t)\big)^2 = -b(t)u^2(t) + \int_t^\infty \big(2c(s) - b'(s)\big)u^2(s)\,ds,$$

that is,

$$\frac{2u''(t)}{u(t)} - \left(\frac{u'(t)}{u(t)}\right)^2 = -b(t) + \int_t^\infty \big(2c(s) - b'(s)\big)\frac{u^2(s)}{u^2(t)}\,ds.$$

Since $u(t)$ is positive and decreasing, we have

$$\frac{2u''(t)}{u(t)} - \left(\frac{u'(t)}{u(t)}\right)^2 \leq -b(t) + \int_t^\infty \big(2c(s) - b'(s)\big)\,ds. \tag{2.43}$$

From Lemma 2.1.3, it follows that

$$x(t) = u(t)v'(t) - u'(t)v(t) = u^2(t)\left(\frac{v(t)}{u(t)}\right)'$$

is an oscillatory solution of

$$\left(\frac{x'(t)}{u(t)}\right)' + \left(\frac{u''(t) + b(t)u(t)}{u^2(t)}\right)x = 0. \tag{2.44}$$

The substitution $x = w(t)u^{1/2}(t)$ transforms (2.44) into

$$w'' + \left[b(t) + \frac{3}{4}\left(\frac{2u''(t)}{u(t)} - \frac{(u'(t))^2}{u^2(t)}\right)\right]w = 0. \tag{2.45}$$

We may note that (2.44) is oscillatory, if and only if (2.45) is oscillatory. Now using (2.43), we get

$$\begin{aligned} b(t) + \frac{3}{4}\left(\frac{2u''(t)}{u(t)} - \frac{(u'(t))^2}{u^2(t)}\right) &\leq \frac{1}{4}b(t) + \frac{3}{4}\int_t^\infty \big(2c(s) - b'(s)\big)\,ds \\ &\leq b(t) + \frac{3}{2}\int_t^\infty c(s)\,ds. \end{aligned} \tag{2.46}$$

From Sturm's comparison theorem, it follows that (2.41) is oscillatory, a contradiction. Hence (2.1) is nonoscillatory. This completes the proof of the theorem. □

Remark 2.1.8 If $\lim_{t\to\infty} b(t) = 0$ is not assumed in Theorem 2.1.9, one may proceed as follows: since

$$F\big[u(t)\big] \geq \big(u'(t)\big)^2 - 2u(t)u''(t),$$

we have $\lim_{t\to\infty} F(u(t)) \geq 0$. Consequently,

$$2u(t)u''(t) - \big(u'(t)\big)^2 \leq -b(t)u^2(t) + \int_t^\infty \big(2c(s) - b'(s)\big)u^2(s)\,ds$$

and hence we obtain (2.43). Proceeding as in Theorem 2.1.9, we get

$$b(t) + \frac{3}{4}\left(\frac{2u''(t)}{u(t)} - \frac{(u'(t))^2}{u^2(t)}\right) \leq \frac{1}{4}b(t) + \frac{3}{4}\int_t^\infty \big(2c(s) - b'(s)\big)\,ds.$$

Thus, one may restate Theorem 2.1.9 as follows:

Theorem 2.1.10 *Consider* (2.1) *with* $a(t) \equiv 0$. *If* $2c(t) - b'(t) \geq 0$, $\int_\sigma^\infty (2c(t) - b'(t))\,dt < \infty$ *and the second-order differential equation*

$$z'' + \left[\frac{1}{4}b(t) + \frac{3}{4}\int_t^\infty \big(2c(s) - b'(s)\big)\,ds\right]z = 0$$

is nonoscillatory, then (2.1) *is nonoscillatory.*

Example 2.1.3 Consider

$$x''' - e^{-t}x' + e^{-t}x = 0, \quad t \geq 0.$$

In this case, Eq. (2.41) takes the form $z'' + \frac{1}{2}e^{-t}z = 0$. From Hille's theorem (p. 45, [38]) it follows that this equation is nonoscillatory. Hence from Theorem 2.1.9, it follows that all solutions of the given third-order equation are nonoscillatory.

Example 2.1.4 Consider

$$x''' - \big(1 + e^{-t}\big)x' + e^{-t}x = 0, \quad t \geq 0.$$

In this equation, $2c(t) - b'(t) = e^{-t}$ and

$$\frac{1}{4}b(t) + \frac{3}{4}\int_t^\infty \big(2c(s) - b'(s)\big)\,ds = -\frac{1}{4} + \frac{1}{2}e^{-t} < 0$$

for large t. Thus the second-order differential equation associated with Theorem 2.1.10 is nonoscillatory. By Theorem 2.1.10, all solutions of the equation are nonoscillatory. In particular, $x(t) = e^t$ is a nonoscillatory solution of the equation. We may note that $\lim_{t\to\infty} b(t) \neq 0$.

Now, we provide some results concerning the asymptotic behaviour of nonoscillatory solutions of (2.1) in the presence of an oscillatory solution.

Theorem 2.1.11 *Suppose that* (2.1) *has an oscillatory solution and* $\lim_{t\to\infty} t^2 b(t) \neq 0$. *If* $u(t)$ *is a nonoscillatory solutions of* (2.1), *then* $\lim_{t\to\infty} u(t) = 0$.

Proof From Lemma 1.5.4, we obtain

$$u(t)u'(t)u''(t) \neq 0, \quad \operatorname{sgn} u(t) = \operatorname{sgn} u''(t) \neq \operatorname{sgn} u'(t) = \operatorname{sgn} u'''(t)$$

for $t \geq \sigma$ and

$$\lim_{t\to\infty} u'(t) = \lim_{t\to\infty} u''(t) = 0, \qquad \lim_{t\to\infty} u(t) = \lambda \neq \pm\infty.$$

Without any loss of generality, we may assume that $u(t) > 0$ for $t \geq \sigma$. So $u'(t) < 0$, $u''(t) > 0$, $u'''(t) < 0$ for $t \geq \sigma$ and $\lim_{t\to\infty} u(t) = \lambda$, where $0 \leq \lambda < \infty$. If possible, let $\lambda \neq 0$. Further, without loss of generality, we may assume that $\lambda = \frac{1}{2}$. Since $u'(t) < 0$, $t \geq \sigma$, it is possible to find $t_0 \geq \sigma$ such that $\frac{1}{2} < u(t) < 1$ for $t \geq t_0$.

Suppose that $v(t)$ is an oscillatory solution of (2.1). Then $v(t)$ is also a solution of (2.2). It may be verified that $W(u, v)(t) = u(t)v'(t) - v(t)u'(t)$ is an oscillatory solution of

$$\left(\frac{r(t)z'}{u(t)}\right)' + \left(\frac{r(t)u''(t) + q(t)u(t)}{u^2(t)}\right)z = 0. \tag{2.47}$$

Proceeding as in Jones [19], one may get $\lim_{t\to\infty} t^2 u''(t) = 0$. Since $r(t)$ is monotonic increasing, we have

$$\frac{r(t)}{u(t)} > r(t) > r(\sigma) = 1 \quad \text{for } t \geq t_0.$$

Further,

$$\lim_{t\to\infty} t^2\big[u''(t) + b(t)u(t)\big] < 0$$

implies that

$$\lim_{t\to\infty} t^2\big[r(t)u''(t) + q(t)u(t)\big] = \lim_{t\to\infty} t^2 r(t)\big[u''(t) + b(t)u(t)\big] < 0.$$

Hence the equation

$$z'' + \left[\frac{r(t)u''(t) + q(t)u(t)}{u^2(t)}\right]z = 0 \tag{2.48}$$

is nonoscillatory (see p. 45, [38]). From Sturm's comparison theorem, it follows that (2.47) is nonoscillatory, a contradiction. Hence $\lambda = 0$. The theorem is proved. □

Remark 2.1.9 The condition $\lim_{t\to\infty} t^2 b(t) \neq 0$ in Theorem 2.1.11 may be replaced by the assumption "$r(t)$ is bounded". Indeed, if $r(t)$ is bounded, then there exists a $k > 0$ such that $r(t) < k$ for $t \geq \sigma$. Now

$$\lim_{t\to\infty} t^2 r(t)\big[u''(t) + b(t)u(t)\big] < k \lim_{t\to\infty} t^2 u''(t) = 0.$$

So (2.48) is nonoscillatory.

Theorem 2.1.12 *Suppose that* (2.1) *has an oscillatory solution and* $\int_\sigma^\infty c(t)\,dt = \infty$. *If* $u(t)$ *is a nonoscillatory solution of* (2.1), *then* $\lim_{t\to\infty} u(t) = 0$.

Proof From Lemma 1.5.4, it follows that

$$u(t)u'(t)u''(t) \neq 0, \quad \operatorname{sgn} u(t) = \operatorname{sgn} u''(t) \neq \operatorname{sgn} u'(t) = \operatorname{sgn} u'''(t)$$

for $t \geq \sigma$ and

$$\lim_{t\to\infty} u'(t) = \lim_{t\to\infty} u''(t) = 0, \qquad \lim_{t\to\infty} u(t) = \lambda \neq \pm\infty.$$

Without any loss of generality, we may assume that $u(t) > 0$ for $t \geq \sigma$. So $u'(t) < 0$, $u''(t) > 0$, and $u'''(t) < 0$ for $t \geq \sigma$ and $\lim_{t\to\infty} u(t) = \lambda$, $0 \leq \lambda < \infty$. We claim that $\lambda = 0$. If not, then $\lambda > 0$. Since $u(t)$ satisfies (2.2), integrating (2.2) from σ to t, we obtain

$$\begin{aligned} r(t)u''(t) &= r(\sigma)u''(\sigma) - \int_\sigma^t q(s)u'(s)\,ds - \int_\sigma^t p(s)u(s)\,ds \\ &\leq r(\sigma)u''(\sigma) - u(t)\int_\sigma^t p(s)\,ds \\ &\leq r(\sigma)u''(\sigma) - r(\sigma)u(t)\int_\sigma^t c(s)\,ds \\ &\leq r(\sigma)u''(\sigma) - \lambda\int_\sigma^t c(s)\,ds. \end{aligned}$$

This, in turn, implies that $u''(t) < 0$ for large t, a contradiction. Hence $\lambda = 0$. This completes the proof of the theorem. □

Theorem 2.1.13 *Suppose that* (2.1) *admits an oscillatory solution. If $x(t)$ is a nonoscillatory solution of* (2.1) *with* $\lim_{t\to\infty} x(t) = 0$, *then every nonoscillatory solution of* (2.1) *is a constant multiple of $x(t)$.*

Proof Let $u_1(t)$ and $u_2(t)$ be two solutions of (2.2) on $[\sigma, \infty)$ with initial conditions

$$u_1(\sigma) = u_1'(\sigma) = 0, \qquad r(\sigma)u_1''(\sigma) = 1$$

and

$$u_2(\sigma) = u_2''(\sigma) = 0, \qquad u_2'(\sigma) = -1.$$

From Lemma 2.1.1 and Lemma 1.5.10, it follows that both $u_1(t)$ and $u_2(t)$ are oscillatory.

Clearly, $W_1(t) = W_1(u_1, u_2)(t) = u_1(t)u_2'(t) - u_1'(t)u_2(t)$ is a solution of the adjoint equation

$$\big((r(t)z')' + q(t)z\big)' - p(t)z = 0, \tag{2.49}$$

which is the adjoint of Eq. (2.2). Observe that

$$W_1(\sigma) = W_1'(\sigma) = 0 \quad \text{and} \quad \big(rW_1'\big)'(\sigma) > 0.$$

It is easy to see that $W_1(t) > 0$ for $t > \sigma$. Indeed, $(rW_1')'(\sigma) > 0$ and the continuity of $(rW_1')'(t)$ imply that $(rW_1')'(t) > 0$ for $t \in [\sigma, \sigma + \delta)$, for some $\delta > 0$. This, in turn, implies that $W_1'(t) > 0$ for $t \in (\sigma, \sigma + \delta)$ and hence $W_1(t) > 0$ for $t \in (\sigma, \sigma + \delta)$. We claim that $W_1(t) > 0$ for $t > \sigma$. If not, then there is a $t_1 > \sigma$ such that $W_1(t_1) = 0$ and $W_1(t) > 0$ for $t \in (\sigma, t_1)$. Since $W_1(t)$ is a solution of (2.49), we have

$$\big((r(t)W_1')' + q(t)W_1(t)\big)' = p(t)W_1(t) > 0, \quad t \in (\sigma, t_1)$$

implies that $(rW_1')'(t)+q(t)W_1(t)$ is nondecreasing on $[\sigma,t_1)$. Hence $(rW_1')'(t)>0$ for $t\in[\sigma,t_1]$. Consequently, $W_1'(t)>0$ for $t\in(\sigma,t_1]$, a contradiction. Hence our claim holds. Further, as $u_1(t)$ and $u_2(t)$ are linearly independent oscillatory solutions of the second-order differential equation, we have

$$\begin{vmatrix} u_1(t) & u_2(t) & y \\ u_1'(t) & u_2'(t) & y' \\ u_1''(t) & u_2''(t) & y'' \end{vmatrix} = 0,$$

that is,

$$\left(\frac{y'}{W_1(t)}\right)' + \left(\frac{(rW_1')'(t)+q(t)W_1(t)}{r(t)W_1^2(t)}\right)y = 0. \tag{2.50}$$

So any nontrivial linear combination of $u_1(t)$ and $u_2(t)$ is oscillatory. Clearly, $\{u_1(t),u_2(t),x(t)\}$ forms a basis of solution space of (2.2).

Without any loss of generality, we may assume that $x(t)>0$ for $t\ge t_0^*\ge\sigma$.

If possible, let $w(t)$ be a nonoscillatory solution of (2.1) on $[\sigma,\infty)$ such that $x(t)$ and $w(t)$ are linearly independent. Then

$$w(t)=\lambda_1u_1(t)+\lambda_2u_2(t)+\lambda_3x(t),$$

where λ_1, λ_2 and λ_3 are constants. $\lambda_3=0$ implies that $w(t)$ is oscillatory. Hence $\lambda_3\neq 0$. Dividing $w(t)$ by λ_3, we get $z(t)=x(t)+c_1u_1(t)+c_2u_2(t)$, where $z(t)=\frac{w(t)}{\lambda_3}$, $c_1=\frac{\lambda_1}{\lambda_3}$ and $c_2=\frac{\lambda_2}{\lambda_3}$. Clearly, c_1 and c_2 cannot be equal to zero simultaneously. Since $c_1u_1(t)+c_2u_2(t)$ is oscillatory, there exists a $t_0\ge t_0^*$ such that $x(t)$ and $z(t)$ are of the same sign for $t\ge t_0$.

Setting $x_1(t)=-c_1u_1(t)-c_2u_2(t)$, we obtain $z(t)=x(t)-x_1(t)$. Clearly, $x_1(t)$ is an oscillatory solution of (2.50) and (2.2). Let $t_1>t_0$ be a zero of $x_1(t)$ such that $x_1'(t_1)>0$. Let $x_2(t)$ be a solution of (2.2) on $[t_1,\infty)$ such that $x_2(t_1)=x_2'(t_1)=0$ and $x_2''(t_1)=1$. From Lemma 2.1.1 and Lemma 1.5.10, it follows that $x_2(t)$ is oscillatory. Clearly, $W(t)=W(x_1,x_2)(t)=x_1(t)x_2'(t)-x_1'(t)x_2(t)$ is a solution of (2.49) with $W(t_1)=0=W'(t_1)$ and $(rW')'(t_1)>0$. Hence $W(t)>0$ for $t>t_1$. Consequently, it follows from (2.49) that $(rW')'+qW$ is increasing in $[t_1,\infty)$. So, for $t>t_1$,

$$\begin{aligned}(rW')'(t) &\ge (rW')'(t)+q(t)W(t)\\ &\ge (rW')'(t_1)+q(t_1)W(t_1)\\ &= (rW')'(t_1)>0.\end{aligned} \tag{2.51}$$

This, in turn, implies that $W'(t)>0$ for $t>t_1$. Clearly, $\{x_1(t),x_2(t),x(t)\}$ forms a basis of solution space of (2.2), because $x_1(t)$ and $x_2(t)$ are linearly independent oscillatory solutions of the second-order equation

$$\left(\frac{x'}{W(t)}\right)' + \left(\frac{(rW')'(t)+q(t)W(t)}{r(t)W^2(t)}\right)x = 0. \tag{2.52}$$

So

$$\begin{vmatrix} x_1(t) & x_2(t) & x(t) \\ x_1'(t) & x_2'(t) & x'(t) \\ r(t)x_1''(t) & r(t)x_2''(t) & r(t)x''(t) \end{vmatrix} = K,$$

a nonzero constant, that is,

$$K = W(t)r(t)x''(t) - W'(t)r(t)x'(t) + \big((rW')'(t) + q(t)W(t)\big)x(t).$$

From Lemma 1.5.4, it follows that $x'(t) < 0$ and $x''(t) > 0$ for $t > t_1$. So (2.51) yields $K > 0$ and, for $t > t_1$,

$$0 < \big((rW')'(t) + q(t)W(t)\big)x(t) < K.$$

Let $\langle\sigma_n\rangle$ be an increasing sequence of maximum points of $x_1(t)$ such that $\sigma_n > t_1$ and $\sigma_n \to \infty$ as $n \to \infty$. So $x_1(\sigma_n) > 0$ and $x_1'(\sigma_n) = 0$. Since $z(t) > 0$ for $t \ge t_0$, $x_1(t) < x(t)$. From (2.51), we obtain $(rW')'(t) + q(t)W(t) > 0$ for $t \ge t_1$. Hence

$$\begin{aligned} 0 &< \big[(rW')'(\sigma_n) + q(\sigma_n)W(\sigma_n)\big]x_1(\sigma_n) \\ &< \big[(rW')'(\sigma_n) + q(\sigma_n)W(\sigma_n)\big]x(\sigma_n) \\ &< K. \end{aligned}$$

Further, since $\lim_{n\to\infty} x(\sigma_n) = 0$, we have $\lim_{n\to\infty} x_1(\sigma_n) = 0$. Consequently,

$$\lim_{n\to\infty}\big[(rW')'(\sigma_n) + q(\sigma_n)W(\sigma_n)\big]x_1^2(\sigma_n) = 0. \tag{2.53}$$

On the other hand, if

$$H(t) = r(t)W(t)\big(x_1'(t)\big)^2 + \big((rW')'(t) + q(t)W(t)\big)x_1^2(t),$$

then

$$\begin{aligned} H'(t) &= 2r(t)W(t)x_1'(t)x_1''(t) + (rW)'(t)\big(x_1'(t)\big)^2 \\ &\quad + p(t)W(t)x_1^2(t) + 2\big((rW')'(t) + q(t)W(t)\big)x_1(t)x_1'(t). \end{aligned}$$

Since $x_1(t)$ is a solution of (2.52), we have

$$r(t)W(t)x_1''(t) = r(t)W'(t)x_1'(t) - \big[(rW')'(t) + q(t)W(t)\big]x_1(t)$$

and hence, for $t > t_1$,

$$H'(t) = 2r(t)W(t)\big(x_1'(t)\big)^2 + (rW)'(t)\big(x_1'(t)\big)^2 + p(t)W(t)x_1^2(t) > 0,$$

because $(rW)'(t) = r'(t)W(t) + r(t)W'(t) > 0$. So $H(t)$ is a positive increasing function. But from (2.53), we obtain $\lim_{n\to\infty} H(\sigma_n) = 0$, a contradiction. So $x(t)$ and $z(t)$ are linearly dependent. Consequently, $x(t)$ and $w(t)$ are linearly dependent. Hence the theorem is proved. □

Corollary 2.1.1 *Suppose that* (2.1) *has an oscillatory solution and* $\lim_{t\to\infty} t^2 b(t) \neq 0$ *or* $\int_\sigma^\infty c(t)\,dt = \infty$. *Then all nonoscillatory solutions of* (2.1) *tend to zero as* $t \to \infty$.

Theorem 2.1.14 *Let* $(2c(t) - a(t)b(t) - b'(t))r(t) \geq d > 0$. *If* (2.1) *is oscillatory, then every nonoscillatory solution* $u(t)$ *of* (2.1) *satisfies the property* $u(t) \to 0$, $u'(t) \to 0$ *and* $u''(t) \to 0$ *as* $t \to \infty$.

Proof Lemma 2.1.1, or Lemma 2.1.2, guarantees the existence of a nonoscillatory solution of (2.1). If $u(t)$ has a zero in $[\sigma, \infty)$, then Lemma 1.5.10 implies that $u(t)$ is oscillatory. Hence $u(t) \neq 0$ for $t \geq \sigma$. Since (2.1) is oscillatory, by Lemma 1.5.4, $u(t)$ satisfies the property

$$\begin{aligned} &u(t)u'(t)u''(t) \neq 0, \\ &\operatorname{sgn} u(t) = \operatorname{sgn} u''(t) \neq \operatorname{sgn} u'(t) = \operatorname{sgn} u'''(t), \quad t \geq \sigma \end{aligned}$$

and

$$\lim_{t\to\infty} u'(t) = \lim_{t\to\infty} u''(t) = 0, \qquad \lim_{t\to\infty} u(t) = \lambda \neq \pm\infty.$$

Let $u(t) > 0$ for $t \geq \sigma$. Suppose that $\lambda \neq 0$. Hence for $0 < \epsilon < \lambda$, there exists $T > \sigma$ such that $u(t) > \lambda - \epsilon$ for $t \geq T$. Since (2.1) is oscillatory, from Lemma 1.5.6, it follows that $F[u(t)] < 0$ for $t \geq \sigma$, where

$$\begin{aligned} F\big[u(t)\big] &= r(t)\big(u'(t)\big)^2 - 2r(t)u(t)u''(t) - q(t)u^2(t) \\ &= F\big[u(\sigma)\big] + \int_\sigma^t r'(s)\big(u'(s)\big)^2\,ds + \int_\sigma^t \big(2p(s) - q'(s)\big)u^2(s)\,ds. \end{aligned} \tag{2.54}$$

Hence

$$\begin{aligned} d\int_\sigma^\infty u^2(t)\,dt &\leq \int_\sigma^\infty \big(2c(t) - a(t)b(t) - b'(t)\big)r(t)u^2(t)\,dt \\ &\leq \int_\sigma^\infty \big(2p(t) - q'(t)\big)u^2(t)\,dt \leq -F\big[u(\sigma)\big] < \infty \end{aligned}$$

implies that

$$\int_\sigma^\infty u^2(t)\,dt < \infty.$$

On the other hand,

$$\int_\sigma^\infty u^2(t)\,dt > \int_T^\infty u^2(t)\,dt = \infty,$$

a contradiction. Hence $\lambda = 0$. This completes the proof of the theorem. □

Theorem 2.1.15 *Assume that $c_2(t)$ does not vanish identically on any subinterval of $[\sigma, \infty)$, and*

$$b_1(t) \le b_2(t) \le 0 \quad and \quad 0 \le c_1(t) \le c_2(t)$$

for all $t \ge \sigma$. If the differential equation

$$x''' + a(t)x'' + b_1(t)x' + c_1(t)x = 0$$

is oscillatory on $[\sigma, \infty)$, then the equation

$$x''' + a(t)x'' + b_2(t)x' + c_2(t)x = 0$$

is oscillatory on $[\sigma, \infty)$.

2.2 Behaviour of Solutions of $x''' + a(t)x'' + b(t)x' + c(t)x = 0$ with $a(t) \le 0$, $b(t) \le 0$ and $c(t) > 0$

This section deals with Eq. (2.1) with $a(t) \le 0$, $b(t) \le 0$ and $c(t) > 0$. We state a sufficient condition for the oscillation of (2.1). Further, the asymptotic behaviour of nonoscillatory solutions of (2.1) is given in the presence of an oscillatory solution of (2.1).

Lemma 2.2.1 *Equation* (2.1) *is of type C_I. Hence* (2.49) *is of type C_{II}.*

Theorem 2.2.1 *Suppose that $b(t) - a'(t) \le 0$. If*

$$\int_\sigma^\infty \left[\frac{2a^3(t)}{27} - \frac{a(t)b(t)}{3} + c(t) - \frac{2}{3\sqrt{3}}\left(\frac{a^2(t)}{3} - b(t) + a'(t)\right)^{3/2}\right] dt = \infty, \quad (2.55)$$

then (2.1) *is oscillatory.*

Proof Suppose that $x(t)$ is a nonoscillatory solution of (2.1). From Lemma 1.5.3, it follows that there exists a $t_0 \in [\sigma, \infty)$ such that $x'(t) \le 0$ or ≥ 0 for $t \in [t_0, \infty)$. In view of Lemma 1.5.4 and the second part of Lemma 1.5.3, it is enough to prove that $x(t)x'(t) \ge 0$ for $t \ge t_0$ does not hold. Setting $u(t) = \frac{x'(t)}{x(t)}$, $t \ge t_0$, one may verify that $u(t) \ge 0$ is a solution of the Riccati equation (2.13). Integrating (2.13) from t_0 to t, we obtain

$$\begin{aligned} u'(t) &= u'(t_0) - \frac{3}{2}u^2(t) + \frac{3}{2}u^2(t_0) - a(t)u(t) + a(t_0)u(t_0) \\ &\quad - \int_{t_0}^t \left[u^3(s) + a(s)u^2(s) + \big(b(s) - a'(s)\big)u(s) + c(s)\right] ds. \end{aligned} \quad (2.56)$$

The minimum of $u^3(s) + a(s)u^2(s) + (b(s) - a'(s))u(s) + c(s)$, for $u(s) \geq 0$, is attained at

$$u(s) = \frac{1}{3}\left[-a(s) + \left(a^2(s) - 3\left(b(s) - a'(s)\right)\right)^{1/2}\right]$$

and its minimum value is given by

$$\frac{2a^3(s)}{27} - \frac{a(s)b(s)}{3} + \frac{a(s)a'(s)}{3} + c(s) - \frac{2}{3\sqrt{3}}\left(\frac{a^2(s)}{3} - b(s) + a'(s)\right)^{3/2}.$$

If $H(u(t), t) = -\frac{3}{2}u^2(t) - a(t)u(t)$, then the maximum of $H(u(t), t)$, for $u(t) \geq 0$, is attained at $u(t) = \frac{-a(t)}{3}$ and the maximum value of $H(u(t), t)$ is given by $\frac{a^2(t)}{6}$. Hence from (2.56), we have

$$\begin{aligned}
u'(t) &\leq u'(t_0) + \frac{3}{2}u^2(t_0) + a(t_0)u(t_0) + \frac{a^2(t)}{6} \\
&\quad - \int_{t_0}^{t}\left[\frac{2a^3(s)}{27} - \frac{a(s)b(s)}{3} + \frac{a(s)a'(s)}{3} + c(s)\right. \\
&\quad \left. - \frac{2}{3\sqrt{3}}\left(\frac{a^2(s)}{3} - b(s) + a'(s)\right)^{3/2}\right] ds \\
&\leq u'(t_0) + \frac{3}{2}u^2(t_0) + a(t_0)u(t_0) + \frac{a^2(t_0)}{6} \\
&\quad - \int_{t_0}^{t}\left[\frac{2a^3(s)}{27} - \frac{a(s)b(s)}{3} + c(s) - \frac{2}{3\sqrt{3}}\left(\frac{a^2(s)}{3} - b(s) + a'(s)\right)^{3/2}\right] ds.
\end{aligned}$$

Hence (2.55) implies that $\lim_{t\to\infty} u'(t) = -\infty$. Consequently, $u(t) < 0$ for large t, a contradiction. Then (2.1) is oscillatory. This completes the proof of the theorem. □

Theorem 2.2.2 *Let* $-\infty < \lim_{t\to\infty} ta(t) \leq 0$ *and* $1 - t^2b(t) + \frac{1}{3}t^2a^2(t) + t^2a'(t) \geq 0$. *If* (2.17) *holds, then* (2.1) *is oscillatory.*

Proof If possible, let (2.1) be nonoscillatory. From Lemma 1.5.3 and Lemma 1.5.4, it follows that Eq. (2.1) admits a nonoscillatory solution $x(t)$ such that $x(t)x'(t) \geq 0$ for $t \geq t_0 \geq \sigma$. Setting $u(t) = \frac{t^2x'(t)}{x(t)}$ for $t \geq t_0$, we observe that $u(t) \geq 0$ is a solution of (2.15), where $G(u(t), t)$ is given by (2.16). A simple calculation shows that $G(u(t), t)$ attains the minimum value

$$\begin{aligned}
&\frac{2t^2a^3(t)}{27} - \frac{t^2a(t)b(t)}{3} + t^2c(t) - \frac{2}{3}a(t) + tb(t) + \frac{t^2a(t)a'(t)}{3} - ta'(t) \\
&\quad - \frac{2}{3\sqrt{3}t}\left(1 - t^2b(t) + \frac{1}{3}t^2a^2(t) + t^2a'(t)\right)^{3/2}
\end{aligned}$$

for $u(t) \geq 0$ at

$$u(t) = t\left[\left(1 - \frac{ta(t)}{3}\right) + \frac{1}{\sqrt{3}}\left(1 - t^2 b(t) + \frac{1}{3}t^2 a^2(t) + t^2 a'(t)\right)^{1/2}\right].$$

Integrating (2.15) from t_0 to t and by using the minimum of $\frac{3}{2}\frac{u^2(t)}{t^2} - \frac{4u(t)}{t} + a(t)u(t)$ as $-\frac{1}{6}(ta(t) - 4)^2$ for $u(t) > 0$ at $U(t) = \frac{t^2}{3}(\frac{4}{t} - a(t))$, we obtain

$$u'(t) \leq k + \frac{1}{6}\big(ta(t) - 4\big)^2 - \int_{t_0}^{t} \min G\big(u(s), s\big)\, ds.$$

From the given hypothesis, it follows that $u'(t) \to -\infty$ as $t \to \infty$ and hence $u(t) < 0$ for large t, a contradiction. Thus, the theorem is proved. □

Remark 2.2.1 Theorem 2.2.2 holds for the Euler equation (2.5) with $a_0 \leq 0$, $b(t) \leq 0$ and $c_0 > 0$. Indeed, if (2.18) holds, then (2.5) is oscillatory.

Theorem 2.2.3 *Suppose that* (2.1) *has an oscillatory solution and a nonoscillatory solution which tends to zero as $t \to \infty$. Then all nonoscillatory solutions of* (2.1) *tend to zero as $t \to \infty$.*

Proof Suppose that $x_0(t)$ is a nonoscillatory solution of (2.1) such that $x_0(t) \to 0$ as $t \to \infty$. Without any loss of generality, we may assume that $x_0(t) > 0$ for $t \geq t_0 \geq \sigma$. Let $x(t)$ be any nonoscillatory solution of (2.1). We claim that $\lim_{t\to\infty} x(t) = 0$. If not, assume that $\lim_{t\to\infty} x(t) = \mu_1 \neq 0$. Clearly, $x_0(t)$ and $x(t)$ are linearly independent.

Define the solutions $u_1(t)$ and $u_2(t)$ of (2.2) with the initial conditions

$$u_1(\sigma) = u_1'(\sigma) = 0, \qquad r(\sigma)u_1''(\sigma) = 1$$

and

$$u_2(\sigma) = u_2''(\sigma) = 0, \qquad u_2'(\sigma) = -1.$$

From Lemma 1.5.10, it follows that both $u_1(t)$ and $u_2(t)$ are oscillatory. Let $W_1(t) = W(u_1, u_2)(t) = u_1(t)u_2'(t) - u_1'(t)u_2(t)$. Proceeding as in Theorem 2.1.13, we may get $W_1(t) > 0$ for $t > \sigma$.

Clearly, $u_1(t)$ and $u_2(t)$ are solutions of the second-order differential equation (2.50). Consequently, $\{u_1(t), u_2(t), x_0(t)\}$ forms a basis for the solution space of (2.1), and hence for (2.2). Let $x(t) = c_1x_0(t) + c_2u_1(t) + c_3u_2(t)$. Clearly, $c_1 \neq 0$. Setting $z(t) = \frac{x(t)}{c_1}$ and $x_1(t) = -[\frac{c_2}{c_1}u_1(t) + \frac{c_3}{c_1}u_2(t)]$, we have $z(t) = x_0(t) - x_1(t)$. Again $x_0(t) > 0$ for large t and $x_1(t)$ oscillatory imply that $z(t)$ cannot eventually be negative. Hence $z(t) > 0$ for large t. Consequently, $\mu > 0$ where $\mu = \frac{\mu_1}{c_1} = \lim_{t\to\infty} z(t) > 0$.

Let $\langle\sigma_n\rangle$ be an increasing sequence of maxima of $x_1(t)$ such that $x_1(\sigma_n) \geq 0$, $x_1'(\sigma_n) = 0$ for $\sigma_n > \sigma$. From Lemma 1.5.14, it follows that $z'(t) < 0$. Clearly,

$\lim_{n\to\infty} z(\sigma_n) = \mu$. Again $z(t) > 0$ implies that $x_0(\sigma_n) > x_1(\sigma_n)$. Now

$$0 < \mu = \lim_{n\to\infty} z(\sigma_n) = \lim_{n\to\infty} \big(x_0(\sigma_n) - x_1(\sigma_n)\big) = 0,$$

a contradiction. Hence $\lim_{t\to\infty} z(t) = 0$. Consequently, $\lim_{t\to\infty} x(t) = 0$. This completes the proof. □

Lemma 2.2.2 *If $z_1(t)$ is an oscillatory solution of* (2.49), *then there exists an oscillatory solution $z_2(t)$ of* (2.49) *such that $W(t) = W(z_1, z_2)(t) = z_1(t)z_2'(t) - z_1'(t)z_2(t) > 0$ for large t.*

Proof Let $\beta \geq \sigma$ be such that $z_1(\beta) \neq 0$. Suppose that $\{t_n\}_{n=1}^{\infty}$ is a sequence of zeros of $z_1(t)$ in (β, ∞) such that $t_n \to \infty$ as $n \to \infty$. Define a sequence $\langle x_n(t)\rangle_{n=1}^{\infty}$ of solutions of (2.49) on $[\beta, \infty)$ with the boundary conditions

$$x_n(\beta) = x_n(t_n) = 0, \qquad x_n'(t_n) > 0.$$

Suppose that $\{u_1(t), u_2(t), u_3(t)\}$ forms a basis for the solution space of (2.49). Then there exist real constants c_{1n}, c_{2n} and c_{3n} such that

$$x_n(t) = c_{1n}u_1(t) + c_{2n}u_2(t) + c_{3n}u_3(t)$$

with $c_{1n}^2 + c_{2n}^2 + c_{3n}^2 = 1$.

We claim that the zeros of $z_1(t)$ and $x_n(t)$ separate in (β, t_n). Let α_1 and α_2 ($\alpha_1 < \alpha_2$) be two consecutive zeros of $z_1(t)$ in (β, t_n). If possible, let $x_n(t) > 0$ or < 0 for $t \in [\alpha_1, \alpha_2]$. From Lemma 1.5.13, it follows that there exists a constant λ such that $z_1(t) - \lambda x_n(t)$ has a double zero in (α_1, α_2). This is a contradiction, because (2.49) is of type C_{II} and $z_1(t) - \lambda x_n(t)$ has a zero at t_n. Hence $x_n(t)$ has a zero in $[\alpha_1, \alpha_2]$. Again $x_n(t_n) = z_1(t_n) = 0$ implies that $x_n(\alpha_1) \neq 0$ and $x_n(\alpha_2) \neq 0$ (see Theorem 2.10, [16]). Similarly, if α_1 and α_2 are consecutive zeros of $x_n(t)$ in (β, t_n), then $z_1(t)$ has a zero in (α_1, α_2). Thus the claim holds, that is, the zeros of $z_1(t)$ and $z_2(t)$ separate each other in (β, t_n).

The bounded sequences $\langle c_{i_n}\rangle_{n=1}^{\infty}$, $i = 1, 2, 3$ admit a convergent subsequence, say $\langle c_{i_{n_j}}\rangle$, $i = 1, 2, 3$, respectively. Let $c_i = \lim_{n_j\to\infty} c_{i_{n_j}}$, $i = 1, 2, 3$. So $\langle x_{n_k}(t)\rangle$ converges uniformly to the solution $z_2(t) = c_1u_1(t) + c_2u_2(t) + c_3u_3(t)$ of (2.49). Thus $z_2(\beta) = 0$ and the zeros of $z_1(t)$ and $z_2(t)$ separate in (β, ∞). Since $z_1(\beta) \neq 0$, it follows that $z_1(t)$ and $z_2(t)$ are linearly independent oscillatory solutions of (2.49).

Next, we show that every linear combination of $z_1(t)$ and $z_2(t)$ is oscillatory. If possible, suppose that $\mu_1 z_1(t) + \mu_2 z_2(t)$ is nonoscillatory, where μ_1 and μ_2 are some constants. Without any loss of generality, we suppose that $\mu_1 z_1(t) + \mu_2 z_2(t) > 0$ for $t \geq t_0 > \beta$. Let t_1, t_2 and t_3 be successive zeros of $z_1(t)$ in $[t_0, \infty)$. So $\mu_2 z_2(t_i) > 0$, $i = 1, 2, 3$. This contradicts the fact that the zeros of $z_1(t)$ and $z_2(t)$ separate in (β, ∞). Hence every linear combination of $z_1(t)$ and $z_2(t)$ is oscillatory. Next, we prove that $W[z_1, z_2](t) \neq 0$ for $t \in (\beta, \infty)$. If not, then there exists a γ, $\gamma > \beta$ such that $W[z_1, z_2](\gamma) = 0$, that is, $z_1(\gamma)z_2'(\gamma) - z_1'(\gamma)z_2(\gamma) = 0$. Since $z_1(t)$ and $z_2(t)$ separate their zeros in (β, ∞), both of $z_1(\gamma)$ and $z_2(\gamma)$ are not equal to zero. Hence $v(t) = z_1(\gamma)z_2(t) - z_1(t)z_2(\gamma)$ is a nontrivial solution of (2.49) with $v(\gamma) = v'(\gamma) = 0$ and $v''(t) \neq 0$. Consequently, $v(t) > 0$ or < 0 for $t > \gamma$, a contradiction to the fact that every linear combination of $z_1(t)$ and $z_2(t)$ is oscillatory. Hence $W[z_1, z_2](t) > 0$ or < 0 for $t > \beta$. If $W[z_1, z_2](t) < 0$ for $t > \beta$, then one may take $-z_2(t)$ in place of $z_2(t)$ to obtain the required result. This completes the proof of the lemma. □

Lemma 2.2.3 *$z_1(t)$ and $z_2(t)$ are linearly independent oscillatory solutions of the second-order differential equation*

$$\left(\frac{r(t)z'}{u(t)}\right)' + \left(\frac{r(t)u''(t) + q(t)u(t)}{u^2(t)}\right)z = 0, \tag{2.57}$$

where $u(t) = r(t)W(t)$.

Proof Clearly, $z_1(t)$ and $z_2(t)$ are solutions of the second-order differential equation

$$\begin{vmatrix} z_1(t) & z_2(t) & y(t) \\ z_1'(t) & z_2'(t) & y'(t) \\ (rz_1')'(t) & (rz_2')'(t) & (ry')'(t) \end{vmatrix} = 0.$$

Expanding the above determinant, we obtain (2.57). □

Remark 2.2.2 $u(t)$ is a positive solution of (2.2).

Lemma 2.2.4 *Suppose that $a(t) \leq \alpha < 0$ for $t \geq \beta \geq \sigma$. If $z(t)$ is a solution of (2.49) with the initial conditions $z(\beta) = z'(\beta) = 0$, $z''(\beta) > 0$, then $z(t)$ has the following properties:*

$$z(t) > 0, \qquad z'(t) > 0, \qquad z''(t) > 0 \quad \text{and} \quad \left(rz'\right)'(t) > 0$$

for $t \geq \beta$. Moreover, for every real number $M > 0$, there exists a real $N > 0$ such that

$$z(t) > -\frac{\alpha M}{2} r(t)\left(\int_\beta^t \frac{ds}{r(s)}\right)^2$$

for $t > N$.

Proof Since (2.49) is of type C_{II}, we have $z(t) > 0$ for $t > \beta$. From (2.49), we have $(rz')'(t) + q(t)z(t) > (rz')'(\beta) + q(\beta)z(\beta) > 0$ for $t > \beta$. Integrating this inequality from β to t, we get $r(t)z'(t) > r(\beta)z''(\beta)(t - \beta)$. Then, $z'(t) > 0$ for $t > \beta$. Moreover, for every real number $M > 0$, there exists an integer $N_1 > 0$ such that $z'(t) > \frac{M}{r(t)}$ for $t > N_1$. Again $(rz')'(t) > r(\beta)z''(\beta)$ implies that $r(t)z''(t) > -r'(t)z'(t) + r(\beta)z''(\beta) > 0$. Thus, $z''(t) > 0$ for $t > \beta$ and

$$z''(t) > -\frac{r'(t)}{r(t)}z'(t) + \frac{r(\beta)z''(\beta)}{r(t)} > -a(t)\frac{M}{r(t)} + z''(\beta) > -\frac{\alpha M}{r(t)} + z''(\beta).$$

Integrating the above inequality from N_1 to t, we have

$$z'(t) > z'(N_1) - \alpha M \int_{N_1}^{t} \frac{ds}{r(s)} + z''(\beta)(t - N_1).$$

But there exists $N_2 > N_1$ such that for $t > N_2$, we have

$$-\alpha M \int_{\beta}^{N_1} \frac{ds}{r(s)} < z''(\beta)(t - N_1).$$

Hence, for $t > N_2$,

$$z'(t) > z'(N_1) - \alpha M \int_{\beta}^{t} \frac{ds}{r(s)}.$$

Dividing both sides by $r(t)$ and integrating the resulting inequality from N_2 to t, we obtain

$$\frac{z(t)}{r(t)} > \frac{z(N_2)}{r(N_2)} + z'(N_1)(t - N_2) - \alpha M \int_{N_2}^{t} \left(\frac{1}{r(s)} \int_{s}^{\beta} \frac{d\theta}{r(\theta)} \right) ds. \tag{2.58}$$

It is possible to choose $N > 0$ such that for $t > N > N_2$,

$$-\alpha M \int_{\beta}^{N_2} \left(\frac{1}{r(s)} \int_{\beta}^{s} \frac{d\theta}{r(\theta)} \right) ds < z'(N_1)(t - N_2).$$

Hence (2.58) yields

$$z(t) > -\alpha M r(t) \int_{\beta}^{t} \left(\frac{1}{r(s)} \int_{s}^{\beta} \frac{d\theta}{r(\theta)} \right) ds = -\frac{\alpha M}{2} r(t) \left(\int_{\beta}^{t} \frac{ds}{r(s)} \right)^2.$$

The proof is complete. □

Theorem 2.2.4 *If $a(t) \le \alpha < 0$ for $t \ge \sigma$ and* (2.1) *is nonoscillatory, then* (2.1) *admits a nonoscillatory solution which tends to zero as $t \to \infty$.*

Proof Clearly (2.1) is oscillatory, if and only if (2.2) is oscillatory. From Lemma 1.5.11, it follows that (2.49) is oscillatory. Suppose that $z_1(t)$ is an oscillatory solution of (2.49). By Lemma 2.2.2, there exists an oscillatory solution $z_2(t)$ of (2.49) such that $W[z_1, z_2](t) > 0$ for $t > \beta$. From Remark 2.2.2, it follows that $u(t) = r(t)W(t)$ is a positive solution of (2.2) and hence of (2.1). So $u'(t) < 0$, $u''(t) > 0$ for $t \ge \beta$, $\lim_{t\to\infty} u'(t) = \lim_{t\to\infty} u''(t) = 0$ and $\lim_{t\to\infty} u(t) = \lambda \ne \infty$. If possible, suppose that $\lambda > 0$.

Let $z(t)$ be a solution of (2.49) with $z(\beta) = z'(\beta) = 0$ and $z''(\beta) > 0$. So, $z(t) > 0$ for $t > \beta$. Clearly, the set $\{z_1(t), z_2(t), z(t)\}$ forms a basis for the solution space of (2.49) and

$$\frac{d}{dt}\left\{ r(t) \begin{vmatrix} z_1(t) & z_2(t) & z(t) \\ z_1'(t) & z_2'(t) & z'(t) \\ (rz_1')'(t) & (rz_2')'(t) & (rz')'(t) \end{vmatrix} \right\} = 0. \tag{2.59}$$

Integrating (2.59) from β to t, we have

$$\begin{vmatrix} z_1(t) & z_2(t) & z(t) \\ z_1'(t) & z_2'(t) & z'(t) \\ (rz_1')'(t) & (rz_2')'(t) & (rz')'(t) \end{vmatrix} = \frac{W(\beta)z''(\beta)}{r(t)}. \tag{2.60}$$

Expanding (2.60), we get

$$\left(\frac{r(t)z'(t)}{u(t)}\right)' + \left(\frac{r(t)u''(t) + q(t)u(t)}{u^2(t)}\right)z(t) = \frac{W(\beta)z''(\beta)}{u^2(t)}. \tag{2.61}$$

Since $u'(t) < 0$ for $t \geq \beta$, from Lemma 2.2.4, it follows that $(\frac{r(t)z'(t)}{u(t)})' > 0$ for $t \geq \beta$. Hence, for $t \geq \beta$, (2.61) yields

$$\big(r(t)u''(t) + q(t)u(t)\big)z(t) < W(\beta)z''(\beta). \tag{2.62}$$

Since $z_1(t)$ and $z_2(t)$ are oscillatory solutions of (2.57), from Theorem 2.3 due to Ohriska [25], we conclude that

$$\limsup_{t\to\infty}\left\{\left(\int_\beta^t \frac{u(s)}{r(s)}\,ds\right)^2 \frac{r(t)}{u(t)}\left(\frac{r(t)u''(t) + q(t)u(t)}{u^2(t)}\right)\right\} = \mu \geq \frac{1}{4}. \tag{2.63}$$

So there exists a sequence $\{t_n\}_{n=1}^{\infty}$, $t_n \to \infty$ as $n \to \infty$, such that

$$\left(\int_\beta^{t_n} \frac{u(s)}{r(s)}\,ds\right)^2 \frac{r(t_n)}{u(t_n)}\left(\frac{ru'' + qu}{u^2}\right)(t_n) > 0, \tag{2.64}$$

for every n and

$$\lim_{n\to\infty}\left\{\left(\int_\beta^{t_n} \frac{u(s)}{r(s)}\,ds\right)^2 \frac{r(t_n)}{u(t_n)}\left(\frac{ru'' + qu}{u^2}\right)(t_n)\right\} = \mu.$$

For $M = -\frac{16u^2(\beta)W(\beta)z''(\beta)}{\alpha\lambda^3} > 0$, there exists a real number $N > \beta$ (by Lemma 2.2.4) such that

$$\left(\int_\beta^{t_n} \frac{u(s)}{r(s)}\,ds\right)^2 \leq u^2(\beta)\left(\int_\beta^{t_n} \frac{ds}{r(s)}\right)^2 \leq -\frac{2u^2(\beta)z(t_n)}{\alpha M r(t_n)}, \quad t_n \geq N.$$

This, in turn, because of (2.64) and (2.62), implies that

$$\left(\int_\beta^{t_n} \frac{u(s)}{r(s)}\,ds\right)^2 \frac{r(t_n)}{u(t_n)}\left(\frac{ru'' + qu}{u^2}\right)(t_n) \leq -\frac{2u^2(\beta)W(\beta)z''(\beta)}{\alpha M\lambda^3} = \frac{1}{8}.$$

Then $\frac{1}{4} \leq \mu \leq \frac{1}{8}$, a contradiction. Consequently, $\lim_{t\to\infty} u(t) = 0$. This completes the proof of the theorem. □

Theorem 2.2.5 *Suppose that* $a(t) \leq \alpha < 0$ *for* $t \geq \sigma$. *Then* (2.1) *is oscillatory, if and only if all nonoscillatory solutions of* (2.1) *tend to zero as* $t \to \infty$.

Proof Suppose that (2.1) is oscillatory. From Theorems 2.2.3 and 2.2.4, it follows that all nonoscillatory solutions of (2.1) tend to zero as $t \to \infty$.

Conversely, suppose that all nonoscillatory solutions of (2.1) tend to zero as $t \to \infty$. Let $u(t)$ be a nonoscillatory solution of (2.1). From Lemma 1.5.2, it follows that $u(t)u'(t) \geq 0$ or ≤ 0 for $t \geq t_0 \geq \sigma$. Since $\lim_{t\to\infty} u(t) = 0$, it follows that $u(t)u'(t) < 0$ for $t \geq t_0$. Then (2.1) is oscillatory, by Lemma 1.5.4. The proof is complete. □

Remark 2.2.3 Theorem 2.2.5 holds good if we replace the condition "$a(t) \leq \alpha < 0$" by "$\int_\sigma^\infty a(s)\,ds \neq -\infty$", which we give here as a theorem:

Theorem 2.2.6 *Suppose that* $\int_\sigma^\infty a(s)\,ds \neq -\infty$. *Then* (2.1) *is oscillatory, if and only if all nonoscillatory solutions of* (2.1) *tend to zero as* $t \to \infty$.

Proof In view of Theorem 2.2.3 and the second part of Theorem 2.2.5, it is enough to show that (2.2) admits a nonoscillatory solution which tends to zero as $t \to \infty$.

From Remark 2.2.2, it follows that $u(t) = r(t)W(t)$ is a positive solution of (2.2) and hence of (2.1). From Lemma 1.5.4, we have $u'(t) < 0$ for $t \geq \beta$. If possible, let $\lim_{t\to\infty} u(t) = \lambda > 0$. Without any loss of generality, we may take $\lambda = \frac{1}{2}$. Let $K = \exp(\int_\sigma^\infty a(s)\,ds)$. Let $M > 0$ be such that $u(t) < 1$ for $t > M$. Hence, for $t > M$, $\frac{r(t)}{u(t)} > K$. From Lemmas 2.2.3 and 1.5.23, it follows that the second-order differential equation

$$z'' + \frac{1}{K}\left(\frac{r(t)u''(t) + q(t)u(t)}{u^2(t)}\right)z = 0 \tag{2.65}$$

is oscillatory. Proceeding as in Jones [19], we may show that (2.65) is nonoscillatory, which is a contradiction. Hence $\lim_{t\to\infty} u(t) = 0$. The proof is complete. □

Example 2.2.1 By Theorem 2.2.1, the equation

$$x''' - \left(1 - \frac{1}{t^2}\right)x'' - \frac{2}{t^3}x' + \left(2 - \frac{1}{t^2} - \frac{2}{t^3}\right)x = 0$$

is oscillatory. Theorem 2.2.5 is applicable to this equation, whereas Theorem 2.2.6 is not applicable to this example. Clearly, $x(t) = e^{-t}$ is a nonoscillatory solution of this equation which tends to zero as $t \to \infty$.

Example 2.2.2 By Theorem 2.2.1, the equation

$$x''' - \frac{1}{t^2}x'' - \frac{2}{t^3}x' + \left(8 + \frac{4}{t^2} - \frac{4}{t^3}\right)x = 0, \quad t \geq 5$$

is oscillatory. Theorem 2.2.6 is applicable to this example, whereas Theorem 2.2.5 cannot be applied to this example. This equation admits a nonoscillatory solution $x(t) = e^{-2t}$, which tends to zero as $t \to \infty$.

2.3 Behaviour of Solutions of $x''' + a(t)x'' + b(t)x' + c(t)x = 0$ with $a(t) \leq 0$, $b(t) \leq 0$ and $c(t) < 0$

In this section, we study the oscillation, nonoscillation and their asymptotic behaviour of Eq. (2.1), when $a(t) \leq 0$, $b(t) \leq 0$ and $c(t) < 0$.

Lemma 2.3.1 *Equation* (2.1) *is of type* C_{II}.

Proof Let $x(t)$ be a solution of (2.1) with $x(t_0) = 0 = x'(t_0)$ and $x''(t_0) > 0$ for $t_0 > \sigma$. From the continuity of $x''(t)$, it follows that $x''(t) > 0$, $t \in (t_0, t_0 + \delta)$ for some $\delta > 0$. We claim that $x''(t) > 0$ for $t > t_0$. If not, there exists a $t_1 > t_0$ such that $x''(t_1) = 0$ and $x(t) > 0$, $x'(t) > 0$ and $x''(t) > 0$ for $t \in (t_0, t_1)$. Since $x(t)$ satisfies (2.2), integrating (2.2) from t_0 to t_1, we obtain

$$0 > -r(t_0)x''(t_0) = -\int_{t_0}^{t_1} \big[q(t)x'(t) + p(t)x(t)\big]\,dt > 0,$$

a contradiction. Hence our claim holds. This in turn implies that $x(t) > 0$ for $t > t_0$. The proof of the lemma is complete. □

Remark 2.3.1 From Lemma 2.3.1, it follows that a solution $x(t)$ of (2.1) with the property $x(t_0) = x'(t_0) = 0$, $x''(t_0) > 0$ satisfies $x(t) > 0$, $x'(t) > 0$, $x''(t) > 0$ for $t > t_0$.

In Lemma 2.3.2, we do not assume that $b(t) \leq 0$ and $c(t) < 0$.

Lemma 2.3.2 *If* $2c(t) - b'(t) - a(t)b(t) \leq 0$ *but* $\not\equiv 0$ *on any subinterval of* $[\sigma, \infty)$, *then* (2.1) *is of type* C_{II}.

Proof Suppose that $x(t)$ is a solution of (2.1) with $x(t_0) = x'(t_0) = 0$ and $x''(t_0) > 0$, $t_0 \geq \sigma$. We claim that $x(t) > 0$ for $t > t_0$. If not, then $x(t_1) = 0$ for some $t_1 > t_0$. Since $x(t)$ is a solution of (2.1), it satisfies (2.2). Multiplying (2.2) through by $x(t)$ and integrating the resulting identity from t_0 to t_1, we obtain

$$r(t_1)\big(x'(t_1)\big)^2 \leq \int_{t_0}^{t_1} \big(2p(t) - q'(t)\big)x^2(t)\,dt,$$

that is,

$$0 < r(t_1)\big(x'(t_1)\big)^2 \le \int_{t_0}^{t_1} \big(2c(t) - b'(t) - a(t)b(t)\big)r(t)x^2(t)\,dt < 0,$$

a contradiction. Hence $x(t) > 0$ for $t > t_0$. This completes the proof of the lemma. □

Lemma 2.3.3 *Equation* (2.49) *is of type* C_I.

Lemma 2.3.4 *Equation* (2.2) *is oscillatory if and only if* (2.49) *is oscillatory.*

Theorem 2.3.1 *Suppose that* $2c(t) - b'(t) - a(t)b(t) \le 0$ *but* $\not\equiv 0$ *on any subinterval of* $[\sigma, \infty)$ *such that*

$$\int_{\sigma}^{\infty} \big(2c(t) - b'(t) - a(t)b(t)\big)\exp\left(\int_{\sigma}^{t} a(s)\,ds\right)dt = -\infty. \tag{2.66}$$

Then Eq. (2.1) *is oscillatory, if and only if all nonoscillatory solutions of the second-order differential equation* (2.11) *are eventually positive.*

Proof Suppose that all nonoscillatory solutions of (2.11) are eventually positive. We have to show that (2.1) has an oscillatory solution. If possible, let all solutions of (2.1) be nonoscillatory. From Lemma 1.5.19, it follows that there exists a nonoscillatory solution $u(t)$ of (2.1) which does not satisfy the conditions

$$\left.\begin{aligned} &u(t)u'(t)u''(t) \ne 0 \quad \text{for } t \ge t_0 \ge \sigma,\\ &\operatorname{sgn} u(t) = \operatorname{sgn} u'(t) = \operatorname{sgn} u''(t), \quad t \ge t_0 \ge \sigma. \end{aligned}\right\} \tag{2.67}$$

Without any loss of generality, we may assume that $u(t) > 0$ for $t \ge t_0 \ge \sigma$. From Lemma 1.5.14, it follows that all solutions of the second-order nonhomogeneous differential equation

$$\big(r(t)x'(t)\big)' + q(t)x(t) = p(t)u(t), \quad t > t_0,$$

are nonoscillatory. Since $-u'(t)$ satisfies the above equation, $-u'(t)$ is nonoscillatory. Let $u'(t) > 0$ or < 0 for $t \ge t_1 \ge t_0$. If $\operatorname{sgn} u(t) = \operatorname{sgn} u'(t)$, then $u'(t) > 0$ for $t \ge t_1$. Since $u(t)$ satisfies (2.2), multiplying (2.2) through by $u(t)$ and integrating the resulting identity from t_1 to t, we have

$$F\big[u(t)\big] = F\big[u(t_1)\big] + \int_{t_1}^{t} \big(2p(s) - q'(s)\big)u^2(s)\,ds + \int_{t_1}^{t} r'(s)\big(u'(s)\big)^2\,ds, \tag{2.68}$$

where $F[u(t)]$ is given in (2.54). Since $r'(t) \le 0$ and $u(t)$ is increasing, we have

$$\begin{aligned} F\big(u(t)\big) &\le F\big(u(t_1)\big) + \int_{t_1}^{t} \big(2p(s) - q'(s)\big)u^2(s)\,ds\\ &\le F\big(u(t_1)\big) + u^2(t_1)\int_{t_1}^{t} \big(2c(s) - b'(s) - a(s)b(s)\big)r(s)\,ds. \end{aligned}$$

Hence $F(u(t)) \to -\infty$ as $t \to \infty$. Let $t_2 > t_1$ such that $F(u(t)) < 0$ for $t \geq t_2$. Hence

$$-2r(t)u(t)u''(t) < -r(t)\big(u'(t)\big)^2 + q(t)u^2(t) < 0, \quad t > t_2.$$

Consequently, $u(t)u''(t) > 0$ for $t > t_2$. This shows that $u(t)$ satisfies (2.67), a contradiction. Hence $\operatorname{sgn} u(t) \neq \operatorname{sgn} u'(t)$. Now taking $z(t) = \frac{u'(t)}{u(t)}$, it may be easily verified that $z(t)$ is a negative nonoscillatory solution of (2.11), a contradiction again. Hence (2.1) admits an oscillatory solution.

Conversely, suppose that (2.1) has an oscillatory solution. We have to prove that all nonoscillatory solutions of (2.11) are eventually positive. If possible, suppose that $z(t)$ is a negative nonoscillatory solution of (2.11) in $[t_3, \infty)$, $t_3 \geq \sigma$. It may easily be shown that $u(t) = \exp(\int_{t_3}^t z(s)\,ds)$ is a positive nonoscillatory solution of (2.1) with $u'(t) < 0$. It contradicts Lemma 1.5.19. Hence all nonoscillatory solutions of (2.11) are eventually positive. This completes the proof of the theorem. □

Theorem 2.3.2 *Suppose that* $a'(t) \geq 0$, $b(t) - 2a'(t) \leq 0$ *and* $c(t) - b'(t) + a''(t) < 0$. *If*

$$\int_\sigma^\infty \bigg[-\frac{2}{27}a^3(t) + \frac{1}{3}a(t)b(t) - c(t) - \frac{2}{3}a(t)a'(t) + b'(t) - a''(t) - \frac{2}{3\sqrt{3}}\left(\frac{a^2(t)}{3} - b(t) + 2a'(t)\right)^{3/2}\bigg]dt = \infty, \tag{2.69}$$

then (2.1) *is oscillatory.*

Proof Since (2.1) is of type C_{II}, from Lemma 1.5.11 it follows that (2.1) is oscillatory if and only if its adjoint (2.3) is oscillatory, that is, if and only if

$$x''' - a(t)x'' + \big(b(t) - 2a'(t)\big)x' + \big(c(t) - b'(t) + a''(t)\big)x = 0$$

is oscillatory. Clearly, Eq. (2.3) satisfies hypotheses of Theorem 2.1.2. So it is oscillatory, and hence (2.1) is oscillatory. The theorem is proved. □

Note that Theorem 2.3.2 does not contain any sign restrictions on $b(t)$ and $c(t)$.

Theorem 2.3.3 *Suppose that* $b(t) - 2a'(t) \leq 0$, $c(t) - b'(t) + a''(t) < 0$, $ta(t) \geq -3$ *and* $1 - t^2 b(t) + \frac{t^2 a^2(t)}{3} + t^2 a'(t) \geq 0$. *If*

$$\int_\sigma^\infty \bigg[-\frac{2t^2a^3(t)}{27} + \frac{t^2a(t)b(t)}{3} - t^2c(t) + \frac{2}{3}a(t) + tb(t) - \frac{t^2a(t)a'(t)}{3} - ta'(t) - t^2a''(t) + t^2b'(t) - \frac{2}{3\sqrt{3}t}\left(1 - t^2b(t) + \frac{1}{3}t^2a^2(t) + t^2a'(t)\right)^{3/2}\bigg]dt = \infty, \tag{2.70}$$

then (2.1) *is oscillatory.*

Theorem 2.3.4 *Suppose that* $b(t) - 2a'(t) \le 0$, $c(t) - b'(t) + a''(t) < 0$, $ta(t) \le -3$ *and* $2ta(t) + t^2 b(t) - t^2 a'(t) < -2$. *If* (2.70) *holds, then* (2.1) *is oscillatory.*

Remark 2.3.2 Applying Theorem 2.3.3 to the Euler equation (2.5), it follows that (2.5) is oscillatory if $-3 \le a_0 < 0$, $b_0 < 0$, $c_0 < 0$ and

$$-\frac{2a_0^3}{27} + \frac{a_0^2}{3} - \frac{a_0}{3} + \frac{a_0 b_0}{3} - b_0 - c_0 - \frac{2}{3\sqrt{3}}\left(1 - b_0 + \frac{a_0^2}{3} - a_0\right)^{3/2} > 0 \quad (2.71)$$

holds. Similarly, applying Theorem 2.3.4 to the Euler equation (2.5), we observe that (2.5) is oscillatory if $a_0 \le -3$, $b_0 < 0$, $c_0 < 0$, $3a_0 + b_0 < -2$ and (2.71) holds.

One may apply Theorem 2.3.2 to Eq. (1.5). We find that (1.5) is oscillatory if (1.9) holds. Thus, one may treat Theorem 2.3.2 as a generalisation of Proposition 1.2.3(i).

Example 2.3.1 By Theorem 2.3.3, the equation

$$x''' - \frac{1}{t}x'' - \frac{1}{t^2}x' - \frac{8}{t^3}x = 0, \quad t \ge 1$$

is oscillatory. In particular, $x_1(t) = \cos(\sqrt{2}\log t)$ and $x_2(t) = \sin(\sqrt{2}\log t)$ are the two oscillatory solutions of this equation.

Example 2.3.2 Consider

$$x''' - \frac{6}{t}x'' - \frac{1}{t^2}x' - \frac{63}{t^3}x = 0, \quad t \ge 1.$$

Since all conditions of Theorem 2.3.4 are satisfied, this equation has an oscillatory solution. One may observe that $x_1(t) = \cos(\sqrt{7}\log t)$ and $x_2(t) = \sin(\sqrt{7}\log t)$ are the oscillatory solutions of the equation.

Theorem 2.3.5 *Suppose that* $2c(t) - b'(t) - a(t)b(t) < 0$ *but* $\not\equiv 0$ *on any subinterval of* $[\sigma, \infty)$ *such that* (2.66) *holds and* (2.1) *has a nonoscillatory solution* $M(t)$ *such that* $M(t) > 0$ *and* $M'(t) > 0$ *for* $t > \sigma$. *Then* (2.1) *is oscillatory if and only if the second-order differential equation* (2.34) *is oscillatory.*

Proof If (2.1) has an oscillatory solution $x(t)$, then it is easy to verify that $(\frac{x(t)}{M(t)})'$ is an oscillatory solution of (2.34). Hence (2.1) being oscillatory implies that (2.34) is oscillatory.

Conversely, suppose that (2.34) is oscillatory. Further, if possible, suppose that all solutions of (2.1) are nonoscillatory. By Lemma 1.5.19, there exists at least one nonoscillatory solution $u(t)$ of (2.1) which does not satisfy the condition

$$u(t)u'(t)u''(t) \ne 0 \quad \text{for } t \ge t_0 \ge \sigma.$$

$$\operatorname{sgn} u(t) = \operatorname{sgn} u'(t) = \operatorname{sgn} u''(t) \quad \text{for } t \ge t_0 \ge \sigma.$$

Without any loss of generality, let us assume that $u(t) > 0$ for $t \ge t_0 \ge \sigma$. Clearly, $u(t)$ is a solution of (2.2) implies that $u'(t)$ is a solution of the second-order differential equation

$$\big(r(t)z'\big)' + q(t)z = -p(t)u, \quad t > t_0. \tag{2.72}$$

From Lemma 1.5.14, it follows that all solutions of (2.72) are nonoscillatory. So $u'(t)$ is nonoscillatory. Proceeding as in Theorem 2.3.1, we obtain $u'(t) < 0$ for $t \ge t_1 \ge t_0$. It is easy to verify that $(\frac{u(t)}{M(t)})'$, $t \ge t_1$, is a solution of (2.34). Since all solutions of (2.34) are oscillatory, $(\frac{u(t)}{M(t)})'$ is oscillatory. But

$$\left(\frac{u(t)}{M(t)}\right)' = \big[M(t)u'(t) - M'(t)u(t)\big]M^{-2}(t) < 0, \quad t \ge t_1,$$

a contradiction. Hence (2.1) is oscillatory. This completes the proof of the theorem. □

Remark 2.3.3 Proceeding as in Lemma 2.3.1, one may show that (2.1) admits a solution $M(t)$ with the initial conditions $M(\sigma) = 0 = M'(\sigma)$ and $M''(\sigma) > 0$ which satisfies conditions $M(t) > 0$ and $M'(t) > 0$ for $t > \sigma$.

The following theorem shows that, if (2.1) has an oscillatory solution, then the set of all oscillatory solutions of (2.1) forms a two-dimensional subspace of its solution space.

Theorem 2.3.6 *Suppose that $b(t) \not\equiv 0$ on any subinterval of $[\sigma, \infty)$. If (2.1) has an oscillatory solution, then there exist two linearly independent oscillatory solutions $x_1(t)$ and $x_2(t)$ of (2.1) whose zeros separate and such that any oscillatory solution of (2.1) can be expressed as a linear combination of $x_1(t)$ and $x_2(t)$.*

In the following, we provide some results which are interesting in themselves and helpful in establishing Theorem 2.3.6.

Theorem 2.3.7 *Equation (2.49) admits a nonoscillatory solution $N(t)$ satisfying $N(t) > 0$, $N'(t) < 0$ and $(rN')'(t) + q(t)N(t) > 0$ for $t \in [\sigma, \infty)$.*

Proof Let $\{u_1(t), u_2(t), u_3(t)\}$ be a basis for the solution space of (2.49). Let $z_n(t)$ be a solution of (2.49) with $z_n(n) = z_n'(n) = 0$, $z_n''(n) > 0$, $n \ge \sigma$. So there exist constants c_{1n}, c_{2n}, c_{3n} such that

$$z_n(t) = c_{1n}u_1(t) + c_{2n}u_2(t) + c_{3n}u_3(t),$$

with $c_{1n}^2 + c_{2n}^2 + c_{3n}^2 = 1$. Since $\langle c_{in}\rangle_{n=1}^{\infty}$, $i = 1, 2, 3$ is a bounded sequence of real numbers, it admits a convergent subsequence. Without any loss of generality, we denote this convergent subsequence by $\langle c_{in_j}\rangle_{n=1}^{\infty}$, $i = 1, 2, 3$. Let $\lim_{n_j\to\infty} c_{in_j} = c_i$ for $i = 1, 2, 3$. Let

$$N(t) = c_1u_1(t) + c_2u_2(t) + c_3u_3(t).$$

Since (2.49) is of type C_I, we have $z_n(t) > 0$ for $t \in [\sigma, n)$. Now from (2.49) it follows that

$$\left[\left(r(t)z_n'(t)\right)' + q(t)z_n(t)\right]' = p(t)z_n(t) < 0, \quad t \in [\sigma, n).$$

Thus, for $\sigma \leq t < n$,

$$\begin{aligned}\left(rz_n'\right)'(t) + q(t)z_n(t) &> \left(rz_n'\right)'(n) + q(n)z_n(n)\\ &= r(n)z_n''(n) + r'(n)z_n'(n) + q(n)z_n(n)\\ &= r(n)z_n''(n) > 0,\end{aligned}$$

that is,

$$\left(rz_n'\right)'(t) > -q(t)z_n(t) > 0.$$

This, in turn, implies that $r(t)z_n'(t)$ is increasing in $[\sigma, n)$. Hence, for $\sigma \leq t < n$, $r(t)z_n'(t) < r(n)z_n'(n) = 0$ implies that $z_n'(t) < 0$.

Since the sequences $\langle z_n \rangle$, $\langle z_n' \rangle$ and $\langle (rz_n')' + qz_n \rangle$ converge uniformly to N, N' and $(rN')' + qN$, respectively, on any compact interval of $[\sigma, \infty)$, it follows that $N(t) \geq 0$, $N'(t) \leq 0$ and $(rN')'(t) + q(t)N(t) \geq 0$ for $t \in [\sigma, \infty)$. If there is a point $t_1 \geq \sigma$ such that $N(t_1) = 0$, then $N(t) \equiv 0$ for $t \geq t_1$, because $N'(t) \leq 0$ for all $t \in [\sigma, \infty)$. But $c_1^2 + c_2^2 + c_3^2 = 1$ implies that $N(t)$ is a nontrivial solution of (2.49), a contradiction. Thus $N(t) > 0$ for $t \in [\sigma, \infty)$. Further, $(rN')'(t) + q(t)N(t) \geq 0$, $t \geq \sigma$ implies that $(rN')'(t) \geq -q(t)N(t) \geq 0$. Consequently, $r(t)N'(t)$ is nondecreasing on $[\sigma, \infty)$. If $(rN')'(t_2) = 0$ for some $t_2 \geq \sigma$, it follows that $r(t)N'(t) = 0$ for $t \geq t_2$. Hence $N'(t) = 0$, $t \geq t_2$. So $N(t)$ is a positive constant. Now $(rN')'(t) + q(t)N(t) \geq 0$ for $t \geq \sigma$ implies that $q(t)N(t) \geq 0$ for $t \geq \sigma$, a contradiction. Hence $N'(t) < 0$ for $t \geq \sigma$. Further, if $[(rN')' + qN](t_3) = 0$ for some $t_3 \geq \sigma$, then $[(rN')' + qN](t) \equiv 0$ for $t \geq t_3$. This in turn implies that $[(rN')'(t) + q(t)N(t)]' = p(t)N(t) \equiv 0$, a contradiction. So $(rN')'(t) + q(t)N(t) > 0$ for all $t \geq \sigma$.

Hence the proof of the theorem is complete. □

Lemma 2.3.5 *The following statements hold.*

(i) $\lim_{t \to \infty} r(t)N'(t) = 0$
(ii) $\lim_{t \to \infty} tr(t)N'(t) = 0$
(iii) $\lim_{t \to \infty} t^2[(rN')'(t) + q(t)N(t)] = 0$.

Theorem 2.3.8 *If* (2.1) *has an oscillatory solution, then there exist two linearly independent oscillatory solutions, $u_1(t)$ and $u_2(t)$, of*

$$\left(\frac{x'(t)}{N(t)}\right)' + \left(\frac{(rN')'(t) + q(t)N(t)}{r(t)N^2(t)}\right)x(t) = 0, \tag{2.73}$$

which satisfy (2.1).

Proof Since (2.1) has an oscillatory solution, (2.2) has an oscillatory solution. From Lemma 2.3.4 it follows that (2.49) is oscillatory. It is clear from Lemmas 2.3.3 and 1.5.10 that a solution of (2.49) which has at least one zero is oscillatory.

Let $z_1(t)$ and $z_2(t)$ be two linearly independent solutions of (2.49) with

$$z_1(\sigma) = 0, \qquad z_1'(\sigma) = 0, \qquad z_1''(\sigma) = 1,$$
$$z_2(\sigma) = 0, \qquad z_2'(\sigma) = 1, \qquad \left(rz_2'\right)'(\sigma) = 0.$$

So $z_1(t)$ and $z_2(t)$ are oscillatory. It is easy to verify that

$$w_1(t) = N(t)z_1'(t) - N'(t)z_1(t) = N^2(t)\left(\frac{z_1}{N}\right)'(t)$$

and

$$w_2(t) = N(t)z_2'(t) - N'(t)z_2(t) = N^2(t)\left(\frac{z_2}{N}\right)'(t)$$

are oscillatory solutions of

$$\left(\frac{r(t)x'(t)}{N(t)}\right)' + \left[\frac{(rN')'(t) + q(t)N(t)}{N^2(t)}\right]x = 0. \tag{2.74}$$

Consequently, $u_1(t) = r(t)w_1(t)$ and $u_2(t) = r(t)w_2(t)$ are oscillatory solutions of (2.73). It may easily be verified that $u_1(t)$ and $u_2(t)$ satisfy (2.2), and hence (2.1).

To complete the proof of the theorem, it remains to show that $u_1(t)$ and $u_2(t)$ are linearly independent. If possible, let $u_1(t)$ and $u_2(t)$ are linearly dependent. So there exist constants c_1 and c_2 not both zero, such that $c_1u_1(t) + c_2u_2(t) = 0$ for $t \in [\sigma, \infty)$, that is, $c_1w_1(t) + c_2w_2(t) = 0$ for $t \in [\sigma, \infty)$. Since $w_1(t)$ and $w_2(t)$ are nontrivial solutions of (2.74), $c_1 = 0$ implies that $c_2 = 0$ and $c_2 = 0$ implies that $c_1 = 0$. Hence $c_1 \neq 0$ and $c_2 \neq 0$. Now $w_1(t) + \lambda w_2(t) = 0$ for $t \in [\sigma, \infty)$, where $\lambda = \frac{c_2}{c_1}$, implies that

$$N(t)\left(z_1'(t) + \lambda z_2'(t)\right) - N'(t)\left(z_1(t) + \lambda z_2(t)\right) = 0,$$

that is,

$$\frac{z_1'(t) + \lambda z_2'(t)}{z_1(t) + \lambda z_2(t)} = \frac{N'(t)}{N(t)},$$

which gives upon integration $N(t) = c(z_1(t) + \lambda z_2(t))$, $c \neq 0$. Consequently, $z_1(t) + \lambda z_2(t)$ is nonoscillatory. Hence there exists a $t_1 \geq \sigma$ such that $z_1(t) + \lambda z_2(t)$ has one sign for $t \geq t_1$. Let t_2 and t_3 $(t_1 < t_2 < t_3)$ be consecutive zeros of $z_1(t)$. From Lemma 1.5.13, it follows that there exists a constant $\mu \neq 0$ such that $(z_1(t) + \lambda z_2(t)) + \mu z_1(t)$ has a double zero in (t_2, t_3) with a zero at $t = \sigma$. This contradicts the fact that (2.49) is of type C_I. Thus $u_1(t)$ and $u_2(t)$ are linearly independent. This completes the proof of the theorem. □

Remark 2.3.4 (i) Any solution of (2.73) is a solution of (2.1).

(ii) It is possible to choose two linearly independent solutions $x_1(t)$ and $x_2(t)$ of (2.73) such that $W(t) > 0$, where $W(t) = x_1(t)x_2'(t) - x_1'(t)x_2(t)$. Since (2.73) is oscillatory, $x_1(t)$ and $x_2(t)$ are oscillatory. Moreover, $x_1(t)$ and $x_2(t)$ are solutions of (2.1). By the Sturm separation theorem, the zeros of $x_1(t)$ and $x_2(t)$ separate each other in $[\sigma, \infty)$. This observation leads us to obtain the following corollary:

Corollary 2.3.1 *If* (2.1) *has an oscillatory solution, then there exist two linearly independent oscillatory solutions u and v of* (2.1) *such that their zeros separate, and any linear combination of u and v is also oscillatory.*

Theorem 2.3.9 *N and W are linearly dependent. In fact, $W(t) = \lambda N(t)$, where $\lambda > 0$ is a constant.*

Proof We have chosen linearly independent solutions $x_1(t)$ and $x_2(t)$ of (2.73) such that $W(t) > 0$ for $t \geq \sigma$. Clearly, $x_1(t)$ and $x_2(t)$ are solutions of

$$\begin{vmatrix} x & x_1(t) & x_2(t) \\ x' & x_1'(t) & x_2'(t) \\ r(t)x'' & r(t)x_1''(t) & r(t)x_2''(t) \end{vmatrix} = 0,$$

$t \in [\sigma, \infty)$, that is, of

$$x'' - \frac{W'(t)}{W(t)}x' + \left(\frac{(rW')'(t) + q(t)W(t)}{r(t)W(t)}\right)x = 0. \tag{2.75}$$

Equation (2.73) may be written as

$$x'' - \frac{N'(t)}{N(t)}x' + \left(\frac{(rN')'(t) + q(t)N(t)}{r(t)N(t)}\right)x = 0. \tag{2.76}$$

Since $x_1(t)$ and $x_2(t)$ are linearly independent solutions of both (2.75) and (2.76), so it is clear that (2.75) and (2.76) have the same solution space. If $u(t)$ is a solution of (2.75), then it is a solution of (2.76) and hence $u(t)$ is a solution of the first-order equation

$$a_1(t)x' + b_1(t)x = 0,$$

where

$$a_1(t) = \frac{N'(t)}{N(t)} - \frac{W'(t)}{W(t)}$$

and

$$b_1(t) = \frac{(rW')'(t) + q(t)W(t)}{r(t)W(t)} - \frac{(rN')'(t) + q(t)N(t)}{r(t)N(t)}.$$

Hence, in particular,

$$a_1(t)x_1'(t) + b_1(t)x_1(t) = 0$$

and

$$a_1(t)x_2'(t) + b_1(t)x_2(t) = 0.$$

Since $W(t) \neq 0$ for $t \geq \sigma$, we have $a_1(t) = 0$ and $b_1(t) = 0$ for $t \geq \sigma$. But $a_1(t) = 0$, $t \geq \sigma$, implies that

$$\frac{N'(t)}{N(t)} = \frac{W'(t)}{W(t)}, \qquad t \geq \sigma.$$

Integrating the above equality, we have $W(t) = \lambda N(t)$, where λ is a constant. Further $W(t) > 0$ and $N(t) > 0$ for $t \geq \sigma$ imply that $\lambda > 0$. Thus, $W = \lambda N(t)$ for $t \geq \sigma$. The theorem is proved. □

Remark 2.3.5 Theorems 2.3.7 and 2.3.8 and Lemma 2.3.5 hold good when $N(t)$ is replaced by $W(t)$ in view of Theorem 2.3.9.

Theorem 2.3.10 *For any solution $x(t)$ of* (2.52), *the function $G(x(t))$ is a decreasing function of t, where*

$$G\big(x(t)\big) = r(t)W(t)\big(x'(t)\big)^2 + \big(\big(rW'\big)'(t) + q(t)W(t)\big)x^2(t).$$

Proof of Theorem 2.3.6 From the Remark 2.3.4, it follows that there exist two linearly independent oscillatory solutions $x_1(t)$ and $x_2(t)$ of (2.1) whose zeros separate. To complete the proof of theorem, it is enough to show that any oscillatory solution of (2.1) can be expressed as a linear combination of $x_1(t)$ and $x_2(t)$.

Let $x_3(t)$ be a solution of (2.1) with $x_3(\sigma) = 0 = x_3'(\sigma)$, $x_3''(\sigma) > 0$. From Lemma 2.3.1, it follows that $x_3(t) > 0$ for $t > \sigma$. Consequently, Remark 2.3.1 implies that $x_3'(t) > 0$ and $x_3''(t) > 0$ for $t > \sigma$. Clearly, $\{x_1(t), x_2(t), x_3(t)\}$ is a linearly independent set of solutions of (2.1). Hence

$$\begin{vmatrix} x_1(t) & x_2(t) & x_3(t) \\ x_1'(t) & x_2'(t) & x_3'(t) \\ r(t)x_1''(t) & r(t)x_2''(t) & r(t)x_3''(t) \end{vmatrix} = k,$$

where $k \neq 0$ is a constant. Thus

$$\begin{vmatrix} x_1(t) & x_2(t) & u(t) \\ x_1'(t) & x_2'(t) & u'(t) \\ r(t)x_1''(t) & r(t)x_2''(t) & r(t)u''(t) \end{vmatrix} = 1,$$

where $u(t) = \frac{x_3(t)}{k}$. Expanding the above determinant,

$$r(t)W(t)u''(t) - r(t)W'(t)u'(t) + \big(\big(rW'\big)'(t) + q(t)W(t)\big)u(t) = 1. \tag{2.77}$$

We may note that $k < 0$ implies that $u(t) < 0$, $u'(t) < 0$ and $u''(t) < 0$. This, in turn, leads to a contradiction in (2.77), where the left-hand side becomes negative and the right-hand side is positive. Thus $k > 0$.

Let $z(t)$ be an oscillatory solution of (2.1). We claim that $z(t)$ can be expressed as a linear combination of $x_1(t)$ and $x_2(t)$. If not, there exist c_1, c_2 and c_3, $c_3 \neq 0$, such that $z(t) = c_1x_1(t) + c_2x_2(t) + c_3u(t)$. We may note that c_1 and c_2 cannot be zero simultaneously. If

$$z_1(t) = \frac{z(t)}{c_3} \quad \text{and} \quad x(t) = -\frac{c_1x_1(t) + c_2x_2(t)}{c_3},$$

then $z_1(t) = u(t) - x(t)$. Clearly $x(t)$ is a nontrivial oscillatory solution of (2.52) and (2.1). Thus $z_1(t)$ is a solution of (2.77). Consequently,

$$\begin{aligned} &r(t)W(t)\big(u(t) - x(t)\big)'' - r(t)W'(t)\big(u(t) - x(t)\big)' \\ &\quad + \big((rW')'(t) + q(t)W(t)\big)\big(u(t) - x(t)\big) = 1. \end{aligned} \tag{2.78}$$

Since $z(t)$ is oscillatory, $u(t) - x(t)$ is oscillatory. From Theorem 2.3.10, it follows that $((rW')'(t) + q(t)W(t))x^2(t)$ is bounded. As

$$\begin{aligned} \big[\big((rW')'(t) + q(t)W(t)\big)x(t)\big]^2 = {}&\big((rW')'(t) + q(t)W(t)\big)x^2(t)t^2\big((rW')'(t) \\ &+ q(t)W(t)\big)t^{-2}, \end{aligned}$$

from Lemma 2.3.5(iii), we obtain

$$\lim_{t\to\infty}\big[\big((rW')'(t) + q(t)W(t)\big)x(t)\big] = 0.$$

Hence there exists a $T > \sigma$ such that

$$\big|\big((rW')'(t) + q(t)W(t)\big)x(t)\big| < \frac{1}{4}$$

for $t \geq T$. From (2.77) we get, for $t \geq \sigma$,

$$0 < \big((rW')'(t) + q(t)W(t)\big)u(t) < 1.$$

Let $t_0 > T$ be a maximum of $u(t) - x(t)$. So $u(t_0) - x(t_0) \geq 0$ and $u'(t_0) - x'(t_0) = 0$. Now multiplying (2.78) through by $u'(t) - x'(t)$ and integrating the resulting identity from t_0 to t, we obtain

$$\begin{aligned} &\frac{1}{2}r(t)W(t)\big(u'(t) - x'(t)\big)^2 - \frac{1}{2}\int_{t_0}^{t}(rW)'(s)\big(u'(s) - x'(s)\big)^2\,ds \\ &\quad - \int_{t_0}^{t} r(s)W'(s)\big(u'(s) - x'(s)\big)^2\,ds \\ &\quad + \frac{1}{2}\big((rW')'(t) + q(t)W(t)\big)\big(u(t) - x(t)\big)^2 \\ &\quad - \frac{1}{2}\big((rW')'(t_0) + q(t_0)W(t_0)\big)\big(u(t_0) - x(t_0)\big)^2 \end{aligned}$$

$$-\frac{1}{2}\int_{t_0}^{t} p(s)W(s)\big(u(s)-x(s)\big)^2\,ds$$
$$= (u-x)(t) - (u-x)(t_0),$$

since $W(t)$ is a solution of (2.49). As $(rW)'(t) = r'(t)W(t) + r(t)W'(t) < 0$, we have

$$(u-x)(t_0)\left[1 - \frac{1}{2}\big((rW')'(t_0) + q(t_0)W(t_0)\big)(u-x)(t_0)\right] < (u-x)(t).$$

But

$$\begin{aligned}\big((rW')'(t_0) + q(t_0)W(t_0)\big)(u-x)(t_0) &= \big((rW')'(t_0) + q(t_0)W(t_0)\big)u(t_0)\\ &\quad - \big((rW')'(t_0) + q(t_0)W(t_0)\big)x(t_0)\\ &< 1 + \frac{1}{4} = \frac{5}{4}.\end{aligned}$$

So, for $t > t_0$, $(u-x)(t) > \frac{3}{8}(u-x)(t_0) \geq 0$, which contradicts the fact that $(u-x)$ is oscillatory. Hence our claim holds. This completes the proof of the theorem. □

Theorem 2.3.11 *If* (2.1) *admits an oscillatory solution, then there exist two linearly independent oscillatory solutions u and v of* (2.1) *such that any nontrivial linear combination of u and v is also oscillatory and their zeros separate.*

2.4 Behaviour of Solutions of $x''' + a(t)x'' + b(t)x' + c(t)x = 0$ with $a(t) \geq 0$, $b(t) \leq 0$ and $c(t) < 0$

In this section, we present some results on the oscillation and asymptotic behaviour of solutions of Eq. (2.1) with $a(t) \geq 0$, $b(t) \leq 0$ and $c(t) < 0$. The results presented in this section generalise Proposition 1.2.4 of Chap. 1 with variable coefficients. Results pertaining to the behaviour of nonoscillatory solutions in the presence of an oscillatory solution are also presented.

Lemma 2.4.1 *Let $a(t)b(t) + b'(t) - c(t) \leq 0$. Then Eq.* (2.1) *is of type C_I.*

Proof Let $x(t)$ be a solution of (2.1) with $x(t_0) = x'(t_0) = 0$ and $x''(t_0) > 0$, where $t_0 > \sigma$. From the continuity of $x''(t)$, it follows that there exists a δ, $\sigma < t_0 - \delta$, such that $x''(t) > 0$ for $t \in [t_0 - \delta, t_0]$. We claim that $x''(t) > 0$ for $t \in [\sigma, t_0]$. If not, then there exists a $t_1 \in [\sigma, t_0 - \delta]$ such that $x''(t_1) = 0$ and $x''(t) > 0$ for $t \in (t_1, t_0]$. Thus, $x'(t) < 0$ and $x(t) > 0$ for $t \in (t_1, t_0)$. Integrating (2.2) from t_1 to t_0, we obtain

$$0 < r(t_0)x''(t_0) = q(t_1)x(t_1) + \int_{t_1}^{t_0} \big(q'(s) - p(s)\big)x(s)\,ds < 0,$$

a contradiction. Hence our claim holds. Consequently, $x(t) > 0$, $x'(t) < 0$ for $t \in [\sigma, t_0)$. The proof of the lemma is complete. □

Lemma 2.4.2 *If $x(t)$ is a solution of* (2.1) *with $x(t_0) \geq 0$, $x'(t_0) \geq 0$ and $x''(t_0) > 0$ for some $t_0 \geq \sigma$, then $x(t) > 0$, $x'(t) > 0$ and $x''(t) > 0$ for some $t > t_0$. Similarly, if $x(t_0) \leq 0$, $x'(t_0) \leq 0$ and $x''(t_0) < 0$ for $t_0 \geq \sigma$, then $x(t) < 0$, $x'(t) < 0$ and $x''(t) < 0$ for $t > t_0$.*

Proof Let $x(t)$ be a solution of (2.1) with $x(t_0) \geq 0$, $x'(t_0) \geq 0$ and $x''(t_0) > 0$ for $t_0 \geq \sigma$. Hence there exists a $\delta > 0$ such that $x''(t) > 0$ for $t \in [t_0, t_0 + \delta)$. If there is a $t_1 \geq t_0 + \delta$ such that $x''(t_1) = 0$ and $x''(t) > 0$ for $t_0 \leq t < t_1$, then $x'(t) > 0$ and $x(t) > 0$ for $t_0 \leq t < t_1$. Multiplying Eq. (2.2) through by $x'(t)$ and integrating the resulting identity from t_0 to t_1, we obtain

$$0 < \int_{t_0}^{t_1} r(t)\big(x''(t)\big)^2\,dt = \int_{t_0}^{t_1} q(t)\big(x'(t)\big)^2\,dt + \int_{t_0}^{t_1} p(t)x(t)x'(t)\,dt < 0,$$

a contradiction. Hence $x''(t) > 0$ for $t \geq t_0$. Then $x(t) > 0$ and $x'(t) > 0$ for $t > t_0$. The other assertion follows similarly. The lemma is proved. □

Corollary 2.4.1 *Equation* (2.1) *is of Class II.*

Theorem 2.4.1 *Equation* (2.1) *admits a positive increasing solution which goes to ∞ as $t \to \infty$. Further, if*

$$\int_{\sigma}^{\infty} \frac{dt}{r(t)} = \infty,$$

then the derivative of the solution tends to ∞ as $t \to \infty$.

Proof If $x(t)$ is a solution of (2.1) with $x(t_0) \geq 0$, $x'(t_0) \geq 0$ and $x''(t_0) > 0$, $t_0 \geq \sigma$, then from Lemma 2.4.2 it follows that $\lim_{t\to\infty} x(t) = \infty$. Since $x(t) > 0$ and $x'(t) > 0$ for $t > t_0$, $r(t)x''(t)$ is monotonically increasing in $[t_0, \infty)$. Thus $\int_{\sigma}^{\infty} \frac{dt}{r(t)} = \infty$ implies that $x'(t) \to \infty$ as $t \to \infty$. This completes the proof of the theorem. □

Theorem 2.4.2 *If $b(t) - a'(t) \leq 0$, $b(t) - 2a'(t) \leq 0$, $c(t) - b'(t) + a''(t) < 0$ and*

$$\begin{aligned}\int_{\sigma}^{\infty} \bigg[&-\frac{2a^3(t)}{27} + \frac{a(t)b(t)}{3} - c(t) + b'(t) - a''(t) - \frac{2a(t)a'(t)}{3} \\ &- \frac{2}{3\sqrt{3}} \left(\frac{a^2(t)}{3} - b(t) + a'(t) \right)^{\frac{3}{2}} \bigg]\,dt = \infty, \end{aligned} \tag{2.79}$$

then (2.1) *is oscillatory.*

Theorem 2.4.3 *Let* $b(t) - 2a'(t) \leq 0$, $c(t) - b'(t) + a''(t) < 0$, $0 \leq \lim_{t\to\infty} ta(t) < \infty$ *and* $1 - t^2b(t) + \frac{1}{3}t^2a^2(t) + t^2a'(t) \geq 0$. *If* (2.70) *holds, then* (2.1) *is oscillatory.*

Remark 2.4.1 Theorem 2.4.3 holds for the Euler equation (2.5) with $a_0 \geq 0$, $b_0 \leq 0$ and $c_0 < 0$. Indeed, if $c_0 + 2b_0 + 2a_0 < 0$, $1 - b_0 + \frac{a_0^2}{3} - a_0 \geq 0$, $b_0 + 2a_0 \leq 0$ and (2.71) hold, then (2.5) is oscillatory.

Example 2.4.1 By Theorem 2.4.3, the equation

$$x''' + \frac{1}{t}x'' - \frac{3}{t^2}x' - \frac{3}{t^3}x = 0, \quad t \geq 1$$

is oscillatory. In particular, $x_1(t) = t^{-1/2}\cos(\sqrt{3}\log t)$ and $x_2(t) = t^{-1/2}\sin(\sqrt{3}\log t)$ are the oscillatory solutions of the equation, and $x_3(t) = t^3$ is a nonoscillatory solution of the equation.

Remark 2.4.2 One may observe in the above example that $x_3(t) \to \infty$ as $t \to \infty$ along with its derivatives, whereas, $x_1(t)$ and $x_2(t) \to 0$ as $t \to \infty$. This indicates that there exists oscillatory solution of Eq. (2.1) converging to zero as $t \to \infty$. However, no such result has been obtained in the literature. This has been left as an open problem to the readers. However, we provide a partial answer to the above problem. See Corollary 2.4.2 and Theorem 2.8.1.

Lemma 2.4.3 *If* $x(t)$ *is a nonoscillatory solution of* (2.1), *then there exists a* $t_0 \geq \sigma$ *such that* $x(t)x'(t) \leq 0$ *or* $x(t)x'(t) > 0$ *for* $t \geq t_0$.

Proof Without any loss of generality, we may assume that $x(t) > 0$ for $t \geq T \geq \sigma$. Let t_1 and t_2 ($T \leq t_1 < t_2$) be two consecutive zeros of $x'(t)$ such that $x'(t) > 0$ for $t \in (t_1, t_2)$. Multiplying (2.1) through by $x'(t)$ and integrating the resulting identity from t_1 to t_2, we obtain the inequality

$$0 < \int_{t_1}^{t_2} r(t)\big(x''(t)\big)^2\,dt = \int_{t_1}^{t_2} q(t)\big(x'(t)\big)^2\,dt + \int_{t_1}^{t_2} p(t)x(t)x'(t)\,dt < 0,$$

a contradiction. Hence there exists a $t_0 > T$ such that $x'(t) > 0$ or ≤ 0 for $t \geq t_0$. Thus the lemma is proved. □

Theorem 2.4.4 *If* (2.1) *has an oscillatory solution, then every nonoscillatory solution* $x(t)$ *of* (2.1) *satisfies the following conditions*:

$$x(t)x'(t) \neq 0, \qquad \operatorname{sgn} x(t) = \operatorname{sgn} x'(t), \quad t \geq t_0 \geq \sigma$$

and

$$\lim_{t\to\infty} |x(t)| = \infty.$$

If, in addition

$$\int_{\sigma}^{\infty} p(t)\,dt = -\infty,$$

then

$$x(t)x'(t)x''(t) \neq 0$$

and

$$\operatorname{sgn} x(t) = \operatorname{sgn} x'(t) = \operatorname{sgn} x''(t), \quad t \geq T_0 \geq t_0.$$

Proof Let $x(t) > 0$ for $t \geq T \geq \sigma$. Let $z(t)$ be an oscillatory solution of (2.1). To prove that $W(t) = x(t)z'(t) - x'(t)z(t)$ must vanish for some value of $t \in [T, \infty)$. If not, then $W(t) \neq 0$ for $t \in [T, \infty)$. If $u(t) = \frac{z(t)}{x(t)}$, then $u'(t) = \frac{W(t)}{x^2(t)} \neq 0$ for $t \geq T$. If t_1 and t_2 ($T \leq t_1 < t_2$) are consecutive zeros of $z(t)$, then $u(t_1) = 0$, $u(t_2) = 0$ and $u(t) \neq 0$ for $t \in (t_1, t_2)$. This is impossible, since $u'(t) \neq 0$ for $t \geq T$. Thus $W(t)$ must vanish for some value of $t \in [\sigma, \infty)$. Let $W(\alpha) = 0$ for some $\alpha \in [T, \infty)$. It is possible to obtain c_1 and c_2, not both zero, such that

$$c_1 x(\alpha) + c_2 z(\alpha) = 0, \qquad c_1 x'(\alpha) + c_2 z'(\alpha) = 0 \quad \text{and} \quad c_1 x''(\alpha) + c_2 z''(\alpha) = 0,$$

because $x(t)$ and $z(t)$ are linearly independent on $[T, \infty)$. Without any loss of generality, we may assume that $c_1 x''(\alpha) + c_2 z''(\alpha) > 0$. If $v(t) = c_1 x(t) + c_2 z(t)$, then $v(t)$ is a solution of (2.1) with $v(\alpha) = 0$, $v'(\alpha) = 0$ and $v''(\alpha) > 0$. Proceeding as in Lemma 2.4.2, we obtain $v(t) \to \infty$ as $t \to \infty$.

From Lemma 2.4.3, it follows that $x'(t) > 0$ or ≤ 0 for $t \geq t_0 \geq T$. If $x'(t) \leq 0$ for $t \geq t_0$, then $\lim_{t\to\infty} x(t) = \lambda$ exists, where $0 \leq \lambda < \infty$. Clearly, $c_2 = 0$ implies that $\lim_{t\to\infty} v(t) = c_1\lambda < \infty$, a contradiction. Thus $c_2 \neq 0$. If $c_2 > 0$, then

$$\begin{aligned}
\liminf_{t\to\infty} c_2 z(t) &= \liminf_{t\to\infty}\bigl[v(t) - c_1 x(t)\bigr] \\
&\geq \liminf_{t\to\infty} v(t) + \liminf_{t\to\infty}\bigl(-c_1 x(t)\bigr) \\
&= \liminf_{t\to\infty} v(t) - c_1 \liminf_{t\to\infty} x(t) \\
&= \infty
\end{aligned}$$

implies that $\liminf_{t\to\infty} z(t) = \infty$, and hence $\lim_{t\to\infty} z(t) = \infty$. If $c_2 < 0$, then $\liminf_{t\to\infty} c_2 z(t) = \infty$ implies that $\limsup_{t\to\infty} z(t) = -\infty$, and hence $\lim_{t\to\infty} z(t) = -\infty$. Thus $\lim_{t\to\infty} z(t) = \infty$ or $-\infty$ according to $c_2 > 0$ or < 0. In either case, it is a contradiction since $z(t)$ is oscillatory. Hence $x'(t) > 0$ for $t \geq t_0$. Clearly, $c_1 \neq 0$ because $c_1 = 0$ implies that $c_2 \neq 0$ and $v(t) = c_2 z(t)$ is oscillatory, a contradiction. If $c_2 = 0$, then $\lim_{t\to\infty} c_1 x(t) = \lim_{t\to\infty} v(t) = \infty$. As $c_1 < 0$ implies that $x(t) < 0$ for large t, then $c_1 > 0$, and hence $\lim_{t\to\infty} x(t) = \infty$. Suppose that $c_2 \neq 0$. If $\lim_{t\to\infty} x(t)$ exists finitely, then $\lim_{t\to\infty} z(t) = \pm\infty$, contradicting the oscillatory nature of $z(t)$. Thus $\lim_{t\to\infty} x(t) = \infty$.

Suppose that $\int_\sigma^\infty p(t)\,dt = -\infty$. Since $x(t) > 0$ and $x'(t) > 0$ for $t \ge t_0$, $r(t)x''(t)$ is monotonic increasing, and hence $x''(t)$ has a constant sign for $t \ge T_0 \ge t_0$. If $x''(t) < 0$ for $t \ge T_0$, then integrating (2.2) from T_0 to t we obtain

$$\begin{aligned} r(t)x''(t) &\ge r(T_0)x''(T_0) - \int_{T_0}^t p(s)x(s)\,ds \\ &\ge r(T_0)x''(T_0) - x(T_0)\int_{T_0}^t p(s)\,ds. \end{aligned}$$

Thus $x''(t) > 0$ for large t, a contradiction. Hence $x''(t) > 0$ for $t \ge T_0$. This completes the proof of the theorem. □

Corollary 2.4.2 *If* (2.1) *has an oscillatory solution, then every bounded solution of* (2.1) *oscillates.*

Theorem 2.4.5 *Let* $\int_\sigma^\infty p(t)\,dt = -\infty$. *Then Eq.* (2.1) *has an oscillatory solution, if and only if every nonoscillatory solution* $x(t)$ *of* (2.1) *satisfies the properties*

$$\left.\begin{aligned} &x(t)x'(t)x''(t) \ne 0, \\ &\operatorname{sgn} x(t) = \operatorname{sgn} x'(t) = \operatorname{sgn} x''(t), \quad t \ge T_0 \ge \sigma, \quad \textit{and} \\ &\lim_{t\to\infty} |x(t)| = \infty. \end{aligned}\right\} \tag{2.80}$$

Proof The necessity follows from Theorem 2.4.4. For sufficiency, assume that (2.80) holds for every nonoscillatory solution $x(t)$ of (2.1). To show that (2.1) admits an oscillatory solution, let z_0, z_1, z_2 be solutions of (2.1) with initial conditions

$$z_k^{(j)}(\sigma) = \begin{cases} 0, & j \ne k, \\ 1, & j = k, \end{cases}$$

$j, k = 0, 1, 2$. Clearly z_0, z_1, z_2 are linearly independent. For each positive integer $n > \sigma$, it is possible to determine the real numbers a_{0n}, a_{2n}, b_{1n} and b_{2n} such that

$$\begin{aligned} a_{0n}z_0(n) + a_{2n}z_2(n) &= 0, \\ b_{1n}z_1(n) + b_{2n}z_2(n) &= 0 \end{aligned}$$

and $a_{0n}^2 + a_{2n}^2 = 1$, $b_{1n}^2 + b_{2n}^2 = 1$. Define, for each positive integer $n > \sigma$,

$$\begin{aligned} u_n &= a_{0n}z_0 + a_{2n}z_2, \\ v_n &= b_{1n}z_1 + b_{2n}z_2. \end{aligned}$$

Thus u_n and v_n are solutions of (2.1) with $u_n(n) = 0$ and $v_n(n) = 0$. Clearly, there exists a sequence $\langle n_j \rangle$ of positive integers $> \sigma$ such that $a_{0n_j} \to a_0$, $a_{2n_j} \to a_2$, $b_{1n_j} \to b_1$ and $b_{2n_j} \to b_2$ as $n_j \to \infty$ and hence $a_0^2 + a_2^2 = 1$ and $b_1^2 + b_2^2 = 1$. If

$u = a_0 z_0 + a_2 z_2$ and $v = b_1 z_1 + b_2 z_2$, then u and v are nontrivial solutions of (2.1) and

$$\lim_{n_j \to \infty} u_{n_j}^{(k)} = u^{(k)}, \qquad \lim_{n_j \to \infty} v_{n_j}^{(k)} = v^{(k)},$$

$k = 0, 1, 2$, uniformly on any compact subinterval of $[\sigma, \infty)$. To show that each of u and v is an oscillatory solution of (2.1). If u is nonoscillatory, then there exists a $T_0 \geq \sigma$ such that

$$u(t)u'(t)u''(t) \neq 0, \qquad \operatorname{sgn} u(t) = \operatorname{sgn} u'(t) = \operatorname{sgn} u''(t), \quad t \geq T_0$$

and $\lim_{t \to \infty} |u(t)| = \infty$. In particular,

$$u(T_0)u'(T_0)u''(T_0) \neq 0, \qquad \operatorname{sgn} u(T_0) = \operatorname{sgn} u'(T_0) = \operatorname{sgn} u''(T_0).$$

Hence there exists a positive integer N such that

$$u_{n_j}(T_0)u'_{n_j}(T_0)u''_{n_j}(T_0) \neq 0,$$
$$\operatorname{sgn} u_{n_j}(T_0) = \operatorname{sgn} u'_{n_j}(T_0) = \operatorname{sgn} u''_{n_j}(T_0)$$

for $n_j \geq N$. From Lemma 2.4.2, it follows that $u_{n_j} \neq 0$ for $n_j \geq N$ and $t \geq T_0$. Thus $u_{n_j}(n_j) \neq 0$ for all $n_j > \max\{N, T_0\}$. This contradicts the fact that $u_n(n) = 0$ for every positive integer $n > \sigma$. Hence $u(t)$ is oscillatory. Similarly, it may be shown that $v(t)$ is oscillatory. Thus the theorem is proved. □

Theorem 2.4.6 *Let $\int_\sigma^\infty p(t)\,dt = -\infty$. If* (2.1) *admits an oscillatory solution, then there exist two linearly independent oscillatory solutions u and v of* (2.1) *such that any nontrivial linear combination of u and v is also oscillatory and the zeros of u and v separate.*

Remark 2.4.3 Theorems 2.4.5 and 2.4.6 are similar to Theorems 6.23 and 6.25, respectively, in [14]. While $\int_\sigma^\infty p(t)\,dt = -\infty$ is assumed in Theorem 6.23, the disconjugacy of $x'' + a(t)x' + b(t)x = 0$ is assumed in Theorem 6.25 in [14]. The proof of Theorems 6.23 and 6.25 may be found in Gera [12]. Note that the assumption $\int_\sigma^\infty p(t)\,dt = -\infty$ is not needed in the sufficiency part of Theorem 2.1.6. Moreover, this condition is satisfied if a, b and c are constants.

Theorem 2.4.7 *Suppose that $q'(t) - p(t) \leq 0$ but $\not\equiv 0$ in any neighbourhood of infinity. Then Eq.* (2.1) *admits a solution $x(t)$ with the following properties*:

$$x(t)x'(t)x''(t) \neq 0,$$
$$\operatorname{sgn} x(t) = \operatorname{sgn} x''(t) \neq \operatorname{sgn} x'(t), \quad t \geq \sigma,$$
$$\lim_{t \to \infty} x'(t) = 0 \quad \text{and} \quad \lim_{t \to \infty} x(t) = \lambda, \quad -\infty < \lambda < \infty.$$

If, in addition, $\int_\sigma^\infty (q'(t) - p(t))\,dt = -\infty$ and $\lim_{t \to \infty} q(t) = k$, $-\infty < k < \infty$, then $\lim_{t \to \infty} x(t) = 0$.

Proof For every positive integer $n > \sigma$, let $x_n(t)$ be a solution of (2.1) with initial conditions

$$x_n(n) = 0, \qquad x_n'(n) = 0, \qquad x_n''(n) > 0.$$

Since $q'(t) - p(t) = (a(t)b(t) + b'(t) - c(t))r(t)$, Lemma 2.4.1 yields $x_n(t) > 0$, $x_n'(t) < 0$ and $x_n''(t) > 0$ for $t \in [\sigma, n)$. We may write

$$x_n(t) = c_{1n}u_1(t) + c_{2n}u_2(t) + c_{3n}u_3(t), \quad t \in [\sigma, n),$$

where $c_{1n}^2 + c_{2n}^2 + c_{3n}^2 = 1$ and $\{u_1(t), u_2(t), u_3(t)\}$ is a basis of the solution space of (2.1). The sequence $\langle c_{in} \rangle$, $i = 1, 2, 3$ has a convergent subsequence $\langle c_{in_j} \rangle$ such that $c_{in_j} \to c_i$ as $n_j \to \infty$ and $i = 1, 2, 3$. Hence $c_1^2 + c_2^2 + c_3^2 = 1$. Setting $x(t) = c_1u_1(t) + c_2u_2(t) + c_3u_3(t)$, we see that $x(t)$ is a solution of (2.1) with

$$\lim_{n_j \to \infty} x_{n_j}^{(k)}(t) = x^{(k)}(t), \quad k = 0, 1, 2$$

uniformly on every compact subinterval of $[\sigma, \infty)$. Thus, $x(t) > 0$, $x'(t) < 0$ and $x''(t) > 0$ for $t \geq \sigma$. As $\lim_{t \to \infty} x'(t) = L$, $-\infty < L < 0$, which implies that $x(t) < 0$ for large t, then $\lim_{t \to \infty} x'(t) = 0$. Clearly $\lim_{t \to \infty} x(t) = \lambda$, $0 \leq \lambda < \infty$. If $\lim_{t \to \infty} x(t) = \lambda$, $\lambda > 0$, then integrating (2.2) from σ to t and by using the additional conditions, we get

$$\begin{aligned} 0 < r(t)x''(t) &= r(\sigma)x''(\sigma) - q(t)x(t) + q(\sigma)x(\sigma) + \int_\sigma^t \big(q'(s) - p(s)\big)x(s)\,ds \\ &\leq r(\sigma)x''(\sigma) - q(t)x(t) + q(\sigma)x(\sigma) + x(t)\int_\sigma^t \big(q'(s) - p(s)\big)\,ds \\ &< 0, \end{aligned}$$

for large t, a contradiction. The theorem is proved. □

Theorem 2.4.8 *If $q'(t) - p(t) \leq 0$ but $\not\equiv 0$ in any neighbourhood of infinity, then* (2.1) *is nonoscillatory.*

Proof If possible, let (2.1) admits an oscillatory solution. From Theorem 2.4.4, it follows that every nonoscillatory solution $x(t)$ of (2.1) satisfies the property $|x(t)| \to \infty$ as $t \to \infty$. On the other hand, Theorem 2.4.7 yields the result that (2.1) has a nonoscillatory solution $u(t)$ such that $\lim_{t \to \infty} u(t) = \lambda$, $-\infty < \lambda < \infty$, a contradiction. This completes the proof of the theorem. □

An alternate proof of Theorem 2.4.8 is given below:

Clearly, (2.1) is of type C_{II}. On the other hand $q'(t) - p(t) \leq 0$ implies that (2.1) is of type C_I. We now prove, since (2.1) is both C_I and C_{II}, (2.1) is nonoscillatory. Since (2.1) is of type C_{II}, the solution $x(t)$ of (2.1) with the initial conditions $x(\sigma) = x'(\sigma) = 0$, $x''(\sigma) > 0$ has the property that $x(t) > 0$ for

$t > \sigma$. If possible, let $z(t)$ be an oscillatory solution of (2.1). Let $\alpha_1, \beta_1, \alpha_2, \beta_2$ ($\sigma < \alpha_1 < \beta_1 < \alpha_2 < \beta_2$) be the successive zeros of $z(t)$ such that $z(t) > 0$ for $t \in (\alpha_1, \beta_1) \cup (\alpha_2, \beta_2)$. By Lemma 1.5.13, there exist nonzero constants λ_1 and λ_2 such that $z_1(t) = z(t) - \lambda_1 x(t)$ has a double zero at $t_1 \in (\alpha_1, \beta_1)$ and $z_2(t) = z(t) - \lambda_2 x(t)$ has a double zero at $t_2 \in (\alpha_2, \beta_2)$. Since $z(t) > 0$ in (α_1, β_1) and (α_2, β_2) and $x(t) > 0$ for $t > \sigma$, we have $\lambda_1 > 0$ and $\lambda_2 > 0$. If $\lambda_1 > \lambda_2$, then $z_2(t_1) = z(t_1) - \lambda_2 x(t_1) > z(t_1) - \lambda_1 x(t_1) = z_1(t_1) = 0$ and $z_2(\beta_1) = z(\beta_1) - \lambda_2 x(\beta_1) = -\lambda_2 x(\beta_1) < 0$. Thus, $z_2(t)$ is a solution of (2.2) with a zero in (t_1, β_1) and a double zero at t_2, which contradicts the assumption that (2.2) is of type C_I. If $\lambda_1 \leq \lambda_2$, then $z_1(t_2) = z(t_2) - \lambda_1 x(t_2) \geq z_2(t_2) = 0$ and $z_1(\beta_1) = z(\beta_1) - \lambda_1 x(\beta_1) = -\lambda_1 x(\beta_1) < 0$. Hence $z_1(t)$ is a solution of (2.1) with a zero in $(\beta_1, t_2]$ and a double zero at t_1, a contradiction to the assumption that (2.1) is of type C_{II}. Hence (2.1) cannot have an oscillatory solution. Thus, (2.1) is nonoscillatory.

The above argument leads us to the following corollary;

Corollary 2.4.3 *If* (2.1) *is of type C_I and C_{II}, then it is nonoscillatory.*

In the literature, one may find several conditions for the nonoscillation of (2.1). Below, we state them: We mention the results for Eq. (2.2) without any sign restrictions on $p(t)$ and $q(t)$. We make the assumption that $r(t)$ is sufficiently smooth so that $r(t) > 0$ and $r(t) \neq 0$ in any neighbourhood of infinity.

Theorem 2.4.9 *Let* (2.1) *be of type C_I. If $r'(t) \leq 0$ and $2p(t) - q'(t) \leq 0$ for $t \geq \sigma$, then* (2.1) *is nonoscillatory.*

Theorem 2.4.10 *Let* (2.1) *be of type C_{II}. If $r'(t) \geq 0$ and $2p(t) - q'(t) \geq 0$ for $t \geq \sigma$, then* (2.1) *is nonoscillatory.*

Remark 2.4.4 One may observe that $r'(t) \leq 0$ and $2p(t) - q'(t) \leq 0$ implies that (2.1) is of type C_{II}. Similarly, $r'(t) \geq 0$ and $2p(t) - q'(t) \geq 0$ for $t \geq \sigma$ implies that (2.1) is of type C_I. Thus, the proof of Theorems 2.4.9 and 2.4.10 follow from Corollary 2.4.3.

On the other hand, when $r'(t) \geq 0$ and $p(t) > 0$, the condition $2p(t) - q'(t) \geq 0$ is better than the condition $p(t) - q'(t) \geq 0$. However, for the case considered in this section, that is, $a(t) \geq 0$, $b(t) \leq 0$ and $c(t) < 0$, the condition $p(t) - q'(t) \geq 2p(t) - q'(t)$ holds. Hence Theorem 2.4.8 provides a better condition than the ones given in Theorem 2.4.10.

Theorem 2.4.11 *Suppose that $q'(t) - p(t) \geq 0$ and*

$$\int_{\sigma}^{\infty} \big(q'(t) - p(t)\big)\, dt = \infty.$$

Then (2.1) *has an oscillatory solution, if and only if every nonoscillatory solution $x(t)$ of* (2.1) *satisfies the properties* (2.80).

Proof The sufficient part is similar to that of Theorem 2.4.5. For necessity, one may proceed, as in Theorem 2.4.4, to obtain $x(t)x'(t) \neq 0$, $\operatorname{sgn} x(t) = \operatorname{sgn} x'(t)$ for $t \geq t_0 \geq \sigma$ and $\lim_{t\to\infty} |x(t)| = \infty$. In order to be definite about the sign of $x''(t)$, we may assume that $x(t) > 0$ for $t \geq t_0$. Hence $x'(t) > 0$ for $t \geq t_0$. Since $r(t)x''(t)$ is monotonically increasing, we have $x''(t) > 0$ or < 0 for $t \geq T_0 \geq t_0$. If $x''(t) < 0$ for $t \geq T_0$, then integration of (2.2) from T_0 to t yields

$$\begin{aligned} r(t)x''(t) &\geq r(T_0)x''(T_0) + q(T_0)x(T_0) + \int_{T_0}^{t} \big(q'(s) - p(s)\big)x(s)\,ds \\ &\geq r(T_0)x''(T_0) + q(T_0)x(T_0) + x(T_0)\int_{T_0}^{t} \big(q'(s) - p(s)\big)\,ds. \end{aligned}$$

Hence $x''(t) > 0$ for large t. This contradiction completes the proof of the theorem. □

Example 2.4.2 Consider

$$x''' + \frac{1}{t^2}x'' - \left(1 + \frac{2}{t^3} - \frac{1}{3t^4}\right)x' - \left(e^t + \frac{2}{3\sqrt{3}}\right)x = 0, \quad t \geq 1.$$

Clearly, the conditions of Theorems 2.4.2, 2.4.5 and 2.4.11 are satisfied. Hence the given equation admits an oscillatory solution, and all nonoscillatory solutions of the equation tend to ∞ as $t \to \infty$.

Theorem 2.4.12 *Suppose that* $q'(t) - p(t) \leq 0$,

$$\int_{\sigma}^{\infty} \big(q'(t) - p(t)\big)\,dt = -\infty$$

and $\lim_{t\to\infty} q(t) = k$, $-\infty < k < 0$. *Then every solution* $x(t)$ *of* (2.1) *satisfies*

$$x(t)x'(t)x''(t) \neq 0, \qquad \operatorname{sgn} x(t) = \operatorname{sgn} x'(t) \neq \operatorname{sgn} x''(t), \quad t \geq t_0 \geq \sigma$$

or

$$\begin{aligned} &x(t)x'(t)x''(t) \neq 0, \qquad \operatorname{sgn} x(t) = \operatorname{sgn} x''(t) \neq \operatorname{sgn} x'(t), \quad t \geq t_0 \geq \sigma, \\ &\lim_{t\to\infty} x(t) = 0 \quad \text{and} \quad \lim_{t\to\infty} x'(t) = 0. \end{aligned}$$

Proof Let $x(t)$ be any solution of (2.1). From Theorem 2.4.8, it follows that (2.1) is nonoscillatory, and hence $x(t)$ is nonoscillatory. We may assume, without any loss of generality, that $x(t) > 0$ for $t \geq T \geq \sigma$. Lemma 2.4.3 yields $x'(t) > 0$ or < 0 for $t \geq T_0 \geq T$. If $x'(t) > 0$ for $t \geq T_0$, then $r(t)x''(t)$ is monotonically increasing and hence $x''(t) > 0$ or < 0 for large t. As $x''(t) > 0$ for large t yields, due to Theorem 2.4.5, the result that (2.1) has an oscillatory solution, the case $x''(t) > 0$ is not possible. Thus $x''(t) < 0$ for $t \geq t_0 \geq T_0$. Thus, we have

$$\operatorname{sgn} x(t) = \operatorname{sgn} x'(t) \neq \operatorname{sgn} x''(t), \quad \text{for } t \geq t_0.$$

Next, suppose that $x'(t) < 0$ for $t \geq T_0$. If possible, let $x''(t)$ be oscillatory with a sequence of zeros $\langle t_n \rangle$ such that $T_0 < t_1 < t_2 < \cdots$ and $t_n \to \infty$ as $n \to \infty$. Clearly, $\lim_{t\to\infty} x(t) = \alpha$ exists. If $\alpha = 0$, then integrating (2.2) from t_1 to t_n, we obtain

$$0 = r(t_n)x''(t_n) - r(t_1)x''(t_1) + q(t_n)x(t_n) - q(t_1)x(t_1) + \int_{t_1}^{t_n} \big(p(t) - q'(t)\big)x(t)\,dt$$

$$= q(t_n)x(t_n) - q(t_1)x(t_1) + \int_{t_1}^{t_n} \big(p(t) - q'(t)\big)x(t)\,dt.$$

If the zeros of $x''(t)$ and $q(t)$ coincide, then we get a contradiction $0 > 0$ from the above identity. Otherwise, taking limit in

$$0 > q(t_n)x(t_n) - q(t_1)x(t_1)$$

as $n \to \infty$, we obtain $0 \geq -q(t_1)x(t_1) > 0$, a contradiction. If $\alpha > 0$, then taking limit as $n \to \infty$ in

$$0 \geq q(t_n)x(t_n) - q(t_1)x(t_1) + x(t_n) \int_{t_1}^{t_n} \big(p(t) - q'(t)\big)\,dt,$$

we get a contradiction. Thus, $x''(t) > 0$ or < 0 for large t. As $x''(t) < 0$ for large t implies that $x(t) < 0$ for large t, then $x''(t) > 0$ for $t \geq t_0 \geq T_0$. If $\lim_{t\to\infty} x'(t) = \lambda$, $-\infty < \lambda < 0$, then $x(t) < 0$ for large t. Thus, $\lambda = 0$. Let $\alpha > 0$. Integrating (2.2) from t_0 to t, we obtain

$$0 < r(t)x''(t) \leq r(t_0)x''(t_0) - q(t)x(t) + \int_{t_0}^{t} \big(q'(s) - p(s)\big)x(s)\,ds$$

$$\leq r(t_0)x''(t_0) - q(t)x(t) + x(t) \int_{t_0}^{t} \big(q'(s) - p(s)\big)\,ds$$

and hence $x''(t) < 0$ for large t, a contradiction. Thus $x(t)x'(t)x''(t) \neq 0$, $\operatorname{sgn} x(t) = \operatorname{sgn} x''(t) \neq \operatorname{sgn} x'(t)$, $t \geq t_0$, and $\lim_{t\to\infty} x(t) = 0$ and $\lim_{t\to\infty} x'(t) = 0$. This completes the proof of the theorem. □

Theorem 2.4.13 *Suppose that $r(t) \equiv 1$, $p(t) > 0$, and $q(t) \leq 0$. If $\int_\alpha^\infty p(t)\,dt < \infty$ for large α, $2p(t) - q'(t) \geq 0$, $\lim_{t\to\infty} q(t) = 0$ and the second-order differential equation*

$$z'' + \left[q(t) + \frac{3}{2}\int_t^\infty p(s)\,ds\right]z = 0$$

is nonoscillatory, then (2.2) *is nonoscillatory.*

Theorem 2.4.14 *Suppose that $r(t) \equiv 1$, $p(t) > 0$, and $q(t) \leq 0$. If $\int_\alpha^\infty (2p(t) - q'(t))\,dt < \infty$ for large α, $2p(t) - q'(t) \geq 0$ and the second-order differential equa-*

tion

$$z'' + \left[\frac{1}{4}q(t) + \frac{3}{4}\int_t^{\infty}\big(2p(s) - q'(s)\big)\,ds\right]z = 0$$

is nonoscillatory, then (2.2) *is nonoscillatory.*

2.5 Behaviour of Solutions of $x''' + a(t)x'' + b(t)x' + c(t)x = 0$ with $a(t) \geq 0$, $b(t) \geq 0$ and $c(t) > 0$

In this section, we consider (2.1) with $a(t) \geq 0$, $b(t) \geq 0$ and $c(t) > 0$. Sufficient conditions are given in terms of the coefficient functions $a(t)$, $b(t)$ and $c(t)$ for the oscillation of (2.1). Asymptotic behaviour of nonoscillatory solutions of (2.1) in the presence of an oscillatory solution are also presented.

For most of the results given in this section, we assume

$$2c(t) - a(t)b(t) - b'(t) \geq 0. \tag{2.81}$$

Lemma 2.5.1 *If* (2.81) *holds, then* (2.1) *is of type* C_I.

Proof Suppose that $x(t)$ is a solution of (2.1) with $x(t_0) = 0$, $x'(t_0) = 0$ and $x''(t_0) > 0$, $t_0 > \sigma$. From the continuity of $x''(t)$ at $t = t_0$, it follows that there exists a $\delta > 0$ such that $x''(t) > 0$ for $t \in (t_0 - \delta, t_0]$. Hence $x'(t) < 0$ and $x(t) > 0$ for $t \in (t_0 - \delta, t_0]$. We claim that $x(t) > 0$ for $t \in [\sigma, t_0)$. If possible, suppose that $x(t_1) = 0$ for some $t_1 \in [\sigma, t_0)$. Multiplying (2.2) throughout by $x(t)$ and integrating the resulting identity from t_1 to t_0, we obtain

$$-r(t_1)\big(x'(t_1)\big)^2 \geq \int_{t_1}^{t_0}\big(2p(t) - q'(t)\big)x^2(t)\,dt,$$

that is,

$$0 \geq -r(t_1)\big(x'(t_1)\big)^2 \geq \int_{t_1}^{t_0}\big(2c(t) - a(t)b(t) - b'(t)\big)r(t)x^2(t)\,dt > 0,$$

a contradiction. Hence $x(t) > 0$ for $t \in [\sigma, t_0)$. This completes the proof of the lemma. □

Theorem 2.5.1 *Let* (2.81) *hold. If* $ta(t) \geq 3$, $2ta(t) - t^2b(t) + t^2a'(t) > 2$ *and* (2.17) *holds, then* (2.1) *is oscillatory.*

Proof We may note that $ta(t) \geq 3$ and $2ta(t) - t^2b(t) + t^2a'(t) > 2$ imply that

$$1 + \frac{t^2a^2(t)}{3} + t^2a'(t) - t^2b(t) > 0$$

and

$$\left(1-\frac{ta(t)}{3}\right)+\frac{1}{\sqrt{3}}\left(1+\frac{t^2a^2(t)}{3}+t^2a'(t)-t^2b(t)\right)^{1/2}>0.$$

If possible, let (2.1) be nonoscillatory. From Lemma 1.5.5, it follows that (2.1) admits a nonoscillatory solution $x(t)$ such that $x(t)x'(t)>0$ for $t\geq t_0>\sigma$. We may assume, without any loss of generality, that $x(t)>0$ for $t\geq t_0$ and $t_0>\sigma$. Setting $u(t)=t^2\frac{x'(t)}{x(t)}$ for $t\geq t_0$, we observe that $u(t)$ is a positive solution of the second-order Riccati equation (2.15), where $G(u(t),t)$ is defined as in (2.16). Then the rest of the proof is similar to Theorem 2.1.4. The proof is complete. □

Theorem 2.5.2 *Let* (2.81) *and* (2.17) *hold. If* $ta(t)\leq 3$ *and*

$$1+\frac{t^2a^2(t)}{3}+t^2a'(t)-t^2b(t)\geq 0$$

holds, then (2.1) *is oscillatory.*

Remark 2.5.1 Applying Theorems 2.5.1 and 2.5.2 to the Euler equation (2.5), we observe the following:

(i) Let $a_0>0$, $b_0\geq 0$ and $c_0>0$ hold. Further, suppose that $a_0\geq 3$, $a_0-b_0>2$, $2c_0-a_0b_0+2b_0\geq 0$ and (2.18) hold. Then (2.5) is oscillatory.
(ii) Let $a_0\geq 0$, $b_0\geq 0$ and $c_0>0$ hold. If $0\leq a_0\leq 3$, $2c_0-a_0b_0+2b_0\geq 0$, $1+\frac{a_0^3}{3}-a_0-b_0\geq 0$ and (2.18) hold, then (2.5) is oscillatory.

Theorem 2.5.3 *Let* (2.81) *hold. If* $\frac{a^2(t)}{3}-b(t)+a'(t)\geq 0$ *and* (2.14) *holds, then* (2.1) *is oscillatory.*

Remark 2.5.2 Theorem 2.5.3 holds for Eq. (1.5). Indeed, if $2c-ab\geq 0$, $a^2-3b\geq 0$ and (1.8) hold, then (1.5) is oscillatory. However, the theorem fails to hold for the Euler equation (2.5). From Proposition 1.2.5(iv) of Chap. 1, it is clear that the assertions $a^2-3b\geq 0$ and (1.8) are superfluous for oscillation of (1.5). Further, $a^2-3b\geq 0$ and $9c\geq ab$ imply that

$$\frac{2a^3}{27}-\frac{ab}{3}+c\geq\frac{2a}{9}\left(\frac{a^2}{3}-b\right)^{3/2}\geq 0$$

and hence (1.8) holds and (1.12) reduces to (1.8). Thus if the condition $2c\geq ab$ is weakened to $9c\geq ab$, then the condition $a^2\geq 3b$ becomes relevant for the oscillation of Eq. (1.5). Indeed, it is clear from Proposition 1.2.5(iii) in Chap. 1 that $a^2-3b\geq 0$ and $9c\geq ab$ imply that (1.5) is oscillatory. This observation is strengthened by Example 2.5.1.

Example 2.5.1 Consider

$$x''' + 3x'' + 2x' + \frac{15}{8}x = 0, \quad t \geq 0.$$

Clearly, $a^2 - 3b = 3 > 0$ and $9c - ab > 0$. However, $2c - ab = -\frac{9}{4} < 0$. The equation is oscillatory because it admits oscillatory solutions

$$x_1(t) = e^{-t/4}\cos\left(\frac{\sqrt{11}}{4}t\right) \quad \text{and} \quad x_2(t) = e^{-t/4}\sin\left(\frac{\sqrt{11}}{4}t\right).$$

The above example suggests that the assumption (2.81) could be weakened to

$$9c(t) - a(t)b(t) - b'(t) \geq 0. \tag{2.82}$$

However, it has not been established yet. Theorem 2.5.3 may be viewed as a generalisation of the observation as in Proposition 1.2.5(iii) in Chap. 1.

Theorem 2.5.4 *Suppose that* (2.81) *holds. If* $a'(t) \leq 0$ *and* $\int_\sigma^\infty c(t)\,dt = \infty$, *then* (2.1) *is oscillatory.*

Proof On the contrary, let (2.1) be nonoscillatory. By Lemma 1.5.5, Eq. (2.1) admits a nonoscillatory solution $x(t)$ such that $x(t)x'(t) > 0$, $t \geq t_0 > \sigma$. If $z(t) = \frac{x'(t)}{x(t)}$, then it satisfies

$$z''(t) + 3z(t)z'(t) + a(t)z'(t) = -\left[z^3(t) + a(t)z^2(t) + b(t)z(t) + c(t)\right] \leq -c(t).$$

Integrating the above inequality from t_0 to t yields

$$\begin{aligned} z'(t) &\leq z'(t) + \frac{3}{2}z^2(t) + a(t)z(t) - \int_{t_0}^t a'(s)z(s)\,ds \\ &\leq -\int_{t_0}^t c(s)\,ds + z'(t_0) + \frac{3}{2}z^2(t_0) + a(t_0)z(t_0). \end{aligned}$$

Then $z'(t) \to -\infty$ as $t \to \infty$ and hence $z(t) < 0$ for large t, a contradiction. This completes the proof of the theorem. □

Theorem 2.5.5 *Let* (2.81) *hold. If*

$$\int_\sigma^\infty \exp\left(-\int_\sigma^t a(s)\,ds\right)dt = \infty$$

and

$$\int_\sigma^\infty \left(2c(t) - a(t)b(t) - b'(t)\right)\exp\left(\int_\sigma^t a(s)\,ds\right)dt = \infty,$$

then (2.1) *is oscillatory.*

Example 2.5.2 Consider

$$x''' + \frac{2}{t}x'' + \frac{1}{t^2}x' + \frac{1}{t}x = 0, \quad t \geq 1.$$

By Theorem 2.5.4, this example has an oscillatory solution. However, $ta(t) = 2 < 3$ implies that Theorem 2.5.1 cannot be applied to this example. Again, $1 + \frac{t^2a^2(t)}{3} + t^2a'(t) - t^2b(t) = -\frac{2}{3} < 0$ implies that Theorem 2.5.2 cannot be applied to this example. Theorem 2.5.3 fails to apply to this example because $\frac{a^2(t)}{3} - b(t) + a'(t) = -\frac{5}{3t^2} < 0$. Further, Theorem 2.5.5 cannot be applied to this example.

Remark 2.5.3 Theorem 2.5.4 fails to hold for the Euler equation (2.5) because

$$\int_\sigma^\infty c(t)\,dt < \infty,$$

where $c(t) = \frac{c_0}{t^3}$, for any $c_0 > 0$. Theorem 2.5.6 can be applied to the Euler equation (2.5).

Theorem 2.5.6 *Suppose that* (2.81) *holds,* $a(t) \geq \frac{3}{t}$, $2 - 2ta(t) - t^2a'(t) + t^2b(t) \geq 0$. *If* $\int_\sigma^\infty t^2c(t)\,dt = \infty$, *then* (2.1) *is oscillatory.*

Proof Suppose that (2.1) is nonoscillatory. Then from Lemma 1.5.5, it follows that (2.1) admits a nonoscillatory solution $x(t)$ such that $x(t)x'(t) > 0$ for $t \geq t_0 > \sigma$. If $z(t) = \frac{t^2x'(t)}{x(t)}$ for $t \geq t_0$, then $z(t) > 0$ is a solution of the second-order Riccati equation (2.15). Since

$$\begin{aligned} G\big(z(t), t\big) &= \frac{z^3(t)}{t^4} + \frac{z^2(t)}{t^2}\left(a(t) - \frac{3}{t}\right) \\ &\quad + \frac{z(t)}{t^2}\big(2 - 2ta(t) - t^2a'(t) + t^2b(t)\big) + t^2c(t) \\ &\geq t^2c(t), \end{aligned}$$

(2.15) reduces to

$$\left[z'(t) - \frac{4}{t}z(t) + \frac{3}{2t^2}z^2(t) + a(t)z(t)\right]' \leq -t^2c(t).$$

Integration of the above inequality from t_0 to t yields

$$\left[z'(t) - \frac{4}{t}z(t) + \frac{3}{2t^2}z^2(t) + a(t)z(t)\right] \leq K - \int_{t_0}^t s^2c(s)\,ds,$$

where $K = z'(t_0) - \frac{4}{t_0}z(t_0) + \frac{3}{2t_0^2}z^2(t_0) + a(t_0)z(t_0)$. Since $\frac{3}{2t^2}z^2(t) - \frac{4}{t}z(t)$ attains the minimum $-\frac{8}{3}$ for $z(t) > 0$ at $z(t) = \frac{4t}{3}$, we have

$$z'(t) \le \frac{8}{3} + K - \int_{t_0}^{t} s^2 c(s)\,ds.$$

Thus $z'(t) \to -\infty$ as $t \to \infty$ and hence $z(t) < 0$ for large t, a contradiction. This completes the proof of the theorem. □

Example 2.5.3 Consider

$$x''' + \frac{3}{t}x'' + \frac{2}{t^2}x' + \frac{2}{t^3}x = 0, \quad t \ge 1.$$

Since all the conditions of Theorem 2.5.6 are satisfied, this example has an oscillatory solution. In particular, $x_1(t) = e^{t/2}\cos(\frac{\sqrt{7}}{2}\log t)$ and $x_2(t) = e^{t/2}\sin(\frac{\sqrt{7}}{2}\log t)$ are the oscillatory solutions of this example. Note that $x_3(t) = \frac{1}{t}$ is a nonoscillatory solution of this example. Since $\int_1^\infty c(t)\,dt < \infty$, Theorem 2.5.4 cannot be applied to this example. Further, $\int_1^\infty \exp(-\int_1^\infty a(s)\,ds)\,dt < \infty$ implies that Theorem 2.5.5 cannot be applied to this example.

Remark 2.5.4 In Theorem 2.5.6, it is assumed that $ta(t) \ge 3$. However, no such result exists for the case $ta(t) < 3$. This has been remain as an open problem.

Theorem 2.5.7 *If* $c(t) - a(t)b(t) - b'(t) \ge 0$,

$$0 \le \int_\sigma^\infty a(t)\,dt < \infty$$

and

$$\int_\sigma^\infty t\big[c(t) - a(t)b(t) - b'(t)\big]\exp\left(\int_\sigma^t a(s)\,ds\right)dt = \infty,$$

then (2.1) *is oscillatory.*

Proof If possible, let (2.1) be nonoscillatory. Since (2.81) holds, it follows from Lemma 1.5.5 that (2.1) admits a nonoscillatory solution $x(t)$ such that $x(t)x'(t) > 0$ for $t \ge t_0 > \sigma$. Without any loss of generality, it may be assumed that $x(t) > 0$ for $t \ge t_0$. Hence $x'(t) > 0$ for $t \ge t_0$. From (2.2), we obtain

$$\big(r(t)x''(t)\big)' = -q(t)x'(t) - p(t)x(t) < 0$$

for $t \ge t_0$. Hence $x''(t) > 0$ or < 0 for $t \ge t_1 > \max\{t_0, \sigma, 0\}$. If $x''(t) < 0$ for $t \ge t_1$, then $r(t)x''(t) < r(t_1)x''(t_1)$ implies that

$$x''(t) < r(t_1)x''(t_1)\exp\left(-\int_\sigma^\infty a(t)\,dt\right).$$

Hence $x'(t) < 0$ for large t, a contradiction. Thus $x''(t) > 0$ for $t \geq t_1$. Clearly, $x'(t) > 0$ and $x''(t) > 0$ for $t \geq t_1$ imply that

$$x(t) > x'(t_1)(t - t_1) \tag{2.83}$$

for $t \geq t_1$. Integrating (2.2) from t_1 to t and using (2.83), we get

$$\begin{aligned}
r(t_1)x''(t_1) + q(t_1)x(t_1) &= r(t)x''(t) + q(t)x(t) + \int_{t_1}^{t} \big(p(s) - q'(s)\big)x(s)\,ds \\
&> \int_{t_1}^{t} \big(c(s) - b'(s) - a(s)b(s)\big)r(s)x(s)\,ds \\
&> x'(t_1)\int_{t_1}^{t} (s - t_1)\big(c(s) - a(s)b(s) - b'(s)\big)r(s)\,ds \\
&> x'(t_1)\int_{2t_1}^{t} (s - t_1)\big(c(s) - b'(s) - a(s)b(s)\big)r(s)\,ds \\
&> \frac{1}{2}x'(t_1)\int_{2t_1}^{t} s\big(c(s) - a(s)b(s) - b'(s)\big)r(s)\,ds.
\end{aligned}$$

Hence

$$\int_{2t_1}^{\infty} t\big(c(t) - a(t)b(t) - b'(t)\big)\exp\left(\int_{\sigma}^{t} a(s)\,ds\right)dt < \infty,$$

a contradiction. Thus the theorem is proved. □

Example 2.5.4 From Theorem 2.5.7, it follows that the equation

$$x''' + \frac{1}{t(t-1)}x'' + \left(1 - \frac{1}{t}\right)x' + \frac{3}{t^2}x = 0, \quad t \geq 2$$

is oscillatory. Since $\int_2^{\infty} c(t)\,dt < \infty$, $\frac{a^2(t)}{3} - b(t) + a'(t) < 0$ for large t,

$$1 + \frac{t^2a^2(t)}{3} + t^2a'(t) - t^2b(t) < 0, \quad t \geq 2$$

and $ta(t) < 3$ for $t \geq 2$, then none of the Theorems 2.5.1–2.5.4 can be applied to this example.

Theorem 2.5.8 *Suppose that* $2a^3(t) - 9a(t)b(t) + 27c(t) - 9a''(t) > 0$, $3b(t) - 3a'(t) - a^2(t) \geq 0$, $2a^3(t) - 9a(t)b(t) + 27c(t) + 18a''(t) - 27b'(t) + 18a(t)a'(t) \geq 0$ *and*

$$\int_{\sigma}^{\infty} t\big[2a^3(t) - 9a(t)b(t) + 27c(t) + 18a''(t) - 27b'(t) + 18a(t)a'(t)\big]\,dt = \infty.$$

Then (2.1) *is oscillatory.*

Remark 2.5.5 Theorem 2.5.8 may be viewed as the generalisation of Proposition 1.2.5(ii) in Chap. 1. Indeed, from Theorem 2.5.8, it follows that Eq. (1.5) is oscillatory if $2a^3 - 9ab + 27c > 0$ and $3b - a^2 \geq 0$. However, $3b = a^2$ and $2a^3 - 9ab + 27c > 0$ imply that $9c > ab$. It confirms the earlier observation that $a^2 \geq 3b$ and $9c \geq ab$ imply that

$$x''' + ax'' + bx' + cx = 0$$

is oscillatory, where $a \geq 0$, $b \geq 0$ and $c > 0$ are constants.

Next theorem follows from Theorem 1.3 in [23].

Theorem 2.5.9 *If $a \in C^2([\sigma, \infty), R)$, $3b(t) - 3a'(t) - a^2(t) \leq 0$, $2a^3(t) - 9a(t)b(t) + 27c(t) - 9a''(t) > 0$ and*

$$\int_\sigma^\infty \left[\frac{2a^3(t)}{27} - \frac{a(t)b(t)}{3} + c(t) - \frac{a''(t)}{3} - \frac{2}{3\sqrt{3}}\left(\frac{a^2(t)}{3} + a'(t) - b(t)\right)^{3/2}\right] dt = \infty,$$

then (2.1) *is oscillatory.*

Remark 2.5.6 If $\frac{2a^3}{27} - \frac{ab}{3} + c > 0$, then Theorem 2.5.9 is the generalisation of the observation from Proposition 1.2.5(iii) in Chap. 1.

Theorem 2.5.10 *If Eq.* (2.1) *is of type C_I or C_{II}, then it admits a nonoscillatory solution.*

Proof Suppose that (2.1) is of type C_I. For every integer $n \geq \sigma$, it is possible to construct a solution $x_n(t)$ of (2.1) such that $x_n(n) = x_n'(n) = 0$, $x_n''(n) > 0$. Hence $x_n(t) > 0$ for $t \in [\sigma, n)$. Let $\{u_1(t), u_2(t), u_3(t)\}$ be a basis for the solution space of (2.1). Hence it is possible to write

$$x_n(t) = c_{1n}u_1(t) + c_{2n}u_2(t) + c_{3n}u_3(t),$$

where

$$c_{1n}^2 + c_{2n}^2 + c_{3n}^2 = 1.$$

Since $\langle c_{in}\rangle$, $i = 1, 2, 3$, is bounded, it admits a convergent subsequence. Without any loss of generality, it may be assumed that $\langle c_{in_j}\rangle$, $i = 1, 2, 3$ is the convergent subsequence. Let $c_{in_j} \to c_i$, $i = 1, 2, 3$, as $n_j \to \infty$. Then $c_1^2 + c_2^2 + c_3^2 = 1$. If $x_0(t) = c_1u_1(t) + c_2u_2(t) + c_3u_3(t)$, $t \geq \sigma$, then $\langle x_n(t)\rangle$ converges to $x_0(t)$ uniformly on every compact subinterval of $[\sigma, \infty)$. Hence $x_0(t) \geq 0$ for $t \in [\sigma, \infty)$. If $x_0(t_0) = 0 = x_0(t_1)$ for $\sigma \leq t_0 < t_1 < \infty$, then $x_0'(t_0) = 0$, $x_0''(t_0) > 0$ and $x_0'(t_1) = 0$, $x_0''(t_1) > 0$. This is impossible, since (2.1) is of type C_I. Hence $x_0(t)$ is nonoscillatory. If (2.1) is of type C_{II}, then obviously it admits a nonoscillatory solution. □

Remark 2.5.7 No sign restrictions on coefficient functions in (2.1) have been used in Theorem 2.5.10.

Corollary 2.5.1 *If* (2.81) *holds, then* (2.1) *admits a nonoscillatory solution.*

Theorem 2.5.11 *Suppose that* $c(t) \geq d > 0$, $c(t) - a(t)b(t) - b'(t) \geq 0$ *and* $\int_\sigma^\infty a(t)\,dt = \infty$. *If* (2.1) *is oscillatory, then every nonoscillatory solution* $x(t)$ *of* (2.1) *satisfies the property*

$$\lim_{t\to\infty} x(t) = \lim_{t\to\infty} x'(t) = \lim_{t\to\infty} x''(t) = 0. \tag{2.84}$$

Proof Since (2.81) holds, from Lemma 1.5.6 it follows that $F[x(t)] < 0$ for $t \geq \sigma$. If $x(t_0) = 0$ for some $t_0 \in [\sigma, \infty)$, then $F[x(t_0)] \geq 0$, a contradiction. Hence $x(t) \neq 0$ for $t \in [\sigma, \infty)$. Without any loss of generality, it may be assumed that $x(t) > 0$ for $t \geq \sigma$. As $r(t) \geq 1$ and $F[x(t)] < 0$ for $t \geq \sigma$, then

$$0 \leq \big(x'(t)\big)^2 \leq r(t)\big(x'(t)\big)^2 < x(t)\big(2r(t)x''(t) + q(t)x(t)\big). \tag{2.85}$$

Then

$$2r(t)x''(t) + q(t)x(t) > 0, \quad t \geq \sigma. \tag{2.86}$$

Consequently, from (2.2) and (2.86), we obtain

$$\begin{aligned} 0 &< r(t)x''(t) + q(t)x(t) \\ &= r(\sigma)x''(\sigma) + q(\sigma)x(\sigma) + \int_\sigma^t \big(q'(s) - p(s)\big)x(s)\,ds \leq K, \end{aligned} \tag{2.87}$$

since $c(t) - a(t)b(t) - b'(t) \geq 0$, where $K = r(\sigma)x''(\sigma) + q(\sigma)x(\sigma)$. Then

$$2r(t)x''(t) + q(t)x(t) \leq 2K. \tag{2.88}$$

Using (2.85) and (2.88), we get

$$\big(x'(t)\big)^2 \leq 2Kx(t). \tag{2.89}$$

Further from (2.54), for $t \geq \sigma$,

$$\begin{aligned} 0 > F\big[x(t)\big] &\geq F\big[x(\sigma)\big] + \int_\sigma^t \big(2p(s) - q'(s)\big)x^2(s)\,ds \\ &\geq F\big[x(\sigma)\big] + \int_\sigma^t \big(2c(s) - a(s)b(s) - b'(s)\big)x^2(s)\,ds \\ &\geq F\big[x(\sigma)\big] + \int_\sigma^t c(s)x^2(s)\,ds \\ &\geq F\big[x(\sigma)\big] + d\int_\sigma^t x^2(s)\,ds. \end{aligned}$$

This inequality implies that $x \in L^2([\sigma, \infty), R)$. By Lemma 1.5.15, $\lim_{t\to\infty} x(t) = 0$. Hence (2.89) yields $\lim_{t\to\infty} x'(t) = 0$. Next, we show that $\lim_{t\to\infty} x''(t) = 0$. From (2.87), we obtain

$$K \ge \int_\sigma^t \big(p(s) - q'(s)\big)x(s)\,ds = \int_\sigma^t \big(c(s) - a(s)b(s) - b'(s)\big)r(s)x(s)\,ds. \quad (2.90)$$

Since

$$\begin{aligned}\big[r(t)x''(t) + q(t)x(t)\big]' &= -\big(p(t) - q'(t)\big)x(t)\\ &= -\big(c(t) - a(t)b(t) - b'(t)\big)r(t)x(t)\\ &\le 0,\end{aligned}$$

then using (2.87), we get

$$\lim_{t\to\infty}\big[r(t)x''(t) + q(t)x(t)\big] = l, \quad (2.91)$$

where $0 \le l < \infty$. Integrating (2.2) from t to s $(\sigma < t < s)$ and then taking limit as $s \to \infty$, we have

$$r(t)x''(t) + q(t)x(t) = l + \int_t^\infty \big(p(\theta) - q'(\theta)\big)x(\theta)\,d\theta,$$

that is, using (2.90),

$$\begin{aligned}x''(t) + b(t)x(t) &\le \left[l + \int_\sigma^\infty \big(c(\theta) - a(\theta)b(\theta) - b'(\theta)\big)r(\theta)x(\theta)\,d\theta\right]r^{-1}(t)\\ &\le [l + K]r^{-1}(t).\end{aligned}$$

From the given hypothesis, it follows that

$$\lim_{t\to\infty}\big[x''(t) + b(t)x(t)\big] = 0.$$

Since by (2.86),

$$0 < x''(t) + \frac{1}{2}b(t)x(t) \le x''(t) + b(t)x(t),$$

we have

$$\lim_{t\to\infty}\left[x''(t) + \frac{1}{2}b(t)x(t)\right] = 0$$

and, hence

$$\lim_{t\to\infty}\frac{1}{2}b(t)x(t) = \lim_{t\to\infty}\left[\big(x''(t) + b(t)x(t)\big) - \left(x''(t) + \frac{1}{2}b(t)x(t)\right)\right] = 0.$$

Consequently, $\lim_{t\to\infty} x''(t) = 0$. Thus the theorem is proved. □

In view of Lemma 1.5.7 and Theorem 2.5.11, we obtain the following corollary:

Corollary 2.5.2 *Let the conditions of Theorem* 2.5.11 *hold. If* $b(t) \geq L > 0$, *then*

$$\lim_{t\to\infty} x(t) = \lim_{t\to\infty} x'(t) = \lim_{t\to\infty} x''(t) = 0$$

for every nonoscillatory solution $x(t)$ *of* (2.1).

Theorem 2.5.12 *Let* $c(t) \geq d > 0$, $c(t) - a(t)b(t) - b'(t) \geq 0$ *and* $b(t)$ *be bounded. If* (2.1) *is oscillatory, then every nonoscillatory solution of* (2.1) *satisfies the property* (2.84).

Proof From Lemma 1.5.6, it follows that $F[x(t)] < 0$ for $t \geq \sigma$. If $x(t_0) = 0$ for some $t_0 \in [\sigma, \infty)$, then $F[x(t_0)] \geq 0$, a contradiction. Hence $x(t) \neq 0$ for $t \in [\sigma, \infty)$. Without any loss of generality, we may assume that $x(t) > 0$ for $t \geq \sigma$. As $r(t) > 1$ and $F[x(t)] < 0$ for $t \geq \sigma$, then proceeding as in the proof of Theorem 2.5.11, we obtain $\lim_{t\to\infty} x(t) = 0$. Now, we show that $\lim_{t\to\infty} x''(t) = 0$. We consider the following two cases:

$$\int_\sigma^\infty a(t)\,dt = \infty$$

and

$$\int_\sigma^\infty a(t)\,dt < \infty. \tag{2.92}$$

In view of Theorem 2.5.11, it is enough to prove $\lim_{t\to\infty} x''(t) = 0$ when (2.92) holds. Let (2.92) hold. Clearly (2.91) holds, too. Since $b(t)$ is bounded, we have $\lim_{t\to\infty} r(t)x''(t) = \alpha$, $0 \leq \alpha < \infty$. Clearly, (2.92) implies that $\int_\sigma^\infty \frac{1}{r(t)}\,dt = \infty$ and hence $\alpha > 0$ yields $x'(t) \to \infty$ as $t \to \infty$, a contradiction. Hence $\alpha = 0$. Consequently, $\lim_{t\to\infty} x''(t) = 0$. This completes the proof of the theorem. □

Theorem 2.5.13 *Suppose that there exists a positive number* m *such that* $m \leq b(t)$ *for* $t \geq \sigma$. *Suppose, furthermore, that there is a function* $h \in C^1([\sigma, \infty), (0, \infty))$ *such that* $h'(t) \leq 0$ *on* $[t_0, \infty) \subset [\sigma, \infty)$,

$$\lim_{t\to\infty} h(t) = 0, \qquad \int_{t_0}^\infty h(t)\,dt = \infty$$

and

$$\big[c(t) - b'(t) - a(t)b(t)\big]r(t) \geq b(t)h(t)$$

hold. Then (2.1) *is oscillatory and every nonoscillatory solution* $x(t)$ *of* (2.1) *satisfies*

$$x(t) \neq 0 \quad \text{for } t \geq \sigma, \quad \text{and} \quad \lim_{t\to\infty} x(t) = \lim_{t\to\infty} x'(t) = \lim_{t\to\infty} x''(t) = 0.$$

Theorem 2.5.14 *Assume that there exist constants* $0 < m < M$ *such that* $m \le b(t) \le M$ *and* h *have the same properties as in Theorem* 2.5.13. *If either*

$$\big[c(t) - b'(t) - a(t)b(t)\big]r(t) \ge h(t) \quad \textit{for } t \in [\sigma, \infty),$$

or

$$c(t) - b'(t) - a(t)b(t) \ge h(t) \quad \textit{for } t \in [\sigma, \infty),$$

then Eq. (2.1) *is oscillatory, and every nonoscillatory solution* $x(t)$ *of* (2.1) *has the properties*

$$x(t) \ne 0 \quad \textit{for } t \ge \sigma \quad \textit{and} \quad \lim_{t\to\infty} x(t) = \lim_{t\to\infty} x'(t) = \lim_{t\to\infty} x''(t) = 0.$$

Theorem 2.5.15 *Assume the hypothesis of Theorem* 2.5.13 *or Theorem* 2.5.14 *hold and* $\int_\sigma^\infty a(t)\,dt < \infty$. *Then every solution of* (2.1) *is either oscillatory except for a solution* x *(unique up to a linear combination) that satisfies*

$$x(t) \ne 0 \quad \textit{for } t \ge \sigma \quad \textit{and} \quad \lim_{t\to\infty} x(t) = \lim_{t\to\infty} x'(t) = \lim_{t\to\infty} x''(t) = 0.$$

Examples of functions $h \in C^1([\sigma, \infty), (0, \infty))$ that satisfy the hypothesis of Theorem 2.5.13 include $h(t) = \frac{d}{t}$, $d > 0$, $t > 0$ and $h(t) = \frac{1}{(t+1)\ln(t+1)}$, $t > 0$.

Example 2.5.5 Consider

$$x''' + \frac{1}{t}x'' + tx' + \left(2 + \frac{1}{t^3}\right)x = 0, \quad t \ge 1.$$

From Corollary 2.5.2, it follows that every nonoscillatory solution $x(t)$ of the equation satisfies $\lim_{t\to\infty} x(t) = \lim_{t\to\infty} x'(t) = \lim_{t\to\infty} x''(t) = 0$. However, Lemma 8 in [15] fails to hold for this example, because one of the assumption of the lemma does not hold.

Remark 2.5.8 The condition $c \ge ab$ implies that $2c \ge ab$ and $\frac{2a^3}{27} - \frac{ab}{3} + c > 0$. Hence Corollary 2.5.2 may be viewed as a partial generalisation of Proposition 1.2.5(v) in Chap. 1. It seems that the condition $c(t) - a(t)b(t) - b'(t) \ge 0$ of Corollary 2.5.2 may be weakened to (2.81). The following example strengthens this observation.

Example 2.5.6 Consider

$$x''' + x'' + \left(1 - \frac{1}{t^2}\right)x' + \left(1 - \frac{1}{t^2}\right)x = 0, \quad t \ge 2. \tag{2.93}$$

Clearly (2.81) holds but $c(t) - a(t)b(t) - b'(t) = -\frac{2}{t^3} < 0$. Further, $b(t) = c(t) = 1 - \frac{1}{t^2} > \frac{1}{2}$ and $\int_2^\infty a(t)\,dt = \infty$. By Theorem 2.5.4, (2.93) is oscillatory. It is expected that every nonoscillatory solution $x(t)$ of (2.93) satisfies $\lim_{t\to\infty} x(t) = \lim_{t\to\infty} x'(t) = \lim_{t\to\infty} x''(t) = 0$. In fact, $x(t) = e^{-t}$ is such a solution of (2.93).

Theorem 2.5.16 *Let* $3b(t) - a^2(t) - 3a'(t) \le 0$ *and* $2a^3(t) - 9a(t)b(t) + 27c(t) - 9a''(t) > 0$. *If* (2.1) *is oscillatory, then nonoscillatory solutions of* (2.1) *form a one-dimensional subspace of the solution space of* (2.1).

Corollary 2.5.3 *Suppose that the conditions of Theorems* 2.5.4 *and* 2.5.16 *hold. Then nonoscillatory solutions of* (2.1) *form a one-dimensional subspace of the solution space of* (2.1).

Corollary 2.5.4 *Let the conditions of Theorem* 2.5.16 *hold. Suppose that* $b(t) \ge L > 0$, $c(t) \ge d > 0$, $c(t) - a(t)b(t) - b'(t) \ge 0$, $a'(t) \le 0$ *and* $\int_\sigma^\infty a(t)\,dt = \infty$. *Then nonoscillatory solutions of* (2.1) *form a one-dimensional subspace of the solution space of* (2.1) *and every nonoscillatory solution of* (2.1) *tends to zero as* $t \to \infty$ *along with its first and second derivatives.*

This follows from Corollary 2.5.2 and Theorems 2.5.4 and 2.5.16.

Theorem 2.5.17 *Suppose that* $3b(t) - a^2(t) - 3a'(t) \le 0$ *and* $2a^3(t) - 9a(t)b(t) + 27c(t) - 9a''(t) < 0$. *If* (2.1) *is oscillatory, then there exist two linearly independent oscillatory solutions of* (2.1) *whose zeros separate and oscillatory solutions of* (2.1) *form a two-dimensional subspace of the solution space of* (2.1).

Corollary 2.5.5 *Let the conditions of Theorems* 2.5.4 *and* 2.5.17 *be satisfied. Then there exist two linearly independent oscillatory solutions of* (2.1) *whose zeros separate and oscillatory solutions of* (2.1) *form a two-dimensional subspace of the solution space of* (2.1).

Now, we consider Eq. (2.10). In [16], Hanan proved the following theorem.

Theorem 2.5.18 (Hanan, [16]) *If* $c(t) > b'(t)$, *the equation*

$$u'' + b(t)u = 0 \tag{2.94}$$

is nonoscillatory, and if

$$\int_{t_0}^\infty t\big[c(t) - b'(t)\big]\,dt = \infty, \qquad t_0 \ge \sigma, \tag{2.95}$$

then (2.10) *is oscillatory.*

On the other hand, Lazer [23] proved the following theorem for the oscillation of (2.10).

Theorem 2.5.19 *If* $2c(t) - b'(t) \ge 0$ *and not identically zero in any subinterval of* $[\sigma, \infty)$, *and there exists a number* $m < \frac{1}{2}$ *such that the second-order differential equation*

$$u'' + \big(b(t) + mtc(t)\big)u = 0, \tag{2.96}$$

is oscillatory, then (2.10) *is also oscillatory. In fact, if x is any nonzero solution of Eq.* (2.10) *with a zero at some $t_0 \geq \sigma$, then x is oscillatory.*

One can interpret the conclusion of the theorem in the following way: If $x(t)$ is a solution of (2.10) with $F[x(t_0)] \geq 0$ for some $t_0 \geq \sigma$, then $x(t)$ is oscillatory, where $F[x(t)]$ is given in (2.54).

Lemma 2.5.2 *If $2c(t) - b'(t) \geq 0$ and not identically zero in any subinterval of $[\sigma, \infty)$, and $x(t)$ is a nonoscillatory solution of* (2.10) *which is eventually nonnegative with $F[x(t_0)] \geq 0$, $t_0 \geq \sigma$, then there exists a number $t_1 \geq t_0$ such that $x(t) > 0$, $x'(t) > 0$, $x''(t) > 0$ and $x'''(t) \leq 0$ for $t \geq t_1$.*

If $a(t) \equiv 0$, then from Theorem 2.5.2 we see that if $2c(t) - b'(t) \geq 0$, $t^2 b(t) \leq 1$ and

$$\int_\sigma^\infty \left[t^2 c(t) + t b(t) - \frac{2}{3\sqrt{3}t} \left(1 - t^2 b(t)\right)^{3/2} \right] dt = \infty, \tag{2.97}$$

then (2.10) is oscillatory. Let the hypothesis of Lemma 2.5.2 hold. Considering additional assumptions, we eliminate positive increasing solutions and so we obtain oscillation criterion. Since $t^2 b(t) > \frac{1}{4}$, $t > \sigma$, (2.96) is oscillatory, by the Sturm comparison theorem and Kneser-criterion, Theorem 2.5.19 is applicable here. Therefore, we shall now concentrate with the case $t^2 b(t) \leq \frac{1}{4}$, $t \geq \sigma$.

It is easy to verify that the following inequality is fulfilled for all $t \geq \sigma$:

$$t b(t) - \frac{2}{3\sqrt{3}t} \left[1 - t^2 b(t)\right]^{3/2} \leq 0 \quad \text{for } t^2 b(t) \leq \frac{1}{4}, \tag{2.98}$$

since $4t^6 b^3(t) + 15t^4 b^2(t) + 12t^2 b(t) - 4 = [4t^2 b(t) - 1][t^2 b(t) + 2]^2$.

Using this inequality, we obtain the assertion needed to prove next Theorem 2.5.20.

Lemma 2.5.3 *Let $0 \leq t^2 b(t) \leq \frac{1}{4}$ for all $t \geq \sigma$. Let G be the polynomial in the variable z,*

$$G(z) = z^3 - 3z^2 + \left(2 + t^2 b(t)\right) z + t^3 c(t), \quad t > \sigma.$$

Then

$$G(z) \geq t^3 c(t) + t^2 b(t) - \frac{2}{3\sqrt{3}} \left(1 - t^2 b(t)\right)^{3/2}, \quad t > \sigma \tag{2.99}$$

for all $z \geq 1 - 2\sqrt{\frac{1 - t^2 b(t)}{3}}$.

We note that the right-hand side of (2.99) is the minimum of G at the point

$$z_0 = 1 + \sqrt{\frac{1 - t^2 b(t)}{3}}.$$

Theorem 2.5.20 *Let the hypothesis of Lemma* 2.5.3 *hold, and in addition* $t^2b(t) \leq \frac{1}{4}$ *for all* $t \geq \sigma$. *If* (2.97) *is satisfied, then* (2.10) *is oscillatory. In fact, any solution* x *of* (2.10) *which satisfies* $F[x(t^*)] \geq 0$ *for some* $t^* \geq \sigma$ *is oscillatory.*

Remark 2.5.9 Theorems 2.5.2 and 2.5.20 give the same conclusion. In the former, the weaker condition $t^2b(t) < 1$ is used whereas the condition $t^2b(t) < \frac{1}{4}$ is used in the later. Thus Theorem 2.5.2 is better than Theorem 2.5.20. However, no results are obtained on the asymptotic behaviour of nonoscillatory solutions of (2.10) except Theorem 2.5.11. The condition $c(t) \geq d > 0$ has been used in Theorem 2.5.11, which in turn, implies that $\int_\sigma^\infty t^2c(t)\,dt = \infty$. In the following, we prove the asymptotic behaviour of nonoscillatory solutions of (2.10) with the price that $\int_\sigma^\infty t^2c(t)\,dt = \infty$ is satisfied.

Theorem 2.5.21 *Let* $t^2b(t) < \frac{1}{4}$ *hold. If* (2.97) *is satisfied, then every nonoscillatory solution of* (2.10) *has the property* $\lim_{t\to\infty} x(t) = 0$.

Proof Since $t^2b(t) \leq \frac{1}{4}$, from Kneser comparison theorem it follows that Eq. (2.94) is nonoscillatory. So by Theorem 3.6 in [18], it follows that there exists $d \geq \sigma$ such that either $x(t)x'(t) \geq 0$ or $x(t)x'(t) < 0$ for all $t \geq d$. Let $x(t)$ be a nonoscillatory solution of (2.10), and suppose that $x(t) > 0, x'(t) > 0$ for all $t \geq d$. We denote $z(t) = \frac{tx'(t)}{x(t)}$, $t \geq d$. Since (2.97) holds, proceeding as in the proof of Theorem 2.1.4, we get the contradiction that $z(t) < 0$ for large t. Hence $x(t) > 0$ and $x'(t) < 0$ for $t \geq d$. Then $\lim_{t\to\infty} x(t) = L \geq 0$ exists. Let $L > 0$. Multiplying (2.10) by t^2 and integration from d to t yields

$$t^2x''(t) - 2tx'(t) + \frac{9}{4}x(t) \leq K - L\int_d^t s^2c(s)\,ds,$$

where K is some constant. By (2.98) and (2.97), we have $\int_\sigma^\infty t^2c(t)\,dt = \infty$. From this and the last inequality, we obtain $x''(t) < 0$ for large t, which contradicts the fact that $x(t) > 0$ and $x'(t) < 0$ for $t \geq d$. The proof is complete. □

Now, we shall show that either Theorem 2.5.2 or Theorem 2.5.20 can be applied even in the case when Theorems 2.5.18 and 2.5.19 are not applicable to the third-order equation

$$x''' + b_0t^\beta x' + c_0t^{-3}x = 0, \quad t > 0, \tag{2.100}$$

where $\beta \leq -2$, $b_0 > 0$, $c_0 > \frac{2}{3\sqrt{3}}$, and $b_0 < \frac{1}{4}$ if $\beta = -2$; β, b_0 and c_0 are some constants.

Directly, we see that Theorem 2.5.18 is not applicable to (2.100). For $\beta = -2$, Eq. (2.100) becomes the Euler equation. The necessary and sufficient condition for the oscillation of Euler equation (2.100) is

$$c_0 + b_0 - \frac{2}{3\sqrt{3}}(1 - b_0)^{3/2} > 0. \tag{2.101}$$

It is easy to check that (2.101) is equivalent to (2.97) of Theorem 2.5.20. Equation (2.96) with $\beta = -2$ and $b_0 < \frac{1}{4}$ becomes the Euler equation

$$v'' + (b_0 + mc_0)t^{-2}v = 0. \tag{2.102}$$

Equation (2.102) is oscillatory if and only if $b_0 + mc_0 > \frac{1}{4}$ for some $m < \frac{1}{2}$, that is, $2b_0 + c_0 > 2b_0 + 2mc_0 > \frac{1}{2}$. So it is easy to check that inequality $c_0 + b_0 > \frac{1}{2} - b_0 > \frac{2}{3\sqrt{3}}(1 - b_0)^{3/2}$ holds for some $c_0 > 0$ and every $0 < b_0 < \frac{1}{4}$. From this it follows that Theorem 2.5.21 is better than Theorem 2.5.19 in this case. For example, if $b_0 = 0.06$, $c_0 = 0.3$, then condition (2.101) is fulfilled, while (2.102) is nonoscillatory.

Let $\beta < -2$. So there is a number $\delta > 0$ such that $\beta = -2 - \delta$, and hence $t^2 b(t) = b_0 t^{-\delta} \le \frac{1}{4}$ for $t \ge (4b_0)^{1/\delta}$. If we denote $y = b_0 t^{-\delta}$ for $t \ge \sigma$, then the function $f(y) = y - \frac{2}{3\sqrt{3}}(1 - y)^{3/2}$ is increasing and so for $0 \le y \le \frac{1}{4}$, we have

$$\int_\sigma^\infty \frac{1}{t}\left(c_0 + b_0 t^{-\delta} - \frac{2}{3\sqrt{3}}\left(1 - b_0 t^{-\delta}\right)^{3/2}\right) dt \ge \left(c_0 - \frac{2}{3\sqrt{3}}\right) \int_\sigma^\infty \frac{dt}{t}.$$

Hence we see by Theorem 2.5.21 that for $b_0 > 0$, $c_0 > \frac{2}{3\sqrt{3}}$ and $\beta < -2$, Eq. (2.100) is oscillatory. On the other hand, by Theorem 2.5.19, Eq. (2.100) is oscillatory only if $c_0 > 0.5$.

One may find several oscillation and asymptotic behaviour of nonoscillatory solutions of the third-order linear differential equation

$$x''' + A(t)x' + \left[A'(t) + b(t)\right]x = 0, \tag{2.103}$$

in [13], where A' and b are continuous functions on $[\sigma, \infty)$ and $b(t) \ge 0$ for $[\sigma, \infty)$ such that $b(t) \not\equiv 0$ on each subinterval. The following theorem is an easy consequence of Corollary 2.3 and Theorem 2.17 in [14]. In fact, the following theorem can be obtained if one proceeds as in Theorem 2.5.11 or Theorem 2.5.12.

Theorem 2.5.22 *Let $A(t) \ge m > 0$, $A'(t) \le 0$, $A'(t) + b(t) \ge \frac{d}{t} > 0$ for $t \in [\sigma, \infty)$, $\sigma > 0$. Then every solution of* (2.103) *is oscillatory in $[\sigma, \infty)$, except a solution $x(t)$ (unique up to linear dependence), which satisfies the property* (2.84).

2.6 Behaviour of Solutions of $x''' + a(t)x'' + b(t)x' + c(t)x = 0$ with $a(t) \le 0$, $b(t) \ge 0$ and $c(t) < 0$

This section is concerned with the study of behaviour of solutions of (2.1) with $a(t) \le 0$, $b(t) \ge 0$ and $c(t) < 0$. An attempt has been made to generalise, as far as possible, Proposition 1.2.8 to the variable coefficients. An explicit sufficient condition in terms of $a(t)$, $b(t)$ and $c(t)$ is given for the strong oscillation of (2.1). An example is also provided to strengthen the result. Throughout this section, we assume that the condition

$$2c(t) - a(t)b(t) - b'(t) \le 0 \tag{2.104}$$

holds.

Theorem 2.6.1 *Let* (2.104) *hold. If* $\int_\sigma^\infty b(t)\,dt = \infty$, *then* (2.1) *is oscillatory.*

Proof If possible, suppose that (2.1) is nonoscillatory. Then by Lemma 1.5.2, there exists a nonoscillatory solution $x(t)$ of (2.1) such that $F[x(t)] > 0$ for $t \geq \sigma$ and $x(t)x'(t) > 0$ for $t \geq t_0 \geq \sigma$, where F is same as in (2.54). Since $F[x(t)] > 0$ for $t \geq \sigma$, we have

$$2\frac{x''(t)}{x(t)} + b(t) \leq \left(\frac{x'(t)}{x(t)}\right)^2, \quad t \geq \sigma. \tag{2.105}$$

If $z(t) = \frac{x'(t)}{x(t)}$, then $z(t) > 0$ for $t \geq t_0$ and

$$\begin{aligned} z'(t) = \frac{x''(t)}{x(t)} - \left(\frac{x'(t)}{x(t)}\right)^2 &= \frac{x''(t)}{x(t)} - \frac{1}{2}\left(\frac{x'(t)}{x(t)}\right)^2 - \frac{1}{2}\left(\frac{x'(t)}{x(t)}\right)^2 \\ &< -\frac{1}{2}\left(\frac{x'(t)}{x(t)}\right)^2 - \frac{b(t)}{2} < -\frac{b(t)}{2}. \end{aligned}$$

Integrating the inequality from t_0 to t, we get $z(t) < 0$ for large t, a contradiction. Hence (2.1) is oscillatory. This completes the proof of the theorem. □

Theorem 2.6.2 *Let* (2.104) *hold. If* $t^2 b(t) > 1$ *for* $t \geq \sigma$ *and* $\int_\sigma^\infty (\frac{t^2 b(t) - 1}{t})\,dt = \infty$, *then* (2.1) *is oscillatory.*

Proof Let (2.1) be nonoscillatory. Then, by Lemma 1.5.2, there exists a nonoscillatory solution $x(t)$ of (2.1) such that $x(t)x'(t) > 0$ for $t \geq t_0 \geq \sigma$ and $F[x(t)] > 0$ for $t \geq \sigma$. $F[x(t)] > 0$ for $t \geq \sigma$ implies that the inequality (2.105) holds. If $z(t) = \frac{tx'(t)}{x(t)}$, then $z(t) > 0$ for $t \geq t_0$ and

$$z'(t) = \frac{x'(t)}{x(t)} + t\left[\frac{x''(t)}{x(t)} - \left(\frac{x'(t)}{x(t)}\right)^2\right].$$

Using (2.105), we obtain

$$z'(t) + \frac{1}{2t}z^2(t) - \frac{1}{t}z(t) \leq -\frac{1}{2}tb(t).$$

Since the function $\frac{1}{2t}z^2(t) - \frac{1}{t}z(t)$ attains the minimum $-\frac{1}{2t}$ for $z(t) > 0$ at $z(t) = 1$, the above inequality yields

$$z'(t) \leq \frac{1}{2t}\left(1 - t^2 b(t)\right).$$

From the given hypotheses, it follows that $z(t) < 0$ for large t, a contradiction. Hence (2.1) is oscillatory, and the theorem is proved. □

Remark 2.6.1 Theorem 2.6.1 generalise the assertion (iv) of Proposition 1.2.8. This theorem cannot be applied to the Euler equation (2.5) because $\int_\sigma^\infty \frac{b_0}{t^2}\,dt < \infty$. On the other hand, Theorem 2.6.2 can be applied to the Euler equation (2.5). Indeed, if $2c_0 - a_0 b_0 + 2b_0 \le 0$ and $b_0 > 1$, then (2.5) is oscillatory. When $a(t) \equiv a$, $b(t) \equiv b$ and $c(t) \equiv c$ are constants, then the conditions of Theorem 2.6.2 are satisfied for $t > b^{-1/2}$ provided that $2c \le ab$. Hence Theorem 2.6.2 may be regarded as a generalisation of the assertion (iv) of Proposition 1.2.8.

Using the transformation

$$x(t) = z(t)\exp\left(-\frac{1}{3}\int_\sigma^t a(s)\,ds\right),$$

one can obtain (2.4) from (2.1).

Theorem 2.6.3 *Let* $b(t) - \frac{a^2(t)}{3} - a'(t) \le 0$ *and* $\frac{2a^3(t)}{27} - \frac{a(t)b(t)}{3} + c(t) - \frac{a''(t)}{3} > 0$. *If*

$$\int_\sigma^\infty \left[\frac{2t^2a^3(t)}{27} - \frac{t^2a(t)b(t)}{3} + t^2c(t) - \frac{t^2a''(t)}{3} + tb(t) - \frac{ta^2(t)}{3} \right.$$
$$\left. - ta'(t) - \frac{2}{3\sqrt{3}t}\left(1 - t^2b(t) + \frac{1}{3}t^2a^2(t) + t^2a'(t)\right)^{3/2}\right] dt = \infty,$$

then (2.1) *is oscillatory.*

Remark 2.6.2 Theorem 2.6.3 generalises Proposition 1.2.8(iii) of Chap. 1. Indeed, $\frac{a^2}{3} - b > 0$ and $\frac{2a^3}{27} - \frac{ab}{3} + c > \frac{2}{3\sqrt{3}}(\frac{a^2}{3} - b)^{3/2}$ imply that Eq. (1.5) is oscillatory because

$$\frac{2a^3}{27} - \frac{ab}{3} + c - \frac{b}{t} - \frac{a^2}{3t} - \frac{2}{3\sqrt{3}}\left(\frac{1}{t^2} - b + \frac{1}{3}a^2\right)^{3/2}$$
$$\to \frac{2a^3}{27} - \frac{ab}{3} + c - \frac{2}{3\sqrt{3}}\left(\frac{a^2}{3} - b\right)^{3/2} > 0$$

as $t \to \infty$. It holds for the Euler equation (2.5). Indeed (2.5) is oscillatory if

$$b_0 - \frac{a_0^2}{3} + a_0 \le 0, \qquad \frac{2a_0^3}{27} - \frac{a_0b_0}{3} + c_0 - \frac{2a_0}{3} > 0$$

and

$$\frac{2a_0^3}{27} - \frac{a_0b_0}{3} + c_0 - \frac{2a_0}{3} - \frac{2}{3\sqrt{3}}\left(1 - b_0 + \frac{a_0^2}{3} - a_0\right)^{3/2} > 0$$

are satisfied.

Theorem 2.6.4 *Let* $b(t) - \frac{a^2(t)}{3} - a'(t) \le 0$ *and* $\frac{2a^3(t)}{27} - \frac{a(t)b(t)}{3} + c(t) - b'(t) + \frac{2a(t)a'(t)}{3} + \frac{2a''(t)}{3} < 0$. *If*

$$\int_\sigma^\infty \Big[-\frac{2t^2a^3(t)}{27} + \frac{t^2a(t)b(t)}{3} - t^2c(t) + t^2b'(t) - \frac{2t^2a(t)a'(t)}{3} - \frac{2t^2a''(t)}{3} + tb(t) - \frac{t^2a(t)}{3} - ta'(t) - \frac{2}{3\sqrt{3}t}\Big(1 - t^2b(t) + \frac{1}{3}t^2a^2(t) + t^2a'(t)\Big)^{3/2}\Big]\,dt = \infty,$$

then (2.1) *is oscillatory.*

Remark 2.6.3 Theorem 2.6.4 can be applied to the Euler equation (2.5). Indeed, (2.5) is oscillatory if

$$b_0 - \frac{a_0^2}{3} + a_0 < 0, \qquad \frac{2a_0^3}{27} - \frac{a_0b_0}{3} + c_0 + 2b_0 - \frac{2a_0^2}{3} + 4a_0 < 0$$

and

$$-\frac{2a_0^3}{27} + \frac{a_0b_0}{3} - \frac{a_0}{3} - b_0 - c_0 + \frac{a_0^2}{3} - \frac{2}{3\sqrt{3}}\Big(1 - b_0 + \frac{a_0^2}{3} - a_0\Big)^{3/2} > 0$$

are satisfied.

Theorem 2.6.5 *Assume that* (2.104) *holds and*

$$\int_\sigma^\infty \big(2c(t) - a(t)b(t) - b'(t)\big)\exp\Big(\int_\sigma^t a(s)\,ds\Big)\,dt = -\infty.$$

Then (2.1) *is oscillatory.*

Example 2.6.1 Consider

$$x''' - \frac{1}{t}x'' + \Big(2 + \frac{2}{t}\Big)x' - \Big(\frac{1}{t} + \frac{4}{t^2}\Big)x = 0, \quad t \ge 1. \tag{2.106}$$

Since $2c(t) - a(t)b(t) - b'(t) = -\frac{4}{t^2} < 0$ and $\int_1^\infty b(t)\,dt = \infty$, by Theorem 2.6.1, (2.106) has an oscillatory solution. Theorem 2.6.2 can be applied to (2.106). However, $\int_1^\infty (2c(t) - a(t)b(t) - b'(t))\exp(\int_\sigma^\infty a(s)\,ds)\,dt = -\int_1^\infty \frac{4}{t^3}\,dt > -\infty$ implies that Theorem 2.6.5 cannot be applied to (2.106).

Example 2.6.2 Consider

$$x''' - \frac{1}{t}x'' + \frac{2}{t^2}x' - \frac{20}{t^3}x = 0, \quad t \ge 1. \tag{2.107}$$

Clearly, $2c(t) - a(t)b(t) - b'(t) = -\frac{34}{t^3} < 0$, $t^2 b(t) = 2 > 1$ and $\int_1^\infty (\frac{t^2 b(t) - 1}{t})\, dt = \int_1^\infty \frac{dt}{t} = \infty$. Hence by Theorem 2.6.2, (2.107) has an oscillatory solution. In particular, $x_1(t) = \cos(\sqrt{5}\log t)$ and $x_2(t) = \sin(\sqrt{5}\log t)$ are oscillatory solutions of (2.107). Theorem 2.6.1 cannot be applied to (2.107), because $\int_1^\infty \frac{dt}{t^2} < \infty$. Further, since $b(t) - \frac{a^2(t)}{3} - a'(t) = \frac{2}{3t^2} > 0$, Theorems 2.6.3 and 2.6.4 fail to hold for (2.107).

Example 2.6.3 Consider

$$x''' - \frac{1}{t}x'' + \frac{1}{t^2}x' - \frac{16}{t^3}x = 0, \quad t \geq 1. \tag{2.108}$$

Since $b(t) - \frac{a^2(t)}{3} - a'(t) = -\frac{1}{3t^2} < 0$,

$$\frac{2a^3(t)}{27} - \frac{a(t)b(t)}{3} + c(t) - b'(t) + \frac{2a(t)a'(t)}{3} + \frac{2}{3}a''(t) = \frac{-425}{27t^3} < 0$$

and

$$\begin{aligned}
&\int_1^\infty \Bigg[-\frac{2t^2 a^3(t)}{27} + \frac{t^2 a(t)b(t)}{3} - t^2 c(t) + t^2 b'(t) - \frac{2t^2 a(t)a'(t)}{3} - \frac{2t^2 a''(t)}{3} \\
&\quad + tb(t) - \frac{ta^2(t)}{3} - ta'(t) - \frac{2}{3\sqrt{3t}}\left(1 - t^2 b(t) + \frac{1}{3}t^2 a^2(t) + t^2 a'(t)\right)^{3/2} \Bigg] dt \\
&= \frac{400}{27}\int_1^\infty \frac{dt}{t} = \infty,
\end{aligned}$$

then, by Theorem 2.6.4, (2.108) is oscillatory. In particular, $x_1(t) = \cos(2\log t)$ and $x_2(t) = \sin(2\log t)$ are oscillatory solutions of (2.108). On the other hand,

$$\frac{2a^3(t)}{27} - \frac{a(t)b(t)}{3} + c(t) - \frac{a''(t)}{3} = \frac{-407}{27t^3} < 0$$

implies that Theorem 2.6.3 cannot be applied to (2.108).

Example 2.6.4 Consider

$$x''' - \frac{3}{t}x'' + \frac{6}{t^2}x' - \frac{33}{8t^3}x = 0, \quad t \geq 1. \tag{2.109}$$

Here $a_0 = -3$, $b_0 = 6$ and $c_0 = \frac{-33}{8}$. Clearly, $b_0 - \frac{a_0^2}{3} - a_0 = 0$, $\frac{2a_0^3}{27} - \frac{a_0 b_0}{3} + c_0 - \frac{2a_0}{3} = \frac{15}{8} > 0$ and

$$\frac{2a_0^3}{27} - \frac{a_0 b_0}{3} + c_0 - \frac{2a_0}{3} - \frac{2}{3\sqrt{3}}\left(1 - b_0 + \frac{a_0^2}{3} - a_0\right)^{3/2} = \frac{15}{8} - \frac{2}{3\sqrt{3}} > 0.$$

Then by Remark 2.6.2, (2.109) is oscillatory. It may be observed that $x_1(t) = t^{11/4}\cos(\frac{\sqrt{407}}{8}\log t)$ and $x_2(t) = t^{11/4}\sin(\frac{\sqrt{407}}{8}\log t)$ are oscillatory solutions of (2.109). Since

$$\frac{2a^3(t)}{27} - \frac{a(t)b(t)}{3} + c(t) - b'(t) + \frac{2a(t)a'(t)}{3} + \frac{2}{3}a''(t) = \frac{1}{t^3}\left(6 - \frac{33}{8}\right) > 0,$$

Theorem 2.6.4 cannot be applied to Eq. (2.109).

Lemma 2.6.1 *If* (2.104) *holds, then* (2.1) *is of type* C_{II}.

Proof Suppose that $x(t)$ is a solution of (2.1) with $x(t_0) = x'(t_0) = 0$ and $x''(t_0) > 0$, $t_0 \geq \sigma$. From the continuity of $x''(t)$, it follows that there exists a $\delta > 0$ such that $x''(t) > 0$ for $t \in [t_0, t_0 + \delta)$. We claim that $x(t) > 0$ for $t > t_0$. If not, then $x(t_1) = 0$ for some $t_1 > t_0$. Now, multiplying (2.2) through by $x(t)$ and integrating the resulting identity from t_0 to t_1, we obtain

$$r(t_1)\big(x'(t_1)\big)^2 \leq \int_{t_0}^{t_1} \big(2p(t) - q'(t)\big)x^2(t)\,dt,$$

that is,

$$0 < r(t_1)\big(x'(t_1)\big)^2 \leq \int_{t_0}^{t_1} \big(2c(t) - a(t)b(t) - b'(t)\big)r(t)x^2(t)\,dt < 0,$$

a contradiction. Hence $x(t) > 0$ for $t > t_0$. This completes the proof of the theorem. □

Lemma 2.6.2 *If the second-order differential equation*

$$lu = u'' + a(t)u' + b(t)u = 0 \tag{2.110}$$

is disconjugate, then (2.1) *is of type* C_{II}.

Theorem 2.6.6 *If* $(2c(t) - a(t)b(t) - b'(t))r(t) \leq -d < 0$, *then all oscillatory solutions of* (2.1) *tend to zero as* $t \to \infty$.

Proof Since $(2c(t) - a(t)b(t) - b'(t))r(t) \leq -d < 0$, by Theorem 2.6.5, (2.1) is oscillatory. Let $u(t)$ be an oscillatory solution of (2.1). Then there exists a sequence $\langle t_n \rangle$ such that $t_n \to \infty$ as $n \to \infty$ and $u(t_n) = 0$. If

$$F\big[u(t)\big] = r(t)\big(u'(t)\big)^2 - 2r(t)u(t)u''(t) - q(t)u^2(t),$$

then

$$F'\big[u(t)\big] = \big(2p(t) - q'(t)\big)u^2(t) + r'(t)\big(u'(t)\big)^2$$

and hence

$$F\big[u(t)\big] = F\big[u(\sigma)\big] - \int_\sigma^t \big(q'(s) - 2p(s)\big)u^2(s)\,ds + \int_\sigma^t r'(s)\big(u'(s)\big)^2\,ds. \quad (2.111)$$

Clearly, $F[u(t)]$ is a decreasing function of t with $F[u(t_n)] \ge 0$. Put $t = t_n$ in (2.111) to get

$$0 \le F\big[u(t_n)\big] = F\big[u(\sigma)\big] - g(t_n)$$

and hence

$$g(t_n) \le F\big[u(\sigma)\big], \quad (2.112)$$

where

$$g(t) = \int_\sigma^t \big(q'(s) - 2p(s)\big)u^2(s)\,ds - \int_\sigma^t r'(s)\big(u'(s)\big)^2\,ds.$$

Clearly, $g(t)$ is a positive, continuous and increasing function of t. We claim that $g(t) \le F[u(\sigma)]$ for large t, say, for $t \ge T$. If not, then, for every T^*, there exists a $t^* > T^*$ such that $g(t^*) > F[u(\sigma)]$. We can find a $t_n > t^*$. Hence $g(t_n) \ge g(t^*)$. Consequently $g(t_n) > F[u(\sigma)]$, a contradiction to (2.112). Hence $g(t) \le F[u(\sigma)]$ for large t, that is,

$$\int_\sigma^t \big(q'(s) - 2p(s)\big)u^2(s)\,ds - \int_\sigma^t r'(s)\big(u'(s)\big)^2\,ds \le F\big[u(\sigma)\big]$$

for large t. Thus

$$\int_\sigma^t u^2(s)\,ds \le \frac{1}{d}\int_\sigma^t \big(q'(s) - 2p(s)\big)u^2(s)\,ds < \frac{F[u(\sigma)]}{d}$$

implies that

$$\int_\sigma^\infty u^2(s)\,ds < \infty,$$

that is, $u \in L^2[\sigma, \infty)$.

To complete the proof of the theorem, it is enough to show that $u'(t)$ is bounded (see Lemma 1.5.15). Since $F[u(t)]$ is a decreasing function of t, for $t \ge \sigma$, $F[u(t)] \le F[u(\sigma)]$. If $t_1 \in [\sigma, \infty)$ is a zero of $u''(t)$, then

$$\big(u'(t_1)\big)^2 \le \frac{[F[u(\sigma)] + q(t_1)u^2(t_1)]}{r(t_1)} < \infty.$$

Thus the values of $u'(t)$ are bounded at its maxima and minima. Hence $u'(t)$ is bounded. Consequently, by Lemma 1.5.15, $u(t) \to 0$ as $t \to \infty$. Hence the theorem is proved. □

Lemma 2.6.3 *If* (2.110) *is nonoscillatory, then every nonoscillatory solution* $x(t)$ *of* (2.1) *satisfies the property* $x(t)x'(t) > 0$ *for* $t \ge T_x \ge \sigma$.

Proof Let $x(t)$ be a nonoscillatory solution of (2.1). Hence $x(t) \neq 0$ for $t \geq t_x \geq \sigma$. Without any loss of generality, it may be assumed that $x(t) > 0$ for $t \geq t_x$. Since $x'(t)$ is a solution of the second-order nonhomogeneous equation

$$\left(r(t)z'\right)' + q(t)z = -p(t)x(t),$$

by Lemma 1.5.29, $x'(t)$ is nonoscillatory. Thus there exists a $T_x \geq t_x$ such that $x'(t) > 0$ or < 0 for $t \geq T_x$. Let $x'(t) < 0$ for $t \geq T_x$. Then from Eq. (2.2), $(r(t)x''(t))' \geq 0$. Hence $x''(t)$ is nonoscillatory for large t. If $x''(t) < 0$ for large t, then $x(t) < 0$ for large t, a contradiction. If $x''(t) > 0$ for large t, then from Eq. (2.1), $x'''(t) > 0$ for large t, which in turn implies that $x'(t) > 0$ for large t, a contradiction. Thus, $x'(t) > 0$ for $t \geq T_x$. This completes the proof of the lemma. □

Lemma 2.6.4 *Let* (2.110) *be nonoscillatory with a solution* $u(t)$ *such that* $u(t)u'(t) > 0$ *for* $t \geq t_0 \geq \sigma$. *If* $x(t)$ *is a solution of* (2.1) *with*

$$x(t_1) \geq 0, \qquad x'(t_1) = 0 \quad \text{and} \quad x''(t_1) > 0, \quad t_1 > t_0,$$

then $x(t) > 0$, $x'(t) > 0$ *and* $x''(t) > 0$ *for* $t > t_1$ *and*

$$\lim_{t\to\infty} x(t) = \infty.$$

Further, if

$$\int_\sigma^\infty p(t)\,dt = -\infty,$$

then

$$\lim_{t\to\infty} x'(t) = \infty.$$

Proof From the continuity of $x''(t)$, it follows that there exists a $\delta > 0$ such that $x''(t) > 0$ for $t \in [t_1, t_1 + \delta)$. We claim that $x''(t) > 0$ for $t > t_1$. If not, then there exists a $t_2 > t_1$ such that $x''(t_2) = 0$ and $x''(t) > 0$ for $t \in [t_1, t_2)$. Consequently, $x'(t) > 0$ and $x(t) > 0$ for $t \in (t_1, t_2]$. Multiplying (2.2) through by $x'(t)$ and integrating the resulting identity from t_1 to t_2 one may obtain, by Lemma 1.5.1,

$$0 = -\int_{t_1}^{t_2} r(t)\left(x''(t)\right)^2 dt + \int_{t_2}^{t_1} q(t)\left(x'(t)\right)^2 dt + \int_{t_1}^{t_2} p(t)x(t)x'(t)\,dt < 0,$$

a contradiction. Hence $x''(t) > 0$ for $t \geq t_1$. Consequently $x'(t) > 0$ and $x(t) > 0$ for $t > t_1$ and hence $\lim_{t\to\infty} x(t) = \infty$. Further, let $\int_\sigma^\infty p(t)\,dt = -\infty$ holds. If $\lim_{t\to\infty} x'(t) = k$, $0 < k < \infty$, then integration of (2.2) from t_3 $(\geq t_1)$ to t yields

$$\begin{aligned} r(t)x''(t) &= r(t_3)x''(t_3) - \int_{t_3}^{t} q(s)x'(s)\,ds - \int_{t_3}^{t} p(s)x(s)\,ds \\ &> r(t_3)x''(t_3) - x'(t)\int_{t_3}^{t} q(s)\,ds - x(t_3)\int_{t_3}^{t} p(s)\,ds. \end{aligned}$$

Since $\int_\sigma^\infty \frac{dt}{r(t)} = \infty$, we have $\int_\sigma^\infty q(t)\,dt < \infty$, because (2.110) is nonoscillatory. Thus from above, it follows that $r(t)x''(t) \to \infty$ as $t \to \infty$, which in turn implies that $x'(t) \to \infty$ as $t \to \infty$, a contradiction. Hence $k = \infty$. The lemma is proved. □

Theorem 2.6.7 *Suppose that* (2.110) *is nonoscillatory. If* (2.1) *has an oscillatory solution, then every nonoscillatory solution* $x(t)$ *of* (2.1) *satisfies*

$$x(t)x'(t) \neq 0, \qquad \operatorname{sgn} x(t) = \operatorname{sgn} x'(t) \quad \text{for } t \geq T_0 \geq \sigma$$

and

$$\lim_{t\to\infty} |x(t)| = \infty.$$

Proof Let $u(t)$ be a solution of (2.110). Hence $u(t)$ is nonoscillatory. If $u(t) > 0$ for $t \geq t_0^* \geq \sigma$, then $u'(t) > 0$ for $t \geq t_0 \geq t_0^*$ since $b(t) \geq 0$. Let $x(t)$ be a nonoscillatory solution of (2.1). Without any loss of generality, we may assume that $x(t) > 0$ for $t \geq T \geq t_0$. From Lemma 2.6.3, it follows that $x'(t) > 0$ for $t \geq T_0 \geq T$. Let $z(t)$ be an oscillatory solution of (2.1). We claim that $W(t) = x(t)z'(t) - x'(t)z(t)$ must vanish for some values of $t \in [T_0, \infty)$. If not, then $w(t) \neq 0$ for $t \in [T_0, \infty)$. Setting $y(t) = \frac{z(t)}{x(t)}$, $t \geq T_0$, we see that $y(t)$ is oscillatory. Hence $y'(t)$ is oscillatory. On the other hand, $y'(t) = \frac{W(t)}{x^2(t)}$ implies that $y'(t)$ is nonoscillatory. This contradiction proves our claim. Let $W(b) = 0$ for some $b \in [T_0, \infty)$. Hence it is possible to obtain constants c_1 and c_2, not both zero, such that

$$c_1x(b) + c_2z(b) = 0 \quad \text{and} \quad c_1x'(b) + c_2z'(b) = 0.$$

Clearly, $c_1x''(b) + c_2z''(b) \neq 0$. Indeed, $c_1x''(b) + c_2z''(b) = 0$ implies that $c_1x(t) + c_2z(t) \equiv 0$, a contradiction to the fact that $x(t)$ and $z(t)$ are linearly independent. Set $v(t) = c_1x(t) + c_2z(t)$. Then, $v(t)$ is a solution of (2.1) with $v(b) = 0$, $v'(b) = 0$ and $v''(b) \neq 0$. We may assume that $v''(b) > 0$. Hence $v(t) > 0$, $v'(t) > 0$ and $v''(t) > 0$ for $t > b$ and $\lim_{t\to\infty} v(t) = \infty$. Clearly, $c_1 \neq 0$ because $c_1 = 0$ implies that $c_2 \neq 0$ and hence $v(t) = c_2z(t)$ is oscillatory, a contradiction. If $c_2 = 0$, then $c_1 \neq 0$ and $\lim_{t\to\infty} c_1x(t) = \lim_{t\to\infty} v(t) = \infty$. Clearly, $c_1 < 0$ implies that $x(t) < 0$ for large t, a contradiction. Thus, $c_1 > 0$ and $\lim_{t\to\infty} x(t) = \infty$. Suppose that $c_2 \neq 0$. If $\lim_{t\to\infty} x(t)$ exists finitely, then $\lim_{t\to\infty} z(t) = \pm\infty$, depending upon $c_2 > 0$ or < 0, a contradiction to the oscillatory nature of $z(t)$. Hence $\lim_{t\to\infty} x(t) = \infty$. This completes the proof of the theorem. □

Theorem 2.6.8 *Suppose that* $(2c(t) - a(t)b(t) - b'(t))r(t) \leq -d < 0$, *and* (2.110) *is nonoscillatory. If* (2.1) *has an oscillatory solution, then there exist two linearly independent oscillatory solutions* $u_1(t)$ *and* $u_2(t)$ *of* (2.1) *whose zeros separate in* $[\sigma, \infty)$ *and any oscillatory solution of* (2.1) *can be expressed as a linear combination of* $u_1(t)$ *and* $u_2(t)$.

Theorem 2.6.9 *If* $b(t) - \frac{a^2(t)}{3} - a'(t) \le 0$, $\frac{2a^3(t)}{27} - \frac{a(t)b(t)}{3} + c(t) - \frac{a''(t)}{3} > 0$ *and* (2.1) *admits an oscillatory solution, then nonoscillatory solution of* (2.1) *forms a one-dimensional subspace of the solution space of* (2.1).

Theorem 2.6.10 *If* $b(t) - \frac{a^2(t)}{3} - a'(t) \le 0$, $\frac{2a^3(t)}{27} - \frac{a(t)b(t)}{3} + c(t) - \frac{a''(t)}{3} < 0$ *and* (2.1) *has an oscillatory solution, then there exist two linearly independent oscillatory solutions whose zeros separate each other in* $[\sigma, \infty)$ *and they form a basis for the solution space of* (2.1), *that is, oscillatory solutions form a two-dimensional subspace of the solution space of* (2.1).

In [10], Dolan established the following results:

Theorem 2.6.11 *If* (2.1) $\{(2.3)\}$ *is weakly oscillatory, then* (2.1) $\{(2.3)\}$ *is oscillatory.*

Theorem 2.6.12 *If* S $\{S^*\}$ *contains a nonoscillatory two-dimensional subspace, then* S $\{S^*\}$ *is either nonoscillatory or strongly oscillatory, where* S *and* S^* *are the solution spaces of* (2.1) *and* (2.3), *respectively.*

Dolan [10] raised the following two questions:

(i) Does there exist an example of a linear third-order differential equation with the property that every two-dimensional subspace of the solution space is weakly oscillatory?
(ii) Does there exist an example of a linear third-order differential equation such that the solution space S and S^* are strongly oscillatory?

In [24], Neumann has provided answers to the above two questions. He has shown that there does not exist a linear third-order differential equation of the form (2.1) with the property that every two-dimensional subspace of its solution space is weakly oscillatory. Further, he has constructed an example of a strongly oscillatory equation (2.1) whose adjoint (2.3) is also strongly oscillatory.

Now we give the following explicit sufficient condition for the strong oscillation of (2.1).

Clearly, (2.110) is equivalent to

$$\left(r(t)z'\right)' + q(t)z = 0, \tag{2.113}$$

where $r(t) = \exp(\int_\sigma^t a(s)\,ds)$ and $q(t) = r(t)b(t)$. Since $r(t) > 0$ and $r'(t) \le 0$, we have

$$\int_\sigma^\infty \frac{dt}{r(t)} = \infty.$$

Hence by Lemma 1.5.25, we have

$$\int_\sigma^\infty q(t)\,dt < \infty.$$

Theorem 2.6.13 *Let* (2.113) *be nonoscillatory. If* $-\infty < \liminf_{t\to\infty} a(t) \le 0$, $\frac{1}{3}a^2(t) - b(t) + a'(t) > 0$ *and*

$$\int_\sigma^\infty \left[\frac{2a^3(t)}{27} - \frac{a(t)b(t)}{3} + \frac{a(t)a'(t)}{3} + c(t) - \frac{2}{3\sqrt{3}}\left(\frac{a^2(t)}{3} - b(t) + a'(t)\right)^{3/2}\right] dt = \infty, \quad \sigma > 0 \tag{2.114}$$

then (2.1) *is strongly oscillatory.*

Proof If possible, let (2.1) admits a nonoscillatory solution $x(t)$. Then $x(t)x'(t) > 0$ for $t \ge t_0 > 0$ by Lemma 2.6.3. Clearly, $z(t) = \frac{x'(t)}{x(t)} > 0$, $t \ge t_0$, is a positive solution of the second-order Riccati equation (2.11). Integrating (2.11) from t_0 to t $(t > t_0)$, we obtain

$$z'(t) = z'(t_0) + \frac{3}{2}z^2(t_0) + a(t_0)z(t_0) - \frac{3}{2}z^2(t) - a(t)z(t) - \int_{t_0}^t \left[z^3(s) + a(s)z^2(s) + \big(b(s) - a'(s)\big)z(s) + c(s)\right] ds. \tag{2.115}$$

If

$$H\big(z(t)\big) = z^3(t) + a(t)z^2(t) + \big(b(t) - a'(t)\big)z(t) + c(t),$$

then $H(z(t))$ attains its minimum value for $z(t) > 0$ at

$$z(t) = \frac{1}{3}\left[-a(t) + \big(a^2(t) - 3b(t) + 3a'(t)\big)^{1/2}\right]$$

and the minimum value of $H(z(t))$ is given by

$$\frac{2}{27}a^3(t) - \frac{1}{3}a(t)b(t) + \frac{1}{3}a(t)a'(t) + c(t) - \frac{2}{3\sqrt{3}}\left(\frac{a^2(t)}{3} - b(t) + a'(t)\right)^{3/2}.$$

Further, if

$$F\big(z(t)\big) = \frac{3}{2}z^2(t) + a(t)z(t),$$

then $F(z(t))$ attains its minimum value for $z(t) > 0$ at $z(t) = -\frac{a(t)}{3}$ and the minimum value of $F(z(t))$ is given by $-\frac{a^2(t)}{6}$. Hence (2.115) yields

$$z'(t) \le z'(t_0) + \frac{3}{2}z^2(t_0) + a(t_0)z(t_0) + \frac{a^2(t)}{6} - \int_{t_0}^t \left[\frac{2a^3(s)}{27} - \frac{a(s)b(s)}{3} + \frac{a(s)a'(s)}{3} + c(s) - \frac{2}{3\sqrt{3}}\left(\frac{a^2(s)}{3} - b(s) + a'(s)\right)^{3/2}\right] ds.$$

From (2.114), it follows that $\lim_{t\to\infty} z'(t) = -\infty$. Thus $z(t) < 0$ for large t, a contradiction. The proof of the theorem is complete. □

Theorem 2.6.14 *Suppose that r and $q \in C^1((0,\infty), R)$ such that r is positive and q is nonnegative in $(0,\infty)$ and*

$$\int_1^\infty \frac{dt}{r(t)} = \infty.$$

If $L = \lim_{t\to\infty} r(t)\{[r(t)q(t)]^{-1/2}\}'$ exists and $L > 2$, then (2.113) is nonoscillatory.

Remark 2.6.4 Theorem 2.6.13 does not hold for (1.5), the third-order differential equation with constant coefficients, with $a < 0$, $b > 0$ and $c < 0$ because

$$L = \lim_{t\to\infty} e^{at}\{[e^{2at}b]^{-1/2}\}' > 2$$

if and only if $a^2 > 4b$. Further,

$$\frac{2a^3}{27} - \frac{ab}{3} - \frac{2}{3\sqrt{3}}\left(\frac{a^2}{3} - b\right)^{3/2} > 0,$$

if and only if $a^2 < 4b$. Thus, the condition (2.114) and $L > 0$ do not hold simultaneously.

Example 2.6.5 Consider

$$x''' - x'' + \left(\frac{1}{4.0000004} + \frac{1}{t}\right)x' - \frac{K}{t^2}x = 0, \quad t \geq 12, \tag{2.116}$$

where $K > 0$ is a constant. In this case $L = 2.0000001 > 2$. Hence, by Theorem 2.6.14, (2.113) is nonoscillatory. Further, the calculation shows that

$$\frac{2a^3(t)}{27} - \frac{a(t)b(t)}{3} + \frac{a(t)a'(t)}{3} + c(t) - \frac{2}{3\sqrt{3}}\left(\frac{a^2(t)}{3} - b(t) + a'(t)\right)^{3/2}$$
$$= 0.00000005 + \frac{1}{3t} + \frac{0.1666664}{t} + \cdots - \frac{K}{t^2}$$

and

$$\frac{a^2(t)}{3} - b(t) + a'(t) = \frac{1.0000004}{12.000001} - \frac{1}{t} > 0$$

for $t \geq 12$. Hence the conditions of Theorem 2.6.13 are satisfied. This, in turn, implies that (2.116) is strongly oscillatory.

Remark 2.6.5 A similar strong oscillation criteria for Eq. (2.1) can be obtained when $a(t) > 0$, $b(t) \geq 0$ and $c(t) < 0$. However, one has to assume the additional condition $\int_\sigma^\infty \frac{dt}{r(t)} = \infty$ in Theorem 2.6.13. One may observe in the rest six cases, viz., (i) $a(t) \geq 0, b(t) \leq 0, c(t) > 0$, (ii) $a(t) \leq 0, b(t) \leq 0, c(t) > 0$, (iii) $a(t) \leq 0, b(t) \leq 0, c(t) < 0$, (iv) $a(t) \geq 0, b(t) \leq 0, c(t) < 0$, (v) $a(t) \geq 0, b(t) \geq 0, c(t) > 0$ and (vi) $a(t) \leq 0, b(t) \geq 0, c(t) > 0$, Eq. (2.1) always admits a nonoscillatory solution. Hence strong oscillation of (2.1) cannot be studied in the above six cases.

2.7 Oscillation and Nonoscillation of Third-Order Linear Differential Equations of the Form $(r_2(t)(r_1(t)x')')' + r_3(t)x = 0$

In this section, we shall study the oscillatory and nonoscillatory behaviour of solutions of the third-order linear differential equation

$$\big(r_2(t)\big(r_1(t)x'\big)'\big)' + r_3(t)x = 0, \tag{2.117}$$

where $r_1 > 0, r_2 > 0$ and $r_3 > 0$ are twice continuously differentiable functions with fixed sign on $[\sigma, \infty)$, that is, $r_2(t) \neq 0, r_1(t) \neq 0$ and $r_3(t) \neq 0$ for $t \in [\sigma, \infty)$.

First of all, we present the property "Correspondence Principle" between the solutions of (2.117) and

$$\left(\frac{1}{r_3(t)}\big(r_2(t)y'\big)'\right)' + \frac{1}{r_1(t)}y = 0 \tag{2.118}$$

and

$$\left(r_1(t)\left(\frac{1}{r_3(t)}z'\right)'\right)' + \frac{1}{r_2(t)}z = 0, \tag{2.119}$$

obtained by means of an ordered cyclic permutation on the coefficients r_1, r_2 and r_3 of (2.117). Such a result plays an important role in the study of the classification of solutions as well as in the theory of disconjugacy.

Denote $\mathcal{S}_i, i = 1, 2, 3$, the linear space of solutions of (2.117), (2.118) and (2.119), respectively, and by $\mathcal{O}_i$ and $\mathcal{N}_i, i = 1, 2, 3$, the subsets of $\mathcal{S}_i$ given by oscillatory and nonoscillatory solutions of (2.117), (2.118) and (2.119), respectively.

Definition 2.7.1 The spaces $\mathcal{S}_i, i = 1, 2, 3$ is said to be isomorphic with respect to the oscillation, if there exists an isomorphism $\mathcal{L}_{ij}: \mathcal{S}_i \to \mathcal{S}_j$ which keeps the oscillatory properties of the solutions, that is,

$$g \in \mathcal{O}_i \Rightarrow \mathcal{L}_{ij}(g) \in \mathcal{O}_j, \qquad g \in \mathcal{N}_i \Rightarrow \mathcal{L}_{ij}(g) \in \mathcal{N}_j.$$

In this case, the operator $\mathcal{L}_{ij}$ is said to be an isomorphism of oscillation.

It is obvious to note that if the space $\mathcal{S}_i, \mathcal{S}_j, i, j = 1, 2, 3$ are isomorphic with respect to the oscillation, then $\mathcal{L}_{ij}$ maps $\mathcal{O}_i$ into $\mathcal{O}_j$ and $\mathcal{N}_i$ into $\mathcal{N}_j$. Hence the existence of an isomorphism of oscillation between the spaces $\mathcal{S}_i$ and $\mathcal{S}_j$ enables us to describe the oscillatory and nonoscillatory behaviour of the solutions of (2.117), (2.118) and (2.119) by the oscillatory and nonoscillatory behaviour of the solutions of (2.118), (2.119) and (2.117), respectively.

Lemma 2.7.1 *If $x(t)$ is a solution of* (2.117), *then $y = r_1(t)x'$ is a solution of* (2.118), *$z = r_2(t)(r_1(t)x')'$ is a solution of* (2.119), *and $w = \frac{1}{r_3(t)}(r_2(t)(r_1(t)x')')'$ is a solution of* (2.117).

Let Γ_1, Γ_2 and Γ_3 be the first-order differential operators

$$\Gamma_1 : \mathcal{S}_1 \to \mathcal{S}_2 \quad \text{by } \Gamma_1 x(t) = r_1(t)x'(t) = y(t),$$

$$\Gamma_2 : \mathcal{S}_2 \to \mathcal{S}_3 \quad \text{by } \Gamma_2 x(t) = r_2(t)y'(t) = z(t),$$

$$\Gamma_3 : \mathcal{S}_3 \to \mathcal{S}_1 \quad \text{by } \Gamma_3 x(t) = \frac{1}{r_3(t)}z'(t).$$

Then we have the following property.

Correspondence Principle The operators $\Gamma_i, i = 1, 2, 3$ are isomorphisms of oscillations.

As a consequence of the "Correspondence Principle", the linear spaces $\mathcal{S}_i, i = 1, 2, 3$ have the same structures with respect to oscillation or nonoscillation. Hence the oscillation (nonoscillation) criteria for Eqs. (2.118)–(2.119) may immediately be transformed into oscillation (nonoscillation) criteria for Eq. (2.117). As we shall see, this fact enables us to obtain easily some new results about oscillation or nonoscillation of (2.117) that may be hard to prove directly.

Lemma 2.7.2 *The following hold*:

(a) *If* (2.117) *is oscillatory and of type C_I, then any solution of* (2.117) *with a zero is oscillatory.*
(b) *If* (2.117) *is of type C_{II}, then any solution of* (2.117) *with a double zero is nonoscillatory.*
(c) *If* (2.117) *is both of type C_I and C_{II}, then* (2.117) *is nonoscillatory.*
(d) *If* (2.117) *is of type C_I, then it admits a nonoscillatory solution.*
(e) *Equation* (2.117) *is of type C_I if and only if its adjoint equation*

$$\big(r_1(t)\big(r_2(t)u'\big)'\big)' - r_3(t)u = 0 \tag{2.120}$$

is of type C_{II}. Similarly, if (2.117) *is of type C_{II}, then* (2.120) *is of type C_I.*
(f) *Equation* (2.117) *is oscillatory if and only if its adjoint equation* (2.120) *is oscillatory.*

We note that, if x_1 and x_2 are two linearly independent solutions of (2.117), then $u(t)=r_1(t)(x_1'(t)x_2(t)-x_1(t)x_2'(t))$ is a solution of (2.120).

Lemma 2.7.3 *If* $r_3(t)>0$, *then* (2.117) *is of type* C_I.

Proof Assume that (2.117) does not belongs to C_I. The for some $a>\sigma$, there is a solution $x(t)$ of (2.117) with $x(a)=0, x'(a)=0$ and $x''(a)>0$ vanishes in $[\sigma,a)$. Let $b, \sigma<b<a$ be such that $x(b)=0, x(t)\neq 0$ for $t\in(b,a)$. Then there exists a $c, b<c<a$ such that $x'(c)=0, x'(t)\neq 0$ for $t\in(c,a)$. By integrating (2.117) from t to a, we obtain

$$r_2(a).\big(r_1(t)x'(t)\big)'\big|_{t=a}+\int_t^a r_3(s)x(s)\,ds=r_2(t)\big(r_1(t)x'(t)\big)'$$

or

$$\frac{1}{r_2(t)}L_2x(a)+\frac{1}{r_2(t)}\int_t^a r_3(s)x(s)\,ds=\big(r_1(t)x'(t)\big)', \tag{2.121}$$

where $L_0x(t)=x(t), L_1x(t)=r_1(t)x'(t), L_2x(t)=r_2(t)(r_1(t)x'(t))'$ and $L_3x(t)=(r_2(t)L_2x(t))'$. Integrating (2.121) from t to a, we have

$$L_2x(a)\int_t^a\frac{1}{r_2(s)}\,ds+\int_t^a\frac{1}{r_2(s)}\int_s^a r_3(u)x(u)\,du\,ds=r_1(a)x'(a)-r_1(t)x'(t).$$

For $t=c$, we get

$$L_2x(a)\int_c^a\frac{1}{r_2(s)}\,ds+\int_c^a\frac{1}{r_2(s)}\int_s^a r_3(u)x(u)\,du\,ds=r_1(a)x'(a)-r_1(c)x'(c). \tag{2.122}$$

Since $x'(a)=0$ and $x''(a)>0$, we have

$$\begin{aligned}L_2x(a)=r_2(a)\big(r_1(t)x'(t)\big)'\big|_{t=a}&=r_2(a).\big(r_1(a)x''(a)+r_1'(a)x'(a)\big)\\&=r_2(a)r_1(a)x''(a)>0.\end{aligned}$$

Then, taking into account that $x(t), r_2(t)$ and $r_3(t)$ are positive functions in (c,a), from (2.122), we obtain a contradiction. The lemma is proved. □

If functions r_1, r_2 and r_3 are positive functions on $[\sigma,\infty)$, then Eqs. (2.118) and (2.119) are of type C_I.

Lemma 2.7.3 shows that Eq. (2.117) can never be strongly oscillatory. This is because (2.117) is of type C_I and hence admits a nonoscillatory solution.

Lemma 2.7.4 *The equation*

$$\big(r_2(t)\big(r_1(t)x'\big)'\big)'-r_3(t)x=0 \tag{2.123}$$

is of type C_{II}.

Theorem 2.7.1 *Let* (2.117) *be oscillatory. Then the space* $\mathcal{S}_1$ *contains a two-dimensional subspace* $\mathcal{W}_1$ *of oscillatory solutions. In addition, there exists a solution* $x \in \mathcal{S}_1$ *such that* $x \in \mathcal{O}_1$ *and* $x \notin \mathcal{W}_1$.

The above conclusion remains true for (2.123).

Proof Clearly (2.117) is of type C_I and so its adjoint equation (2.120) is of type C_{II}. From Lemma 2.7.2(f), it follows that (2.120) is oscillatory. By Dolan [10], the linear space of solutions of the adjoint equation (2.120) contains a two-dimensional subspace $\mathcal{W}_1$ of oscillatory solutions. Hence the assertion follows by taking into account that the adjoint of (2.120) is (2.117).

Now, we consider (2.123). Clearly the adjoint of (2.123) given by

$$\big(r_1(t)\big(r_2(t)u'\big)'\big)' + r_3(t)u = 0 \tag{2.124}$$

is of type C_I. Hence from Lemma 2.7.2(f), Eq. (2.124) is oscillatory. By the same argument as above, the first assertion follows.

In order to complete the proof, it is enough to show that there exists $x \in \mathcal{S}_1$ such that $x \in \mathcal{O}_1$ and $x \notin \mathcal{W}_1$ for (2.117). Let $x_1 = x_1(t)$ and $x_2 = x_2(t)$ be a basis for the subspace $\mathcal{W}_1$, and let $\overline{x} = \overline{x}(t)$ be a nonoscillatory solution of (2.117), $\overline{x}(t) \neq 0$ for $t \geq a \geq \sigma$, and let $b \in (a, \infty)$ be such that $x_1(b) \neq 0$. Now, we consider the solution $x_3(t)$ of (2.117) is given by

$$x_3(t) = x_1(t) + \frac{-x_1(b)}{\overline{x}(b)}\overline{x}(t). \tag{2.125}$$

Since $x_3(b) = 0$, $x_3(t)$ is oscillatory. Let us prove, by contradiction, that $x_3(t) \notin \mathcal{W}_1$. If $x_3(t) \in \mathcal{W}_1$, then there exist λ_1 and $\lambda_2 \in R$, $\lambda_1^2 + \lambda_2^2 = 1$, such that $x_3(t) = \lambda_1 x_1(t) + \lambda_2 + x_2(t)$. Then from (2.125), we obtain

$$(\lambda_1 - 1)x_1(t) + \lambda_2 x_2(t) = \frac{-x_1(b)}{\overline{x}(b)}\overline{x}(t), \tag{2.126}$$

which is a contradiction because the right-hand side of (2.126) is a nonoscillatory solution of (2.117) and the left-hand side is an oscillatory solution. The proof is now complete. □

Remark 2.7.1 Consider (2.123). Here, the set of oscillatory solutions $\mathcal{O}_1$ may be exactly a two-dimensional subspace of $\mathcal{S}_1$. That is, it may happen that there do not exist solutions $x(t)$ of (2.117) such that $x \in \mathcal{O}_1$, $x \notin \mathcal{W}_1$, as equations with constant coefficients show.

Theorem 2.7.2 *If* (2.117) *is oscillatory, then* $\mathcal{S}_1$ *cannot contain a two-dimensional subspace of nonoscillatory solutions.*

Remark 2.7.2 Lemmas's 2.7.1–2.7.4, and Theorems 2.7.1–2.7.2 are still valid for Eqs. (2.118) and (2.119).

Theorem 2.7.3 *If*

$$\int_{\sigma}^{\infty} r_3(t) \int_{\sigma}^{t} \frac{1}{r_1(s)} \int_{\sigma}^{s} \frac{1}{r_2(u)}\, du\, ds\, dt < \infty, \tag{2.127}$$

then (2.117) *is nonoscillatory.*

Proof Let $a > \sigma$ be such that

$$\int_{a}^{\infty} r_3(t) \int_{a}^{t} \frac{1}{r_1(s)} \int_{a}^{s} \frac{1}{r_2(u)}\, du\, ds\, dt < 1,$$

and let $x(t)$ be a solution of (2.117) such that $x(a) = 0$, $L_1x(a) = 0$ and $L_2x(a) = 1$, where $L_1x(t) = r_1(t)x'(t)$ and $L_2x(t) = r_2(t)(r_1(t)x'(t))'$. Integrating (2.117) three times from a to t, we obtain

$$\begin{aligned} x(t) = x(a) + L_1x(a) \int_{a}^{t} \frac{1}{r_1(s)}\, ds + L_2x(a) \int_{a}^{t} \frac{1}{r_1(s)} \int_{a}^{s} \frac{1}{r_2(u)}\, du\, ds \\ - \int_{a}^{t} \frac{1}{r_1(s)} \int_{a}^{s} \frac{1}{r_2(u)} \int_{a}^{u} r_3(v)x(v)\, dv\, du\, ds. \end{aligned} \tag{2.128}$$

Then from (2.128), we have

$$x(t) = \int_{a}^{t} \frac{1}{r_1(s)} \int_{a}^{s} \frac{1}{r_2(u)}\, du\, ds - \int_{a}^{t} \frac{1}{r_1(s)} \int_{a}^{s} \frac{1}{r_2(u)} \int_{a}^{u} r_3(v)x(v)\, dv\, du\, ds. \tag{2.129}$$

Assume that (2.117) is oscillatory. Since (2.117) is of type C_I, $x(t)$ is oscillatory. Let $b > a$ be such that $x(b) = 0$ and $x(t) > 0$ for $t \in (a, b)$. Hence from (2.129), we obtain for $t \in (a, b)$ that

$$x(t) \le \int_{a}^{t} \frac{1}{r_1(s)} \int_{a}^{s} \frac{1}{r_2(u)}\, du\, ds. \tag{2.130}$$

Since $x(b) = 0$, from (2.129) and (2.130), we have

$$\begin{aligned} 0 = x(b) &= \int_{a}^{b} \frac{1}{r_1(s)} \int_{a}^{s} \frac{1}{r_2(u)}\, du\, ds - \int_{a}^{b} \frac{1}{r_1(s)} \int_{a}^{s} \frac{1}{r_2(u)} \int_{a}^{u} r_3(v)x(v)\, dv\, du\, ds \\ &= \int_{a}^{b} \frac{1}{r_1(s)} \int_{a}^{s} \frac{1}{r_2(u)} \left(1 - \int_{a}^{u} r_3(v)x(v)\, dv\right) du\, ds \\ &\ge \int_{a}^{b} \frac{1}{r_1(s)} \int_{a}^{s} \frac{1}{r_2(u)} \left(1 - \int_{a}^{u} r_3(v) \int_{a}^{v} \frac{1}{r_1(\theta)} \int_{a}^{\theta} \frac{1}{r_2(\tau)}\, d\tau\, d\theta\, dv\right) du\, ds \end{aligned}$$

which is a contradiction, since

$$\int_{a}^{b} r_3(t) \int_{a}^{t} \frac{1}{r_1(s)} \int_{a}^{s} \frac{1}{r_2(u)}\, du\, ds\, dt < \int_{a}^{\infty} r_3(t) \int_{a}^{t} \frac{1}{r_1(s)} \int_{a}^{s} \frac{1}{r_2(u)}\, du\, ds\, dt < 1.$$

Hence (2.117) is nonoscillatory. The theorem is proved. □

Remark 2.7.3 Let $r_1(t) \equiv 1$ and $r_2(t) \equiv 1$ and $r_3(t) = c(t)$. Then (2.117) reduces to (2.6). As an application of Theorem 2.7.3, we see that if $\int^\infty t^2 c(t)\,dt < \infty$, then (2.6) is nonoscillatory, which we may restate as

Corollary 2.7.1 *If Eq.* (2.6) *is oscillatory, then* $\int_\sigma^\infty t^2 c(t)\,dt = \infty$.

Remark 2.7.4 The converse of Corollary 2.7.1 is not necessarily true, that is, $\int^\infty t^2 c(t)\,dt = \infty$ need not imply that (2.6) is oscillatory. The following example strengthens this remark.

Example 2.7.1 Let $c(t) = \frac{1}{t^2}$. then $\int_t^\infty c(s)\,ds = -\frac{1}{s}$. Then the equation $z'' - \frac{3}{2t}z = 0$ is nonoscillatory. Thus, by Theorem 2.1.9, the equation $x''' + \frac{1}{t^2}x = 0, t \geq 2$ is nonoscillatory. Observe that $\int^\infty t^2 c(t)\,dt = \infty$.

On the other hand, the equation $x''' + \frac{3}{t^2}x = 0$, $t \geq 2$ is oscillatory by Theorem 2.5.7, and the property $\int^\infty t^2 c(t)\,dt = \infty$ holds.

Theorem 2.7.4 *If*

$$\int_\sigma^\infty r_3(t) \int_\sigma^t \frac{1}{r_2(s)} \int_\sigma^s \frac{1}{r_1(u)}\,du\,ds\,dt < \infty, \tag{2.131}$$

then (2.123) *is nonoscillatory.*

Applying the "Correspondence Principle" to (2.117), we obtain the following theorem:

Theorem 2.7.5 (i) *Equation* (2.117) *is nonoscillatory in the case any of the following conditions is satisfied*:

(a) $\int_\sigma^\infty \frac{1}{r_2(t)} \int_\sigma^t r_3(s) \int_\sigma^s \frac{1}{r_1(u)}\,du\,ds\,dt < \infty$,
(b) $\int_\sigma^\infty \frac{1}{r_1(t)} \int_\sigma^t \frac{1}{r_2(s)} \int_\sigma^s r_3(u)\,du\,ds\,dt < \infty$,

(ii) *Eq.* (2.123) *is nonoscillatory in the case any of the following is satisfied*:

(a) $\int_\sigma^\infty \frac{1}{r_2(t)} \int_\sigma^t \frac{1}{r_1(s)} \int_\sigma^s r_3(u)\,du\,ds\,dt < \infty$,
(b) $\int_\sigma^\infty \frac{1}{r_1(t)} \int_\sigma^t r_3(s) \int_\sigma^s \frac{1}{r_2(u)}\,du\,ds\,dt < \infty$.

Definition 2.7.2 A nonoscillatory solution $x(t)$ of (2.117) is said to be a Kneser solution or a completely monotone solution if there exists a $t_0 \geq \sigma$ such that

$$(-1)^i x(t) L_i x(t) > 0, \quad i = 0, 1, 2, \ t \geq t_0, \tag{2.132}$$

where $L_i x(t)$ denotes the ith quasiderivative of x defined as

$$\left.\begin{aligned} L_0x(t) &= x(t), \\ L_1x(t) &= r_1(t)\big(L_0x(t)\big)' = r_1(t)x'(t), \\ L_2x(t) &= r_2(t)\frac{d}{dt}\big(L_1x(t)\big)' = r_2(t)\big(r_1(t)x'(t)\big)', \\ L_3x(t) &= \frac{d}{dt}\big(L_2x(t)\big) = \big(r_2(t)\big(r_1(t)x'(t)\big)'\big)'. \end{aligned}\right\} \tag{2.133}$$

Assuming the conditions

$$\int_\sigma^\infty \frac{1}{r_2(t)}\,dt = \int_\sigma^\infty \frac{1}{r_1(t)}\,dt = \infty, \tag{2.134}$$

Kiguradze [22] obtained the following classical result:

Theorem 2.7.6 (i) *Every nonoscillatory solution of* (2.117) *is either a Kneser solution or satisfies*

$$x(t)L_ix(t) > 0, \quad i = 0, 1, 2 \tag{2.135}$$

for $t \ge t_0 \ge \sigma$.

(ii) *Equation* (2.123) *does not have any Kneser solution.*

Corollary 2.7.2 *The set of all Kneser solutions of* (2.117) *is a nonempty subset of* $\mathcal{S}_1$ *in case any of the following conditions are satisfied*:

(i) $\int_\sigma^\infty \frac{1}{r_2(t)}\,dt = \int_\sigma^\infty \frac{1}{r_1(t)}\,dt = \infty$, $\int_\sigma^\infty r_3(t)\int_\sigma^t \frac{1}{r_1(s)}\int_\sigma^s \frac{1}{r_2(u)}\,du\,ds\,dt < \infty$.
(ii) $\int_\sigma^\infty \frac{1}{r_2(t)}\,dt = \int_\sigma^\infty r_3(t)\,dt = \infty$, $\int_\sigma^\infty \frac{1}{r_1(t)}\int_\sigma^t \frac{1}{r_2(s)}\int_\sigma^s r_3(u)\,du\,ds\,dt < \infty$.
(iii) $\int_\sigma^\infty r_3(t)\,dt = \int_\sigma^\infty \frac{1}{r_1(t)}\,dt = \infty$, $\int_\sigma^\infty \frac{1}{r_2(t)}\int_\sigma^t r_3(s)\int_\sigma^s \frac{1}{r_1(u)}\,du\,ds\,dt < \infty$.

Theorem 2.7.7 *If*

$$\int_\sigma^\infty \frac{1}{r_2(t)}\,dt = \int_\sigma^\infty \frac{1}{r_1(t)}\,dt = \infty, \qquad \int_\sigma^\infty r_3(t)\int_\sigma^t \frac{1}{r_1(s)}\,ds\,dt = \infty,$$

then (2.117) *is oscillatory.*

Proof If (2.117) is nonoscillatory, then by [8], there exists a nonoscillatory solution $x(t)$ of (2.117) such that $x(t)$ satisfies (2.135) for large t. Without any loss of generality, we may assume that $x(t) > 0$, $L_1x(t) > 0$ and $L_2x(t) > 0$ for some $t \ge t_0 \ge \sigma$, where $L_ix(t)$ is defined in (2.133). Integrating (2.117) from t_0 to t, we obtain

$$r_2(t)\big(r_1(t)x'(t)\big)' - r_2(t)\big(r_1(t)x'(t)\big)'\big|_{t=t_0} + \int_{t_0}^t r_3(s)x(s)\,ds = 0.$$

Since $L_2x(t) > 0$, we have

$$r_2(t)\big(r_1(t)x'(t)\big)'\big|_{t=t_0} > \int_{t_0}^{t} r_3(s)x(s)\,ds. \tag{2.136}$$

Since $L_1x(t)$ is a positive increasing function, we have

$$x(t) > x(t_0) + r_1(t_0)x'(t_0)\int_{t_0}^{t}\frac{1}{r_1(u)}\,du.$$

Hence (2.136) gives

$$\begin{aligned}r_2(t)\big(r_1(t)x'(t)\big)'\big|_{t=t_0} &> \int_{t_0}^{t} r_3(s)\left(x(t_0) + r_1(t_0)x'(t_0)\int_{t_0}^{s}\frac{1}{r_1(u)}\,du\right)ds\\ &> r_1(t_0)x'(t_0)\int_{t_0}^{t} r_3(s)\int_{t_0}^{s}\frac{1}{r_1(u)}\,du\,ds.\end{aligned}$$

The above inequality yields a contradiction by taking the limit as $t \to \infty$. The theorem is proved. □

Theorem 2.7.8 *If*

$$\int_{\sigma}^{\infty}\frac{1}{r_2(t)}\,dt = \int_{\sigma}^{\infty}\frac{1}{r_1(t)}\,dt = \infty \quad \text{and} \quad \int_{\sigma}^{\infty} r_3(t)\int_{\sigma}^{t}\frac{1}{r_2(s)}\,ds\,dt = \infty,$$

then (2.123) *is oscillatory.*

Applying the "Correspondence Principle" to (2.117), we obtain the following:

Theorem 2.7.9 (i) *Equation* (2.117) *is oscillatory in the case any one of the following conditions is satisfied*:

(a) $\int_{\sigma}^{\infty} r_3(t)\,dt = \int_{\sigma}^{\infty}\frac{1}{r_2(t)}\,dt = \int_{\sigma}^{\infty}\frac{1}{r_1(t)}\int_{\sigma}^{t}\frac{1}{r_2(s)}\,ds\,dt = \infty,$
(b) $\int_{\sigma}^{\infty}\frac{1}{r_1(t)}\,dt = \int_{\sigma}^{\infty} r_3(t)\,dt = \int_{\sigma}^{\infty}\frac{1}{r_2(t)}\int_{\sigma}^{t} r_3(s)\,ds\,dt = \infty.$

(ii) *Equation* (2.123) *is oscillatory in the case any one of the following conditions is satisfied*:

(a) $\int_{\sigma}^{\infty} r_3(t)\,dt = \int_{\sigma}^{\infty}\frac{1}{r_2(t)}\,dt = \int_{\sigma}^{\infty}\frac{1}{r_1(t)}\int_{\sigma}^{t} r_3(s)\,ds\,dt = \infty,$
(b) $\int_{\sigma}^{\infty}\frac{1}{r_1(t)}\,dt = \int_{\sigma}^{\infty} r_3(t)\,dt = \int_{\sigma}^{\infty}\frac{1}{r_2(t)}\int_{\sigma}^{t}\frac{1}{r_1(s)}\,ds\,dt = \infty.$

Corollary 2.7.3 *If any one of the following conditions*:

(i) $\int_{\sigma}^{\infty}\frac{1}{r_2(t)}\,dt = \int_{\sigma}^{\infty}\frac{1}{r_1(t)}\,dt = \int_{\sigma}^{\infty} r_3(t)\int_{\sigma}^{t}\frac{1}{r_1(s)}\,ds\,dt = \infty,$
(ii) $\int_{\sigma}^{\infty} r_3(t)\,dt = \int_{\sigma}^{\infty}\frac{1}{r_2(t)}\,dt = \int_{\sigma}^{\infty}\frac{1}{r_1(t)}\int_{\sigma}^{t}\frac{1}{r_2(s)}\,ds\,dt = \infty,$
(iii) $\int_{\sigma}^{\infty}\frac{1}{r_1(t)}\,dt = \int_{\sigma}^{\infty} r_3(t)\,dt = \int_{\sigma}^{\infty}\frac{1}{r_2(t)}\int_{\sigma}^{t} r_3(s)\,ds\,dt = \infty$

is satisfied, then

(a) *any solution* $x(t)$ *of* (2.117) *which vanishes* (*at some* $t_0 \geq \sigma$)

$$x(t_0) = 0 \quad or \quad x'(t_0) = 0 \quad or \quad \left(r_1(t)x'(t)\right)'\Big|_{t=t_0} = 0$$

is oscillatory;

(b) *every nonoscillatory solution of* (2.117) *is a Kneser solution on* $[\sigma, \infty)$.

Proof (a) From Theorems 2.7.7 and 2.7.9, it follows that Eqs. (2.117)–(2.119) are oscillatory. Let $x(t)$ be a solution of (2.117) such that $x(t_0) = 0$. Since (2.117) is of type C_I, by Lemma 2.7.2(a), $x(t)$ is oscillatory.

Assume that $x(t)$ is a solution of (2.117) with $x'(t_0) = 0$. Then $y(t) = r_1(t)x'(t)$ is a solution of (2.118). Clearly, $y(t_0) = 0$. Since $r_1(t) > 0$, (2.118) is of type C_I. Hence the "Correspondence Principle" implies that (2.118) is oscillatory. This in turn implies that $y(t)$ is oscillatory. Applying the "Correspondence Principle", once again we obtain the required result. A similar argument may be applied when $(r_1(t)x'(t))'|_{t=t_0} = 0$ holds.

(b) Assume the condition (i) and let $x(t)$ be a nonoscillatory solution of (2.117). Since (2.117) is oscillatory, by [8], there exists a $t_0 \geq \sigma$ such that $x(t)$ is the Kneser solution on $[t_0, \infty)$. If, by contradiction, $x(t_1) = 0$, $t_0 \geq t_1 \geq \sigma$, then $x(t)$ is oscillatory, a contradiction. Hence $x(t)$ is a Kneser solution on $[\sigma, \infty)$.

If the conditions (ii) or (iii) is valid, then the assertion follows from the "Correspondence Principle" and by using arguments similar to that above. The proof is complete. □

It is well known that (2.117) is oscillatory, if and only if its adjoint (2.120) is oscillatory. The same situation does not occur for Eqs. (2.117) and (2.123) even if $\int_\sigma^\infty \frac{1}{r_2(t)}\,dt = \int_\sigma^\infty \frac{1}{r_1(t)}\,dt = \infty$, as the following example shows:

Example 2.7.2 Let $\epsilon \in (0, 1)$ and $T > 1$. Consider the equations

$$\left(t \ln t \left(\frac{x'(t)}{\ln t}\right)'\right)' + \frac{1}{t^2(\ln t)^{1+\epsilon}} x(t) = 0, \quad t \in [T, \infty), \tag{2.137}$$

and

$$\left(t \ln t \left(\frac{x'(t)}{\ln t}\right)'\right)' - \frac{1}{t^2(\ln t)^{1+\epsilon}} x(t) = 0, \quad t \in [T, \infty). \tag{2.138}$$

By Theorem 2.7.7, (2.137) is oscillatory, and by Theorem 2.7.4, (2.138) is nonoscillatory.

Lemma 2.7.1 shows that (2.117) is closely related to (2.118) and (2.119) obtained by means of an ordered cyclic permutations of the functions r_2, r_1 and r_3. Similarly, if $u(t)$ is a solution of (2.120), then $v(t) = r_1(t)u'(t)$ is a solution of

$$\left(\frac{1}{r_3(t)}\left(r_1(t)v'\right)'\right)' - \frac{1}{r_2(t)} v = 0, \tag{2.139}$$

and $w(t) = r_2(t)(r_1(t)v'(t))'$ is a solution of

$$\left(r_2(t)\left(\frac{1}{r_3(t)}w'\right)'\right)' - \frac{1}{r_1(t)}w = 0. \tag{2.140}$$

Denote $\aleph$ and $\aleph(a)$ as the set of all nontrivial nonoscillatory solutions of (2.117) and its adjoint equation (2.120). If $x(t)$ is nonoscillatory, then $L_1x(t)$, $L_2x(t)$ and $L_3x(t)$ are nonoscillatory. In view of this, one can divide the set $\aleph$ into the following four classes:

(i) $\aleph_0 = \{x \in \aleph, \exists T_x;\ x(t)L_1x(t) < 0, x(t)L_2x(t) > 0 \text{ for } t \geq T_x\}$,
(ii) $\aleph_1 = \{x \in \aleph, \exists T_x;\ x(t)L_1x(t) > 0, x(t)L_2x(t) < 0 \text{ for } t \geq T_x\}$,
(iii) $\aleph_2 = \{x \in \aleph, \exists T_x;\ x(t)L_1x(t) > 0, x(t)L_2x(t) > 0 \text{ for } t \geq T_x\}$,
(iv) $\aleph_3 = \{x \in \aleph, \exists T_x;\ x(t)L_1x(t) < 0, x(t)L_2x(t) < 0 \text{ for } t \geq T_x\}$,

and the set $\aleph(a)$ into the following four classes:

(i) $\aleph_0(a) = \{u \in \aleph(a), \exists T_u;\ u(t)L_1u(t) < 0, u(t)L_2u(t) > 0 \text{ for } t \geq T_u\}$,
(ii) $\aleph_1(a) = \{u \in \aleph(a), \exists T_u;\ u(t)L_1u(t) > 0, u(t)L_2u(t) < 0 \text{ for } t \geq T_u\}$,
(iii) $\aleph_2(a) = \{u \in \aleph(a), \exists T_u;\ u(t)L_1u(t) < 0, u(t)L_2u(t) < 0 \text{ for } t \geq T_u\}$,
(iv) $\aleph_3(a) = \{u \in \aleph(a), \exists T_u;\ u(t)L_1u(t) > 0, u(t)L_2u(t) > 0 \text{ for } t \geq T_u\}$.

Obviously, if every nonoscillatory solution $x(t)$ of (2.117) satisfies the property $\lim_{t\to\infty} |x^{(i)}(t)| = 0, i = 0, 1, 2$, then $x(t)$ belongs to the class $\aleph_0$. Similarly, if $u \in \aleph(a)$ and satisfies the property $\lim_{t\to\infty} |u^{(i)}(t)| = \infty, i = 0, 1, 2$, then u belongs to the class $\aleph_3(a)$. This means that if (2.117) has property A, then $\aleph = \aleph_0$ and if (2.120) has property B, then $\aleph(a) = \aleph_3(a)$. In addition, if $x \in \aleph_0$, then the quasiderivatives $L_ix(t)$, $i = 0, 1, 2, 3$ have eventually an alternate sign which we call a Kneser solution. Similarly, if $u \in \aleph_3(a)$, then the quasiderivatives $L_iu(t)$, $i = 0, 1, 2, 3$ have eventually same sign, which we call a strongly monotone solution. Their existence is ensured by the following lemma.

Lemma 2.7.5 *Equation* (2.117) *has always a Kneser solution, and* (2.120) *has always a strongly monotone solution.*

Remark 2.7.5 It is easy to see that if $x \in \aleph_0$, then $x(t)L_1x(t) < 0$ and $x(t)L_2x(t) > 0$ not only eventually but also for all $t \geq \sigma$. Indeed, let $x(t) > 0, L_1x(t) < 0, L_2x(t) > 0$ for $t \geq T \geq \sigma$ and suppose that there exists $t_1 \in [\sigma, T)$ such that $x'(t_1) = 0$ and $x(t) > 0$ for $t \in (t_1, T)$. Then $L_2x(t)$ is decreasing on (t_1, T). Because $L_2x(T) > 0$, we have $L_2x(t) > 0$ for $t \in (t_1, T)$, which implies that $L_1x(t)$ is decreasing on (t_1, T). Again, since $L_1x(T) < 0$, we obtain $L_1x(t_1) = r_1(t_1)x'(t_1) < 0$, which is a contradiction. Thus $x(t)L_1x(t) < 0$ for $t \geq \sigma$. By a similar argument, Lemma 2.7.1 implies that $x(t)L_2x(t) > 0$ for $t \geq \sigma$.

Lemma 2.7.6 *Let x and y be two linearly independent solutions of* (2.117) [(2.120)], *then*

$$w = L_1x(t) \cdot y(t) - x(t) \cdot L_1y(t) \tag{2.141}$$

is a solution of (2.120) [(2.117)] *and its quasiderivatives satisfy*

$$L_1 w(t) = L_2 x(t) \cdot y(t) - x(t) \cdot L_2 y(t)$$

and

$$L_2 w(t) = L_2 x(t) \cdot L_1 y(t) - L_1 x(t) \cdot L_2 y(t).$$

Lemma 2.7.7 *The following conditions are equivalent*:

(i) $\aleph = \aleph_0$
(ii) $\aleph(a) = \aleph_3(a)$.

Proof First, we prove that (ii) $\Rightarrow$ (i). By Lemma 2.7.5, we have $\aleph_0 \neq \phi$. Assume that there exists $j \in \{1, 2, 3\}$ such that $\aleph_j \neq \phi$. Let $x \in \aleph_0$ and $y \in \aleph_j$. Without any loss of generality, we may assume that $x(t) > 0$ and $y(t) > 0$, for large t. Then the function w defined by (2.141) is a solution of (2.120) and satisfies the properties

$$w(t) < 0, \qquad L_1 w(t) > 0 \quad \text{if } j = 1;$$

$$w(t) < 0, \qquad L_2 w(t) > 0 \quad \text{if } j = 2;$$

and

$$L_1 w(t) > 0, \qquad L_2 w(t) < 0 \quad \text{if } j = 3$$

for large t, which contradicts the fact that all nonoscillatory solutions of (2.120) are strongly monotone.

The claim (i) $\Rightarrow$ (ii) can be proved by using a similar argument as given in the first part. □

Lemma 2.7.8 *If there exists $x \in \aleph_0$ such that*

$$\lim_{t\to\infty} L_i x(t) = 0, \quad i = 0, 1, 2, \tag{2.142}$$

then

$$\int_\sigma^\infty r_3(t) \int_\sigma^t \frac{1}{r_2(s)} \int_\sigma^s \frac{1}{r_1(\tau)} d\tau \, ds \, dt = \infty.$$

Proof Suppose that

$$\int_\sigma^\infty r_3(t) \int_\sigma^t \frac{1}{r_2(s)} \int_\sigma^s \frac{1}{r_1(\tau)} d\tau \, ds \, dt < \infty.$$

Then there exists a $t_0 > \sigma$ such that

$$\int_{t_0}^\infty r_3(t) \int_{t_0}^t \frac{1}{r_2(s)} \int_{t_0}^s \frac{1}{r_1(\tau)} d\tau \, ds \, dt < \frac{1}{2} \tag{2.143}$$

holds. Let $x(t)$ be an eventually positive solution of (2.117), which belongs to the class $\aleph_0$ such that $x(t)$ satisfies (2.142) for $t \geq t_0$. From repeated integration of (2.117) from t to ∞, $t \geq t_0$, we obtain

$$x(t) = \int_t^\infty \frac{1}{r_1(s)} \int_s^\infty \frac{1}{r_2(\tau)} \int_\tau^\infty r_3(\theta)x(\theta)\, d\theta\, d\tau\, ds$$
$$\leq x(t) \int_t^\infty \frac{1}{r_1(s)} \int_s^\infty \frac{1}{r_2(\tau)} \int_\tau^\infty r_3(\theta)\, d\theta\, d\tau\, ds.$$

Thus,

$$1 \leq \int_t^\infty \frac{1}{r_1(s)} \int_s^\infty \frac{1}{r_2(\tau)} \int_\tau^\infty r_3(\theta)\, d\theta\, d\tau\, ds.$$

Then, by interchanging the order of integration, we obtain a contradiction to (2.143). The lemma is proved. □

In a similar way, it can be proved that, if $x \in \aleph_0$ satisfies (2.142), then

$$\int_\sigma^\infty \frac{1}{r_1(t)} \int_\sigma^t r_3(s) \int_\sigma^s \frac{1}{r_2(\tau)}\, d\tau\, ds\, dt = \infty$$

and

$$\int_\sigma^\infty \frac{1}{r_2(t)} \int_\sigma^t \frac{1}{r_1(s)} \int_\sigma^s r_3(\tau)\, d\tau\, ds\, dt = \infty$$

hold.

Lemma 2.7.9 *The following hold*:

(i) *if there exists* $x \in \aleph_0$ *such that* $\lim_{t\to\infty} x(t) \neq 0$, *then* (2.131) *holds*;
(ii) *if there exists* $x \in \aleph_0$ *such that* $\lim_{t\to\infty} L_1 x(t) \neq 0$, *then*

$$\int_\sigma^\infty \frac{1}{r_1(t)} \int_\sigma^t r_3(s) \int_\sigma^s \frac{1}{r_2(\tau)}\, d\tau\, ds\, dt < \infty; \tag{2.144}$$

(iii) *if there exists a* $x \in \aleph_0$ *such that* $\lim_{t\to\infty} L_2 x(t) \neq 0$, *then*

$$\int_\sigma^\infty \frac{1}{r_2(t)} \int_\sigma^t \frac{1}{r_1(s)} \int_\sigma^s r_3(\tau)\, d\tau\, ds\, dt < \infty. \tag{2.145}$$

Proof (i) Let $x(t)$ be an eventually positive solution of (2.117) in the class $\aleph_0$ such that $\lim_{t\to\infty} x(t) = \lambda > 0$. Hence there exists a $T \geq \sigma$ such that $x(t) > 0$, $L_1 x(t) < 0$ and $L_2 x(t) > 0$ for $t \geq T$. In view of the signs of $\int_\sigma^\infty \frac{1}{r_2(t)}\, dt$ and $\int_\sigma^\infty \frac{1}{r_1(t)}\, dt$, three possible cases may arise:

(I) $\int_\sigma^\infty \frac{1}{r_2(t)}\, dt < \infty$, $\int_\sigma^\infty \frac{1}{r_1(t)}\, dt < \infty$,
(II) $\int_\sigma^\infty \frac{1}{r_2(t)}\, dt = \infty$, $\int_\sigma^\infty \frac{1}{r_1(t)}\, dt < \infty$, and

(III) $\int_\sigma^\infty \frac{1}{r_1(t)}\,dt = \infty$.

First consider the case (I), that is, $\int_\sigma^\infty \frac{1}{r_2(t)}\,dt < \infty$ and $\int_\sigma^\infty \frac{1}{r_1(t)}\,dt < \infty$. Integrating (2.117) from t to ∞, $t \geq T$, we obtain

$$\begin{aligned} L_2x(t) &= L_2x(\infty) + \int_t^\infty r_3(s)x(s)\,ds \\ &\geq L_2x(\infty) + x(\infty)\int_t^\infty r_3(s)\,ds. \end{aligned}$$

Then $\int_\sigma^\infty r_3(s)\,ds < \infty$, and so (2.131) holds.

Now assume the case (II), that is, $\int_\sigma^\infty \frac{1}{r_2(s)}\,ds = \infty$ and $\int_\sigma^\infty \frac{1}{r_1(s)}\,ds < \infty$. Clearly, $\int_\sigma^\infty \frac{1}{r_2(s)}\,ds = \infty$ implies that $L_2x(\infty) = 0$. Integrating (2.117) twice from t to ∞, $t > T$, we obtain

$$\begin{aligned} L_1x(t) &= L_1x(\infty) - \int_t^\infty \frac{1}{r_2(s)}\int_s^\infty r_3(\tau)x(\tau)\,d\tau\,ds \\ &\leq L_1x(\infty) - x(\infty)\int_t^\infty \frac{1}{r_2(s)}\int_s^\infty r_3(\tau)\,d\tau\,ds \\ &\leq L_1x(\infty) - x(\infty)\int_t^\infty r_3(s)\int_t^s \frac{1}{r_2(\tau)}\,d\tau\,ds. \end{aligned}$$

Then $\int_\sigma^\infty r_3(t)\int_\sigma^t \frac{1}{r_2(s)}\,ds\,dt < \infty$, and hence (2.131) holds.

Finally, suppose that the case (III) holds, that is, $\int_\sigma^\infty \frac{1}{r_1(s)}\,ds = \infty$ holds. Proceeding as in case (II), we see that $\int_\sigma^\infty \frac{1}{r_1(s)}\,ds = \infty$ implies that $L_1x(\infty) = 0$. Integrating (2.117) three times from t to ∞, $t > T$, we obtain

$$\begin{aligned} x(t) \geq x(\infty) &+ L_2x(\infty)\int_t^\infty \frac{1}{r_1(s)}\int_s^\infty \frac{1}{r_2(\tau)}\,d\tau\,ds \\ &+ x(\infty)\int_t^\infty \frac{1}{r_1(s)}\int_s^\infty \frac{1}{r_2(\tau)}\int_\tau^\infty r_3(\theta)\,d\theta\,d\tau\,ds \end{aligned}$$

and then by interchanging the order of integration, we get (2.131).

The claims (ii) and (iii) may be proved by applying similar arguments as above by using (2.118) and (2.119) instead of (2.117). □

Theorem 2.7.10 *Equation* (2.117) *has property* A, *if and only if* (2.120) *has property* B.

Proof Assume that (2.117) has property A. Then $\aleph = \aleph_0$ and, by Lemma 2.7.7, $\aleph(a) = \aleph_3(a)$. By Lemma 2.7.5, $\aleph_3(a) \neq \phi$. Let $u \in \aleph_3(a)$, that is, there exists a $T \geq \sigma$ such that $L_iu(t) > 0$ for $t > T$ and $i = 0, 1, 2$. Suppose that (2.120) does not have property B, that is, there exists $i \in \{0, 1, 2\}$ such that $L_iu(t)$ is bounded.

First, let $L_2u(t)$ be bounded. Observe that $w(t) = L_2u(t)$ is a solution of (2.140) and $L_1w(t) = u(t)$, $L_2w(t) = L_1u(t)$. Then $L_iw(t) > 0$ for $t \geq T$, $i = 0, 1, 2$, and $w(t)$ is bounded. Thus, there exists $k_i > 0$, $i = 1, 2$ such that

$$0 < k_1 \leq w(t) \leq k_2 \quad \text{for } t \geq T. \tag{2.146}$$

Integrating (2.140) three times from T to t, we have

$$\begin{aligned} w(t) &= w(T) + L_1w(T)\int_T^t r_3(s)\,ds + L_2w(T)\int_T^t r_3(s)\int_T^s \frac{1}{r_2(\tau)}\,d\tau\,ds \\ &+ \int_T^t r_3(s)\int_T^s \frac{1}{r_2(\tau)}\int_T^\tau \frac{1}{r_1(\theta)}w(\theta)\,d\theta\,d\tau\,ds. \end{aligned} \tag{2.147}$$

Since $L_iw(T) > 0$ for $i = 0, 1, 2$, from (2.146) and (2.147), we obtain

$$k_1\int_T^t r_3(s)\int_T^s \frac{1}{r_2(\tau)}\int_T^\tau \frac{1}{r_1(\theta)}\,d\theta\,d\tau\,ds \leq w(t) \leq k_2. \tag{2.148}$$

By Lemma 2.7.5, (2.117) has a Kneser solution. Since (2.117) has property A, any Kneser solution of (2.117) satisfies (2.142). Thus, by Lemma 2.7.8, the integral on the left-hand side of (2.148) is divergent, which is a contradiction.

If $L_1u(t)$ is bounded, we consider (2.139) instead of (2.140). Using a similar argument as above and Lemma 2.7.8, we again get a contradiction. Hence property A of (2.117) implies the property B of (2.120).

Now, assume that (2.120) has property B. Then $\aleph(a) = \aleph_3(a)$. From Lemmas 2.7.5 and 2.7.7, we have $\aleph = \aleph_0 \neq \phi$. Assume that there exists a Kneser solution $x(t)$ of (2.117) such that for some $i \in \{0, 1, 2\}$, $\lim_{t\to\infty} L_ix(t) = c \neq 0$. First, suppose that $\lim_{t\to\infty} x(t) = c \neq 0$. Then by Lemma 2.7.9, we obtain (2.131), so also $\int_\sigma^\infty r_3(t)\,dt < \infty$ and $\int_\sigma^\infty r_3(t)\int_\sigma^t \frac{1}{r_2(s)}\,ds\,dt < \infty$.

Let $w(t)$ be a nonoscillatory solution of (2.140). Without any loss of generality, we may assume that $w(t)$ is eventually positive. Clearly, there exists a solution $u(t)$ of (2.120) such that $w(t) = L_2u(t)$. Since (2.120) has property B, $u(t)$ satisfies the property (2.135). Hence $w(t) \to \infty$ as $t \to \infty$ and there exists a $T \geq \sigma$ such that $L_iw(t) > 0$ for all $t \geq T$. In view of (2.131), we can find T large enough such that

$$\int_T^\infty r_3(t)\int_T^t \frac{1}{r_2(s)}\int_T^s \frac{1}{r_1(\tau)}\,d\tau\,ds\,dt < 1. \tag{2.149}$$

Repeated integration of (2.140) from T to t and by using the property w is a nondecreasing function, it follows from (2.147) that

$$w(t) \leq f(t) + w(t)g(t),$$

where

$$f(t) = w(T) + L_1w(T)\int_T^t r_3(s)\,ds + L_2w(T)\int_T^t r_3(s)\int_T^s \frac{1}{r_2(\tau)}\,d\tau\,ds$$

and

$$g(t) = \int_T^t r_3(s) \int_T^s \frac{1}{r_2(\tau)} \int_T^\tau \frac{1}{r_1(\theta)} \, d\theta \, d\tau \, ds.$$

Then

$$w(t) \le \frac{f(t)}{1 - g(t)}$$

which implies that $w(t)$ is bounded, a contradiction to the property B of (2.120).

Similar contradictions can be obtained if

$$\lim_{t\to\infty} L_1 x(t) = c_1 \neq 0 \quad \text{and} \quad \lim_{t\to\infty} L_2 x(t) = c_2 \neq 0.$$

Here one has to use Lemma 2.7.9. The proof is complete. □

Corollary 2.7.4 *If* (2.117) *has property* A*, then* (2.118) *and* (2.119) *have property* A*, and* (2.120), (2.139) *and* (2.140) *has property* B*.*

Corollary 2.7.5 *Every Kneser solution* x *of* (2.117) *satisfies* $\lim_{t\to\infty} L_i x(t) = 0$ *for* $i = 0, 1, 2$ *if and only if every strongly monotone solution* u *of* (2.120) *satisfies* $\lim_{t\to\infty} L_i u(t) = \infty$ *for* $i = 0, 1, 2$.

Theorem 2.7.11 *If* (2.117) *has property* A *or* (2.120) *has property* B*, then it is oscillatory.*

Theorem 2.7.12 *If any one of the following conditions*:

(i) $\int_\sigma^\infty \frac{1}{r_1(t)} \, dt = \int_\sigma^\infty \frac{1}{r_2(t)} \, dt = \int_\sigma^\infty r_3(t) \int_\sigma^t \frac{1}{r_1(s)} \, ds \, dt = \int_\sigma^\infty r_3(t) \int_\sigma^t \frac{1}{r_2(s)} \, ds \, dt = \infty,$

(ii) $\int_\sigma^\infty \frac{1}{r_2(t)} \, dt = \int_\sigma^\infty r_3(t) \, dt = \int_\sigma^\infty \frac{1}{r_1(t)} \int_\sigma^t \frac{1}{r_2(s)} \, ds \, dt = \int_\sigma^\infty \frac{1}{r_1(t)} \int_\sigma^t r_3(s) \, ds \, dt = \infty,$

(iii) $\int_\sigma^\infty \frac{1}{r_1(t)} \, dt = \int_\sigma^\infty r_3(t) \, dt = \int_\sigma^\infty \frac{1}{r_2(t)} \int_\sigma^t r_3(s) \, ds \, dt = \int_\sigma^\infty \frac{1}{r_2(t)} \int_\sigma^t \frac{1}{r_1(s)} ds \, dt = \infty,$

(iv) $\int_\sigma^\infty \frac{1}{r_2(t)} \, dt = \infty$, $\int_\sigma^\infty r_3(t) \int_\sigma^t \frac{1}{r_2(s)} \, ds \, dt < \infty$, *and*

$$\int_\sigma^\infty \frac{1}{r_1(t)} \left(\int_t^\infty r_3(s) \, ds \right) \left(\int_t^\infty \frac{1}{r_2(s)} \int_s^\infty r_3(\tau) \, d\tau \, ds \right) dt = \infty,$$

(v) $\int_\sigma^\infty r_3(t) \, dt = \infty$, $\int_\sigma^\infty \frac{1}{r_1(t)} \int_\sigma^t r_3(s) \, ds \, dt < \infty$, *and*

$$\int_\sigma^\infty \frac{1}{r_2(t)} \left(\int_t^\infty \frac{1}{r_1(s)} ds \right) \left(\int_t^\infty r_3(s) \int_s^\infty \frac{1}{r_1(\tau)} d\tau \, ds \right) dt = \infty,$$

is satisfied, then (2.120) *has property* B*. Consequently,* (2.139) *and* (2.140) *also have property* B*.*

Theorem 2.7.13 *If any one of the following conditions*

(i) $\int_\sigma^\infty \frac{1}{r_1(t)}\,dt = \int_\sigma^\infty r_3(t)\,dt = \int_\sigma^\infty \frac{1}{r_2(t)} \int_\sigma^t r_3(s)\,ds\,dt = \int_\sigma^\infty \frac{1}{r_2(t)} \int_\sigma^t \frac{1}{r_1(s)}\,ds\,dt = \infty$,
(ii) $\int_\sigma^\infty \frac{1}{r_1(t)}\,dt = \int_\sigma^\infty \frac{1}{r_2(t)}\,dt = \int_\sigma^\infty r_3(t) \int_\sigma^t \frac{1}{r_1(s)}\,ds\,dt = \int_\sigma^\infty r_3(t) \int_\sigma^t \frac{1}{r_2(s)}\,ds\,dt = \infty$,
(iii) $\int_\sigma^\infty \frac{1}{r_2(t)}\,dt = \int_\sigma^\infty r_3(t)\,dt = \int_\sigma^\infty \frac{1}{r_1(t)} \int_\sigma^t \frac{1}{r_2(s)}\,ds\,dt = \int_\sigma^\infty \frac{1}{r_1(t)} \int_\sigma^t r_3(s)ds\,dt = \infty$,
(iv) $\int_\sigma^\infty r_3(t)\,dt = \infty$, $\int_\sigma^\infty \frac{1}{r_1(t)} \int_\sigma^t r_3(s)\,ds\,dt < \infty$, *and*

$$\int_\sigma^\infty \frac{1}{r_2(t)} \left(\int_t^\infty \frac{1}{r_1(s)}\,ds \right) \left(\int_t^\infty r_3(s) \int_s^\infty \frac{1}{r_1(\tau)}\,d\tau\,ds \right) dt = \infty,$$

(v) $\int_\sigma^\infty \frac{1}{r_2(t)}\,dt = \infty$, $\int_\sigma^\infty r_3(t) \int_\sigma^t \frac{1}{r_2(s)}\,ds\,dt < \infty$, *and*

$$\int_\sigma^\infty \frac{1}{r_1(t)} \left(\int_t^\infty r_3(s)\,ds \right) \left(\int_t^\infty \frac{1}{r_2(s)} \int_s^\infty r_3(\tau)\,d\tau\,ds \right) dt = \infty$$

is satisfied, then (2.117)–(2.119) *are oscillatory and have property A.*

Theorem 2.7.14 *If* $\int_\sigma^\infty \frac{1}{r_1(t)}\,dt = \infty$, $\int_\sigma^\infty \frac{1}{r_2(t)} \int_\sigma^t \frac{1}{r_1(s)}\,ds\,dt < \infty$, *and*

$$\int_\sigma^\infty r_3(t) \left(\int_t^\infty \frac{1}{r_2(s)}\,ds \right) \left(\int_t^\infty \frac{1}{r_1(s)} \int_s^\infty \frac{1}{r_2(\tau)}\,d\tau\,ds \right) dt = \infty,$$

then

(i) (2.117)–(2.119) *are oscillatory and have property A.*
(ii) (2.120), (2.139) *and* (2.140) *have property B.*

Theorem 2.7.15 *The following assertions are equivalent*:

(a) (2.117) *has property A.*
(b) (2.117) *is oscillatory and*

$$\int_\sigma^\infty r_3(t) \int_\sigma^t \frac{1}{r_2(s)} \int_\sigma^s \frac{1}{r_1(u)}\,du\,ds\,dt = \infty,$$

$$\int_\sigma^\infty \frac{1}{r_1(t)} \int_\sigma^t r_3(s) \int_\sigma^s \frac{1}{r_2(u)}\,du\,ds\,dt = \infty,$$

$$\int_\sigma^\infty \frac{1}{r_2(t)} \int_\sigma^t \frac{1}{r_1(s)} \int_\sigma^s r_3(u)\,du\,ds\,dt = \infty.$$

Corollary 2.7.6 *If the functions* $r_1(t)$, $r_2(t)$ *and* $r_3(t)$ *satisfy* (2.131), *and*

$$\int_\sigma^\infty \frac{1}{r_1(t)}\,dt = \int_\sigma^\infty \frac{1}{r_2(t)}\,dt = \int_\sigma^\infty r_3(t) \int_\sigma^t \frac{1}{r_1(s)}\,ds\,dt = \infty$$

holds, then there exists an unbounded oscillatory solution of (2.117).

Corollary 2.7.7 *Let* $\int_{\sigma}^{\infty}\frac{1}{r_1(t)}\,dt=\int_{\sigma}^{\infty}\frac{1}{r_2(t)}\,dt=\infty$. *If* (2.117) *is oscillatory and has no property* A, *then* (2.123) *is nonoscillatory.*

Corollaries 2.7.6 and 2.7.7 are illustrated by the following example.

Example 2.7.3 Let $\epsilon\in(0,1)$ and $T>1$. Consider Eqs. (2.137) and (2.138). Clearly

$$\int_T^t \frac{1}{r_1(s)}\,ds=t(\ln t-1)+c_1\to\infty,$$

$$\int_T^t \frac{1}{r_2(s)}\,ds=\ln\ln t+c_2\to\infty \quad \text{as } t\to\infty,$$

$$\int_T^t \frac{1}{r_2(s)}\int_T^s \frac{1}{r_1(s)}\,ds\,dt=t-\int_T^t\frac{ds}{\ln s}-T-\int_T^t\frac{c\,ds}{s\ln s}\to\infty \quad \text{as } t\to\infty,$$

$$\int_T^\infty r_3(t)\int_T^t \frac{1}{r_2(s)}\int_T^s\frac{1}{r_1(\tau)}\,d\tau\,ds\,dt\le\int_T^\infty\frac{dt}{t(\ln t)^{1+\epsilon}}=\int_{\ln T}^\infty\frac{du}{u^{1+\epsilon}}<\infty$$

and

$$\int_T^\infty r_3(t)\int_T^t\frac{1}{r_1(s)}\,ds\,dt\ge\int_T^\infty\frac{\ln t-1}{t(\ln t)^{1+\epsilon}}\,dt=\int_T^\infty\frac{dt}{t(\ln t)^{\epsilon}}=\int_{\ln T}^\infty\frac{ds}{s^{\epsilon}}=\infty$$

implies that Corollary 2.7.6 can be applied to (2.137). By Corollary 2.7.6, Eq. (2.137) has an unbounded oscillatory solution and a Kneser solution tending to a nonzero constant. Similarly, by Corollary 2.7.7, Eq. (2.138) is nonoscillatory.

Theorem 2.7.16 *Let one of the following two conditions hold:*

(i) $\int_\sigma^\infty\frac{1}{r_2(t)}\,dt=\infty$, $\int_\sigma^\infty\frac{1}{r_1(t)}\int_\sigma^t\frac{1}{r_2(s)}\,ds\,dt<\infty$ *and*

$$\int_\sigma^\infty r_3(t)\left(\int_t^\infty\frac{1}{r_2(s)}\,ds\right)\int_t^\infty\frac{1}{r_1(s)}\int_s^\infty\frac{1}{r_2(\tau)}\,d\tau\,ds\,dt=\infty,$$

(ii) $\int_\sigma^\infty\frac{1}{r_1(t)}\,dt<\infty$, $\int_\sigma^\infty\frac{1}{r_2(t)}\,dt<\infty$ *and*

$$\int_\sigma^\infty r_3(t)\left(\int_t^\infty\frac{1}{r_1(s)}\,ds\right)\left(\int_t^\infty\mu(s)\,ds\right)\int_t^\infty\frac{1}{r_2(s)}\int_s^\infty\frac{1}{r_1(\tau)}\,d\tau\,ds\,dt=\infty,$$

where $\mu(t)=\left(\frac{1}{\int_t^\infty\frac{1}{r_1(s)}\,ds}\right)'\int_t^\infty\frac{1}{r_2(s)}\int_s^\infty\frac{1}{r_1(\tau)}\,d\tau\,ds$. *Then* (2.117)–(2.119) *are oscillatory and their adjoint equations* (2.120), (2.139) *and* (2.140) *are also oscillatory. In addition, every nonoscillatory solution of* (2.117) *is a Kneser solution or satisfies the property* $x(t)L_1x(t)<0$ *and* $x(t)L_2x(t)<0$ *for large* t *and tends to zero as* $t\to\infty$.

Theorem 2.7.17 *Equation* (2.117) *is oscillatory, if and only if every nonoscillatory solution* $x(t)$ *of* (2.117) *satisfies* $x(t)L_1x(t)<0$ *and* $x(t)L_2x(t)>0$ *for* $t\ge\sigma$.

Theorem 2.7.18 *Equation* (2.123) *is oscillatory, if and only if every nonoscillatory solution* $u(t)$ *of* (2.123) *satisfies* $u(t)L_1u(t) > 0$ *and* $u(t)L_2u(t) > 0$ *for* $t \geq \sigma$.

The following corollary is a consequence of Theorem 2.7.9.

Corollary 2.7.8 *If*

$$\int_{\sigma}^{\infty} r_3(t) \int_{\sigma}^{t} \frac{1}{r_2(s)} \int_{\sigma}^{s} \frac{1}{r_1(\tau)} d\tau\, ds\, dt = \infty,$$

then any solution $x(t)$ *of* (2.117) *such that* $x(t)L_1x(t) < 0$ *and* $x(t)L_2x(t) > 0$ *for* $t \geq \sigma$ *satisfies* $\lim_{t\to\infty} x(t) = 0$.

Theorems 2.7.3–2.7.5 yield the following corollary:

Corollary 2.7.9 *If* (2.127) *holds, then* (2.120) *is nonoscillatory.*

Corollary 2.7.10 *If* (2.131) *holds, then* (2.123) *is nonoscillatory and Eq.* (2.124) *is nonoscillatory.*

Now, we consider the second-order linear homogeneous equation (2.94), where b is a continuous function for $t \geq \sigma$ with either $b(t) > 0$ or $b(t) < 0$ for $t \geq \sigma$. Let $h(t)$ be a positive solution of (2.94) on $[t_0, \infty)$, $t_0 \geq \sigma$. Then the third-order differential equation (2.10) can be written in the disconjugate form

$$\left(h^2(t)\left(\frac{1}{h(t)}x'\right)'\right)' + c(t)h(t)x = 0, \tag{2.150}$$

which is a special case of the linear disconjugate equation (2.117).

In the following, we shall make an attempt to study the oscillation, nonoscillation, property A and property B of (2.10) with the help of Eq. (2.150).

Remark 2.7.6 Assume that $b(t) \leq 0$ for $t \geq \sigma$. Hartman [17] proved that (2.94) is nonoscillatory and every nonoscillatory solution $h(t)$ of (2.94) satisfying either

$$h(t)h'(t) > 0 \quad \text{for large } t \quad \text{and} \quad \lim_{t\to\infty} |h(t)| = \infty$$

or

$$h(t)h'(t) \leq 0 \quad \text{for } t \geq \sigma.$$

Further, (2.94) admits a nondecreasing nonoscillatory solution and a nonincreasing nonoscillatory solution $h(t)$, such that $\lim_{t\to\infty} h(t) = c_h \neq 0$ if and only if $\int_{\sigma}^{\infty} t|b(t)|\, dt < \infty$.

Lemma 1.5.3 shows that (2.10) is oscillatory, if and only if every nonoscillatory solution of the equation is a Kneser solution. However, it does not guarantee that this equation has property A. The following result proves the equivalence between the oscillation and property A for (2.10).

Theorem 2.7.19 *Let $b(t) \le 0$ and $c(t) > 0$. Then* (2.10) *is oscillatory if and only if* (2.10) *has property A.*

Proof By Lemma 1.5.4, it is sufficient to prove that if (2.10) is oscillatory, then every Kneser solution of (2.10) tends to zero as $t \to \infty$. Remark 2.7.6 implies that (2.94) has a nonincreasing solution. Let $h(t)$ be the nonincreasing solution of (2.94) for $t \ge \sigma$. Then (2.10) can be written in the disconjugate form (2.150). Clearly, (2.150) is oscillatory and hence by Theorem 2.7.3, we have

$$\int_\sigma^\infty c(t)h(t) \int_\sigma^t h(s) \int_\sigma^s \frac{1}{h^2(\tau)}\, d\tau\, ds\, dt = \infty. \tag{2.151}$$

Since $h(t)$ is a nonincreasing function for $t \ge \sigma$, the function $\frac{1}{h^2}$ is nondecreasing for $t \ge \sigma$. Then

$$\int_\sigma^t \frac{1}{h^2(s)} \int_\sigma^s h(u)\, du\, ds \ge \int_\sigma^t \frac{1}{h^2(s)} \int_\sigma^s h(s)\, du\, ds = \int_\sigma^t \frac{s-\sigma}{h(s)}\, ds,$$

and

$$\int_\sigma^t h(s) \int_\sigma^s \frac{1}{h^2(u)}\, du\, ds \le \int_\sigma^t h(s) \int_\sigma^s \frac{1}{h^2(s)}\, du\, ds \le \int_\sigma^t \frac{s-\sigma}{h(s)}\, ds.$$

The above two inequalities together with (2.151) implies that

$$\int_\sigma^\infty c(t)h(t) \int_\sigma^t \frac{1}{h^2(s)} \int_\sigma^s h(\tau)\, d\tau\, ds\, dt = \infty.$$

Since (2.150) is oscillatory, Corollary 2.7.8 implies that every Kneser solution of the equation tends to zero as $t \to \infty$. This completes the proof of the theorem. □

Remark 2.7.7 Let $b(t) \le 0$ and $c(t) < 0$. Lazer [23] proved that if (2.10) is oscillatory, then every nonoscillatory solution $x(t)$ of (2.10) is a strongly monotone solution and

$$\lim_{t\to\infty} |x(t)| = \lim_{t\to\infty} |x''(t)| = \infty.$$

Gera in [12] proved the converse of the above statement as follows: if every nonoscillatory solution of (2.10) is strongly monotone, then (2.10) is oscillatory. Thus we have the following corollary:

Corollary 2.7.11 *Equation* (2.10) *is oscillatory, if and only if every nonoscillatory solution of* (2.10) *is strongly monotone.*

Theorem 2.4.4 partially improves the above quoted Lazer's result to equation (2.1) with $a(t) \geq 0$. On the other hand, Theorem 2.4.5 extends Corollary 2.7.11. Applying Theorem 2.4.5 to (1.5), we obtain the following corollary:

Corollary 2.7.12 *Let $a \geq 0$, $b \leq 0$ and $c < 0$. Equation* (1.5) *admits an oscillatory solution, if and only if every nonoscillatory solution of* (1.5) *satisfies the property* (2.80).

Now, we give a stronger result which establishes the equivalence between oscillation and property B of (2.10).

Theorem 2.7.20 *Let $b(t) \leq 0$ and $c(t) < 0$. Then* (2.10) *is oscillatory if and only if* (2.10) *has property B.*

Proof In view of Corollary 2.7.11, it is sufficient to show that if (2.10) is oscillatory, then (2.10) has property B. Let $x(t)$ be a nonoscillatory solution of (2.10). Since (2.10) is oscillatory, $x(t)$ is strongly monotone. Without any loss of generality, we may assume that there exists a $t_0 \geq \sigma$ such that $x(t) > 0, x'(t) > 0$ and $x''(t) > 0$ for $t \geq t_0$. Then $x'''(t) > 0$ for $t \geq t_0$. Repeated integration of $x'''(t) > 0$ from t_0 to t gives

$$x'(t) > x''(t_0)(t - t_0)$$

and

$$x(t) \geq x(t_0) + \frac{x''(t_0)}{2}(t - t_0)^2 > \frac{x''(t_0)}{2}(t - t_0)^2$$

for $t \geq t_0$. Integrating (2.10) from t_0 to t, we get

$$\begin{aligned} x''(t) &= x''(t_0) - \int_{t_0}^{t} b(s)x'(s)\,ds - \int_{t_0}^{t} c(s)x(s)\,ds \\ &\geq x''(t_0)\left[1 - \int_{t_0}^{t} (s - t_0)b(s)\,ds - \frac{1}{2}\int_{t_0}^{t} (s - t_0)^2 c(s)\,ds\right]. \end{aligned}$$

Since $x(t)$ is an eventually positive strongly monotone solution of (2.10), we have $\lim_{t\to\infty} x(t) = \lim_{t\to\infty} x'(t) = \infty$. Hence, it is sufficient to show that $\lim_{t\to\infty} x''(t) = \infty$. Assume that $\lim_{t\to\infty} x''(t) < \infty$. From the above inequality, we have

$$-\int_{\sigma}^{\infty} t^2 c(t)\,dt < \infty \quad \text{and} \quad -\int_{\sigma}^{\infty} tc(t) < \infty. \tag{2.152}$$

By Remark 2.7.6, there exists a positive nonincreasing solution $h(t)$ of (2.94) such that $\lim_{t\to\infty} h(t) = l_h > 0$. Thus $h(\sigma) \geq h(t) \geq l_h$ and

$$\int_{\sigma}^{t} \frac{1}{h^2(s)} \int_{\sigma}^{s} h(u)\,du\,ds \leq kt^2,$$

where $k = h(\sigma)/2l_h^2$. Consequently, by using (2.152), we obtain

$$\int_\sigma^\infty c(t)h(t)\int_\sigma^t \frac{1}{h^2(s)}\int_\sigma^s h(u)\,du\,ds\,dt$$
$$\le k\int_\sigma^\infty t^2 c(t)h(t)\,dt \le k\,h(\sigma)\int_\sigma^\infty t^2 c(t)\,dt < \infty,$$

and by Corollary 2.7.10, (2.150) is nonoscillatory, which is a contradiction, because (2.150) is the disconjugate form of (2.10). The theorem is proved. □

Finally, we consider the linear differential equation with quasiderivatives

$$L_3x(t) + q(t)x'(t) + p(t)x(t) = 0, \tag{2.153}$$

where $p(t)$ and $q(t)$ are continuous functions on $[\sigma,\infty)$, $L_3x = (L_2x)'$, $L_2x = r_2(t)(L_1x)'$, $L_1x = r_1(t)x'$, $L_0x = x$, and r_1 and r_2 are continuous and positive functions on $[\sigma,\infty)$.

Lemma 2.7.10 *Let $2p(t) - q'(t) \ge 0$ and $\frac{r_2(t)}{r_1(t)}$ be differentiable on $[\sigma,\infty)$, and $(\frac{r_2(t)}{r_1(t)})' \ge 0$ on $[\sigma,\infty)$. Then every solution of* (2.153) *with a double zero at some point $t_0 > \sigma$ has no simple zero in $[\sigma, t_0)$.*

Proof At first, we shall say that the solution $x(t)$ of (2.153) has a double zero at a point t_0 if $x(t_0) = 0$, $L_1x(t_0) = 0$ and $L_2x(t_0) \ne 0$. Define a function $w(t, x(t))$ by

$$w\big(t, x(t)\big) = \frac{1}{2}\left[\big(2p(t) - q'(t)\big)x^2(t) + \left(\frac{r_2(t)}{r_1(t)}\right)'\big(r_1(t)x'(t)\big)^2\right].$$

Multiplying (2.153) by $x(t)$, we obtain the identity

$$\left[r_2(t)\big(r_1(t)x'(t)\big)'x(t) - \frac{1}{2}\frac{r_2(t)}{r_1(t)}\big(r_1(t)x'(t)\big)^2 + \frac{1}{2}q(t)x^2(t)\right]' = -w\big(t, x(t)\big). \tag{2.154}$$

Without any loss of generality, we may suppose that $x(t_0) = 0$, $L_1x(t_0) = 0$ and $L_2x(t_0) > 0$. If there exists a $t_1 < t_0$ such that $x(t_1) = 0$, $L_1x(t_1) \ne 0$, then from $L_2x(t_0) > 0$, it follows that $x(t)$ has a minimum at t_0. Suppose that $x(t) > 0$ on (t_1, t_0). Then integrating the identity (2.154) over the interval $[t_1, t_0]$, we get

$$0 < \frac{1}{2}\frac{r_2(t_1)}{r_1(t_1)}\big(r_1(t_1)x'(t_1)\big)^2 = -\int_{t_0}^{t_1} w\big(t, x(t)\big)\,dt \le 0,$$

which is a contradiction. Therefore $x(t) > 0$ for $t \le t_0$. The proof is complete. □

Lemma 2.7.11 *Let the conditions of Lemma* 2.7.10 *hold. Then there exists a nonnegative solution of* (2.153).

Lemma 2.7.12 *Let the conditions of Lemma* 2.7.10 *hold. If* (2.153) *is oscillatory, then every solution of* (2.153), *which vanishes at least once is oscillatory.*

Theorem 2.7.21 *Let the conditions of Lemma* 2.7.10 *hold and* $b(t) \le 0$. *Then there exists a Kneser solution of* (2.153).

Lemma 2.7.13 *If* $q(t) \le 0$, $p(t) > 0$ *and* $x(t)$ *is any solution of* (2.153) *satisfying the condition* $x(t_0) \ge 0$, $L_1 x(t_0) \le 0$ *and* $L_2 x(t_0) > 0$ *for* $t_0 > \sigma$, *then* $x(t) > 0$, $L_1 x(t) < 0$ *and* $L_2 x(t) > 0$ *for* $\sigma \le t < t_0$.

Lemma 2.7.14 *If* $q(t) \le 0$, $p(t) > 0$ *and* (2.153) *is oscillatory, then every solution of* (2.153), *which vanishes at least once is oscillatory.*

Lemma 2.7.15 *If* $q(t) \le 0$, $p(t) > 0$, *then there exists a Kneser solution of* (2.153).

From Theorem 2.7.21, it follows that the equation $x''' + p(t)x = 0$ has Kneser-solution if $p(t) \ge 0$. This condition is sufficient for the uniqueness of a Kneser solution of $x''' + p(t)x = 0$ with $x(\sigma) = 1$. However, the conditions of Theorem 2.7.21 are not sufficient for the uniqueness of Kneser solution for (2.153). This is evident from the equation $x''' - 3x' + 2x = 0$, which has two Kneser solutions $x_1(t) = e^{-t}$ and $x_2(t) = te^{-t}$.

Theorem 2.7.22 *Let the conditions of Lemma* 2.7.10 *hold,* $b(t) \le 0$ *and*

$$\int^{\infty} \frac{1}{r_2(t)}\, dt = \infty \tag{2.155}$$

hold, and (2.153) *be oscillatory. Then there exists a unique Kneser solution of* (2.153) *such that*

$$x(\sigma) = 1, \qquad x(t) > 0, \qquad L_1 x(t) < 0 \quad \text{and} \quad L_2 x(t) \ge 0 \quad \text{for } t \ge \sigma. \tag{2.156}$$

Proof On the contrary, suppose that there exist two solutions $x_1(t)$ and $x_2(t)$ of (2.153) that satisfy (2.156). Then the solution $v(t) = x_1(t) - x_2(t)$ of (2.153) satisfies the condition $v(\sigma) = 0$. Since (2.153) is oscillatory, by Lemma 2.7.12, $v(t)$ is an oscillatory solution of (2.153). Hence there exists a sequence $\{t_n\}$ of zeros of $v(t)$ such that $\lim_{n\to\infty} t_n = \infty$. Integrating (2.154) over $[t_n, t_{n+1}]$, we get

$$r_2(t_{n+1}) r_1(t_{n+1}) \big(v'(t_{n+1})\big)^2 \ge r_2(t_n) r_1(t_n) \big(v'(t_n)\big)^2. \tag{2.157}$$

Now two cases may occur: (i) there exists n_0 such that $v'(n_0) \ne 0$ (and so $v'(t_n) \ne 0$ for $n > n_0$) or (ii) $v'(t_n) = 0$ for all n. Consider the case (i). If $v'(n_0) \ne 0$, then (2.157) implies that

$$\liminf_{t\to\infty} r_2(t_n) r_1(t_n) \big(v'(t_n)\big)^2 \ge c > 0. \tag{2.158}$$

Since $L_1 x_1 \le 0$, then $x_1'(t) \le 0$. Hence $\lim_{t\to\infty} r_2(t)x_1'(t) = 0$. Similarly

$$\lim_{t\to\infty} r_2(t)x_2'(t) = 0 \quad \text{and so} \quad \lim_{t\to\infty} r_2(t)v'(t) = 0.$$

By the second assumption of Lemma 2.7.10, we see that $r_1(t)/r_2(t)$ is nonincreasing. Then

$$\lim_{t\to\infty} r_2(t)r_1(t)\big(v'(t)\big)^2 = \lim_{t\to\infty} \frac{r_1(t)}{r_2(t)}\big(r_2(t)v'(t)\big)^2 = 0,$$

which contradicts (2.158). Now, we consider the case (ii). If $v'(t_n) = 0$ for all $n = 1, 2, 3, \ldots$, then with respect to the initial data, we have $v(t) \ge 0$ or ≤ 0 on $[\sigma, \infty)$. Let us suppose that $v(t) \ge 0$. Let $\{\tau_n\}$ be a sequence of zeros of $v'(t)$ such that $\tau_n < t_n < \tau_{n+1} < t_{n+1}$ and $v'(t) \ne 0$ on (τ_n, t_n). Since $v'(\tau_n) = v'(t_n) = 0$, there exists a number $\xi \in (\tau_n, t_n)$ such that $(r_1(t)v'(t))' = 0$ for $t = \xi$. Now if we integrate the equality (2.154), we shall obtain a contradiction. Therefore $v(t) \equiv 0$, that is, $x_1(t) = x_2(t)$ for $t \ge \sigma$. The theorem is proved. □

Corollary 2.7.13 *Let $r_1(t) = r_2(t) = r(t)$ and (2.155) hold. Let either*

(i) *$q(t) \ge k$ and $2p(t) - q'(t) \ge \frac{\alpha}{r(t)}$, where $\beta \le 0$ and $\alpha > \frac{4}{3\sqrt{3}}(-\beta)^{3/2}$, α, β are constants, or*

(ii) *$q(t) \ge \frac{\beta}{R^2(t)}$ and $2p(t) - q'(t) \ge \frac{\epsilon}{r(t)R^3(t)}$, where $\beta < 1, \epsilon > \frac{4}{3\sqrt{3}}(1-\beta)^3$ are constants and $R(t) = \int^t \frac{1}{r(s)}\,ds$.*

Then there exists a unique Kneser solution of (2.153).

Example 2.7.4 By Corollary 2.7.13(i), the equation

$$\big(t\big(tx'\big)'\big)' - 2x' + \frac{4}{t}x = 0$$

admits a unique Kneser solution. Note that $x(t) = t^{-2}$ on $[1, \infty)$ is the unique Kneser solution of the equation.

Lemma 2.7.16 *If $q(t) \le 0$, $p(t) > 0$ and*

$$\int^{\infty} \frac{1}{r_1(t)}\,dt = \infty. \tag{2.159}$$

Let $x(t)$ be a nonoscillatory solution of (2.153). Then either

(i) *$x(t)x'(t) \le 0$, $L_2x(t) \ge 0$ and $x(t) \ne 0$ on $[\sigma, \infty)$, or*
(ii) *there exists a number $t_0 \in [\sigma, \infty)$ such that $x(t)x'(t) \ge 0$ and $x(t) \ne 0$ for $t \ge t_0$.*

Theorem 2.7.23 *Let $q(t) \le 0$, $p(t) \ge 0$ and (2.155), (2.159) hold. Let m be the positive stationary point of the function*

$$F\big(x(t)\big) = h(t)\frac{1}{r_1(t)}x^3(t) - \frac{1}{2}h'(t)x^2(t) + q(t)\frac{1}{r_1(t)}x(t), \tag{2.160}$$

where $h(t) = \frac{r_1(t)}{r_2(t)}$ and

$$\int^{\infty}\big[p(t) + F\big(m(t)\big)\big]\,dt = \infty.$$

Then (2.153) has an oscillatory solution.

Proof Suppose that $u(t)$ is any nonoscillatory solution of (2.153). By Lemma 2.7.16, there exists a number T such that either

$$x(t) = r_1(t)\frac{u'(t)}{u(t)} \ge 0 \quad \text{for } t \ge T \tag{2.161}$$

or

$$x(t) = r_1(t)\frac{u'(t)}{u(t)} \le 0 \quad \text{for } t \ge T. \tag{2.162}$$

We assert that (2.161) is impossible. On the contrary, assume that (2.161) holds. The function $x(t)$ satisfies the second-order nonlinear differential equation

$$\left(r_2(t)x' + \frac{3}{2}h(t)x^2\right)' = -\big[F\big(x(t)\big) + p(t)\big], \tag{2.163}$$

where $F(x(t))$ is defined in (2.160). The function $F(x(t)) + p(t)$ has a minimum on the interval $[0, \infty)$ at the point

$$m(t) = \frac{1}{6h(t)}\left(r_1(t)h'(t) + \left(r_1^2(t)\big(h'(t)\big)^2 - 12h(t)q(t)\right)^{1/2}\right)$$

and

$$F\big(m(t)\big) = \frac{2}{3}m(t)\frac{q(t)}{r_1(t)} + \frac{1}{18}\frac{h'(t)}{h(t)}\left(q(t) - m(t)r_1(t)\frac{h'(t)}{h(t)}\right). \tag{2.164}$$

Then (2.163) gives

$$\left(r_2(t)x'(t) + \frac{3}{2}h(t)x^2(t)\right)' = -\big[F\big(x(t)\big) + p(t)\big] \le -\big[F\big(m(t)\big) + p(t)\big], \quad t \ge T.$$

Integrating the last inequality from T to t, we get

$$r_2(t)x'(t) \le r_2(t_0)x'(t_0) + \frac{3}{2}h(t_0)x^2(t_0) - \frac{3}{2}h(t)x^2(t) - \int_{t_0}^{t}\big(p(s) + F\big(m(s)\big)\big)\,ds,$$

which tends to ∞. Therefore,

$$\lim_{t\to\infty} r_2(t)\left(r_1(t)\frac{u'(t)}{u(t)}\right)' = -\infty \quad \text{as } t\to\infty$$

and

$$r_1(t)\frac{u'(t)}{u(t)} < -k^2\frac{1}{r_2(t)}, \qquad h \neq 0$$

eventually. Since $\int^\infty \frac{1}{r_2(t)}\,dt = \infty$, (2.161) is impossible and hence $u(t)u'(t) \le 0$ and $u(t) \neq 0$ on $[0,\infty)$. Since $u(t)$ was taken to be any nonoscillatory solution, it follows that (2.153) has an oscillatory solution. The proof is complete. □

Theorem 2.7.24 *Let $q(t) \le 0$, $p(t) \ge 0$, the conditions of Lemma* 2.7.10 *and the integral condition* (2.155) *hold, and*

$$\int^\infty \big(p(t) + F\big(m(t)\big)\big)\,dt = \infty$$

holds. Then (2.153) *has an unique Kneser solution.*

Corollary 2.7.14 *Let $q(t) \le 0$, $p(t) \ge 0$, $2p(t) - q'(t) \ge 0$, $\int^\infty \frac{1}{r(t)}\,dt = \infty$ and*

$$\int^\infty \left(p(t) - \frac{2}{3\sqrt{3}}\frac{(-q(t))^{3/2}}{r(t)}\right)dt = \infty.$$

Then there exists a unique Kneser solution of the differential equation

$$\big(r(t)\big(r(t)x'\big)'\big)' + q(t)x'(t) + p(t)x = 0.$$

2.8 Open Problems and Discussions

- Theorem 2.1.2 gives the sufficient condition that if (2.12) holds, then (2.1) is oscillatory. In view of Proposition 1.2.1(i), we propose the following: If (2.1) is oscillatory, then prove that (2.12) holds when $a(t) \ge 0$, $b(t) \le 0$ and $c(t) > 0$. In a similar way, can we prove that (2.55) holds if (2.1) is oscillatory, when $a(t) \le 0$, $b(t) \le 0$ and $c(t) > 0$? This problem is proposed in view of Proposition 1.2.2(i) and Theorem 2.2.1.
- Let $a(t) \le 0, b(t) \le 0$ and $c(t) < 0$. In view of Theorem 2.3.2 and Proposition 1.2.3(i), it would be interesting to prove the following open problem: If (2.1) is oscillatory, then (2.69) holds.
- Let $a(t) \ge 0, b(t) \le 0$ and $c(t) < 0$. The following assertions are yet to be established:

  (A) If (2.1) admits an oscillatory solution, then (2.79) holds.

(B) If (2.1) admits an oscillatory solution, then all oscillatory solutions of (2.1) tend to zero as $t \to \infty$.

Corollary 2.4.2 provides an indication in this direction.

(C) If (2.1) has an oscillatory solution, then all oscillatory solutions of (2.1) form a two-dimensional subspace of the solution space of (2.1).

We can prove (C) with the assumption that (B) holds.

Theorem 2.8.1 *Suppose that the existence of an oscillatory solution of* (2.1) *implies that all oscillatory solutions of* (2.1) *tends to zero as* $t \to \infty$. *If* $\int_\sigma^\infty p(t)\,dt = -\infty$, *and* (2.1) *admits an oscillatory solution, then all oscillatory solutions of* (2.1) *form a two-dimensional subspace of the solution space of* (2.1).

Proof From Theorem 2.4.6, it follows that (2.1) admits two linearly independent oscillatory solutions u and v whose linear combination is an oscillatory solution of (2.1). Let $x(t)$ be any oscillatory solution of (2.1). Theorem 2.4.1 yields the result that (2.1) admits a positive solution $x_0(t)$ such that $x_0(t) \to \infty$ as $t \to \infty$. Clearly $\{u, v, x_0\}$ is a basis of the solution space of (2.1). If possible, let $x(t) = c_1u(t) + c_2v(t) + c_3x_0(t)$, where c_1, c_2 and c_3 are reals such that $c_3 \neq 0$. Thus $x(t) \to \infty$ or $-\infty$ as $t \to \infty$ according to $c_3 > 0$ or < 0. In either case, we get a contradiction because $x(t)$ is oscillatory. Thus, $x(t)$ can be expressed as a linear combination of u and v, and hence the theorem is proved. □

- Let $a(t) \geq 0, b(t) \geq 0$ and $c(t) > 0$. The following assertions are yet to be established:
  - (i) Let $9c(t) - a(t)b(t) - b'(t) \geq 0$. If (2.1) is oscillatory, then it admits a nonoscillatory solution $x(t)$ such that $x(t)x'(t) > 0$ for $t \geq t_0 \geq \sigma$.
  - (ii) Let $9c(t) - a(t)b(t) - b'(t) \geq 0$, that is, (2.82) holds. If $\frac{a^2(t)}{3} - b(t) + a'(t) \geq 0$, then (2.1) is oscillatory.
  - (iii) Suppose that $c(t) \geq d > 0$, $2c(t) - a(t)b(t) - b'(t) \geq 0$ and $\int_\sigma^\infty a(t)\,dt = \infty$. If (2.1) is oscillatory, then every nonoscillatory solution of (2.1) satisfies
$$\lim_{t\to\infty} x(t) = \lim_{t\to\infty} x'(t) = \lim_{t\to\infty} x''(t) = 0.$$
  - (iv) Every oscillatory solution of (2.1) tends to zero as $t \to \infty$ under the assumption of Corollary 2.5.5.
  - (v) To generalise the observation in Proposition 1.2.5(vii).
  - (vi) To obtain a result similar to Theorem 2.5.6 for $ta(t) < 3$.
- To generalise the observations in Proposition 1.2.6.
- To generalise the observations in Proposition 1.2.7.
- Let $a(t) \leq 0, b(t) \geq 0$ and $c(t) < 0$. The following assertions are yet to be established:

  To generalise the assertions (i), (ii), (iii), (v), (vi) and (vii) of Proposition 1.2.8 to Eq. (2.1).

2.9 Notes

Lemmas 2.1.1 and 2.1.2 are the easy consequences of the definitions of Class I. Theorems 2.1.1, 2.1.2, 2.1.11–2.1.13 are due to Parhi and Das [28]. Theorems 2.1.4 and 2.1.5 are taken from Padhi [26], whereas Remarks 2.1.8 and 2.1.9 are adopted from Škerlik [36]. Theorems 2.1.7–2.1.9 are proved in Das [9]. Theorem 2.2.2 is taken from Padhi [26], whereas the rest part of Sect. 2.2 is brought from Parhi and Das [29]. Lemma 2.3.4 follows from Theorem 4.1 due to Hanan [16]. Theorems 2.3.3 and 2.3.4 are taken from Padhi [26]. The proof of Theorem 2.3.5 is as in proof of a theorem due to Jones [21], Theorem 2.3.11 is due to Ahmad and Lazer [1], and the rest part of Sect. 2.3 is taken from Parhi and Das [31]. Theorems 2.4.2 and 2.4.3 are due to Padhi [26], whereas Theorems 2.4.4–2.4.8, 2.4.11 and 2.4.12 are due to Parhi and Padhi [32]. Theorems 2.4.9–2.4.10, 2.8.1 and 2.4.13 are taken from Parhi and Das [30]. Theorems 2.5.1–2.5.11 and Lemma 2.5.1 are due to Parhi and Padhi [34]. Theorem 2.5.12 is taken from Padhi [27]. Theorem 2.5.16 follows from Theorem 4 due to Jones [20], whereas Theorem 2.5.17 follows from Theorem 1 by Jones [21]. Theorems 2.5.20 and 2.5.21 are due to Škerlik [37]. Theorem 2.6.5 is due to Gera [11]. Theorems 2.6.11 and 2.6.12 are taken from Dolan [10], and Theorem 2.6.13 is due to Parhi and Padhi [33], while Theorem 2.6.14 is adopted from Swanson [38]. The rest of Sect. 2.6 are due to Padhi [26]. Some of the results of Sect. 2.7 are adopted from [3–6] and Rovder [35].

References

1. S. Ahmad and A. C. Lazer; *On the oscillatory behavior of a class of linear third order differential equations*, Journal of Mathematical Analysis and Applications, 28(3) (1970), 681–689.
2. M. Cecchi and M. Marini; *On the oscillatory behavior of a third order nonlinear differential equation*, Nonlinear Analysis; Theory Methods and Applications, 15(2) (1990), 141–153.
3. M. Cecchi, Z. Doslá, M. Marini and G. Villari; *On the qualitative behavior of solutions of third order differential equations*, Journal of Mathematical Analysis and Applications, 197(3) (1996), 749–766.
4. M. Cecchi, Z. Doslá and M. Marini; *An equivalence theorem of properties A, B for third order differential equations*, Annali di Matematica Pura ed Applicata, 173(1) (1997), 373–389.
5. M. Cecchi, Z. Doslá and M. Marini; *Some properties of third order differential operators*, Czechoslovak Mathematical Journal, 47(122) (1997), 729–748.
6. M. Cecchi, Z. Doslá and M. Marini; *On third order differential equations with Property A and B*, Journal of Mathematical Analysis and Applications, 231(2) (1999), 509–525.
7. T. A. Chanturiya; *Integral tests for oscillation of solutions of higher order linear differential equations I* (In Russian), Differentsial'nye Uravneniya, 16(3) (1980) 470–482.
8. T. A. Chanturiya; *Integral tests for oscillation of solutions of higher order linear differential equations II* (In Russian), Differentsial'nye Uravneniya, 16(4) (1980) 635–644.
9. P. Das; *Contributions to the Study of Qualitative Behaviour of Solutions of Third-Order Differential Equations*, Ph.D. Thesis, Berhampur University, India, 1991.
10. J. M. Dolan; *On the relationship between the oscillatory behaviour of a linear third-order differential equation and its adjoint*, Journal of Differential Equations, 7 (1970), 367–388.
11. M. Gera; *Some oscillation conditions for a linear third order differential equations*, Soviet Mathematics. Doklady, 36(2) (1988), 277–280.

12. M. Gera; *On the behaviour of solutions of the differential equation* $x''' + a(t)x'' + b(t)x' + c(t)x = 0$, Habilitation Thesis (in Slovak), Faculty of Mathematics and Physics, Comenius Univ., Bratislava.
13. M. Greguš; *Oscillatory properties of solutions of a third order differential equation of the type* $y''' + 2A(x)y' + [A'(x) + b(x)]y = 0$, Acta Facultatis Rerum Naturalium Universitatis Comenianae, Mathematica, 6 (1961), 275–300.
14. M. Greguš; *Third Order Linear Differential Equations*, D. Reidel Publishing Company, Boston, 1987.
15. M. Greguš, J.R. Graef and M. Gera; *Oscillating nonlinear third order differential equations*, Nonlinear Analysis; Theory, Methods and Applications, 28(10) (1997), 1611–1622.
16. M. Hanan; *Oscillation criteria for third order linear differential equations*, Pacific Journal of Mathematics, 11 (1961), 919–944.
17. P. Hartman; *Ordinary Differential Equations*, Wiley, New York, 1964, and Birkhäuser, Boston, 1982.
18. J. W. Heidel; *Qualitative behaviour of solutions of a third order nonlinear differential equation*, Pacific Journal of Mathematics, 27 (1968), 507–526.
19. G. D. Jones; *An asymptotic property of solutions of* $y''' + py' + qy = 0$, Pacific Journal of Mathematics, 48(1) (1973), 135–138.
20. G. D. Jones; *Oscillatory behaviour of third order differential equations*, Proceedings of American Mathematical Society, 43 (1974), 133–136.
21. G. D. Jones; *Properties of solutions of a class of third order differential equations*, Journal of Mathematical Analysis and Applications, 48(1) (1974), 165–169.
22. I. T. Kiguradze and T. A. Chanturia; *Asymptotic Properties of Solutions of Nonautonomous Ordinary Differential Equations*, Kluwer Academic Publishers, Dordrecht, 1993.
23. A. C. Lazer; *The behaviour of solutions of the differential equation* $y''' + p(x)y' + q(x)y = 0$, Pacific Journal of Mathematics, 17 (1966), 435–466.
24. F. Neumann; *On two problems about oscillation of linear differential equations of third order*, Journal of Differential Equations, 15 (1974), 589–596.
25. J. Ohriska; *On the oscillation of a linear differential equation of second order*, Czechoslovak Mathematical Journal, 39 (114) (1989), 16–23.
26. S. Padhi; *Contributions to the Oscillation Theory of Ordinary and Delay Differential Equations of Third Order*, Ph.D. Thesis, Berhampur University, India, 1998.
27. S. Padhi; *On the asymptotic behaviour of solutions of third order delay differential equations*, Georgian Mathematical Journal, 12 (2005), 369–376.
28. N. Parhi and P. Das; *Asymptotic property of solutions of a class of third order differential equations*, Proceedings of American Mathematical Society, 110 (1990), 387–393.
29. N. Parhi and P. Das; *On asymptotic property of solutions of linear homogeneous third order differential equations*, Bollettino dell'Unione Matematica Italiana) 7-B (1993), 775–786.
30. N. Parhi and P. Das; *On nonoscillation of third order differential equations*, Bulletin of the Institute of Mathematics Academia Sinica, 22 (1994), 267–274.
31. N. Parhi and P. Das; *On the oscillation of a class of linear homogeneous third order differential equations*, Archivum Mathematicum, 34(4) (1998), 435–443.
32. N. Parhi and S. Padhi; *On oscillation and asymptotic property of a class of third order differential equations*, Czechoslovak Mathematical Journal, 49(124) (1999), 21–33.
33. N. Parhi and S. Padhi; *On oscillatory linear differential equations of third order*, Archivum Mathematicum, 37(1) (2001), 33–38.
34. N. Parhi and S. Padhi; *On oscillatory linear third order differential equations*, Journal of the Indian Mathematical Society, 69 (2002), 113–128.
35. J. Rovder; *On monotone solution of the third-order differential equation*, Journal of Computational and Applied Mathematics, 66(1–2) (1996), 421–432.
36. T. L. Sherman; *Conjugate points and simple zeros for ordinary linear differential equations*, Transactions of the American Mathematical Society, 146 (1969), 397–411.
37. A. Škerlik; *An integral condition of oscillation for equation* $y''' + p(t)y' + q(t)y = 0$ *with nonnegative coefficients*, Archivum Mathematicum, 31(2) (1995), 155–161.
38. C. A. Swanson; *Comparison and Oscillation Theory of Linear Differential Equations*, Academic Press, New York, 1968.

Chapter 3
Oscillation of Solutions of Linear Nonhomogeneous Differential Equations of Third Order

This chapter is concerned with the oscillatory and nonoscillatory behaviour of solutions of linear nonhomogeneous third-order differential equations of the form

$$x''' + a(t)x'' + b(t)x' + c(t)x = f(t), \tag{3.1}$$

where a, b and c are defined as in Chap. 2, $f \in C([\sigma, \infty), R)$.

Clearly, (3.1) is equivalent to

$$\big(r(t)x''\big)' + q(t)x' + p(t)x = F(t), \tag{3.2}$$

where $r(t) = \exp(\int_\sigma^t a(s)\,ds)$, $q(t) = r(t)b(t)$, $p(t) = r(t)c(t)$ and $F(t) = r(t)f(t)$.

For the rest of this chapter, we assume that $p(t) \neq 0$, $F(t) \neq 0$ and r is sufficiently smooth. Note that the homogeneous equations associated with Eqs. (3.1) and (3.2) are given by (2.1) and (2.2), respectively.

Let $\{u_1(t), u_2(t), u_3(t)\}$ be a solution basis of (2.2) and

$$W(t) = W(u_1, u_2, u_3)(t) = \begin{vmatrix} u_1(t) & u_2(t) & u_3(t) \\ u_1'(t) & u_2'(t) & u_3'(t) \\ r(t)u_1''(t) & r(t)u_2''(t) & r(t)u_3''(t) \end{vmatrix} = 1.$$

Then the general solution of (3.2) is given by

$$x(t) = c_1u_1(t) + c_2u_2(t) + c_3u_3(t) + x_p(t),$$

where c_1, c_2 and c_3 are constants and $x_p(t)$ is a particular solution of (3.2), and is given by

$$x_p(t) = \int_\sigma^t \frac{1}{W(s)} \begin{vmatrix} u_1(t) & u_2(t) & u_3(t) \\ u_1(s) & u_2(s) & u_3(s) \\ u_1'(s) & u_2'(s) & u_3'(s) \end{vmatrix} f(s)\,ds.$$

Clearly, $x_p(\sigma) = 0$, $x_p'(\sigma) = 0$ and $x_p''(\sigma) = 0$. This shows that (3.2) admits a solution with three zeros, counting multiplicity.

S. Padhi, S. Pati, *Theory of Third-Order Differential Equations*,
DOI 10.1007/978-81-322-1614-8_3, © Springer India 2014

3.1 Nonoscillatory Behaviour of Solutions of Nonhomogeneous Linear Differential Equations

This section is concerned with Eq. (3.1) and its corresponding homogeneous equation (2.1). It has been proved under certain conditions that the nonoscillation of (2.1) implies the nonoscillation of (3.1). If $W(t) = 1$, then

$$x_p(t) = \int_\sigma^t \begin{vmatrix} u_1(t) & u_2(t) & u_3(t) \\ u_1(s) & u_2(s) & u_3(s) \\ u_1'(s) & u_2'(s) & u_3'(s) \end{vmatrix} f(s)\,ds.$$

If $W_i(t)$ denotes the determinant obtained from $W(u_1, u_2, u_3)(t)$ by replacing the ith column with the vector $(0, 0, 1)^T$, $i = 1, 2, 3$, then

$$x_p(t) = \sum_{i=1}^{3} u_i(t) \int_\sigma^t f(s) W_i(s)\,ds$$

and

$$x(t) = \sum_{i=1}^{3} u_i(t) \left[c_i + \int_\sigma^t f(s) W_i(s)\,ds \right].$$

Lemma 3.1.1 *If $x(t)$ is a solution of (3.1) such that*

$$x(t) = \sum_{i=1}^{3} c_i u_i(t) + x_p(t),$$

then

$$r(t)W_i(t)x''(t) - r(t)W_i'(t)x'(t) + \left[q(t)W_i(t) + \left(r(t)W_i'(t)\right)'\right]x(t)$$
$$= c_i + \int_\sigma^t f(s) W_i(s)\,ds, \quad i = 1, 2, 3. \tag{3.3}$$

Proof One may see that $S_i(t) = c_i + \int_\sigma^t f(s) W_i(s)\,ds$ can be obtained by replacing the ith column of $W(u_1, u_2, u_3)(t)$ with the vector $(x(t), x'(t), r(t)x''(t))^T$. Indeed, $x_p(\sigma) = 0$, $x_p'(\sigma) = 0$, and $x_p''(\sigma) = 0$. For $i = 1$, we write

$$S_1(t) = \begin{vmatrix} x(t) & u_2(t) & u_3(t) \\ x'(t) & u_2'(t) & u_3'(t) \\ r(t)x''(t) & r(t)u_2''(t) & r(t)u_3''(t) \end{vmatrix}.$$

Then

$$S_1'(t) = \begin{vmatrix} x(t) & u_2(t) & u_3(t) \\ x'(t) & u_2'(t) & u_3'(t) \\ (r(t)x''(t))' & (r(t)u_2''(t))' & (r(t)u_3''(t))' \end{vmatrix}$$

$$= \begin{vmatrix} x(t) & u_2(t) & u_3(t) \\ x'(t) & u_2'(t) & u_3'(t) \\ f(t) - q(t)x'(t) - p(t)x(t) & -q(t)u_2'(t) - p(t)u_2(t) & -q(t)u_3'(t) - p(t)u_3(t) \end{vmatrix}$$

$$= \begin{vmatrix} x(t) & u_2(t) & u_3(t) \\ x'(t) & u_2'(t) & u_3'(t) \\ f(t) & 0 & 0 \end{vmatrix} = f(t)W_1(t)$$

implies that

$$S_1(t) = S_1(\sigma) + \int_\sigma^t f(s)W_1(s)\,ds.$$

But $S_1(\sigma) = c_1 W(\sigma) = c_1$. Consequently,

$$S_1(t) = c_1 + \int_\sigma^t f(s)W_1(s)\,ds.$$

Now expanding $S_1(t)$, we obtain (3.3) for $i = 1$. Similarly, one may obtain (3.3) for $i = 2$ and 3. This completes the proof of lemma. □

Theorem 3.1.1 *Suppose that $W_1(t)$ is nonoscillatory. If $x(t)$ is a solution of* (3.1) *such that $x(t) = \sum_{i=1}^{3} c_i u_i(t) + x_p(t)$, then it is a solution of the second-order nonhomogeneous differential equation*

$$\big(R(t)x'\big)' + Q(t)x = F(t), \tag{3.4}$$

where

$$R(t) = \frac{1}{W_1(t)}, \qquad Q(t) = \frac{q(t)W_1(t) + (r(t)W_1'(t))'}{r(t)W_1^2(t)}$$

and

$$F(t) = \frac{1}{r(t)W_1^2(t)}\left[c_1 + \int_\sigma^t f(s)W_1(s)\,ds\right].$$

Theorem 3.1.2 *If $W_1(t)$ is nonoscillatory, then $u_2(t)$ and $u_3(t)$ satisfy*

$$\big(R(t)x'\big)' + Q(t)x = 0. \tag{3.5}$$

Proof Clearly, $u_2(t)$ and $u_3(t)$ are solutions of the second-order differential equation

$$\begin{vmatrix} u_2(t) & u_3(t) & x \\ u_2'(t) & u_3'(t) & x' \\ r(t)u_2''(t) & r(t)u_3''(t) & r(t)x'' \end{vmatrix} = 0.$$

Expanding the determinant, we obtain (3.4). This completes the proof of the theorem. □

Theorem 3.1.3 *If $W_1(t)$ is nonoscillatory and*

$$\big(R(t)x'\big)' + Q(t)x = F_c(t) \tag{3.6}$$

is nonoscillatory for every constant c, where $R(t)$ and $Q(t)$ are same as in Theorem 3.1.1, *and*

$$F_c(t) = \frac{1}{r(t)W_1^2(t)}\left[c + \int_\sigma^t f(s)W_1(s)\,ds\right], \tag{3.7}$$

then (3.1) *is nonoscillatory.*

Remark 3.1.1 The adjoint of (2.2) is given by (2.49). Since $q(t)$ is differentiable, (2.49) takes the form

$$\big(r(t)x'\big)'' + q(t)x' + \big(q'(t) - p(t)\big)x = 0. \tag{3.8}$$

It is easy to verify that $W_1(t)$ satisfies (3.8) and

$$\big(r(t)z'\big)' + q(t)z = g(t),$$

where $g(t) = r(t)(u_2'(t)u_3''(t) - u_3'(t)u_2''(t))$.

Theorem 3.1.4 *Suppose that* (2.2) *is of type C_I or C_{II}, and $f(t)$ does not change sign for large t. If* (2.2) *is nonoscillatory, then* (3.1) *is nonoscillatory.*

Proof From Lemma 1.5.11 it follows that (3.8) is nonoscillatory. So $W_1(t)$ is nonoscillatory. Suppose that $W_1(t) \neq 0$ for $t > t_0 \geq \sigma$. Further, from Theorem 3.1.2 it is clear that (3.5) is nonoscillatory. Suppose that $f(t) > 0$ or < 0 for $t \geq t_1 \geq t_0$. For any constant c,

$$h(t) = c + \int_\sigma^t f(s)W_1(s)\,ds = \lambda + \int_{t_1}^t f(s)W_1(s)\,ds$$

is nonoscillatory, where

$$\lambda = c + \int_\sigma^{t_1} f(s)W_1(s)\,ds.$$

Indeed, if t_2 and t_3 ($t_1 \leq t_2 < t_3$) are consecutive zeros of $h(t)$, then $h'(\theta) = 0$ for some $\theta \in (t_2, t_3)$, that is, $f(\theta)W_1(\theta) = 0$, a contradiction. From Theorem 3 due to Keener [6], it is clear that (3.6) is nonoscillatory. Hence the theorem follows from Theorem 3.1.3. □

Example 3.1.1 Consider

$$x''' + e^{-t}x' + 2e^{-t}x = 3 + e^t, \quad t \geq 0. \tag{3.9}$$

From Lemma 2.5.1, it follows that the homogeneous equation associated with Eq. (3.9)

$$x''' + e^{-t}x' + 2e^{-t}x = 0 \tag{3.10}$$

is of type C_I and by Theorem 3.5 due to Lazer [7], all solutions of the homogeneous equation (3.10) are nonoscillatory. Hence by Theorem 3.1.4, all solutions of Eq. (3.9) are nonoscillatory. In particular, $x(t) = e^t$ is a nonoscillatory solution of (3.9).

Remark 3.1.2 (i) Hanan [3] and Lazer [7] have obtained various sufficient conditions for nonoscillation of (2.2).

(ii) Although Keener [6] has proved his Theorem 3 for $r(t) \equiv 1$, $p(t) \geq 0$ and $f(t) \geq 0$, his result holds good for

$$\big(r(t)x'\big)' + p(t)x = f(t),$$

where $r(t) > 0$ and $f(t)$ does not change sign for large t. There is no sign restriction on $p(t)$.

If $f(t)$ is allowed to change sign, then (3.6) is nonoscillatory provided that (3.5) is nonoscillatory and

$$\int_\sigma^t F_c(s)u(s)\,ds \tag{3.11}$$

is nonoscillatory, where $u(t)$ is a solution of (3.5). Hence we have the following result:

Theorem 3.1.5 *Suppose that* (2.2) *is of type C_I or C_{II} and* (3.11) *is nonoscillatory, where $u(t)$ is a solution of* (3.5). *Then* (2.2) *is nonoscillatory implies that* (3.2) *is nonoscillatory.*

Lemma 3.1.2 *If $u(t)$ is a nonoscillatory solution of* (2.2) *and $v(t)$ is a solution of* (3.2), *then $(u(t)v'(t) - u'(t)v(t))$ is a solution of*

$$\left(\frac{r(t)y'}{u(t)}\right)' + \left(\frac{r(t)u''(t) + q(t)u(t)}{u^2(t)}\right)y = f(t). \tag{3.12}$$

Lemma 3.1.3 (Jones [5]) *If $u(t)$ is a nonoscillatory solution of* (2.2), *then there exist two linearly independent solutions of* (3.8) *which satisfy*

$$\left(\frac{r(t)y'}{u(t)}\right)' + \left(\frac{r(t)u''(t) + q(t)u(t)}{u^2(t)}\right)y = 0, \tag{3.13}$$

the homogeneous equation associated with (3.12).

Theorem 3.1.6 *If $u(t)$ is a nonoscillatory solution of* (2.2) *and Eq.* (3.12) *is nonoscillatory, then* (3.2) *is nonoscillatory.*

Proof If possible, let $x(t)$ be an oscillatory solution of (3.2). From Lemma 3.1.2, it follows that $y(t) = u(t)x'(t) - u'(t)x(t)$ is a solution of (3.12). Since $y(t) = u^2(t)(\frac{x(t)}{u(t)})'$ is oscillatory, it contradicts our assumption that (3.12) is nonoscillatory. Thus Eq. (3.2) is nonoscillatory. Hence the theorem is completed. □

Remark 3.1.3 The existence of a nonoscillatory solution of (2.2) has been considered by Lazer [7] for three cases, viz., (i) $p(t) > 0$ and $q(t) \leq 0$, (ii) $p(t) \leq 0$ and $q(t) \leq 0$ and (iii) $p(t) \geq 0$ and $q(t) \geq 0$ such that $2p(t) - q'(t) \geq 0$. The existence problem for the fourth case, viz., $p(t) \leq 0$ and $q(t) \geq 0$ has been considered by Parhi [18]. Further, one may refer to Chap. 2 for the existence of a nonoscillatory solution of (2.2).

Theorem 3.1.7 *Suppose that* (2.2) *is of type C_I or C_{II} and $f(t)$ does not change sign for large t. If* (2.2) *is nonoscillatory, then* (3.2) *is nonoscillatory.*

Proof From Lemma 1.5.11, it follows that (3.8) is nonoscillatory. Suppose that $u(t)$ is a nonoscillatory solution of (3.3). Using Lemma 3.1.3, (3.13) is nonoscillatory. This, in turn, implies that (3.12) is nonoscillatory (see Remark 3.1.2(ii)). The conclusion of the theorem follows from Theorem 3.1.6. □

Example 3.1.2 Consider

$$x''' - e^{-t}x' + e^{-t}x = e^t, \quad t \geq 0.$$

From Theorem 2.1.9, it follows that $x''' - e^{-t}x' + e^{-t}x = 0$ is nonoscillatory. Clearly, this equation is of type C_I. Hence all solutions of the nonhomogeneous equation are nonoscillatory. In particular, $x(t) = e^t$ is a nonoscillatory solution of the equation.

Parhi and Parhi [16] have obtained several nonoscillation criteria for the nonoscillation of third-order nonlinear and nonhomogeneous differential equations which, in particular, are applicable to third-order linear nonhomogeneous differential equations of the form (3.1) or (3.2). Some simple calculations show that the theorem in [16] cannot be applied to the nonhomogeneous equation considered in Example 3.1.2.

If $f(t)$ changes sign, then a theorem similar to Theorem 3.1.5 holds.

Theorem 3.1.8 *Suppose that* (2.2) *is of type C_I or C_{II} and*

$$\int_\sigma^t f(s)u(s)\,ds$$

is nonoscillatory, where $u(t)$ is a solution of (3.13). *Then the nonoscillation of* (2.2) *implies the nonoscillation of* (3.2).

Next, we deal with the distribution of zeros of solutions of (3.1) or (3.2).

By saying that Eq. (3.1) or (3.2) does not admit a solution having $(2,2)$-distribution of zeros, we mean that if $x(t)$ is a solution of (3.1) or (3.2) on $[\sigma,\infty)$, then there exist no t_1 and $t_2 \in [\sigma,\infty)$, $t_1 < t_2$, such that $x(t_1) = 0 = x'(t_1)$, $x(t_2) = 0 = x'(t_2)$ and $x(t) > 0$ or < 0 for $t \in (t_1, t_2)$.

Lemma 3.1.4 *Suppose that x and $z \in C^1((a,b), R)$ such that z is of constant sign in (a,b). If α and β $(a < \alpha < \beta < b)$ are consecutive zeros of x, then there exists a constant $\mu \neq 0$ such that the function $\mu z - x$ has a double zero at $t^* \in (\alpha,\beta)$ and is of constant sign in $(t^*, \beta]$.*

Proof Let $x(t) > 0$ for $t \in (\alpha,\beta)$ and $z(t) > 0$ for $t \in (a,b)$. From Lemma 1.5.13, it follows that there exists a constant $\lambda \neq 0$ such that $\lambda z(t) - x(t)$ has a double zero at $t_0 \in (\alpha,\beta)$. So $\lambda = \frac{x(t_0)}{z(t_0)} > 0$. If $\lambda z(t) - x(t) > 0$ for $t \in (t_0,\beta]$, then we take $\mu = \lambda$ and $t^* = t_0$. Otherwise, since $\lambda z(\beta) - y(\beta) > 0$, there exists $t_0' \in (t_0,\beta)$ such that $\lambda z(t_0') - x(t_0') = 0$, $\lambda z(t) - x(t) > 0$ for $t \in (t_0',\beta)$ and there exists $t_1 \in [t_0, t_0')$ such that $\lambda z(t_1) - x(t_1) = 0$ and $\lambda z(t) - x(t) < 0$ for $t \in (t_1, t_0')$. It is possible to find $\gamma \geq \lambda$ and $t_\gamma \in (t_1, t_0']$ such that $\gamma z(t_\gamma) - x(t_\gamma) = 0$ and $\gamma z(t) - x(t) > 0$ for $t \in (t_\gamma, \beta]$. Set $t^* = \inf t_\gamma$ and $\mu = x(t^*)/z(t^*)$. So $\mu z(t) - x(t) = 0$ for $t = t^*$. Clearly, there exists a sequence $\langle t_{\gamma_n}\rangle \subset (t_1, t_0']$ such that $t_{\gamma_n} \to t^*$ as $n \to \infty$, $\gamma_n \geq \lambda$, $\gamma_n z(t_{\gamma_n}) - x(t_{\gamma_n}) = 0$ and $\gamma_n z(t) - x(t) > 0$ for $t \in (t_{\gamma_n}, \beta]$. So $\gamma_n \to \mu$ as $n \to \infty$, $\mu \geq \lambda$ and $\mu z(t) - x(t) > 0$ for $t \in (t^*, \beta]$. Next, we show that $\mu z'(t^*) - x'(t^*) = 0$. If possible, let $\mu z'(t^*) - x'(t^*) < 0$. So there exists a $\delta > 0$ such that $\mu z'(t) - x'(t) < 0$ for $t \in [t^*, t^* + \delta)$. Hence, for $t \in (t^*, t^* + \delta)$, $\mu z(t) - x(t) < \mu z(t^*) - x(t^*) = 0$, a contradiction. Suppose that $\mu z'(t^*) - x'(t^*) > 0$. So there exists a $\delta > 0$ such that $\mu z'(t) - x'(t) > 0$ for $t \in (t^* - \delta, t^*]$. This in turn implies that $\mu z(t) - x(t) < 0$ for $t \in (t^* - \delta, t^*)$. It is possible to choose $\epsilon > 0$ such that $\mu z(\tilde{t}) - x(\tilde{t}) + \epsilon z(\tilde{t}) < 0$, where $\tilde{t} \in (t^* - \delta, t^*)$. Now $\mu z(t^*) - x(t^*) + \epsilon z(t^*) > 0$ implies that there exists a $t' \in (\tilde{t}, t^*)$ such that $\mu z(t') - x(t') + \epsilon z(t') = 0$ and $\mu z(t) - x(t) + \epsilon z(t) > 0$ for $t \in (t', t^*)$, that is, $(\mu + \epsilon)z(t') - x(t') = 0$ and $(\mu + \epsilon)z(t) - x(t) > 0$ for $t \in [t', t^*]$. Since $(\mu + \epsilon)z(t) - x(t) > 0$ for $t \in [t^*, \beta]$, we have $(\mu + \epsilon)z(t) - x(t) > 0$ for $t \in (t', \beta]$. This contradicts the fact that t^* is the inf of $t_\gamma \in (t_1, t_0']$ with the prescribed property. Hence $\mu z'(t^*) - x'(t^*) = 0$.

If $x(t) < 0$ for $t \in (\alpha,\beta)$ and $z(t) < 0$ for $t \in (a,b)$, then set $u(t) = -x(t)$ and $v(t) = -z(t)$.

From the above discussion, it follows that there exists a constant $\mu > 0$ such that $\mu v(t) - u(t)$ has a double zero at $t^* \in (\alpha,\beta)$ and $\mu v(t) - u(t) > 0$ for $t \in (t^*, \beta]$, that is, $\mu z(t) - x(t)$ has a double zero at $t^* \in (\alpha,\beta)$ and $\mu z(t) - x(t) < 0$ for $t \in (t^*, \beta]$. Other two cases may be dealt with similarly. This completes the proof of the lemma. □

Remark 3.1.4 Note that $\mu > 0$ when both $x(t)$ and $z(t)$ have the same sign in respective intervals and $\mu < 0$ when $x(t)$ and $z(t)$ have opposite signs in respective intervals.

Lemma 3.1.5 *Suppose that* (2.2) *is of type* C_I *and* (3.2) *does not admit a solution with a* $(2, 2)$*-distribution of zeros. Then Eq.* (3.2) *does not admit a solution with a* $(1, 1, 2)$*-distribution of zeros.*

Proof If possible, let $x(t)$ be a solution of (3.2) such that $x(\alpha_1) = x(\alpha_2) = 0$, $x(\alpha_3) = x'(\alpha_3) = 0$, where $\sigma < \alpha_1 < \alpha_2 < \alpha_3$, and $x(t) \neq 0$ for $t \in (\alpha_1, \alpha_2)$ and $t \in (\alpha_2, \alpha_3)$. Let $z(t)$ be a solution of (2.2) with $z(\alpha_3) = 0 = z'(\alpha_3)$ and $z''(\alpha_3) > 0$. Since (2.2) is of type C_I, we have $z(t) > 0$ for $t \in (\sigma, \alpha_3)$. From Lemma 3.1.4, it follows that there exists a constant $\mu \neq 0$ such that $\mu z(t) - x(t)$ has a double zero at $t^* \in (\alpha_1, \alpha_2)$ and is of constant sign in $(t^*, \alpha_2]$. Clearly $x(t) - \mu z(t)$ is a solution of (3.2) with double zeros at $t = t^*$ and $t = \alpha_3$. From Remark 3.1.4, it is clear that $\mu > 0$ if $x(t) > 0$ for $t \in (\alpha_1, \alpha_2)$, and $\mu < 0$ if $x(t) < 0$ for $t \in (\alpha_1, \alpha_2)$. Further, $x(t) < 0$ for $t \in (\alpha_2, \alpha_3)$ if $x(t) > 0$ for $t \in (\alpha_1, \alpha_2)$, and $x(t) > 0$ for $t \in (\alpha_2, \alpha_3)$ if $x(t) < 0$ for $t \in (\alpha_1, \alpha_2)$. Since $z(t) > 0$ for $t \in (\sigma, \alpha_3)$, it is clear that $x(t) - \mu z(t) \neq 0$ for $t \in (\alpha_2, \alpha_3)$. So (3.2) admits a solution $x(t) - \mu z(t)$, having $(2, 2)$-distribution of zeros, a contradiction. Hence the lemma is proved. □

Theorem 3.1.9 *Suppose that* (2.2) *is of type* C_I *and* (3.2) *does not admit a solution with a* $(2, 2)$*-distribution of zeros. If* (2.2) *is oscillatory, then* (3.2) *is oscillatory.*

Proof If possible, suppose that (3.2) is nonoscillatory. Let $x(t)$ be an oscillatory solution of (2.2) and $z(t)$ be a nonoscillatory solution of (3.2). So there exists a $t_0 > \sigma$ such that $z(t) > 0$ or $z(t) < 0$ for $t \geq t_0$. Suppose that $z(t) > 0$ for $t \geq t_0$. The case $z(t) < 0$ for $t \geq t_0$ may be treated in a similar way.

Let α and β $(t_0 < \alpha < \beta)$ be consecutive zeros of $x(t)$ such that $x(t) > 0$ for $t \in (\alpha, \beta)$. From Lemma 1.5.13, it follows that there exists a $\lambda_1 > 0$ such that $z(t) - \lambda_1 x(t)$ has a double zero at $t_1 \in (\alpha, \beta)$. Clearly, $z(t) - \lambda_1 x(t)$ is a solution of (3.2) and hence is nonoscillatory. We claim that there exists a point $t_2 > t_1$ such that $z(t) - \lambda_1 x(t) > 0$ for $t \geq t_2$. Indeed, if $z(t) - \lambda_1 x(t) < 0$ for $t \geq t_2$, then $0 < z(t) < \lambda_1 x(t)$ for $t \geq t_2$. This contradicts the fact that $x(t)$ is oscillatory. Let α_1 and β_1 $(t_2 < \alpha_1 < \beta_1)$ be two consecutive zeros of $x(t)$ such that $x(t) > 0$ for $t \in (\alpha_1, \beta_1)$. Hence there exists a $\lambda_2 > 0$ such that $z(t) - \lambda_2 x(t)$ has a double zero at $t_3 \in (\alpha_1, \beta_1)$. Now $z(t_3) - \lambda_1 x(t_3) > 0$ and $z(t_3) - \lambda_2 x(t_3) = 0$ imply that $\lambda_2 > \lambda_1$. This in turn implies that $z(t_1) - \lambda_2 x(t_1) < 0$. However, $z(t) - \lambda_2 x(t)$ is positive at $t = \alpha$ and $t = \beta$. Since $z(t) - \lambda_2 x(t)$ is continuous, it has at least two zeros in (α, β). So (3.2) admits a solution $z(t) - \lambda_2 x(t)$ which has $(1, 1, 2)$-distribution of zeros, contradicting Lemma 3.1.5. Thus the theorem is proved. □

Corollary 3.1.1 *Suppose that* $f(t)$ *does not change sign for large* t*, Eq.* (2.2) *is of type* C_I *and* (3.2) *does not admit a solution with* $(2, 2)$*-distribution of zeros. Then* (2.2) *is oscillatory, if and only if* (3.2) *is oscillatory.*

Now we give some sufficient conditions in terms of coefficient functions and the forcing term so that (2.2) is of type C_I and (3.2) does not admit a solution with a $(2, 2)$-distribution of zeros.

Theorem 3.1.10 *Suppose that* $p(t) \geq 0$, $p'(t) \geq 0$, $f(t) \geq 0$ *and* $f'(t) \leq 0$. *If*

$$\big(r(t)z'\big)' + q(t)z = 0 \tag{3.14}$$

is nonoscillatory, then (2.2) *is of type* C_I *and* (3.2) *does not admit a solution with* $(2,2)$*-distribution of zeros.*

Proof Let $x(t)$ be a solution of (2.2) with $x(\alpha) = x'(\alpha) = 0$ and $x''(\alpha) > 0$, $\alpha > \sigma$. We claim that $x(t) > 0$ for $t \in [\sigma, \alpha)$. If not, there exists a point $\beta \in [\sigma, \alpha)$ such that $x(\beta) = 0$ and $x(t) > 0$ for $t \in (\beta, \alpha)$. So, there is a point $\alpha_1 \in (\beta, \alpha)$ such that $x'(\alpha_1) = 0$ and $x'(t) < 0$ for $t \in (\alpha_1, \alpha)$. Now multiplying (2.2) through by $x'(t)$ and integrating the resulting identity from α_1 to α, we obtain

$$\begin{aligned}
0 &= \big[r(t)x'(t)x''(t)\big]_{\alpha_1}^{\alpha} \\
&= \int_{\alpha_1}^{\alpha} \big[r(t)\big(x''(t)\big)^2 - q(t)\big(x'(t)\big)^2\big]\,dt - \int_{\alpha_1}^{\alpha} p(t)x(t)x'(t)\,dt \\
&> \int_{\alpha_1}^{\alpha} \big[r(t)\big(x''(t)\big)^2 - q(t)\big(x'(t)\big)^2\big]\,dt > 0,
\end{aligned}$$

since (3.14) is nonoscillatory (see [18]), a contradiction. Hence (2.2) is of type C_I.

If possible, let $x(t)$ be a solution of (3.2) with $x(\alpha) = 0 = x'(\alpha)$, $x(\beta) = 0 = x'(\beta)$ and $x(t) \neq 0$ for $t \in (\alpha, \beta)$, $\sigma < \alpha < \beta$. Let $x(t) > 0$ for $t \in (\alpha, \beta)$. Now multiplying (3.2) through by $x'(t)$, we get

$$\big[r(t)x'(t)x''(t)\big]' = r(t)\big(x''(t)\big)^2 - q(t)\big(x'(t)\big)^2 + f(t)x'(t) - p(t)x(t)x'(t). \tag{3.15}$$

Integrating (3.15) from α to β,

$$\begin{aligned}
0 &= \big[r(t)x'(t)x''(t)\big]_{\alpha}^{\beta} \\
&= \int_{\alpha}^{\beta} \big[r(t)\big(x''(t)\big)^2 - q(t)\big(x'(t)\big)^2\big]\,dt + \int_{\alpha}^{\beta} f(t)x'(t)\,dt - \int_{\alpha}^{\beta} p(t)x(t)x'(t)\,dt \\
&> \big[f(t)x(t)\big]_{\alpha}^{\beta} - \int_{\alpha}^{\beta} f'(t)x(t)\,dt - \frac{1}{2}\big[p(t)x^2(t)\big]_{\alpha}^{\beta} + \frac{1}{2}\int_{\alpha}^{\beta} p'(t)x^2(t)\,dt \\
&> 0,
\end{aligned}$$

a contradiction. If $x(t) < 0$ for $t \in (\alpha, \beta)$, then there exists a $\alpha_1 \in (\alpha, \beta)$ such that $x'(\alpha_1) = 0$ and $x'(t) > 0$ for $t \in (\alpha_1, \beta)$. Now, integrating (3.15) from α_1 to β, we get

$$\begin{aligned}
0 &= \big[r(t)x'(t)x''(t)\big]_{\alpha_1}^{\beta} \\
&= \int_{\alpha_1}^{\beta} \big[r(t)\big(x''(t)\big)^2 - q(t)\big(x'(t)\big)^2\big]\,dt + \int_{\alpha_1}^{\beta} f(t)x'(t)\,dt - \int_{\alpha_1}^{\beta} p(t)x(t)x'(t)\,dt \\
&> 0,
\end{aligned}$$

a contradiction, which completes the proof of the theorem. □

Remark 3.1.5 Theorem 3.1.10 holds if $f(t) \geq 0$ is continuous and $f'(t)$ exists almost everywhere with $f'(t) \leq 0$ whenever it exists.

The following result is analogous to a result due to Skidmore and Leighton [20] in the second-order case.

Theorem 3.1.11 *Suppose that* $p(t) \geq 0$, $p'(t) \geq 0$, $f(t) \geq 0$ *and* $f'(t) \leq 0$. *Then*

$$x''' + p(t)x = f(t) \tag{3.16}$$

is oscillatory.

Proof From Theorem 3.1.10, it follows that $x''' + p(t)x = 0$ is of type C_I and (3.16) does not admit a solution with a (2, 2)-distribution of zeros. Since $\int_\sigma^\infty p(t)\,dt = \infty$, from Theorem 1.3 due to Lazer [7], it is clear that the equation $x''' + p(t)x = 0$ is oscillatory. Hence (3.16) is oscillatory, by Theorem 3.1.9. □

The following examples illustrate the above results.

Example 3.1.3 Consider

$$x''' - \frac{1}{t^2}x' + t^2 x = e^{-t}, \quad t \geq 1. \tag{3.17}$$

Clearly, $x''' - \frac{1}{t^2}x' + t^2 x = 0$ is oscillatory (Theorem 1.3, [7]). So (3.17) is oscillatory.

Example 3.1.4 Consider

$$x''' + \frac{1}{t^3}x' + tx = e^{-t}, \quad t \geq 2. \tag{3.18}$$

Clearly, $z'' + \frac{1}{t^3}z = 0$ is nonoscillatory (p. 45, [21]). Since $\int_2^\infty t(t + \frac{3}{t^4})\,dt = \infty$, from Theorem 5.12 due to Hanan [3], it follows that $x''' + \frac{1}{t^3}x' + tx = 0$ is oscillatory. Consequently, (3.18) is oscillatory.

Theorem 3.1.12 *Suppose that* $p(t) \leq 0$, $p'(t) \geq 0$, $f(t) \geq 0$, $f'(t) \geq 0$, $r'(t) \geq 0$ *and* $2p(t) - q'(t) > 0$. *If* (3.14) *is nonoscillatory, then* (2.2) *is of type* C_I *and* (3.2) *does not admit a solution with a* (2, 2)*-distribution of zeros.*

Proof The proof that (2.2) is of type C_I is easy. Let $x(t)$ be a solution of (3.2) with $x(\alpha) = 0 = x'(\alpha)$, $x(\beta) = 0 = x'(\beta)$ and $x(t) \neq 0$ for $t \in (\alpha, \beta)$, $\sigma \leq \alpha < \beta$. Let $x(t) > 0$ for $t \in (\alpha, \beta)$. So there exists a $\alpha_1 \in (\alpha, \beta)$ such that $x'(\alpha_1) = 0$ and $x'(t) > 0$ for $t \in (\alpha, \alpha_1)$. Now integrating (3.15) from α to α_1, we get

$$\begin{aligned}
0 &= \left[r(t)x'(t)x''(t)\right]_{\alpha}^{\alpha_1} \\
&= \int_{\alpha}^{\alpha_1} \left[r(t)\left(x''(t)\right)^2 - q(t)\left(x'(t)\right)^2\right] dt + \int_{\alpha}^{\alpha_1} f(t)x'(t)\, dt \\
&\quad - \int_{\alpha}^{\alpha_1} p(t)x(t)x'(t)\, dt \\
&> 0,
\end{aligned}$$

a contradiction. If $x(t) < 0$ for $t \in (\alpha, \beta)$, then we integrate the identity (3.15) from α to β and we obtain

$$\begin{aligned}
0 &= \left[r(t)x'(t)x''(t)\right]_{\alpha}^{\beta} \\
&= \int_{\alpha}^{\beta} \left[r(t)\left(x''(t)\right)^2 - q(t)\left(x'(t)\right)^2\right] dt + \int_{\alpha}^{\beta} f(t)x'(t)\, dt - \int_{\alpha}^{\beta} p(t)x(t)x'(t)\, dt \\
&> \left[f(t)x(t)\right]_{\alpha}^{\beta} - \int_{\alpha}^{\beta} f'(t)x(t)\, dt - \frac{1}{2}\left[p(t)x^2(t)\right]_{\alpha}^{\beta} + \frac{1}{2}\int_{\alpha}^{\beta} p'(t)x^2(t)\, dt \\
&> 0,
\end{aligned}$$

a contradiction. which completes the proof of the theorem. □

Example 3.1.5 Consider

$$x''' + \frac{2}{t^3}x' - \frac{1}{t^4}x = e^t, \quad t \geq 1. \tag{3.19}$$

Clearly, the equation $z'' + \frac{2}{t^3}z = 0$ is nonoscillatory (p. 45, [21]) and $2p(t) - q'(t) = \frac{4}{t^4} > 0$ for $t \geq 1$. So, the homogeneous equation associated with (3.19) is of type C_I, and (3.19) does not admit a solution with a $(2, 2)$-distribution of zeros.

Theorem 3.1.13 *Suppose that* (2.2) *is of type* C_I. *Then Eq.* (3.2) *admits a solution* $x(t)$ *on* $[\sigma, \infty)$ *satisfying* $x(\alpha) = 0$, $x(\beta) = x'(\beta) = 0$, $\sigma \leq \alpha < \beta$.

Proof Let $x(t)$ be a solution of (3.2) on $[\sigma, \infty)$ satisfying $x(\beta) = x'(\beta) = 0$, $\beta > \sigma$. If $x(\alpha) = 0$, $\sigma \leq \alpha < \beta$, then $x(t)$ is the required solution. Suppose that $x(\alpha) \neq 0$. Let $z(t)$ be a solution of (2.2) satisfying $z(\beta) = z'(\beta) = 0$, $z''(\beta) > 0$. Since (2.2) is of type C_I, we have $z(t) > 0$ for $t \in [\sigma, \beta)$. Setting $\lambda = \frac{x(\alpha)}{z(\alpha)}$ and $y(t) = x(t) - \lambda z(t)$, we see that $y(t)$ is a solution of (3.2) with $y(\alpha) = 0$ and $y(\beta) = y'(\beta) = 0$. Thus $y(t)$ is the required solution. Hence the theorem is proved. □

Theorem 3.1.14 *Suppose that* (2.2) *is of type* C_I, *and* (3.2) *does not admit a solution with* $(2, 2)$*-distribution of zeros. If* (2.2) *is oscillatory, then every solution of* (3.2) *with two zeros* (*counting multiplicities*) *is oscillatory. In particular,* $x_p(t)$ *is oscillatory.*

Proof Let $x(t)$ be a solution of (3.2) on $[\sigma, \infty)$ such that $x(\alpha) = 0 = x(\beta)$, $\sigma \le \alpha < \beta$. If possible, let $x(t)$ be nonoscillatory. So, there exists a $\alpha_1 > \beta$ such that $x(t) \neq 0$ for $t \ge \alpha_1$. Let $z(t)$ be a nontrivial solution of (2.2) on $[\sigma, \infty)$ with $z(\alpha) = 0 = z(\beta)$. From Lemma 1.5.10, it is clear that $z(t)$ is oscillatory. Let t_1 and t_2 $(\alpha_1 < t_1 < t_2)$ be two consecutive zeros of $z(t)$. So, there exists a constant $\lambda \neq 0$ such that $x(t) - \lambda z(t)$ has a double zero at $\alpha_2 \in (t_1, t_2)$. Thus (3.2) admits a solution $x(t) - \lambda z(t)$ with $(1, 1, 2)$-distribution of zeros, contradicting Lemma 3.1.5.

If $x(t)$ is a solution of (3.2) on $[\sigma, \infty)$ with $x(\alpha) = 0 = x'(\alpha)$, $\alpha \ge \sigma$ and $x(t) \neq 0$ for $t \ge \alpha_1 > \alpha$, then we choose $z(t)$ to be a nontrivial solution of (2.2) with $z(\alpha) = 0 = z'(\alpha)$. In this case, (3.2) admits a solution $x(t) - \lambda z(t)$ with double zeros at α and α_2. If $x(t) - \lambda z(t) \neq 0$ for $t \in (\alpha, \alpha_2)$, then (3.2) admits a solution with a $(2, 2)$-distribution of zeros. If $x(t) - \lambda z(t)$ admits zeros in (α, α_2), then $x(t) - \lambda z(t)$ has $(1, 1, 2)$-distribution of zeros. In either case, there is a contradiction. Hence $x(t)$ is oscillatory. This completes the proof of the theorem. □

Theorem 3.1.15 *Suppose that* (2.2) *is of type* C_I *and* (3.2) *does not admit a solution with a* $(2, 2)$*-distribution of zeros. If* (2.2) *is oscillatory, then* (3.2) *admits a nonoscillatory solution with a single zero.*

Proof By Theorem 3.1.13, it is possible to define a sequence $\langle x_n(t) \rangle$ of solutions of (3.2) on $[\sigma, \infty)$ satisfying $x_n(\sigma) = 0$ and $x_n(n) = 0 = x_n'(n)$, where n is a positive integer such that $\sigma < n$. By Lemma 3.1.5, it follows that (3.2) does not admit a solution having $(2, 2)$ or $(1, 1, 2)$-distribution of zeros. Hence $x_n'(\sigma) \neq 0$ and $x_n(t) \neq 0$ for $t \in (\sigma, n)$.

We may write, for $t \in [\sigma, \infty)$,

$$x_n(t) = x_p(t) + \lambda_{1n} u_1(t) + \lambda_{2n} u_2(t) + \lambda_{3n} u_3(t),$$

where λ_{1n}, λ_{2n} and λ_{3n} are constants and $\{u_1(t), u_2(t), u_3(t)\}$ is a solution basis for (2.2). Choose a constant $\mu_n > 0$ such that

$$\mu_n = \left(1 + \lambda_{1n}^2 + \lambda_{2n}^2 + \lambda_{3n}^2\right)^{-1/2}.$$

If we set $y_n(t) = \mu_n x_n(t)$, then

$$y_n(t) = \mu_n x_p(t) + c_{1n} u_1(t) + c_{2n} u_2(t) + c_{3n} u_3(t), \tag{3.20}$$

where $c_{in} = \mu_n \lambda_{in}$, $i = 1, 2, 3$. Clearly, $0 < \mu_n \le 1$ and $|c_{in}| \le 1$, $i = 1, 2, 3$, with

$$\mu_n^2 + c_{1n}^2 + c_{2n}^2 + c_{3n}^2 = 1.$$

Note that $y_n(\sigma) = 0$, $y_n'(\sigma) \neq 0$ and $y_n(t) \neq 0$ for $t \in (\sigma, n)$. Clearly, each of the sequences $\langle \mu_n \rangle$ and $\langle c_{in} \rangle$, $i = 1, 2, 3$, has a convergent subsequence. Thus $\langle y_n \rangle$ has a subsequence which converges uniformly in $[\sigma, \infty)$. If $y(t) = \lim_{n_k \to \infty} y_{n_k}(t)$, then from (3.20),

$$y(t) = \mu x_p(t) + c_1 u_1(t) + c_2 u_2(t) + c_3 u_3(t)$$

with $y(\sigma) = 0$, where μ is the limit of μ_{n_k}, c_i is the limit of $\langle c_{in_k} \rangle$ and $\mu^2 + c_1^2 + c_2^2 + c_3^2 = 1$. Moreover, $\langle y'_{n_k} \rangle$ converges uniformly to y'.

Here $y(t) \neq 0$. Indeed, if $y(t) \equiv 0$ and $\mu = 0$, then from linear independence of u_1, u_2, u_3, it is clear that $c_1 = 0, c_2 = 0$ and $c_3 = 0$, a contradiction. If $y(t) \equiv 0$ and $\mu \neq 0$, then $x_p(t)$ is a solution of (2.2) and hence $f(t) \equiv 0$, a contradiction again.

If possible suppose that $y(t)$ is oscillatory. Let α and β $(\sigma < \alpha < \beta)$ be two consecutive zeros of $y(t)$ such that $y(t) > 0$ for $t \in (\alpha, \beta)$. We claim that $y'(\beta) \neq 0$. If not, let $y'(\beta) = 0$. If $\mu = 0$, then $y(t)$ is a solution of (2.2) with $y(\beta) = 0 = y'(\beta)$. Since (2.2) is of type C_I, we have $y(t) \neq 0$ for $t < \beta$, a contradiction. If $\mu \neq 0$, then $\frac{y(t)}{\mu}$ is a solution of (3.2) with $(1, 1, 2)$-distribution of zeros, a contradiction. So our claim holds. In fact, $y'(\beta) < 0$. Consequently, it is possible to find t_1 and t_2 $(t_1 < \beta < t_2)$ such that $y(t_1) > 0$ and $y(t_2) < 0$. From the uniform convergence of the sequence $\langle y_{n_k} \rangle$ to y it follows that there exists an integer $N > \max\{0, t_2\}$ such that $y_{n_k}(t_1) > 0$ and $y_{n_k}(t_2) < 0$ for $n_k > N$. So $y_{n_k}(t)$ has a zero in (t_1, t_2). Thus $y_{n_k}(t)$ is a solution of (3.2) with $(1, 1, 2)$-distribution of zeros, a contradiction. Hence $y(t)$ is nonoscillatory.

If $\mu = 0$, then $y(t)$ is a solution of (2.2) with $y(\sigma) = 0$. Hence $y(t)$ is oscillatory, a contradiction. Consequently, $\mu \neq 0$ and $\frac{y(t)}{\mu}$ is a nonoscillatory solution of (3.1). If $y(t)$ has two zeros, counting multiplicities, then from Theorem 3.1.14 it follows that $y(t)$ is oscillatory, a contradiction. Hence $\frac{y(t)}{\mu}$ is the required nonoscillatory solution of (3.2) with a single zero at $t = \sigma$. This completes the proof of the theorem. □

Example 3.1.6 Consider

$$x''' + x = 3e^{-t}, \quad t \geq 0. \tag{3.21}$$

Let $\alpha > 0$. Clearly, $x(t) = (t - \alpha)e^{-t}$ is a nonoscillatory solution of (3.21) with a single zero at $t = \alpha$. From Theorem 3.1.10, it follows that (3.21) does not admit a solution with a $(2, 2)$-distribution of zeros and $x''' + x = 0$ is of type C_I. Further, $x''' + x = 0$ is oscillatory (see the condition (2.9) in Chap. 2).

Next, we assume that f is a continuous function of bounded variation which is allowed to change its sign. The object here is to show that the oscillation of (2.2) implies the oscillation of (3.2). We begin with the following lemma:

Lemma 3.1.6 *Suppose that* $p(t) \geq 0$ *and* $q(t) \leq 0$. *If* $x(t)$ *is a solution of* (2.49) *with* $x(\alpha) = 0 = x'(\alpha)$ *and* $(rx')'(\alpha) > 0$, $\alpha \in [\sigma, \infty)$, *then* $x(t) > 0$, $x'(t) > 0$ *and* $(rx')'(t) > 0$ *for* $t > \alpha$.

Proof $(rx')'(t)$ is continuous and $(rx')'(\alpha) > 0$ imply that there exists a $\delta > 0$ such that $(rx')'(t) > 0$ for $t \in [\alpha, \alpha + \delta)$. This, in turn, implies that $x'(t) > 0$ in $(\alpha, \alpha + \delta)$ and hence $x(t) > 0$ for $t \in (\alpha, \alpha + \delta)$.

We claim that $x(t) > 0$ for $t \in (\alpha, \infty)$. If not, there is a $\beta > \alpha$ such that $x(\beta) = 0$ and $x(t) > 0$ for $t \in (\alpha, \beta)$. So $(r(t)x'(t))' + q(t)x(t)$ is nondecreasing in $[\alpha, \beta)$. Hence $(r(t)x'(t))' > 0$ for $t \in [\alpha, \beta]$. On the other hand, there is a $t_1 \in (\alpha, \beta)$ such that $x'(t_1) = 0$. Consequently, there exists a $t_2 \in (\alpha, t_1)$ such that $(rx')'(t_2) = 0$, a contradiction. Hence $x(t) > 0$ for $t \in (\alpha, \infty)$ holds. This, in turn, implies that $x'(t) > 0$ and $(rx')'(t) > 0$ for $t \in (\alpha, \infty)$. The lemma is proved.

It is assumed that $\{u_1(t), u_2(t), u_3(t)\}$ to be a solution basis of (2.2). Suppose that

$$\begin{array}{lll} u_1(\sigma) = 1, & u_1'(\sigma) = 0, & u_1''(\sigma) = 0, \\ u_2(\sigma) = 0, & u_2'(\sigma) = 1, & u_2''(\sigma) = 0, \\ u_3(\sigma) = 0, & u_3'(\sigma) = 0, & r(\sigma)u_3''(\sigma) = 1. \end{array}$$

So $W(t) \equiv 1$. Suppose that (2.2) is of type C_I and oscillatory. From Lemma 1.5.10, it follows that $u_2(t)$ and $u_3(t)$ are oscillatory solutions of (2.2). Clearly, $W_1(t) = u_2(t)u_3'(t) - u_3(t)u_2'(t)$ is a solution of (2.49) with $W_1(\sigma) = 0$, $W_1'(\sigma) = 0$ and $(rW_1')'(\sigma) = 1$. From Lemma 3.1.6, it follows that $W_1(t) > 0$ and $W_1'(t) > 0$ for $t > \sigma$. If $x(t)$ is a solution of (3.2) such that $x(t) = x_p(t) + \sum_{i=1}^{3} c_i u_i(t)$, where c_1, c_2 and c_3 are constants, then from Theorem 3.1.1, $x(t)$ is a solution of (3.4). Further, from Theorem 3.1.2, it follows that $u_2(t)$ and $u_3(t)$ are oscillatory solutions of (3.5). Hence (3.5) is oscillatory. □

Theorem 3.1.16 *Suppose that $p(t) \geq 0$, $q(t) \leq 0$, $f(t) \geq 0$ and $\int_\sigma^\infty f(t)\,dt = \infty$. If* (2.2) *is oscillatory, then every nonoscillatory solution of* (3.2) *is positive for large t.*

Proof Let $x(t)$ be a nonoscillatory solution of (3.2) on $[\sigma, \infty)$. If possible, let $x(t) < 0$ for $t \geq t_0 > \sigma$. If $x(t) = x_p(t) + \sum_{i=1}^{3} c_i u_i(t)$, where c_1, c_2, c_3 are constants, then $x(t)$ is a solution of (3.4). Hence $x(t)$ satisfies

$$\big(R(t)y'\big)' + \left[Q(t) - \frac{F(t)}{x(t)}\right] y = 0 \tag{3.22}$$

for $t \geq t_0$. Now, for $t > t_0$,

$$\int_\sigma^t f(s)W_1(s)\,ds > \int_\sigma^{t_0} f(s)W_1(s)\,ds + W_1(t_0)\int_{t_0}^t f(s)\,ds,$$

because $W_1'(t) > 0$ for $t > \sigma$. So, from the hypothesis it follows that

$$c_1 + \int_\sigma^t f(s)W_1(s)\,ds > 0$$

for large t. Since

$$Q(t) < Q(t) - \frac{F(t)}{x(t)}$$

for large t and from the above discussions, it follows that (3.5) is oscillatory. Then by Sturm's comparison theorem, (3.22) is oscillatory for large t. Consequently, $x(t)$ is oscillatory for large t. This contradiction completes the proof of the theorem. □

Remark 3.1.6 If f is a function of bounded variation, then it may be written as $f(t) = g(t) - h(t) + f(\sigma)$, $t \in [\sigma, \infty)$, where $g(t) = \frac{1}{2}[v_f(t) + f(t) - f(\sigma)]$ and $h(t) = \frac{1}{2}[v_f(t) - f(t) + f(\sigma)]$ are monotonic increasing functions and

$$v_f(t) = lub \sum_{i=1}^{n} \left| f(t_i) - f(t_{i-1}) \right|,$$

the least upper bound is taken over all partitions on $[\sigma, t]$. Further, if f is continuous, then v_f is continuous and hence each of g and h is continuous. But

$$\begin{aligned} v_f(t) &= lub \sum_{i=1}^{n} \left| f(t_i) - f(t_{i-1}) \right| \\ &= lub \sum_{i=1}^{n} \left| \int_{t_{i-1}}^{t_i} f'(s)\, ds \right| \\ &\le lub \sum_{i=1}^{n} \int_{t_{i-1}}^{t_i} \left| f'(s) \right| ds \\ &= \int_{\sigma}^{t} \left| f'(s) \right| ds \end{aligned}$$

and

$$f(t) - f(\sigma) = \int_{\sigma}^{t} f'(s)\, ds \le \left| \int_{\sigma}^{t} f'(s)\, ds \right| \le \int_{\sigma}^{t} \left| f'(s) \right| ds.$$

So

$$2g(t) = v_f(t) + f(t) - f(\sigma) \le 2 \int_{\sigma}^{t} \left| f'(s) \right| ds,$$

that is,

$$g(t) \le \int_{\sigma}^{t} \left| f'(s) \right| ds \le \int_{\sigma}^{\infty} \left| f'(s) \right| ds.$$

Similarly,

$$h(t) \le \int_{\sigma}^{\infty} \left| f'(s) \right| ds.$$

Hence

$$f(t) = \left[\int_{\sigma}^{\infty} \left| f'(s) \right| ds - h(t) \right] - \left[\int_{\sigma}^{\infty} \left| f'(s) \right| ds - g(t) \right] + f(\sigma).$$

If $f(\sigma) \ge 0$, then we take

$$g_1(t) = \left[\int_{\sigma}^{\infty} \left| f'(s) \right| ds - h(t) \right] + f(\sigma)$$

and

$$g_2(t) = \left[\int_\sigma^\infty |f'(s)|\,ds - g(t)\right].$$

Hence $f(t) = g_1(t) - g_2(t)$, where $g_1(t)$ and $g_2(t)$ are positive monotonically decreasing functions. If $f(\sigma) < 0$, then we take

$$g_1(t) = \left[\int_\sigma^\infty |f'(s)|\,ds - h(t)\right]$$

and

$$g_2(t) = \left[\int_\sigma^\infty |f'(s)|\,ds - g(t)\right] - f(\sigma).$$

Clearly, $f(t) = g_1(t) - g_2(t)$, where $g_1(t)$ and $g_2(t)$ are positive monotonically decreasing functions. Thus, if f is a function of bounded variation and

$$\int_\sigma^\infty |f'(s)|\,ds < \infty$$

then it can be expressed as a difference of two positive monotonically decreasing functions.

Theorem 3.1.17 *Suppose that $p(t) \geq 0$, $p'(t) \geq 0$ and $q(t) \leq 0$. Let $f(t)$ be a continuous function of bounded variation such that $\int_\sigma^\infty |f'(s)| < \infty$. If* (2.2) *is oscillatory, then* (3.2) *is oscillatory.*

Proof Since $f(t)$ is a continuous function of bounded variation such that

$$\int_\sigma^\infty |f'(s)|\,ds < \infty,$$

from Remark 3.1.6, it is possible to write $f(t) = g_1(t) - g_2(t)$, where g_1 and g_2 are nonnegative, continuous and monotonically decreasing functions. If $f_1(t) = 1 + g_1(t)$ and $f_2(t) = 1 + g_2(t)$, then $f_1(t)$ and $f_2(t)$ are positive monotonic decreasing and continuous functions such that

$$f(t) = f_1(t) - f_2(t), \qquad \int_\sigma^\infty f_1(t)\,dt = \infty \quad \text{and} \quad \int_\sigma^\infty f_2(t)\,dt = \infty.$$

Next, we consider two equations, viz,

$$\left(r(t)x''\right)' + q(t)x' + p(t)x = f_1(t) \tag{3.23}$$

and

$$\left(r(t)x''\right)' + q(t)x' + p(t)x = f_2(t). \tag{3.24}$$

From Theorem 3.1.10, it is clear that neither (3.23) nor (3.24) admits a solution with $(2, 2)$-distribution of zeros. Further, each of (3.23) and (3.24) admits a nonoscillatory solution (see Theorem 3.1.15), and these nonoscillatory solutions are positive for large t (see Theorem 3.1.16). Let these solutions be $x_1(t)$ and $x_2(t)$, respectively. Let $x_1(t) > 0$ and $x_2(t) > 0$, for $t \geq t_0 > \sigma$.

Let $u(t)$ be an oscillatory solution of (2.2). Let α_1 and β_1 $(t_0 < \alpha_1 < \beta_1)$ be two consecutive zeros of $u(t)$ such that $u(t) > 0$ for $t \in (\alpha_1, \beta_1)$. So there exists a $\lambda_1 > 0$ such that the function $x_1(t) - \lambda_1 u(t)$ has a double zero at $t_1 \in (\alpha_1, \beta_1)$. Further, let α_2 and β_2 $(t_0 < \alpha_2 < \beta_2)$ be two consecutive zeros of $u(t)$ such that $u(t) < 0$ for $t \in (\alpha_2, \beta_2)$. So there exists a $\lambda_2 > 0$ such that the function $x_2(t) + \lambda_2 u(t)$ has a double zero at $t_2 \in (\alpha_2, \beta_2)$. Let $\lambda = \max\{\lambda_1, \lambda_2\}$. Now $x_1(t_1) - \lambda u(t_1) \leq x_1(t_1) - \lambda_1 u(t_1) = 0$, $x_1(\alpha_1) - \lambda u(\alpha_1) = x_1(\alpha_1) > 0$ and $x_1(\beta_1) - \lambda u(\beta_1) = x_1(\beta_1) > 0$ imply that the function $x_1(t) - \lambda u(t)$ has at least two zeros (counting multiplicities) in (α_1, β_1). Similarly, the function $x_2(t) + \lambda u(t)$ has at least two zeros (counting multiplicities) in (α_2, β_2). From Theorem 3.1.14, it is clear that each of $x_1(t) - \lambda u(t)$ and $x_2(t) - \lambda u(t)$ is oscillatory.

Clearly, $x_1(t) - x_2(t)$ is a solution of (3.2) and hence $x_1(t) - x_2(t) - \lambda u(t)$ is a solution of (3.2). We claim that $x_1(t) - x_2(t) - \lambda u(t)$ is oscillatory. If not, assume that $x_1(t) - x_2(t) - \lambda u(t) > 0$ for large t. Since $x_2(t) > 0$ for large t, we have $x_1(t) - \lambda u(t) > x_2(t) > 0$ for large t, a contradiction to the fact that $x_1(t) - \lambda u(t)$ is oscillatory. If $x_1(t) - x_2(t) - \lambda u(t) < 0$ for large t, then $0 < x_1(t) < x_2(t) + \lambda u(t)$ for large t, which contradicts the oscillatory nature of the function $x_2(t) + \lambda u(t)$. Hence our claim holds and the theorem is proved. □

Remark 3.1.7 In Theorem 3.1.17, $f(t)$ is allowed to change sign.

The following example illustrates the above theorem.

Example 3.1.7 Consider

$$x''' - \frac{1}{t^2}x' + t^2 x = \frac{\sin t}{t^3}, \quad t \geq \sigma, \tag{3.25}$$

where $\sigma = 2\pi + \frac{\pi}{2}$. Clearly,

$$f'(t) = \frac{\cos t}{t^3} - \frac{3\sin t}{t^4} \quad \text{and} \quad \left|f'(t)\right| \leq \frac{1}{t^3} + \frac{3}{t^4}.$$

Further, $f(t) = f_1(t) - f_2(t)$, where $f_1(t) = \frac{\sin t}{t^3} + \frac{1}{t}$ and $f_2(t) = \frac{1}{t}$ with $f_1'(t) \leq \frac{1}{t^3} + \frac{3}{t^4} - \frac{1}{t^2} < 0$ and $f_2'(t) < 0$ for $t \geq \sigma$. Hence, by Theorem 4 (p. 100, [19]), f is of bounded variation. Consequently, $f(t)$ is a continuous function of bounded variation with

$$\int_\sigma^\infty \left|f'(t)\right| dt < \infty.$$

From Theorem 1.3 due to Lazer [7], it is clear that the homogeneous equation

$$x''' - \frac{1}{t^2}x' + t^2 x = 0$$

is oscillatory. Hence (3.25) is oscillatory, by Theorem 3.1.17.

Lemma 3.1.7 *If $W_1(t)$ is nonoscillatory, then every solution of* (3.6) *is a solution of* (3.2).

Proof Let $x_c(t) = cu_1(t) + x_p(t)$. From Theorem 3.1.1, it follows that $x_c(t)$ is a solution of (3.6). Since $\{u_2, u_3\}$ is a basis of solution space of (3.5) by Theorem 3.1.2, a general solution of (3.6) is written as

$$\begin{aligned} z(t) &= \lambda_2 u_2(t) + \lambda_3 u_3(t) + x_c(t) \\ &= cu_1(t) + \lambda_2 u_2(t) + \lambda_3 u_3(t) + x_p(t). \end{aligned}$$

Clearly, $z(t)$ is a solution of (3.2), and this completes the proof. □

Theorem 3.1.18 *If $W_1(t)$ is nonoscillatory and Eq.* (3.6) *is oscillatory for some constant c, then* (3.2) *is oscillatory.*

The proof follows from Lemma 3.1.7.

Theorem 3.1.19 *Suppose that $p(t) \geq 0$, $q(t) \leq 0$, $f(t) \geq 0$ and*

$$z'' + g(t)z = 0$$

is oscillatory implies that

$$z'' + g(t)z = h(t)$$

is oscillatory, where g and h are real-valued continuous functions on $[\sigma, \infty)$ such that $h(t) \geq 0$. If (2.2) *is oscillatory, then* (3.2) *is oscillatory.*

Proof Let $u_1(t)$, $u_2(t)$ and $u_3(t)$ be a linearly independent set of solutions of (2.2) with

$$\begin{aligned} &u_1(\sigma) = 1, \qquad u_1'(\sigma) = 0, \qquad u_1''(\sigma) = 0, \\ &u_2(\sigma) = 0, \qquad u_2'(\sigma) = 1, \qquad u_2''(\sigma) = 0 \end{aligned}$$

and

$$u_3(\sigma) = 0, \qquad u_3'(\sigma) = 0, \qquad r(\sigma)u_3''(\sigma) = 1.$$

So $W(u_1, u_2, u_3)(t) \equiv 1$. Now, $W_1(t) = u_2(t)u_3'(t) - u_3(t)u_2'(t)$ is a solution of (3.8). Clearly, $W_1(\sigma) = 0 = W_1'(\sigma)$, $(rW_1')'(\sigma) = 1$ and (2.2) is of type C_I. Indeed, if $x(t)$ is a solution of (2.2) with $x(t_1) = 0 = x'(t_1)$, $x''(t_1) > 0$, $t_1 > \sigma$, then

$x''(t) > 0$ for $t \in [\sigma, t_1]$. If not, it is possible to find a $t_2 \in [\sigma, t_1)$ such that $x''(t_2) = 0$, $x'(t) < 0$ and $x(t) > 0$ for $t \in [t_2, t_1)$. Now, integrating (2.2) from t_2 to t_1, we get a contradiction. This, in turn, implies that $x(t) > 0$ for $t \in [\sigma, t_1)$.

From Lemma 3.1.6, it follows that $W_1(t) > 0$ for $t > \sigma$. Since (2.2) is oscillatory and is of type C_I, Lemma 1.5.10 implies that (3.8) is oscillatory. If $v(t)$ is an oscillatory solution of (3.8), then

$$W_1(t)v'(t) - W_1'(t)v(t) = W_1^2(t)\left(\frac{v(t)}{W_1(t)}\right)'$$

is an oscillatory solution of

$$\left(\frac{(r(t)\theta)'}{W_1(t)}\right)' + \left(\frac{(r(t)W_1'(t))' + q(t)W_1(t)}{W_1^2(t)}\right)\theta = 0.$$

Putting $y = r(t)\theta$, we get

$$\left(\frac{y'}{W_1(t)}\right)' + \left(\frac{(r(t)W_1'(t))' + q(t)W_1(t)}{r(t)W_1^2(t)}\right)y = 0. \tag{3.26}$$

The transformation $y = zW_1^{1/2}(t)$ transforms (3.26) and

$$\left(\frac{y'}{W_1(t)}\right)' + \left(\frac{(r(t)W_1'(t))' + q(t)W_1(t)}{r(t)W_1^2(t)}\right)y = \frac{1}{r(t)W_1^2(t)}\int_\sigma^t f(s)W_1(s)\,ds \tag{3.27}$$

respectively, to

$$z'' + \left[\frac{q(t)}{r(t)} + \frac{(r(t)W_1'(t))'}{r(t)W_1(t)} + \frac{1}{2}\frac{W_1''(t)}{W_1(t)} - \frac{3}{4}\frac{(W_1'(t))^2}{W_1^2(t)}\right]z = 0 \tag{3.28}$$

and

$$z'' + \left[\frac{q(t)}{r(t)} + \frac{(r(t)W_1'(t))'}{r(t)W_1(t)} + \frac{1}{2}\frac{W_1''(t)}{W_1(t)} - \frac{3}{4}\frac{(W_1'(t))^2}{W_1^2(t)}\right]z$$
$$= \frac{1}{r(t)(W_1(t))^{3/2}}\int_\sigma^t f(s)W_1(s)\,ds. \tag{3.29}$$

So (3.26) is oscillatory, if and only if (3.28) is oscillatory and (3.27) is oscillatory if and only if (3.29) is oscillatory. From the hypothesis, it follows that (3.29) is oscillatory. Hence (3.27) is oscillatory. Consequently, (3.2) is oscillatory by Theorem 3.1.18. This completes the proof of the theorem. □

Theorem 3.1.20 *Let $p(t) \geq 0$, $q(t) \geq 0$, and $f(t) \geq 0$. Let r and q be differentiable such that $r'(t) \geq 0$ and $2p(t) - q'(t) \geq 0$. Suppose that*

$$z'' + g(t)z = 0$$

is oscillatory implies that

$$z'' + g(t)z = h(t)$$

is oscillatory, where g and h are real-valued continuous functions on $[\sigma, \infty)$ such that $h(t) \geq 0$. If (2.2) *is oscillatory, then* (3.2) *is oscillatory.*

Proof If $x(t)$ is a solution of (2.2) with $x(t_1) = 0 = x'(t_1)$ and $x''(t_1) > 0$, $t_1 > \sigma$, then $x(t) > 0$ for $t \in [\sigma, t_1)$. This is because otherwise there exists a $t_2 \in [\sigma, t_1)$ such that $x(t_2) = 0$ and $x(t) > 0$ for $t \in (t_2, t_1)$. Multiplying (2.2) through by $x(t)$ and integrating the resulting identity from t_2 to t_1, we get

$$\begin{aligned} 0 &= r(t_1)x(t_1)x''(t_1) - r(t_2)x(t_2)x''(t_2) \\ &= \frac{1}{2}r(t_1)\big(x'(t_1)\big)^2 - \frac{1}{2}r(t_2)\big(x'(t_2)\big)^2 - \frac{1}{2}\int_{t_2}^{t_1} r'(t)\big(x'(t)\big)^2\,dt \\ &\quad - \frac{1}{2}q(t_1)x^2(t_1) + \frac{1}{2}q(t_2)x^2(t_2) - \frac{1}{2}\int_{t_2}^{t_1} \big(2p(t) - q'(t)\big)x^2(t)\,dt < 0, \end{aligned}$$

a contradiction. So (2.2) is of type C_I. Here $r(\sigma)W_1''(\sigma) = 1$. Since (3.8) is of type C_{II} (see Lemma 1.5.12), we have $W_1(t) > 0$ for $t > \sigma$. The rest of the proof is similar to that of Theorem 3.1.19. □

Remark 3.1.8 From Theorems 3.1.19 and 3.1.20,

$$g(t) = \frac{q(t)}{r(t)} + \frac{(r(t)W_1'(t))'}{r(t)W_1(t)} + \frac{1}{2}\frac{W_1''(t)}{W_1(t)} - \frac{3}{4}\frac{(W_1'(t))^2}{W_1^2(t)}$$

and

$$h(t) = \frac{1}{r(t)(W_1(t))^{\frac{3}{2}}}\int_\sigma^t f(s)W_1(s)\,ds > 0.$$

Remark 3.1.9 The statement "$z' + g(t)z = 0$ is oscillatory implies that $z'' + g(t)z = h(t)$ is oscillatory" in Theorems 3.1.19 and 3.1.20 holds in a natural way when p, q, r and f are real constants such that $r > 0$, $q \leq 0$ or >0, $p > 0$ and $f > 0$.

In this case (2.2), (3.2) and (3.8) are reduced, respectively, to

$$x''' + q_1 x' + p_1 x = 0, \tag{3.30}$$

$$x''' + q_1 x' + p_1 x = f_1, \tag{3.31}$$

$$x''' + q_1 x' - p_1 x = 0, \tag{3.32}$$

$t \geq \sigma$, where $p_1 = \frac{p}{r} > 0$, $q_1 = \frac{q}{r} \leq 0$ or ≥ 0 and $f_1 = \frac{f}{r} > 0$. The characteristic equations of (3.30) and (3.32) are

$$m^3 + q_1 m + p_1 = 0 \tag{3.33}$$

and

$$n^3 + q_1 n - p_1 = 0. \tag{3.34}$$

Since (3.30) is assumed to be oscillatory, (3.33) admits two imaginary roots, say $\alpha + i\beta$ and $\alpha - i\beta$ and one real root γ, say. Without any loss of generality, we may assume that $\beta > 0$. Clearly,

$$\{e^{\gamma t}, e^{(\alpha+i\beta)t}, e^{(\alpha-i\beta)t}\}$$

and hence

$$\left\{\frac{1}{k}e^{\gamma t}, e^{\alpha t}\cos\beta t, e^{\alpha t}\sin\beta t\right\}$$

forms a basis for solution space of (3.30), where $k \neq 0$ is the value of the Wronskian of

$$\{e^{\gamma t}, e^{\alpha t}\cos\beta t, e^{\alpha t}\sin\beta t\}.$$

If $u_1(t) = \frac{1}{k}e^{\gamma t}$, $u_2(t) = e^{\alpha t}\cos\beta t$ and $u_3(t) = e^{\alpha t}\sin\beta t$, then $W(u_1, u_2, u_3)(t) \equiv 1$ and $W_1(t) = u_2(t)u_3'(t) - u_2'(t)u_3(t) = \beta e^{2\alpha t}$. Since $W_1(t)$ is a solution of (3.32), $e^{2\alpha t}$ satisfies (3.32). Further, since (3.30) is oscillatory implies that (3.32) is oscillatory, (3.34) admits two imaginary roots and a real root. As 2α satisfies (3.34) and $p_1 > 0$, then $2\alpha > 0$, that is, $\alpha > 0$. Now,

$$g(t) = q_1 + \frac{3}{4}\left(\frac{2W_1(t)W_1''(t) - (W_1'(t))^2}{W_1^2(t)}\right) = q_1 + 3\alpha^2$$

and

$$h(t) = \left(W_1^{-3/2}(t) f_1\right)\int_\sigma^t W_1(s)\,ds > 0.$$

For sufficiently large t,

$$h'(t) = \frac{f_1}{2\beta^{1/2}}\left[3e^{\alpha(2\sigma-3t)} - e^{-\alpha t}\right] < 0.$$

Clearly, $u_2(t)$ and $u_3(t)$ are linearly independent oscillatory solutions of

$$\begin{vmatrix} u_2(t) & u_3(t) & y \\ u_2'(t) & u_3'(t) & y' \\ u_2''(t) & u_3''(t) & y'' \end{vmatrix} = 0,$$

that is, of

$$\left(\frac{y'}{W_1(t)}\right)' + \left(\frac{W_1''(t) + q_1 W_1(t)}{W_1^2(t)}\right)y = 0. \tag{3.35}$$

The transformation $y = zW_1^{1/2}(t)$ transforms Eq. (3.35) into

$$z'' + \left[q_1 + \frac{3}{4}\left(\frac{2W_1(t)W_1''(t) - (W_1'(t))^2}{W_1^2(t)}\right)\right]z = 0, \tag{3.36}$$

that is,

$$z'' + g(t)z = 0. \tag{3.37}$$

Since (3.36) is oscillatory if and only if (3.35) is oscillatory, (3.37) is oscillatory and consequently $g(t) = q_1 + 3\alpha^2 > 0$. From Lemma 1.5.29, it follows that

$$z'' + g(t)z = h(t)$$

is oscillatory. Hence (3.30) is oscillatory implies that (3.31) is oscillatory. Consequently, we have the following theorem:

Theorem 3.1.21 *Consider* (2.2) *and* (3.2) *with* p, q, r *and* f *constants such that* $r > 0$, $q \geq 0$ *or* <0, $p > 0$ *and* $f > 0$. *Then* (2.2) *is oscillatory implies that* (3.2) *is oscillatory.*

Remark 3.1.10 Complete equations

$$y''' + r_1 y'' + q_1 y' + p_1 y = 0 \tag{3.38}$$

and

$$y''' + r_1 y'' + q_1 y' + p_1 y = f e^{-r_1 t/3}, \tag{3.39}$$

where p_1, q_1, r_1 and f are constants such that $p_1 \geq 0$, $q_1 \leq 0$, $r_1 > 0$ and $f > 0$, are reduced to

$$x''' + qx' + px = 0 \tag{3.40}$$

and

$$x''' + qx' + px = f, \tag{3.41}$$

by the transformation $y = xe^{-r_1 t/3}$, where

$$p = \frac{2r_1^3}{27} - \frac{r_1 q_1}{3} + p_1 \quad \text{and} \quad q = q_1 - \frac{r_1^2}{3}.$$

Clearly, $p > 0$ and $q < 0$ are constants. We may note that (3.38) is oscillatory, if and only if (3.40) is oscillatory, and (3.39) is oscillatory if and only if (3.41) is oscillatory. Consequently, from Theorem 3.1.21, it follows that (3.39) is oscillatory if (3.38) is oscillatory. In fact, this conclusion holds if p_1, q_1 and r_1 in (3.38) are such that

$$\frac{2r_1^3}{27} - \frac{r_1 q_1}{3} + p_1 > 0 \quad \text{and} \quad q_1 - \frac{r_1^2}{3} < 0 \text{ or } \geq 0.$$

3.2 Oscillatory Behaviour of Solutions of Nonhomogeneous Linear Differential Equations

This section is concerned with the oscillatory behaviour of solutions of (3.1), satisfying

$$\left.\begin{array}{lll} a \in C^{(2)}\big([\sigma,\infty),(0,\infty)\big), & b \in C^{(1)}\big([\sigma,\infty),(-\infty,0)\big), & \\ c \in C^{(1)}\big([\sigma,\infty),(0,\infty)\big), & f \in C^{(1)}\big([\sigma,\infty),(0,\infty)\big), & b(t)-a'(t) \le 0, \\ c'(t) \ge 0, \quad f'(t) \le 0, & t \ge \sigma \text{ for some real } \sigma \ge 0. & \end{array}\right\} \tag{3.42}$$

Theorem 3.2.1 *If* (3.42) *and* (2.14) *hold, then* (3.1) *is oscillatory* (*that is*, (3.1) *admits an oscillatory solution*).

Proof By Theorem 2.1.3, (2.1) is oscillatory. Again by Lemma 2.1.1, (2.1) is of type C_I. In the proof, we shall construct a solution of (3.1) through the solutions of (2.1) which is oscillatory.

Let $\{u_1(t), u_2(t), u_3(t)\}$ be a set of linearly independent solutions of (2.1) defined by the initial conditions,

$$\begin{array}{lll} u_1(\alpha)=1, & u_1'(\alpha)=0, & u_1''(\alpha)=0, \\ u_2(\alpha)=0, & u_2'(\alpha)=1, & u_2''(\alpha)=0, \\ u_3(\alpha)=0, & u_3'(\alpha)=0, & u_3''(\alpha)=1, \end{array}$$

for some $\alpha \in [\sigma,\infty)$. It may be verified that the function $x_p(t)$ defined by

$$x_p(t)=\int_\alpha^t \frac{1}{W(s)} \begin{vmatrix} u_1(t) & u_2(t) & u_3(t) \\ u_1(s) & u_2(s) & u_3(s) \\ u_1'(s) & u_2'(s) & u_3'(s) \end{vmatrix} f(s)\,ds,$$

where

$$W(s)=\begin{vmatrix} u_1(s) & u_2(s) & u_3(s) \\ u_1'(s) & u_2'(s) & u_3'(s) \\ u_1''(s) & u_2''(s) & u_3''(s) \end{vmatrix}$$

is a solution of (3.1) satisfying the conditions

$$x_p(\alpha)=x_p'(\alpha)=x_p''(\alpha)=0 \quad \text{and} \quad x_p'''(\alpha)=f(\alpha)>0.$$

Indeed,

$$x_p'(t)=\int_\alpha^t \frac{1}{W(s)} \begin{vmatrix} u_1'(t) & u_2'(t) & u_3'(t) \\ u_1(s) & u_2(s) & u_3(s) \\ u_1'(s) & u_2'(s) & u_3'(s) \end{vmatrix} f(s)\,ds,$$

$$x_p''(t) = \int_\alpha^t \frac{1}{W(s)} \begin{vmatrix} u_1''(t) & u_2''(t) & u_3''(t) \\ u_1(s) & u_2(s) & u_3(s) \\ u_1'(s) & u_2'(s) & u_3'(s) \end{vmatrix} f(s)\,ds,$$

and

$$x_p'''(t) = \int_\alpha^t \frac{1}{W(s)} \begin{vmatrix} u_1'''(t) & u_2'''(t) & u_3'''(t) \\ u_1(s) & u_2(s) & u_3(s) \\ u_1'(s) & u_2'(s) & u_3'(s) \end{vmatrix} f(s)\,ds + f(t).$$

Since $u_i(t)$, $i = 1, 2, 3$, are solutions of (2.1), putting $u_i'''(t) = -a(t)u_i''(t) - b(t)u_i'(t) - c(t)u_i(t)$, $i = 1, 2, 3$, in $x_p'''(t)$, we obtain

$$x_p'''(t) = -a(t)x_p''(t) - b(t)x_p'(t) - c(t)x_p(t) + f(t),$$

that is, $x_p(t)$ is a solution of (3.1) with the prescribed properties.

To complete the proof of the theorem, it is enough to show that $x_p(t)$ is oscillatory. For the sake of contradiction, assume that $x_p(t)$ is nonoscillatory. In the following, we prove that this assumption leads us to the conclusion that (3.1) admits a solution having consecutive double zeros at some point in $[\sigma, \infty)$, which leads us to the final contradiction.

By definition, there exists a $t_0 > \alpha \geq \sigma$ such that $x_p(t) > 0$ or $x_p(t) < 0$ for $t \geq t_0$. From Lemma 2.1.1, (2.1) is of type C_I and it admits an oscillatory solution. Further, by Lemma 1.5.10 the solution of (2.1) which admits a zero is oscillatory. In particular, $u_3(t)$ is oscillatory. Let t_1, t_2 and t_3 be successive zeros of $u_3(t)$ in (t_0, ∞). Then by Lemma 1.5.13, there exists a μ such that $\mu x_p(t) + u_3(t)$ has a double zero at some point $t_4 \in (t_2, t_3)$. Clearly, $\mu \neq 0$ (otherwise $u_3(t_4) = 0$ leads to a contradiction). Hence $\mu(x_p(t) + \lambda u_3(t))$, where $\lambda = \frac{1}{\mu}$, has a double zero at $t = t_4$ implies that $x_p(t) + \lambda u_3(t)$ has a double zero at $t = t_4$. It may be observed that zeros of $\mu x_p(t) + u_3(t)$ are the zeros of $x_p(t) + \lambda u_3(t)$. Now, we have two cases at hand; these are: $x_p(t) + \lambda u_3(t)$ has no zero in (α, t_4) and $x_p(t) + \lambda u_3(t)$ has at least one zero in (α, t_4). In the former case, we see that $x_p(t) + \lambda u_3(t)$ has double zeros at $t = \alpha$ and $t = t_4$. Further, $x_p(t)$ is a solution of (3.1) and $u_3(t)$ is a solution of (2.1) implies that $x_p(t) + \lambda u_3(t)$ is a solution of (3.1). Hence $x_p(t) + \lambda u_3(t)$ itself is a solution of (3.1) with consecutive double zeros at $t = \alpha$ and $t = t_4$. In the latter case, if $x_p(t) + \lambda u_3(t)$ has a zero in (α, t_4) then we shall show that (3.1) still admits a solution with consecutive double zeros. Suppose that there exists $t_5 \in (\alpha, t_4)$ such that $x_p(t) + \lambda u_3(t)$ vanishes at $t = t_5$ and $x_p(t) + \lambda u_3(t)$ is of one sign (either positive or negative) in (t_5, t_4). Here, two subcases arise, that is, $t = t_5$ is a double zero of $x_p(t) + \lambda u_3(t)$ and $t = t_5$ is a single zero of $x_p(t) + \lambda u_3(t)$. Again in the former subcase, $x_p(t) + \lambda u_3(t)$ is again a solution of (3.1) which has consecutive double zeros at $t = t_5$ and $t = t_4$. In the latter subcase, we shall show that (3.1) also admits a solution with consecutive double zeros as follows. Let $t = t_5$ be a single zero of $w(t) = x_p(t) + \lambda u_3(t)$, that is, $w(t_5) = 0$ and $w'(t_5) \neq 0$. Since $w(\alpha) = w(t_5) = 0$, there exists $t_6 \in [\alpha, t_5)$ such that $w(t_6) = 0$, and $w(t)$ has opposite signs in (t_6, t_5) and (t_5, t_4). Indeed, if $w(t)$ has no zero in (α, t_5), then α will serve the purpose of t_6. Suppose that $w(t) > 0$ in (t_5, t_4) and $w(t) < 0$ in (t_6, t_5). (The case when $w(t) < 0$ in (t_5, t_4) and $w(t) > 0$ in (t_6, t_5) may be treated similarly.)

Now, consider the solution $x(t)$ of (2.1) with initial conditions:

$$x(t_4) = x'(t_4) = 0, \qquad x''(t_4) = -1.$$

From Lemma 2.1.1, it follows that $x(t) < 0$ for $\sigma \le t < t_4$. Since $w(t_6) = w(t_5) = 0$ and $x(t) < 0$ for $t \in [t_6, t_5]$, by Lemma 1.5.13, there exists a real number μ_1 such that $w(t) - \mu_1 x(t)$ has a double zero at some point $t_7 \in (t_6, t_5)$. Set $\mu_1 = \frac{w(t_7)}{x(t_7)}$. Since $w(t) < 0$ and $x(t) < 0$ for $t \in (t_6, t_5)$, it follows that $\mu_1 > 0$. Consider a set D consisting of all μ_1 for which $w(t) - \mu_1 x(t)$ has at least a double zero in (t_6, t_5). Since $t_7 \in D$, it is nonempty. Let μ_m be the maximum of D or supremum of D according as D is finite or infinite. If D is finite, then $w(t) - \mu_m x(t)$ has a double zero at some point in (t_6, t_5). If D is infinite, then there exists a sequence of real numbers $\langle \sigma_n \rangle_{n=1}^{\infty}$ such that $\sigma_n \to \mu_m$ as $n \to \infty$. Consequently, the sequence of functions $\langle w(t) - \sigma_n x(t) \rangle_{n=1}^{\infty}$ converges uniformly to $w(t) - \mu_m x(t)$ in (t_6, t_5). Hence $w(t) - \mu_m x(t)$ has a double zero at some point in $[t_6, t_5]$. Let $w(t) - \mu_m x(t)$ has a double zero at $t_8 \in [t_6, t_5]$. If $w(t) - \mu_m x(t)$ has more than one double zero in $[t_6, t_5]$, without any loss of generality, assume that t_8 is the largest among all such points. Since $w(t) > 0$, $x(t) < 0$ in $[t_5, t_4)$, and $\mu_1 > 0$, it is obvious that $w(t) - \mu_m x(t) \ge 0$ in $[t_5, t_4)$. Therefore, to show that $w(t) - \mu_m x(t)$ has consecutive double zeros, in particular, at $t = t_8$ and $t = t_4$, we claim that $w(t) - \mu_m x(t) \ge 0$ in $t_8 < t < t_5$. On the contrary, suppose that there exists $t_9 \in (t_8, t_5)$ such that $w(t_9) - \mu_m x(t_9) < 0$. Since $x(t) < 0$ for $\sigma \le t < t_4$,

$$\frac{w(t_9) - \mu_m x(t_9)}{x(t_9)} > 0.$$

Clearly,

$$\big(w(t_9) - \mu_m x(t_9)\big) - \left(\frac{w(t_9) - \mu_m x(t_9)}{x(t_9)}\right) x(t_9) = 0,$$

that is,

$$w(t_9) - \left[\mu_m + \frac{w(t_9) - \mu_m x(t_9)}{x(t_9)}\right] x(t_9) = 0,$$

which is a contradiction to the fact that μ_m is the supremum of D. Hence $w(t) - \mu_m x(t) \ge 0$ in (t_8, t_4) and $w(t) - \mu_m x(t)$ has double zeros at $t = t_8$ and $t = t_4$. If $w(t) - \mu_m x(t)$ has zeros in (t_8, t_4), then these zeros will also be double zero due to the fact that $w(t) - \mu_m x(t) \ge 0$ in (t_8, t_4). Hence $w(t) - \mu_m x(t)$ is a function with consecutive double zeros at some points in $[\sigma, \infty)$. Further

$$w(t) - \mu_m x(t) = x_p(t) + \lambda u_3(t) - \mu_m x(t)$$

and $x_p(t)$ is a solution of (3.1), $u_3(t) - \mu_m x(t)$ is a solution of (2.1) (since $u_3(t)$, $x(t)$ are solutions of (2.1)) implies that $x_p(t) + \lambda u_3(t) - \mu_m x(t)$ is a solution of (3.1). Hence (3.1) admits a solution $w(t) - \mu_m x(t)$, which has consecutive double zeros.

Thus, the assumption $x_p(t)$ is nonoscillatory leads us to the conclusion that (3.1) admits a solution with consecutive double zeros. This conclusion will lead to the following final contradiction.

For our convenience, let $X(t)$ be a solution of (3.1) having consecutive double zeros, say at $t = \alpha$ and $t = \beta$, that is,

$$X(\alpha) = X(\beta) = 0 = X'(\alpha) = X'(\beta),$$

and either

$$X(t) > 0 \quad \text{for } \alpha < t < \beta$$

or

$$X(t) < 0 \quad \text{for } \alpha < t < \beta.$$

If the first case holds, multiplying (3.1) throughout by $X'(t)$ and integrating the resulting identity from α to β, we get

$$\begin{aligned} 0 &> -\int_\alpha^\beta \left(X''(t)\right)^2 dt + \int_\alpha^\beta \left(\frac{2b(t) - a'(t)}{2}\right)\left(X'(t)\right)^2 dt - \int_\alpha^\beta c'(t)\frac{(X(t))^2}{2}\, dt \\ &= -\int_\alpha^\beta f'(t)X(t)\, dt > 0, \end{aligned}$$

which is a contradiction.

If the second case holds, using the arguments employed in Lemma 2.1.1 by replacing $x(t)$ by $-X(t)$, we can find a $\gamma \in (\alpha, \beta)$ such that $X''(\gamma) = 0$ and $X(t) < 0$, $X'(t) > 0$, $X''(t) < 0$ for $t \in (\gamma, \beta)$. Next integrating (3.1) from γ to β, we obtain

$$0 = X''(\beta) - X''(\gamma) = \int_\gamma^\beta \left[f(t) - a(t)X''(t) - b(t)X'(t) - c(t)X(t)\right] dt > 0,$$

which, again is a contradiction. Hence $x_p(t)$ must oscillate. The theorem is proved. □

Clearly, the condition $a(t) \geq 0$, $b(t) \leq 0$, $c(t) > 0$, $f(t) > 0$, $b(t) - a'(t) \leq 0$, $c'(t) \geq 0$ and $f'(t) \leq 0$ is required in the statement of Theorem 3.2.1. Further, the condition (2.14) is required in the statement of Theorem 3.2.1 to show that (2.1) is oscillatory. One may use several different condition stated in Sect. 2.1 of Chap. 2 for the oscillation of (2.1), which ultimately yields the oscillation of (3.1).

We now relax the above norms and provide a simple condition under which the oscillation of (2.1) implies the oscillation of (3.1).

Theorem 3.2.2 *Let $a(t) \geq (\leq)0$, $b(t) \leq 0$, $c(t) > 0$, $F(t) \geq 0$ and $F'(t) \leq 0$ hold, where $F(t) = f(t)\exp(\int_\sigma^t a(s)\, ds)$. If (2.1) is oscillatory, then (3.1) is oscillatory.*

Proof Let $\{u_1(t), u_2(t), u_3(t)\}$ be a solution basis for (2.1) with

$$\begin{aligned} u_1(\alpha) &= 1, & u_1'(\alpha) &= 0, & u_1''(\alpha) &= 0, \\ u_2(\alpha) &= 0, & u_2'(\alpha) &= 1, & u_2''(\alpha) &= 0, \\ u_3(\alpha) &= 0, & u_3'(\alpha) &= 0, & u_3''(\alpha) &= 1. \end{aligned}$$

So

$$W(t) = W\big(u_1(t), u_2(t), u_3(t)\big) = \begin{vmatrix} u_1(t) & u_2(t) & u_3(t) \\ u_1'(t) & u_2'(t) & u_3'(t) \\ u_1''(t) & u_2''(t) & u_3''(t) \end{vmatrix}.$$

Since (2.1) is of type C_I, by Lemma 1.5.10, $u_3(t)$ is oscillatory. By the method of variation of parameters,

$$x(t) = \mu u_3(t) + \int_a^t f(s) W(t,s)\, ds$$

is a solution of (3.1) for any real constant μ, where

$$W(t,s) = \begin{vmatrix} u_1(t) & u_2(t) & u_3(t) \\ u_1(s) & u_2(s) & u_3(s) \\ u_1'(s) & u_2'(s) & u_3'(s) \end{vmatrix}$$

and $\int_\alpha^t f(s) W(t,s)\, ds$ is a particular solution of (3.1). If s is fixed, then $W(t,s)$ is a solution of (2.1) with double zeros at s and $W''(t,s) = 1$. Indeed, if s is fixed, then

$$W'(t,s) = \begin{vmatrix} u_1'(t) & u_2'(t) & u_3'(t) \\ u_1(s) & u_2(s) & u_3(s) \\ u_1'(s) & u_2'(s) & u_3'(s) \end{vmatrix}$$

and

$$W''(t,s) = \begin{vmatrix} u_1''(t) & u_2''(t) & u_3''(t) \\ u_1(s) & u_2(s) & u_3(s) \\ u_1'(s) & u_2'(s) & u_3'(s) \end{vmatrix}$$

implies that $W(s,s) = 0$, $W'(s,s) = 0$ and $W''(s,s) = W(u_1, u_2, u_3)(s) \equiv 1$. We shall show that $x(t)$ is oscillatory. From the continuity of $u_3(t)$, it follows that there exists a $\delta > 0$ such that $u_3(t) > 0$ for $t \in (\alpha, \alpha + \delta)$. Let t_1 be the next zero of $u_3(t)$ to the right of α. Hence $t_1 \geq \alpha + \delta$ and $u_3(t) > 0$ for $t \in (\alpha, t_1)$. We claim that $x(t)$ has a zero in $(\alpha, \alpha + \delta)$. If $x(t)$ has no zero in $(\alpha, \alpha + \delta)$, from Lemma 1.5.13, it follows that there exists a constant $\lambda \neq 0$ such that $x(t) + \lambda u_3(t)$ has a double zero at $t_2 \in (\alpha, \alpha + \delta)$. Setting $X(t) = x(t) + \lambda u_3(t)$, we see that $X(t)$ is a solution of (3.1) with $X(\alpha) = 0 = X'(\alpha)$ and $X(t_2) = 0 = X'(t_2)$ and $X(t) > 0$ or $X(t) < 0$ in (α, t_2). Equation (3.1) can be written in the form of (3.2). Let $X(t) > 0$ in (α, t_2).

Now, multiplying (3.2) through by $X'(t)$, we get

$$\big[r(t)X'(t)X''(t)\big]' = r(t)\big(X''(t)\big)^2 - q(t)\big(X'(t)\big)^2 + F(t)X'(t) - p(t)X(t)X'(t). \tag{3.43}$$

Integrating (3.43) from α to t_2, we have

$$\begin{aligned} 0 = \big[r(t)X'(t)X''(t)\big]_\alpha^{t_2} &= \int_\alpha^{t_2} r(t)\big(X''(t)\big)^2\,dt - \int_\alpha^{t_2} q(t)\big(X'(t)\big)^2\,dt \\ &\quad + \int_\alpha^{t_2} F(t)\big(X'(t)\big)^2\,dt - \int_\alpha^{t_2} p(t)X(t)X'(t)\,dt \\ &\geq - \int_\alpha^{t_2} F'(t)X(t)\,dt - \int_\alpha^{t_2} p(t)X(t)X'(t)\,dt \\ &\geq - \int_\alpha^{t_2} p(t)X(t)X'(t)\,dt, \end{aligned}$$

that is, $\int_\alpha^{t_2} p(t)X(t)X'(t)\,dt > 0$, which in turn implies that $X(t)X'(t) > 0$ in (α, t_2), a contradiction, because $X(t)$ has double zeros at α and t_2. If $X(t) < 0$ in (α, t_2), then there exists a $t_3 \in (\alpha, t_2)$ such that $X'(t_3) = 0$ and $X'(t) > 0$ for $t \in (t_3, t_2)$. Integrating (3.43) from t_3 to t_2,

$$\begin{aligned} 0 = \big[r(t)X'(t)X''(t)\big]_{t_3}^{t_2} &= \int_{t_3}^{t_2} r(t)\big(X''(t)\big)^2\,dt - \int_{t_3}^{t_2} q(t)\big(X'(t)\big)^2\,dt \\ &\quad + \int_{t_3}^{t_2} F(t)\big(X'(t)\big)^2\,dt - \int_{t_3}^{t_2} p(t)X(t)X'(t)\,dt \\ &> 0, \end{aligned}$$

a contradiction. Hence our claim holds, that is, $x(t)$ has a zero in $(\alpha, \alpha + \delta)$, that is, in (α, t_1). Proceeding as above, one can prove that between any two consecutive zeros of $u_3(t)$, $x(t)$ has one and only one zero. Consequently, $x(t)$ is oscillatory. This completes the proof of the theorem. □

Corollary 3.2.1 *Let $F(t) \geq 0$ and $F'(t) \leq 0$ hold, where $F(t) = f(t)\exp(\int_\sigma^t a(s)\,ds)$. Then* (2.1) *is oscillatory, if and only if* (3.1) *is oscillatory.*

Theorem 3.2.3 *Let $2b(t) - a'(t) \leq 0$, $f(t) \geq 0$ and $f'(t) \leq 0$ hold. If* (2.1) *is oscillatory, then* (3.1) *is oscillatory.*

Proof Proceeding as in the proof of Theorem 3.2.2, one may find that a solution $X_1(t)$ of (3.1) exists with double zeros at α_1 and α_2 with $X_1(t) \neq 0$ in (α_1, α_2), that is, $X_1(\alpha) = 0 = X_1'(\alpha)$, $X_1(\beta) = 0 = X_1'(\beta)$ and $X_1(t) \neq 0$ for $t \in (\alpha, \beta)$. Let $X_1(t) > 0$ for $t \in (\alpha, \beta)$. Multiplying (3.1) through by $X_1'(t)$ and integrating the resulting identity from α to β, we get

$$-\int_\alpha^\beta \left(X_1'(t)\right)^2 dt + \int_\alpha^\beta \left(2b(t)-a'(t)\right)\left(X_1'(t)\right)^2 dt + \int_\alpha^\beta c(t)X_1(t)X_1'(t)\,dt$$
$$= -\int_\alpha^\beta f'(t)X_1(t)\,dt$$

which, in turn, implies that $\int_\alpha^\beta c(t)X_1(t)X_1'(t)\,dt > 0$. Consequently, $X_1'(t) > 0$ for $t \in (\alpha, \beta)$, a contradiction, because $X_1(\alpha) = 0 = X_1'(\alpha)$, $X_1(\beta) = 0 = X_1'(\beta)$ and $X_1(t) \neq 0$ for $t \in (\alpha, \beta)$. Let $X_1(t) < 0$ for $t \in (\alpha, \beta)$. We can find a $\gamma \in (\alpha, \beta)$ such that $X_1'(\gamma) = 0 = X''(\gamma)$ and $X_1(t) < 0$, $X_1'(t) > 0$ and $X_1''(t) < 0$ for $t \in (\gamma, \beta)$. Integrating (3.1) from γ to β, we obtain

$$0 = X_1''(\beta) - X_1''(\gamma) = \int_\gamma^\beta \left[f(t) - a(t)X_1''(t) - b(t)X_1'(t) - c(t)X_1(t)\right]dt > 0,$$

a contradiction. Consequently, (3.1) is oscillatory. This completes the proof. □

Corollary 3.2.2 *Let* $2b(t) - a'(t) \leq 0$, $f(t) \geq 0$ *and* $f'(t) \leq 0$ *hold. Then* (3.1) *is oscillatory, if and only if* (2.1) *is oscillatory.*

Example 3.2.1 Consider the equations

$$x''' + \frac{4}{t}x'' - \frac{6}{t^2}x' + \frac{16}{t^3}x = 0 \tag{3.44}$$

and

$$y''' + \frac{4}{t}y'' - \frac{6}{t^2}y' + \frac{16}{t^3}y = \frac{10}{t^2}, \quad t \geq 1. \tag{3.45}$$

Since $2b(t) - a'(t) = -\frac{8}{t^2} \leq 0$, $f(t) = \frac{10}{t^2} \geq 0$ and $f'(t) = -\frac{20}{t^3} < 0$, by Corollary 3.2.2, (3.44) is oscillatory if and only if (3.45) is oscillatory. In particular, $x(t) = t^{3/2}\cos(\frac{\sqrt{7}}{2}\log t)$ and $y(t) = t[1 + t^{1/2}\cos(\frac{\sqrt{7}}{2}\log t)]$ are the oscillatory solutions of (3.44) and (3.45), respectively. On the other hand, $r(t) = \exp(\int_1^t a(s)\,ds) = t^4$, $F(t) = 10t^2 > 0$ and $F'(t) = 20t > 0$ imply that Corollary 3.2.1 cannot be applied to this example.

Example 3.2.2 Consider the equations

$$x''' + \frac{3}{t}x'' - \frac{1}{t^2}x' + \frac{4}{t^3}x = 0 \tag{3.46}$$

and

$$y''' + \frac{3}{t}y'' - \frac{1}{t^2}y' + \frac{4}{t^3}y = \frac{5}{t^4}, \quad t \geq 1. \tag{3.47}$$

Since $2b(t) - a'(t) = \frac{1}{t^2} > 0$, Corollary 3.2.2 cannot be applied to this example. Here $r(t) = t^3$ and $f(t) = \frac{5}{t^4}$ implies that $F(t) = \frac{5}{t}$ and hence $F'(t) = -\frac{5}{t^2} < 0$ for $t \geq 1$. Thus, by Corollary 3.2.1, (3.46) is oscillatory if and only if (3.47) is oscillatory. In particular, $x(t) = t\sin(\log t)$ and $y(t) = \frac{1}{t} + t\sin(\log t)$ are the oscillatory solutions of (3.46) and (3.47), respectively.

Remark 3.2.1 Theorem 3.2.2 also holds when $a(t) \leq 0$, $b(t) \leq 0$ and $c(t) > 0$ and $f(t) \geq 0$, and hence Corollary 3.2.2 remains true for this case.

Remark 3.2.2 Theorem 3.2.3 also holds when $a(t) \leq 0$, $b(t) \leq 0$ and $c(t) > 0$ and $f(t) \geq 0$. However, in this case, one has to proceed as in Theorem 3.2.2 to obtain a contradiction when $X_1(t) < 0$ for $t \in (\alpha, \beta)$. Thus Corollary 3.2.2 also holds for this case.

Remark 3.2.3 Corollaries 3.2.1 and 3.2.2 generalise Proposition 1.2.10 when $a(t)(\geq) \leq 0$, $b(t) \leq 0$, $c(t) > 0$ and $f(t) \geq 0$. Indeed, if $a(t) \equiv a \leq 0$, $b(t) \equiv b \leq 0$, $c(t) \equiv c > 0$ and $f(t) \equiv f > 0$ are constants, then $H \leq 0$. (1.8) implies that $G > 0$ and hence $G^2 + 4H^3 > 0$. Consequently, (1.5) is oscillatory. On the other hand, if (1.5) admits an oscillatory solution, then (1.6) has two imaginary roots and a real root. So, (1.7) has two imaginary roots and a real root. This, in turn, implies that $G^2 + 4H^3 > 0$. Since $c > 0$, the real root of (1.6) is negative. Hence the real root γ of (1.7) is negative. Consequently, $G > 0$ and (1.8) holds. It can be easily verified that (1.5) is oscillatory, if and only if (1.13) is oscillatory.

Remark 3.2.4 Clearly $a(t) \leq 0$, $f(t) \geq 0$ and $f'(t) \leq 0$ implies that $r(t) \geq 0$ and $F'(t) = r(t)(a(t)f(t) + f'(t)) \leq 0$. Hence Corollary 3.2.1 gives a stronger result concerning the oscillation and nonoscillation of (3.2).

Example 3.2.3 Consider the equations

$$x''' - \frac{1}{t}x'' - \frac{1}{t^2}x' + \frac{7}{t^3}x = 0 \tag{3.48}$$

and

$$y''' - \frac{1}{t}y'' - \frac{1}{t^2}y' + \frac{7}{t^3}y = \frac{6}{t^2}, \quad t \geq 1. \tag{3.49}$$

By Corollary 3.2.1, (3.48) is oscillatory if and only if (3.49) is oscillatory. In particular, $x(t) = t^{5/2}\cos(\frac{\sqrt{3}}{2}\log t)$ and $y(t) = t + t^{5/2}\cos(\frac{\sqrt{3}}{2}\log t)$ are the oscillatory solutions of (3.48) and (3.49), respectively.

Here, we have considered only two cases, viz., $a(t) \geq 0 (\leq 0)$, $b(t) \leq 0$ and $c(t) > 0$. One may obtain similar results when the coefficients $a(t)$, $b(t)$ and $c(t)$ satisfies $a(t) \geq 0$, $b(t) \geq 0$ and $c(t) > 0$. Note that the condition (2.1) is of type C_I and being oscillatory is needed in the proof of Theorems 3.2.1 and 3.2.2. One may find sufficient conditions for the property C_I and oscillation of (2.1) in Sect. 2.5 of Chap. 2. The following theorem can be obtained.

Theorem 3.2.4 *Let* $a(t) \geq 0$, $b(t) \geq 0$, $c(t) > 0$ *and* $2c(t) - a(t)b(t) - b'(t) \geq 0$ *hold. If* (2.1) *is oscillatory, then* (3.1) *is oscillatory.*

3.3 Nonoscillation of Nonhomogeneous Linear Differential Equations

In Sects. 3.1 and 3.2, we have obtained several comparison results for the oscillation and nonoscillation of (3.1) or (3.2). Note that we have used either (3.1) or (3.2) according to our convenience. In this section, we shall make a direct study on the nonoscillation of (3.1). The method we follow here is same as the methods used by Parhi and Parhi in [14–17]. We shall introduce a definition given by Parhi in [18].

Definition 3.3.1 A solution of a third-order differential equation is said to be of z-type oscillatory solution, if it has arbitrarily large zeros, but is ultimately nonnegative or nonpositive. If it is nonnegative, we say that the solution is nonnegative z-type oscillatory solution. Similarly, nonpositive z-type oscillatory solution may be defined.

Theorem 3.3.1 *Let* $p(t) \geq 0$ *and* $q(t) \leq 0$. *Suppose that* $F(t) + p(t) > 0$ *for large* t, *and* $\lim_{t\to\infty} \frac{F(t)}{p(t)} = \infty$. *Then all bounded solutions of* (3.2) *are nonoscillatory.*

Proof Let $x(t)$ be a solution of (3.2) on $[\sigma, \infty)$ such that $|x(t)| \leq K$ for $t \geq T \geq \sigma$ and $K > 0$ is a constant. From the given condition, it follows that there exists a $t_0 \geq T$ such that $F(t) \geq Kp(t)$ for $t \geq t_0$.

If possible, let $x(t)$ be nonnegative z-type. Let α and β $(t_0 \leq \alpha < \beta)$ be consecutive double zeros of $x(t)$. So, there exists a $\gamma \in (\alpha, \beta)$ such that $x'(\gamma) = 0$ and $x'(t) > 0$ for $t \in (\alpha, \gamma)$. Multiplying (3.2) through by $x'(t)$, we get

$$\big[r(t)x''(t)x'(t)\big]' = r(t)\big(x''(t)\big)^2 - q(t)\big(x'(t)\big)^2 - p(t)x(t)x'(t) + F(t)x'(t). \tag{3.50}$$

Integration of (3.50) from α to γ yields

$$\begin{aligned} 0 &= \big[r(t)x'(t)x''(t)\big]_\alpha^\gamma \\ &\geq \int_\alpha^\gamma \big[F(t) - p(t)x(t)\big]x'(t)\,dt \\ &\geq \int_\alpha^\gamma \big[F(t) - Kp(t)\big]x'(t)\,dt \\ &> 0, \end{aligned}$$

a contradiction. Next, suppose that $x(t)$ is nonpositive z-type with consecutive double zeros at α and β $(t_0 \leq \alpha < \beta)$. Thus there exists a $\gamma \in (\alpha, \beta)$ such that $x'(\gamma) = 0$

and $x'(t) > 0$ for $t \in (\gamma, \beta)$. Integrating (3.50) from γ to β, we get

$$0 = \int_{\gamma}^{\beta} \left[r(t)\left(x''(t)\right)^2 - q(t)\left(x'(t)\right)^2 - p(t)x(t)x'(t) + F(t)x'(t)\right] dt > 0,$$

a contradiction.

Finally, suppose that $x(t)$ is oscillatory. Let α, β and γ $(t_0 \le \alpha < \beta < \gamma)$ be three consecutive zeros of $x(t)$ such that $x'(\alpha) \le 0$, $x'(\beta) \ge 0$, $x'(\gamma) \le 0$, $x(t) < 0$ for $t \in (\alpha, \beta)$ and $x(t) > 0$ for $t \in (\beta, \gamma)$. So there exist points $\alpha' \in (\alpha, \beta)$ and $\beta' \in (\beta, \gamma)$ such that $x'(\alpha') = 0$, $x'(\beta') = 0$ and $x'(t) > 0$ for $t \in (\alpha', \beta')$. We shall consider two possible cases, viz., (I) $x''(\beta) \le 0$ and (II) $x''(\beta) > 0$. First, suppose that $x''(\beta) \le 0$. Integrating (3.50) from α' to β, we obtain

$$\begin{aligned} 0 &\ge r(\beta)x'(\beta)x''(\beta) \\ &= \int_{\alpha'}^{\beta} \left[r(t)\left(x''(t)\right)^2 - q(t)\left(x'(t)\right)^2 - p(t)x(t)x'(t) + F(t)x'(t)\right] dt \\ &> 0, \end{aligned}$$

a contradiction. Let $x''(\beta) > 0$. Integrating (3.50) from β to β', we get

$$0 \ge -r(\beta)x'(\beta)x''(\beta) = \int_{\beta}^{\beta'} \left[F(t) - Kp(t)\right]x'(t)\, dt > 0$$

a contradiction. Hence all bounded solutions of (3.2) are nonoscillatory. This completes the proof of the theorem. □

Theorem 3.3.2 *If $q(t) \le 0$, $p(t) \ge 0$, $p(t)$ and $F(t)$ are once continuously differentiable such that $p'(t) \ge 0$ and $F'(t) \le 0$, then all solutions of* (3.2) *are nonoscillatory.*

Proof Let $x(t)$ be a solution of (3.2) on $[\sigma, \infty)$. If possible, let $x(t)$ be of nonnegative z-type with consecutive double zeros at α and β $(\sigma \le \alpha < \beta)$. Integrating (3.50) from α to β, we get

$$\begin{aligned} 0 &= \int_{\alpha}^{\beta} \left[r(t)\left(x''(t)\right)^2 - q(t)\left(x'(t)\right)^2 - p(t)x(t)x'(t) + F(t)x'(t)\right] dt \\ &= \int_{\alpha}^{\beta} \left[r(t)\left(x''(t)\right)^2 - q(t)\left(x'(t)\right)^2\right] dt + \frac{1}{2}\int_{\alpha}^{\beta} p'(t)x^2(t)\, dt - \int_{\alpha}^{\beta} F'(t)x(t)\, dt \\ &> 0, \end{aligned}$$

a contradiction.

Let $x(t)$ be of nonpositive z-type with consecutive double zeros at α and β $(\sigma \le \alpha < \beta)$. So there exists a $\gamma \in (\alpha, \beta)$, such that $x'(\gamma) = 0$ and $x'(t) > 0$ for $t \in (\gamma, \beta)$.

Integrating (3.50) from γ to β yields

$$\begin{aligned} 0 &= \int_{\gamma}^{\beta} \left[r(t)\left(x''(t)\right)^2 - q(t)\left(x'(t)\right)^2 - p(t)x(t)x'(t) + F(t)x'(t)\right] dt \\ &> 0, \end{aligned}$$

a contradiction.

The last part of the proof of the theorem is similar to the last part of the proof of Theorem 3.3.1. This is left to the readers. □

Suppose that the second-order linear differential equation (2.110) is nonoscillatory. Then as an application of Lemma 1.5.1, any solution $x(t)$ of (3.1) with $x(t) > 0$ on (α, β), the following condition:

$$\int_{\alpha}^{\beta} r(t)\left(x''(t)\right)^2 dt \geq \int_{\alpha}^{\beta} q(t)\left(x'(t)\right)^2 dt$$

is satisfied. Using the above inequality and the methods used in Theorems 3.3.1 and 3.3.2, the following theorem can be proved.

Theorem 3.3.3 *If* (2.110) *is nonoscillatory,* $p(t) \leq 0$ *and* $\lim_{t\to\infty} \frac{F(t)}{-p(t)} = \infty$, *then all bounded solutions of* (3.2) *are nonoscillatory.*

We note that (2.110) is nonoscillatory, if $q(t) \leq 0$. Hence we have the following result.

Theorem 3.3.4 *If* $p(t) \leq 0$, $q(t) \leq 0$ *and* $\lim_{t\to\infty} \frac{F(t)}{-p(t)} = \infty$, *then all bounded solutions of* (3.2) *are nonoscillatory.*

Theorem 3.3.5 *Let* $p(t) \geq 0$, $p'(t) \geq 0$ *be such that* $F(t) \geq p(t)$. *If* (2.110) *is nonoscillatory, then all solutions of* (3.2) *which are bounded above by* 1 *are nonoscillatory.*

Theorem 3.3.6 *Let* $p(t) \leq 0$ *be such that* $p(t) + F(t) \geq 0$. *If* (2.110) *is nonoscillatory, then all solutions of* (3.2) *which are bounded above by* 1 *are nonoscillatory.*

In [1], Barrett considered homogeneous third-order linear differential equations of the form

$$\left[r_2(t)\left\{\left(r_1(t)x'\right)' + q_1(t)x\right\}\right]' + q_2(t)\left(r_1(t)x'\right) = 0, \tag{3.51}$$

where r_1, r_2, q_1 and $q_2 \in C([\sigma, \infty), R)$, $\sigma \in R$, $r_1(t) > 0$ and $r_2(t) > 0$. By a solution of (3.51) on $[\sigma, \infty)$, we mean a function $x \in C^1([\sigma, \infty), R)$ such that $r_1(t)x'(t)$ and $r_2(t)\{(r_1(t)x'(t))' + q_1(t)x(t)\} \in C^1([\sigma, \infty), R)$ and (3.51) is satisfied identically. Equation (3.51) is known as the canonical form. The adjoint of (3.51) is given by

$$\left[r_1(t)\left\{\left(r_2(t)x'\right)' + q_2(t)x\right\}\right]' + q_1(t)\left(r_2(t)x'\right) = 0. \tag{3.52}$$

The nonhomogeneous equations associated with (3.51) and (3.52) are given, respectively, by

$$\left[r_2(t)\left\{\left(r_1(t)x'\right)' + q_1(t)x\right\}\right]' + q_2(t)\left(r_1(t)x'\right) = f_1(t) \tag{3.53}$$

and

$$\left[r_1(t)\left\{\left(r_2(t)x'\right)' + q_2(t)x\right\}\right]' + q_1(t)\left(r_2(t)x'\right) = g_1(t) \tag{3.54}$$

with f_1 and $g_1 \in C([\sigma,\infty), R)$ such that $f_1(t) \geq 0$ and $g_1(t) \geq 0$.

Suppose that $\int_\sigma^\infty \frac{dt}{r_1(t)} = \infty$. The Liouville transformation $s = R(t)$, $y(s) = x(t)$, where $R(t) = \int_\sigma^t \frac{du}{r_1(u)}$, transforms (3.53) into

$$\frac{d}{ds}\left[\frac{r_2(t)}{r_1(t)}\frac{d^2y}{ds^2} + r_2(t)q_1(t)y\right] + r_1(t)q_2(t)\frac{dy}{ds} = r_1(t)f_1(t) \tag{3.55}$$

with $t = R^{-1}(s)$. If $\int_\sigma^\infty \frac{dt}{r_1(t)} < \infty$, then the Kummer transformation $s = \frac{1}{\rho(t)}$, $y(s) = sx(t)$, where $\rho(t) = \int_t^\infty \frac{du}{r_1(u)}$, transforms (3.53) into

$$\frac{d}{ds}\left[\frac{r_2(t)}{r_1(t)}s^3\frac{d^2y}{ds^2} + \frac{r_2(t)}{s}q_1(t)y\right] + \frac{r_1(t)q_2(t)}{s}\frac{dy}{ds} - \frac{r_1(t)q_2(t)}{s^2}y = \frac{r_1(t)}{s^2}f_1(t) \tag{3.56}$$

with $t = \rho^{-1}(1/s)$. However, Eq. (3.56) may be written as

$$\frac{d}{ds}\left[\psi(s)\frac{d^2y}{ds^2} + \left(\lambda(s) - \int_\sigma^s \nu(u)\,du\right)y\right] + \left[\phi(s) + \int_\sigma^s \nu(u)\,du\right]\frac{dy}{ds} = \frac{r_1(t)}{s^2}f_1(t), \tag{3.57}$$

where $\psi(s) = r_2(t)s^2/r_1(t)$, $\lambda(s) = r_2(t)q_1(t)/s$, $\phi(s) = r_1(t)q_2(t)/s$ and $\nu(s) = r_1(t)q_2(t)/s^2$.

Note that $y(s)$ is oscillatory, if and only if $x(t)$ is nonoscillatory. Furthermore, Eqs. (3.55) and (3.57) have the same general form. If $\int_\sigma^\infty \frac{dt}{r_2(t)} = \infty$ or $\int_\sigma^\infty \frac{dt}{r_2(t)} < \infty$, then (3.54) is transformed to an equation of the type (3.55) or (3.57), which is obtained by interchanging r_1 with r_2 and q_1 with q_2. Hence we shall emphasise the study of solutions of

$$\left(r(t)x'' + p(t)x\right)' + q(t)x' = f(t), \tag{3.58}$$

where p, q, r and $f \in C([\sigma,\infty), R)$, $r(t) > 0$ and $f(t) \geq 0$.

Linear differential equations of the type (3.2) occur in the study of entry-flow phenomenon in hydrodynamics [4]. Note that (3.2) is a particular case of (3.58). Indeed, one may write Eq. (3.2) as

$$\left[r(t)x'' + \left(\int_\sigma^t p(s)\,ds\right)x\right]' + \left(q(t) - \int_\sigma^t p(s)\,ds\right)x' = f(t).$$

Theorem 3.3.7 *If $p(t) \leq 0$ and $q(t) \leq 0$ for large t, then (3.58) is nonoscillatory.*

Proof Let $x(t)$ be a solution of (3.58) on $[\sigma, \infty)$. Let $p(t) \le 0$ and $q(t) \le 0$ for $t \ge t_0 \ge \sigma$. Let $x(t)$ be a nonnegative z-type solution with consecutive double zeros at t_1 and t_2 ($t_0 < t_1 < t_2$). So there exists a $b \in (t_1, t_2)$ such that $x'(b) = 0$, $x''(b) \le 0$ and $x'(t) > 0$ for $t \in (t_1, b)$. Integrating (3.58) from t_1 to b, we get

$$\begin{aligned} 0 &\ge r(b)x''(b) + p(b)x(b) - p(t_1)x''(t_1) \\ &= \int_{t_1}^{b} f(t)\,dt - \int_{t_1}^{b} q(t)x'(t)\,dt \\ &> 0, \end{aligned}$$

because $x''(t_1) \ge 0$. Suppose that $x(t)$ is a nonpositive z-type solution with consecutive double zeros at t_1 and t_2 ($t_0 < t_1 < t_2$). Then there exists a $b \in (t_1, t_2)$, such that $x'(b) = 0$ and $x'(t) > 0$ for $t \in (b, t_2)$. We note that $x''(b) \ge 0$ and $x''(t_2) \le 0$. Now integrating (3.58) from b to t_2 yields

$$\begin{aligned} 0 &\ge r(t_2)x''(t_2) + r(b)x''(b) - p(b)x(b) \\ &= \int_{b}^{t_2} f(t)\,dt - \int_{b}^{t_2} q(t)x'(t)\,dt \\ &> 0, \end{aligned}$$

a contradiction. Hence, $x(t)$ cannot be of z-type.

Suppose that $x(t)$ is an oscillatory solution with consecutive zeros at t_1, t_2 and t_3 ($t_0 < t_1 < t_2 < t_3$) such that $x(t) < 0$ for $t \in (t_1, t_2)$ and $x(t) > 0$ for $t \in (t_2, t_3)$. So there exist $b \in (t_1, t_2)$ and $c \in (t_2, t_3)$ such that $x'(b) = 0$, $x'(c) = 0$, $x'(t) > 0$ for $t \in (b, t_2)$ and $x'(t) > 0$ for $t \in (t_2, c)$. If $x''(t_2) \ge 0$, then integrating (3.58) from t_2 to c, we obtain

$$0 \ge r(c)x''(c) + p(c)x(c) - r(t_2)x''(t_2) = \int_{t_2}^{c} f(t)\,dt - \int_{t_2}^{c} q(t)x'(t)\,dt > 0,$$

a contradiction, because $x''(c) \le 0$. Furthermore, if $x''(t_2) < 0$, then integrating (3.58) from b to t_2 yields

$$0 > r(t_2)x''(t_2) - r(b)x''(b) - p(b)x(b) = \int_{b}^{t_2} f(t)\,dt - \int_{b}^{t_2} q(t)x'(t)\,dt > 0,$$

a contradiction, because $x''(b) \ge 0$. Hence $x(t)$ cannot be oscillatory. This completes the proof of the theorem. □

Theorem 3.3.8 *If* $\int_\sigma^t p(\theta)\,d\theta \le 0$ *and* $q(t) \le \int_\sigma^t p(\theta)\,d\theta$ *for large* t, *then* (3.58) *is nonoscillatory.*

Theorem 3.3.9 *If* $p(t) \ge 0$, $q(t) \le 0$ *and* $p(s) + q(t) \le 0$ *for* t *and* $s \in [\sigma, \infty)$ *and* $p(s) + q(t) \not\equiv 0$ *on any subinterval of* $[\sigma, \infty)$, *then* (3.58) *is nonoscillatory.*

Proof Let $x(t)$ be a solution of (3.58) on $[a,\infty)$. If $x(t)$ is of nonnegative z-type with consecutive double zeros at t_1 and t_2 $(\sigma < t_1 < t_2)$, then there exists a point $b \in (t_1, t_2)$ such that $x'(b) = 0$ and $x'(t) > 0$ for $t \in (t_1, b)$. Since $x''(t) \geq 0$ and $x''(b) \leq 0$, integrating (3.58) from t_1 to b, we obtain

$$\left.\begin{aligned} 0 &\geq r(b)x''(b) - r(t_1)x''(t_1) \\ &\geq -p(b)x(b) - \int_{t_1}^{b} q(t)x'(t)\,dt \\ &\geq -\int_{t_1}^{b} \big[q(t) + p(b)\big]x'(t)\,dt \\ &> 0, \end{aligned}\right\} \tag{3.59}$$

a contradiction. If $x(t)$ is of nonpositive z-type with consecutive double zeros at t_1 and t_2 $(a < t_1 < t_2)$, then there exists a point $b \in (t_1, t_2)$ such that $x'(b) = 0$ and $x'(t) > 0$ for $t \in (b, t_2)$. Clearly, $x''(b) \geq 0$ and $x''(t_2) \leq 0$. So integrating (3.58) from b to t_2 yields

$$\begin{aligned} 0 &\geq r(t_2)x''(t_2) - r(b)x''(b) \\ &\geq p(b)x(b) - \int_{b}^{t_2} q(t)x'(t)\,dt \\ &\geq -\int_{b}^{t_2} \big[q(t) + p(b)\big]x'(t)\,dt \\ &> 0, \end{aligned}$$

a contradiction. Hence $x(t)$ cannot be of z-type.

Suppose that $x(t)$ is oscillatory. Let t_1, t_2, t_3 $(a < t_1 < t_2 < t_3)$ be consecutive zeros of $x(t)$ such that $x'(t_1) \leq 0$ and $x'(t_2) \geq 0$ and $x'(t_3) \leq 0$. So there exist $b \in (t_1, t_2)$ and $c \in (t_2, t_3)$ such that $x'(t) > 0$ for $t \in (b, t_2)$ and $t \in (t_2, c)$. Clearly, $x''(b) \geq 0$ and $x''(c) \leq 0$. If $x''(t_2) \geq 0$, then integrating (3.58) from t_2 to c, we obtain

$$\begin{aligned} 0 &\geq r(c)x''(c) - r(t_2)x''(t_2) \\ &\geq -p(c)x(c) - \int_{t_2}^{c} q(t)x'(t)\,dt \\ &\geq -\int_{t_2}^{c} \big[q(t) + p(c)\big]x'(t)\,dt \\ &> 0, \end{aligned}$$

a contradiction. If $x''(t_2) < 0$, then integrating (3.58) from b to t_2, we get

$$0 \geq r(t_2)x''(t_2) - r(b)x''(b)$$

$$\begin{aligned}&\geq p(b)x(b) - \int_b^{t_2} q(t)x'(t)\,dt\\ &\geq \int_b^{t_2} \bigl[q(t) + p(b)\bigr]x'(t)\,dt\\ &> 0.\end{aligned}$$

This contradiction completes the proof of the theorem. □

Theorem 3.3.10 *If* $\int_\sigma^t p(u)\,du \geq 0$, $q(t) \leq \int_\sigma^t p(u)\,du$ *and* $\int_\sigma^s p(u)\,du \leq \int_\sigma^t p(u)\,du - q(t)$, *then* (3.2) *is nonoscillatory.*

Example 3.3.1 Consider

$$\left(2t^3x'' + \frac{1}{t+2}x\right)' - 4tx' = 4t^2 + \frac{t(t+4)}{(t+2)^2}, \quad t \geq 1. \tag{3.60}$$

Clearly, $p(s) = 1/(s+2) \leq 1/3 < 4t = |q(t)|$ for $s, t \geq 1$. From Theorem 3.3.9, it follows that Eq. (3.60) is nonoscillatory. In particular, $x(t) = t^2$ is a nonoscillatory solution of the equation. Note that (3.60) may be written as

$$\left(2t^3x''\right)' - \left(4t - \frac{1}{t+2}\right)x' - \frac{1}{(t+2)^2}x = 4t^2 + \frac{t(t+4)}{(t+2)^2}, \quad t \geq 1.$$

Clearly, Theorem 3.3.10 cannot be applied to (3.60). We note that

$$\int_1^t -\left[\frac{1}{(t+2)^2}\right]dt = \frac{1}{t+2} - \frac{1}{3}.$$

However, Theorems 3.3.9 and 3.3.10 can be applied to the equation

$$\left(5t^4x'' + 2x\right)' - 8x' = 40t^3 - 12t, \quad t \geq 0,$$

which admits the nonoscillatory solution $x(t) = t^2$.

Theorem 3.3.11 *If* $p(t) \leq 0$, $q(t) \geq 0$ *and* $p(t) + q(s) \leq 0$ *for* t *and* $s \in [\sigma, \infty)$ *such that* $p(t) + q(s) \not\equiv 0$ *on any subinterval of* $[\sigma, \infty)$, *then* (3.58) *is nonoscillatory.*

Theorem 3.3.12 *If* $\int_\sigma^t p(u)\,du \leq 0$, $q(t) \geq \int_\sigma^t p(u)\,du$ *and* $\int_\sigma^t p(u)\,du \leq \int_\sigma^s p(u)\,du - q(s)$ *for* t *and* $s \in [\sigma, \infty)$, *then* (3.58) *is nonoscillatory.*

Theorem 3.3.13 *Let* $p(t) \geq 0$ *and* $q(t) \geq 0$. *If* $\lim_{t\to\infty} \frac{f(t)}{(p(s)+q(t))} = \infty$ *uniformly for* $s \geq \sigma$, *then every solution of* (3.58) *whose first derivative is bounded is nonoscillatory.*

3.4 Asymptotic Behaviour of Oscillatory Solutions of Third-Order Linear Nonhomogeneous Differential Equations

Theorem 2.6.6 gives a sufficient condition under which all oscillatory solutions of (2.1) tend to zero as $t \to \infty$. However, no such result has been obtained for the other cases studied in Chap. 2. In this section, we shall give a new sufficient condition under which every oscillatory solution of the third-order linear nonhomogeneous equation

$$\big(r(t)x''\big)' = p_0(t)x + p_1(t)x' + f(t) \tag{3.61}$$

tends to zero as $t \to \infty$, where $r : [0, \infty) \to [0, \infty)$, $p_i : [0, \infty) \to R$, $i = 0, 1$ and $f : [0, \infty) \to R$ are continuous functions. We remark that no sign restriction is imposed on the coefficient functions to prove the following theorem.

Theorem 3.4.1 *Let*

$$\begin{aligned} &\int_0^\infty \frac{t}{r(t)}\,dt < \infty, \\ &\int_0^\infty \big|p_i(t)\big|\,dt < \infty, \quad i = 0, 1, \qquad \int_0^\infty \big|f(t)\big|\,dt < \infty. \end{aligned} \tag{3.62}$$

Then an arbitrary oscillatory solution of (3.61) *satisfies the conditions*

$$\lim_{t\to\infty} x(t) = \lim_{t\to\infty} x'(t) = \lim_{t\to\infty} r(t)x''(t) = 0. \tag{3.63}$$

Proof Let the sequence $\{t_K\}_{K=1}^\infty$ be such that

$$x(t_K) = 0, \quad 1 < t_K < t_{K+1} \ (K = 1, 2, \ldots).$$

Then for each natural K, there exists $\tilde{t}_K \in (t_K, t_{K+2})$ such that

$$x''(\tilde{t}_K) = 0.$$

Hence (3.61) gives

$$x''(t) = \frac{1}{r(t)} \int_{\tilde{t}_K}^t \big[p_0(s)x(s) + p_1(s)x'(s) + f(s)\big]\,ds.$$

If now we set

$$\begin{aligned} \rho_{iK} &= \max\big\{\big|x^{(i)}(t)\big|; t_K \le t \le t_{K+2}\big\}, \quad i = 0, 1, \\ \rho_{2K} &= \max\big\{r(t)\big|x''(t)\big|; t_K \le t \le t_{K+2}\big\} \end{aligned}$$

and

$$\epsilon_K = \int_{t_K}^{t_{K+2}} \left(\frac{t}{r(t)} + \big|p_0(t)\big| + \big|p_1(t)\big| + \big|f(t)\big|\right) dt,$$

then from the latter identity, we find that

$$|x''(t)| \le \frac{1}{r(t)} \epsilon_K(\rho_{0K} + \rho_{1K} + 1) \quad \text{for } t_K \le t \le t_{K+2}, \tag{3.64}$$

and

$$\rho_{2K} \le \epsilon_K(\rho_{0K} + \rho_{1K} + 1) \quad \text{for } t_K \le t \le t_{K+2}. \tag{3.65}$$

On the other hand, by virtue of conditions (3.62), we have

$$\lim_{K\to\infty} \epsilon_K = 0. \tag{3.66}$$

Therefore, without any loss of generality, it can be assumed that

$$\epsilon_K < \frac{1}{2}, \quad K = 1, 2, \ldots. \tag{3.67}$$

By the Green formula, for each natural K, we have

$$\left.\begin{aligned} x(t) &= \int_{t_K}^{t_{K+2}} G_K(t,s)x''(s)\,ds, \\ x'(t) &= \int_{t_K}^{t_{K+2}} \frac{\partial G_K(t,s)}{\partial t} x'(s)\,ds \end{aligned}\right\} \tag{3.68}$$

for $t_K \le t \le t_{K+2}$, where

$$G_K(t,s) = \begin{cases} \frac{(t-t_K)(s-t_{K+2})}{t_{K+2}-t_K} & \text{for } t < s, \\ \frac{(t-t_{K+2})(s-t_K)}{t_{K+2}-t_K} & \text{for } t > s \end{cases}$$

is the Green function. Moreover,

$$|G_K(t,s)| \le s - t_K < s$$

and

$$\left|\frac{\partial G_K(t,s)}{\partial t}\right| \le 1 \quad \text{for } t_K \le t,\ s \le t_{K+2}.$$

By virtue of these estimates and the inequalities (3.64), we find from (3.68) that

$$\rho_{0K} \le \epsilon_K(\rho_{0K} + \rho_{1K} + 1) \int_{t_K}^{t_{K+2}} \frac{s}{r(s)}\,ds \le \epsilon_K^2(\rho_{0K} + \rho_{1K} + 1),$$

$$\rho_{1K} \le \epsilon_K(\rho_{0K} + \rho_{1K} + 1) \int_{t_K}^{t_{K+2}} \frac{1}{r(s)}\,ds$$

$$\leq \epsilon_K(\rho_{0K} + \rho_{1K} + 1)\int_{t_K}^{t_{K+2}} \frac{s}{r(s)}\,ds$$

$$\leq \epsilon_K^2(\rho_{0K} + \rho_{1K} + 1).$$

Therefore,

$$\rho_{0K} + \rho_{1K} \leq 2\epsilon_K^2(\rho_{0K} + \rho_{1K}) + 2\epsilon_K^2.$$

Hence, by (3.65), (3.66) and (3.67), it follows that

$$\rho_{0K} + \rho_{1K} \leq 4\epsilon_K^2, \qquad \rho_{2K} \leq 2\epsilon_K, \quad K = 0, 1, 2, \ldots$$

and

$$\lim_{K\to\infty} \rho_{iK} = 0, \quad i = 0, 1, 2.$$

Therefore, the properties (3.63) is satisfied. This completes the proof of the theorem. □

In the literature many authors have studied the asymptotic behaviour of oscillatory solutions for second- and nth-order linear differential equations. The method given in Theorem 3.4.1 is completely different from the earlier ones.

3.5 Open Problems and Discussions

The results in Sects. 3.1 and 3.2 are on the oscillation and nonoscillation of solutions of (3.1) or (3.2). Theorem 3.3.1 gives a sufficient condition under which all bounded solutions of (3.1) or (3.2) are nonoscillatory. Further, Theorem 3.4.1 gives a sufficient condition under which all oscillatory solutions of (3.4) converges to zero eventually. In view of the above discussion, it would be interesting to obtain the following results:

(i) Obtain a sufficient condition under which all solutions of (3.1) oscillate.

(ii) Study the asymptotic behaviour of nonoscillatory solutions of (3.1) or (3.2).

(iii) Obtain a sufficient condition under which all solutions of (3.1) or (3.2) tend to zero as $t \to \infty$.

(iv) Study the asymptotic behaviour of nonoscillatory solutions of (3.1) or (3.2) in the presence of an oscillatory solution of (3.1) or (3.2).

(v) Obtain results similar to Theorems 3.2.1 and 3.2.2 for the case $a(t) \leq 0$, $b(t) \geq 0$ and $c(t) > 0$. Note that some sufficient conditions are required for Eq. (2.1) to be of type C_I and oscillatory.

(vi) Consider the general fourth-order linear differential equations of the form

$$\big(r(t)y'''\big)' = a(t)y + b(t)y' + c(t)y'' + f(t), \tag{3.69}$$

where $r : [0, \infty) \to (0, \infty)$, a, b, c and $f : [0, \infty) \to R$ are continuous functions. Then Padhi and Qian [10] proved the following theorem.

Theorem 3.5.1 *Let $\int_{t_0}^{\infty} |a(t)|\,dt < \infty$ and $\int_{t_0}^{\infty} |f(t)|\,dt < \infty$ hold. Further suppose that*

$$\int_0^{\infty} \frac{t^2}{r(t)}\,dt < \infty, \qquad \int_0^{\infty} \big|b(t)\big|\,dt < \infty, \qquad \int_0^{\infty} \big|c(t)\big|\,dt < \infty. \tag{3.70}$$

Then every oscillatory solution of Eq. (3.69) *satisfies*

$$\lim_{t\to\infty} y(t) = \lim_{t\to\infty} y'(t) = \lim_{t\to\infty} y''(t) = \lim_{t\to\infty} r(t)y'''(t) = 0. \tag{3.71}$$

Proof Let $\{t_k\}_{k=1}^{\infty}$, $1 < t_k < t_{k+1}$ $(k = 1, 2, 3, \dots)$ be such that $y(t_k) = 0$. Then for each natural k, there exists $t'''_{k+1} \in (t_{k+1}, t_{k+3})$ such that $y'''(t'''_{k+1}) = 0$. Hence (3.69) implies

$$y'''(t) = \frac{1}{r(t)} \int_{t'''_{k+1}}^{t} \big[a(s)y(s) + b(s)y'(s) + c(s)y''(s) + f(s)\big]\,ds. \tag{3.72}$$

We can find t'_{k+1} and $t''_{k+1} (t_{k+1} < t'_{k+1} < t''_{k+1})$ such that $y(t_{k+1}) = 0$, $y'(t'_{k+1}) = 0$ and $y''(t''_{k+1}) = 0$. We note that $t'''_{k+1} \le t''_{k+1}$.

If now we set

$$\begin{aligned}
\rho_{ik} &= \max\big\{\big|y^{(i)}(t)\big| : t_{k+1} \le t \le t''_{k+1}\big\}, \quad i = 0, 1, 2, \\
\rho_{3k} &= \max\big\{r(t)\big|y'''(t)\big| : t_{k+1} \le t \le t''_{k+1}\big\}, \\
\epsilon_k &= \int_{t_k}^{t''_{k+1}} \left(\frac{t^2}{r(t)} + \big|a(t)\big| + \big|b(t)\big| + \big|c(t)\big| + \big|f(t)\big|\right) dt,
\end{aligned}$$

then it follows that

$$\big|y'''(t)\big| \le \frac{1}{r(t)} \epsilon_k (\rho_{0k} + \rho_{1k} + \rho_{2k} + 1), \quad t_{k+1} \le t \le t''_{k+1}, \tag{3.73}$$

$$\rho_{3k} \le \epsilon_k (\rho_{0k} + \rho_{1k} + \rho_{2k} + 1), \quad t_{k+1} \le t \le t''_{k+1}. \tag{3.74}$$

On the other hand, by conditions $\int_{t_0}^{\infty} |a(t)|\,dt < \infty$, $\int_{t_0}^{\infty} |f(t)|\,dt < \infty$ and (3.70), we have

$$\lim_{k\to\infty} \epsilon_k = 0. \tag{3.75}$$

Therefore, without any loss of generality, it can be assumed that

$$\epsilon_k < \frac{1}{\sqrt{7}} \quad (k = 1, 2, 3, \dots). \tag{3.76}$$

By Green's formula, for each natural k, we have

$$\left.\begin{aligned} y(t) &= \int_{t_{k+1}}^{t''_{k+1}} G_k(t,s)\,y'''(s)\,ds, \\ y'(t) &= \int_{t_{k+1}}^{t''_{k+1}} \frac{\delta G_k(t,s)}{\delta s}\,y'''(s)\,ds, \\ y''(t) &= \int_{t_{k+1}}^{t''_{k+1}} \frac{\delta^2 G_k(t,s)}{\delta s^2}\,y'''(s)\,ds \end{aligned}\right\} \tag{3.77}$$

where

$$G_k(t,s) = \begin{cases} s \in [t_{k+1}, t'_{k+1}] : \begin{cases} \frac{(t-t_k)}{2}(2s - t - t_k), & t \le s, \\ \frac{(s-t_{k+1})^2}{2}, & s \le t. \end{cases} \\ s \in [t'_{k+1}, t''_{k+1}] : \begin{cases} \frac{(t-t_{k+1})}{2}(2t'_{k+1} - t - t_k), & t \le s, \\ \frac{(t-s)^2}{2} + \frac{(t-t_{k+1})}{2}(2t'_{k+1} - t - t_{k+1}), & s \le t \end{cases} \end{cases}$$

is the Green function for $y'''(t) = 0$, $y(t_{k+1}) = 0$, $y'(t'_{k+1}) = 0$, $y''(t''_{k+1}) = 0$. Moreover,

$$\left|G_k(t,s)\right| < \frac{3s^2}{2}, \qquad \left|\frac{\delta G_k(t,s)}{\delta t}\right| < s, \qquad \left|\frac{\delta^2 G_k(t,s)}{\delta t^2}\right| < 1,$$

for $t_{k+1} \le s \le t''_{k+1}$. By these estimates and inequalities (3.73) and (3.77), we have

$$\rho_{0k} \le \frac{3}{2}\epsilon_k(\rho_{0k} + \rho_{1k} + \rho_{2k} + 1)\int_{t_{k+1}}^{t''_{k+1}} \frac{s^2}{r(s)}\,ds \le \frac{3}{2}\epsilon_k^2(\rho_{0k} + \rho_{1k} + \rho_{2k} + 1),$$

$$\rho_{1k} \le \epsilon_k(\rho_{0k} + \rho_{1k} + \rho_{2k} + 1)\int_{t_{k+1}}^{t''_{k+1}} \frac{s}{r(s)}\,ds \le \epsilon_k^2(\rho_{0k} + \rho_{1k} + \rho_{2k} + 1),$$

$$\rho_{2k} \le \epsilon_k(\rho_{0k} + \rho_{1k} + \rho_{2k} + 1)\int_{t_{k+1}}^{t''_{k+1}} \frac{1}{r(s)}\,ds \le \epsilon_k^2(\rho_{0k} + \rho_{1k} + \rho_{2k} + 1).$$

Thus

$$\begin{aligned} \rho_{0k} + \rho_{1k} + \rho_{2k} &\le \frac{7}{2}\epsilon_k^2(\rho_{0k} + \rho_{1k} + \rho_{2k}) + \frac{7}{2}\epsilon_k^2 \\ &\le \frac{7}{2}\cdot\frac{1}{7}(\rho_{0k} + \rho_{1k} + \rho_{2k}) + \frac{7}{2}\epsilon_k^2 \\ &\le \frac{1}{2}(\rho_{0k} + \rho_{1k} + \rho_{2k}) + \frac{7}{2}\epsilon_k^2 \end{aligned}$$

which, in turn, implies

$$\rho_{0k} + \rho_{1k} + \rho_{2k} \le 7\epsilon_k^2$$

and from (3.74),

$$\rho_{3k} \le \epsilon_k(\rho_{0k} + \rho_{1k} + \rho_{2k} + 1) \le \epsilon_k\bigl(7\epsilon_k^2 + 1\bigr).$$

Since (3.75) holds, the above inequality yields $\rho_{ik} \to 0$ as $k \to \infty, i = 0, 1, 2, 3$. This completes the proof of the theorem. □

Example 3.5.1 Consider the equation

$$\bigl(t^4 y'''(t)\bigr)' = \frac{1}{t^2} y(t) + \frac{1}{t^2} y'(t) + \frac{1}{t^2} y''(t) + f(t), \quad t \ge 1, \tag{3.78}$$

where

$$f(t) = \frac{\sin t}{t^2} + \frac{20\cos t}{t^3} - \frac{190\sin t}{t^4} - \frac{896\cos t}{t^5} + \frac{1760\sin t}{t^6} - \frac{\cos t}{t^8} - \frac{6\sin t}{t^9} - \frac{12\cos t}{t^9} + \frac{42\sin t}{t^{10}}.$$

All the conditions of Theorem 3.5.1 are satisfied. We note that $y(t) = \frac{\sin t}{t^6}$ is an oscillatory solution of Eq. (3.78) satisfying the property (3.71).

It is clear that the conclusion of Theorem 3.5.1 holds for the homogeneous equation

$$\bigl(r(t)y'''(t)\bigr)' = a(t)y(t) + b(t)y'(t) + c(t)y''(t). \tag{3.79}$$

However, from Example 3.5.1, it seems that the forcing term $f(t)$ plays a crucial role in constructing the example. Thus, it would be interesting to obtain an example for the homogeneous equation (3.79), satisfying the conclusions of Theorem 3.5.1 under the conditions $\int_{t_0}^{\infty} |a(t)|\,dt < \infty$ and (3.70).

It would be interesting to obtain sufficient conditions on the coefficient functions using the techniques of Theorem 3.4.1, so that an arbitrary oscillatory solution $x(t)$ of the general nth-order differential equations of the form

$$\bigl(r(t)x^{(n-1)}(t)\bigr)' = \sum_{i=0}^{n-2} p_i(s)x^{(i)}(s) + f(t) \tag{3.80}$$

satisfies

$$\lim_{t\to\infty} x(t) = \lim_{t\to\infty} x'(t) = \cdots = \lim_{t\to\infty} x^{(n-2)}(t) = \lim_{t\to\infty} r(t)x^{(n-1)}(t) = 0, \tag{3.81}$$

where r and f are defined as in Sect. 3.4 and $p_i : [0, \infty) \to R, i = 0, 1, \ldots, n-2$.

It seems that the following open problem is yet to be established:

Open Problem Under the conditions $\int_0^{\infty} |f(t)|\,dt < \infty$, $\int_0^{\infty} \frac{t^{n-2}}{r(t)}\,dt < \infty$ and $\int_0^{\infty} |p_i(t)|\,dt < \infty$, $i = 0, 1, 2, \ldots, n-2$, does every oscillatory solution of (3.80) satisfy the property (3.81)?

To prove the above problem, one needs to find a suitable Green function and its estimates.

3.6 Notes

The results of Sect. 3.1 are taken from [12, 13], whereas the results of Sect. 3.2 are due to Das [2] and Padhi [9]. Theorems 3.3.5–3.3.6 are new, whereas Theorem 3.4.1 is taken from Padhi [8]. Theorems 3.3.7 to 3.3.13 are due to Parhi [11]. Theorem 3.5.1 is taken from Padhi and Qian [10].

References

1. J. H. Barrett; *Oscillation theory of ordinary linear differential equations*, Advances in Mathematics, 3(4) (1969), 415–509. (Reprinted in *Lectures in Ordinary Differential Equations*, Edited by R. McKelvev, Academic Press, New York, 1970).
2. P. Das; *On oscillation of third order forced equations*, Journal of Mathematical Analysis and Applications, 196(2) (1995), 502–513.
3. M. Hanan; *Oscillation criteria for third order linear differential equations*, Pacific Journal of Mathematics, 11 (1961), 919–944.
4. G. Jayaraman, N. Padmanabhan and R. Mehrotra; *Entry flow into a circular tube of slowly varying cross section*, Fluid Dynamics Research, 1(2) (1986), 131–144.
5. G. D. Jones; *Oscillation criteria for third order differential equations*, SIAM Journal of Mathematical Analysis, 7(1) (1976), 13–15.
6. M. S. Keener; *On the solutions of certain linear nonhomogeneous second order differential equations*, Applicable Analysis, 1 (1971), 57–63.
7. A. C. Lazer; *The behaviour of solutions of the differential equation* $y''' + p(x)y' + q(x)y = 0$, Pacific Journal of Mathematics, 17 (1966), 435–466.
8. S. Padhi; *On oscillatory solutions of third order differential equations*, Memoires on Differential Equations and Mathematical Physics, 31 (2004), 109–111.
9. S. Padhi; *On oscillatory linear third order forced differential equations*, Differential Equations and Dynamical Systems, 13 (2005), 343–358.
10. S. Padhi and C. Qian; *On asymptotic behaviour of oscillatory solutions of fourth order differential equations*, Electronic Journal of Differential Equations, 22(2007) (2007), 1–5.
11. N. Parhi; *On non-homogeneous canonical third-order linear differential equations*, Journal of the Australian Mathematical Society (Series A), 57 (1994), 138–148.
12. N. Parhi and P. Das; *On the zeros of solutions of nonhomogeneous third order differential equations*, Czechoslovak Mathematical Journal, 41 (116) (1991), 641–652.
13. N. Parhi and P. Das; *Oscillation and nonoscillation of nonhomogeneous third order differential equations*, Czechoslovak Mathematical Journal, 44(119) (1994), 443–459.
14. N. Parhi and S. Parhi; *Oscillation and nonoscillation theorems for nonhomogeneous third order differential equations*, Bulletin of the Institute of Mathematics Academia Sinica, 11 (1983), 125–139.
15. N. Parhi and S. Parhi; *Nonoscillation and asymptotic behaviour for forced nonlinear third order differential equations*, Bulletin of the Institute of Mathematics Academia Sinica, 13 (1985), 367–384.
16. N. Parhi and S. Parhi; *On the behaviour of solutions of the differential equation* $(r(t)y'')' + q(t)(y')^{\beta} + p(t)y^{\alpha} = f(t)$, Annales Polonici Mathematici, XLVII (1986), 137–148.

17. N. Parhi and S. Parhi; *Qualitative behaviour of solutions of forced nonlinear third order differential equations*, Rivista di Mathematics della University do Parma, 13 (1987), 137–148.
18. S. Parhi; *On Qualitative Behaviour of Solutions of Third Order Differential Equations*, Ph.D. Thesis, Berhampur University, India, 1984.
19. H. L. Royden; *Real Analysis*, Second Edition, McMillan Publishing Co. Inc., New York, 1968.
20. A. Skidmore and W. Leighton; *On the differential equation* $y'' + p(x)y = f(x)$, J. Math. Anal. Appl., 43 (1973), 46–55.
21. C. A. Swanson; *Comparison and Oscillation Theory of Linear Differential equations*, Academic Press, New York, 1968.

Chapter 4
Oscillation and Nonoscillation of Homogeneous Third-Order Nonlinear Differential Equations

This chapter is concerned with the study of oscillatory and asymptotic behaviour of nonoscillatory solutions of homogeneous third-order nonlinear differential equations.

In 1966, Waltman provided two oscillation theorems, namely, Theorems 1 and 2 in [39], for the nonlinear equations

$$x''' + b(t)x' + c(t)x^{\alpha} = 0, \quad t \geq \sigma \tag{4.1}$$

and

$$x''' + b(t)x' + c(t)f(x) = 0, \quad t \geq \sigma, \tag{4.2}$$

where $b \in C^1([\sigma, \infty), R)$ and $c \in C([\sigma, \infty), R)$. Both Theorems 1 and 2 in [39], involve integral criterions for the oscillation of solutions of (4.1) and (4.2) and are of the form that any solution of (4.1) or (4.2) which has a zero is oscillatory. For Eq. (4.1), Waltman [39], in his Theorem 1, proved that a solution of (4.1) which has a zero is oscillatory provided that $c(t)$ and $b(t)$ are nonnegative continuous functions and $b'(t) < 0$, α is a ratio of odd integers and

$$A + Bt - \int_{\sigma}^{t} P(s)\,ds < 0, \tag{4.3}$$

for t sufficiently large, where A and B are any constants and $P(t) = \int_{\sigma}^{t} c(s)\,ds$. In a similar way, in Theorem 2, Waltman [39] proved that any solution of (4.2) which has a zero is oscillatory provided that $c(t)$ and $b(t)$ are positive continuous functions, $Kc(t) - b'(t) > 0$, $\frac{f(u)}{u} \geq K$ for some constant $K > 0$ and $t \geq \sigma$ and

$$\int_{\sigma}^{\infty} t\big[Kc(t) - b'(t)\big]\,dt = \infty$$

holds (see Theorem 2 in [39]). However, the above results fails to apply when $c(t) = \frac{r_1}{t^3}$ and $b(t) = \frac{r_2}{t^2}$, where r_1 and r_2 are constants. In 1968, Heidel [19] studied the existence of an oscillatory solution and asymptotic behaviour of nonoscillatory

S. Padhi, S. Pati, *Theory of Third-Order Differential Equations*,
DOI 10.1007/978-81-322-1614-8_4, © Springer India 2014

solutions of (4.1) for the two different cases: (i) $b(t) \le 0$, $c(t) < 0$ and (ii) $b(t) \ge 0$, $c(t) \ge 0$. First, suppose that (i) holds. Heidel [19] proved that if either $c(t) < 0$ with $0 < \alpha < 1$ or $c(t) \le 0$ with $\alpha \ge 1$ holds and $x(t)$ is a nonoscillatory solution of (4.1), then there exists a number $t_0 \ge \sigma$ such that either

(I) $x(t)x'(t) < 0$ or
(II) $x(t)x'(t) > 0$

holds for $t \ge t_0$. Further, if (I) holds, and $-\infty < -M \le b(t)t^\alpha$ in $[\sigma, \infty)$ and

$$\int_\sigma^\infty s^\alpha c(s)\,ds = -\infty, \qquad \text{for some } \alpha \le 2,$$

then every nonoscillatory solution of (4.1) satisfying (I) tends to zero as $t \to \infty$. On the other hand, if

$$\int_\sigma^\infty sb(s)\,ds > -\infty,$$

then every nonoscillatory solution of (4.1) satisfy the property (II). Similarly, if either $c(t) < 0$ with $0 < \alpha < 1$ or $c(t) \le 0$ with $\alpha \ge 1$ holds and $-\frac{2}{t^2} \le b(t) \le 0$, then every nonoscillatory solution of (4.1) satisfies the property (II). Further, if $\int_\sigma^\infty s^2 c(s)\,ds = -\infty$ and $x(t)$ is a nonoscillatory solution of (4.1) such that (II) holds, then $\lim_{t\to\infty} |x(t)| = \infty$. In addition, if $\alpha > 1$ and (II) holds, then $x'(t)x''(t) \ge 0$ eventually and

$$\lim_{t\to\infty} \left|x(t)\right| = \lim_{t\to\infty} \left|x'(t)\right| = \lim_{t\to\infty} \left|x''(t)\right| = \infty$$

holds. Now, suppose that

$$\int_\sigma^\infty t^4\big(b'(s) - 2c(s)\big)\,ds = \infty,$$

then Eq. (4.1) with $\alpha = 1$ has an oscillatory solution. This clearly generalises and improves a theorem due to Hanan [16].

Next, if

$$\int_{t_0}^\infty c(t)u(t)\,dt = -\infty, \qquad t_0 > \max\{1, \sigma\}, \tag{4.4}$$

where $u(t)$ is one of the functions

$$t^{2-\gamma}, t^2(\log t)^{-1-\gamma}, t^2(\log t)^{-1}\big(\log(\log t)\big)^{-1-\gamma}$$

for some $0 < \gamma < 1$, then every nonoscillatory solution $x(t)$ of (4.1) with $\alpha = 1$ such that $x(t)x'(t) > 0$ eventually satisfies $x'(t)x''(t) \ge 0$ eventually and

$$\lim_{t\to\infty} \left|x(t)\right| = \lim_{t\to\infty} \left|x'(t)\right| = \lim_{t\to\infty} \left|x''(t)\right| = \infty.$$

The above conclusion may be obtained if one replaces (4.4) by

$$\int_{\sigma}^{\infty} s^{2\alpha} c(s)\, ds = -\infty, \quad 0 < \alpha < 1.$$

Next, suppose that (ii) holds, that is, $b(t) \geq 0$ and $c(t) \geq 0$. Heidel [19] proved that, if

(i) $\int_{\sigma}^{\infty} s^{2\alpha} c(s)\, ds = \infty$, if $0 < \alpha < 1$;
(ii) $\int_{\sigma}^{\infty} u(s)c(s)\, ds = \infty$, if $\alpha = 1$, where $u(t)$ is one of the functions $t^{2-\gamma}$, $t^2(\log t)^{-1-\gamma}, t^2(\log t)^{-1}, \log(\log t)^{-1-\gamma}, \ldots$ for some $\gamma > 0$;
(iii) $\int_{\sigma}^{\infty} s^{1+\alpha} c(s)\, ds = \infty$ if $\alpha > 1$,

then, every nonoscillatory solution $x(t)$ of (4.1) satisfies the property that $|x(t)|$ is not eventually nondecreasing.

In 1970, Singh [33] considered the third-order nonlinear equation

$$x''' + b(t)x' + f(t,x) = 0, \tag{4.5}$$

where $b(t)$ is a continuous function in R and does not change sign in $[\sigma, \infty)$, $\sigma \geq 0$, and $f(t,x) \in C([\sigma,\infty) \times (-\infty,\infty), R)$ with $a(t)\phi(x) \geq f(t,x) \geq h(t)\psi(x)$ for $(t,x) \in [\sigma,\infty) \times (-\infty,\infty)$, $a(t)$ and $h(t)$ are locally integrable functions and

$$\frac{\phi(x)}{x}, \frac{\psi(x)}{x} \geq \alpha > 0.$$

Singh [33] showed that if $b(t)$ is nonpositive and $a(t)$, $h(t)$ are nonnegative, and

$$\int_{\sigma}^{\infty} \left[\alpha a(t) - \frac{2}{3\sqrt{3}}\big(-b(t)\big)^{3/2}\right] dt = \infty$$

and

$$\int_{\sigma}^{\infty} \left[\alpha h(t) - \frac{2}{3\sqrt{3}}\big(-b(t)\big)^{3/2}\right] dt = \infty,$$

then every continuable solution of (4.5) which has a zero is oscillatory. On the other hand, if $b(t)$, $a(t)$ and $h(t)$ are nonnegative, $\alpha a(t) - b'(t)$ and $\alpha h(t) - b'(t)$ are positive and

$$\int_{\sigma}^{t} t\big[\alpha a(t) - b'(t)\big]\, dt = \infty \quad \text{and} \quad \int_{\sigma}^{t} t\big[\alpha a(t) - b'(t)\big]\, dt = \infty$$

hold, then every continuable solution of (4.5) which has a zero is oscillatory. Similarly, asymptotic behaviour of nonoscillatory solutions of (4.5), when all coefficients are positive, has also been discussed in [33].

Let $a(t)$, $b(t)$ and $h(t)$ be nonpositive. If

$$\int_{\sigma}^{\infty} \big[b'(t) - 2\alpha a(t)\big]\, dt = \infty,$$

$$\int_{\sigma}^{\infty} \left[b'(t) - 2\alpha h(t)\right] dt = \infty$$

and

$$\int_{\sigma}^{\infty} tb(t)\, dt > -\infty,$$

then Singh [33] proved that every continuable nontrivial solution $x(t)$ of (4.5) is either nonoscillatory such that

$$\lim_{t\to\infty} \left|x(t)\right| = \lim_{t\to\infty} \left|x'(t)\right| = \infty$$

or oscillatory if and only if $F[x(t)] < 0$ for all $t \geq \sigma$, where $F[x(t)]$ is given by

$$F\left[x(t)\right] = 2x(t)x''(t) - \left(x'(t)\right)^2 + b(t)x^2(t).$$

A solution of (4.1) or (4.2) is called a Kneser solution if for every nonoscillatory solution $x(t)$ of the equation there exists $t_0 \geq \sigma$ such that either $x(t)x'(t) < 0$ and $x(t)x''(t) > 0$ for $t \geq t_0$.

Bartušek et al. [6] obtained several sufficient conditions for the existence of Kneser solutions of equation of the form (4.2). Kura [23] was the first to initiate the work of nonoscillation of the equation

$$x''' + f(t, x) = 0, \tag{4.6}$$

a prototype of which is

$$x''' + c(t)|x|^{\alpha} \operatorname{sgn} x = 0, \tag{4.7}$$

where $f(t, x)$ and $c(t)$ are continuous functions on $[\sigma, \infty) \times R$ and $[\sigma, \infty)$, respectively, and α is a positive constant. He considered two different cases, where (4.6) is a generalisation of (4.7) with $c(t) < 0$ and $\alpha > 0$ and the case where $c(t) > 0$ with $0 < \alpha < 1$. The results were sharp. Theorem 1 due to Kura [23] is a nonlinear analogue of Theorem 3.5 due to Lazer [24], whereas Theorem 3 of Kura [23] is a third-order analogue of a theorem of Kiguradze [21, Theorem 2] for the second-order equation

$$x'' + c(t)|x|^{\alpha} \operatorname{sgn} x = 0, \quad \alpha > 1.$$

A special case of (4.6) is the equation

$$x''' - c(t)|x|^{\alpha} \operatorname{sgn} x = 0, \quad \alpha > 1, \tag{4.8}$$

where $c(t)$ is a positive continuous function on $[\sigma, \infty)$. If there exists a constant $\epsilon > 0$ such that

$$\frac{d}{dt}\left[c(t)t^{3+h(\alpha-1)+\epsilon}\right] \leq 0 \quad \text{for } t \geq t_0 > \sigma,$$

where $h = 1 - \frac{1}{\sqrt{3}}$, then (4.8) is oscillatory.

Hanan [16] proved that in the case where $c(t) \geq 0$, the equation $x''' - c(t)x = 0$ has an oscillatory solution if and only if its corresponding adjoint equation $x''' + c(t)x = 0$ has an oscillatory solution. By means of the above result on (4.8) and Heidel's theorem [19], it can be shown that no such remarkable relation exists between the nonlinear equations (4.7) and (4.8) with $\alpha \neq 1$. For example, $x''' + t^{-2-\alpha}|x|^\alpha \operatorname{sgn} x = 0$, $\alpha > 1$, has an oscillatory solution, whereas $x''' - 2^{-2-\alpha}|x|^\alpha \operatorname{sgn} x = 0$ is nonoscillatory.

Later, Erbe and Rao [11] studied the nonoscillation results of Eqs. (4.6) and (4.7). Theorem 2.1 of Erbe and Rao [11] extends and improves Theorem 3 due to Kura [23] with a simpler proof, whereas Theorem 3.2 due to Erbe and Rao [11] complements Theorem 6 of Kura [23]. On the other hand, Theorems 2.2 and 3.1 in [11] give nonoscillation criteria for the cases which were not considered by Kura [23].

In this chapter, we have made an attempt to provide all latest results available in the literature on the oscillation and asymptotic behaviour of nonoscillatory solutions of (4.1), (4.2) and the general equation

$$x''' + a(t)x'' + b(t)x' + c(t)x^\alpha = 0 \tag{4.9}$$

and

$$x''' + a(t)x'' + b(t)x' + c(t)f(x) = 0, \tag{4.10}$$

where a, b and $c \in C([\sigma, \infty), R)$, $\alpha > 0$ is a ratio of odd integers, $f \in C(R, R)$ such that $\frac{f(x)}{x} \geq \beta > 0$ for $x \neq 0$.

Note that we restrict our attention to those solutions of the above considered equations which exist on some ray $[T_x, \infty)$, $T_x \geq \sigma$ and are nontrivial in any neighbourhood of infinity. Further, such a solution is called "oscillatory", if it has arbitrarily large zeros, otherwise, it is called "nonoscillatory".

4.1 Behaviour of Solutions of $x''' + b(t)x' + c(t)x^\alpha = 0$ and $x''' + b(t)x' + c(t)f(x) = 0$

In this section, we present some results on the oscillation and nonoscillation of Eqs. (4.1) and (4.2) for four different possible cases on the signs of $b(t)$ and $c(t)$. We use the following notations for our use in the sequel:

(H_1) $b(t) \geq 0$ and $c(t) > 0$,
(H_2) $b(t) \leq 0$ and $c(t) > 0$,
(H_3) $b(t) \leq 0$ and $c(t) < 0$,
(H_4) $b(t) \geq 0$ and $c(t) < 0$.

Theorem 4.1.1 *Suppose that* (H_1) *holds and* $b'(t) \leq 0$. *If either*

$$\int_\sigma^\infty b(t)\,dt = \infty \quad \text{or} \quad \int_\sigma^\infty c(t)\,dt = \infty,$$

then a solution $x(t)$ *of* (4.1) *on* $[T_x, \infty)$, $T_x \geq \sigma$, *for which* $F[x(t_0)] \geq 0$ *for some* $t_0 \geq T_x$ *is oscillatory, where* $F[x(t)]$ *is given by*

$$F\big[x(t)\big] = \big(x'(t)\big)^2 - 2x(t)x'(t) - b(t)x^2(t). \tag{4.11}$$

Remark 4.1.1 Clearly, $x(t_0) = 0$ implies that $F[x(t_0)] > 0$. The following corollary follows from Theorem 4.1.1.

Corollary 4.1.1 *Suppose that the conditions of Theorem* 4.1.1 *are satisfied. Then any solution of* (4.1) *which vanishes at least once is oscillatory.*

Proof of Theorem 4.1.1 If $x(t)$ is nonoscillatory, then there exists a $t_1 \geq t_0$ such that $F[x(t_1)] \geq 0$ and $x(t) \neq 0$ for $t > t_1$. Without any loss of generality, we may assume that $x(t) > 0$ for $t > t_1$. Multiplying (4.1) through by $x(t)$ and integrating the resulting identity from t_1 to t, we obtain

$$F\big[x(t)\big] = F\big[x(t_1)\big] + \int_{t_1}^{t} \big(2c(s)x^{\alpha+1}(s) - b'(s)x^2(s)\big)\, ds. \tag{4.12}$$

If possible, let t_2 and t_3 ($t_1 \leq t_2 < t_3$) be two consecutive zeros of $x'(t)$. Then from (4.12), it follows that $F[x(t_2)] > 0$, that is,

$$-2x(t_2)x''(t_2) > b(t_2)x^2(t_2) > 0,$$

that is, $x''(t_2) < 0$. Similarly, $F[x(t_3)] > 0$ implies that $x''(t_3) < 0$. This is impossible, because t_2 and t_3 are two consecutive zeros of $x'(t)$. So $x'(t)$ has at most one zero in $[t_1, \infty)$. Consequently, there exists a $t_4 \geq t_1$ such that $x'(t) > 0$ or < 0 for $t \geq t_4$.

If possible, let $x'(t) > 0$ for $t \geq t_4$. Clearly, $-x'(t)$ is a nonoscillatory solution of the second-order nonhomogeneous equation

$$z'' + b(t)z = c(t)x^{\alpha}(t), \quad t \geq t_4.$$

If $\int_\sigma^\infty b(t) = \infty$, then some simple calculations show that $-x'(t) > 0$ for large t, that is, $x'(t) < 0$ for large t, a contradiction. If $\int_\sigma^\infty c(t)\, dt = \infty$, then

$$\int_\sigma^\infty c(t)x^\alpha(t)\, dt > \int_\sigma^{t_4} c(t)x^\alpha(t)\, dt + x^\alpha(t_4)\int_{t_4}^\infty c(t)\, dt$$

implies that $\int_\sigma^\infty c(t)x^\alpha(t)\, dt = \infty$. Hence $x'(t) < 0$ for large t, a contradiction.

Suppose that $x'(t) < 0$ for $t \geq t_4$. If $x''(t) < 0$ for large t, then $x(t) < 0$ eventually, a contradiction. If $x''(t) > 0$ for large t, then $\lim_{t\to\infty} x'(t)$ exists and nonpositive. From (4.12) we obtain

$$\big(x'(t)\big)^2 \geq \int_{t_1}^{t} \big(2c(s)x^{\alpha+1}(s) - b'(s)x^2(s)\big)\, ds.$$

Hence $\lim_{t\to\infty}(x'(t))^2$ is positive, that is, $\lim_{t\to\infty} x'(t) = \beta < 0$. This, in turn, implies that $x(t) < 0$ for large t, a contradiction. Hence $x''(t)$ changes sign for large t. So $x'(t)$ has maxima for large t. We claim that $\limsup_{t\to\infty} x'(t) = 0$. If not, then $\limsup_{t\to\infty} x'(t) = \lambda < 0$ yields the result that $x(t) < 0$ for large t, a contradiction. Hence our claim holds. Let $\{t_n\}$ be the sequence of maxima of $x'(t)$. So $\limsup_{n\to\infty} x'(t_n) = 0$. Clearly, the sequence $\langle t_n\rangle$ contains a subsequence $\langle s_n\rangle$ such that $\lim_{n\to\infty} x'(s_n) = 0$. However, since $x''(s_n) = 0$, from (4.12), we find that $(x'(s_n))^2 > 0$, a contradiction. Hence $x(t)$ is oscillatory, and this completes the proof. □

Remark 4.1.2 Following statements are equivalent:

(i) (4.3) holds for all real numbers A and B, and
(ii) $\int_\sigma^\infty c(t)\,dt = \infty$.

(ii) implies (i). Indeed, if $\int_\sigma^\infty c(t)\,dt = \infty$, then for $M > \max\{0, B\}$, there exists a $T > \sigma$ such that $\int_\sigma^t c(s)\,ds > M$ for $t \geq T$. Set

$$T_1 > \max\left\{\frac{A - \lambda + MT}{M - B}, T\right\},$$

where $\lambda = \int_\sigma^T (\int_\sigma^s c(u)\,du)\,ds$. Now, for $t \geq T_1$,

$$\begin{aligned}\int_\sigma^t \left(\int_\sigma^s c(u)\,du\right) ds &= \int_\sigma^T \left(\int_\sigma^s c(u)\,du\right) ds + \int_T^t \left(\int_\sigma^s c(u)\,du\right) ds \\ &> \lambda + M(t - T).\end{aligned}$$

Hence, for $t \geq T_1$,

$$\begin{aligned}A + BT &- \int_\sigma^t \left(\int_\sigma^s c(u)\,du\right) ds \\ &< A + Bt - [\lambda + M(t - T)] \\ &= (A - \lambda + MT) - (M - B)t \\ &< 0.\end{aligned}$$

Thus, (ii) implies (i). Next, suppose that for any real numbers A and B, (4.3) holds for large t. If possible, suppose that $L = \int_\sigma^\infty c(t)\,dt < \infty$. So $L > \int_\sigma^t c(s)\,ds$ for $t \geq \sigma$. Now, integrating this inequality from σ to t, we get

$$\int_\sigma^t \left(\int_\sigma^s c(u)\,du\right) ds < L(t - \sigma),$$

that is,

$$-\sigma L + Lt - \int_\sigma^t \left(\int_\sigma^s c(u)\,du\right) ds > 0,$$

a contradiction for $A = -\sigma L$ and $B = L$. Hence, $\int_\sigma^\infty c(t)\,dt = \infty$. This proves the assertion.

In view of the above remark, it is clear that Theorem 4.1.1 is more general than Theorem 1 due to Waltman [39]. The following example further strengthens this remark.

Example 4.1.1 Consider

$$x''' + \left(1 + \frac{1}{t^2}\right)x' + \frac{1}{t^3}x^\alpha = 0, \quad t \ge 1,$$

where α is a quotient of odd integers. Waltman's result cannot be applied to this example, because

$$\int_1^\infty c(t)\,dt < \infty.$$

However, Theorem 4.1.1 can be applied to this example.

Lemma 4.1.1 *Let $b(t) \ge 0$. If the second-order linear homogeneous differential equation* (2.94) *is oscillatory, then all nonoscillatory solutions of*

$$x'' + b(t)x = g(t) \tag{4.13}$$

are eventually positive, where $g \in C([\sigma, \infty), R)$ is such that $g(t) \ge 0$.

Proof Let $x(t)$ be a nonoscillatory solution of (4.13). If possible, let $x(t) < 0$ for $t \ge t_0 \ge \sigma$. Clearly, $x(t)$ is a nonoscillatory solution of the second-order linear homogeneous equation

$$z'' + \left(b(t) - \frac{g(t)}{x(t)}\right)z = 0, \quad t \ge t_0. \tag{4.14}$$

Since $b(t) < b(t) - \frac{g(t)}{x(t)}$ for $t \ge t_0$, from Sturm's comparison theorem, that is, Lemma 1.5.22, it follows that (4.14) is oscillatory, a contradiction. The proof is complete. □

Theorem 4.1.2 *Suppose that* (H_1) *holds and $b'(t) \le 0$. If either $\int_\sigma^\infty c(t)\,dt = \infty$ or* (2.94) *is oscillatory, then a solution $x(t)$ of* (4.1) *on $[T_x, \infty)$, $T_x \ge \sigma$, for which $F[x(t_0)] \ge 0$ for some $t_0 \ge T_x$ is oscillatory.*

Corollary 4.1.2 *Let* (H_1) *hold and $b'(t) \le 0$. If either $\int_\sigma^\infty c(t)\,dt = \infty$ or* (2.94) *is oscillatory, then any solution of* (4.1) *which has a zero is oscillatory.*

Example 4.1.2 One may observe that neither Theorem 4.1.1 nor Waltman's Theorem 1 can be applied to

$$x''' + \frac{1}{t^2}x' + \frac{1}{t^2}x^\alpha = 0, \quad t \ge 1.$$

On the other hand, $\lim_{t\to\infty} t^2 b(t) = 1 > \frac{1}{4}$ implies that $y'' + \frac{1}{t^2} y = 0$ is oscillatory. Hence, Theorem 4.1.2 can be applied to this example.

For $f(x) = x^\alpha$, when $\alpha > 0 (\neq 1)$ is a quotient of odd integers, the condition $\frac{f(x)}{x} \geq \beta \geq 0$ for $x \neq 0$ and some β takes the form $\frac{f(x)}{x} > \beta > 0$ and for $\alpha = 1$, $\beta = 1$. In view of this, the following theorem holds for both (4.1) and (4.2).

Theorem 4.1.3 *Let* (H_1) *hold. Suppose that* $\frac{f(x)}{x} \geq \beta$ *for some* $\beta > 0$ *and* $2\beta c(t) - b'(t) \geq 0$ *but not identically equal to zero on any subinterval of* $[\sigma, \infty)$. *If the second-order differential equation*

$$z'' + \big(b(t) + m\beta t c(t)\big)z = 0, \tag{4.15}$$

$0 < m < \frac{1}{2}$, *is oscillatory, then a solution* $x(t)$ *of* (4.2) *on* $[T_x, \infty)$, $T_x \geq \sigma$ *for which* $F[x(t_0)] \geq 0$ *for some* $t_0 \geq T_x$, *is oscillatory. That is, any solution of* (4.2) *which has a zero is oscillatory.*

Proof Suppose that $x(t)$ is nonoscillatory. So there exists a $t_1 \geq t_0$ such that $F[x(t_1)] \geq 0$ and $x(t) \neq 0$ for $t \geq t_1$. Let $x(t) > 0$ for $t > t_1$. The case when $x(t) < 0$ for $t \geq t_1$ may be treated in a similar way.

Multiplying (4.2) through by $x(t)$ and integrating the resulting identity from t_1 to t, we obtain

$$F\big[x(t)\big] \geq \int_{t_1}^{t} \big(2\beta c(s) - b'(s)\big)x^2(s)\,ds > 0.$$

Proceeding as in Theorem 4.1.1, we obtain $x'(t) > 0$ or < 0 for some $t > t_2 > t_1$. If $x'(t) < 0$, for $t > t_2$, then we proceed as in Theorem 4.1.1 to get a contradiction.

So $x'(t) > 0$ for $t > t_2$. From (4.2), it is clear that $x'''(t) < 0$ for $t > t_2$. Consider $t_3 > t_2$. From Lemma 3.2 due to Lazer [24], it follows that

$$\frac{x(t)}{x'(t)} > \frac{1}{2}(t - t_3)$$

for $t > t_3$. Thus, for $t > t_3$,

$$\frac{f(x(t))}{x'(t)} = \frac{f(x(t))}{x(t)} \frac{x(t)}{x'(t)} > \frac{\beta}{2}(t - t_3).$$

Consequently, $\liminf_{t\to\infty} \frac{f(x(t))}{tx'(t)} \geq \frac{\beta}{2}$. If $\liminf_{t\to\infty} \frac{f(x(t))}{tx'(t)} = \lambda$, $0 < \lambda < \infty$, then it is possible to choose m, $0 < m < \frac{1}{2}$ such that $\lambda > m\beta$. For $0 < \epsilon < \lambda - m\beta$, there exists a $T > t_3$ such that $f(x(t)) > (\lambda - \epsilon)tx'(t)$, that is, $f(x(t)) > m\beta t x'(t)$ for $t > T$. If $\liminf_{t\to\infty} \frac{f(x(t))}{tx'(t)} = \infty$, then $f(x(t)) > m\beta t x'(t)$ for $t > T$. Hence, in any case,

$$\frac{f(x(t))}{tx'(t)} > m\beta$$

for $t > T$. Clearly, $x'(t)$ is a nonoscillatory solution of the second-order differential equation

$$z'' + \left[b(t) + c(t)\frac{f(x(t))}{x'(t)}\right]z = 0$$

for $t > T$. From Sturm's comparison theorem, that is, Lemma 1.5.22, it follows that (4.15) is nonoscillatory, a contradiction. Hence $x(t) > 0$ for $t > t_1$ is not possible. A similar contradiction may be obtained when $x(t) < 0$ for $t > t_1$. Thus, $x(t)$ is oscillatory. The theorem is proved. □

Remark 4.1.3 Theorem 4.1.3 is more general than Theorem 2 due to Waltman [39]. In the following, we show that the assumptions in Theorem 2 in [39] imply the assumptions of Theorem 4.1.3. Clearly, $\beta c(t) - b'(t) > 0$ implies that $2\beta c(t) - b'(t) > 0$. Further, $\int_\sigma^\infty t(\beta c(t) - b'(t))\,dt = \infty$ implies that $\int_\sigma^\infty tc(t)\,dt = \infty$ or $-\int_\sigma^\infty tb'(t)\,dt = \infty$. But $-\int_\sigma^t sb'(s)\,ds \le \sigma b(\sigma) + \int_\sigma^t b(s)\,ds$ implies that $\int_\sigma^\infty b(t)\,dt = \infty$, when $-\int_\sigma^\infty tb'(t)\,dt = \infty$. Hence $\int_\sigma^\infty t(\beta c(t) - b'(t))\,dt = \infty$ implies that $\int_\sigma^\infty (b(t) + m\beta tc(t))\,dt = \infty$, and consequently (4.15) is oscillatory.

In the following, we give an example to which Waltman's theorem cannot be applied but to which Theorem 4.1.3 can be applied.

Example 4.1.3 Consider

$$x''' + \left(1 + \frac{1}{t^2}\right)x' + \frac{1}{t^3}(\beta + e^x)x = 0, \quad t \ge 1,$$

where $\beta > 0$. Clearly, $\beta c(t) - b'(t) > 0$ but

$$\int_1^\infty t\big(\beta c(t) - b'(t)\big)\,dt = (\beta + 2)\int_1^\infty \frac{dt}{t^2} < \infty.$$

So Waltman's theorem cannot be applied to this example. Now, for $0 < m < \frac{1}{2}$,

$$\int_1^\infty \big(b(t) + m\beta tc(t)\big)\,dt = \int_1^\infty \left(1 + \frac{1}{t^2}\right)dt + m\beta\int_1^\infty \frac{dt}{t^2} = \infty.$$

Thus, (4.15) is oscillatory. Consequently, Theorem 4.1.3 can be applied to this example.

Theorem 4.1.3 generalises the following theorem due to Lazer [24].

Theorem 4.1.4 *If $c(t) \ge 0$, $b(t) \ge 0$, $2c(t) - b'(t) \ge 0$ and not identically zero in any interval, and there exists a number $m < \frac{1}{2}$ such that the second-order differential equation*

$$z'' + \big[b(t) + mtc(t)\big]z = 0$$

is oscillatory, then the third-order linear differential equation (2.10) *admits oscillatory solutions.*

Lemma 4.1.2 *Let* (H_1) *hold,* $b'(t) \leq 0$ *for* $t \geq \sigma$*. Let* $x(t)$ *be a solution of* (4.1) *or* (4.2) *with* $F[x(t_0)] \geq 0$ *for some* $t_0 \geq \sigma$*, where* $F[x(t)]$ *is given in* (4.11)*. Then* $x(t)$ *is either oscillatory or there exists a number* $T \geq t_0$ *such that*

$$x(t) > 0, \qquad x'(t) > 0, \qquad x''(t) > 0 \quad \text{and} \quad x'''(t) \leq 0 \quad \text{for } t \geq T$$

or

$$x(t) < 0, \qquad x'(t) < 0, \qquad x''(t) < 0 \quad \text{and} \quad x'''(t) \geq 0 \quad \text{for } t \geq T$$

holds.

Assume that f satisfies the condition $\frac{f(x)}{x} \geq \beta > 0$ for $x \neq 0$ and $\alpha > 0$ is a ratio of odd integers. Thus, for Eq. (4.2), the condition $b'(t) \leq 0$ for $t \geq \sigma$ in Lemma 4.1.2 may be replaced by a weaker condition

$$2\beta c(t) - b'(t) \geq 0 \quad \text{for } t \geq \sigma \tag{4.16}$$

and $2\beta c(t) - b'(t)$ is not identically zero on any subinterval of $[\sigma, \infty)$. Further, from Lemma 4.1.2, it follows that any nonoscillatory solution $x(t)$ of (4.1) with $\alpha \geq 1$ and (4.2) with $F[x(t_0)] \geq 0$ for some $t_0 \geq \sigma$ is unbounded.

Theorem 4.1.5 *Let the hypothesis of Lemma* 4.1.2 *hold, and* $t^2 b(t) \leq \frac{1}{4}$ *for all* $t > 0$ *and* $\alpha \geq 1$*. If*

$$\int_\sigma^\infty \left[\mu t^2 c(t) + t b(t) - \frac{2}{3\sqrt{3}t}\left(1 - t^2 b(t)\right)^{3/2}\right] dt = \infty \tag{4.17}$$

for all $\mu > 0$*, then every solution of* (4.1) *which has a zero is oscillatory.*

If $c_0 > \frac{2}{3\sqrt{3}}$ and $\delta < -2$ are constants, then by Theorem 4.1.5, every solution of the equation

$$x''' + b_0 t^\delta x' + c_0 t^{-3} x = 0, \tag{4.18}$$

which has a zero, is oscillatory.

Theorem 4.1.6 *Let the hypothesis of Lemma* 4.1.2 *with* (4.16) *hold,* $t^2 b(t) \leq \frac{1}{4}$ *for all* $t > 0$*. If*

$$\int_\sigma^\infty \left[t^2 \beta c(t) + t b(t) - \frac{2}{3\sqrt{3}t}\left(1 - t^2 b(t)\right)^{3/2}\right] dt = \infty, \tag{4.19}$$

then every solution of (4.2) *which has a zero is oscillatory.*

Theorem 4.1.7 *Let the hypothesis of Lemma* 4.1.2 *hold*, $t^2b(t) \le \frac{1}{4}$ *for all* $t > 0$. *Suppose that* $\alpha < 1$. *If*

$$\int_\sigma^\infty \left[\mu t^2 c(t) + tb(t) - \frac{2}{3\sqrt{3}t}\left(1 - t^2 b(t)\right)^{3/2}\right] dt = \infty$$

for every $\mu > 0$, *then every bounded solution of* (4.1) *which has a zero is oscillatory.*

Lemma 4.1.3 *Suppose that the second-order linear differential equation* (2.94) *is nonoscillatory. If* $x(t)$ *is a nonoscillatory solution of* (4.1) *with* $c(t) > 0$, *then there exists a* $t_0 \in [\sigma, \infty)$ *such that* $x(t)x'(t) > 0$ *or* $x(t)x'(t) < 0$ *for* $t \ge t_0$.

Remark 4.1.4 In Theorems 4.1.5–4.1.7, if the hypotheses of Lemma 4.1.2 are replaced by the hypotheses of Lemma 4.1.3, then the same conclusions hold.

Theorem 4.1.8 *Suppose that* $\alpha \ge 1$. *If* (4.17) *holds, then every solution of* (4.1) *which has a zero is oscillatory.*

Let $\delta < -2$ and $c_0 > \frac{2}{3\sqrt{3}}$. By Theorem 4.1.8, every solution of (4.18) which has a zero is oscillatory.

Theorem 4.1.9 *Let* $\alpha < 1$. *If* (4.17) *holds for every* $\mu > 0$, *then every solution of* (4.2) *which has a zero is oscillatory.*

Theorem 4.1.10 *Let* $\frac{f(x)}{x} \ge \beta > 0$ *for* $x \ne 0$. *Suppose that* (4.19) *holds. Then every solution of* (4.2) *which has a zero is oscillatory.*

By Theorem 4.1.10, every solution of (4.18) which has a zero is oscillatory if $c_0 > \frac{2}{3\sqrt{3}\beta}$ and $\delta < -2$.

Theorem 4.1.11 *Let* (H_2) *hold*, $b'(t) \le 0$ *and* $b(t)$ *be bounded. Let* $\alpha \ge 1$. *If*

$$\int_\sigma^\infty \left[c(t) - \frac{2}{3\sqrt{3}}\left(-b(t)\right)^{3/2}\right] dt = \infty, \tag{4.20}$$

then a solution of (4.1) *which has a zero is oscillatory.*

Proof Let $|b(t)| \le M$, where M is a positive real number. Let $x(t)$ be a solution of (4.1) such that $x(t_0) = 0$ for some $t_0 \ge T_x \ge \sigma$. We claim that $x(t)$ is oscillatory. If not, then there exists a $t_0' \ge t_0$ such that $x(t_0') = 0$ and $x(t) \ne 0$ for $t > t_0'$. Without any loss of generality, we may assume that $x(t) > 0$ for $t > t_0'$. Clearly, $-x'(t)$ is a solution of the second-order differential equation

$$z''(t) + b(t)z = c(t)x^\alpha(t).$$

From Lemma 1.5.14, it follows that the above equation is nonoscillatory. So there exists a $t_1 > t_0'$ such that $x'(t) \ne 0$ for $t > t_1$.

Suppose that $x'(t) < 0$ for $t > t_1$. Since $x(t_0') = 0$ and $x(t) > 0$ for $t > t_0'$, there exists a t_2, $t_0' < t_2 \le t_1$, such that $x'(t_2) = 0$ and $x'(t) < 0$ for $t > t_2$. Now multiplying (4.1) through by $x'(t)$ and integrating the resulting identity from t_2 to t ($t_2 < t$), we obtain

$$x'(t)x''(t) = \int_{t_2}^{t} \left[\left(x''(s)\right)^2 - b(s)\left(x'(s)\right)^2 - c(s)x^{\alpha}(s)x'(s)\right] ds > 0.$$

Thus, $x''(t) < 0$ for $t > t_2$. This, in turn, implies that $x(t) < 0$ for large t, a contradiction.

So $x'(t) > 0$ for $t > t_1$. Suppose that $x(t)$ is bounded. Integrating (4.1) from t_1 to t, we obtain

$$x''(t) \le x''(t_1) - b(t)x(t) - x^{\alpha}(t_1) \int_{t_1}^{t} c(s)\, ds.$$

Clearly, (4.20) implies that $\int_{\sigma}^{\infty} c(t)\, dt = \infty$. Hence $\lim_{t\to\infty} x''(t) = -\infty$. This, in turn, implies that $x'(t) < 0$ for large t, a contradiction. Next, let $x(t)$ be unbounded. So there exists a $t_3 > t_1$ such that $x(t) > 1$ for $t \ge t_3$. Then $z(t) = \frac{x'(t)}{x(t)}$, for $t \ge t_3$, satisfies

$$z''(t) + 3z(t)z'(t) \le -\left[z^3(t) + b(t)z(t) + c(t)\right]. \tag{4.21}$$

The minimum of $z^3(t) + b(t)z(t) + c(t)$ over all positive $z(t)$ is

$$c(t) - \frac{2}{3\sqrt{3}}\left(-b(t)\right)^{3/2}.$$

Thus,

$$z''(t) + 3z(t)z'(t) \le -\left[c(t) - \frac{2}{3\sqrt{3}}\left(-b(t)\right)^{3/2}\right]$$

for $t \ge t_2$. Integrating the above inequality from t_3 to t and making the use of (4.20), we obtain $z'(t) \to -\infty$ as $t \to \infty$, which implies that $z(t) < 0$ for large t, a contradiction. This completes the proof of the theorem. □

Further, Theorem 4.1.11 may be put in the following form which may be viewed as a stability theorem for nonoscillatory solutions of (4.1).

Theorem 4.1.12 *Suppose that the conditions of Theorem* 4.1.11 *hold. Then for any nonoscillatory solution* $x(t)$ *of* (4.1), *the following properties hold for large* t:

$$\left.\begin{array}{ll}\text{(i)} & \operatorname{sgn} x(t) = \operatorname{sgn} x''(t) \ne \operatorname{sgn} x'(t) = \operatorname{sgn} x'''(t), \\ \text{(ii)} & \lim_{t\to\infty} x(t) = \lim_{t\to\infty} x'(t) = \lim_{t\to\infty} x''(t) = 0.\end{array}\right\} \tag{4.22}$$

Proof From Theorem 4.1.11, it follows that $x(t) \ne 0$ for $t \ge \sigma$. Without any loss of generality, we may assume that $x(t) > 0$ for $t \ge \sigma$. Proceeding as in Theorem 4.1.11, we may show that $x'(t) < 0$ for large t. Consequently, from (4.1) we obtain $x'''(t) < 0$ for large t, and hence $x''(t) > 0$ or < 0 for large t. But $x''(t) < 0$ for large t gives $x(t) < 0$ for large t, a contradiction. Thus, $x''(t) > 0$ eventually. Hence (i) holds. Clearly, (ii) follows from (i) and the observation that (4.20) implies $\int_\sigma^\infty p(t)\,dt = \infty$. Thus the theorem is proved. □

Nelson [28] obtained the following result:

Theorem 4.1.13 *Let $b'(t)$ and $c(t)$ be continuous and $c(t) \ge 0$, $b(t) < 0$ with $b'(t) \ge 0$. For any A and B, suppose that* (4.3) *holds for large t. Then any nonoscillatory solution $x(t)$ of the equation*

$$x''' + b(t)x' + c(t)x^{2n+1} = 0, \quad n = 1, 2, 3, \ldots$$

has the following properties

$$\operatorname{sgn} x(t) = \operatorname{sgn} x''(t) \ne \operatorname{sgn} x'(t),$$
$$\lim_{t\to\infty} x''(t) = \lim_{t\to\infty} x'(t) = 0, \qquad \lim_{t\to\infty} |x(t)| = L \ge 0.$$

Further, if

$$c(t) > \epsilon > 0,$$

then

$$\lim_{t\to\infty} y(t) = 0.$$

The following Theorem 4.1.14 improves Theorem 4.1.13.

Theorem 4.1.14 *Suppose that $c(t) \ge 0$, $b(t) \le 0$ and $b'(t) \ge 0$. Let $\alpha > 1$. If $\int_\sigma^\infty c(t)\,dt = \infty$, then for any nonoscillatory solution $x(t)$ of* (4.1), *the property* (4.22) *holds for large t.*

In view of Remark 4.1.2, the proof of Theorem 4.1.14 is the same as that of Theorem 4.1.13, except that we have to show that $\lim_{t\to\infty} x(t) = 0$ without the extra condition $c(t) > \epsilon > 0$.

Let $x(t) > 0$ for large t. So $x'(t) < 0$ and $x''(t) > 0$ for large t. If possible, let $\lim_{t\to\infty} x(t) = \lambda$, $\lambda > 0$. So $x(t) > \lambda$ for large t. Now integrating (4.1) from t_1 to t, t_1 sufficiently large, we get

$$x''(t) \le x''(t_1) - \lambda^\alpha \int_{t_1}^t c(s)\,ds.$$

Hence $x''(t) < 0$ for large t, a contradiction. Then $\lim_{t\to\infty} x(t) = 0$.

One may view Theorem 4.1.14 as an oscillation criteria for (4.1) as follows:

Theorem 4.1.15 *Suppose that the conditions of Theorem* 4.1.14 *hold. Then a solution of* (4.1) *which has a zero is oscillatory.*

Proof If possible, suppose that $x(t)$ is a nonoscillatory solution of (4.1) with $x(t_0) = 0$ for some $t_0 \geq \sigma$ and $x(t) \neq 0$ for $t > t_0$. Without any loss of generality, we may assume that $x(t) > 0$ for $t > t_0$. Consequently, from Theorem 4.1.14, it follows that $x'(t) < 0$ for $t \geq t_1 > t_0$. Hence there exists a $t_2 \in (t_0, t_1)$ such that $x'(t_2) = 0$ and $x'(t) < 0$ for $t > t_2$. Now, multiplying (4.1) through by $x'(t)$ and integrating the resulting identity from t_2 to t ($t_2 < t$), we get

$$x'(t)x''(t) = \int_{t_2}^{t} \left[\left(x''(s)\right)^2 - b(s)\left(x'(s)\right)^2 - c(s)x^\alpha(s)x'(s)\right] ds > 0.$$

Thus $x''(t) < 0$ for $t > t_2$, a contradiction, which completes the proof of the theorem. □

According to the classical results of I.T. Kiguradze [21] (see Definition 1.1, Definition 1.2 and Lemma 1.1 in [21]), the following results are often accomplished by introducing the concepts of "equation with property A" and/or "equation with property B". More precisely,

Definition 4.1.1 Equation (4.1) or (4.2) is said to have property A if every solution of (4.1) or (4.2) is either oscillatory or satisfies the conditions

$$x(t)x'(t) < 0, \qquad x(t)x''(t) > 0 \quad \text{for } t \geq t_0 \geq \sigma \tag{4.23}$$

and

$$\lim_{t\to\infty} x(t) = \lim_{t\to\infty} x'(t) = \lim_{t\to\infty} x''(t) = 0. \tag{4.24}$$

Equation (4.1) or (4.2) is said to have property B if every solution of (4.1) or (4.2) is either oscillatory or satisfies the conditions

$$x(t)x'(t) > 0, \qquad x(t)x''(t) > 0 \quad \text{for all large } t, \tag{4.25}$$

and

$$\lim_{t\to\infty} \left|x(t)\right| = \lim_{t\to\infty} \left|x'(t)\right| = \lim_{t\to\infty} \left|x''(t)\right| = \infty. \tag{4.26}$$

Now, we consider a superlinear case. In the case of Eq. (4.1), this means that $\alpha > 1$, that is,

$$\int_\sigma^\infty \frac{du}{u^\alpha} < \infty \quad \text{for any } \sigma > 0 \tag{4.27}$$

and for Eq. (4.2), the condition

$$\int_{\pm\sigma}^{\pm\infty} \frac{du}{f(u)} < \infty \quad \text{for any } \sigma > 0 \tag{4.28}$$

is required.

A more general nonlinear equation than (4.1) is given by

$$x''' + b(t)x' + c(t)|x|^{\alpha} \operatorname{sgn} x = 0, \quad \alpha > 0. \tag{4.29}$$

The following lemmas are easy to prove:

Lemma 4.1.4 *If* (H_2) *holds, there exists a solution* $x(t)$ *of* (4.29) ((4.2)) *with* $x(t) \not\equiv 0$ *and* $x(t) \geq 0$, $x'(t) \leq 0$, $x''(t) > 0$, $t \geq T \geq \sigma$. *If* $\alpha \geq 1$, *then* $x(t) > 0$, $x'(t) < 0$ *and* $x''(t) > 0$, $t \geq T \geq \sigma$.

Lemma 4.1.5 *Let* (H_2) *hold. Suppose that* $x(t)$ *is a nonoscillatory solution of* (4.29) ((4.2)), *which exists on* $[\sigma, \infty)$. *Then either*

$$x(t) > 0, \qquad x'(t) < 0, \qquad x''(t) > 0, \qquad x'''(t) < 0 \quad \text{on } [\sigma, \infty)$$

and

$$\lim_{t\to\infty} x'(t) = \lim_{t\to\infty} x''(t) = 0, \ \lim_{t\to\infty} x(t) = L < \infty$$

or there exists $T \geq \sigma$ *such that*

$$x(t) > 0 \quad \text{and} \quad x'(t) > 0 \quad \text{for all } t \geq T.$$

In the following, we give two different criteria of property A for Eq. (4.29). We consider the superlinear case, that is, $\alpha > 1$. So, the condition (4.27) holds.

Theorem 4.1.16 *Let* (H_2) *hold and* $0 \leq \delta \leq 2$ *be a real number, and suppose that for some real constant* $M > 0$,

$$t^{\delta} b(t) \geq -M > -\infty \quad \text{on } [\sigma, \infty) \tag{4.30}$$

and

$$\int^{\infty} t^{\delta} c(t)\, dt = \infty \tag{4.31}$$

holds. Then (4.29) *has property* A.

Proof Let $x(t)$ be a nonoscillatory solution of (4.29). Without any loss of generality, we may assume that $x(t) > 0$ for $t \geq t_0 \geq \sigma$. By Lemma 4.1.5, $x'(t) > 0$ or $x'(t) < 0$ for $t \geq T \geq t_0$. First, we show that $x'(t) > 0$ for $t \geq t_0$ does not hold. Multiplying (4.29) by $t^{\delta} x^{-\alpha}(t)$ and integrating the resulting identity from T to t, we obtain

$$\frac{t^{\delta} x''(t)}{x^{\alpha}(t)} - \delta \frac{t^{\delta-1} x'(t)}{x^{\alpha}(t)} + \frac{\alpha}{2} \frac{t^{\delta}(x'(t))^2}{x^{\alpha+1}(t)} + \delta(\delta-1) \int_T^t \frac{s^{\delta-2} x'(s)}{x^{\alpha}(s)}\, ds$$
$$+ \int_T^t \frac{s^{\delta} b(s) x'(s)}{x^{\alpha}(s)}\, ds + \frac{1}{2}\alpha(\alpha+1) \int_T^t \frac{s^{\delta}(x'(s))^3}{x^{\alpha+2}(s)}\, ds$$

$$- \frac{3}{2}\alpha\delta \int_T^t \frac{s^{\delta-1}(x'(s))^2}{x^{\alpha+1}(s)}\,ds$$

$$= K - \int_T^t s^\delta c(s)\,ds, \tag{4.32}$$

where K is a constant.

From $0 \le \delta \le 2$, (4.27) and (4.30), it follows that the first two integrals on the left-hand side of (4.32) are bounded from below. Using the well-known inequality $Az^2 - Bz \ge -B^2/4A$, $A > 0$, we get

$$\int_T^t \left[\frac{\alpha+1}{2} s^\delta \left(\frac{x'(s)}{x(s)}\right)^2 - \frac{3}{2}\delta s^{\delta-1}\frac{x'(s)}{x(s)}\right]\frac{x'(s)}{x^\alpha(s)}\,ds$$

$$\ge -\frac{9\delta^2}{8(\alpha+1)}\int_T^t s^{\delta-2}\frac{x'(s)}{x^\alpha(s)}\,ds$$

$$> -\infty.$$

So

$$\frac{t^\delta x''(t)}{x^\alpha(t)} - \delta\frac{t^{\delta-1}x'(t)}{x^\alpha(t)} + \frac{\alpha}{2}\frac{t^\delta(x'(t))^2}{x^{\alpha+1}(t)} \le K_1 - \int_T^t s^\delta c(s)\,ds,$$

where K_1 is a constant. Integrating the above inequality from T to t, we have

$$\frac{t^\delta x'(t)}{x^\alpha(t)} + \frac{3\alpha}{2}\int_T^t \frac{s^\delta(x'(s))^2}{x^{\alpha+1}(s)}\,ds - 2\delta\int_T^t \frac{s^{\delta-1}x'(s)}{x^\alpha(s)}\,ds$$

$$\le K_2 + K_1 t - \int_T^t\int_T^s u^\delta c(u)\,du\,ds. \tag{4.33}$$

Since

$$\int_T^t \left[\frac{3\alpha}{2} s^\delta\left(\frac{x'(s)}{x(s)}\right)^2 - 2\delta s^{\delta-1}\frac{x'(s)}{x(s)}\right]x^{1-\alpha}(s)\,ds$$

$$\ge -\frac{2\delta^2}{3\alpha}\int_T^t s^{\delta-2}x^{1-\alpha}(s)\,ds$$

$$\ge -\frac{2\delta^2}{3\alpha}T^{\delta-2}x^{1-\alpha}(T)\int_T^t ds,$$

then from (4.33) we get

$$t^\delta\frac{x'(t)}{x^\alpha(t)} \le K_4 + K_3 t - \int_T^t\int_T^s u^\delta c(u)\,du\,ds.$$

So, it follows from (4.31) that $x'(t) < 0$ for large t, a contradiction. Hence $x'(t) > 0$ does not hold for $t \ge t_0$. To complete the proof of the theorem, it is enough to show, in view of Lemma 4.1.5, that $\lim_{t\to\infty} x(t) = 0$. From (4.29), we have

$$t^\delta x'''(t) \le -L^\alpha t^\delta c(t), \quad t > T_1 \ge \sigma.$$

Integrating the above inequality from T_1 to t, we get

$$t^2 x''(t) - \delta t^{\delta-1} x'(t) + \delta(\delta-1)\int_{T_1}^t s^{\delta-2} x'(s)\,ds \le K_5 - L^\alpha \int_{T_1}^t s^\delta c(s)\,ds.$$

By the integral condition (4.31), the right-hand side of the above inequality tends to $-\infty$ as $t \to \infty$, while all the terms on the left-hand side are either positive or bounded. This contradiction proves the theorem. □

Theorem 4.1.17 ([35]) *Let* (H_2) *hold and* $b(t) \equiv 0$ *on* $[\sigma, \infty)$. *A necessary and sufficient condition for Eq.* (4.29) *to have property A is that*

$$\int^\infty t^2 c(t)\,dt = \infty. \tag{4.34}$$

Remark 4.1.5 Let $b(t) \equiv$ on $[\sigma, \infty)$. Then it is sufficient to require $\delta = 2$ in Theorem 4.1.16, that is, in (4.34). Although Theorem 4.1.17 gives a necessary and sufficient condition to have the property A of (4.29) with $b(t) \equiv 0$, Theorem 4.1.16 provides a better sufficient condition for Eq. (4.29) to have property A with the price that this theorem provides only a sufficient condition.

Erbe [10] and Nelson [28] have studied Eq. (4.29) with $\alpha > 1$, where α is a quotient of odd integers. Their result is presented in the following Theorem 4.1.18.

Theorem 4.1.18 *Let*

$$\begin{aligned} &b(t) \le 0, \qquad b'(t) \ge 0 \quad \text{and} \\ &c(t) > 0 \quad \big(\text{or, } b(t) < 0, b'(t) \ge 0 \text{ and } c(t) \ge 0\big) \text{ on } [0, \infty) \end{aligned} \tag{4.35}$$

and $\int^\infty c(t)\,dt = \infty$. *Then* (4.29) *has property A.*

Example 4.1.4 Consider the equation

$$x''' - t^{-2}x' + \beta\big[1 - (\beta-1)(\beta-2)\big]t^{\beta(1-\alpha)-3}|x|^\alpha \operatorname{sgn} x = 0, \tag{4.36}$$

$t \ge a > 0, \alpha > 1$ and $\beta < 0$ or $\beta \in (\frac{3-\sqrt{5}}{2}, \frac{3+\sqrt{5}}{2})$. The conditions (4.35) of Theorem 4.1.18 and (4.30) of Theorem 4.1.16 are satisfied. First, we consider the case $\beta < 0$. The condition (4.31) of Theorem 4.1.16 is satisfied for any $\alpha > 1$ with $\delta = 2$, and hence Eq. (4.36) has property A. Note that $x(t) = t^\beta$ is a solution of (4.36) and $\beta < 0$. Theorem 4.1.18 can be applied only in the case when $\alpha \ge \frac{\beta-2}{\beta}$. Next, we consider the case $\beta \in (\frac{3-\sqrt{5}}{2}, \frac{3+\sqrt{5}}{2})$. The condition (4.31) of Theorem 4.1.16 is not satisfied. In this case, a solution of (4.36) is $x(t) = t^\beta$ shows that the equation does not have property A.

Lemma 4.1.6 *Let Q be a polynomial, $Q(z) = Az^3 - Bz + C$, $A > 0$, $B \geq 0$ and $z \in R$. Then*

$$Q(z) \geq c - \frac{2}{3}\left(\frac{B^3}{3A}\right)^{1/2}, \quad \text{for all } z \geq 0.$$

The following theorem concerns with the case when the condition (4.30) fails.

Theorem 4.1.19 *Assume* (H$_2$) *and* $0 \leq \delta \leq 2$. *If*

$$\int^\infty \left[c(t) - D\left(-b(t)\right)^{3/2}\right] t^\delta \, dt = \infty \tag{4.37}$$

for any positive constant D, then (4.29) *has property A.*

Proof Let $x(t)$ be a nonoscillatory solution of (4.29), and suppose that $x(t)$ satisfies the property $x(t)x'(t) > 0$ for some $t \geq T \geq \sigma$. The identity (4.32) may be rewritten as

$$\begin{aligned} & t^\delta \frac{x''(t)}{x^\alpha(t)} - \delta t^{\delta-1} \frac{x'(t)}{x^\alpha(t)} + \frac{\alpha}{2} t^\delta \frac{(x'(t))^2}{x^{\alpha+1}(t)} \\ & + \int_T^t \left[\frac{\alpha^2}{2}\left(x^{\alpha-1}(s)\right)^2 \left(\frac{x'(s)}{x^\alpha(s)}\right)^3 + b(s)\frac{x'(s)}{x^\alpha(s)} + c(s)\right] s^\delta \, ds \\ & + \delta(\delta-1) \int_T^t s^{\delta-2} \frac{x'(s)}{x^\alpha(s)} \, ds \\ & + \frac{\alpha}{2} \int_T^t \left[s^\delta \left(\frac{x'(s)}{x(s)}\right)^2 - 3\delta s^{\delta-1} \frac{x'(s)}{x(s)}\right] \frac{x'(s)}{x^\alpha(s)} \, ds = K. \end{aligned} \tag{4.38}$$

Similarly, as in Theorem 4.1.16, we can easily prove that the last two integrals on the left-hand side of (4.38) are bounded from below. Now, denote $z(t) = \frac{x'(t)}{x(t)}$ for $t \geq T$. Then by Lemma 4.1.6, we have

$$\begin{aligned} \frac{\alpha^2}{2}\left(x^{\alpha-1}(t)\right)^2 z^3 - \left(-b(t)\right)z + c(t) &\geq c(t) - \left(\frac{2}{3}\right)^{3/2} \frac{x^{1-\alpha}(t)}{\alpha}\left(-b(t)\right)^{3/2} \\ &\geq c(t) - \left(\frac{2}{3}\right)^{3/2} \frac{x^{1-\alpha}(T)}{\alpha}\left(-b(t)\right)^{3/2} \\ &= c(t) - D_0\left(-b(t)\right)^{3/2} \quad \text{for all } t \geq T. \end{aligned}$$

After substituting this estimate in (4.38), we get

$$t^\delta \frac{x''(t)}{x^\alpha(t)} - \delta t^{\delta-1} \frac{x'(t)}{x^\alpha(t)} + \frac{\alpha}{2} t^\delta \frac{(x'(t))^2}{x^{\alpha+1}(t)} \leq K_6 - \int_T^t \left[c(s) - D_0\left(-b(s)\right)^{3/2}\right] s^\delta \, ds.$$

The rest of the proof is similar to Theorem 4.1.16 and hence omitted. ☐

Example 4.1.5 By Theorem 4.1.19, the equation

$$x''' + (3-2t^2)x' + (4t^3-6t)e^{(\alpha-1)t^2}|x|^\alpha \operatorname{sgn} x = 0, \quad \alpha > 1,\ t > (3/2)^{1/2} \tag{4.39}$$

has property A. An example of such a solution is $x(t) = e^{-t^2}$. Theorem 4.1.18 cannot be applied to (4.39) since $b'(t) < 0$ for $t > (3/2)^{1/2}$.

Note that Theorem 4.1.19 cannot be applied to Eq. (4.36).

In order to obtain results for the superlinear equation (4.2) similar to Theorems 4.1.16 and 4.1.19, we assume that $f \in C^2(R, R)$ and that for all $|u_0| > 0$, there exist constants $K_0 = K_0(u_0) > 0$, $k_0 = k_0(u_0) > 0$ such that

$$\begin{aligned} &f'(u_0) \ge k_0, \qquad 2\big(f'(u)\big)^2 - f(u)f''(u) > 0, \quad \text{and} \\ &\big(f'(u)\big)^2/\big[2\big(f'(u)\big)^2 - f(u)f''(u)\big] \le K_0 \quad \text{for all } |u| \ge |u_0|. \end{aligned} \tag{4.40}$$

Some example of functions f_1, f_2 and $f_3 : R \to R$, satisfying the conditions (4.28) and (4.40), are

$$\left.\begin{aligned} f_1(u) &= |u|^\alpha \operatorname{sgn} u, \quad \alpha > 1 \\ f_2(u) &= \frac{|u|^{2\alpha} \operatorname{sgn} u}{1+|u|^\alpha}, \quad \alpha > 1, \\ f_3(u) &= \sinh u. \end{aligned}\right\} \tag{4.41}$$

Theorem 4.1.20 *Suppose that* (H$_2$) *is satisfied. Let the conditions* (4.28), (4.30), (4.31) *and* (4.40) *hold. Then* (4.2) *has property A.*

Proof The proof is similar to the proof of Theorem 4.1.16. Let $x(t)$ be a nonoscillatory solution of (4.2), and suppose that $x(t)$ satisfies the property $x(t)x'(t) > 0$ for some $t \ge T \ge \sigma$. Multiplying (4.2) by $t^\delta/f(x(t))$ and integrating the resulting identity by parts from T to t, we obtain

$$\begin{aligned} &\frac{t^\delta x''(t)}{f(x(t))} - \delta\frac{t^{\delta-1}x'(t)}{f(x(t))} + \frac{t^\delta f'(x(t))(x'(t))^2}{2f^2(x(t))} + \delta(\delta-1)\int_T^t \frac{s^{\delta-2}x'(s)}{f(x(s))}\,ds \\ &\quad + \int_T^t \frac{s^\delta b(s)x'(s)}{f(x(s))}\,ds + \frac{1}{2}\int_T^t \bigg[\frac{2(f'(x))^2 - f(x)f''(x)}{f^3(x(s))} s^\delta \big(x'(s)\big)^3 \\ &\quad - 3\delta\frac{s^{\delta-1}f'(x)}{f^2(x(s))}\big(x'(s)\big)^2\bigg]\,ds \\ &= K - \int_T^t s^\delta c(s)\,ds, \end{aligned} \tag{4.42}$$

where K is a constant. Since

$$\begin{aligned}
&\int_T^t \left[\left(2\left(f'(x)\right)^2 - f(x)f''(x)\right)s^\delta \left(\frac{x'(s)}{f(x)}\right)^2 - 3\delta s^{\delta-1} f'(x)\frac{x'(s)}{f(x)}\right]\frac{x'(s)}{f(x)}\,ds \\
&\geq -\frac{9}{4}\delta^2 \int_T^t \frac{(f'(x))^2}{2(f'(x))^2 - f(x)f''(x)} s^{\delta-2}\frac{x'(s)}{f(x)}\,ds \\
&\geq -\frac{9}{4}\delta^2 K_0 \int_T^t s^{\delta-2}\frac{x'(s)}{f(x(s))}\,ds \\
&> -\infty,
\end{aligned}$$

it follows from (4.28), (4.30) and (4.40) that all the integrals on the left-hand side of (4.42) are bounded from below, and hence we have

$$\frac{t^\delta x''(t)}{f(x(t))} - \delta\frac{t^{\delta-1}x'(t)}{f(x(t))} + \frac{t^\delta f'(x(t))(x'(t))^2}{2f^2(x(t))} \leq K_1 - \int_T^t s^\delta c(s)\,ds.$$

Integrating the above inequality from T to t, we obtain

$$\begin{aligned}
&\frac{t^\delta x'(t)}{f(x(t))} + \int_T^t \left[\frac{3s^\delta f(x(s))(x'(s))^2}{2f^2(x(s))} - 2\delta\frac{s^{\delta-1}x'(s)}{f(x(s))}\right]ds \\
&\quad \leq K_2 + K_1 t - \int_T^t \int_T^s u^\delta c(u)\,du\,ds.
\end{aligned} \tag{4.43}$$

Since

$$\begin{aligned}
\int_T^t \left[\frac{3}{2}s^\delta f'(x)\left(\frac{x'(s)}{f(x)}\right)^2 2\delta\frac{s^{\delta-1}x'(s)}{f(x(s))}\right]ds &\geq -\frac{2}{3}\delta^2 \int_T^t \frac{s^{\delta-2}}{f'(x)}\,ds \\
&\geq -\frac{2}{3}\delta^2 \frac{T^{\delta-2}}{k_0}\int_T^t ds,
\end{aligned}$$

from (4.43), we get

$$\frac{t^\delta x'(t)}{f(x(t))} \leq K_4 + K_3 t - \int_T^t \int_T^s u^\delta c(u)\,du\,ds.$$

The rest of the proof is similar to that of Theorem 4.1.16, and hence is omitted. □

Example 4.1.6 By Theorem 4.1.20, the equation

$$x''' - \frac{1}{t^2}x' + \frac{5}{t^4 \sinh\frac{1}{t}}\sinh x = 0, \quad t \geq a > 0 \tag{4.44}$$

has property A with $\delta = 2$. In particular, $x(t) = \frac{1}{t}$ is a solution of (4.44) satisfying the conditions of property A.

Remark 4.1.6 If $b(t) \equiv 0$, the condition (4.34) is sufficient for Eq. (4.29) to have property A. But, in the case of (4.2), the function f need not be twice differentiable. For example, if f is nondecreasing and there exists a constant $C > 0$ such that $|f(uv)| \geq Cf(u)|f(v)|$ for $u \geq 0$, $v \in R$, then the condition (4.34) and

$$\int^{\infty} tb(t)f(t)\,dt = \infty$$

is sufficient for the superlinear equation (4.2) to have property A, one may refer to Corollary 1 in [34].

Suppose that $f \in C^2(R, R)$ and that for all $|u_0| > 0$, there exist constants $K_0 = K_0(u_0) > 0$, $k_0 = k_0(u_0) > 0$ and $k_0' = k_0'(u_0) > 0$ such that

$$\left.\begin{aligned} &f'(u) \geq k_0, \qquad A\big(f'(u)\big)^2 - f(u)f''(u) > 0 \\ &\text{and} \\ &\big(f'(u)\big)^2/\big[A\big(f'(u)\big)^2 - f(u)f''(u)\big] \leq K_0 \quad \text{or} \quad A\big(f'(u)\big)^2 - f(u)f''(u) \geq k_0' \end{aligned}\right\} \tag{4.45}$$

for all $|u| \geq |u_0|$, where A is some constant with $0 < A < 2$.

An example of functions that satisfies the conditions in (4.45) are given in (4.41).

The next theorem concerns with the case when the condition (4.30) fails to satisfy Eq. (4.2).

Theorem 4.1.21 *Let the conditions* (4.28), (4.37) *and* (4.45) *hold. Then* (4.2) *has property* A.

Proof Suppose that $x(t)$ is a nonoscillatory solution of (4.2) that satisfies the property $x(t)x'(t) > 0$ for $t \geq T \geq \sigma$. Then the identity (4.42) may be rewritten as

$$\begin{aligned} &t^{\delta}\frac{x''(t)}{f(x(t))} - \delta\frac{t^{\delta-1}x'(t)}{f(x(t))} + t^{\delta}\frac{f'(x(t))(x'(t))^2}{2f^2(x(t))} \\ &+ \int_T^t \left[\frac{A(f'(x))^2 - f(x)f''(x)}{2}s^{\delta}\left(\frac{x'(s)}{f(x(s))}\right)^2 - \frac{3}{2}\delta s^{\delta-1}f'(x)\frac{x'(s)}{f(x(s))}\right] \\ &\times \frac{x'(s)}{f(x(s))}\,ds + \delta(\delta-1)\int_T^t s^{\delta-2}\frac{x'(s)}{f(x(s))}\,ds \\ &+ \int_T^t \left[\frac{B}{2}\big(f'(x)\big)^2\left(\frac{x'(s)}{f(x(s))}\right)^3 + b(s)\frac{x'(s)}{f(x(s))} + c(s)\right]s^{\delta}\,ds = K, \end{aligned} \tag{4.46}$$

or

$$t^\delta \frac{x''(t)}{f(x(t))} - \delta \frac{t^{\delta-1}x'(t)}{f(x(t))} + t^\delta \frac{f'(x(t))(x'(t))^2}{2f^2(x(t))}$$
$$+ \int_T^t \left[\frac{B}{2} f'^2(x)s^\delta \left(\frac{x'(s)}{f(x(s))} \right)^2 - \frac{3}{2}\delta s^{\delta-1} f'(x)\frac{x'(s)}{f(x(s))} \right] \frac{x'(s)}{f(x(s))}\, ds$$
$$+ \delta(\delta-1) \int_T^t s^{\delta-2} \frac{x'(s)}{f(x(s))}\, ds$$
$$+ \int_T^t \left[\frac{A(f'(x))^2 - f(x)f''(x)}{2} \left(\frac{x'(s)}{f(x(s))} \right)^3 + b(s)\frac{x'(s)}{f(x(s))} + c(s) \right] s^\delta\, ds = K, \tag{4.47}$$

where $B = 2 - A$ is a constant.

From $0 \le \delta \le 2$, (4.28) and (4.45), it follows that the first two integrals on the left-hand side of (4.46) or (4.47) are bounded from below. If we denote $z(t) = \frac{x'(t)}{f(x(t))}$ for $t \ge T$ and we use Lemma 4.1.6, then from (4.46) or (4.47), we get

$$\frac{t^\delta x''(t)}{f(x(t))} - \delta \frac{t^{\delta-1}x'(t)}{f(x(t))} + t^\delta \frac{f'(x(t))(x'(t))^2}{2f^2(x(t))} \le K_7 - \int_T^t \left[c(s) - D_0\big(-b(s)\big)^{3/2} \right] ds.$$

The rest of the proof is similar to Theorem 4.1.16, and hence omitted. □

Example 4.1.7 By Theorem 4.1.21, the equation

$$x''' + (3 - 2t^2)x' + (4t^3 - 6t)\big(e^{5t^2} + e^{2t^2}\big)\frac{x^6 \operatorname{sgn} x}{1 + |x|^3} = 0, \qquad t > \frac{\sqrt{3}}{2}$$

has property A. Note that $x(t) = e^{-t^2}$ is a solution of this example.

Theorem 4.1.21 cannot be applied to (4.44), because the condition (4.37) fails to hold.

Next, we give results on the oscillation of (4.1) or (4.2) under the assumption (H_3).

Theorem 4.1.22 *Suppose that* (H_3) *holds and* $b'(t) \ge 0$. *If*

$$\int_\sigma^\infty c(t)\, dt = -\infty, \tag{4.48}$$

then every bounded solution of (4.1) *is either oscillatory or tends to zero as* $t \to \infty$.

Proof Let $x(t)$ be a bounded solution of (4.1) and suppose that $x(t)$ is nonoscillatory. Without any loss of generality, we assume that $x(t) > 0$ for $t \ge t_0 \ge T_x \ge \sigma$. We have to show that $\lim_{t\to\infty} x(t) = 0$.

Clearly, $x'(t)$ is a solution of the second-order nonhomogeneous equation

$$z''(t) + b(t)z(t) = -c(t)x^{\alpha}(t).$$

By Lemma 1.5.14, $x'(t)$ is nonoscillatory. So there exists a $t_1 > t_0$ such that $x'(t) > 0$ or $x'(t) < 0$ for $t \geq t_1$. Suppose that $x'(t) > 0$ for $t \geq t_1$. Then $x'''(t) \geq 0$ for $t \geq t_1$ from (4.1), Thus, $x''(t)$ is nonoscillatory. We note that $x''(t) \equiv 0$ implies that $x(t)$ is unbounded. If $x''(t) > 0$ for large t, then $x(t)$ is unbounded. So, $x''(t) < 0$ for $t \geq t_2 \geq t_1$. Now multiplying (4.1) through by $x(t)$ and integrating the resulting identity from t_2 to t, we obtain

$$\begin{aligned}(x'(t))^2 &\leq (x'(t))^2 - 2x(t)x''(t) - b(t)x^2(t)\\ &\leq (x'(t_2))^2 - b(t_2)x^2(t_2) - 2x(t_2)x''(t_2) + 2x^{\alpha+1}(t_2)\int_{t_2}^{t} c(s)\,ds.\end{aligned}$$

Thus $(x'(t))^2 < 0$ for large t, a contradiction.

Hence $x'(t) < 0$ for $t \geq t_1$. Consequently, $\lim_{t\to\infty} x(t)$ exists. If possible, let $\lim_{t\to\infty} x(t) = \lambda$ and $\lambda > 0$. Integration of (4.1) from t_1 to t gives

$$x''(t) \geq x''(t_1) + b(t_1)x(t_1) - x^{\alpha}(t)\int_{t_1}^{t} c(s)\,ds,$$

which further yields $\lim_{t\to\infty} x''(t) = \infty$ and so $x'(t) > 0$ for large t, a contradiction. Thus $\lim_{t\to\infty} x(t) = 0$. This completes the proof of the theorem. □

Let $\alpha > 1$ be a ratio of odd integers. Suppose that $b(t) \equiv 0$ for $t \geq \sigma$. Then Parhi and Parhi [31] proved that if $c(t) < 0$ for $t \geq \sigma$ and (4.48) holds, then every bounded nontrivial solution of (4.1) is oscillatory. Further, a necessary and sufficient condition for a solution $x(t)$ of (4.1) to be oscillatory for $t \geq t_0 \geq \sigma$ is that $F[x(t)] > 0$ for $t \geq t_0$, where $F[x(t)]$ is given in (4.11).

One may verify that the following statements are true.

(a) Let $x(t) \in C^2([t_0, \infty))$. Then $x(t) > 0$, $x'(t) < 0$ and $x''(t) < 0$ cannot hold for large t.
(b) Let $x(t) \in C^3([t_0, \infty))$. Then $x(t) > 0$, $x'(t) < 0$ and $x'''(t) \geq 0$ cannot hold for large t.

Lemma 4.1.7 *Let* (H$_3$) *holds and* $x(t)$ *be a solution of* (4.1) *with the properties* $x(t_0) \geq 0$, $x'(t_0) \geq 0$ *and* $x''(t_0) > 0$ *for some* $t_0 \geq \sigma$. *Then* $x(t) > 0$, $x'(t) > 0$, $x''(t) > 0$ *for* $t > t_0$ *and* $x(t) \to \infty$ *and* $x'(t) \to \infty$ *as* $t \to \infty$.

Lemma 4.1.8 *Let* (H$_3$) *holds and* $c(t)$ *satisfy the property* (4.48). *If* $x(t)$ *is a solution of* (4.1) *with* $x(t) > 0$ *for some* $t \geq t_0 \geq \sigma$, *then there exists a* $t_1 \geq t_0$ *such that either*

$$x(t) > 0, \qquad x'(t) > 0 \quad \textit{and} \quad x''(t) > 0 \quad \textit{for } t \geq t_1$$

or

$$x(t) > 0, \qquad x'(t) < 0 \quad \textit{for } t > t_1,$$

and

$$\lim_{t\to\infty} x(t) = 0, \qquad \limsup_{t\to\infty} x'(t) = 0.$$

Lemma 4.1.9 *Let* (H$_3$) *and* (4.48) *hold. Then for any solution* $x(t)$ *of* (4.1) *which converges monotonically to zero with the property* $x(t)x'(t) < 0$ *for* $t \ge t_0 \ge \sigma$, *there exists* $T \ge t_0$ *such that* $F[x(t)] > 0$ *for* $t \ge t_0$, *where* $F[x(t)]$ *is given in* (4.11).

Theorem 4.1.23 *Let* (H$_3$), (4.48) *and* $b'(t) > 0$. *Then a necessary and sufficient condition for a solution* $x(t)$ *of* (4.1) *to be oscillatory or to be monotonically converging to zero is* $F[x(t)] < 0$ *for* $t \ge \sigma$.

Theorem 4.1.24 *Let* (H$_4$) *hold,* $b'(t) \ge 0$, $b(t)$ *be bounded and* (4.48) *hold. Then every bounded solution of* (4.1) *oscillates.*

Proof Let $x(t)$ be a bounded solution of (4.1). We claim that $x(t)$ is oscillatory. If not, then $x(t)$ is nonoscillatory for $t \ge t_0 \ge \sigma$. Let $x(t) > 0$ for $t \ge t_0$. Then $x'(t) > 0$ or < 0 or oscillatory for $t \ge T \ge t_0$. First suppose that $x'(t) > 0$ for $t \ge T$. Then the integral identity

$$x''(t) + b(t)x(t) + \int_T^t \big(c(s)x^{\alpha-1}(s) - b'(s)\big)x(s)\,ds = k_0, \tag{4.49}$$

implies that

$$x''(t) > -b(t)x(t) + k_0 - x^\alpha(T)\int_T^t c(s)\,ds$$

and hence $x''(t) > 0$ for large t, where k_0 is a constant. Consequently, $x(t)$ is unbounded, a contradiction. Next, suppose that $x'(t) < 0$ for $\ge T$. Then from (4.1), $x'''(t) > 0$ for $t \ge T$. So there exists a $t_1 \ge t_0$ such that either $x''(t) > 0$ or < 0 for $t \ge t_1$. If $x''(t) > 0$ for $t \ge t_1$, then $x'''(t) > 0$ for $t \ge t_1$ which further implies that $x'(t) > 0$ for large t, a contradiction. If $x''(t) < 0$ for $t \ge t_1$, then $x'(t) < 0$ for $t \ge t_1$ implies that $x(t) < 0$ for large t, a contradiction. Finally, suppose that $x'(t)$ is oscillatory. Then it has infinitely many zeros where it changes its signs. Clearly, $\liminf_{t\to\infty} x(t) = 0$. Indeed, if $x(t) > k > 0$ for $t \ge t_0$, then from (4.49), we obtain $x''(t) > 0$ for $t \ge T_1 > t_0$ and hence $x'(t)$ is increasing for $t \ge T_1$, a contradiction. Hence $\liminf_{t\to\infty} x(t) = 0$. We consider the following two cases: (i) $-\int_T^\infty (c(s)x^{\alpha-1}(s) - b'(s))x(s)\,ds = \infty$ and (ii) $-\int_T^\infty (c(s)x^{\alpha-1}(s) - b'(s))x(s)\,ds < \infty$. Let (i) holds. Then from (4.49), we obtain $x''(t) > 0$ for large t, which contradicts the oscillatory nature of $x(t)$. Hence (ii) holds. In this case, let $\{t_n\}_{n=1}^\infty$, $t_n \to \infty$ as $n \to \infty$ be a sequence of zeros of

$x'(t)$ and $x''(t_n) > 0$. Clearly, $x(t_i) \to 0$ as $i \to \infty$. Then it follows from (4.49) that $\{x''(t_i)\}$ is bounded on $[t_0, \infty)$. On the other hand, if we rewrite the integral identity

$$F\big[x(t)\big] = F\big[x(t_1)\big] + \int_{t_1}^{t} \big(2c(s)x^{\alpha-1}(s) - b'(s)\big)x^2(s)\,ds \tag{4.50}$$

by

$$x(t)x''(t) - \frac{1}{2}\big(x'(t)\big)^2 + \frac{1}{2}b(t)x^2(t) + \int_{t_1}^{t}\left(c(s)x^{\alpha-1}(s) - \frac{1}{2}b'(s)\right)x^2(s)\,ds = k, \tag{4.51}$$

where $k = x(t_1)x''(t_1) - \frac{1}{2}(x'(t_1))^2 + \frac{1}{2}b(t_1)x^2(t_1) > 0$, we obtain for $x''(t_i), i = 1, 2, 3, \ldots$, the equality

$$x''(t_i) = \frac{k}{x(t_i)} - \frac{1}{2}b(t_i)x(t_i) - \frac{1}{x(t_i)}\int_{t_1}^{t_i}\left(c(s)x^{\alpha-1}(s) - \frac{1}{2}b'(s)\right)x^2(s)\,ds,$$

because $x'(t_i) = 0, i = 1, 2, 3, \ldots$. Hence $x''(t_i) \to \infty$ as $t_i \to \infty$, which is a contradiction to the boundedness of $x''(t)$. This completes the proof of the theorem. □

Theorem 4.1.25 *Suppose that the conditions of Theorem* 4.1.24 *are satisfied. Then a necessary and sufficient condition for a solution $x(t)$ of* (4.1) *to be oscillatory is that $F[x(t)] < 0$ for $t \geq \sigma$, where $F[x(t)]$ is given in* (4.11).

One may verify from the identity (4.51) that if the conditions of Theorem 4.1.24 or Theorem 4.1.25 are satisfied and $x(t)$ is a solution of (4.1) with $x(t_0) = x'(t_0) = 0$ and $x''(t_0) > 0$, $t_0 \geq \sigma$, then $x(t) > 0$ for $t \geq t_0$. This shows the existence of a nonoscillatory solution of (4.1).

Theorem 4.1.26 *Let* (H_4) *hold, $b'(t) \leq 0, \lim_{t\to\infty} b(t) = 0$ and $c(t)$ satisfies the property* (4.48). *Then every bounded solution of* (4.1) *is either oscillatory or* $\liminf_{t\to\infty} |x(t)| = 0$.

Proof Let $x(t)$ be a bounded solution of (4.1). If $x(t)$ is oscillatory, then there is nothing to prove. Let $x(t)$ be nonoscillatory. Suppose that $x(t) > 0$ for $t \geq t_0 \geq \sigma$.

Integrating (4.1) term by term from t_0 to t, we get

$$x''(t) + \int_{t_0}^{t} b(s)x'(s)\,ds + \int_{t_0}^{t} c(s)x^{\alpha}(s)\,ds = x''(t_0). \tag{4.52}$$

Let $|x(t)| \leq k, k > 0$. The function $b(t)$ is nonincreasing and $b(t) \to 0$ as $t \to \infty$. Hence, for a given $\epsilon/4k > 0$, there exists $T \geq t_0$ such that $0 \leq b(t) \leq \epsilon/4k$ for $t \geq T_0$. Let T_1 and T_2 be such that $T_0 < T_1 < T_2$. By the second mean value theorem, there exists T_3, $T_1 \leq T_3 \leq T_2$ such that

$$\begin{aligned}\left|\int_{T_1}^{T_2} b(s)x'(s)\,ds\right| &= \left|b(T_1)\int_{T_1}^{T_3} x'(s)\,ds + b(T_2)\int_{T_3}^{T_2} x'(s)\,ds\right| \\ &\le |b(T_1)||x(T_3) - x(T_1)| + |b(T_2)||x(T_2) - x(T_3)| \\ &\le \frac{\epsilon}{4k}\cdot 4k = \epsilon.\end{aligned}$$

Then, by the Cauchy–Bolzano criterion, $\int_{t_0}^{\infty} b(s)x'(s)\,ds$ converges. Now, we consider three different cases, viz., (i) $x'(t) > 0$ for $t \ge t_1 \ge t_0$, (ii) $x'(t) \le 0$ for $t \ge t_1 \ge t_0$ and (iii) $x'(t)$ is oscillatory for $t \ge t_1 \ge t_0$. First suppose that $x'(t) > 0$ for $t \ge t_1$. Then it follows from (4.52) that $x''(t) \to \infty$ as $t \to \infty$, which, in turn, implies that $x(t)$ is unbounded, a contradiction. Now, suppose that $x'(t) \le 0$ for $t \ge t_1$. Then (4.1) implies that $x'''(t) > 0$ for $t \ge t_1$. Thus, $x''(t) > 0$ or < 0 for $t \ge t_2 \ge t_1$. If $x''(t) > 0$ for $t \ge t_2$, then $x'''(t) > 0$ for $t \ge t_1$ implies that $x'(t) > 0$ for large t, a contradiction. Again, if $x''(t) < 0$ for $t \ge t_2$, then $x'(t) \le 0$ for $t \ge t_1$ implies that $x(t) < 0$ for large t, a contradiction. Hence $x'(t)$ is oscillatory on (t_1, ∞), that is, $x'(t)$ changes its sign infinitely many times in $[t_1, \infty)$. In this case, if $x(t) > k_1$ for $t \ge t_1$, then $\int_{t_1}^{\infty} c(t)x^\alpha(t)\,dt = -\infty$ and (4.52) yields $x''(t) \to \infty$ as $t \to \infty$, which is a contradiction with the boundedness of $x(t)$. Therefore $\liminf_{t\to\infty} x(t) = 0$. □

Theorem 4.1.27 *Let* (H$_4$) *hold,* $b'(t) \le 0$, $\lim_{t\to\infty} b(t) = 0$ *and* $-c(t) + \frac{1}{2}b'(t) \ge k > 0$. *If* $x(t)$ *is a solution of* (4.1) *such that* $F[x(t)] > 0$ *for* $t \ge \sigma$, *then* $x(t)$ *is oscillatory. Moreover, any solution of* (4.1) *which has a zero in* $[\sigma, \infty)$ *is oscillatory.*

Proof The conditions $2c(t) - b'(t) \le k < 0$ and $b'(t) \le 0$ imply the condition (4.48).

Now suppose that $F[x(t)] > 0$ for $t \ge \sigma$ and $x(t)$ is nonoscillatory. Let $x(t) > 0$ for $t \ge t_0 \ge \sigma$. If $x'(t) < 0$ for $t \ge t_1 \ge t_0$, then we obtain a contradiction as in Theorem 4.1.26. Suppose that $x'(t) > 0$ for $t \ge t_1 \ge t_0$. If $x(t)$ is bounded from above, then it is oscillatory or $\liminf_{t\to\infty} x(t) = 0$ by Theorem 4.1.26. But $\liminf_{t\to\infty} x(t) = 0$ is a contradiction to the fact that $x(t) > 0$ and $x'(t) > 0$ for $t \ge t_1$. If $\lim_{t\to\infty} x(t) = \infty$, then there exists a $T_1 \ge t_1$ such that $x(t) > 1$ for $t \ge T_1$ and

$$-\left[c(t)x^{\alpha-1}(t) - \frac{1}{2}b'(t)\right]x^2(t) \ge kx^2(t) > k > 0.$$

Then the integral identity (4.51) implies that $F[x(t)] < 0$ for large t, which is a contradiction. Finally, suppose that $x'(t)$ is oscillatory. Then there exists a sequence $\{t_n\}_{n=1}^{\infty}$, $t_n \to \infty$ as $n \to \infty$ of zeros of $x'(t)$ such that $x''(t_n) \ge 0$ for $n = 1, 2, \ldots$ which yields a contradiction to $F[x(t)] > 0$ at the points $t = t_n$, that is, $F[x(t_n)] \le 0$. This contradiction completes the proof of the theorem. □

Theorem 4.1.28 *Let* (H$_3$) *hold,* $b'(t) \le 0$, $b(t)$ *be bounded from below and* (4.48) *hold. Then every bounded solution of* (4.1) *is either oscillatory or converges monotonically to zero as* $t \to \infty$.

Proof Let $x(t)$ be a bounded solution of (4.1). The proof is complete, if $x(t)$ is oscillatory. Let $x(t)$ be nonoscillatory. Suppose that $x(t) > 0$ for $t \geq t_0 \geq \sigma$. If $x'(t) > 0$ for large t, then $x'''(t) > 0$ for large t. Thus, $x''(t) > 0$ or < 0 for large t. If $x''(t) > 0$ for large t, then $x(t)$ is unbounded, a contradiction. Hence $x''(t) < 0$ for $t \geq t_1 \geq t_0$. Integration of (4.1), term by term, we obtain (4.52) with $t_0 = t_1$. Clearly, $\lim_{t\to\infty} x(t) = k < \infty$, $\int_{t_1}^{\infty} c(s)x^{\alpha}(s)\,ds = -\infty$ and $m \leq b(t) \leq 0$ imply that

$$\int_{t_1}^{t} b(s)x'(s)\,ds \geq m\int_{t_1}^{t} x'(s)\,ds \geq m\big[k - x(t_1)\big],$$

and hence $\int_{t_1}^{t} b(s)x'(s)\,ds \to l$, $0 > l > -\infty$. This, in turn, implies from (4.52) that $x''(t) \to \infty$ as $t \to \infty$, a contradiction to the boundedness of $x(t)$. Let $x'(t) \leq 0$ for $t \geq t_1 \geq t_0$. Then $x(t)$ is nonincreasing. If $\lim_{t\to\infty} x(t) = k \geq 0$, then from (4.52) it follows that $x''(t) \to \infty$ as $t \to \infty$, again a contradiction. Hence $\lim_{t\to\infty} x(t) = 0$. Finally, suppose that $x'(t)$ is oscillatory. Then $x'(t)$ changes its sign infinitely many times and there exists a point $T_1 \geq t_0$ such that $x(T_1) > 0$, $x'(T_1) = 0$ and $x''(T_1) \geq 0$. Then by Lemma 4.1.7, $x(t) \to \infty$ as $t \to \infty$, a contradiction. This completes the proof of the theorem. □

Theorem 4.1.29 *Let* (H_3) *hold,* $-c(t) > k > 0$, $b'(t) \leq 0$ *and* $b'(t) \to 0$ *as* $t \to \infty$. *Let* $x(t)$ *be a solution of* (4.1) *such that* $F[x(t)] > 0$ *for* $t \geq t_0 \geq \sigma$. *Then* $x(t)$ *is either oscillatory or converges monotonically to zero as* $t \to \infty$.

Greguš and Greguš Jr. [14] obtained some sufficient conditions for oscillation and nonoscillation of (4.1) and (4.2) with the use of oscillation and nonoscillation of the linear third-order differential equation

$$x''' + 2A(t)x' + \big[A'(t) + B(t)\big]x = 0, \tag{4.53}$$

where A', B are continuous functions on $[\sigma, \infty)$ and $B(t) \geq 0$ for $t \geq \sigma$ with the condition that $B(t) \equiv 0$ does not hold on any subinterval of $[\sigma, \infty)$.

The adjoint equation to (4.53) is

$$z''' + 2A(t)z' + \big[A'(t) - B(t)\big]z = 0. \tag{4.54}$$

If $w(t) > 0$ for $t \geq t_0 \geq \sigma$ is a solution of (4.54), then there exists a two-parameter family of solutions $x(t)$ of Eq. (4.53) that satisfies the equation

$$\left(\frac{1}{w(t)}x'\right)' + \frac{w''(t) + 2A(t)w(t)}{w^2(t)}x = 0. \tag{4.55}$$

For $A(t) \leq 0$ and $A'(t) + B(t) \geq 0$ for $t \geq \sigma$, the solution $w(t)$ of the differential equation (4.54) with the property $w(t_0) = w''(t_0) = 0$, $w'(t_0) > 0$, $t_0 \geq \sigma$ has the property $w(t) > 0$ and $w''(t) + 2A(t)w(t) > 0$ for $t \geq t_0$. Further, if we substitute this solution $w(t)$ into (4.55), then the solution $x(t)$ of (4.55) has the property

$x'(t_0) = 0$. Differentiating (4.55), term by term, we obtain (4.53). Hence all solutions of (4.55) are at the same time solutions of (4.53).

Let $A(t) \le 0$ and $A'(t) + B(t) \ge 0$ for $t \ge \sigma$ with $B(t) \not\equiv 0$ on any subinterval of $[\sigma, \infty)$. Then Greguš [12] showed that if

$$\int_{t_0}^{\infty} \left[A'(t) + B(t) - \frac{4}{3}\left(-\frac{2}{3}A^3(t)\right)^{1/2} \right] dt = \infty, \qquad t_0 \ge \sigma,$$

then (4.53) is oscillatory. In fact, it is proved that any solution of (4.53) which has a zero in $[\sigma, \infty)$ is oscillatory. On the other hand, if $A(t) \le 0$ and $B(t) \ge 0$ for $t \ge \sigma$ and $B(t) \not\equiv 0$ on any subinterval of $[\sigma, \infty)$, then there exists at least one nontrivial solution $x(t)$ of (4.53) such that $x(t) \ne 0$ for $t \ge \sigma$. Moreover, $x(t)$ and $x'(t)$ are monotone functions in $[\sigma, \infty)$ and

$$\operatorname{sgn} x(t) = \operatorname{sgn} x''(t) \ne \operatorname{sgn} x'(t) \quad \text{for } t \ge \sigma.$$

Observations on (4.53) and (4.54) are used in [14] to obtain the following results on (4.1) and (4.2).

Theorem 4.1.30 *Let* (H_2) *hold and* $b'(t) \le 0$. *Further, assume that* $-k^2 < b(t), k \ne 0$ *and* $\lim_{t\to\infty} c(t) = \infty$. *Then each solution* $x(t)$ *of* (4.1) *defined on* (t_0, ∞), $t_0 \ge \sigma$ *with the property* $x(t) \ne 0$ *for* $t \ge t_0$ *has also the property: There exists a* $T \ge t_0$ *such that for* $t \ge T$, *we have* $\operatorname{sgn} x(t) \ne \operatorname{sgn} x'(t)$.

Theorem 4.1.31 *Let the coefficients* $b(t)$ *and* $c(t)$ *satisfy the assumptions of Theorem* 4.1.30. *Let* $f(x)$ *be continuous for* $x \in (-\infty, \infty)$ *and* $\frac{f(x)}{x} > k > 0$ *for* $x \ne 0$. *Then each solution* $x(t)$ *of* (4.2) *defined on* (t_0, ∞) *with* $x(t) \ne 0$ *for* $t \in (t_0, \infty)$, $t_0 \ge \sigma$ *has also the property: There exists* $T \ge t_0$ *such that for* $t \ge T$, *we have* $\operatorname{sgn} x(t) \ne \operatorname{sgn} x'(t)$. *Moreover, any solution of* (4.2) *which has a zero in* $[\sigma, \infty)$ *is oscillatory.*

Now, we consider the equation

$$y''' + 2A(t)y' + \big(A'(t) + B_1(t)\big)y = 0, \tag{4.56}$$

where $B_1(t) \ge 0$ is continuous in $[\sigma, \infty)$. Then Greguš [12] gave the following classical comparison result, which we shall use in the sequel.

Theorem 4.1.32 (Theorem 2.5, [12]) *Let* $B(t) \ge 0, t \ge \sigma$ *be such that* $B(t) \not\equiv 0$ *on any subinterval of* $[\sigma, \infty)$ *and* $B(t) \le B_1(t)$ *for* $t \ge \sigma$. *If the differential equation* (4.53) *is oscillatory in* $[\sigma, \infty)$, *then the differential equation* (4.56) *is also oscillatory. If* (4.56) *is nonoscillatory, then* (4.53) *is nonoscillatory in* $[\sigma, \infty)$.

This classical comparison theorem plays a crucial role in obtaining oscillation and nonoscillation of several third-order nonlinear differential equations, which we state here.

Remark 4.1.7 Theorem 3.1 in [12] states that if $A(t) \geq m > 0$, $A'(t) + B(t) \geq 0$ for every $t \geq \sigma$, then every solution of Eq. (4.53) is oscillatory in $[\sigma, \infty)$ except one solution x (up to the linear dependence) for which $\lim_{t\to\infty} x(t) = \lim_{t\to\infty} x'(t) = 0$ and $x(t)$ is in the class L^2, that is, $\int_{t_0}^{\infty} x^2(s)\,ds < \infty$, $t_0 \geq \sigma$. However, $x \in L^2$ and $\lim_{t\to\infty} x'(t) = 0$ imply, by Lemma 1.5.15, that $\lim_{t\to\infty} x(t) = 0$.

Next, suppose that $B(t) \geq 0$, $B(t) \not\equiv 0$ on any subinterval of $[\sigma, \infty)$ and the second-order linear differential equation

$$z'' + \frac{A(t)}{2} z = 0$$

is oscillatory. Then Greguš, in his monograph ([12], Theorem 2.17 and Corollary 2.3), shows that a necessary and sufficient condition for a nontrivial solution of (4.53) to be nonoscillatory is that $F[x(t)] < 0$ for $t \geq \sigma$, where $F[x(t)]$ is given in (4.11).

Let $b'(t)$ and $c(t)$ be continuous functions of $t \in [\sigma, \infty)$, $xf(x) > 0$ for $x \neq 0$ and

$$\lim_{x\to 0} \frac{f(x)}{x} = \lambda, \quad 0 \leq \lambda < \infty \tag{4.57}$$

holds. If $x_1(t)$ is a nontrivial solution of (4.2), then $x_1(t)$ is a solution of the linear differential equation

$$x''' + b(t)x' + c(t)H(x_1)x = 0, \tag{4.58}$$

where

$$H(x_1) = \begin{cases} \frac{f(x_1)}{x_1} & \text{for } x_1 \neq 0, \\ \lambda & \text{for } x_1 = 0. \end{cases} \tag{4.59}$$

The adjoint to the differential equation (4.58) is

$$y''' + b(t)y' + \left[b'(t) - c(t)H(x_1)\right]y = 0. \tag{4.60}$$

If we multiply (4.58) by $x(t)$ and (4.60) by $y(t)$, and integrating them from t_0 to t, $t_0 \geq \sigma$, we obtain the identities

$$x(t)x''(t) - \frac{1}{2}\left(x'(t)\right)^2 + \frac{1}{2}b(t)x^2(t) + \int_{t_0}^{t} \left[c(s)H\left(x_1(s)\right) - \frac{1}{2}b'(s)\right]x^2(s)\,ds = \text{const.} \tag{4.61}$$

and

$$y(t)y''(t) - \frac{1}{2}\left(y'(t)\right)^2 + \frac{1}{2}b(t)y^2(t) + \int_{t_0}^{t} \left[c(s)H\left(x_1(s)\right) - \frac{1}{2}b'(s)\right]y^2(s)\,ds = \text{const.} \tag{4.62}$$

The following lemma is an immediate consequence of the identities (4.61) and (4.62).

Lemma 4.1.10 *Let $c(t) > 0$ and $b'(t) \leq 0$ for $t \geq \sigma$. Then every solution $x(t)$ of (4.58) defined on $[T, \infty)$, $T > \sigma$, with the property that $x(t_0) = x'(t_0) = 0$ and $x''(t_0) > 0$, $t_0 \geq T$, has no zeros to the left of t_0, and every solution $y(t)$ of (4.60), with the property $y(t_0) = y'(t_0) = 0$ and $y''(t_0) > 0$, $t_0 \geq T$, has no zeros to the right of t_0. Moreover, every solution $x(t)$ of (4.58) and $y(t)$ of (4.60) has at most one double zero.*

Theorem 4.1.33 *Let $xf(x) > 0$ for $x \neq 0$ and (4.57) hold, and let $f(x) \geq \lambda > 0$ for every $x \in R$. Further, let $b(t) \geq M > 0$, $c(t) > M$ and $\lambda c(t) - b'(t) \geq 0$ for $t \geq \sigma$. Then every solution $x(t)$ of (4.2) defined in the interval $[t_0, \infty)$, $t_0 \geq \sigma$ is either oscillatory on $[t_0, \infty)$ or it has no zeros in $[t_0, \infty)$ and then $x(t) \to 0$, $x'(t) \to 0$ for $t \to \infty$ and $x(t) \in L^2([t_0, \infty))$.*

Theorem 4.1.34 *Let $xf(x) > 0$ for $x \neq 0$ and (4.57) hold. Let $b'(t) \leq 0$, $c(t) > 0$ for $t \geq \sigma$ and let the second-order linear differential equation*

$$u'' + \frac{1}{4}b(t)u = 0$$

be oscillatory in $[\sigma, \infty)$. Then, a necessary and sufficient condition for a nontrivial solution $x(t)$ of (4.2) defined on $[t_0, \infty) \subset [\sigma, \infty)$ to be nonoscillatory in $[t_0, \infty)$ is that $F[x(t)] < 0$ for $t \geq t_0$.

We recall that a solution $x(t)$ of (4.1) or (4.2) that satisfies (4.23) is said to be a *Kneser-solution*. Similarly, a solution $x(t)$ of (4.1) or (4.2) which satisfies (4.25) is said to be a *strongly monotone solution*.

Now, we consider (2.10). If (H_2) holds, then Lazer [24] proved that (2.10) has always a Kneser solution and if (H_3) holds, then (2.10) has a strongly monotone solution. Further, Lazer [24] proved that if (H_2) holds, then (2.10) is oscillatory if and only if every nonoscillatory solution of (2.10) is a Kneser solution and $\lim_{t\to\infty} x'(t) = \lim_{t\to\infty} x''(t) = 0$. Similarly, if ($H_3$) holds and (2.10) is oscillatory then every nonoscillatory solution of (2.10) is a strongly monotone solution and $\lim_{t\to\infty} |x(t)| = \lim_{t\to\infty} |x'(t)| = \infty$. This property has been improved by Erbe [10] (Lemma 1.5.4) and Ahmad and Lazer [1] (Lemma 1.5.19).

The asymptotic behaviour of solutions of second-order linear differential equation (2.94) plays a crucial role in the theory of third-order differential equations. Assume that (2.94) is nonoscillatory. If $b(t) \leq 0$, then every nonoscillatory solution $v(t)$ of (2.94) satisfies either

$$v(t)v'(t) > 0 \quad \text{for large } t, \quad \text{and} \quad \lim_{t\to\infty} \left|v(t)\right| = \infty \tag{4.63}$$

or

$$v(t)v'(t) \leq 0 \quad \text{for all } t \geq \sigma. \tag{4.64}$$

Solutions satisfying (4.64) are called *principal solutions at infinity*. They are uniquely determined up to a constant factor and verify the conditions

$$\int_\sigma^\infty \frac{dt}{v^2(t)} = \infty$$

and

$$\lim_{t\to\infty} v(t) = c_v \neq 0 \quad \text{if and only if} \quad \int_\sigma^\infty t\left|b(t)\right| dt < \infty.$$

One may refer to Hartman [17] for the above observation. If $v(t)$ is a solution of (2.94) on $[T,\infty)$, $T \ge \sigma$, then (4.2) can be expressed in the disconjugate form

$$\left(v^2(t)\left(\frac{1}{v(t)}x'\right)'\right)' + c(t)v(t)f(x) = 0. \tag{4.65}$$

One may refer to Sect. 1.4 of Chap. 1 and Remark 2.7.6 of Sect. 2.7 of Chap. 2 to get some basic knowledge on the solutions of the linear equation corresponding to Eq. (4.65).

Equation (4.65) is a special case of the following nonlinear equation

$$\left(r_2(t)\left(r_1(t)x'\right)'\right)' + r_3(t)f(x) = 0, \tag{4.66}$$

where r_1, r_2 and $r_3 \in C([\sigma,\infty), R)$ and are positive for $t \ge \sigma$.

Lemma 4.1.11 *Suppose that either* (H$_2$) *or* (H$_3$) *holds. If $x(t)$ is a nonoscillatory solution of* (4.2), *then either $x(t)x'(t) > 0$ or $x(t)x'(t) < 0$ holds for large t.*

Lemma 4.1.12 *Let* (H$_2$) *hold. If $x(t)$ is a nonoscillatory solution of* (4.2), *then either $x(t)$ is a Kneser solution and*

$$\lim_{t\to\infty} x'(t) = \lim_{t\to\infty} x''(t) = 0$$

or there exists $T_x \ge \sigma$ such that $x(t)x'(t) > 0$ for $t \ge T_x$.

Lemma 4.1.13 *Let* (H$_3$) *hold and $uf(u) > 0$ for $u \neq 0$. Further, assume that there exists a positive solution $v(t)$ of* (2.94), *principal at infinity, such that $\int^\infty v(t)\,dt = \infty$. Let $x(t)$ be a nonoscillatory solution of* (4.2). *Then either $x(t)$ is a strongly monotone solution and*

$$\lim_{t\to\infty} \left|x(t)\right| = \lim_{t\to\infty} \left|x'(t)\right| = \infty$$

or there exists a $T_x \ge \sigma$ such that

$$x(t)x'(t) > 0, \qquad x(t)x''(t) < 0 \quad \text{for } t \ge T_x.$$

Proof Without any loss of generality, we assume that $x(t)$ is eventually positive. First, we claim that $x'(t)$ is eventually positive. If this is not true, then by Lemma 4.1.11, there exists a $t_0 \geq \sigma$ such that $x(t) > 0$ and $x'(t) < 0$ for $t \geq t_0$.

Consider the function G_v defined by

$$G_v(t) = v(t)x''(t) - v'(t)x'(t),$$

where $v(t)$ is the principal solution of (2.94) at infinity satisfying $\int^\infty v(t)\,dt = \infty$. Clearly, $G_v'(t) = c(t)v(t)x(t)f(x(t)) > 0$ implies that G_v is increasing for $t > t_0$. Then two cases are possible, viz., (a) $G_v(t) > 0$ for $t \geq t_0$ and (b) $G_v(t) < 0$ for $t \geq t_0$. First suppose that $G_v(t) > 0$ for $t \geq t_0$. Then from $G_v(t) = v^2(t)\frac{d}{dt}(\frac{x'(t)}{v(t)}) \geq G_v(t_0) > 0$, we obtain

$$\frac{x'(t)}{v(t)} \geq \frac{x'(t_0)}{v(t_0)} + G_v(t_0)\int_{t_0}^t \frac{1}{v^2(s)}\,ds,$$

which gives a contradiction as $t \to \infty$. Next, suppose that $G_v(t) < 0$ for $t \geq t_0$. Since $\frac{d}{dt}(\frac{x'(t)}{v(t)}) < 0$, we obtain $x(t) < \frac{x'(t_0)}{v(t_0)}\int_{t_0}^t v(s)\,ds + x(t_0)$, which, in turn, implies that $x(t) < 0$ for large t, a contradiction. Hence our claim holds and $x'(t)$ is eventually positive.

Clearly $x'''(t)$ is eventually positive. This implies that $x''(t)$ is eventually of one sign. Then the rest part of the proof is easy and left to the reader. □

Theorem 4.1.35 *Let* (H_2) *hold and* $xf(x) > 0$ *for* $x \neq 0$, *and there exists a positive solution* $v(t)$ *of* (2.94), *principal at infinity, such that* $\int_\sigma^\infty v(t)\,dt = \infty$, *and there exists a constant* $k_0 > 0$ *such that* $\liminf_{|u|\to\infty} \frac{f(u)}{u} > k_0$ *and the linear equation*

$$y''' + b(t)y' + k_0c(t)y = 0 \tag{4.67}$$

is oscillatory. Then (4.2) *has property* A.

Proof Assume that (4.2) does not have property A. By Lemma 4.1.12, there exist a nonoscillatory solution $x(t)$ of (4.2) and $T_x \geq \sigma$ such that either (i) $x(t)x'(t) > 0$ for $t \geq T_x$ or (ii) $x(t)x'(t) < 0$, $x(t)x''(t) > 0$ for $t \geq T_x$ and $x'(\infty) = x''(\infty) = 0$ and $x(\infty) \neq 0$. Assume that $x(t)x'(t) > 0$ for $t \geq T_x$. Consider the linearised equation

$$h''' + b(t)h' + c(t)\frac{f(x(t))}{x(t)}h = 0, \quad t \geq T_x. \tag{4.68}$$

Since $b(t) \leq 0$, (2.94) is nonoscillatory. Then (4.68) can be written in the disconjugate form

$$\left(v^2(t)\left(\frac{1}{v(t)}h'(t)\right)'\right)' + c(t)v(t)\frac{f(x(t))}{x(t)}h(t) = 0. \tag{4.69}$$

Taking into the assumption that $\int_\sigma^\infty v(t)\,dt = \infty$ and $v(t)$ is a positive principal solution of (2.94) at infinity, we observe that (4.69) is in canonical form. Hence, every eventually positive increasing solution h satisfies the property $(\frac{1}{v(t)}h'(t))' > 0$ for large t, and hence for $t \geq t_0$, t_0 large enough

$$h(t) > h(t_0) + \frac{1}{v(t_0)}h'(t_0)\int_{t_0}^{t} v(s)\,ds,$$

which implies that $h(\infty) = \infty$. Because $x(t)$ is an eventually positive increasing solution of (4.69), we get $x(\infty) = \infty$ and so there exists $t_1 \geq T_x$ such that $\frac{f(x(t))}{x(t)} > k_0$ for $t \geq t_1$. Since (4.67) is oscillatory, by the classical comparison result, Theorem 4.1.32, (4.68) is oscillatory. This contradicts Lemma 1.5.4, because $x(t)$ is its nonoscillatory solution. Next, suppose that $x(t)x'(t) < 0$ and $x(t)x''(t) > 0$ and $x'(\infty) = x''(\infty) = 0$ and $x(\infty) \neq 0$. Since $x(\infty) \neq 0$, by Lemma 2.7.9(i),

$$\int_{\sigma}^{\infty} c(t)v(t)\int_{\sigma}^{t} \frac{1}{v^2(s)}\int_{\sigma}^{s} v(\tau)\,d\tau\,ds\,dt < \infty \tag{4.70}$$

holds. Because $v(t)$ is a nonincreasing function for $t \geq \sigma$, the function $\frac{1}{v^2(t)}$ is a nondecreasing function for $t \geq \sigma$. Then

$$\begin{aligned}\int_{\sigma}^{t} \frac{1}{v^2(s)}\int_{\sigma}^{s} v(u)\,du\,ds &\geq \int_{\sigma}^{t} \frac{1}{v^2(s)}\int_{\sigma}^{s} v(s)\,du\,ds\\ &\geq \int_{\sigma}^{t} (s-\sigma)\frac{1}{v(s)}\,ds\end{aligned}$$

and

$$\begin{aligned}\int_{\sigma}^{t} v(s)\int_{\sigma}^{s} \frac{1}{v^2(u)}\,du\,ds &\leq \int_{\sigma}^{t} v(s)\int_{\sigma}^{s} \frac{1}{v^2(s)}\,du\,ds\\ &\leq \int_{\sigma}^{t} (s-\sigma)\frac{1}{v(s)}\,ds\end{aligned}$$

implies that

$$\int_{\sigma}^{t} \frac{1}{v^2(s)}\int_{\sigma}^{s} v(u)\,du\,ds \geq \int_{\sigma}^{t} v(s)\int_{\sigma}^{s} \frac{1}{v^2(u)}\,du\,ds.$$

Then, from (4.70), we obtain

$$\int_{\sigma}^{\infty} c(t)v(t)\int_{\sigma}^{t} v(s)\int_{\sigma}^{s} \frac{1}{v^2(\tau)}\,d\tau\,ds\,dt < \infty.$$

This, in turn, implies by Theorem 2.7.3 that the equation

$$\left(v^2(t)\left(\frac{1}{v(t)}y'\right)'\right)' + k_0 c(t)v(t)y = 0 \tag{4.71}$$

is nonoscillatory, which is a contradiction, because (4.71) is the disconjugate form of (4.67). The theorem is proved. □

Theorem 4.1.36 *Let* (H$_2$) *hold,* $uf(u) > 0$ *for* $u \neq 0$ *and*

(i) $\liminf_{|u|\to\infty} \frac{f(u)}{u} > 0$ *and*
(ii) *the linear equation* $y''' + b(t)y' + kc(t)y = 0$ *is oscillatory for every* $k > 0$.

Then (4.2) *has property* A.

Remark 4.1.8 Theorem 4.1.36 is an improvement of Theorem 3.4 due to Erbe [10], where Erbe used the additional condition $\liminf_{t\to\infty} t^3 c(t) > 0$.

Example 4.1.8 By Theorem 4.1.35, the nonlinear equation

$$x''' - \frac{2}{(t+1)^2}x' + \frac{k}{(t+1)^3}|x|^\gamma \operatorname{sgn} x = 0, \quad \gamma > 1$$

has property A for $k > 0$. Theorem 4.1.36 cannot be applied to this example, because the linear equation

$$x''' - \frac{2}{(t+1)^2}x' + \frac{k}{(t+1)^3}x = 0 \tag{4.72}$$

is nonoscillatory for $k < 4$. Indeed, $b'(t) - c(t) < 0$ if and only if $k \leq 4$ and $b'(t) - c(t) \leq 0$ implies that (4.72) is of type C_{II}. On the other hand, (4.72) is of type C_I. Hence Corollary 2.4.3 implies that (4.72) is nonoscillatory.

Example 4.1.9 The nonlinear equation

$$x''' - x' + t\big|x \ln |x|\big|^\gamma \operatorname{sgn} x = 0, \quad \gamma \geq 1$$

has property A, by Theorem 4.1.36. Theorem 4.1.35 cannot be applied to this example.

Theorem 4.1.37 *Let* (H$_3$) *hold,* $uf(u) > 0$ *for* $u \neq 0$, *there exists a positive solution* $v(t)$ *of* (2.94), *principal at infinity, such that* $\int_\sigma^\infty v(t)\,dt = \infty$, *and*

(i) $\liminf_{|u|\to\infty} \frac{f(u)}{u} > 0$ *and*
(ii) *the linear equation* $y''' + b(t)y' - kc(t)y = 0$ *is oscillatory for every* $k > 0$.

Then (4.2) *has property* B.

Proof Assume that (4.2) does not have property B. Let $x(t)$ be a nonoscillatory solution of (4.2) defined on $[T_x, \infty) \subset [\sigma, \infty)$. By Lemma 4.1.13, there exist a nonoscillatory solution $x(t)$ of (4.2) and $T_x \geq \sigma$ such that either

(a) $x(t)x'(t) > 0$, $x(t)x''(t) < 0$ for $t \geq T_x$ or
(b) $x(t)x'(t) > 0$, $x(t)x''(t) > 0$ for $t \geq T_x$ and $x(\infty) = x'(\infty) = \infty$ and $x''(\infty) < \infty$.

Without any loss of generality, we may assume that $x(t) > 0$ for $t \geq T_x$. Now, we consider the linearised equation

$$z'' + b(t)z' - c(t)\frac{f(x(t))}{x(t)}z = 0, \qquad t \geq T_x. \tag{4.73}$$

Since $x(t)$ is an eventually positive increasing solution of (4.73), there exist a $k > 0$ and $t_x \geq T_x$ such that $f(x(t))/x(t) > k$ for $t \geq t_x$. In view of the assumption (ii), we find from a classical comparison result ([12], Theorem 7.5), that (4.73) is oscillatory. This contradicts Theorem 2.7.20 of Chap. 2, because $x(t)$ is its nonoscillatory solution. This completes the proof of the theorem. □

Example 4.1.10 The nonlinear equation

$$x''' - \frac{2}{(t+1)^2}x' - \frac{\lambda}{(t+1)^2}|x|^\gamma \operatorname{sgn} x = 0, \qquad \gamma \geq 1$$

has property B for $\lambda > 0$ arbitrarily by Theorem 4.1.37. Indeed, $v(t) = \frac{1}{t+1}$ satisfying $\int^\infty v(t)\,dt = \infty$ is a solution of $v'' - \frac{2}{(t+1)^2}v = 0$ and (i) of Theorem 4.1.37 holds. Now, we show that the assumption (ii) of Theorem 4.1.37 holds, that is,

$$y''' - \frac{2}{(t+1)^2}y' - \frac{\lambda}{(t+1)^2}y = 0, \tag{4.74}$$

is oscillatory for every $\lambda > 0$. Let k and t_0 be sufficiently large such that

$$\frac{k}{(t+1)^3} < \frac{\lambda}{(t+1)^2}$$

for every $\lambda > 0$ and $t \geq t_0$. Then by applying the classical comparison result, that is, Theorem 4.1.32 on (4.72), we get

$$y''' - \frac{2}{(t+1)^2}y' + \frac{\lambda}{(t+1)^2}y = 0$$

is oscillatory for every $\lambda > 0$. This implies that (4.74) is oscillatory.

Let (H_2) hold. It is well-known (see [21]) that (4.2) has a solution $x(t)$ such that $x(\sigma) > 0$ and

$$x(t) \geq 0, \qquad x'(t) \leq 0 \quad \text{and} \quad x''(t) \geq 0. \tag{4.75}$$

From this, a Kneser solution of (4.2) exists if there exist $\epsilon > 0$ and a continuous, nondecreasing function $g : [0, \epsilon] \to [0, \infty)$ such that

$$f(x) \leq g(x) \quad \text{for } x \in [0, \epsilon] \quad \text{and} \quad \int_0^\epsilon \frac{du}{g(u)} = \infty. \tag{4.76}$$

More precisely, the following holds.

Theorem 4.1.38 *Let* (H$_2$) *hold,*

$$f(x) > 0 \quad \textit{for } x > 0, \qquad f(0) = 0, \tag{4.77}$$

and let (4.76) *hold for large* t. *Then there exists a Kneser solution of* (4.2).

Note that if (4.76) is not valid, then the solution x guaranteed by Theorem 4.1.38 may be trivial in a neighbourhood of infinity. The following example demonstrates this fact.

Example 4.1.11 Let $\lambda \in (0, 1)$. Then every solution of

$$x''' - x' + |x|^{\lambda} \operatorname{sgn} x = 0$$

satisfying (4.75) is trivial in some neighbourhood of infinity, and thus there exist no Kneser solutions (see [21], Corollary 11.1).

Theorem 4.1.39 *Assume that* (H$_1$), (4.76) *and* (4.77) *hold. Then there exists at least one Kneser solution of* (4.2) *if the followings hold.*

(i) (2.94) *is nonoscillatory, and*
(ii) *there exists a positive constant* α *such that*

$$\left|f(x)\right| \geq \alpha|x| \quad \textit{for } x \in R, \qquad b^{3/2}(t) \leq \alpha c(t) \quad \textit{and} \quad b'(t) < 2\alpha c(t)$$

eventually.

Theorem 4.1.39 requires that the function $c(t)$ to be positive. Note that if $b \equiv 0$ and $c \equiv 0$, then (4.2) has no Kneser solutions. Thus the existence of Kneser solutions of (4.1) or (4.2) is more interesting in the case (H$_2$). In the opposite case, that is, when (H$_1$) holds, or $b(t)$ changes sign, the situation is different. Thus, an attention to the existence of Kneser solution of (4.2) would be more interesting when (H$_1$) holds.

We have seen in our previous results on the existence of Kneser solutions of (4.2) that the investigations are easier when the second-order differential equation (2.94) is nonoscillatory. If v is a positive solution of (2.94) on $[T, \infty)$, $T \geq \sigma$, then (4.2) can be expressed in the disconjugate form (4.65). Cecchi and Marini [8] have obtained some properties of solutions of (4.65). Kneser solutions have an interesting property that when (2.94) is nonoscillatory, then under very weak assumptions on $b(t), c(t)$ and $f(t)$ every nonoscillatory solution $x(t)$ of (4.2) is a Kneser solution tending to zero as $t \to \infty$, that is, $\lim_{t\to\infty} x(t) = 0$. This problem has been solved under the name property A, as we have seen in our earlier results. On the other hand, if (2.94) is oscillatory, then (4.2) cannot be expressed in the disconjugate form (4.65) and the above quoted results cannot be applied.

In the following, we shall give some results on (4.2) when (2.94) is oscillatory. In particular, we shall investigate the problem of existence of Kneser solutions as well as conditions assuring that these solutions tend to zero as $t \to \infty$.

Sometimes, the more restrictive assumption

$$xf(x) > 0 \quad \text{for } x \neq 0 \tag{4.78}$$

will also be assumed. (This condition has been assumed in some earlier results.)

In the following, we shall extend the conclusion of Theorem 4.1.39 to the case when (2.94) is oscillatory. The existence of Kneser solution of (4.2) will also be given in the case where f is quasilinear near the origin, that is, there exist α_1, α_2 and $\epsilon \in R$ such that

$$\epsilon > 0, \qquad 0 < \alpha_1 \le \alpha_2 < \infty, \qquad \alpha_1 x \le f(x) \le \alpha_2 x \quad \text{for } x \in [0, \epsilon]. \tag{4.79}$$

Lemma 4.1.14 *Consider the boundary-value problem*

$$x''' + b(t)x' + c(t)f(x) = 0, \quad x \in S, \tag{4.80}$$

where S is a nonempty subset of the Frechet space $C^0(R_+)$ of continuous functions on R_+. Assume that there exists a function $H : R_+ \times R^2 \to R$ such that $H(t, d, d) = c(t)f(d)$ for all $d \in R$. Then the problem (4.80) *admits at least one solution provided that the following conditions are satisfied*:

(i) *There exists a nonempty closed convex subset Ω of $C^2(R_+)$ such that*

  (a) *for any $u \in \Omega$, $H(t, u(t), x)$ is a Caratheodory function on $R_+ \times R$,*
  (b) *for any $u \in \Omega$, there exists a nonempty convex set of solutions Tu of the problem*

$$y''' + b(t)y' + H\big(t, u(t), y\big) = 0, \quad t \in R_+, \tag{4.81}$$

$$y \in S \cap \Omega; \tag{4.82}$$

(ii) *there exists a positive continuous function m on R_+ such that*

$$\big|y(t)\big| \le m(t) \quad \textit{for all } t \in R_+ \textit{ and any } y \in T(\Omega);$$

(iii) *there exists a locally integrable function α on R_+ such that*

$$\big|b(t)\big| \le \alpha(t), \qquad \big|H\big(t, u(t), y\big)\big| \le \alpha(t)$$
$$\textit{for } t \in R_+ \textit{ and } (u, y) \in \Omega \times C^2(R_+)$$

satisfying (4.81) *and* (4.82); *and*

(iv) *if, for some $u \in \Omega$, x is a solution of* (4.80) *which lies in the C^2 closure of $T(\Omega)$, then $x \in S$.*

Lemma 4.1.15 *Let $k \in N$, and let u_k be a solution of*

$$\begin{aligned} &x''' + b(t)x' + \overline{c}(t)x = 0, \quad \overline{c} \in Loc(R_+), \\ &\lim_{s \to t^+} \overline{c}(s) \quad \textit{exists for } t \in R_+ \end{aligned} \tag{4.83}$$

defined on $[0, k]$. *Let* $\psi : R_+ \to R_+$ *be a nondecreasing function such that*

$$\sum_{i=0}^{2} \left|u_k^{(i)}(t)\right| \le \psi(t) \quad \text{for } t \in [0, k], k = 1, 2, \ldots.$$

Then there exists a subsequence $\{u_{k_j}^{(i)}\}_{j=1}^{\infty}$ *such that* $\{u_{k_j}^{(i)}\}_{j=1}^{\infty}$, $i = 0, 1, 2$ *converges uniformly to* $u^{(i)}(t)$ *on every finite interval of* R_+, *that is,*

$$\lim_{j\to\infty} u_{k_j}^{(i)} = u^{(i)}(t), \quad t \in R_+, i \in \{0, 1, 2\}.$$

Moreover, $u(t)$ *is a solution of* (4.83).

Lemma 4.1.16 *Let* $u(t)$ *be a solution of* (4.83) *such that* $|u(t)| \le M_1$ *for* $t \in I$, I *is a bounded interval of* R_+. *Then there exists a constant* M_2 *(depending only on* b, $\overline{c}$, M_1, *and* I*) such that*

$$\left|u'(t)\right| \le M_2, \qquad \left|u''(t)\right| \le M_2 \quad \text{on } I.$$

Proof Let $I = [a, b]$, $v_i = \max_{t\in I} |u^{(i)}(t)|$, $i = 0, 1, 2$, and $v_3 \ge |u'''(t)|$ for almost all $t \in I$. As $v_0 \le M_1$, then by Lemma 5.2 due to Kiguradze and Chanturia [21], we have

$$v_1 \le C_1 + C_2 v_3^{1/3} \le C_1 + C_3(v_1 + v_0)^{1/3} \le C_1 + C_3(v_1 + M_1)^{1/3},$$

where $C_1 = 81M_1(b-a)^{-1}$, $C_2 = 24M_1^{2/3}$ and $C_3 = C_2[\max_{t\in I}(|b(t)|+|c(t)|)]^{1/3}$. From this, it follows that $v_1 \le C_4$ and by integrating (4.83), we easily obtain $v_2 \le C_5$, where C_4 and C_5 are suitable constants independent of the solution u. By choosing $M_2 = \max\{C_4, C_5\}$, we get the assertion. □

Lemma 4.1.17 *Let* $\overline{c} \in Loc(R_+)$ *be nonnegative and* $\overline{c}_+(t) = \lim_{s\to t^+} \overline{c}(s)$ *exists on* R_+. *Then for every* $x_0 > 0$, *there exists a solution* x *of* (4.83) *satisfying*

$$x(\sigma) = x_0, \qquad (-1)^j x^{(j)}(t) \ge 0, \quad t \in R_+, \ j = 0, 1, 2 \tag{4.84}$$

if one of the following assumptions holds:

(i) $K \in R_+$, $b(t) \le K$, *and* $\overline{c}_+(t) > K^{3/2}$ *on* R_+;
(ii) $b \in C^1(R_+)$, $b'(t) \le 2\overline{c}_+(t)$, *and* $b_+^{3/2}(t) < \overline{c}_+(t)$ *for* $t \in R_+$, *where* $b_+(t) = \max\{b(t), 0\}$.

Proof (i) It is enough to prove the result for $x_0 = 1$. Let $k \in \{1, 2, \ldots\}$ and let $x_k : [\sigma, k] \to R$ be a solution of the Cauchy problem

$$x_k(k) = x_k'(k) = 0, \qquad x_k''(k) = 1$$

associated to (4.83). Then there exists a left neighbourhood of J_k at $t=k$ such that

$$x_k(t)>0, \qquad x_k'(t)<0, \qquad x_k''(t)>0 \quad \text{on } J_k. \tag{4.85}$$

Now, we prove that (4.85) is valid not only in J_k but also on $[\sigma,k)$, too.

Suppose to the contrary that there exists $\tau\in[\sigma,k)$ such that $x_k''(\tau)=0$ and $x_k''(t)>0$ for $t\in(\tau,k)$. Then (4.85) yields

$$x_k(t)>0, \qquad x_k'(t)<0, \qquad x_k''(t)>0 \quad \text{on } (\tau,k) \quad \text{and} \quad x_k''(\tau)=0. \tag{4.86}$$

Assume $b(\tau)x_k'(\tau)+\overline{c}_+(\tau)x_k(\tau)>0$. Then $b(t)x_k'(t)+\overline{c}_+(t)x_k(t)>0$ almost everywhere in a suitable right neighbourhood of I_τ of τ, $I_\tau\subset[\sigma,k)$. Hence $x_k'''(t)<0$ almost everywhere on I_τ and

$$x_k''(t_1)=\int_\tau^{t_1} x_k'''(s)\,ds<0, \qquad t_1\in I_\tau,$$

which is a contradiction. Then

$$b(\tau)x_k'(\tau)+\overline{c}_+(\tau)x_k(\tau)\le 0. \tag{4.87}$$

If $b(\tau)<0$, then (4.87) contradicts (4.86) because $\overline{c}_k(\tau)\ge 0$. Hence $b(\tau)\ge 0$. Define

$$F\big[x_k(t)\big]=-2x_k(t)x_k''(t)+\big(x_k'(t)\big)^2-kx_k^2(t), \qquad t\in[\tau,k]. \tag{4.88}$$

Using (4.83) and (4.86), we get

$$F'\big[x_k(t)\big]=2\overline{c}(t)x_k^2(t)+2x_k(t)x_k'(t)\big[b(t)-k\big]\ge 0 \quad \text{almost everywhere on } [\tau,k).$$

From this, $F[x_k(k)]=0$ and F is increasing on $[\tau,k)$, we obtain

$$\big|x_k'(\tau)\big|\le\sqrt{k}\cdot x_k(\tau). \tag{4.89}$$

Then (4.87) and (4.89) yield

$$0\le -b(\tau)x_k'(\tau)-\overline{c}_+(\tau)x_k(\tau)\le\big(b(\tau)\sqrt{k}-\overline{c}_+(\tau)\big)x_k(\tau),$$

which, in view of assumption (i), contradicts (4.86). Thus, (4.85) is valid on $[\sigma,k)$ too. That is,

$$x_k(t)>0, \qquad x_k'(t)<0, \qquad x_k''(t)>0 \quad \text{on } [\sigma,k).$$

Because (4.83) is linear, then a solution $\overline{x}_k(t)$ of (4.83) exists such that

$$\overline{x}_k(\sigma)=1, \qquad \overline{x}_k(t)>0, \qquad \overline{x}_k'(t)<0, \qquad \overline{x}_k''(t)>0 \quad \text{on } [\sigma,k). \tag{4.90}$$

If we put $s\in\{1,2,\dots\}$, $I=[s-1,s]$, $M_1=1$, $M=\max_{t\in I}(|b(t)|+\overline{c}(t))$, denoting by N_s the constant M_2 guaranteed by Lemma 4.1.16, then from Lemma 4.1.16 and (4.90), we obtain

$$\begin{aligned}&\left|\overline{x}_k(t)\right| \le 1, \qquad \left|\overline{x}_k^{(i)}(t)\right| \le N_s,\\ &t \in [s-1, s],\ i = 1, 2,\ s \in \{1, 2, \ldots\},\ k \in \{1, 2, \ldots\}.\end{aligned} \tag{4.91}$$

Let $\psi(t)$, $t \in R_+$ be an arbitrary nondecreasing continuous function such that

$$\psi(t) \ge 1 + 2N_s, \quad t \in [s-1, s],\ s = 1, 2, \ldots.$$

Then (4.91) yields

$$\sum_{i=0}^{2} \left|\overline{x}_k^{(i)}(t)\right| \le \psi(t) \quad \text{for } t \in [\sigma, k],\ k = 1, 2, \ldots,$$

and, by Lemma 4.1.15, there exists a solution x of (4.83) such that (4.84) with $x_0 = 1$ holds.

(ii) The proof is similar, where we have to use

$$F\big[x(t)\big] = -2x_k(t)x_k''(t) + \big(x_k'(t)\big)^2 - b(t)x_k^2(t)$$

in place of (4.88). The proof is left. □

Theorem 4.1.40 *Assume the conditions* (4.77) *and* (4.79). *If any one of the following assumptions holds*:

(i) *there exist* $K, \alpha \in R_+$ *such that* $b(t) \le K, \alpha(t) > K^{3/2}$ *on* R_+;
(ii) $b \in C^1(R_+)$ *and there exists* $\alpha \in R_+$ *such that* $b'(t) \le 2\alpha c(t)$, $[b_+(t)]^{3/2} < \alpha c(t)$ *on* R_+,

then for an arbitrary $x_0 \in (0, \epsilon]$, *then exists a Kneser solution* x *of* (4.2) *such that*

$$x(\sigma) = x_0, \qquad (-1)^j x^{(j)}(t) > 0, \quad t \ge \sigma,\ j = 0, 1, 2.$$

Proof We apply Lemma 4.1.14 with

$$\Omega = S = \big\{x \in C^2(R_+); x(\sigma) = x_0, x(t) \ge 0, x'(t) \le 0, x''(t) \ge 0\big\},$$

$$H(t, u, x) = \begin{cases} c(t)\frac{f(u)}{u}x & \text{for } u \ne 0;\\ \alpha c(t)x & \text{for } u = 0,\end{cases}$$

$m(t) = x_0$ and $\alpha(t) = \alpha_1 x_0 c(t)$ for $t \in R_+$.

Let $u \in \Omega$. Put $\widetilde{c}(t) = c(t)\frac{f(u(t))}{u(t)}$ for $u(t) \ne 0$ and $\overline{c}(t) = \alpha c(t)$ for $u(t) = 0$. Because $t_0 \in (\sigma, \infty)$, $u(t_0) = 0$ implies that $u(t) \equiv 0$ on $[t_0, \infty)$, then $\overline{c} \in L_{\text{loc}}(R_+)$, $\widetilde{c} \ge 0$, and $\lim_{s \to t^+} \widetilde{c}(t)$ exists for $t \in R_+$. Thus, the assumptions of Lemma 4.1.17 are satisfied. Moreover, H satisfies all assumptions given in Lemma 4.1.14 and, hence for any $u \in \Omega$, there exists a nonempty set $T(u) \subset C^2(R_+)$ of solutions of (4.81) and (4.82) that is convex. Hence, by Lemma 4.1.14, (4.2) has at least one solution satisfying (4.84).

As $x_0 > 0$, it follows from (4.79) that $x \not\equiv 0$ in any neighbourhood of ∞ ([21], Theorem 11.5). From this, by using (4.2) and positiveness of c, we obtain

$$x(\sigma) = x_0, \qquad x(t) > 0, \qquad x'(t) < 0, \qquad x''(t) \geq 0 \quad \text{on } R_+. \tag{4.92}$$

Now, we claim that $x''(t) \neq 0$, that is, $x''(t) > 0$ on $[\sigma, \infty)$. If not, then there exists $\tau \in [\sigma, \infty)$ such that $x''(\tau) = 0$. Then (4.92) yields $x'''(\tau) = 0$, that is,

$$0 = -b(\tau)x'(\tau) - c(\tau)f\big(x(\tau)\big) \leq -b(\tau)x'(\tau) - c(\tau)x(\tau). \tag{4.93}$$

Define

$$E(t) = -2x(t)x''(t) + \big(x'(t)\big)^2 - Kx^2(t), \quad t \in R_+.$$

Using (4.92), we have

$$E'(t) = 2x(t)c(t)f\big(x(t)\big) - 2x(t)x'(t)\big[K - b(t)\big] \geq 0.$$

Because $\lim_{t\to\infty} x'(t) = 0$, we can conclude that $\lim_{t\to\infty} E(t) \leq 0$ and thus $E(t) \leq 0$ on $[\sigma, \infty)$. Especially,

$$E(\tau) = x'^2(\tau) - Kx^2(\tau) \leq 0. \tag{4.94}$$

If $b(\tau) \leq 0$, then by virtue of the positiveness of c, (4.93) contradicts (4.92). If $b(\tau) > 0$, then (4.93) and (4.94) yield

$$\sqrt{K}b(\tau) \geq \alpha c(\tau),$$

that contradicts the assumption (i). Hence our claim holds, that is, $x''(t) \neq 0$ on R_+. Consequently, (4.92) implies the existence of a Kneser solution of (4.2).

The proof in the case (ii) is similar. The theorem is proved. □

Theorem 4.1.41 *Assume that* (H_2) *and* (4.78) *holds. Further, assume that* $\int_\sigma^\infty t^2 c(t)\,dt = \infty$, *and* $\liminf_{|u|\to\infty} |f(u)| > 0$. *Then every Kneser solution of* (4.2) *tends to zero as t tends to* ∞.

Theorem 4.1.42 *Assume* (4.78). *If any one of the following conditions holds*:

(i) $\int_\sigma^\infty tb_-(t)\,dt = -K > -\infty$, *the equation* $w'' + e^{-2k}b_+(t)w = 0$ *is nonoscillatory, and* $\int_\sigma^\infty c(t)\,dt = \infty$;
(ii) $\int_\sigma^\infty t|b(t)|\,dt < \infty$ *and* $\int_\sigma^\infty t^2 c(t)\,dt = \infty$;
(iii) $b(t) \leq 0$ *and the linear equation* $z''' + b(t)z' + \lambda c(t)z = 0$ *is oscillatory for every* $\lambda > 0$;
(iv) $b(t) \leq 0$ *and* $\int_\sigma^\infty t^2 c(t)\,dt = \infty$,

then every Kneser solution of (4.2) *tends to zero as* $t \to \infty$.

Proof (i) According to [8], there exists a positive solution $v(t)$ of (2.94) with nonzero limit as $t \to \infty$, and (4.2) can be transformed into the disconjugate form (4.65). Then the proof follows from Theorems 2.7.12–2.7.13.

(ii) From [17], there exists a solution $v(t)$ of (2.94) such that $\lim_{t\to\infty} v(t) = v_0 > 0$. Then the argument is similar to the proof of (i).

(iii) Assume that $x(t)$ is the Kneser solution of (4.2) such that $\lim_{t\to\infty} x(t) > 0$. Consider, for t large, the linearised equation

$$w''' + b(t)w' + c(t)\frac{f(x(t))}{x(t)}w = 0. \tag{4.95}$$

Since $x(t)$ is its solution, we see that (4.95) is nonoscillatory. Then taking into account that there exists a positive constant k such that, for t large, $k \le \frac{f(x(t))}{x(t)}$, the comparison result Theorem 4.1.32 yields a contradiction.

(iv) Let $x(t)$ be a Kneser solution of (4.2). Then there exists a constant $M > 0$ such that $|x(t)| < M$ for $t \ge \sigma$. Now considering the auxiliary equation

$$X''' + b(t)X' + c(t)F(X) = 0 \tag{4.96}$$

with

$$F(x) = \begin{cases} f(x) & \text{if } |x| \le M; \\ f(M) & \text{if } x > M; \\ f(-M) & \text{if } x < -M, \end{cases}$$

we see that $x(t)$ is the Kneser solution of (4.96). Moreover, as $f(x) \ne 0$ for $x \ne 0$, we have $F(x) \ne 0$ for $x \ne 0$, $\lim_{x\to\pm\infty} F(x) = M > 0$. Hence applying Theorem 4.1.41 to (4.96), we conclude that $\lim_{t\to\infty} x(t) = 0$. The theorem is proved. □

The following result, Theorem 4.1.43, gives mainly the case when (2.94) is oscillatory and enlarges the conclusion of Theorem 4.1.42(i).

Theorem 4.1.43 *Assume* (4.78). *If any one of the following conditions holds*:

(i) $\int_\sigma^\infty c(t)\,dt = \infty$ *on* $[\sigma, \infty)$ *and there exists* $M > 0$ *such that* $b(t) \le M$ *on* $[\sigma, \infty)$;

(ii) $b \in C^1([\sigma, \infty), R_+)$, $b(t) > 0$, *and* $b'(t) \ge 0$ *on* R_+, $\int_\sigma^\infty \frac{c(t)}{b(t)}\,dt = \infty$;

(iii) $c \in C^1([\sigma, \infty), R_+)$, $c'(t) \ge 0$ *on* R_+ *and there exists* $M > 0$ *such that* $\frac{b(t)}{c(t)} \le M$ *on* R_+,

then every Kneser solution of (4.2) *tends to zero as* $t \to \infty$.

Proof Let $x(t)$ be a Kneser solution of (4.2) and suppose that $x(t) > 0$, $x'(t) < 0$, $x''(t) > 0$ on $[t_x, \infty)$ and $\lim_{t\to\infty} x(t) = N, t_x \ge \sigma$. Denote $M_1 = \min_{N\le s\le x(t_x)} sf(s)$. Hence $M_1 > 0$.

First, suppose that (i) holds. Put

$$F[x(t)] = -2x(t)x''(t) - Mx^2(t), \quad t \ge t_x.$$

Then

$$F'[x(t)] = -2x'(t)x''(t) + 2c(t)x(t)f(x(t)) - 2x(t)x'(t)[M - b(t)] \geq 0.$$

Thus, F is nondecreasing. Further, since $F[x(t)] \leq 0$ for $t \geq t_x$, then

$$\begin{aligned}\infty > F[x(\infty)] - F[x(t_x)] = \int_{t_x}^{\infty} F'[x(t)]\,dt &\geq 2\int_{t_x}^{\infty} c(t)x(t)f(x(t))\,dt \\ &\geq 2M_1\int_{t_x}^{\infty} c(t)\,dt = \infty,\end{aligned}$$

a contradiction. Hence $x(t) \to 0$ as $t \to \infty$.

Let (ii) be satisfied. Set $F[x(t)] = -(2x(t)x''(t)/b(t)) - x^2(t)$, $t \geq t_x$. Then

$$\begin{aligned}F'[x(t)] &= -2\frac{x'(t)x''(t)}{b(t)} + 2\frac{c(t)}{b(t)}x(t)f(x(t)) + 2\frac{b'(t)}{b^2(t)}x(t)x''(t) \\ &\geq 0, \quad t \in [t_x, \infty).\end{aligned}$$

Thus, F is nondecreasing and $\lim_{t\to\infty} F[x(t)] = F_0 \in (-\infty, 0]$. Hence

$$\begin{aligned}-\infty > F[x(\infty)] - F[x(t_x)] = \int_{t_x}^{\infty} F'[x(t)]\,dt &\geq 2\int_{t_x}^{\infty} \frac{c(t)}{b(t)}x(t)f(x(t))\,dt \\ &\geq 2M_1\int_{t_x}^{\infty} \frac{c(t)}{b(t)}\,dt = \infty,\end{aligned}$$

a contradiction. Hence $x(t) \to 0$ as $t \to \infty$.

The argument is similar to that given in (ii) for the case (iii) by choosing $F[x(t)] = -(2x(t)x''(t)/c(t)) - Mx^2(t)$, $t \geq t_x$. The proof is complete. □

Now, we give some sufficient conditions for the nonoscillation and asymptotic behaviour of solutions for the following third-order nonlinear differential equation

$$x''' + b(t)(x')^{\gamma} + c(t)f(x) = 0, \tag{4.97}$$

where the coefficients $b(t)$ and $c(t)$ satisfy the condition (H_1), $xf(x) > 0$ for $x \neq 0$ and γ is a ratio of odd integers.

In Chap. 3, we introduced the definition of z-type oscillatory solutions. We shall use this definition to study some sufficient conditions for the nonoscillation of (4.97).

Theorem 4.1.44 *Let* (H_1) *hold. Further, suppose that* $b(t) \leq c(t)$ *and* $b(t) \not\equiv 0$ *on any subinterval of* $[\sigma, \infty)$. *If* $x(t)$ *is a solution of* (4.97) *defined on* $[T, \infty)$, $T \geq \sigma$, *such that it satisfies* $(z')^{\gamma} + f(z) < 0$ *on the subinterval of* $[T, \infty)$ *on which* $x'(t) > 0$, *while* $(z')^{\gamma} + f(z) > 0$ *on the subinterval of* $[T, \infty)$ *on which* $x'(t) < 0$, *then* $x(t)$ *is nonoscillatory.*

Proof Let $x(t)$ be a solution of (4.97) on $[T, \infty), T \geq \sigma$. Multiplying (4.97) by $x'(t)$, we get

$$x'(t)x'''(t) = -b(t)\big(x'(t)\big)^{\gamma+1} - c(t)f\big(x(t)\big)x'(t). \tag{4.98}$$

Assume that $x'(t) > 0$ on an interval $[T, \infty)$. Then the condition $(x')^\gamma + f(x) < 0$ yields $f(x) < 0$, and hence $x(t) < 0$. Thus, we cannot have simultaneously $x(t) > 0$ and $x'(t) > 0$. Therefore, an oscillatory or nonnegative z-type solution cannot exist. Hence, in the proof, it is sufficient to exclude nonpositive z-type solutions.

Let $x(t)$ be a nonpositive z-type solution with consecutive double zeros at t_1 and t_2 $(t_1 < t_2)$. So there exists a $t_3 \in (t_1, t_2)$ such that $x'(t_3) = 0$, and hence $x'(t) > 0$ for $t \in (t_3, t_2)$. Now, integration of (4.98) from t_3 to t_2 yields

$$0 = x'(t)x''(t)\big|_{t_3}^{t_2} \geq \int_{t_3}^{t_2} \big(x''(t)\big)^2\,dt - \int_{t_3}^{t_2} b(t)\big[\big(x'(t)\big)^\gamma + f\big(x(t)\big)\big]x'(t)\,dt > 0,$$

which is a contradiction.

When $x'(t) < 0$, one may repeat the above argument to obtain a contradiction. Hence $x(t)$ must be nonoscillatory. The proof is complete. □

Example 4.1.12 By Theorem 4.1.44, the equation

$$x''' + \frac{1}{2}\big(x'\big)^3 + \frac{2e^{3t} + e^t}{2(e^{3t} + e^t)}\big(x + x^3\big) = 0, \quad t \geq 0$$

is nonoscillatory. Note that $x_1(t) = e^{-t}$ and $x_2(t) = -e^{-t}$ are the nonoscillatory solutions of the above equation.

Theorem 4.1.45 *Let* (H_1) *hold and* $b(t) \geq 1$. *If* $x(t)$ *is a solution of* (4.97) *on an interval on which* $x(t)$ *satisfies* $(x''(t))^2 - (x'(t))^{\gamma+1} < 0$, *then* $x(t)$ *is nonoscillatory.*

Proof Let $x(t)$ be a nonnegative z-type with consecutive double zeros at t_1 and t_2 $(t_1 < t_2)$. Thus there exists a $t_3 \in (t_1, t_2)$ such that $x'(t_3) = 0$. Integrating (4.98) from t_1 to t_3, we get

$$\begin{aligned}0 &= x'(t)x''(t)\big|_{t_1}^{t_3}\\ &= \int_{t_1}^{t_3} b(t)\big[\big(x''(t)\big)^2 - \big(x'(t)\big)^{\gamma+1}\big]\,dt - \int_{t_1}^{t_3} c(t)x'(t)f\big(x(t)\big)\,dt.\end{aligned}$$

Since $f(x) > 0$ for $x > 0$, the above inequality yields

$$0 \leq \int_{t_1}^{t_3} b(t)\big[\big(x''(t)\big)^2 - \big(x'(t)\big)^{\gamma+1}\big]\,dt - \int_{t_1}^{t_3} c(t)x'(t)f\big(x(t)\big)\,dt < 0,$$

which is a contradiction.

Next, suppose that $x(t)$ is of nonpositive z-type with consecutive double zeros at t_1 and t_2 ($t_1 < t_2$). So there exists a $t_3 \in (t_1, t_2)$ such that $x'(t_3) = 0$ and $x'(t) < 0$ for $t \in (t_1, t_3)$. Integration of (4.98) from t_1 to t_3 yields

$$\begin{aligned} 0 &= x'(t)x''(t)\big|_{t_1}^{t_3} \\ &\le \int_{t_1}^{t_3} b(t)\big[\big(x''(t)\big)^2 - \big(x'(t)\big)^{\gamma+1}\big]\,dt - \int_{t_1}^{t_3} c(t) f\big(x(t)\big)x'(t)\,dt. \end{aligned}$$

Proceeding as in the previous case, we arrive at a contradiction.

Finally, suppose that $x(t)$ is oscillatory. Let t_1, t_2 and t_3 ($t_1 < t_2 < t_3$) be any three consecutive zeros of $x(t)$ such that $x'(t_1) \le 0$, $x'(t_2) \ge 0$ and $x'(t_3) \le 0$. So, $x(t) < 0$ for $t \in (t_1, t_2)$ and $x(t) > 0$ for $t \in (t_2, t_3)$. Hence there exist $t_4 \in (t_1, t_2)$ and $t_5 \in (t_2, t_3)$ such that $x'(t_4) = 0 = x'(t_5)$ and $x'(t) > 0$ for $t \in (t_4, t_2)$ and $t \in (t_2, t_5)$. Now, we consider two cases, namely, $x''(t_2) \le 0$ and $x''(t_2) > 0$. Let $x''(t_2) \le 0$. Integrating (4.98) from t_2 to t_5, we obtain

$$\begin{aligned} 0 &= x'(t)x''(t)\big|_{t_2}^{t_5} \\ &\le \int_{t_2}^{t_5} b(t)\big[\big(x''(t)\big)^2 - \big(x'(t)\big)^{\gamma+1}\big]\,dt - \int_{t_2}^{t_5} c(t) f\big(x(t)\big)x'(t)\,dt < 0, \end{aligned}$$

a contradiction. Hence $x''(t_2) > 0$. Since $x''(t)$ is continuous, we have $x''(t) > 0$ for $t \in [t_2, t_2 + \delta_1)$, $0 < \delta_1 < t_5 - t_2$. So $x'(t)$ is increasing in $[t_2, t_2 + \delta_1)$. Again $x'(t_5) = 0$ and $x'(t) > 0$ for $t \in (t_2, t_5)$ imply that $x'(t)$ is decreasing in $[t_5 - \delta_2, t_5]$, where $0 < \delta_2 < t_5 - t_2$. This, in turn, implies that $x''(t) < 0$ for $t \in [t_5 - \delta_2, t_5]$. Hence $x''(t_6) = 0$ for some $t_6 \in (t_2, t_5)$. An integration of (4.98) from t_6 to t_5 yields a contradiction. Hence $x(t)$ must be nonoscillatory. This completes the proof of the theorem. □

Example 4.1.13 Clearly $x(t) = 1 + \frac{1}{t}$ is a nonoscillatory solution of the equation

$$x''' + t^{10/3}\big(x'\big)^{5/3} + \frac{t^4 + 6}{2t^4 + 4t^3 + 3t^2 + t}\big(x + x^3\big) = 0, \qquad t \ge 8.$$

Theorem 4.1.45 can be applied to this example.

Theorem 4.1.46 *Any solution $x(t)$ of* (4.97) *which satisfies the inequality*

$$\big(z''\big)^2 - b(t)\big(z'\big)^{\gamma+1} - c(t) f(z) z' > 0$$

on an interval on which $x'(t) > 0$ is nonoscillatory.

Example 4.1.14 By Theorem 4.1.46, the equation

$$x''' + \frac{1}{27}t^{-3}\big(x'\big)^3 + \frac{1}{4}t^{-6}x^3 = 0, \qquad t > 0$$

is nonoscillatory. In particular, $x(t) = t^{3/2}$ is a nonoscillatory solution of this example.

Lemma 4.1.18 *Let $\int_{\sigma}^{\infty} c(t)\,dt = \infty$, and let $f(x)$ be nondecreasing. If $x(t)$ is a nonoscillatory solution of* (4.97) *and $x(t) \neq 0$ for $t \geq t_0 \geq \sigma$, then $|x(t)|$ is nondecreasing for $t \geq t_0$.*

This can be proved by integrating (4.97) from t_0 to t and letting $t \to \infty$.

Theorem 4.1.47 *Let* (H_1) *hold and $\gamma = 1$ in* (4.97). *Let $b(t) \leq M$, $M > 0$, $\int_{\sigma}^{\infty} c(t)\,dt = \infty$ and $f(x)$ be nondecreasing. If $x(t)$ is a nonoscillatory solution of* (4.97) *such that $x(t)x'(t) < 0$ for some $t \geq t_0 \geq \sigma$, then* $\lim_{t\to\infty} x(t) = 0$.

Proof If $x(t) > 0$ and $x'(t) < 0$ for $t \geq t_0$, then $\lim_{t\to\infty} x(t)$ exists. Then, using the nondecreasing nature of $f(x)$, the boundedness of $b(t)$ and $\int_{\sigma}^{\infty} c(t)\,dt = \infty$, we obtain a contradiction by the integration of (4.97) from t_0 to t and taking the limit as $t \to \infty$. □

Theorem 4.1.48 *Let* (H_1) *hold and $\gamma = 1$ in* (4.97). *Let $\int_{\sigma}^{\infty} c(t)\,dt = \infty$ and $f(x)$ be nondecreasing. Suppose that the second-order linear differential equation* (2.94) *is nonoscillatory. If $x(t)$ is a nonoscillatory solution of* (4.97) *and $x(t) \neq 0$ on any subinterval of* $[\sigma, \infty)$, *then $x(t)x'(t) < 0$ for some $t \geq t_0 \geq \sigma$.*

Theorem 4.1.49 *Let the hypothesis of Lemma* 4.1.18 *and the hypothesis of Lemma* 1.5.1 *with $r(t) \equiv 1$ and $q(t) = b(t)$ hold. If $x(t)$ is a solution of* (4.97) *with the properties given in Theorem* 4.1.47, *then* $\lim_{t\to\infty} x(t) = 0$.

Remark 4.1.9 The results obtained above, that is, from Theorems 4.1.44–4.1.49 can be extended to the equation

$$\big(r(t)x''\big)' + q(t)\big(x'\big)^{\alpha} + p(t)f(x) = 0,$$

where $r(t)$ is a continuous and positive function. However, the only minor assumption needed for the analysis is $\int_{\sigma}^{\infty} \frac{1}{r(t)}\,dt = \infty$ to hold, whenever necessary.

4.2 Behaviour of Solutions of $x''' + a(t)x'' + b(t)x' + c(t)x^{\alpha} = 0$ and $x''' + a(t)x'' + b(t)x' + c(t)f(x) = 0$

This section deals with the oscillatory and nonoscillatory behaviour of solutions of nonlinear differential equations of the form (4.9) and (4.10), where $a, b, c \in C([\sigma, \infty), R)$, $\sigma \in R$, $f \in C(R, R)$ such that $\frac{f(x)}{x} \geq \beta > 0$ for $x \neq 0$ and $a > 0$ is a quotient of odd integers.

Consider (1.5) with $b \leq 0$ and $c > 0$. Then, as discussed in Propositions 1.2.1 and 1.2.2, we find that (1.5) is oscillatory if and only if (1.8) holds. Moreover, all nonoscillatory solutions (1.5) tend to zero, if and only if (1.5) has an oscillatory solution. It has been observed in Sects. 2.1 and 2.2 of Chap. 2 that the above observations are found to be true to a large extent for more general equations of the form (2.1), where $a, b, c \in C([\sigma, \infty), R)$ such that $a(t)$ does not change sign, $b(t) \leq 0$ and $c(t) > 0$. In this section, we shall obtain similar results for nonlinear equations of the form (4.9) and (4.10).

We consider the following cases:

(H_5) $a(t) \geq 0, b(t) \leq 0$ and $c(t) > 0$;

and

(H_6) $a(t) \leq 0, b(t) \leq 0$ and $c(t) > 0$.

Equations (4.9) and (4.10) may be written, respectively, as

$$\big(r(t)x''\big)' + q(t)x' + p(t)x^{\alpha} = 0 \tag{4.99}$$

and

$$\big(r(t)x''\big)' + q(t)x' + p(t)f(x) = 0, \tag{4.100}$$

where $r(t) = e^{\int_{\sigma}^{t} a(s)\,ds}$, $q(t) = r(t)b(t)$ and $p(t) = r(t)c(t)$.

Lemma 4.2.1 *Let* (H_5) *hold. If $x(t)$ is a nonoscillatory solution of* (4.9) *or* (4.10) *on $[T_x, \infty)$, $T_x \geq \sigma$, then there exists a $t_0 \in (T_x, \infty)$ such that $x(t)x'(t) > 0$ or $x(t)x'(t) < 0$ for $t \geq t_0$.*

Proof Suppose that $x(t)$ is a nonoscillatory solution of (4.9) on $[T_x, \infty)$. Without any loss of generality, we may assume $x(t) > 0$ for some $t \geq t_0 \geq T_x$. Clearly, $-x'(t)$ is a solution of the second-order nonhomogeneous differential equation

$$\big(r(t)z'\big)' + q(t)z = p(t)x^{\alpha}(t), \quad t \geq t_0. \tag{4.101}$$

From Lemma 1.5.14, it follows that all solutions of (4.101) are nonoscillatory. Consequently, $x'(t)$ is nonoscillatory. Hence the conclusion of the lemma holds for (4.9).

If $x(t)$ is a nonoscillatory solution of (4.10) on $[T_x, \infty)$ and $x(t) > 0$ for $t \geq t_0 \geq T_x$, then proceeding as above, we get the result.

If $x(t) < 0$ for $t \geq t_0 \geq T_x$, we arrive at the conclusion of the lemma in a similar way. □

Theorem 4.2.1 *Suppose that* (H_5) *holds, $a'(t) \leq 0$ and $\alpha \geq 1$. If*

$$\int_{\sigma}^{\infty} \left[\frac{2a^3(t)}{27} - \frac{a(t)b(t)}{3} + \gamma c(t) - \frac{2}{3\sqrt{3}} \left(\frac{a^2(t)}{3} - b(t) \right)^{3/2} \right] dt = \infty \tag{4.102}$$

for every $\gamma > 0$, then a solution of (4.9) *which has a zero is oscillatory.*

Proof Suppose that $x(t)$ is a solution of (4.9) on $[T_x, \infty)$, $T_x \geq \sigma$, with $x(t_1) = 0$ for some $t_1 \geq T_x$. Suppose that $x(t)$ is nonoscillatory. Hence $x(t) > 0$ or < 0 for large t. As $x(t)$ is a solution of (4.9) implies that $-x(t)$ is a solution of that equation, without any loss of generality, we may take $x(t) > 0$ for $t > t_2$, where $t_2 \geq t_1$, is the last zero of $x(t)$. From Lemma 4.2.1, it follows that $x'(t) < 0$ or > 0 for $t \geq t_0 \geq t_2$. Let $x'(t) < 0$ for $t \geq t_0$. Since $x(t_2) = 0$ and $x(t) > 0$ for $t > t_2$, $x'(t) > 0$ for $t \in (t_2, t_2 + \delta)$, $\delta > 0$. So there exists a $t_3 \in (t_2, t_0)$ such that $x'(t_3) = 0$ and $x'(t) < 0$ for $t > t_3$. Now multiplying (4.99) through by $x'(t)$ and integrating the resulting identity from t_3 to t $(t_3 < t)$, we obtain

$$r(t)x'(t)x''(t) = \int_{t_3}^{t} \left[r(s)\left(x''(s)\right)^2 - q(s)\left(x'(s)\right)^2 - p(s)x^\gamma(s)x'(s)\right] ds > 0.$$

Thus, $x''(t) < 0$ for $t > t_3$. This, in turn, implies that $x(t) < 0$ for large t, a contradiction.

If $x'(t) > 0$ for $t \geq t_0$, then setting $z(t) = \frac{x'(t)}{x(t)} > 0$ for $t \geq t_0$, we obtain the Riccati equation

$$z'' + 3zz' + a(t)z' = -\left[z^3 + a(t)z^2 + b(t)z + c(t)x^{\alpha-1}(t)\right]. \tag{4.103}$$

Since $x'(t) > 0$ for $t \geq t_0$, there exists a $\gamma > 0$ such that $x^{\alpha-1}(t) > \gamma$ for $t \geq t_0$. Hence (4.103) reduces to

$$z'' + 3zz' + a(t)z' \leq -\left[z^3 + a(t)z^2 + b(t)z + \gamma c(t)\right]. \tag{4.104}$$

The minimum of $z^3 + a(t)z^2 + b(t)z + \gamma c(t)$ for positive z is given by

$$\frac{2a^3(t)}{27} - \frac{a(t)b(t)}{3} + \gamma c(t) - \frac{2}{3\sqrt{3}}\left(\frac{a^2(t)}{3} - b(t)\right)^{3/2}.$$

Thus, (4.104) yields

$$\begin{aligned} &z'' + 3zz' + a(t)z' \\ &\quad \leq -\left[\frac{2a^3(t)}{27} - \frac{a(t)b(t)}{3} + \gamma c(t) - \frac{2}{3\sqrt{3}}\left(\frac{a^3(t)}{3} - b(t)\right)^{3/2}\right], \quad t \geq t_0. \end{aligned} \tag{4.105}$$

Integrating both sides of (4.105) from t_0 to t, we have

$$\begin{aligned} z'(t) &\leq z'(t_0) + \frac{3}{2}z^2(t_0) + a(t_0)z(t_0) \\ &\quad - \int_{t_0}^{t}\left[\frac{2a^3(s)}{27} - \frac{a(s)b(s)}{3} + \gamma c(s) - \frac{2}{3\sqrt{3}}\left(\frac{a^2(s)}{3} - b(s)\right)^{3/2}\right] ds. \end{aligned}$$

Then from (4.102), it follows that $\lim_{t\to\infty} z'(t) = -\infty$. This, in turn, implies that $z(t) < 0$ for large t, a contradiction. Hence $x(t)$ is oscillatory. This completes the proof of the theorem. □

Theorem 4.2.2 *Let* (H$_5$) *hold. Suppose that* $a'(t) \le 0$, $\alpha < 1$ *and* (4.102) *holds for every* $\gamma > 1$. *Then a bounded solution of* (4.9) *which has a zero is oscillatory.*

Theorem 4.2.3 is a partial generalisation of Theorem 1.3 due to Lazer [24].

Theorem 4.2.3 *Suppose* (H$_5$) *holds,* $a'(t) \le 0$, $b'(t) \le 0$ *and* $b(t)$ *is bounded. If* $\alpha \ge 1$ *and* (2.12) *holds, then a solution of* (4.9) *which has a zero is oscillatory.*

Proof Suppose that $x(t)$ is a nonoscillatory solution of (4.9) with $x(t_1) = 0$ for some $t_1 \ge T_x \ge \sigma$. Without any loss of generality, we may assume that $x(t) > 0$ for $t > t_2$, where $t_2 \ge t_1$ is the last zero of $x(t)$. From Lemma 4.2.1, it follows that $x'(t) > 0$ or < 0 for $t \ge t_0 > t_2$. Proceeding as in Theorem 4.2.1, we may show that $x'(t) < 0$ for $t > t_0$ is not possible. So $x'(t) > 0$ for $t \ge t_0$. Taking

$$A(t) = \frac{2a^3(t)}{27} - \frac{a(t)b(t)}{3}, \qquad B(t) = \frac{2}{3\sqrt{3}}\left(\frac{a^2(t)}{3} - b(t)\right)^{3/2},$$

we see that $A(t) \ge 0$, $B(t) \ge 0$ and $A^2(t) - B^2(t) \le 0$. So $A(t) - B(t) \le 0$ and hence

$$\frac{2a^3(t)}{27} - \frac{a(t)b(t)}{3} - \frac{2}{3\sqrt{3}}\left(\frac{a^2(t)}{3} - b(t)\right)^{3/2} \le 0. \tag{4.106}$$

From (2.12) and (4.106), it follows that $\int_\sigma^\infty c(t)\,dt = \infty$.

If $x(t)$ is bounded, then integrating (4.9) from t_0 to t, we have

$$-\int_{t_0}^t c(s)x^\alpha(s)\,ds = x''(t) - x''(t_0) + a(t)x'(t) - a(t_0)x'(t_0) - b(t_0)x(t_0) + b(t)x(t) - \int_{t_0}^t \big(a'(s)x'(s) + b'(s)x(s)\big)\,ds.$$

That is,

$$-x^\alpha(t_0)\int_{t_0}^t c(s)\,ds \ge x''(t) - x''(t_0) - a(t_0)x'(t_0) + b(t)x(t).$$

Since $b(t)x(t)$ is bounded, we see that $\lim_{t\to\infty} x''(t) = -\infty$ and consequently, $x'(t)$ is negative for large t, a contradiction.

If $x(t)$ is not bounded, then we may assume that $x(t) > 1$ for $t \ge t_3 > t_0$. Clearly, $z(t) = \frac{x'(t)}{x(t)}$ satisfies the inequality

$$z'' + 3zz' + a(t)z' \le -\big[z^3 + a(t)z^2 + b(t)z + c(t)\big],$$

for $t \ge t_3$. Now, finding the minimum of $z^3 + a(t)z^2 + b(t)z + c(t)$ for a positive z and integrating the resulting identity from t_3 to t, we obtain $\lim_{t\to\infty} z'(t) = -\infty$. So $z(t) < 0$ for large t, a contradiction. Hence $x(t)$ is oscillatory. This completes the proof of the theorem. □

Theorem 4.2.4 *Let* (H$_5$) *hold. Suppose that* $a'(t) \le 0, b'(t) \le 0$ *and* $b(t)$ *is bounded. If* $\alpha < 1$ *and* (2.12) *holds, then a bounded solution of* (4.9) *which has a zero is oscillatory.*

Example 4.2.1 Consider

$$x''' + t^{-2/3}x'' - \left(1 + \frac{1}{3}t^{-4/3}\right)x' + e^t x^{3/2} = 0, \quad t \ge 1.$$

Theorem 4.2.1 is applicable to this example, whereas Theorem 4.2.3 cannot be applied to this example.

Example 4.2.2 Consider

$$x''' + t^{-2/3}x'' - \left(1 - \frac{1}{3}t^{-4/3}\right)x' + \frac{2}{3\sqrt{3}}x^{3/2} = 0, \quad t \ge 1.$$

This example satisfies all the conditions of Theorem 4.2.3. But Theorem 4.2.1 is not applicable to this example, because the integral

$$\int_1^\infty \left(\frac{2}{27t^2} + \frac{1}{3}t^{-2/3} - \frac{1}{9t^2} + \frac{2\gamma}{3\sqrt{3}} - \frac{2}{3\sqrt{3}}\right) dt = -\infty$$

for $0 \le \gamma \le 3/4$.

Theorem 4.2.5 *If the hypothesis of either Theorem* 4.2.1 *or Theorem* 4.2.3 *is satisfied, than a nonoscillatory solution* $u(t)$ *of* (4.9) *has the following properties*:

$$\left.\begin{aligned} &u(t) \ne 0, \quad t \ge T_u \ge \sigma, \\ &\operatorname{sgn} u(t) = \operatorname{sgn} u''(t) \ne \operatorname{sgn} u'(t) = \operatorname{sgn} u'''(t), \quad t \ge T_u \ge \sigma, \\ &\lim_{t\to\infty} u(t) = \lim_{t\to\infty} u'(t) = \lim_{t\to\infty} u''(t) = 0. \end{aligned}\right\} \tag{4.107}$$

Proof Suppose that $u(t)$ is a nonoscillatory solution of (4.9) on $[T_u, \infty)$, $T_u \ge \sigma$. If $u(t)$ has a zero in $[T_u, \infty)$, then from Theorem 4.2.1 or Theorem 4.2.3, it follows that $u(t)$ is oscillatory, a contradiction. Thus, $u(t)$ has no zero in $[T_u, \infty)$. Without any loss of generality, we may assume that $u(t) > 0$ for $t \ge T_u$. Proceeding as in the proof of Theorem 4.2.1 or Theorem 4.2.3, it may be shown that $u'(t) \not> 0$ for $t \ge t_0$. So $u'(t) < 0$ for $t \ge t_0$.

Since $(r(t)u''(t))' = -q(t)u'(t) - p(t)u^\alpha(t) < 0$ for $t \ge t_0 \ge T_u$, $u''(t)$ is nonoscillatory. Thus, there exists a $t_1 \ge t_0$ such that $u''(t) > 0$ or < 0 for $t \ge t_1$. If $u''(t) < 0$ for $t \ge t_1$, then $u(t)$ becomes negative for large t, a contradiction. Hence $u''(t) > 0$ for $t \ge t_1$. Now, we claim that neither $u'(t)$ nor $u''(t)$ has a zero in $[T_u, \infty)$. If not, then suppose that $t_2 \in [T_u, \infty)$ is a zero of $u'(t)$ or $u''(t)$. Without any loss of generality, we may assume that $u'(t_2)u''(t_2) = 0$, $u'(t) < 0$ and $u''(t) > 0$ for $t \ge t_2$. Multiplying

$$\big(r(t)u''(t)\big)' + q(t)u'(t) + p(t)u^\alpha(t) = 0 \tag{4.108}$$

throughout by $u'(t)$ and integrating the resulting identity from t_2 to t $(t_2 < t)$, we get

$$r(t)u'(t)u''(t) \geq -\int_{t_2}^{t} \left(q(s)\left(u'(s)\right)^2 + p(s)u'(s)u^{\alpha}(s)\right) ds,$$

a contradiction. Hence $u'(t) < 0$ and $u''(t) > 0$ for $t \geq T_u$. Consequently, $\lim_{t\to\infty} u'(t)$ exists and is zero. Similarly, $\lim_{t\to\infty} u''(t) = 0$. If possible suppose that $\lim_{t\to\infty} u(t) = \mu$, $\mu > 0$. Then $u(t) \geq \mu$, $\mu > 0$ for $t \geq T_u$. Integrating (4.108) from T_u to t, we obtain

$$\begin{aligned} r(t)u''(t) &= r(T_u)u''(T_u) - \int_{T_u}^{t} q(s)u'(s)\,ds - \int_{T_u}^{t} p(s)u^{\alpha}(s)\,ds \\ &\leq r(T_u)u''(T_u) - \int_{T_u}^{t} r(s)c(s)u^{\alpha}(s)\,ds \\ &\leq r(T_u)u''(T_u) - r(T_u)\mu^{\alpha} \int_{T_u}^{t} c(s)\,ds. \end{aligned}$$

Since (4.102) or (2.12) implies that $\int_{\sigma}^{\infty} c(s)\,ds = \infty$, we have $u''(t) < 0$ for large t, a contradiction. Hence $\lim_{t\to\infty} u(t) = 0$. This completes the proof of the theorem. □

Theorem 4.2.6 *Suppose that the hypothesis of either Theorem* 4.2.2 *or Theorem* 4.2.4 *is satisfied. Then a bounded nonoscillatory solution of* (4.9) *has the properties* (4.107).

Example 4.2.3 Consider

$$x''' + t^{-4/3}x'' - \left(\frac{1}{2} - t^{-4/3}\right)x' + \frac{1}{2}e^{2t}x^3 = 0, \quad t \geq 2.$$

From Theorem 4.2.5, it follows that a nonoscillatory solution of this equation has the properties (4.107). In particular, $x(t) = e^{-t}$ is a nonoscillatory solution of the equation which satisfies the properties (4.107).

Corollary 4.2.1 *If the hypothesis of either Theorem* 4.2.1 *or Theorem* 4.2.3 *is satisfied, then a solution $u(t)$ of* (4.9), *whose derivative or second derivative has a zero on $[T_u, \infty)$, $T_u \geq \sigma$ is oscillatory.*

The proof follows from Theorem 4.2.5.

Corollary 4.2.2 *If the hypothesis of either Theorem* 4.2.2 *or Theorem* 4.2.4 *is satisfied, then a bounded solution $u(t)$ of* (4.9), *whose derivative or second derivative has a zero, is oscillatory.*

The proof follows from Theorem 4.2.6.

Theorem 4.2.7 *Let (H_5) hold. Suppose that $a'(t) \le 0, b'(t) \ge 0$ and $\alpha > 1$. If $\int_\sigma^\infty c(t)\,dt = \infty$, then for any nonoscillatory solution $u(t)$ of (4.9) on $[T_u, \infty)$, $T_u \ge \sigma$, the following properties hold for large t:*

$$\operatorname{sgn} u(t) = \operatorname{sgn} u''(t) \ne \operatorname{sgn} u'(t) = \operatorname{sgn} u'''(t), \quad t \ge T_u,$$
$$\lim_{t\to\infty} u(t) = \lim_{t\to\infty} u'(t) = \lim_{t\to\infty} u''(t) = 0.$$

Proof Suppose that $u(t) > 0$ for $t \ge t_0 \ge T_u$. From Lemma 4.2.1, it follows that $u'(t) > 0$ or < 0 for $t \ge t_1 \ge t_0$. Suppose that $u'(t) > 0$ for $t \ge t_1$. Dividing (4.9) through by $u^\alpha(t)$, $t \ge t_1$, we have

$$-c(t) = \frac{u'''(t)}{u^\alpha(t)} + a(t)\frac{u''(t)}{u^\alpha(t)} + b(t)\frac{u'(t)}{u^\alpha(t)}. \tag{4.109}$$

Integrating (4.109) twice from t_1 to t ($t_1 < t$), we obtain

$$\begin{aligned}
&A + Bt - \int_{t_1}^t \left(\int_{t_1}^s c(\theta)\,d\theta\right) ds \\
&= \frac{u'(t)}{u^\alpha(t)} + \frac{3\alpha}{2}\int_{t_1}^t \frac{(u'(s))^2}{u^{\alpha+1}(s)}\,ds + \frac{\alpha(\alpha+1)}{2}\int_{t_1}^t \frac{(t-s)(u'(s))^2}{u^{\alpha+2}(s)}\,ds \\
&\quad - \frac{1}{\alpha-1}\int_{t_1}^t \frac{b(s)}{u^{\alpha-1}(s)}\,ds + \frac{1}{\alpha-1}\int_{t_1}^t \frac{(t-s)b'(s)}{u^{\alpha+1}(s)}\,ds + \int_{t_1}^t \frac{u(s)u'(s)}{u^\alpha(s)}\,ds \\
&\quad + \alpha\int_{t_1}^t \left(\int_{t_1}^\theta a(s)\frac{(u'(s))^2}{u^\alpha(s)}\,ds\right) d\theta.
\end{aligned}$$

From Remark 4.1.2, it follows that the left-hand side of the above identity is negative, whereas the right-hand side is positive for large t, a contradiction. Hence $u'(t) < 0$ for $t \ge t_1$. Thus, from (4.108), we have

$$\big(r(t)u''(t)\big)' = -q(t)u'(t) - p(t)u^\alpha(t) < 0 \quad \text{for } t \ge t_1.$$

Consequently, $u''(t)$ is nonoscillatory. If $u''(t) < 0$ for large t, then $u(t)$ is eventually negative, a contradiction. So $u''(t) > 0$ for large t, say, for $t \ge t_2 \ge t_1$. Hence from (4.9) we have $u'''(t) < 0$ for $t \ge t_2 \ge t_1$. Consequently, $\lim_{t\to\infty} u''(t) = \lim_{t\to\infty} u'(t) = 0$. If possible, suppose that $\lim_{t\to\infty} u(t) = \mu > 0$. Integrating (4.9) from t_2 to t, we have

$$u''(t) - u''(t_2) < -\mu^\alpha \int_{t_2}^t c(s)\,ds.$$

Letting $t \to \infty$, we have $\lim_{t\to\infty} u''(t) = -\infty$, a contradiction. Hence $\lim_{t\to\infty} u(t) = 0$. The theorem is proved. □

One may view Theorem 4.2.7 as follows:

Theorem 4.2.8 *Suppose that the conditions of Theorem* 4.2.7 *are satisfied. Then a solution of* (4.9) *which has a zero is oscillatory.*

Proof If possible, suppose that $x(t)$ is a nonoscillatory solution of (4.9) with $x(t_1) = 0$ for some $t_1 \geq T_x$. Without any loss of generality, we may assume that $x(t) > 0$ for $t > t_2$, where $t_2 \geq t_1$ is the last zero of $x(t)$. So $x'(t) > 0$, $t_2 < t \leq t_2 + \delta$, for some $\delta > 0$. By Lemma 4.2.1, we have $x'(t) > 0$ or < 0 for $t \geq t_0 > t_2$. Proceeding as in Theorem 4.2.7, it may be shown that $x'(t) > 0$ for $t \geq t_0$ is not possible. So $x'(t) < 0$, $t \geq t_0$. Then there exists a point $t_3 \in (t_2, t_0)$ such that $x'(t_3) = 0$ and $x'(t) < 0$ for $t > t_3$. Multiplying (4.99) throughout by $x'(t)$ and integrating the resulting identity from t_3 to t, we get

$$r(t)x'(t)x''(t) = \int_{t_3}^{t} \left[r(s)\left(x''(s)\right)^2 - q(s)\left(x'(s)\right)^2 - p(s)x^{\alpha}(s)x'(s)\right] ds > 0.$$

This implies that $x''(t) < 0$ for $t > t_3$. Consequently, $x(t) < 0$ for large t, a contradiction. Hence $x(t)$ is oscillatory. This completes the proof of the theorem. □

Note that Theorem 4.2.8 is a generalisation of Theorem 1 due to Nelson [28].

Theorem 4.2.9 *Let* (H_5) *hold and* $\frac{f(x)}{x} \geq \beta > 0$ *for* $x \neq 0$. *Suppose that* $a'(t) \leq 0$ *and*

$$\int_{\sigma}^{\infty} \left[\frac{2a^3(t)}{27} - \frac{a(t)b(t)}{3} + \beta c(t) - \frac{2}{3\sqrt{3}}\left(\frac{a^2(t)}{3} - b(t)\right)^{3/2}\right] dt = \infty, \tag{4.110}$$

then every solution of (4.10) *which has a zero is oscillatory.*

Proof Suppose that $x(t)$ is a nonoscillatory solution of (4.10) on $[T_x, \infty)$ with $x(t_1) = 0$ for $t_1 \geq T_x \geq \sigma$. Suppose that $x(t) > 0$ for $t \geq t_2$, where $t_2 \geq t_1$ is the last zero of $x(t)$. From Lemma 4.2.1, it follows that $x'(t) > 0$ or < 0 for $t \geq t_0 > t_1$. If $x'(t) < 0$ for $t \geq t_0$, then we proceed as in Theorem 4.2.1 to obtain a contradiction. If $x'(t) > 0$ for $t \geq t_0$, then setting $z(t) = \frac{x'(t)}{x(t)}$ for $t \geq t_0$, we see that $z(t)$ is a solution of the second-order inequality

$$\begin{aligned} z'' + 3zz' + a(t)z' &\leq -\left[z^3 + a(t)z^2 + b(t)z + \beta c(t)\right] \\ &\leq -\left[\frac{2a^3(t)}{27} - \frac{a(t)b(t)}{3} + \beta c(t) - \frac{2}{3\sqrt{3}}\left(\frac{a^2(t)}{3} - b(t)\right)^{3/2}\right]. \end{aligned} \tag{4.111}$$

Integrating the above inequality from t_0 to t, we obtain

$$z'(t) \le z'(t_0) + \frac{3}{2}z^2(t_0) + a(t_0)z(t_0)$$
$$- \int_{t_0}^{t} \left[\frac{2a^3(s)}{27} - \frac{a(s)b(s)}{3} + \beta c(s) - \frac{2}{3\sqrt{3}} \left(\frac{a^2(s)}{3} - b(s) \right)^{3/2} \right] ds.$$

Hence $\lim_{t\to\infty} z'(t) = -\infty$. So $z(t) < 0$ for large t, a contradiction. Hence $x(t)$ cannot be positive for large t. Proceeding as above, it may be shown that $x(t) < 0$ for large t is not possible. Hence $x(t)$ is oscillatory. This completes the proof of the theorem. □

We may note that Theorem 4.2.1 may be applied to (4.10). But the condition (4.110) is weaker than (4.102). However, Theorem 4.2.3 cannot be applied to (4.10). Further, Theorem 4.2.9 generalises Theorem 1.3 due to Lazer [24].

Example 4.2.4 Consider

$$x''' + \frac{3}{t}x'' - \frac{1}{t^2}x' + e^t(1 + e^x)x = 0, \quad t \ge 1.$$

Here $f(x)/x = 1 + e^x > 1$ for $x \neq 0$. All the conditions of Theorem 4.2.9 are satisfied. So every solution of the equation having a zero is oscillatory.

Theorem 4.2.10 *If the hypothesis of Theorem 4.2.9 is satisfied, then for any nonoscillatory solution $u(t)$ of (4.10), the property (4.107) holds.*

Theorem 4.2.11 *If the hypothesis of Theorem 4.2.9 is satisfied, then a solution of (4.10), whose derivative or second derivative has a zero, is oscillatory.*

Lemma 4.2.2 *Let* (H_6) *hold. If $x(t)$ is a nonoscillatory solution of (4.9) or (4.10) on $[T_x, \infty)$, $T_x \ge \sigma$, then there exists a $t_0 \in [T_x, \infty)$ such that $x(t)x'(t) > 0$ or $x(t)x'(t) < 0$ for $t \ge t_0$.*

Theorem 4.2.12 *Let* (H_6) *hold. Suppose that $b(t) - a'(t) \le 0$ and $\alpha \ge 1$. If*

$$\int_{\sigma}^{\infty} \left[\frac{2a^3(t)}{27} - \frac{a(t)b(t)}{3} + \gamma c(t) - \frac{2}{3\sqrt{3}} \left(\frac{a^2(t)}{3} - b(t) + a'(t) \right)^{3/2} \right] dt = \infty \tag{4.112}$$

for every $\gamma > 0$, then a solution of (4.9) which has a zero is oscillatory.

Proof Suppose that $x(t)$ is a nonoscillatory solution of (4.9) on $[T_x, \infty)$ with $x(t_1) = 0$ for $t_1 \ge T_x \ge \sigma$. Suppose that $x(t) > 0$ for $t \ge t_2$; where $t_2 \ge t_1$ is the last zero of $x(t)$. From Lemma 4.2.2, it follows that $x'(t) > 0$ or < 0 for $t \ge t_3 > t_2$. Proceeding as in Theorem 4.2.1, we may show that $x'(t) < 0$ for $t \ge t_3$ leads us to a contradiction. Hence $x'(t) > 0$ for $t \ge t_3$. Clearly, $z(t) = \frac{x'(t)}{x(t)}$ for $t \ge t_3$ is a positive solution of Eq. (4.103). Since $x(t)$ is increasing for $t \ge t_3$, it is possible to find a $\gamma > 0$ such that $x^{\alpha-1}(t) > \gamma$ for $t \geq t_4 \geq t_3$. Hence, for $t \geq t_4$, (4.103) reduces to (4.104). Integrating (4.104) from t_4 to t, we have

$$\begin{aligned} z'(t) \leq z'(t_4) + \frac{3}{2}z^2(t_4) + a(t_4)z(t_4) - \frac{3}{2}z^2(t) - a(t)z(t) \\ - \int_{t_4}^{t} \left[z^3(s) + a(s)z^2(s) + \big(b(s) - a'(s)\big)z(s) + \gamma c(s)\right] ds. \end{aligned} \tag{4.113}$$

Setting $G(z,t) = -\frac{3}{4}z^2 - a(t)z$, $H(t,z) = z^3 + a(t)z^2 + (b(t) - a'(t))z + \gamma c(t)$, the maximum of $G(z,t)$ for $z > 0$ is given by $\frac{a^2(t)}{6}$, and the minimum of $H(z,t)$ for $z > 0$ is given by

$$\frac{2a^3(t)}{27} - \frac{a(t)(b(t) - a'(t))}{3} + \gamma c(t) - \frac{2}{3\sqrt{3}}\left(\frac{a^2(t)}{3} - b(t) + a'(t)\right)^{3/2}.$$

Hence, from (4.113) we obtain

$$\begin{aligned} z'(t) &\leq z'(t_4) + \frac{3}{2}z^2(t_4) + a(t_4)z(t_4) + \frac{a^2(t)}{6} \\ &\quad - \int_{t_4}^{t}\left[\frac{2a^3(s)}{27} - \frac{a(s)(b(s) - a'(s))}{3} + \gamma c(s) \right. \\ &\quad \left. - \frac{2}{3\sqrt{3}}\left(\frac{a^2(s)}{3} - b(s) + a'(s)\right)^{3/2}\right] ds \\ &\leq z'(t_4) + \frac{3}{2}z^2(t_4) + a(t_4)z(t_4) + \frac{a^2(t)}{6} \\ &\quad - \int_{t_4}^{t}\left[\frac{2a^3(s)}{27} - \frac{a(s)b(s)}{3} + \gamma c(s) - \frac{2}{3\sqrt{3}}\left(\frac{a^2(s)}{3} - b(s) + a'(s)\right)^{3/2}\right] ds. \end{aligned} \tag{4.114}$$

Taking limit as $t \to \infty$ in (4.114), we get $z'(t) \to -\infty$ as $t \to \infty$. Consequently, $z(t) < 0$ for large t, a contradiction. Hence $x(t)$ is oscillatory. The theorem is proved. □

Theorem 4.2.13 *Let* (H$_6$) *hold. Suppose that* $b(t) - a'(t) \leq 0$ *and* $\alpha < 1$. *If* (4.112) *holds for every* $\gamma > 0$, *then a bounded solution of* (4.9) *which has a zero is oscillatory.*

Theorem 4.2.14 *If the hypothesis of Theorem* 4.2.12 *holds, then a nonoscillatory solution* $u(t)$ *of* (4.9) *has the following properties*:

$$\left.\begin{aligned} &u(t) \neq 0, \quad t \geq T_u \geq \sigma, \\ &\operatorname{sgn} u(t) = \operatorname{sgn} u''(t) \neq \operatorname{sgn} u'(t), \quad t \geq T_u \geq \sigma, \\ &\lim_{t\to\infty} u'(t) = 0, \qquad \lim_{t\to\infty} u(t) = \lambda \neq \pm\infty. \end{aligned}\right\} \tag{4.115}$$

Further, if $\int_{\sigma}^{\infty} a(t)\,dt > -\infty$, *then* $\lim_{t\to\infty} u(t) = 0$.

Example 4.2.5 Consider

$$x''' - e^{-t}x'' - e^{-t}x' + e^{2t}x^3 = 0, \quad t \geq 0.$$

From Theorem 4.2.13, it follows that a nonoscillatory solution of the equation has the properties (4.115). In particular, $x(t) = e^{-t}$ is a nonoscillatory solution of this example.

Theorem 4.2.15 *Suppose that the conditions of Theorem* 4.2.13 *hold. Then a bounded nonoscillatory solution* (4.9) *satisfies the property* (4.115).

Theorem 4.2.16 *Let* (H_6) *hold. Suppose that* $b(t) - a'(t) \leq 0$ *and*

$$\int_{\sigma}^{\infty}\left[\frac{2a^3(t)}{27} - \frac{a(t)b(t)}{3} + \beta c(t) - \frac{2}{3\sqrt{3}}\left(\frac{a^2(t)}{3} - b(t) + a'(t)\right)^{3/2}\right]dt = \infty.$$

Then every solution of (4.10) *which has a zero is oscillatory.*

Theorem 4.2.17 *If the hypothesis of Theorem* 4.2.16 *is satisfied, then every nonoscillatory solution* $u(t)$ *of* (4.10) *satisfies the property* (4.115).

Example 4.2.6 Consider

$$x''' - e^{-t}x'' - e^{-t}x' + e^{2t}\left(1 + e^{x}\right)x = 0, \quad t \geq 1.$$

From Theorem 4.2.15, it follows that every nonoscillatory solution of this equation satisfies the property (4.115).

One may find several results concerning the oscillation of (4.9) when $a(t) \geq 0$, $b(t) \geq 0$ and $c(t) > 0$ and $\alpha > 0$ is a quotient of odd positive integers. For example, Erbe [10] proved the following theorem:

Theorem 4.2.18 *Let* $a(t) \geq 0$, $b(t) \geq 0$, $c(t) > 0$ *and* $a(t)b(t) + b'(t) \leq 0$. *Let* $x(t)$ *be a nontrivial solution of* (4.9) *with* $F[x(t_0)] \geq 0$ *for some* $t_0 \geq \sigma$, *where* $F[x(t)]$ *is defined in* (2.54). *Assume further that the equation*

$$\left(r(t)z'\right)' + q(t)z + \lambda^{\alpha}t^{\alpha}p(t)z^{\alpha} = 0$$

is oscillatory for some $0 < \lambda < 1/2$. *Then* (4.9) *is oscillatory.*

4.3 Behaviour of Solutions of $x''' + c(t)x^\alpha = 0$ and $x''' + c(t)f(x) = 0$

In this section, we shall concentrate in providing some necessary and sufficient conditions on the oscillation and nonoscillation of the third-order nonlinear differential equations of the forms

$$x''' + c(t)x^\alpha = 0 \tag{4.116}$$

and

$$x''' + c(t)f(x) = 0, \tag{4.117}$$

where α and f are defined as earlier and $c \in ([\sigma, \infty), R)$.

Theorem 4.3.1 *Let $c(t) > 0$. If*

$$\int_\sigma^\infty tc(t)\,dt = \infty, \tag{4.118}$$

then every nonoscillatory solution $x(t)$ of (4.116) *with $\alpha \geq 1$ or* (4.117) *has the following properties*:

$$\left.\begin{aligned} &x(t) \neq 0, \quad t \geq T_x \geq \sigma, \\ &\operatorname{sgn} x(t) = \operatorname{sgn} x''(t) \neq \operatorname{sgn} x'(t) = \operatorname{sgn} x'''(t), \quad t \geq T_x \geq \sigma, \\ &\lim_{t\to\infty} x(t) = \lim_{t\to\infty} x'(t) = \lim_{t\to\infty} x''(t) = 0. \end{aligned}\right\} \tag{4.119}$$

If

$$\int_\sigma^\infty t^\alpha c(t)\,dt = \infty, \tag{4.120}$$

then every nonoscillatory solution of (4.116) *with $\alpha < 1$ satisfies the above property* (4.119).

Proof Suppose that $x(t)$ is a nonoscillatory solution of (4.116) in $[T_x, \infty)$. Without any loss of generality, we may assume that $x(t) > 0$ for $t \geq t_1 \geq T_x$. Consequently, $x'''(t) < 0$ for $t \geq t_1$. Hence, either $x''(t) < 0$ or > 0 for large t. But the former case implies that $x(t) < 0$ for large t, a contradiction. Thus, $x''(t) > 0$ for $t \geq t_2 \geq t_1$. This, in turn, implies that $x'(t) > 0$ or $x'(t) < 0$ for large t. If $x'(t) > 0$ for large t, then we may have $x(t) > \gamma t$ for large t, say for $t > t_3 > |t_2|$, where $\gamma > 0$ is a constant. Integrating (4.116) from t_3 to t, we have

$$x''(t) - x''(t_3) = -\int_{t_3}^t c(s)x^\alpha(s)\,ds < -\gamma^\alpha \int_{t_3}^t s^\alpha c(s)\,ds. \tag{4.121}$$

As (4.118) implies (4.120) for $\alpha \geq 1$, from (4.121) we have $\lim_{t\to\infty} x''(t) = -\infty$, a contradiction. Hence $x(t) > 0, x'(t) < 0, x''(t) > 0$ and $x'''(t) < 0$ for $t \geq t_3$.

Clearly, $\lim_{t\to\infty} x''(t) = 0$, $\lim_{t\to\infty} x'(t) = 0$ and $\lim_{t\to\infty} x(t) = \lambda, 0 \le \lambda < \infty$ exist. Let $\lambda > 0$. Integrating (4.116) from t to s for $s > t > t_3$, and letting s tends to infinity, we obtain

$$x''(t) \ge \int_t^\infty c(\theta)x^\alpha(\theta)\,d\theta.$$

Integrating the above inequality from t_3 to t, we have

$$x'(t) - x'(t_3) \ge \int_{t_3}^t \int_s^\infty c(\theta)x^\alpha(\theta)\,d\theta\,ds.$$

That is,

$$\begin{aligned} x'(t) - x'(t_3) &\ge (t - t_3)\int_t^\infty c(s)x^\alpha(s)\,ds + \int_{t_3}^t (s - t_3)c(s)x^\alpha(s)\,ds \\ &\ge \int_{t_3}^t (s - t_3)c(s)x^\alpha(s)\,ds \\ &\ge \lambda^\alpha \int_{t_3}^t (s - t_3)c(s)\,ds. \end{aligned} \tag{4.122}$$

We may note that (4.118) holds, if and only if $\int_\sigma^\infty (t - t_3)c(t)\,dt = \infty$, and for $\alpha < 1$, (4.120) implies (4.118). Thus, from (4.122), it follows that $\lim_{t\to\infty} x'(t) = \infty$, a contradiction. Hence $\lim_{t\to\infty} x(t) = 0$.

If $x(t)$ is nonoscillatory solution of (4.117), then proceeding exactly as above, we obtain the stated properties of the solution. Hence the proof of the theorem is completed. □

Theorem 4.3.2 *Suppose that the conditions of Theorem* 4.3.1 *is satisfied. If a solution of* (4.116) *or* (4.117) *has a zero, or its first derivative or second derivative has a zero, then the solution is oscillatory.*

Remark 4.3.1 We may note that for $a(t) \equiv 0$ and $b(t) \equiv 0$, Theorem 4.3.1 is better than Theorems 4.2.5, 4.2.7 and 4.2.10, as it is proved under weaker assumptions.

Now, we assume that $c(t) \ge 0$ for $t \ge \sigma$, $f(x)$ is continuous on $(0, \infty)$ and $f(x) > 0$ for $x > 0$. Now, we give a structure of positive solutions of (4.117) in the case where f has a general nonlinearity.

If $x(t)$ is a positive solution of (4.117) defined on $t \ge t_1 \ge \sigma$, then $x'''(t) < 0$ for $t \ge t_1$, and thus there exists a $t_2 \ge t_1$ such that either of the following properties is satisfied for $t \ge t_2 \ge t_1$:

$$x(t) > 0, \qquad x'(t) < 0, \qquad x''(t) > 0 \quad\text{and}\quad x'''(t) < 0 \tag{4.123a}$$

$$x(t) > 0, \qquad x'(t) > 0, \qquad x''(t) > 0 \quad\text{and}\quad x'''(t) < 0. \tag{4.123b}$$

If $\aleph$ is the set of all nonoscillatory solutions of (4.117), $\aleph_0$ is the set of all nonoscillatory solutions of (4.117), satisfying (4.123a), and $\aleph_2$ is the set of all nonoscillatory solutions of (4.117), satisfying (4.123b), then one can denote $\aleph = \aleph_0 \cup \aleph_2$. By (4.123a) and (4.123b), it is easy to see that if $x \in \aleph_2$, then there exist $a_1 > 0$ and $a_2 > 0$ such that

$$a_1 t \leq x(t) \leq a_2 t^2$$

for large t. More precisely, exactly either of the following cases holds:

$$\lim_{t\to\infty} \frac{x(t)}{t^2} \text{ exists and is a finite value,} \tag{4.124}$$

$$\lim_{t\to\infty} \frac{x(t)}{t^2} = 0 \quad \text{and} \quad \lim_{t\to\infty} \frac{x(t)}{t} = \infty \tag{4.125}$$

and

$$\lim_{t\to\infty} \frac{x(t)}{t} \text{ exists and is a finite value.} \tag{4.126}$$

Now, we give some appropriate class of functions for which $f(x)$ is superlinear or sublinear.

Equation (4.117) is said to be superlinear if

$$x^{-1} f(x) \text{ is nondecreasing in } (0, \infty); \tag{4.127}$$

and (4.117) is said to be sublinear

$$x^{-1} f(x) \text{ is nonincreasing in } (0, \infty). \tag{4.128}$$

If there is an $\epsilon > 0$ such that

$$x^{-1-\epsilon} f(x) \text{ is nondecreasing in } (0, \infty), \tag{4.129}$$

then (4.117) is said to be strongly superlinear. If there is an $\epsilon > 0$ such that

$$x^{-1+\epsilon} f(x) \text{ is nonincreasing in } (0, \infty), \tag{4.130}$$

then (4.117) is said to be strongly sublinear. It is easy to see that (4.117) is strongly superlinear (respectively, strongly sublinear), then it is superlinear (respectively, sublinear).

Theorem 4.3.3 *Let $c(t) > 0$ and $i = 0, 1, 2$. For the case $i = 1$ and 2, suppose that the followings hold:*

$$\left.\begin{aligned} &c(t) \text{ is decomposable in the following way:}\\ &c(t) = g(t)h(t), \quad g(t) \text{ and } h(t) \text{ are continuous}\\ &\text{on } [\sigma, \infty), \sigma > 0 \text{ and } g(t) > 0, h(t) > 0,\ t \geq t_0 \geq \sigma. \end{aligned}\right\} \tag{4.131}$$

Further, $g(t)$ and $h(t)$ in (4.131) *are assumed to satisfy*

$$0 \le \liminf_{t\to\infty} g(t) \le \limsup_{t\to\infty} g(t) < \infty \tag{4.132}$$

and

$$\left.\begin{array}{l}\textit{there is a } \rho \in R \textit{ such that } t^\rho h(t) \textit{ is either}\\ \textit{nondecreasing or nonincreasing on } [t_0,\infty),\ t_0 \ge \sigma.\end{array}\right\} \tag{4.133}$$

Then (4.117) *has a nonoscillatory solution $x(t)$ with the asymptotic condition*

$$\lim_{t\to\infty} \frac{x(t)}{t^i} \text{ exists and is a positive finite value} \tag{4.134}$$

if and only if

$$\int_{t_0}^{\infty} t^{2-i} c(t) f\left(\lambda t^i\right) dt < \infty \quad \text{for some } \lambda > 0. \tag{4.135}$$

We note that for the case $i = 0$ of Theorem 4.3.3, the conditions (4.131), (4.132) and (4.133) are not assumed. As a direct consequence of Theorem 4.3.3, we obtain the following corollary:

Corollary 4.3.1 *Suppose that* (4.131)–(4.133) *hold. Then*

(i) *Equation* (4.117) *has a positive solution satisfying the property* (4.124) *if and only if*

$$\int_{t_0}^{\infty} c(t) f\left(\lambda t^2\right) dt < \infty \quad \text{for some } \lambda > 0, \tag{4.136}$$

(ii) *Equation* (4.117) *has a positive solution satisfying the property* (4.126) *if and only if*

$$\int_{t_0}^{\infty} t c(t) f(\lambda t) \, dt < \infty \quad \text{for some } \lambda > 0. \tag{4.137}$$

Proof of Theorem 4.3.3 First, we prove the "only if" part. Let $i = 0, 1, 2$ and suppose that (4.117) has a positive solution $x(t)$ which satisfies the asymptotic condition (4.134). Let

$$\lim_{t\to\infty} \frac{x(t)}{t^i} = \alpha \in (0, \infty). \tag{4.138}$$

Then it is obvious that

$$\lim_{t\to\infty} \frac{x^{(j)}(t)}{t^{i-j}} = \frac{i!}{(i-j)!}\alpha \quad \text{for } j = 0, 1, 2, \ldots, i \tag{4.139}$$

and

$$\lim_{t\to\infty} x^{(j)}(t) = 0 \quad \text{for } j = i+1, \ldots, 2. \tag{4.140}$$

Taking (4.139) and (4.140) into account and integrating (4.117) repeatedly from t to ∞, we get

$$x^{(i)}(t) = i!\alpha + (-1)^{2-i} \int_t^\infty \frac{(s-t)^{2-i}}{(2-i)!} c(s) f\big(x(s)\big)\, ds$$

for $t \geq T$, where $T > t_0$ is chosen sufficiently large. In particular, we have

$$\int_T^\infty s^{2-i} c(s) f\big(x(s)\big)\, ds < \infty. \tag{4.141}$$

It follows from (4.138) that

$$\frac{\alpha}{2} t^i \leq x(t) \leq 2\alpha t^i \tag{4.142}$$

for all sufficiently large t. Moreover, if $i = 1, 2$, then it follows from (4.139) with $j = 1$ that

$$\frac{\alpha}{2} i t^{i-1} \leq x'(t) \leq 2\alpha i t^{i-1} \tag{4.143}$$

for all sufficiently large t. We may suppose that (4.142) and (4.143) are satisfied for $t \geq T$.

To prove the theorem, we must distinguish between the case $i = 0$ and $i = 1, 2$. First, we consider the case $i = 0$. Then (4.142) with $i = 0$ gives

$$\frac{\alpha}{2} \leq x(t) \leq 2\alpha$$

for $t \geq T$. Therefore, it follows from (4.141) with $i = 0$ that

$$f_0 \int_T^\infty s^2 c(s)\, ds < \infty,$$

where $f_0 = \min\{f(x);\ \frac{\alpha}{2} \leq x \leq 2\alpha\}$ is a positive constant. Thus (4.135) with $i = 0$ is satisfied. Note that the conditions (4.131)–(4.133) are not used in the case $i = 0$.

Next, we consider the case $i = 1, 2$. In this case, we use (4.131)–(4.133). By the condition (4.132), there exist constants $c_1 > 0$ and $c_2 > 0$ such that

$$c_1 \leq g(t) \leq c_2, \quad t \geq t_0. \tag{4.144}$$

By (4.133), $t^\rho h(t)$ is either nondecreasing or nonincreasing on $[t_0, \infty)$ for some $\rho \in R$. We consider the case $t^\rho h(t)$ is nondecreasing on $[t_0, \infty)$. The case where $t^\rho h(t)$ is nonincreasing can similarly be treated. We may assume, without any loss of generality, that $\rho > 0$, since we consider the case where $t^\rho h(t)$ is nondecreasing.

Put $\beta_1 = (2\alpha)^{-1/i}$ and $\beta_2 = (\alpha/2)^{-1/i}$. Then by (4.142), we have

$$\beta_1 \big[x(t)\big]^{1/i} \leq t \leq \beta_2 \big[x(t)\big]^{1/i}, \quad t \geq T. \tag{4.145}$$

The condition (4.138) with $i = 1, 2$ implies that $\lim_{t\to\infty} x(t) = \infty$, and so we may suppose that

$$t_0 \le \beta_1 \big[x(t)\big]^{1/i}, \quad t \ge T. \tag{4.146}$$

Using the inequalities (4.144)–(4.146) and the condition $c(t) = g(t)h(t)$, we obtain

$$c(t) \ge \mu c\big(\beta_1 \big[x(t)\big]^{1/i}\big), \quad t \ge T, \tag{4.147}$$

where $\mu = c_1 c_2^{-1} \beta_1^\rho \beta_2^{-\rho}$ is a positive constant. In fact, we compute as follows:

$$\begin{aligned} c(t) &= g(t)t^{-\rho}t^\rho h(t) \\ &\ge c_1 \big\{\beta_2 \big[x(t)\big]^{1/i}\big\}^{-\rho} \big\{\beta_1 \big[x(t)\big]^{1/i}\big\}^\rho h\big(\beta_1 \big[x(t)\big]^{1/i}\big) \\ &= c_1 \beta_1^\rho \beta_2^{-\rho} h\big(\beta_1 \big[x(t)\big]^{1/i}\big) \\ &\ge c_1 c_2^{-1} \beta_1^\rho \beta_2^{-\rho} c\big(\beta_1 \big[x(t)\big]^{1/i}\big), \quad t \ge T. \end{aligned}$$

By (4.142) and (4.143), there exists a constant $M > 0$ such that

$$\frac{d}{dt}\big\{\beta_1 \big[x(t)\big]^{1/i}\big\} \le M, \quad t \ge T. \tag{4.148}$$

Hence it follows from (4.141) and (4.145)–(4.148) that

$$\int_T^\infty \big\{\beta_1 \big[x(t)\big]^{1/i}\big\}^{2-i} c\big(\beta_1 \big[x(t)\big]^{1/i}\big) f\big(x(t)\big) \frac{d}{dt}\big\{\beta_1 \big[x(t)\big]^{1/i}\big\}\, dt < \infty.$$

Then, letting $r = \beta_1 [x(t)]^{1/i}$ and noting that $r \to \infty$ as $t \to \infty$, we arrive at

$$\int_R^\infty r^{2-i} c(r) f\big(\beta_1^{-1} r^i\big)\, dr < \infty$$

with $R = \beta_1 [x(T)]^{1/i}$. Thus, the integral condition (4.135) is satisfied.

Now, we prove the "if" part of the theorem. Assume that (4.135) is satisfied. First, we suppose that $i = 0$. Then (4.135) reduces to

$$\int_t^\infty t^2 c(t)\, dt < \infty.$$

Let λ be an arbitrary positive number. It is possible to choose $T > t_0$ so large that

$$f_1 \int_T^\infty t^2 c(t)\, dt < 2\lambda,$$

where $f_1 = \max\{f(x); \lambda \le x \le 2\lambda\} > 0$. Denote $C[T, \infty)$ be a Frechet space of all continuous functions on $[T, \infty)$ with the topology of uniform convergence on compact subinterval of $[T, \infty)$, and let X be the closed convex subset of $C[T, \infty)$ defined by

$$X = \big\{x \in C[T, \infty); \lambda \le x(t) \le 2\lambda \text{ for } t \ge T\big\}.$$

Define an operator Φ on X by

$$\big(\Phi(x)\big)(t) = \lambda + \int_t^\infty \frac{(s-t)^2}{2!} c(s) f\big(x(s)\big)\,ds, \quad t \geq T.$$

Then it can be shown, without any difficulty, that Φ is well defined for $x \in X$, Φ maps X into X, Φ is continuous on X and $\Phi(X)$ is relatively compact. The verification is left as a task to the research students.

We shall apply Schauder-Tychonoff fixed-point theorem to the operator Φ. Let $x \in X$ be a fixed point of Φ. Then it can easily be seen that $x(t)$ is a solution of (4.117) for $t \geq T$ and has the property that $\lim_{t\to\infty} x(t) = \lambda > 0$. This proves the "if" part of the theorem in the case $i = 0$. We remark that the conditions (4.131)–(4.133) are not used in the case $i = 0$.

Next, we consider the case $i = 1, 2$. In this case, (4.131)–(4.133) are used. We have (4.144). Moreover, we may suppose that $t^\rho h(t)$ is nondecreasing on $[t_0, \infty)$ for some positive number $\rho > 0$. Put $\beta_1 = (2\lambda)^{-1/i}$, $\beta_2 = \lambda^{-1/i}$ and $\mu = c_1^{-1} c_2 \beta_1^{-\rho} \beta_2^{\rho}$. Here λ is a positive constant appearing in (4.135). Choose $T > t_0$ so large that

$$2^{\frac{i-1}{i}} \mu \int_T^\infty t^{2-i} c(t) f\big(\lambda t^i\big)\,dt \leq \lambda i!(2-i)!. \tag{4.149}$$

Let $C[T,\infty)$ denote the Frechet space of all C^1-functions on $[T,\infty)$ with the usual metric topology. We have the convergence $x_j \to x$ as $j \to \infty$ in the topology of $C^1[T,\infty)$ if and only if $x_j(t) \to x(t)$ as $j \to \infty$ and $x_j'(t) \to x'(t)$ as $j \to \infty$ uniformly on every compact subinterval of $[T,\infty)$. Let X be the subset of all functions $x \in C^1[T,\infty)$ such that

$$\lambda t^i \leq x(t) \leq 2\lambda t^i \quad \text{and} \quad \lambda i t^{i-1} \leq x'(t) \leq 2i\lambda t^{i-1} \quad \text{for } t \geq T.$$

The set X is a closed convex subset of $C[T,\infty)$. Let λ be an arbitrary positive number. We define an operator Φ on X by

$$(\Phi x)(t) = \lambda t^i + (-1)^{2-i} \int_T^t \frac{(t-s)^{i-1}}{(i-1)!} \int_s^\infty \frac{(r-s)^{2-i}}{(2-i)!} c(r) f\big(x(r)\big)\,dr\,ds, \quad t \geq T.$$

(i) *Φ is well defined on X*. We will show that

$$\int_T^\infty t^{2-i} c(t) f\big(x(t)\big)\,dt < \infty \tag{4.150}$$

for each $x \in X$. Let $x \in X$. Set $\lambda t^i \leq x(t) \leq 2\lambda t^i$ for $t \geq T$; then we have

$$\beta_1 \big[x(t)\big]^{1/i} \leq t \leq \beta_2 \big[x(t)\big]^{1/i}, \quad t \geq T \tag{4.151}$$

and we may suppose that

$$t_0 \leq \beta_1 \big[x(t)\big]^{1/i}, \quad t \geq T. \tag{4.152}$$

As in the previous argument, it can be verified that

$$c(t) \le \mu\, c\big(\beta_2\big[x(t)\big]^{1/i}\big), \quad t \ge T. \tag{4.153}$$

Further, it is easy to show that

$$\frac{d}{dt}\big\{\beta_2\big[x(t)\big]^{1/i}\big\} \ge 2^{\frac{1-i}{i}}, \quad t \ge T. \tag{4.154}$$

From (4.151)–(4.154), it follows that

$$\begin{aligned}\int_T^{\infty} t^{2-i} c(t) f\big(x(t)\big)\, dt &\le 2^{\frac{1-i}{i}} \mu \int_T^{\infty} \big\{\beta_2\big[x(t)\big]^{1/i}\big\}^{2-i} c\big(\beta_2\big[x(t)\big]^{1/i}\big) f\big(x(t)\big) \\ &\quad \times \frac{d}{dt}\big\{\beta_2\big[x(t)\big]^{1/i}\big\}\, dt \\ &= 2^{\frac{i-1}{i}} \mu \int_R^{\infty} r^{2-i} c(r) f\big(\lambda r^i\big)\, dr,\end{aligned}$$

where $R = \beta_2[x(T)]^{1/i}$. Then by using (4.151), we get

$$\int_T^{\infty} t^{2-i} c(t) f\big(x(t)\big)\, dt \le 2^{\frac{i-1}{i}} \mu \int_T^{\infty} r^{2-i} c(r) f\big(\lambda r^i\big)\, dr. \tag{4.155}$$

Thus, (4.149) and (4.155) yields

$$\int_T^{\infty} t^{2-i} c(t) f\big(x(t)\big)\, dt \le \lambda i!(2-i)!, \tag{4.156}$$

and hence, in particular, (4.150) holds for $x \in X$.

(ii) *Φ maps X into X*. Let $x \in X$. Differentiating $(\Phi x)(t)$, we obtain

$$(\Phi x)'(t) = 2\lambda t + \int_T^t \int_s^{\infty} c(r) f\big(x(r)\big)\, dr\, ds, \quad t \ge T,$$

for the case $i = 2$; and

$$(\Phi x)'(t) = \lambda - \int_t^{\infty} (s-t) c(s) f\big(x(s)\big)\, ds, \quad t \ge T$$

for the case $i = 1$. Then it is clear that $\Phi x \in C^1[T, \infty)$. Further, we find that

$$\begin{aligned} 0 &\le \int_T^t \frac{(t-s)^{i-1}}{(i-1)!} \left(\int_s^{\infty} \frac{(r-s)^{2-i}}{(2-i)!} c(r) f\big(x(r)\big)\, dr \right) ds \\ &\le \int_T^t \frac{(t-s)^{i-1}}{(i-1)!}\, ds \int_T^{\infty} \frac{(r-T)^{2-i}}{(2-i)!} c(r) f\big(x(r)\big)\, dr \\ &\le \frac{t^i}{i!(2-i)!} \int_T^{\infty} r^{2-i} c(r) f\big(x(r)\big)\, dr \\ &\le \lambda t^i \ \big(\text{using (4.156)}\big) \end{aligned} \tag{4.157}$$

for $t \geq T$. Thus, we see that $\lambda t^i \leq (\Phi x)(t) \leq 2\lambda t^i$ for $t \geq T$. Similarly, we can prove that $\lambda i t^{i-1} \leq (\Phi x)'(t) \leq 2\lambda i t^{i-1}$ for $t \geq T$. Thus, if $x \in X$ then $\Phi x \in X$.

(iii) *Φ is continuous on X*. Let x_j, $j \in N$ and x be functions in X such that $\{x_j\}$ converges to x in the topology $C^1[T, \infty)$. Then it is enough to show that $(\Phi x_j)(t) \to (\Phi x)(t)$ as $j \to \infty$ uniformly on every compact subinterval of $[T, \infty)$. Using the definitions of Φ, we have

$$\left|(\Phi x_j)(t) - (\Phi x)(t)\right| \leq \frac{t^i}{i!(2-i)!}\int_T^\infty s^{2-i}c(s)\left|f\left(x_j(s)\right) - f\left(x(s)\right)\right| ds$$

and

$$\left|(\Phi x_j)'(t) - (\Phi x)'(t)\right| \leq \frac{t^{i-1}}{(i-1)!(2-i)!}\int_T^\infty s^{2-i}c(s)\left|f\left(x_j(s)\right) - f\left(x(s)\right)\right| ds$$

for $t \geq T$ and $j \in N$. Hence, to prove the continuity of Φ, it is enough to claim that

$$\int_T^\infty s^{2-i}c(s)\left|f\left(x_j(s)\right) - f\left(x(s)\right)\right| ds \to 0 \quad \text{as } j \to \infty. \tag{4.158}$$

As in the proof of the inequality (4.155), we have

$$\int_T^\infty t^{2-i}c(s)f\left(x(s)\right) ds \leq 2^{\frac{i-1}{i}}\mu\int_{T_1}^\infty r^{2-i}c(r)f\left(\lambda r^i\right) dr$$

and

$$\int_T^\infty t^{2-i}c(s)f\left(x_j(s)\right) ds \leq 2^{\frac{i-1}{i}}\mu\int_{T_1}^\infty r^{2-i}c(r)f\left(\lambda r^i\right) dr$$

for any $T_1 \geq T$ and $j \in N$. Let $\epsilon > 0$ be given. There is a $T_1 = T_1(\epsilon) > T$ such that

$$2^{(i-1)/i}\mu\int_{T_1}^\infty r^{2-i}c(r)f\left(\lambda r^i\right) dr < \epsilon/4.$$

Then, we have

$$\int_{T_1}^\infty t^{2-i}c(t)f\left(x(t)\right) dt < \epsilon/4$$

and

$$\int_{T_1}^\infty t^{2-i}c(t)f\left(x_j(t)\right) dt < \epsilon/4$$

for all $j \in N$. Note that the number $T_1 (> T)$ is chosen independently of $j \in N$. It is easy to check that

$$\int_T^{T_1} s^{2-i}c(s)\left|f\left(x_j(s)\right) - f\left(x(s)\right)\right| ds \to 0 \quad \text{as } j \to \infty,$$

and hence there is a $j_1 \in N$ such that if $j \geq j_1$, then

$$\int_T^{T_1} s^{2-i} c(s) \big| f\big(x_j(s)\big) - f\big(x(s)\big) \big| \, ds < \epsilon/2.$$

Therefore, if $j \geq j_1$, then

$$\begin{aligned} \int_T^{\infty} s^{2-i} c(s) \big| f\big(x_j(s)\big) - f\big(x(s)\big) \big| \, ds &\leq \int_T^{T_1} s^{2-i} c(s) \big| f\big(x_j(s)\big) - f\big(x(s)\big) \big| \, ds \\ &\quad + \int_{T_1}^{T} s^{2-i} c(s) \big| f\big(x_j(s)\big) - f\big(x(s)\big) \big| \, ds \\ &\quad + \int_{T_1}^{\infty} s^{2-i} c(s) f\big(x(s)\big) \, ds \\ &< \epsilon/2 + \epsilon/4 + \epsilon/4 = \epsilon. \end{aligned}$$

This means that (4.158) holds.

(iv) Φ *is relatively compact*. By the Ascoli–Arzela theorem, it is sufficient to verify that $\{\Phi(x); x \in X\}$ and $\{(\Phi x)'; x \in X\}$ are uniformly bounded and equicontinuous on every compact subinterval of $[T, \infty)$. Let $x \in X$. As shown in (ii), we have $\lambda t^i \leq (\Phi x)(t) \leq 2\lambda t^i$ and $\lambda i t^{i-1} \leq (\Phi x)'(t) \leq 2\lambda i t^{i-1}$ for $t \geq T$. By a similar argument, we obtain

$$2\lambda \leq (\Phi x)''(t) \leq 4\lambda, \quad t \geq T,$$

for $i = 2$; and

$$\big|(\Phi x)''(t)\big| \leq \frac{\lambda}{t} \quad \text{for } t \geq T$$

for $i = 1$. Thus, for any $S > T$, we have

$$\big|(\Phi x)(t)\big| \leq 2\lambda S^i \quad \text{and} \quad \big|(\Phi x)'(t)\big| \leq 2\lambda i S^{i-1}$$

on the compact interval $[T, S] \subset [T, \infty)$. Further, we have

$$\big|(\Phi x)''(t)\big| \leq 2\lambda i(i-1) S^{i-2} \quad \text{for } i = 2$$

and

$$\big|(\Phi x)''(t)\big| \leq \frac{\lambda}{t} \quad \text{for } i = 1$$

on the same interval $[T, S]$. These facts show that $\{\Phi x; x \in X\}$ and $\{(\Phi x)'; x \in X\}$ are uniformly bounded and equicontinuous on every compact subinterval of $[T, \infty)$.

Hence, with the foregoing considerations, the Schauder-Tychonoff fixed-point theorem can be applied to the operator Φ. Thus, we conclude that there exists a function $x \in X$ such that $x(t) = (\Phi x)(t)$, $t \geq T$. It is easy to see that this function

$x(t)$ is a positive solution of (4.117) on $[T, \infty)$ and satisfies

$$\lim_{t\to\infty} \frac{x(t)}{t^j} = \lambda \in (0, \infty).$$

This proves the theorem. □

Lemma 4.3.1 *Suppose that $x(t)$, $y(t)$ and $P(t)$ are continuous functions such that $x(t) > 0$, $y(t) \geq 0$, $P(t) \geq 0$ on $[T, \infty)$, and $P(t) \not\equiv 0$ on $[\tau, \infty)$ for any $\tau \geq T$.*

(i) *If $0 < \gamma < 1$ and*

$$x(t) \geq y(t) \int_t^\infty P(s)\big[x(s)\big]^\gamma \, ds, \quad t \geq T, \tag{4.159}$$

then we have

$$\int_T^\infty \big[y(t)\big]^\gamma P(t)\, dt < \infty; \tag{4.160}$$

(ii) *If $\gamma > 1$ and*

$$x(t) \geq y(t) \int_T^t P(s)\big[x(s)\big]^\gamma \, ds, \quad t \geq T, \tag{4.161}$$

then we also have (4.160).

Proof (i) Let $I(t) = \int_t^\infty P(s)[x(s)]^\gamma \, ds$. Then

$$\begin{aligned} -\frac{d}{dt}\big[I(t)\big]^{1-\gamma} &= (1-\gamma)\big[I(t)\big]^{-\gamma} P(t)\big[x(t)\big]^\gamma \\ &\geq (1-\gamma)\big[I(t)\big]^{-\gamma} P(t)\big[y(t)I(t)\big]^\gamma \\ &\geq (1-\gamma)\big[y(t)\big]^\gamma P(t) \end{aligned} \tag{4.162}$$

for $t \geq T$. An integration of this inequality yields (4.160).

(ii) Let $J(t) = \int_T^t P(s)[x(s)]^\gamma \, ds$. Then there is a $T_1 > T$ such that $J(t) > 0$ for $t \geq T_1$. Then it is easy to see that

$$-\frac{d}{dt}\big[J(t)\big]^{1-\gamma} \geq (\gamma - 1)\big[y(t)\big]^\gamma P(t)$$

for $t \geq T_1$. Integrating this inequality, we find that (4.160) holds. The proof is complete. □

Lemma 4.3.2 *Let $x(t)$ be a positive solution of* (4.117) *satisfying the property* (4.123b). *Then there are constants $\mu > 0$, $\gamma > 0$, and $T > t_0$ such that*

$$x(t) \geq \mu(t-T) \int_T^t (s-T)c(s) f\big(x(s)\big)\, ds + \mu(t-T)^2 \int_T^\infty c(s) f\big(x(s)\big)\, ds, \quad t \geq T, \tag{4.163}$$

and

$$x(t) \geq \gamma t x'(t) > 0, \quad t \geq T. \tag{4.164}$$

Proof Clearly, for $t \geq T_0 \geq T$, we have

$$x'(t) = x'(T_0) + \int_{T_0}^{t} x''(s)\, ds, \quad t \geq T_0.$$

Since $x'(T_0) > 0$ and $x''(t)$ is nonincreasing on $[T_0, \infty)$, we get

$$x'(t) \geq x''(t)(t - T_0), \quad t \geq T_0.$$

Integrating this inequality by parts, we see that

$$x(t) \geq x(T_0) + x'(t)(t - T_0) - x(t) + x(T_0), \quad t \geq T_0,$$

which gives

$$x(t) \geq \frac{1}{2} x'(t)(t - T_0), \quad t \geq T_0.$$

Let $0 < \gamma < 1/2$ and choose $T > T_0$ be such that $(t - T_0)/2 \geq \gamma t$ for $t \geq T$. Then (4.164) is clear.

It is easy to see that $\lim_{t\to\infty} x''(t)$ exists and is a nonnegative finite value. Integrating (4.117) from t $(\geq T)$ to τ, and letting $\tau \to \infty$, we have

$$x''(t) = x''(\infty) + \int_{t}^{\infty} c(s) f\big(x(s)\big)\, ds, \quad t \geq T.$$

Further, integrating this inequality two times over the finite interval $[T, t]$, we get

$$\begin{aligned} x(t) &= x(T) + (t - T)x'(T) + \frac{x''(\infty)}{2!}(t - T)^2 \\ &\quad + \int_{T}^{t} (t - s) \int_{s}^{\infty} c(r) f\big(x(r)\big)\, dr\, ds \\ &= x(T) + (t - T)x'(T) + \frac{x''(\infty)}{2!}(t - T)^2 \\ &\quad + \int_{T}^{t} \left(\int_{T}^{r} (t - s)\, ds \right) c(r) f\big(x(r)\big)\, dr \\ &\quad + \int_{t}^{\infty} \left(\int_{T}^{t} (t - s)\, ds \right) c(r) f\big(x(r)\big)\, dr \end{aligned}$$

for $t \geq T$ and, in particular,

$$x(t) \geq \int_{T}^{t} \left(\int_{T}^{r} (t - s)\, ds \right) c(r) f\big(x(r)\big)\, dr + \int_{t}^{\infty} \left(\int_{T}^{t} (t - s)\, ds \right) c(r) f\big(x(r)\big)\, dr \tag{4.165}$$

for $t \geq T$. Note here that the inequality

$$\int_T^u (u-s)^{i-1}(v-s)^{j-1}\,ds \geq \frac{1}{i+j-1}(u-T)^i(v-T)^{j-1}, \quad T \leq u \leq v,$$

where i and j are positive integers, holds. Then we find that (4.165) gives (4.163) with $\mu = 1/2$. The proof of the lemma is complete. □

Theorem 4.3.4 *Suppose that $c(t) \geq 0$. Let* (4.131) *and* (4.132) *hold. Further, suppose that there exists an $\epsilon > 0$ that*

$$t^{3-\epsilon}h(t) \text{ is nondecreasing on } [t_0, \infty). \tag{4.166}$$

Then (4.117) *has a positive solution $x(t)$ satisfying the property* (4.123b) *if and only if*

$$\int_{t_0}^{\infty} c(t) f\left(\lambda t^2\right) dt < \infty \quad \text{for some } \lambda > 0. \tag{4.167}$$

Proof The "if" part follows from Corollary 4.3.1. It remains to prove the "only if" part. Suppose that (4.117) has a positive solution $x(t)$ satisfying the property (4.123b). Now, we shall show that (4.167) holds. If $c(t) \equiv 0$ on $[\tau, \infty)$ for some $\tau \geq t_0$, then (4.167) is trivially satisfied. Therefore, to prove (4.167), it is enough to consider the case where $c(t) \not\equiv 0$ on $[\tau, \infty)$ for any $\tau \geq t_0$. By the condition (4.166), there is an $\epsilon > 0$ such that $t^{2-\epsilon}h(t)$ is nondecreasing on $[t_0, \infty)$. Note that $\epsilon > 0$ may be chosen sufficiently small so that $0 < \epsilon < 2$. By (4.163) of Lemma 4.3.2, we have

$$x(t) \geq \mu(t-T)^2 \int_t^{\infty} c(s) f\big(x(s)\big)\,ds, \quad t \geq T,$$

for some $\mu > 0$. Applying (i) of Lemma 4.3.1 to the case $\gamma = 1 - \frac{\epsilon}{k}$, $y(t) = \mu(t - T)^2$ and $P(t) = c(t) f(x(t))[x(t)]^{-1+\epsilon/2}$, we conclude that

$$\int_T^{\infty} t^{2-\epsilon} c(t) f\big(x(t)\big)\big[x(t)\big]^{-1+\frac{\epsilon}{2}}\,dt < \infty \tag{4.168}$$

holds. By the conditions (4.131), (4.132) and (4.166), on $c(t) = g(t)h(t)$, there are $c_1 > 0$ and $c_2 > 0$ such that (4.144) holds. Further, since we consider the case $c(t) \not\equiv 0$ on $[\tau, \infty)$ for any $\tau \geq t_0$, there is a $t_1 \geq t_0$ such that $h(t_1) > 0$. Then the nondecreasing property of $t^{3-\epsilon}h(t)$ implies

$$c(t) = g(t)t^{-3+\epsilon}t^{3-\epsilon}h(t) \geq c_1 t_1^{3-\epsilon}h(t_1)t^{-3+\epsilon}$$

for $t \geq t_1$. Thus, we have

$$\int_{t_1}^{\infty} t^2 c(t)\,dt = \infty.$$

By (ii) of Corollary 4.3.1, we find that $x(t)$ does not have a positive finite limit as $t \to \infty$. Consequently, we have

$$\lim_{t \to \infty} x(t) = \infty. \tag{4.169}$$

As is easily seen, there exist positive constants $a_1 > 0$ and $a_2 > 0$ such that

$$a_1 t \le x(t) \le a_2 t^2, \quad t \ge T. \tag{4.170}$$

Since (4.169) holds, we may suppose that $x(t) \ge a_2 t_0^k$ for $t \ge T$. Then

$$t_0 \le \beta\big[x(t)\big]^{1/2} \le t, \quad t \ge T,$$

where $\beta = a_2^{-1/2}$. Consequently, we have

$$\begin{aligned} c(t) &= g(t)t^{-3+\epsilon}t^{3-\epsilon}h(t) \\ &\ge c_1 t^{-3+\epsilon}\big\{\beta\big[x(t)\big]^{1/2}\big\}^{3-\epsilon}h\big(\beta\big[x(t)\big]^{1/2}\big) \\ &\ge \frac{c_1}{c_2}t^{-3+\epsilon}\big\{\beta\big[x(t)\big]^{1/2}\big\}^{3-\epsilon}c\big(\beta\big[x(t)\big]^{1/2}\big) \\ &= dt^{-3+\epsilon}\big[x(t)\big]^{(3-\epsilon)/2}c\big(\beta\big[x(t)\big]^{1/2}\big), \quad t \ge T, \end{aligned} \tag{4.171}$$

where $d = c_1 c_2^{-1}\beta^{3-\epsilon} > 0$. Then it is easy to verify that (4.164) of Lemma 4.3.2, (4.168) and (4.171) give

$$\int_T^{\infty} c\big(\beta\big[x(t)\big]^{1/2}\big) f\big(x(t)\big)\frac{d}{dt}\big\{\beta\big[x(t)\big]^{1/2}\big\}\,dt < \infty.$$

Letting $r = \beta[x(t)]^{1/2}$ and by (4.169), $r \to \infty$ as $t \to \infty$, we see that (4.167) holds with $\lambda = \beta^{-2}$. This completes the proof of the theorem. □

Theorem 4.3.5 *Let* $c(t) \ge 0$. *Suppose that* (4.131), (4.132) *and there is an* $\epsilon > 0$ *such that*

$$t^{3+\epsilon}h(t) \textit{ is nonincreasing on } [t_0, \infty). \tag{4.172}$$

Then (4.117) *has a positive solution* $x(t)$ *satisfying* (4.123b) *if and only if*

$$\int_{t_0}^{\infty} tc(t)f(\lambda t)\,dt < \infty \quad \textit{for some } \lambda > 0. \tag{4.173}$$

Proof The "if" part follows from part (ii) of Corollary 4.3.1. It remains to prove the "only if" part of the theorem. Suppose that (4.117) has a positive solution $x(t)$ satisfying the property (4.123b). We shall show that (4.173) holds. Clearly, we have (4.169). From (4.163) of Lemma 4.3.2, it follows that

$$x(t) \ge \mu\,(t - T)\int_T^t (s - T)c(s)f\big(x(s)\big)\,ds, \quad t \ge T,$$

for some $\mu > 0$ and $T > t_0$. We apply (ii) of Lemma 4.3.1 to the case

$$\gamma = 1 + \frac{\epsilon}{2}, \qquad y(t) = \mu(t - T) \quad \text{and} \quad P(t) = (t - T)c(t) f\big(x(t)\big)\big[x(t)\big]^{-1-\epsilon},$$

we obtain

$$\int_T^\infty t^{2+\epsilon} c(t) f\big(x(t)\big)\big[x(t)\big]^{-1-\epsilon}\, dt < \infty. \tag{4.174}$$

Since (4.123b) holds, there exist positive constants $a_1 > 0$ and $a_2 > 0$ satisfying (4.170). Therefore, we have

$$t \le \beta\big[x(t)\big], \qquad t \ge T,$$

where $\beta = a_1^{-1}$. Hence

$$\begin{aligned} c(t) &= g(t) t^{-2-\epsilon} t^{2+\epsilon} h(t) \\ &\ge c_1 t^{-3-\epsilon} \big\{\beta\big[x(t)\big]\big\}^{3+\epsilon} h\big(\beta\big[x(t)\big]\big) \\ &\ge \frac{c_1}{2} t^{-3-\epsilon} \big\{\beta\big[x(t)\big]\big\}^{3+\epsilon} c\big(\beta\big[x(t)\big]\big) \\ &= d t^{-3-\epsilon} \big[x(t)\big]^{3+\epsilon} c\big(\beta\big[x(t)\big]\big), \quad t \ge T, \end{aligned} \tag{4.175}$$

where $d = c_1 c_2^{-1} \beta^{3+\epsilon} > 0$. Then it is easy to verify that (4.164) of Lemma 4.3.2, (4.174) and (4.175) give

$$\int_T^\infty \big[x(t)\big] c\big(\beta\big[x(t)\big]\big) f\big(x(t)\big) \frac{d}{dt} \big\{\beta\big[x(t)\big]\big\}\, dt < \infty.$$

Putting $r = \beta[x(t)]$ and by using (4.169), we find that (4.173) holds with $\lambda = \beta^{-1}$. The theorem is proved. □

Remark 4.3.2 Consider the Euler equation

$$x''' + \delta t^{-3} x = 0, \quad t \ge 1, \tag{4.176}$$

where δ is a positive constant. Suppose that, in this case, $c(t) = \delta t^{-3}$, $g(t) = \delta$, $h(t) = t^{-3}$ and $f(x) = x$. Denote

$$\eta_2^* = \max\big\{-\eta(\eta - 1)(\eta - 2)\big\}$$

on the interval $[1, 2]$. It is easy to see that $\eta_2^* > 0$ and that if δ satisfies $0 < \delta \le \eta_2^*$, then the indicial equation for (4.176)

$$\eta(\eta - 1)(\eta - 2) + \delta = 0$$

has a root η_2 in the open interval $(1, 2)$. Thus the function $x(t) = t^{\eta_2}$ is a positive root of (4.176) and satisfies the property (4.123b). In this example, while $t^3 h(t) = 1$

is nondecreasing and nonincreasing on $[1, \infty)$, neither (4.167) nor (4.173) is satisfied. This shows that the condition (4.167) (respectively (4.172)) of Theorem 4.3.4 (respectively Theorem 4.3.5), it is impossible to take $\epsilon = 0$.

Now, we consider the case $c(t) < 0$. If $x(t)$ is a nonoscillatory solution of (4.117) such that $x(t) > 0$ for $t \geq t_1 \geq t_0$, then $x'''(t) > 0$ for $t \geq t_1$, then either of the following holds for large t, say for $t \geq t_2 \geq t_1$:

$$x(t) > 0, \qquad x'(t) > 0, \qquad x''(t) < 0 \quad \text{and} \quad x'''(t) > 0 \tag{4.177}$$

or

$$x(t) > 0, \qquad x'(t) > 0, \qquad x''(t) > 0 \quad \text{and} \quad x'''(t) > 0. \tag{4.178}$$

If N is the set of all nonoscillatory solutions of (4.117), N_1 is the set of all nonoscillatory solutions of (4.117) satisfying (4.177), and N_3 is the set of all nonoscillatory solutions of (4.117) satisfying (4.178), then N can be classified as $N = N_1 \cup N_3$. If we proceed as in Theorems 4.3.3–4.3.5, we can prove the following theorem.

Theorem 4.3.6 *Let* $c(t) = -e(t)$, $e(t) > 0$ *for* $t \geq \sigma$. *Let* $i = 0, 1, 2$. *For the case* $i = 1$ *and* 2, *suppose that*

$$\left.\begin{array}{l} e(t) \text{ is decomposable in the following way:} \\ e(t) = g(t)h(t),\ \ g(t) \text{ and } h(t) \text{ are continuous on} \\ [t_0, \infty),\ \ t_0 > \sigma, \text{ and } g(t) > 0,\ \ h(t) \geq 0 \text{ for } t \geq t_0, \end{array}\right\} \tag{4.179}$$

and $g(t)$ *and* $h(t)$ *satisfy the condition* (4.132) *and* (4.133). *Then* (4.117) *has a positive solution* $x(t)$ *with the asymptotic condition* $\lim_{t\to\infty} \frac{x(t)}{t^i}$ *exists and it has a positive finite value if and only if*

$$\int_{t_0}^{\infty} t^{2-i} e(t) f\left(\lambda t^i\right) dt < \infty \quad \text{for some } \lambda > 0.$$

Corollary 4.3.2 *Suppose that the conditions of Theorem* 4.3.6 *are satisfied. Then*

(i) *Equation* (4.117) *has a positive nonoscillatory solution* $x(t)$ *which satisfies the property* $\lim_{t\to\infty} \frac{x(t)}{t}$, *and it has a positive finite value if and only if*

$$\int_{t_0}^{\infty} t e(t) f(\lambda t)\, dt < \infty \quad \text{for some } \lambda > 0. \tag{4.180}$$

(ii) *Equation* (4.117) *has a positive nonoscillatory solution* $x(t)$ *satisfying the property* $\lim_{t\to\infty} x(t)$, *and it has a positive finite value if and only if*

$$\int_{t_0}^{\infty} t^2 e(t)\, dt < \infty. \tag{4.181}$$

Theorem 4.3.7 *Let $c(t) = -e(t)$, $e(t) > 0$ for $t \geq \sigma$. Further suppose that* (4.179), (4.132) *and* (4.166) *hold. Then* (4.117) *has a positive solution $x(t)$ satisfying* (4.177) *if and only if* (4.180) *holds.*

Theorem 4.3.8 *Let $c(t) = -e(t)$, $e(t) > 0$ for $t \geq \sigma$. Suppose that* (4.179), (4.132) *and* (4.172) *hold. Then* (4.117) *has a positive solution $x(t)$ satisfying* (4.177) *if and only if* (4.181) *holds.*

We recall that a solution $x(t)$ of (4.116) or (4.117) or any of the above considered third-order differential equation is called a Kneser solution vanishing at infinity if $(-1)^i x^{(i)}(t) > 0$ and $|x^{(i)}(t)| \to 0$ as $t \to \infty$, $i = 0, 1, 2$.

In [3], Astashova obtained sufficient conditions for the existence of a Kneser solution vanishing at infinity of the following equation:

$$x''' + c(t)|x|^{\lambda-1}x = 0, \quad 0 < \lambda < 1, \ t \geq \sigma. \tag{4.182}$$

Theorem 4.3.9 *Let $0 < \lambda < 1$. Then for any $\mu > \frac{1}{2\lambda+1}$ and any continuous positive function $\phi(x)$, $x \geq 0$, if*

$$\int_\sigma^\infty c^\mu(t)\phi(t)\,dt = \infty,$$

then (4.182) *has a Kneser solution vanishing at infinity.*

Moreover, Astashova [3] proved that, if $0 < \lambda < 1$ and $\phi(t)$ is a continuous positive function for $t \geq \sigma$, $\mu > \frac{1}{2\lambda+1}$, then there exists a C^∞ function $x(t)$, $t \geq \sigma$ such that

$$\left|x^{(i)}(t)\right| \to 0, \quad t \to \infty, \ i = 0, 1, 2$$

and

$$\int_\sigma^\infty \left|x'''(t)\right|^\mu \left|x(t)\right|^{-\mu\lambda} \phi(t)\,dt = \infty.$$

Now, suppose that $c(t) > 0$, (4.78) holds and $f(x)$ satisfies the property

$$\lim_{x \to 0} \frac{f(x)}{x} = \theta, \quad 0 \leq \theta < \infty. \tag{4.183}$$

Note that the assumption (4.183) is sufficient to ensure the uniqueness of trivial solution with respect to the initial data.

Lemma 4.3.3 *Let $x(t)$ be a nontrivial solution of* (4.117). *Then $x(t)$ and $x'(t)$ have at most one double zero in $[\sigma, \infty)$.*

Lemma 4.3.4 *Let $e : [\sigma, \infty) \to R$ be a nonnegative continuous function. Then the boundary-value problems*

$$x'''(t) + e(t)x(t) = 0, \tag{4.184}$$

$$x(\sigma) = 1, \qquad x(t) \geq 0, \qquad x'(t) \leq 0 \tag{4.185}$$

and

$$x'''(t) + e(t)x(t) = 0, \tag{4.186}$$

$$x(\sigma) = -1, \qquad x(t) \leq 0, \qquad x'(t) \geq 0 \tag{4.187}$$

admit a unique solution.

Proof Without any loss of generality, we consider the boundary-value problems (4.184)–(4.185). The existence of solutions satisfying (4.184)–(4.185) follows from the result of Hartman and Wintner [18]. In order to prove the uniqueness, let u_1 and u_2 be solutions of (4.184)–(4.185), $u_1 \neq u_2$, and let $v = u_1 - u_2$. Obviously, $v(\sigma) = 0$. We claim that there exists a $T \geq \sigma$ such that $v(T) = 0$ and v has a constant sign for $t \geq T$. Suppose it is not true. Then there exists a sequence $\{t_k\}_{k=1}^{\infty}$, $\lim_{k\to\infty} t_k = \infty$ such that $v(t_k) = 0$ and v changes sign in t_k. Taking into account the uniqueness of trivial solution with respect to the initial data, we get $v'(t_k) = 0$. Consider the function

$$G[v(t)] = \frac{1}{2}\big(v'(t)\big)^2 - v(t)v''(t).$$

Since the function $G[v(t)]$ is a decreasing function of t, we have

$$G[v(t_k)] \geq F[v(t_1)]$$

or

$$[v'(t_k)]^2 \geq [v'(t_1)]^2 > 0$$

which contradicts $\lim_{t\to\infty} v'(t) = 0$. Hence our claim holds. Without any loss of generality, suppose that $v(t) \geq 0$ for $t \geq T$. Then $v'''(t) = -c(t)v(t) \leq 0$ for $t \geq T$ and $v''(t)$ is nonincreasing for $t \geq T$. As $\lim_{t\to\infty} v''(t) = 0$, we have $v''(t) \geq 0$ for $t \geq T$. Again $v'(t) \leq 0$ for $t \geq T$. Since $v(T) = 0$, we have $v(t) \leq 0$ for $t \geq T$, and so $v(t) = 0$ for $t \geq T$. Hence the uniqueness of the trivial solution of (4.184) with respect to the initial data implies that $v(t) = 0$ for $t \geq \sigma$, that is, $u_1(t) = u_2(t)$ for $t \geq \sigma$. The lemma is proved. □

Theorem 4.3.10 *Equation* (4.117) *admits at least both a positive decreasing and a negative increasing solution.*

Proof We shall prove that (4.117) has at least a positive decreasing solution. A similar argument may be used to prove that (4.117) admits a negative increasing solution. Thus, we leave it.

The assumption (4.183) implies that there exists a positive constant k such that

$$\left|\frac{f(u)}{u}\right| \leq k, \quad |x| \leq 1. \tag{4.188}$$

Let $C[\sigma,\infty)$ be the Frechet space of real continuous functions defined on $[\sigma,\infty)$, $\sigma\in R$ with the topology of uniform convergence on compact subinterval of $[\sigma,\infty)$, and let

$$\Omega=\big\{u\in C[\sigma,\infty);\,0\le u(t)\le 1\big\}.$$

For every $u\in\Omega$, consider the linear differential equation

$$y'''+c(t)f_u(t)y=0, \tag{4.189}$$

where

$$f_u(t)=\begin{cases}\frac{f(u(t))}{u(t)} & \text{if } u(t)\neq 0,\\ \theta & \text{if } u(t)=0.\end{cases}$$

Then from Lemma 4.3.4, (4.189) has a unique solution y_u which is nonnegative nonincreasing such that $y_u(\sigma)=1$. Therefore, we may define an operator T as

$$T:\Omega\to T(\Omega);\quad T(u)=y_u.$$

Since $0\le (Tu)(t)\le y_u(t)\le 1$, we have $T(\Omega)\subset\Omega$. It is easy to verify that the operator T is continuous in Ω and $T(\Omega)$ is relatively compact in $C[\sigma,\infty)$. From Schauder–Tychonoff theorem, T has at least a fixed point x in Ω. Obviously, x is a nonnegative nonincreasing solution of (4.117). We claim that if x is a solution of (4.117) on $[T_x,\infty)$ such that $x(t)x'(t)<0$ eventually, then $x(t)x'(t)<0$ for every $t\in(T_x,\infty)$. In fact, if $x(t)>0$, $x'(t)<0$ for $t\ge T$, and there exists $t_1, T_x<t_1\le T$ such that $x'(t_1)=0$, $x(t)<0$ on (t_1,T), then $x(t)>0$ on (t_1,∞). Integrating (4.117) on (t_1,t), we obtain

$$x''(t)\le-\int_{t_1}^{t}c(s)f\big(x(s)\big)\,ds<0.$$

Since $x'''(t)\le 0$, $x''(t)$ is nonincreasing, so $\lim_{t\to\infty}x''(t)=L_x$, $-\infty\le L_x<0$, which gives a contradiction. Hence our claim holds. In view of this, it is enough to prove that $x(t)x'(t)<0$ eventually. First let us show that $x'(t)$ is eventually negative. If it is not true, then there exists a sequence $\{\tau_k\}$ of zeros of $x'(t)$ such that $\tau_k\to\infty$ as $k\to\infty$. From Lemma 4.3.3, $x(t)$ has at most one double zero on $[\sigma,\infty)$, and thus we may suppose $x(\tau_k)>0$. Since $x'(t)$ takes in τ_k a maximum, $x''(t)$ vanishes for every τ_k. From (4.117), $x''(t)$ is nonincreasing and so $x'(t)$ and $x''(t)$ are eventually vanishing. This implies that $x(t)$ is eventually a positive constant which gives a contradiction since $x(t)$ is a solution of (4.117). A similar argument proves that x is eventually positive. The proof is now complete. □

Lemma 4.3.5 *For every positive decreasing (negative increasing) solution $x(t)$ of* (4.117), *we have*

$$\int^{\infty}c(s)x(s)f\big(x(s)\big)\,ds<\infty.$$

Theorem 4.3.11 (a) *If*

$$\int^{\infty} t^2 c(t)\, dt < \infty, \tag{4.190}$$

then every positive decreasing or negative increasing solution of (4.117) *tends to a nonzero limit as* $t \to \infty$.

(b) *If Eq.* (4.117) *has at least a positive decreasing or negative increasing solution tending to a nonzero limit as* $t \to \infty$, *then* (4.190) *is satisfied.*

Proof (a) Suppose that there exists a positive decreasing solution $x(t)$ of (4.117) such that $\lim_{t\to\infty} x(t) = 0$. Choose a large $t_0 > \sigma$ such that

$$0 \le x(t) \le 1, \quad t \ge t_0 \tag{4.191}$$

and $T \ge t_0$ such that

$$k \int_T^{\infty} (s-T)^2 c(s)\, ds < 1, \tag{4.192}$$

where k is defined in (4.188). Integrating (4.117) in (t, ∞), $t \ge T$ and by using (4.188) and (4.191), we obtain

$$x''(t) \le kx(T) \int_T^{\infty} c(s)\, ds.$$

Integrating again in (t, ∞), $t \ge T$ we get

$$x(t) - x(\infty) = x(t) \le \frac{1}{2} k\, x(T) \int_t^{\infty} (s-t)^2 c(s)\, ds,$$

which implies

$$x(T) \le \frac{1}{2} k\, x(T) \int_T^{\infty} (s-T)^2 c(s)\, ds$$

and, hence from (4.192), we obtain

$$x(T) \le \frac{1}{2} x(T),$$

a contradiction. A similar argument holds, if we assume that there exists a negative increasing solution of (4.117) which tends to zero as $t \to \infty$. Hence (a) is now proved.

(b) Without any loss of generality, we suppose that there exists a positive decreasing solution $x(t)$ of (4.117) such that $x(\infty) = \lim_{t\to\infty} x(t) > 0$. Let $M = \min_{t\in[\sigma,\infty)} f(x(t))$. Since f is continuous and $x(t) > x(\infty) > 0$, we have $M > 0$. Thus for $t \ge \sigma$, $f(x(t)) \ge M > 0$. Integrating (4.117) three times from t to ∞, we obtain

$$x(t) - x(\infty) \ge \frac{1}{2} M \int_t^{\infty} (s-t)^2 c(s)\, ds,$$

which implies

$$x(\sigma) - x(\infty) \geq \frac{1}{2} M \int_{\sigma}^{\infty} s^2 c(s)\, ds$$

and so the condition (4.190) is satisfied. The theorem is proved. □

The above proof suggests the following stronger result.

Corollary 4.3.3 *Every positive decreasing or negative increasing solution of* (4.117) *tends to zero as $t \to \infty$ if and only if* (4.34) *holds, that is,*

$$\int^{\infty} t^2 c(t)\, dt = \infty.$$

Theorem 4.3.12 *Let $\phi : [\sigma, \infty) \to R$ be a positive function such that*

$$\int^{\infty} t^2 c(t) \phi(t)\, dt < \infty.$$

If $x(t)$ is a solution of (4.117) *such that*

$$\big|f\big(x(t)\big)\big| \leq \big|x(t)\big|\big|\phi(t)\big| \quad \textit{eventually},$$

then $x(t)$ is nonoscillatory.

Corollary 4.3.4 *If* (4.190) *holds, then Eq.* (4.117) *does not have bounded oscillatory solutions.*

Corollary 4.3.5 *Let* (4.190) *holds, and there exists $M > 0$ such that $\frac{f(u)}{u} \leq M$, $u \neq 0$, then* (4.117) *is nonoscillatory.*

Now we consider Eq. (4.7) *with $\alpha > 1$, that is,*

$$x''' + c(t)|x|^{\alpha} \operatorname{sgn} x = 0, \quad \alpha > 1$$

which is a particular case of (4.116) *or* (4.117).

Theorem 4.3.13 *If there exists an $n > 0$ such that*

$$\int^{\infty} t^{n(\alpha-1)+2} c(t)\, dt < \infty,$$

then (4.7) *does not have oscillatory solution $x(t)$ such that $|x(t)| \leq t^n$ eventually. In particular, for $n = 2$, if*

$$\int^{\alpha} t^{2\alpha} c(t)\, dt < \infty,$$

then (4.7) *does not have oscillatory solutions $x(t)$ such that $|x(t)| \leq t^2$.*

The proof easily follows from Theorem 4.3.12 with $\phi(t) = t^{n(\alpha-1)}$, since

$$\frac{f(x(t))}{x(t)} = \frac{|x(t)|^\alpha \operatorname{sgn} x}{x(t)} = |x(t)|^{\alpha-1} \quad (x(t) \neq 0).$$

Theorem 4.3.14 *If the assumption* (4.190) *and*

$$\int^\infty t^{1+\alpha} c(t)\, dt = +\infty$$

are satisfied, then every oscillatory solution of (4.7) *is unbounded.*

Theorem 4.3.15 *Suppose that* (4.117) *has a continuable bounded solution vanishing at some* $t_0 \geq \sigma$. *If there exists a* $\lambda > 0$ *such that*

$$\int^\infty t^{2-\lambda} c(t)\, dt = +\infty, \tag{4.193}$$

then (4.117) *is oscillatory.*

Theorem 4.3.16 *Suppose that there exists a* $k > 0$ *such that*

$$\frac{f(u)}{u} \geq k \quad \text{for large } |u|. \tag{4.194}$$

If the assumption (4.193) *is satisfied, then any continuable solution of* (4.117) *which vanishes at least once in* $[\sigma, \infty)$ *is oscillatory.*

Heidel [19] and Erbe [10] proved that, under certain conditions, (4.117) may have a noncontinuable solution with infinitely many zeros in a finite interval. In view of this, we must suppose that the continuability of the solution $x(t)$ in Theorems 4.3.15 and 4.3.16.

Some of the results given above may be extended, with minor changes, to Eq. (4.6). That is,

$$x''' + F(t, x) = 0,$$

where $F : [\sigma, \infty) \times R \to R$ is a continuous function such that $F(t, x)x > 0$ for $x \neq 0$, $\lim_{x\to\infty} \frac{F(t,x)}{x} = \gamma(t)$, $0 \leq \gamma(t)$ and there exist a $h > 0$ and a continuous function $g : [\sigma, \infty) \to (0, \infty)$ such that $\frac{F(t,x)}{x} \leq g(t)$, $0 < |x| \leq h$. In view of this, some results due to Kura [23] cannot be ignored. Below, we state some interesting results due to Kura [23].

First, we consider the equation

$$x''' - F(t, x) = 0, \tag{4.195}$$

where $F : [\sigma, \infty) \times R \to R$ is a continuous function such that the following hold:

$$F(t, -x) = -F(t, x) \quad \text{on } [\sigma, \infty) \times R; \tag{4.196}$$

$$xF(t,x) \geq 0 \quad \text{on } [\sigma,\infty) \times R; \tag{4.197}$$

$$\frac{1}{x}F(t,x) \text{ is continuous on } [\sigma,\infty) \times R; \tag{4.198}$$

$$\frac{1}{x}F(t,x) \text{ is nondecreasing in } x \text{ for } x > 0. \tag{4.199}$$

It is easy to see that the equation

$$x''' - c(t)|x|^{\alpha} \operatorname{sgn} x = 0, \quad \alpha > 1, \tag{4.200}$$

where $c(t)$ is a positive continuous function on $[\sigma,\infty)$, is a special case of (4.195) satisfying (4.196)–(4.199).

Theorem 4.3.17 *Suppose that the conditions* (4.196)–(4.199) *are satisfied and*

$$\int_{\sigma}^{\infty} \frac{1}{t}\big|F(t,\lambda t)\big|\,dt < \infty \quad \text{for any } \lambda > 0.$$

If the second-order linear equation

$$z'' + \lambda_1\left(\int_t^{\infty} \frac{1}{s}\big|F(s,\lambda_2 s)\big|\,ds\right)z = 0 \tag{4.201}$$

is nonoscillatory for some $\lambda_1 > 0$ *and* $\lambda_2 > 0$, *then* (4.195) *is nonoscillatory.*

Proof It suffices to prove that if (4.195) has an oscillatory solution, then (4.201) is oscillatory for some $\lambda_1 > 0$ and $\lambda_2 > 0$.

Let $x(t)$ be an oscillatory solution of (4.195) on $[T,\infty)$, $t \geq \sigma$. Let $\{t_n\}_{n=1}^{\infty}$ be an increasing sequence of zeros of $x'(t)$, $t_n \to \infty$ as $n \to \infty$. Choose, for each n, $\tau_n \in (t_n, t_{n+1})$ such that $|x'(\tau_n)| = \max\{|x'(t)|; t_n \leq t \leq t_{n+1}\}$. Now consider the function

$$G\big(x(t)\big) = -x''(t)x(t) + \frac{1}{2}\big(x'(t)\big)^2. \tag{4.202}$$

Then $G'(x(t)) = -x(t)F(t,x(t)) \leq 0$ for $t \geq T$. Since $x''(\tau_n) = 0$ for all n, $G(x(\tau_n)) = \frac{1}{2}(x'(\tau_n))^2$ is nonincreasing in n, and so there exists $M > 0$ such that $|x'(\tau_n)| \leq M$ for all n. From the choice of τ_n it follows that $|x'(t)| \leq M$ for $t \geq t_1$. Consequently, $|x(t)| < Mt$, $t \geq T_1$, $T_1 > t_1$ is sufficiently large. From the assumptions (4.196)–(4.199), we have

$$\left|\frac{1}{x(t)}F\big(t,x(t)\big)\right| \leq \frac{1}{Mt}\big|f(t,Mt)\big| \quad \text{for } t \geq T_1. \tag{4.203}$$

On the other hand, since $x(t)$ is an oscillatory solution of

$$z''' - \left|\frac{F(t,x(t))}{x(t)}\right|z = 0,$$

then Lemma 1.5.11 guarantees that the equation

$$z''' + \left|\frac{F(t, x(t))}{x(t)}\right| z = 0$$

has an oscillatory solution. Then Theorem 2.19, due to Barrett [5], shows that the second-order equation

$$z'' + \frac{3}{2}\left(\int_t^\infty \left|\frac{1}{x(s)} F\big(s, x(s)\big)\right| ds\right) z = 0$$

has an oscillatory solution. In view of (4.203) and Sturm comparison theorem, that is, Lemma 1.5.22, it follows that the equation

$$z'' + \frac{3}{2M}\left(\int_t^\infty \frac{1}{s}\big|F(s, Ms)\big| ds\right) z = 0$$

has an oscillatory solution. This completes the proof of the theorem. □

Corollary 4.3.6 *Equation* (4.200) *is nonoscillatory if*

$$\lim_{t\to\infty} t^2 \int_t^\infty s^{\alpha-1} c(s)\, ds = 0. \tag{4.204}$$

Theorem 4.3.18 *If there exists a constant $\epsilon > 0$ such that*

$$\frac{d}{dt}\big[c(t)t^{3+h(\alpha-1)+\epsilon}\big] \le 0 \quad \textit{for } t \ge t_0 > \sigma,$$

where $h = 1 - \frac{1}{\sqrt{3}}$, then (4.200) *is nonoscillatory.*

Hanan [16] has proved that the case where $c(t) > 0$, the equation $x''' - c(t)x = 0$ has an oscillatory solution if and only if so does the equation $x''' + c(t)x = 0$. However, Theorem 4.3.18 shows that no such remarkable relation exists between the nonlinear equations $x''' - c(t)|x|^\alpha \operatorname{sgn} x = 0$ and $x''' + c(t)|x|^\alpha \times \operatorname{sgn} x = 0$, $\alpha \ne 1$. For example, $x''' + t^{-2-\alpha}|x|^\alpha \operatorname{sgn} x = 0$, $\alpha > 1$ has an oscillatory solution. But from Theorem 4.3.18, the equation $x''' - t^{-2-\alpha}|x|^\alpha \operatorname{sgn} x = 0$ is nonoscillatory.

Example 4.3.1 Consider the equation

$$x''' - t^\beta |x|^\alpha \operatorname{sgn} x = 0, \quad \alpha > 1, \tag{4.205}$$

where β is a constant. By Corollary 4.3.6, (4.205) is nonoscillatory if $\alpha + \beta + 2 < 0$. Applying Theorem 4.3.18, we see that (4.205) is nonoscillatory if $\beta + (1 - \frac{1}{\sqrt{3}})(\alpha - 1) + 3 < 0$. Obviously, the latter condition is better than the former.

We have obtained nonoscillation criteria of superlinear equation in Theorems 4.3.13–4.3.18. Now, we provide some nonoscillation criteria for the solutions of (4.6) when it is sublinear.

Equation (4.6) includes Eq. (4.7). Let $0 < \alpha < 1$. The conditions we assume for (4.6) are

$$xF(t,x) \geq 0 \quad \text{on } [\sigma,\infty) \times R; \tag{4.206a}$$

$$\left|F(t,x)\right| \leq w\big(t,|x|\big) \quad \text{on } [\sigma,\infty) \times R; \tag{4.206b}$$

$$\frac{1}{r} w(t,r) \text{ is nonincreasing in } r \text{ for } r > 0, \tag{4.206c}$$

where $w(t,r)$ is continuous on $[\sigma,\infty) \times R_+$, $R_+ = [0,\infty)$ and nondecreasing in r. It is clear that (4.206a)–(4.206c) are satisfied for (4.7) with $0 < \alpha < 1$.

Lemma 4.3.6 *Suppose that* (4.206b), (4.206c) *and*

$$\int_{t_0}^{\infty} t w(t,\lambda t)\,dt < \infty \quad \textit{for any } \lambda > 0,\ t_0 \geq \sigma \tag{4.207}$$

are satisfied. Let $x(t)$ be an oscillatory solution of (4.6). *Then* $\lim_{t\to\infty} x'(t) = 0$.

Proof Condition (4.206c) implies that

$$w(t,rs) \leq s w(t,r) \quad \text{for } s \geq 1, r > 0 \text{ and } t \in [t_0,\infty). \tag{4.208}$$

Let $T_0 > t_0 + 1$ be such that $x(T_0) \neq 0$. Define

$$\nu(t) = \sup\left\{ \frac{|x(s)|}{s^2};\ T_0 \leq s \leq t \right\}.$$

Integrating (4.6) three times over $[T_0,t]$, $t \geq T_0$, yields

$$\left|x(t)\right| \leq \lambda t^2 + \int_{T_0}^{t} (t-s)^2 w\big(s,\left|x(s)\right|\big)\,ds \quad \text{for } t \geq T_0,$$

where $\lambda > 0$ is a constant. Dividing the above expression by t^2, taking the supremum on $[T_0,t]$ and by using (4.208), we obtain

$$\nu(t) \leq \lambda + \int_{T_0}^{t} w\big(s,s^2\nu(s)\big)\,ds \leq \lambda + \int_{T_0}^{t} s w\big(s,\nu(T_0)s\big)\frac{\nu(s)}{\nu(T_0)}\,ds$$

for $t \geq T_0$. By Gronwall's inequality, we have

$$\nu(t) \leq \lambda \exp\left(\frac{1}{\nu(T_0)} \int_{T_0}^{t} s w\big(s,\nu(T_0)s\big)\,ds \right), \quad t \geq T_0.$$

Hence from (4.207), we see that $\nu(t)$ is bounded on $[T_0, \infty)$. Integrating (4.6) over $[t_1, t_2]$, $T_0 \le t_1 \le t_2$, we have

$$\left|x''(t_2) - x''(t_1)\right| \le \int_{t_1}^{t_2} w\left(s, \left|x(s)\right|\right) ds \le \int_{t_1}^{t_2} s w(s, Ms)\, ds \to 0$$

as $t_1, t_2 \to \infty$, where $M = \sup\{\nu(t); t \ge T_0\}$. This shows that $\lim_{t\to\infty} x''(t)$ exists. Since $x(t)$ is oscillatory, $\lim_{t\to\infty} x''(t) = 0$ and we have

$$\left|x''(t)\right| \le \int_{t}^{\infty} w\left(s, \left|x(s)\right|\right) ds, \quad t \ge T_0. \tag{4.209}$$

Since $|x(t)| \le Mt^2, t \ge T_0$, the right-hand side of (4.209) is finite.

Now, we show that $\frac{x(t)}{t}$ is bounded. On the contrary, suppose that $\frac{x(t)}{t}$ is not bounded. By (4.207), there exists $T_1 > T_0$ such that

$$\int_{T_1}^{\infty} t w(t, t)\, dt < \frac{1}{2} \quad \text{and} \quad x'(T_1) = 0. \tag{4.210}$$

Define $m(t) = \sup\{\frac{|x(s)|}{s}; T_1 \le s \le t\}$. Then there exists $T_2 > T_1$ such that $m(t) \ge 1$ and

$$\frac{|x(T_1)|}{T_1} \le \frac{m(t)}{2}, \quad t \ge T_2. \tag{4.211}$$

Dividing the inequality

$$\begin{aligned}\left|x(t)\right| &\le \left|x(T_1)\right| + \int_{T_1}^{t} \left|x'(s)\right| ds \\ &\le \left|x(T_1)\right| + (t - T_1) \sup\left\{\left|x'(s)\right|; T_1 \le s \le T\right\}, \quad t \ge T_1,\end{aligned}$$

by t and taking the supremum over $[T_1, t]$, we obtain

$$m(t) \le \frac{|x(T_1)|}{T_1} + \sup\left\{\left|x'(s)\right|; T_1 \le s \le t\right\}, \quad t \ge T_1,$$

which, in view of (4.211), yields

$$\frac{1}{2} m(t) \le \sup\left\{\left|x'(s)\right|; T_1 \le s \le t\right\}, \quad t \ge T_2. \tag{4.212}$$

Integrating (4.209) over $[T_1, t]$, $t \ge T_1$, and by using $x'(T_1) = 0$, we have

$$\left|x'(t)\right| \le \int_{T_1}^{t} \int_{s}^{\infty} w\left(\tau, \left|x(\tau)\right|\right) d\tau\, ds, \quad t \ge T_1. \tag{4.213}$$

It follows from (4.212) and (4.213) that, for $t \ge T_2$,

$$\begin{aligned}
\frac{1}{2} m(t) &\le \int_{T_1}^{t} \int_{s}^{\infty} w\left(\tau, \left|x(\tau)\right|\right) d\tau\, ds \\
&= (t - T_1) \int_{t}^{\infty} w\left(s, \left|x(s)\right|\right) ds + \int_{T_1}^{t} (s - T_1) w\left(s, \left|x(s)\right|\right) ds \\
&\le (t - T_1) \int_{t}^{\infty} w\left(s, s m(s)\right) ds + \int_{T_1}^{t} (s - T_1) w(s, s) m(s)\, ds \\
&\le t \int_{t}^{\infty} m(s)\, w(s, s)\, ds + m(t) \int_{T_1}^{t} s w(s, s)\, ds.
\end{aligned} \tag{4.214}$$

For each $t \ge T_1$, there exists a $s \in [T_1, t]$ such that

$$\frac{m(t)}{t} = \frac{|x(s)|}{st} \le \frac{|x(s)|}{s^2} \le M.$$

Therefore, $\frac{m(t)}{t} \le M$ for $t \ge T_1$. Let $M_1 = \sup\{\frac{m(t)}{t}; t \ge T_2\}$. Then by (4.214) and (4.210), we obtain

$$\begin{aligned}
\frac{1}{2t} m(t) &\le M_1 \int_{t_1}^{\infty} s w(s, s)\, ds + M_1 \int_{T_1}^{t} s w(s, s)\, ds \\
&= M_1 \int_{T_1}^{\infty} s w(s, s)\, ds < \frac{M_1}{4}, \quad t \ge T_2.
\end{aligned}$$

This implies that $M_1 \le M_1/2$. This contradiction proves the boundedness of $\frac{x(t)}{t}$ as claimed. Let $L > 0$ be such that $|x(t)| \le Lt$ for $t \ge T_0$. Integrating (4.209) over $[t_1, t_2]$, $T_0 \le t_1 \le t_2$, we get by using (4.207)

$$\begin{aligned}
\left|x'(t_1) - x'(t_2)\right| &\le \int_{t_1}^{t_2} \int_{s}^{\infty} w(\tau, L\tau)\, d\tau\, ds \\
&= (t_2 - t_1) \int_{t_2}^{\infty} w(s, Ls)\, ds + \int_{t_1}^{t_2} (s - t_1) w(s, Ls)\, ds \\
&\le 2 \int_{t_1}^{\infty} s w(s, Ls)\, ds \to 0 \quad \text{as } t_1 \to \infty.
\end{aligned}$$

This shows that $x'(t)$ tends to a finite limit as $t \to \infty$. Since $x(t)$ is oscillatory, it follows that $\lim_{t\to\infty} x'(t) = 0$. This completes the proof of the lemma. □

Theorem 4.3.19 *Suppose that the conditions* (4.206a)–(4.206c) *and* (4.207) *are satisfied. Then* (4.6) *is nonoscillatory.*

Proof Suppose to the contrary that there exists an oscillatory solution $x(t)$ of (4.6) on $[T, \infty)$, $T > \sigma$. Let $\{t_n\}_{n=1}^{\infty}$ be an increasing sequence of zeros of $x'(t)$ such that $t_n \to \infty$, as $n \to \infty$. Choose, for each n, $\tau_n \in (t_n, t_{n+1})$ such that $|x'(\tau_n)| = \max\{|x'(t)|; t_n \le t \le t_{n+1}\}$. There exists a $N > 1$ such that $|x'(t_N)| > 0$. Consider the function $G(x(t))$ defined by (4.202). Then $G'(x(t)) = x(t)F(t, x(t)) \ge 0$ for $t \ge T$. We have $\lim_{n\to\infty} G(x(\tau_n)) = 0$, since $x''(\tau_n) = 0$ for all n and $\lim_{t\to\infty} x'(t) = 0$ by Lemma 4.3.6. Since $G(x(t))$ is nondecreasing, we finally obtain $0 < G(x(\tau_n)) = x'^2(\tau_N)/2 \le \lim_{n\to\infty} G(x(\tau_n)) = 0$. This contradiction proves that (4.6) is nonoscillatory. The theorem is proved. □

Applying Theorem 4.3.19 to (4.7) with $0 < \alpha < 1$, we obtain the following corollary.

Corollary 4.3.7 *Equation* (4.7) *with* $0 < \alpha < 1$ *is nonoscillatory, if*

$$\int_\sigma^\infty t^{1+\alpha} c(t)\, dt < \infty.$$

Theorem 4.3.20 *Suppose that there exists a constant* ϵ *with* $0 < \epsilon < \frac{1-\alpha}{\sqrt{3}}$ *such that* $c(t)t^{3+k(\alpha-1)+\epsilon}$ *is bounded and*

$$\frac{d}{dt}\left[c(t)t^{3+k(\alpha-1)+\epsilon}\right] \ge 0 \quad \text{for } t \ge t_0 \ge \sigma,$$

where $k = 1 + \frac{1}{\sqrt{3}}$. *Then* (4.7) *with* $0 < \alpha < 1$ *is nonoscillatory.*

Example 4.3.2 By Corollary 4.3.7, the equation

$$x''' + t^\beta |x|^\alpha \operatorname{sgn} x = 0, \quad 0 < \alpha < 1,$$

where β is a constant, is nonoscillatory if $\alpha + \beta + 2 < 0$. On the other hand, by Theorem 4.3.20, this example is nonoscillatory if $\alpha + \beta + 2 > 0$ and $\beta + (1 + \frac{1}{\sqrt{3}})(\alpha - 1) + 3 < 0$.

4.4 Behaviour of Solutions of $(r_2(t)(r_1(t)x')')' + q(t)x' + p(t)x^\alpha = 0$ and $(r_2(t)(r_1(t)x')')' + q(t)x' + p(t)f(x) = 0$

This section deals with the oscillation, nonoscillation of solutions and their asymptotic behaviour of the nonlinear differential equations of the form

$$\big(r_2(t)\big(r_1(t)x'\big)'\big)' + q(t)x' + p(t)x^\alpha = 0, \quad t \ge \sigma \tag{4.215}$$

and

$$\big(r_2(t)\big(r_1(t)x'\big)'\big)' + q(t)x' + p(t)f(x) = 0, \quad t \ge \sigma \tag{4.216}$$

and some of their particular cases, where r_1, r_2, p and $q : [\sigma, \infty) \to R$ are continuous functions with $r_2(t) > 0$, $r_1(t) > 0$, $f : R \to R$ is a continuous function with the property that

$$\frac{f(x)}{x} \geq \beta \quad \text{for some } \beta > 0 \text{ and for all } x \neq 0. \tag{4.217}$$

Note that if $x(t)$ is a solution of (4.216), then $-x(t)$ is a solution of

$$\big(r(t)\big(r_1(t)x'\big)'\big)' + q(t)x' + p(t)f^*(x) = 0,$$

where $f^*(x) = -f(-x)$. Since f and f^* are of the same class, we may restrict our attention only to a positive solution of (4.216), whenever a nonoscillatory solution of the equation is concerned. A similar property also holds for (4.215).

If we denote $l_0x = x$, $L_ix = r_i(t)(L_{i-1}x)'$, $i = 1, 2$ and $L_3x = (L_2x)'$, then Eqs. (4.215) and (4.216) may be rewritten, respectively, as

$$L_3x + q(t)x' + p(t)x^\alpha = 0 \tag{4.218}$$

and

$$L_3x + q(t)x' + p(t)f(x) = 0. \tag{4.219}$$

Škerlik [34] introduced the definition of property V_2 in the following way:

Definition 4.4.1 Let $x(t)$ be a solution of (4.218) or (4.219). We say that the solution $x(t)$ has property V_2 on $[T, \infty)$, $T \geq \sigma$, if and only if

$$L_0x(t)L_ix(t) > 0, \quad i = 0, 1, 2; \qquad L_0x(t)L_3x(t) \leq 0 \tag{4.220}$$

for every $t \geq T$.

Define the functions

$$R_2(t, T) = \int_T^t \frac{ds}{r_2(s)}$$

and

$$R_{12}(t, T) = \int_T^t \frac{R_2(s, T)}{r_1(s)}\, ds = \int_T^t \frac{1}{r_1(s)} \int_T^s \frac{1}{r_2(\theta)}\, d\theta\, ds, \quad \sigma \leq T \leq t < \infty.$$

Lemma 4.4.1 *Let $q(t) \geq 0$, $p(t) \geq 0$ and $R_2(t, \sigma) \to \infty$ as $t \to \infty$ and $x(t)$ be a nonoscillatory solution of* (4.216) *such that $x(t)L_1x(t) \geq 0$ for every $t \geq T \geq \sigma$. Then $x(t)$ has property V_2 for all large t.*

Theorem 4.4.1 *Let $q(t) \geq 0$, $p(t) \geq 0$ and $R_2(t, \sigma) \to \infty$ as $t \to \infty$. Suppose that $\frac{r_2}{r_1}, q \in C^1([\sigma, \infty), R)$ and*

$$q'(t) \leq 0, \qquad \left(\frac{r_2(t)}{r_1(t)}\right)' \geq 0 \quad \textit{for } t \geq \sigma. \tag{4.221}$$

Let $x(t)$ be a solution of (4.216), *and assume further that there exists a $t_0 \in [\sigma, \infty)$ such that $F_1[x(t_0)] \le 0$, where*

$$F_1\big[x(t)\big] = 2x(t)L_2x(t) - \frac{r_2(t)}{r_1(t)}\big(L_1x(t)\big)^2 + q(t)x^2(t). \tag{4.222}$$

Then either $x(t)$ is oscillatory or $x(t)$ has property V_2 for large t.

Proof Let $x(t)$ be a solution of (4.216) satisfying the condition $F_1[x(t_0)] \le 0$ for some $t_0 \ge \sigma$. If $x(t)$ is oscillatory, then there is nothing to prove. Suppose that $x(t)$ is nonoscillatory. Without any loss of generality, we may assume that $x(t) > 0$ for $t \ge T > t_0$. Then

$$F_1'\big[x(t)\big] = -\left[\frac{r_2(t)}{r_1(t)}\right]'\big(L_1x(t)\big)^2 + q'(t)x^2(t) - 2p(t)x(t)f\big(x(t)\big) \le 0 \quad \text{for } t \ge T.$$

So, there exists a $t_0^* \ge T$ such that $F_1[x(t)] < 0$ for $t \ge t_0^*$, and $\lim_{t\to\infty} F_1[x(t)] = F_0 < 0$ exists (finite or infinite). Then from (4.222), we obtain

$$\begin{aligned}
2r_2(t)\left(\frac{L_1x(t)}{x(t)}\right)' &= \frac{2}{x^2(t)}\left(x(t)L_2x(t) - \frac{r_2(t)}{r_1(t)}\big(L_1x(t)\big)^2\right)\\
&\le \left[2x(t)L_2x(t) - \frac{r_2(t)}{r_1(t)}\big(L_1x(t)\big)^2\right]x^{-2}(t)\\
&< -q(t) \le 0,
\end{aligned} \tag{4.223}$$

for $t \ge t_0^*$. Hence the function $\frac{L_1x}{x}$ is decreasing on $[t_0^*, \infty)$. Then, either $L_1x(t) > 0$, $t \ge t_0^*$, and by Lemma 4.4.1, $x(t)$ has property V_2 or there exists a $T_1 \ge t_0^*$ such that $L_1x(t) < 0$ for $t \ge T_1$. Let $L_1x(t) < 0$ for $t \ge T_1$. Then from (4.221) with $R_2(t,\sigma) \to \infty$ as $t \to \infty$, we have $r_2(t) \ge A\, r_1(t)$, $t \ge \sigma$, where $A = \frac{r_2(\sigma)}{r_1(\sigma)}$ and so

$$\left.\begin{aligned}
&\lim_{t\to\infty} R_1(t,\sigma) = \lim_{t\to\infty}\int_\sigma^t \frac{ds}{r_1(s)} = \infty,\\
&r_2(t)x'(t) \le A\, L_1x(t) < 0, \quad t \ge T_1.
\end{aligned}\right\} \tag{4.224}$$

We consider the function $L_2x(t)$. Since $L_2x(t) < 0$ for large t implies that $x(t) < 0$ for large t, a contradiction, we have $L_2x(t) > 0$ for large t. Assume that $L_2x(t) \ge 0$ for $t \ge T_3 \ge T_1$. We assert that $\lim_{t\to\infty} L_1x(t) = \limsup_{t\to\infty} r_2(t)x'(t) = 0$. (In fact, $\limsup_{t\to\infty} r_2(t)x'(t) < 0$ implies that $x(t) < 0$ for large t, a contradiction.) Otherwise, a calculation shows that

$$\begin{aligned}
0 > F_0 = \lim_{t\to\infty} F_1\big[x(t)\big] &= \limsup_{t\to\infty}\big[2x(t)L_2x(t) - r_2(t)x'(t)L_1x(t) + q(t)x^2(t)\big]\\
&= \limsup_{t\to\infty}\big[2x(t)L_2x(t) + q(t)x^2(t)\big] \ge 0,
\end{aligned}$$

a contradiction.

Finally, $x(t) > 0$, $L_1x(t) < 0$ for $t \geq T_1$ and $L_2x(t)$ changes sign for arbitrarily large t. Denote

$$G(t) = \big(r_2(t)x'(t)\big)' = \left(\frac{r_2(t)}{r_1(t)}\right)' L_1x(t) + \frac{L_2x(t)}{r_1(t)}, \qquad t \geq T_1.$$

The function G cannot be nonpositive on $[T_2, \infty)$ for some $T_2 \geq T_1$, since there is $r_2(t)x'(t) \leq r_2(T_2)x'(T_2) < 0$, $t \geq T_2$, and from $R_2(t, \sigma) \to \infty$ as $t \to \infty$, we obtain the contradiction that $x(t) < 0$ for large t. Let G change sign. Hence there exists an unbounded sequence of zeros of the function G. Let it be $\{t_n\}_{n=1}^{\infty}$, $t_n \to \infty$ as $n \to \infty$ such that $a_1 \leq a_2 \leq a_3 \cdots \leq a_n \leq \cdots$, where $a_n = r_2(t_n)x'(t_n)$, $n = 1, 2, \ldots$ are nondecreasing relative maxima of $r_2(t)x'(t)$. Clearly $\lim_{t\to\infty} a_n = 0$. From (4.224), it follows that $\lim_{n\to\infty} L_1x(t_n) = 0$. We see that $L_2x(t_n) \geq 0$. So a simple calculation shows that $0 > F_0 = \lim_{t\to\infty} F_1[x(t)] \geq 0$, a contradiction, too. This completes the proof of the lemma. □

Remark 4.4.1 Equation (4.216) is called weak superlinear if the function f has the property: for any $x \neq 0$, there exists a number $m > 0$ such that $uf(u) \geq mu^2$. Let us note that any linear equation is weak superlinear with $m = 1$. If Eq. (4.216) is weak superlinear, then the condition $q'(t) \leq 0$ in (4.221) in Lemma 4.4.2 may be replaced by a weaker condition $2mp(t) - q'(t) \geq 0$ for $t \geq \sigma$ and not identically zero in any subinterval of $[\sigma, \infty)$. Again, from Theorem 4.4.1, it follows that any nonoscillatory solution $x(t)$ of (4.216) with $F_1[x(t_0)] \leq 0$ for some $t_0 \in [\sigma, \infty)$ is unbounded.

Example 4.4.1 Theorem 4.4.1 can be applied to

$$\left(t^{\frac{1}{2}}\left(t^{-\frac{1}{2}}x'\right)'\right)' + \left(\frac{1}{36t^2}\right)x' + \frac{5}{108}t^{-((5\alpha+4)/3)}|x|^{\alpha}\operatorname{sgn} x = 0,$$

$\alpha > 0$ and $t > 0$, which has an unbounded nonoscillatory solution $x(t) = t^{5/3}$.

Now, we assume that the function f satisfies the condition f is nondecreasing, there exists a constant $\mu > 0$ such that

$$\big|f(uv)\big| \geq \mu f(u)\big|f(v)\big| \quad \text{for } u \geq 0,\ v \in R. \tag{4.225}$$

Further, suppose that $R_{12}(t, \sigma) \to \infty$ as $t \to \infty$, $R_2(t, \sigma) \to \infty$ as $t \to \infty$ and $x(t)$ is a nonoscillatory solution of (4.216), say $x(t) > 0$, with property V_2 for $t \geq t_0 \geq \sigma$. Then some simple calculations shows that

$$x(t) \geq \frac{R_{12}(t, t_0)}{R_2(t, t_0)} L_1x(t), \quad t \geq t_1 > t_0$$

holds. Thus, for every $\lambda \in (0, 1)$, there exists a number T, $T = t_\lambda \geq t_1$ such that

$$\frac{R_{12}(t, t_0)}{R_2(t, t_0)} \geq \lambda \frac{R_{12}(t, \sigma)}{R_2(t, \sigma)}, \quad t \geq T,$$

since $\lim_{t\to\infty} \frac{R_{12}(t,t_0)}{R_2(t,t_0)} \frac{R_2(t,\sigma)}{R_{12}(t,\sigma)} = 1$. Then by using the condition f is nondecreasing and (4.225), we obtain

$$f\big(x(t)\big) \geq f\left[\lambda \frac{R_{12}(t,\sigma)}{R_2(t,\sigma)} L_1 x(t)\right] \geq \mu f(\lambda) f\left[\frac{R_{12}(t,\sigma)}{R_2(t,\sigma)}\right] f\big(L_1 x(t)\big)$$

for some $\mu > 0$ and $t \geq T$. Then from (4.216), we obtain

$$L_1 x(t)\left\{L_3 x(t) + \frac{q(t)}{r_1(t)} L_1 x(t) + \mu f(\lambda) f\left[\frac{R_{12}(t,\sigma)}{R_2(t,\sigma)}\right] p(t) f\big(L_1 x(t)\big)\right\} \leq 0 \tag{4.226}$$

for every $t \geq T$. It is clear that the inequality (4.226) for a negative function $x(t)$ of (4.216) with property V_2 holds, too.

Let the condition $R_2(t,\sigma) \to \infty$ and $R_{12}(t,\sigma) \to \infty$ as $t \to \infty$ hold and let Eq. (4.216) be weak superlinear.

Let $x(t)$ be a nonoscillatory solution of (4.216). Similarly as above, we can derive the inequality

$$L_1 x(t)\left\{L_3 x(t) + \left[\frac{q(t)}{r_1(t)} + m\lambda \frac{R_{12}(t,\sigma)}{R_2(t,\sigma)} p(t)\right] L_1 x(t)\right\} \leq 0 \tag{4.227}$$

for every $t \geq T > \sigma$.

An application of Theorem 4.4.1 is the following theorem.

Theorem 4.4.2 *Let $R_2(t,\sigma) \to \infty$ and $R_{12}(t,\sigma) \to \infty$ as $t \to \infty$, f be nondecreasing, and let there exist a constant $\mu > 0$ such that* (4.225) *holds, and assume that the equation*

$$\big(r_2(t) z'\big)' + \frac{q(t)}{r_1(t)} z + \mu f(\lambda) f\left[\frac{R_{12}(t,\sigma)}{R_2(t,\sigma)}\right] p(t) f(z) = 0 \tag{4.228}$$

is oscillatory for some $0 < \lambda < 1$ and $\mu > 0$. Then no nonoscillatory solution $x(t)$ of (4.216) *has the property V_2 for all large t.*

Corollary 4.4.1 *Let $R_2(t,\sigma) \to \infty$ and $R_{12}(t,\sigma) \to \infty$ as $t \to \infty$, f is nondecreasing, and let* (4.225) *hold. Further assume that*

$$\int^\infty f\big(R_2(t,\sigma)\big) f\left[\frac{R_{12}(t,\sigma)}{R_2(t,\sigma)}\right] p(t)\, dt = \infty, \tag{4.229}$$

if (4.216) *is sublinear, or*

$$\int^\infty R_2(t,\sigma) f\left[\frac{R_{12}(t,\sigma)}{R_2(t,\sigma)}\right] p(t)\, dt = \infty,$$

if (4.216) *is superlinear. Then no nonoscillatory solution $x(t)$ of* (4.216) *has property V_2 for all large t.*

Theorem 4.4.3 *Let $R_2(t,\sigma) \to 2\infty$, $R_{12}(t,\sigma) \to \infty$ as $t \to \infty$, f is nondecreasing and there exists a constant $\mu > 0$ such that* (4.225) *holds, and* (4.216) *is sublinear. If*

$$\int^{\infty} f\big(R_{12}(t,\sigma)\big)p(t)\,dt = \infty \tag{4.230}$$

holds, then no nonoscillatory solution $x(t)$ of (4.216) *has property V_2 for all large t.*

Since

$$f\big(R_{12}(t,\sigma)\big) = f\left(R_2(t,\sigma)\frac{R_{12}(t,\sigma)}{R_2(t,\sigma)}\right) \ge \lambda f\big(R_2(t,\sigma)\big) f\left(\frac{R_{12}(t,\sigma)}{R_2(t,\sigma)}\right),$$

the condition (4.230) is weaker than the condition (4.229).

Theorem 4.4.4 *Let $R_2(t,\sigma) \to \infty$ and $R_{12}(t,\sigma) \to \infty$ as $t \to \infty$, and let Eq.* (4.216) *be weak superlinear. If the equation*

$$\big(r_2(t)z'\big)' + \left[\frac{q(t)}{r_1(t)} + m\lambda \frac{R_{12}(t,\sigma)}{R_2(t,\sigma)}p(t)\right]z = 0$$

is oscillatory for some $m > 0$, $0 < \lambda < 1$, then no nonoscillatory solution $x(t)$ of (4.216) *has property V_2 for large t.*

One of the simple criterion for the conclusion, same as Theorem 4.4.4, is given below in Theorem 4.4.5.

Theorem 4.4.5 *Let the function f satisfy the condition*

$$\liminf_{|x|\to\infty} \big|f(x)\big| > 0.$$

If

$$\int^{\infty} p(t)\,dt = \infty,$$

then no nonoscillatory solution $x(t)$ of (4.216) *has the property V_2 for large t.*

Theorem 4.4.6 *Let $p(t) \ge 0$, $q(t) \ge 0$, $(\frac{r_2(t)}{r_1(t)})' \ge 0$, $q'(t) \le 0$ and $R_2(t,\sigma) \to \infty$ as $t \to \infty$. In addition, assume that the hypothesis of any of the Theorems* 4.4.2–4.4.5 *is fulfilled. Let $x(t)$ be a solution of* (4.216) *which exists on the interval $[T,\infty)$, $T \ge \sigma$. Then $x(t)$ is oscillatory, if and only if there exists a point $t_0 \ge T$ such that $F_1[x(t_0)] \le 0$.*

Lemma 4.4.2 *Let $g \in C([\sigma,\infty), R)$ be such that $g(t) \le t$ and $g(t) \to \infty$ as $t \to \infty$. If $x(t)$ is a solution of* (4.216) *with property V_2, then there exists a constant $t_0 \ge \sigma$ such that*

$$L_1x(t) \ge R_2\big(t, g(t)\big)L_2x(t) \quad \textit{for } t \ge t_0.$$

Proof If $x(t)$ is a solution of (4.216) with the property $L_0x(t) > 0$, $L_1x(t) > 0$, $L_2x(t) > 0$ and $L_3x(t) < 0$ for $t \geq t_1 \geq T$, then

$$L_1x(t) \geq \int_{g(t)}^{t} \big(L_1x(s)\big)' \, ds = \int_{g(t)}^{t} \frac{1}{r_2(s)} L_2x(s) \, ds \geq R_2\big(t, g(t)\big) L_2x(t).$$

The lemma is proved. □

Lemma 4.4.3 *Suppose that the second-order linear equation*

$$\big(r_2(t)z'\big)' + \frac{q(t)}{r_1(t)} z = 0 \tag{4.231}$$

is nonoscillatory. If $x(t)$ is a nonoscillatory solution of (4.216), *then there exists a $t_1 \geq \sigma$ such that $x(t)L_1x(t) > 0$ or $x(t)L_1x(t) < 0$ for $t \geq t_1$.*

Lemma 4.4.4 *Let $R_2(t, \sigma) \to \infty$ as $t \to \infty$ and let* (4.231) *be disconjugate on $[\sigma, \infty)$. Then for any function $v \in c^1[b, c]$, $\sigma \leq b < c$ satisfying $v(b) = 0$ and $v(t) \not\equiv 0$ on (b, c), the strict inequality*

$$\int_b^c \left[r_2(s) v'^2(s) - \frac{q(s)}{r_1(s)} v^2(s) \right] ds > 0$$

holds.

Theorem 4.4.7 *Let $R_2(t, \sigma) \to \infty$ as $t \to \infty$, $(\frac{r_2(t)}{r_1(t)})' \geq 0$ for $t \geq \sigma$, $2\beta p(t) - q'(t) \geq 0$ and not identically zero on any subinterval of $[\sigma, \infty)$, where β is given in* (4.217). *Further, suppose that $g(t) \to \infty$ as $t \to \infty$ where $0 < g(t) \leq t$ is a real-valued function. If there exists an eventually positive function $\rho(t) \in C^1[\sigma, \infty)$ such that*

$$\limsup_{t \to \infty} \int_T^t \left[\beta \rho(s) p(s) - \frac{B^2(s)}{4A(s)} \right] ds = \infty, \tag{4.232}$$

for some $T \geq \sigma$, where $A(t) = \frac{R_2(t, g(t))}{r_1(t)\rho(t)}$ and $B(t) = \frac{\rho'(t)}{\rho(t)} - q(t)\frac{R_2(t, g(t))}{r_1(t)}$, then (4.216) *is oscillatory. In fact, any solution $x(t)$ satisfying $F_1[x(t_0)] \leq 0$ for some $t_0 \geq \sigma$ is oscillatory, where the function F_1 is given in* (4.222).

Proof Let $x(t)$ be a solution of (4.216) with $F_1[x(t_0)] \leq 0$ for some $t_0 \geq \sigma$. By Remark 4.4.1, $x(t)$ is either oscillatory or nonoscillatory with property V_2 for all large t. Let $x(t)$ be nonoscillatory with property V_2. Then, there exists a $t_1 \geq t_0$ such that $L_0x(t) > 0$, $L_1x(t) > 0$, $L_2x(t) > 0$ and $L_3x(t) \leq 0$ for $t \geq t_1$. Defining $w(t) = \rho(t)\frac{L_2x(t)}{x(t)}$ for $t \geq t_1$, we obtain, by using Lemma 4.4.2, that

$$w'(t) < -\beta \rho(t) p(t) + \frac{B^2(t)}{4A(t)}.$$

Integrating the above inequality from t_1 to t, we obtain a contradiction to (4.232). Hence $x(t)$ cannot be nonoscillatory. The theorem is proved. □

Example 4.4.2 From Theorem 4.4.7, the equation

$$\left(t\left(tx'\right)'\right)' + \left(t^2 - 1\right)x' + \frac{3t}{2 + \sin 2t}\left(x + x^3\right) = 0, \quad t \geq 1$$

is oscillatory. Indeed, $x(t) = \sin t + \cos t$ is a solution of the equation. Here, one is required to assume $\rho(t) = 1$ and $g(t) = te^{-1/t^2}$.

Now, let $\rho(t) = t^2$, $f(x) = x$ and $r_1(t) = r_2(t) = 1$. Then the following corollary is an easy consequence of Theorem 4.4.7.

Corollary 4.4.2 *Let $p(t) \geq 0, q(t) \geq 0, 2p(t) - q'(t) \geq 0, g : [\sigma, \infty) \to R$ be such that $g(t) \leq t$ and $g(t) \to \infty$ as $t \to \infty$. If*

$$\limsup_{t\to\infty} \int_\sigma^t \left[s^2 p(s) + sq(s) - \frac{1}{s - g(s)} - \left(s - g(s)\right)\frac{(sq(s))^2}{4}\right] ds = \infty,$$

then the third-order linear differential equation

$$x''' + q(t)x' + p(t)x = 0$$

is oscillatory.

Example 4.4.3 By Corollary 4.4.2, the equation

$$x''' + \frac{1}{4t^2}x' + \frac{25}{4t^3}x = 0, \quad t \geq 1$$

with $g(t) = \frac{t}{2}$ is oscillatory. Note that $x(t) = t^3 \sin(\frac{3}{2}\sin t)$ is an oscillatory solution of the equation. On the other hand, Theorem 2.5.18 cannot be applied to this example because

$$\int_1^\infty t\left(\frac{25}{4t^3} + \frac{1}{2t^3}\right) dt < \infty.$$

This implies that (2.95) fails to hold. Similarly, with $m = 0$, some easy calculations show that Theorem 2.5.19 cannot be applied to this example.

Now, we give some new oscillation results for (4.216), by using integral averages condition of Philos-type. First, we introduce a class of function $\wp$. Let

$$D_0 = \left\{(t, s); t > s \geq \sigma\right\} \quad \text{and} \quad D = \left\{(t, s); t \geq s \geq \sigma\right\}.$$

A function $H \in C(D, R)$ is said to belong to a class $\wp$ if

(i) $H(t, t) = 0$ for all $t \geq \sigma$, $H(t, s) > 0$ for all $(t, s) \in D_0$.

(ii) $H(t,s)$ has a continuous nonpositive partial derivative on D_0 with respect to the second variable, and for a positive continuous function $h(t,s)$,

$$-\frac{\partial H(t,s)}{\partial s} = h(t,s)\sqrt{H(t,s)} \quad \text{for all } (t,s) \in D_0.$$

For the choice $H(t,s) = (t-s)^n$, $n \geq 1$, the Philos-type condition reduces to the Kamenev-type one. Other choices of H include

$$\left(\ln\frac{t}{s}\right)^n, \qquad (\sqrt{t}-\sqrt{s})^n, \qquad \left(t^3 - s^3\right)^n, \qquad \left(e^{t-s} - e^{s-t}\right)^n.$$

One may find several examples of $H(t,s)$ with appropriate $h(t,s)$.

Theorem 4.4.8 *Let $R_2(t,\sigma) \to \infty$ as $t \to \infty$, $(\frac{r_2(t)}{r_1(t)})' \geq 0$, $2\beta p(t) - q'(t) \geq 0$ and not identically zero on any subinterval of $[\sigma,\infty)$, $g \in C([\sigma,\infty), R)$ with $0 < g(t) \leq t$ and $g(t) \to \infty$ as $t \to \infty$. If there exists a positive function $\rho \in C^1[\sigma,\infty)$ and a function $H \in \wp$ such that*

$$\limsup_{t\to\infty} \frac{1}{H(t,T)} \int_T^t \left[\beta\rho(s)p(s)H(t,s) - \frac{Q^2(t,s)}{4A(s)}\right] ds = \infty \tag{4.233}$$

for every $T \geq \sigma$, where $A(t) = \frac{R_2(t,g(t))}{r_1(t)\rho(t)}$ and

$$Q(t,s) = h(t,s) - \sqrt{H(t,s)}\left[\frac{\rho'(s)}{\rho(s)} - q(s)\frac{R_2(s,g(s))}{r_1(s)}\right],$$

then (4.216) *is oscillatory.*

Proof Proceeding as in Theorem 4.4.7 and setting $w(t) = \rho(t)\frac{L_2x(t)}{x(t)}$, we obtain

$$w'(t) \leq -\beta\rho(t)p(t) - \left[w^2(t)A(t) - w(t)B(t)\right],$$

where $A(t)$ and $B(t)$ are given in the statement of Theorem 4.4.7. Then some simple calculation shows that, for $t_2 \geq t_1$,

$$\int_{t_2}^t \rho(s)p(s)H(t,s)\,ds \leq H(t,t_2)w(t_2) + \int_{t_2}^t \frac{Q^2(t,s)}{4A(s)}\,ds,$$

which, in turn, implies that

$$\frac{1}{H(t,t_2)} \int_{t_2}^t \left[s^2 p(s)H(t,s) - \frac{Q^2(t,s)}{4A(s)}\right] ds \leq w(t_2),$$

a contradiction to (4.233). Thus the theorem is proved. □

Corollary 4.4.3 *Suppose that the condition of Theorem* 4.4.8 *is satisfied except the condition* (4.233) *is replaced by*

$$\limsup_{t\to\infty}\frac{1}{H(t,T)}\int_T^t \rho(s)p(s)H(t,s)\,ds=\infty$$

and

$$\limsup_{t\to\infty}\frac{1}{H(t,T)}\int_T^t \frac{Q^2(t,s)}{A(s)}\,ds<\infty.$$

Then (4.216) *is oscillatory.*

Theorem 4.4.9 *Let the assumption, except* (4.233), *of Theorem* 4.4.8 *hold. Suppose that for every $t\ge\sigma$,*

$$0<\inf_{s\ge T}\left[\liminf_{t\to\infty}\frac{H(t,s)}{H(t,T)}\right]\le\infty,$$

and

$$\limsup_{t\to\infty}\frac{1}{H(t,T)}\int_T^t \frac{Q^2(t,s)}{A(s)}\,ds<\infty,$$

and that there exists a function $\psi\in C[\sigma,\infty]$ such that

$$\int_T^\infty \psi_+^2(s)A(s)\,ds=\infty,\qquad \psi_+(s)=\max\{\psi(t),0\};$$

and

$$\limsup_{t\to\infty}\frac{1}{H(t,T)}\int_T^t\left[\beta\rho(s)p(s)H(t,s)-\frac{Q^2(t,s)}{4A(s)}\right]ds\ge\sup_{t\ge T}\psi(t).$$

Then (4.216) *is oscillatory.*

Example 4.4.4 Consider the third-order differential equation

$$\left(\frac{1}{t^2}x'\right)''+\frac{1}{t^2}x=0,\qquad t>1.$$

Assume $H(t,s)=(t-s)^2$, $\rho(s)\equiv 1$, $g(s)=s-1$ and $\psi(t)=\frac{1}{t}$. From Theorem 4.4.9, this example has an oscillatory solution.

Theorem 4.4.10 *Suppose that* (4.231) *is nonoscillatory, and that $R_1(t,\sigma)\to\infty$ and $R_2(t,\sigma)\to\infty$ as $t\to\infty$, $g\in C[\sigma,\infty)$ such that $0<g(t)\le t$ and $g(t)\to\infty$ as $t\to\infty$. If there exists a positive function $\rho\in C^1[\sigma,\infty)$ such that* (4.232) *holds, then* (4.216) *is oscillatory.*

Roveder [32] used Ky Fan fixed-point theorem to find the existence and uniqueness of a Kneser-type solution of Eq. (4.216). The result, due to Roveder [32], is given below:

Theorem 4.4.11 *Let $q(t) \le 0$, $\frac{r_2(t)}{r_1(t)}$ be differentiable on $[\sigma, \infty)$ and $(\frac{r_2(t)}{r_1(t)})' \ge 0$ on $[\sigma, \infty)$, $xf(x) > 0$ for $x \ne 0$ and $\lim_{x\to 0} \frac{f(x)}{x} = \theta < \infty$ hold, and for every*

$$u \in Q = \{u \in C_p^2([\sigma, \infty)); 0 \le u(t) \le c_0\},$$

C_p^2 denoting the Frechet space of all functions $x : [\sigma, \infty) \to R$ with the topology of uniform convergence on compact subinterval of $[\sigma, \infty)$ of all the quasiderivatives up to the order 2. Let

$$2p(t)f_u(t) - q'(t) \ge 0,$$

where

$$f_u(t) = \begin{cases} \frac{f(u(t))}{u(t)}, & \text{if } u(t) \ne 0, \\ \theta & \text{if } u(t) = 0. \end{cases}$$

Then (4.216) with $x(\sigma) = c_0$, $x(t) > 0$, $L_1x(t) \le 0$, $L_2x(t) \ge 0$ on $[\sigma, \infty)$ admits a unique Kneser solution.

A particular case of (4.216) is the nonlinear equation

$$\big(r_2(t)\big(r_1(t)x'\big)'\big)' + r_3(t)f(x) = 0, \quad t \ge \sigma, \tag{4.234}$$

where $r_1, r_2, r_3 \in C([\sigma, \infty), R)$, $r_1(t) > 0$, $r_2(t) > 0$, $r_3(t) > 0$ on $[\sigma, \infty)$ and $f \in C(R, R)$ with $xf(x) > 0$ for $x \ne 0$.

The special case of (4.234) is the linear equation (2.117), where oscillatory and asymptotic properties of nonoscillatory solutions are given in Sect. 2.7 of Chap. 2. As is customary, we shall say that Eq. (4.234) or Eq. (2.117) is in the canonical form if $\int^\infty \frac{1}{r_1(t)}\,dt = \int^\infty \frac{1}{r_2(t)}\,dt = \infty$, and noncanonical form, if either $\int^\infty \frac{1}{r_1(t)}\,dt < \infty$ or $\int^\infty \frac{dt}{r_2(t)} < \infty$ or both $\int^\infty \frac{1}{r_1(t)}\,dt < \infty$ and $\int^\infty \frac{dt}{r_2(t)}\,dt < \infty$.

If (4.234) is of canonical form, then the use of $\int^\infty \frac{dt}{r_1(t)} = \infty$ and $\int^\infty \frac{dt}{r_2(t)} = \infty$ helps to study the oscillation and nonoscillation of the equation and their asymptotic behaviour. Thus, it would be interesting to study (4.234) in the noncanonical case.

Let $\aleph(N)$ and $\aleph$ denote the set of all proper nonoscillatory solutions of (4.234) and (2.117), respectively. As given in Sect. 2.7 of Chap. 2, the sets $\aleph(N)$ and $\aleph$ can be divided into the following four classes:

$$\aleph_0 = \{x \in \aleph(N)[x \in \aleph], \exists T_x; x(t)L_1x(t) < 0, x(t)L_2x(t) > 0, t \ge T_x\},$$

$$\aleph_1 = \{x \in \aleph(N)[x \in \aleph], \exists T_x; x(t)L_1x(t) > 0, x(t)L_2x(t) < 0, t \ge T_x\},$$

$$\aleph_2 = \{x \in \aleph(N)[x \in \aleph], \exists T_x; x(t)L_1x(t) > 0, x(t)L_2x(t) > 0, t \ge T_x\},$$

and

$$\aleph_3 = \{x \in \aleph(N)[x \in \aleph], \exists T_x; x(t)L_1x(t) < 0, x(t)L_2x(t) < 0, t \geq T_x\}.$$

Note that if $x \in \aleph_0$, then the quasiderivative satisfies the inequality $L_ix(t)L_{i+1}x(t) < 0$, $i = 0, 1, 2$ for sufficiently large t, which we call a Kneser solution.

Denote $\aleph_0^0 = \{x \in \aleph_0 : \lim_{t\to\infty} x(t) = 0\}$ and $\aleph_3^0 = \{x \in \aleph_3 : \lim_{t\to\infty} x(t) = 0\}$. Now, we shall give some comparison results to find the existence of Kneser solution of (4.234) by using Lemma 2.7.9 and Theorem 2.7.17.

Theorem 4.4.12 *Assume that*

$$\liminf_{|x|\to\infty} \frac{f(x)}{x} > 0, \qquad \liminf_{x\to 0} \frac{f(x)}{x} > 0, \qquad \int_\sigma^\infty \frac{dt}{r_1(t)} < \infty,$$

and

$$\int_\sigma^\infty \frac{dt}{r_2(t)} = \int_\sigma^\infty r_3(t)\,dt = \int_\sigma^\infty \frac{1}{r_1(t)} \int_\sigma^t \frac{1}{r_2(s)}\,ds\,dt = \infty.$$

If (2.117) *has property A, then* (4.234) *has property A.*

Proof Let $x(t)$ be a nonoscillatory solution of (4.234). We assume that there exists a constant $T > \sigma$ such that $x(t) > 0$ for all $t \geq T$. If (4.234) does not have property A, then from the above classifications of nonoscillatory solutions of (4.234), the following four cases may arise: (i) $x \in \aleph_1$, (ii) $x \in \aleph_2$, (iii) $x \in \aleph_0$ such that $\lim_{t\to\infty} L_ix(t) \neq 0$ for some $i \in \{0, 1, 2\}$ and (iv) $x \in \aleph_3$. First suppose that $x(t)$ is a positive nonoscillatory solution of (4.234) in the class $\aleph_1$. Then there exists a $T_1 \geq T$ such that $x(t) > 0$, $L_1x(t) > 0$, $L_2x(t) < 0$ for $t \geq T_1$. Moreover, since $x(t)$ is a positive increasing function, then either $\lim_{t\to\infty} x(t) = \alpha < \infty$ or $\lim_{t\to\infty} x(t) = \infty$ holds. In the case $\lim_{t\to\infty} x(t) = \alpha < \infty$, the continuity of the function $f(x)$ in the interval $[x(T_2), \alpha]$, where $T_2 \geq T_1$, ensures the existence of a positive constant K such that

$$\frac{f(x(t))}{x(t)} \geq K, \quad t \geq T_2. \tag{4.235}$$

In the case $\lim_{t\to\infty} x(t) = \infty$, the corresponding inequality (4.235) holds for some positive constant K, for $t \geq T_3 \geq T_2$ as $\liminf_{t\to\infty} \frac{f(x)}{x} > 0$ holds. Hence in either case, we see that there exists a $T_4 \geq T_1$ such that

$$\frac{f(x(t))}{x(t)} \geq K_1, \quad \text{for } t \geq T_4. \tag{4.236}$$

We note that $L_1x(t)$ is bounded. Now, integrating (4.234) twice in the interval $[T_4, t]$, we obtain

$$L_1x(t) < L_1x(T_4) - \int_{T_4}^t \frac{1}{r_2(s)} \int_{T_4}^{s_4} r_3(u) f(x(u))\,du\,ds.$$

Using the inequality (4.236), we obtain

$$L_1x(T_4) - L_1x(t) > K_1x(T_4)\int_{T_4}^t \frac{1}{r_2(s)}\int_{T_4}^s r_3(\theta)\,d\theta\,ds.$$

When $t \to \infty$, we have

$$\int_\sigma^\infty \frac{1}{r_2(s)}\int_\sigma^s r_3(\theta)\,d\theta\,ds < \infty,$$

which, taken together with $\int_\sigma^\infty \frac{1}{r_1(t)}\,dt < \infty$, leads to

$$\int_\sigma^\infty \frac{1}{r_1(t)}\int_\sigma^t \frac{1}{r_2(s)}\int_\sigma^s r_3(\theta)\,d\theta\,ds\,dt < \infty.$$

On the other hand, since (2.117) has property A, then Theorem 2.7.15 implies that

$$\int_\sigma^\infty \frac{1}{r_1(t)}\int_\sigma^t \frac{1}{r_2(s)}\int_\sigma^s r_3(\theta)\,d\theta\,ds\,dt = \infty,$$

a contradiction. Now, suppose that $x \in \aleph_2$. Then there exists a $T_1 \geq T$ such that $x(t) > 0$, $L_1x(t) > 0$ and $L_2x(t) > 0$ for $t \geq T_1$. Since $(L_2x(t))' = -r_3(t)f(x(t)) < 0$ for $t \geq T_1$, $L_2x(t)$ is a positive decreasing function. Hence $0 \leq L_2x(\infty) < \infty$. As done in the case $x \in \aleph_1$, by using the positive increasing nature of $x(t)$, we obtain (4.236). Integrating (4.234) in $[T_1, \infty)$, by using (4.236), we observe that there exists a constant $c > 0$ such that

$$c = \int_{T_4}^\infty r_3(t)f\big(x(t)\big)\,dt \geq K_1\int_{T_4}^\infty r_3(t)x(t)\,dt \geq K_1x(T_4)\int_{T_4}^\infty r_3(t)\,dt,$$

which contradicts the assumption $\int_\sigma^\infty r_3(t)\,dt = \infty$.

Next, assume that $x \in \aleph_0$ with $\lim_{t\to\infty} L_ix(t) \neq 0$ for some $i \in \{0, 1, 2\}$. Consider the linearised equation

$$\big(r_2(t)\big(r_1(t)w'\big)'\big)' + r_3(t)F(t)w = 0, \tag{4.237}$$

where $F(t) = \frac{f(x(t))}{x(t)}$. For its nonoscillatory solution $w = x$, Eq. (4.237) has a Kneser solution such that $\lim_{t\to\infty} L_iw(t) \neq 0$ for some $i \in \{0, 1, 2\}$. By Lemma 2.7.9, at least one of the following integrals:

$$\int_\sigma^\infty r_3(t)F(t)\int_\sigma^t \frac{1}{r_2(s)}\int_\sigma^s \frac{1}{r_1(u)}\,du\,ds\,dt,$$

$$\int_\sigma^\infty \frac{1}{r_1(t)}\int_\sigma^t r_3(s)F(s)\int_\sigma^s \frac{1}{r_2(u)}\,du\,ds\,dt,$$

$$\int_\sigma^\infty \frac{1}{r_2(t)}\int_\sigma^t \frac{1}{r_1(s)}\int_\sigma^s r_3(u)F(u)\,du\,ds\,dt$$

is convergent. Since $x(t)$ is positive decreasing, either (i) $\lim_{t\to\infty} x(t) = \beta > 0$ or (ii) $\lim_{t\to\infty} x(t) = 0$. In case (i), the continuity of the function $f(u)$ on the interval $[\beta, x(T)]$, where $T \geq \sigma$, ensures the existence of a positive constant M such that

$$F(t) = \frac{f(x(t))}{x(t)} \geq M \tag{4.238}$$

for t sufficiently large.

In case (ii), the condition $\liminf_{x\to 0} \frac{f(x)}{x} > 0$, then the inequality of the form (4.238) holds for some positive constant M_1, for sufficiently large t. We see, then, that there exists a positive number M_2 such that

$$\frac{f(x(t))}{x(t)} \geq M_2 \quad \text{for sufficiently large } t,$$

which implies that

$$\begin{aligned}
&M_2 \int_\sigma^\infty r_3(t) \int_\sigma^t \frac{1}{r_2(s)} \int_\sigma^s \frac{1}{r_1(u)}\, du\, ds\, dt \\
&\quad \leq \int_\sigma^\infty r_3(t)F(t) \int_\sigma^t \frac{1}{r_2(s)} \int_\sigma^s \frac{1}{r_1(u)}\, du\, ds\, dt, \\
&M_2 \int_\sigma^\infty \frac{1}{r_1(t)} \int_\sigma^t r_3(s) \int_\sigma^s \frac{1}{r_2(u)}\, du\, ds\, dt \\
&\quad \leq \int_\sigma^\infty \frac{1}{r_1(t)} \int_\sigma^t r_3(s)F(s) \int_\sigma^s \frac{1}{r_2(u)}\, du\, ds\, dt, \\
&M_2 \int_\sigma^\infty \frac{1}{r_2(t)} \int_\sigma^t \frac{1}{r_1(s)} \int_\sigma^s r_3(u)\, du\, ds\, dt \\
&\quad \leq \int_\sigma^\infty \frac{1}{r_2(t)} \int_\sigma^t \frac{1}{r_1(s)} \int_\sigma^s r_3(u)F(u)\, du\, ds\, dt
\end{aligned}$$

and so, at least one of the integral on the left of the above three inequalities is convergent. On the other hand, since (2.117) has property A, then it follows from Theorem 2.7.15 that the following integrals hold:

$$\begin{aligned}
&\int_\sigma^\infty r_3(t) \int_\sigma^t \frac{1}{r_2(s)} \int_\sigma^s \frac{1}{r_1(u)}\, du\, ds\, dt = \infty, \\
&\int_\sigma^\infty \frac{1}{r_1(t)} \int_\sigma^t r_3(s) \int_\sigma^s \frac{1}{r_2(u)}\, du\, ds\, dt = \infty, \\
&\int_\sigma^\infty \frac{1}{r_2(t)} \int_\sigma^t \frac{1}{r_1(s)} \int_\sigma^s r_3(u)\, du\, ds\, dt = \infty,
\end{aligned}$$

which is a contradiction.

Finally, suppose that $x \in \aleph_3$. Then there exists a $T_1 \geq T$ such that $x(t) > 0$, $L_1x(t) < 0$, $L_2x(t) < 0$ for $t \geq T_1$. Integrating the inequality $-L_2x(t) \geq -L_2x(T_1) > 0$ for $t \geq T_1$ twice from T_1 to t, we obtain

$$x(t) < x(T_1) + L_2x(T_1)\int_{T_1}^{t}\frac{1}{r_1(s)}\int_{T_1}^{s}\frac{1}{r_2(u)}\,du\,ds,$$

which yields a contradiction as $t \to \infty$.

The case $x(t) < 0$ for $t \geq T$ may be treated similarly. Thus, we have proved that any proper solution $x(t)$ of (4.234) is either oscillatory or belongs to the case $\aleph_0$ such that $\lim_{t\to\infty} L_ix(t) = 0$, $i \in \{0, 1, 2\}$. This completes the proof of the theorem. □

Corollary 4.4.4 *Assume that* $\liminf_{|x|\to\infty}\frac{f(x)}{x} > 0$ *and* $\liminf_{x\to 0}\frac{f(x)}{x} > 0$ *hold. Assume that one of the following conditions is satisfied*:

(i) $\int_\sigma^\infty \frac{dt}{r_1(t)} < \infty$, $\int_\sigma^\infty \frac{1}{r_2(t)}dt = \infty$,

$$\int_\sigma^\infty \frac{1}{r_1(t)}\int_\sigma^t \frac{1}{r_2(s)}\,ds\,dt = \int_\sigma^\infty r_3(t)\,dt = \int_\sigma^\infty \frac{1}{r_1(t)}\int_\sigma^t r_3(s)\,ds\,dt = \infty,$$

(ii) $\int_\sigma^\infty \frac{1}{r_1(t)}\int_\sigma^t r_3(s)\,ds\,dt < \infty$,

$$\int_\sigma^\infty \frac{1}{r_2(t)}dt = \int_\sigma^\infty \frac{1}{r_1(t)}\int_\sigma^t \frac{1}{r_2(s)}ds = \int_\sigma^\infty r_3(t)\,dt = \infty$$

and

$$\int_\sigma^\infty \frac{1}{r_2(t)}\left(\int_t^\infty \frac{1}{r_1(s)}ds\right)\left(\int_t^\infty r_3(s)\int_s^\infty \frac{1}{r_1(u)}\,du\,ds\right)dt = \infty.$$

Then (4.234) *has property A*.

Example 4.4.5 Consider

$$\left(\frac{1}{t+1}\left((t+1)^3x'(t)\right)'\right)' + (t+1)^2\left(x^3(t) + x(t)\right) = 0, \quad t \geq 0.$$

Here $r_1(t) = (t+1)^3$ and $\int_0^\infty \frac{1}{r_1(t)}dt < \infty$. Hence this example is in the non-canonical form. Theorem 4.4.12 can be applied to this example.

Theorem 4.4.13 *Assume that*

$$\liminf_{t\to\infty}\frac{f(x)}{x} > 0, \qquad \liminf_{x\to 0}\frac{f(x)}{x} > 0, \qquad \int_\sigma^\infty \frac{dt}{r_2(t)} < \infty$$

and

$$\int_\sigma^\infty \frac{dt}{r_1(t)} = \int_\sigma^\infty \frac{1}{r_2(t)}\int_\sigma^t r_3(s)\,ds\,dt = \int_\sigma^\infty r_3(t)\,dt = \infty.$$

If (2.117) *has property A*, *then* (4.234) *has property A*.

Corollary 4.4.5 *Assume that* $\liminf_{|x|\to\infty}\frac{f(x)}{x}>0$, $\liminf_{x\to 0}\frac{f(x)}{x}>0$ *and one of the following conditions is satisfied*:

(i) $\int_\sigma^\infty \frac{dt}{r_2(t)}<\infty$, $\int_\sigma^\infty \frac{dt}{r_1(t)}=\infty$,

$$\int_\sigma^\infty \frac{1}{r_2(t)}\int_\sigma^t r_3(s)\,ds\,dt=\int_\sigma^\infty r_3(t)\,dt=\int_\sigma^\infty \frac{1}{r_2(t)}\int_\sigma^t \frac{1}{r_1(s)}\,ds\,dt=\infty,$$

(ii) $\int_\sigma^\infty \frac{1}{r_2(t)}\int_\sigma^t \frac{1}{r_1(s)}\,ds\,dt<\infty$,

$$\int_\sigma^\infty \frac{dt}{r_1(t)}=\int_\sigma^\infty \frac{1}{r_2(t)}\int_\sigma^t r_3(s)\,ds\,dt=\int_\sigma^\infty r_3(t)\,dt=\infty$$

and

$$\int_\sigma^\infty r_3(t)\left(\int_t^\infty \frac{1}{r_2(s)}\,ds\right)\left(\int_t^\infty \frac{1}{r_1(s)}\int_s^\infty \frac{1}{r_2(u)}\,du\,ds\right)dt=\infty.$$

Then (4.234) *has property* A.

Theorem 4.4.14 *Assume that* $\liminf_{|x|\to\infty}\frac{f(x)}{x}>0$, $\liminf_{x\to 0}\frac{f(x)}{x}>0$, $\int_\sigma^\infty \frac{dt}{r_2(t)}<\infty$ *and* $\int_\sigma^\infty \frac{dt}{r_1(t)}<\infty$. *If* (2.117) *has property* A, *then* (4.234) *has property* A *or any solution* $x(t)$ *of* (4.234) *from the class* $\aleph_3$ *tends to zero as* $t\to\infty$.

Proof Let $x(t)$ be a nonoscillatory solution of (4.234). Then $x\in\aleph_0\cup\aleph_1\cup\aleph_2\cup\aleph_3$. We assume that there exists a $T\geq\sigma$ such that $x(t)>0$ for all $t\geq T$. Suppose that (4.234) does not have property A, and that there exists a solution $x(t)$ of (4.234) from the class $\aleph_3$ that tends to a positive constant as $t\to\infty$. Then, there are four possibilities: (i) $x\in\aleph_1$, (ii) $x\in\aleph_2$, (iii) $x\in\aleph_0$ such that $\lim_{x\to\infty}L_i x(t)\neq 0$ for some $i\in\{0,1,2\}$ and (iv) $x\in\aleph_3$ such that $\lim_{t\to\infty}x(t)=c>0$. In each of the cases (i) and (iii), we get a contradiction as in Theorem 4.4.12.

Let $x\in\aleph_2$. Proceeding as in the proof of Theorem 4.4.12, we get $\int_\sigma^\infty r_3(t)\,dt<\infty$, which, taken together with $\int_\sigma^\infty \frac{dt}{r_1(t)}<\infty$ and $\int_\sigma^\infty \frac{dt}{r_2(t)}<\infty$, leads to

$$\int_\sigma^\infty r_3(t)\int_\sigma^t \frac{1}{r_2(s)}\int_\sigma^s \frac{1}{r_1(u)}\,du\,ds\,dt<\infty.$$

On the other hand, since (2.117) has property A, from Theorem 2.7.15, it follows that

$$\int_\sigma^\infty r_3(t)\int_\sigma^t \frac{1}{r_2(s)}\int_\sigma^s \frac{1}{r_1(u)}\,du\,ds\,dt=\infty,$$

a contradiction.

Finally, suppose that $x\in\aleph_3$ such that $\lim_{t\to\infty}x(t)=c>0$. Then there exists a $T_1\geq T$ such that $x(t)>0$, $L_1x(t)<0$ and $L_2x(t)<0$ for $t\geq T_1$. Integrating

(4.234) three times from T_1 to t, we obtain

$$x(t) < x(T_1) - \int_{T_1}^{t} \frac{1}{r_1(s)} \int_{T_1}^{s} \frac{1}{r_2(\theta)} \int_{T_1}^{\theta} r_3(u) f\big(x(u)\big)\, du\, d\theta\, ds, \qquad t \geq T_1.$$

Since x is positive and decreasing such that $\lim_{t\to\infty} x(t) = c > 0$, we have $0 < c \leq x(t) \leq x(T_2)$ for all $t \geq T_2 \geq T_1$. The continuity of $\frac{f(x)}{x}$ on the interval $[c, x(T_2)]$ ensures the existence of a positive constant K_1 such that $f(x(t)) \geq K_1 x(t)$ for $t \geq T_2$. Then the above integral inequality yields

$$x(t)\left(1 + K_1 \int_{T_2}^{t} \frac{1}{r_1(s)} \int_{T_2}^{s} \frac{1}{r_2(\theta)} \int_{T_2}^{\theta} r_3(u)\, du\, d\theta\, ds\right) \leq x(T_1), \qquad t \geq T_2.$$

Letting $t \to \infty$, we obtain

$$\int_{T_2}^{\infty} \frac{1}{r_1(t)} \int_{T_2}^{t} \frac{1}{r_2(s)} \int_{T_2}^{s} r_3(u)\, du\, ds\, dt < \infty.$$

However, from Theorem 2.7.15, Eq. (2.117) is oscillatory because (2.117) has property A. Hence from Theorem 2.7.5, we obtain

$$\int_{\sigma}^{\infty} \frac{1}{r_1(t)} \int_{\sigma}^{t} \frac{1}{r_2(s)} \int_{\sigma}^{s} r_3(u)\, du\, ds\, dt = \infty,$$

a contradiction.

The case $x(t) < 0$ for $t \geq T$ may be treated similarly. We have shown that any solution $x(t)$ of (4.234) is either oscillatory or belongs to the class $\aleph_0$ such that $\lim_{t\to\infty} L_i x(t) = 0$ for all $i \in \{0, 1, 2\}$, or belongs to the class $\aleph_3$ such that $\lim_{t\to\infty} x(t) = 0$. This completes the proof of the theorem. □

Theorem 4.4.15 *If* $\int_{\sigma}^{\infty} \frac{dt}{r_1(t)} = \int_{\sigma}^{\infty} \frac{dt}{r_2(t)} = \infty$, *then any solution* $x(t)$ *of* (4.234) *in the class* $\aleph_0$ *satisfies the property* $\lim_{t\to\infty} L_i x(t) = 0$, $i = 1, 2$.

Theorem 4.4.16 *Assume that* $\limsup_{x\to 0} \frac{f(x)}{x} < \infty$ *and* $\int_{\sigma}^{\infty} \frac{dt}{r_1(t)} = \int_{\sigma}^{\infty} \frac{dt}{r_2(t)} = \infty$. *If there exists a solution* $x(t)$ *of* (4.234) *in the class* $\aleph_0$ *such that* $\lim_{t\to\infty} L_i x(t) = 0$, $i = 0, 1, 2$, *then*

$$\int_{\sigma}^{\infty} r_3(t) \int_{\sigma}^{t} \frac{1}{r_2(s)} \int_{\sigma}^{s} \frac{1}{r_1(u)}\, du\, ds\, dt = \infty.$$

Theorem 4.4.17 *Let* $\int_{\sigma}^{\infty} \frac{dt}{r_1(t)} < \infty$, $\int_{\sigma}^{\infty} \frac{1}{r_2(t)} \int_{\sigma}^{t} r_3(s)\, ds\, dt < \infty$ *and assume that the function* f *satisfies the Lipschitz condition on the interval* $[0, 2\int_{\sigma}^{\infty} \frac{dt}{r_1(t)}]$. *Then* (4.234) *has a solution* $x(t)$ *in the class* $\aleph_3$ *such that*

$$\lim_{t\to\infty} x(t) = 0, \quad \text{i.e.} \quad \aleph_3^0 \neq \emptyset.$$

Proof Here, we prove the existence of a positive solution of (4.234) in the class $\aleph_3$ which tends to zero as $t \to \infty$.

Let L denote the Lipschitz constant of the function f on the interval $[0, 2\int_\sigma^\infty \frac{dt}{r_1(t)}]$ and let $t_0 \geq \sigma$ be such that

$$\int_{t_0}^{\sigma} \frac{1}{r_1(t)}\, dt \leq \frac{1}{L+1}$$

and

$$\int_{t_0}^{\infty} \frac{1}{r_2(t)} \int_{t_0}^{t} r_3(s)\, ds\, dt \leq \min\left\{\frac{1}{K}, 1\right\},$$

where

$$K = \max\left\{ f(x); x \in \left[0, 2\int_\sigma^\infty \frac{dt}{r_1(t)}\right]\right\}.$$

Denote $H_1(t) = \int_t^\infty \frac{1}{r_1(s)}\, ds$, $t \geq t_0$. Let us define the set

$$\Delta_1 = \left\{u \in C\left([t_0, \infty), R\right); H_1(t) \leq u(t) \leq 2H_1(t)\right\},$$

where $C([t_0, \infty), R)$ denotes the Banach space of all continuous and bounded functions defined on the interval with the sup. norm

$$\|u\| = \sup\left\{\left|u(t)\right|,\ t \geq t_0\right\}.$$

Clearly, Δ_1 is a nonempty closed subset of space $C([t_0, \infty), R)$ and so Δ_1 is a nonempty complete metric space. For every $u \in \Delta_1$, we consider a mapping

$$T_1 : \Delta_1 \to C\left([t_0, \infty), R\right)$$

given by

$$\begin{aligned} x_u(t) &= (T_{1u})(t) \\ &= H_1(t) + \int_t^\infty \frac{1}{r_1(\tau)} \int_{t_0}^{\tau} \frac{1}{r_2(s)} \int_{t_0}^{s} r_3(z) f\left(u(z)\right) dz\, ds\, d\tau, \quad t \geq t_0. \end{aligned}$$

Now, $H_1(t) \leq x_u(t)$ and the estimation

$$\begin{aligned} x_u(t) &= H_1(t) + \int_t^\infty \frac{1}{r_1(\tau)} \int_{t_0}^{\tau} \frac{1}{r_2(s)} \int_{t_0}^{s} r_3(z) f\left(u(z)\right) dz\, ds\, d\tau \\ &\leq H_1(t) + K \int_t^\infty \frac{1}{r_1(\tau)} \int_{t_0}^{\tau} \frac{1}{r_2(s)} \int_{t_0}^{s} r_3(z)\, dz\, ds\, d\tau \\ &\leq H_1(t) + K\left(\int_{t_0}^\infty \frac{1}{r_2(s)} \int_{t_0}^{s} r_3(z)\, dz\, ds\right)\left(\int_t^\infty \frac{1}{r_1(\tau)}\, d\tau\right) \\ &\leq H_1(t) + H_1(t) = 2H_1(t) \end{aligned}$$

implies that T_1 maps Δ_1 to Δ_1.

Let u_1 and $u_2 \in \Delta_1$ and $t \geq t_0$. Since f is Lipschitzian,

$$\begin{aligned}
&\left|(T_1u_1)(t) - (T_1u_2)(t)\right| \\
&\leq \int_t^\infty \frac{1}{r_1(\tau)} \int_{t_0}^\tau \frac{1}{r_2(s)} \int_{t_0}^s r_3(z)\left|f\big(u_1(z)\big) - f\big(u_2(z)\big)\right| dz\,ds\,d\tau \\
&\leq \int_{t_0}^\infty \frac{1}{r_1(\tau)} \int_{t_0}^\tau \frac{1}{r_2(s)} \int_{t_0}^s r_3(z)\left|f\big(u_1(z)\big) - f\big(u_2(z)\big)\right| dz\,ds\,d\tau \\
&\leq L \int_{t_0}^\infty \frac{1}{r_1(\tau)} \int_{t_0}^\tau \frac{1}{r_2(s)} \int_{t_0}^s r_3(z)|u_1(z) - u_2(z)|\,dz\,ds\,d\tau \\
&\leq L\|u_1 - u_2\| \left(\int_{t_0}^\infty \frac{1}{r_1(s)}\,ds\right)\left(\int_{t_0}^\infty \frac{1}{r_2(s)} \int_{t_0}^s r_3(z)\,dz\,ds\right) \\
&\leq \frac{L}{L+1}\|u_1 - u_2\| = Q_1\|u_1 - u_2\|, \quad 0 < Q_1 < 1,
\end{aligned}$$

that is, T_1 is a contraction mapping in Δ_1. Hence, by the Banach fixed-point theorem, there exists a unique fixed point $x \in \Delta_1$ such that

$$x(t) = H_1(t) + \int_t^\infty \frac{1}{r_1(\tau)} \int_{t_0}^\tau \frac{1}{r_2(s)} \int_{t_0}^s r_3(z) f\big(x(z)\big)\,dz\,ds\,d\tau, \quad t \geq t_0.$$

Observe that $L_1x(t) < 0$ and $L_2x(t) < 0$ and $x(t)$ is a positive solution of (4.234) in the class $\aleph_3$ which tends to zero as $t \to \infty$, that is, $x \in \aleph_3^0$. □

Theorem 4.4.18 *If* $\int_\sigma^\infty \frac{dt}{r_1(t)} = \infty$ *or* $\int_\sigma^\infty \frac{1}{r_1(t)} \int_\sigma^t \frac{1}{r_2(s)}\,ds\,dt = \infty$, *then* $\aleph_3 = \emptyset$.

Corollary 4.4.6 *Let f satisfy the Lipschitz condition on the interval* $[0, 2\int_\sigma^\infty \frac{1}{r_1(t)}\,dt]$ *and*

$$\int_\sigma^\infty \frac{1}{r_2(t)} \int_\sigma^t r_3(s)\,ds\,dt < \infty.$$

Then a necessary and sufficient condition for (4.234) *to have a solution $x(t)$ in the class $\aleph_3^0$ is that* $\int_\sigma^\infty \frac{dt}{r_1(t)} < \infty$.

Evidently, the following Corollary 4.4.7 holds.

Corollary 4.4.7 *Let f satisfy the Lipschitz condition on the interval* $[0, 2\int_\sigma^\infty \frac{dt}{r_1(t)}]$ *and*

$$\int_\sigma^\infty \frac{1}{r_2(t)} \int_\sigma^t r_3(s)\,ds\,dt < \infty.$$

Then a necessary and sufficient condition for (4.234) *to have a solution $x(t)$ in the class $\aleph_3$ is that* $\int_\sigma^\infty \frac{dt}{r_1(t)} < \infty$.

Theorem 4.4.19 *Let* $\int_\sigma^\infty \frac{1}{r_2(t)} \int_\sigma^t \frac{1}{r_1(s)}\, ds\, dt < \infty$, $\int_\sigma^\infty r_3(t)\, dt < \infty$ *and assume that the function* f *satisfies the Lipschitz condition on the interval*

$$\left[0, 2\int_\sigma^\infty \frac{1}{r_2(t)} \int_\sigma^t \frac{1}{r_1(s)}\, ds\, dt\right].$$

Then (4.234) *has a solution* $x(t)$ *in the class* $\aleph_0$ *such that* $\lim_{t\to\infty} x(t) = 0$, *that is,* $\aleph_0^0 \neq \emptyset$.

Proof Let N denote the Lipschitz constant of f on the interval $[0, 2\int_\sigma^\infty \frac{1}{r_2(t)} \times \int_\sigma^t \frac{1}{r_1(s)}\, ds\, dt]$ and let $t_0 \geq \sigma$ be such that

$$\int_{t_0}^\infty \frac{1}{r_2(t)} \int_{t_0}^t \frac{1}{r_1(s)}\, ds\, dt \leq \frac{1}{L+1}$$

and

$$\int_{t_0}^\infty r_3(t)\, dt \leq \min\left\{\frac{1}{K}, 1\right\},$$

where

$$K = \max\left\{f(x); x \in \left[0, 2\int_{t_0}^\infty \frac{1}{r_1(t)} \int_t^\infty \frac{1}{r_2(s)}\, ds\, dt\right]\right\}.$$

We observe that

$$\int_{t_0}^\infty \frac{1}{r_1(t)} \int_t^\infty \frac{1}{r_2(s)}\, ds\, dt = \int_{t_0}^\infty \frac{1}{r_2(t)} \int_{t_0}^t \frac{1}{r_1(s)}\, ds\, dt.$$

Denote $H_2(t) = \int_t^\infty \frac{1}{r_1(t)} \int_t^\infty \frac{1}{r_2(s)}\, ds\, dt$. Define a set Δ_2 as

$$\Delta_2 = \left\{u \in C\left([t_0, \infty), R\right); H_2(t) \leq u(t) \leq 2H_2(t)\right\},$$

where $C([t_0, \infty), R)$ denotes the Banach space of all continuous and bounded functions defined on $[t_0, \infty)$ under the sup. norm

$$\|u\| = \sup\left\{\left|u(t)\right|; t \geq t_0\right\}.$$

Clearly, Δ_2 is a nonempty closed subset of the space $C([t_0, \infty), R)$ and so Δ_2 is a nonempty complete metric space. For every $u \in \Delta_2$, define a mapping $T_2 : \Delta_2 \to C([\sigma, \infty), R)$ by

$$\begin{aligned} x_u(t) &= (T_2u)(t) \\ &= H_2(t) + \int_t^\infty \frac{1}{r_1(\tau)} \int_\tau^\infty \frac{1}{r_2(s)} \int_s^\infty r_3(z) f\left(u(z)\right) dz\, ds\, d\tau, \quad t \geq t_0. \end{aligned}$$

One may verify that T_2 maps Δ_2 into itself and is a contraction mapping in Δ_2. Hence Banach fixed-point theorem ensures the existence of the unique fixed-point $x \in \Delta_2$ such that

$$x(t) = H_2(t) + \int_t^\infty \frac{1}{r_1(\tau)} \int_\tau^\infty \frac{1}{r_2(s)} \int_s^\infty r_3(z) f\bigl(x(z)\bigr)\, dz\, ds\, d\tau, \quad t \ge t_0.$$

Note that this $x(t)$ is a solution of (4.234) with $x \in \aleph_0$ which tends to zero as $t \to \infty$, that is, $x \in \aleph_0^0$. This completes the proof of the theorem. □

Theorem 4.4.20 *Assume that* $\limsup_{x\to 0} \frac{f(x)}{x} < \infty$. *If*

$$\int_\sigma^\infty r_3(t) \int_\sigma^t \frac{1}{r_2(s)} \int_\sigma^s \frac{1}{r_1(u)}\, du\, ds\, dt < \infty \quad \textit{and}$$

$$\int_\sigma^\infty \frac{1}{r_1(t)}\, dt = \int_\sigma^\infty \frac{1}{r_2(t)}\, dt = \infty,$$

then $\aleph_0^0 \ne \emptyset$.

Bakulikova et al. [4] obtained some sufficient conditions and also conditions of Kamenev-type and Philos-type for the oscillation and asymptotic properties of nonoscillatory solutions of

$$\bigl(r_2(t)\bigl(\bigl(r_1(t)x'\bigr)'\bigr)^\gamma\bigr)' + r_3(t)x^\gamma = 0, \quad t \ge \sigma, \tag{4.239}$$

where r_1, r_2 and r_3 are real-valued continuous function, γ is a quotient of odd positive integers and

$$\int_\sigma^\infty \left(\frac{1}{r_2(t)}\right)^\gamma dt = \infty, \qquad \int_\sigma^\infty \frac{1}{r_1(t)}\, dt = \infty. \tag{4.240}$$

Let (4.240) hold, then one may verify that, if $x(t)$ is an eventually positive solution of (4.239), then there exists a $t_1 \ge \sigma$ such that either

$$x(t) > 0, \qquad x'(t) > 0, \qquad \bigl(r_1(t)x'(t)\bigr)' > 0 \quad \text{for } t \ge t_1$$

or

$$x(t) > 0, \qquad x'(t) < 0, \qquad \bigl(r_1(t)x'(t)\bigr)' > 0 \quad \text{for } t \ge t_1.$$

If the former inequalities hold, then there exists a $t_2 \ge t_0$ such that

$$x'(t) \ge \frac{\delta(t)(r_2(t))^{1/\gamma}}{r_1(t)} \bigl(r_1(t)x'(t)\bigr)', \quad t \ge t_1,$$

holds, where $\delta(t) = \int_{t_2}^t \frac{ds}{(r_2(s))^{1/\gamma}}$. In fact, the calculation

$$r_1(t)x'(t) \ge \bigl(r_2(t)\bigr)^{1/\gamma} \bigl(r_1(t)x'(t)\bigr)' \int_{t_2}^t \frac{ds}{(r_2(s))^{1/\gamma}}, \quad t \ge t_2$$

leads to the above conclusion.

Using the above calculations, Bakulikova [4] proved Theorem 4.4.21.

Theorem 4.4.21 *Assume that* (4.240) *hold. Further, assume that there exists a* $\rho \in C'([\sigma, \infty), [0, \infty))$ *such that*

$$\limsup_{t\to\infty} \int_\sigma^t \left(\rho(s) r_3(s) - \frac{r_1^\gamma(s)(\rho'(s))_+^{\gamma+1}}{(\gamma+1)^{\gamma+1}(\delta(s))^\gamma \rho^\gamma(s)} \right) ds = \infty$$

and

$$\int_\sigma^\infty \frac{1}{r_1(t)} \int_t^\infty \left(\frac{1}{r_2(s)} \int_s^\infty r_3(\theta)\, d\theta \right)^{1/\gamma} ds\, dt = \infty, \tag{4.241}$$

where $(\rho'(s))_+ = \max\{0, \rho'(s)\}$. *Then every solution* $x(t)$ *of* (4.239) *is either oscillatory or* $x(t) \to 0$ *as* $t \to \infty$.

Suitable choices of $\rho(t)$ leads to various oscillation criteria for (4.239). For instance, if $\rho(t) = 1$ and $\rho(t) = t^{1+\gamma}$, $t \ge \sigma$, then Theorem 4.4.21 give the following corollaries.

Corollary 4.4.8 *Assume that* (4.240) *and* (4.241) *hold. Furthermore, assume that*

$$\limsup_{t\to\infty} \int_\sigma^t \left[s r_3(s) - \frac{r_1^\gamma(s)}{(\gamma+1)^{\gamma+1}(\delta(s))^\gamma s^\gamma} \right] ds = \infty,$$

then every solution $x(t)$ *of* (4.239) *is either oscillatory or* $x(t) \to 0$ *as* $t \to \infty$.

Corollary 4.4.9 *Assume that* (4.240) *and* (4.241) *hold. Furthermore, assume that*

$$\limsup_{t\to\infty} \int_\sigma^t \left(s^{\gamma+1} r_3(s) - \frac{r_1^\gamma(s)}{(\delta(s))^\gamma} \right) ds = \infty,$$

then every solution $x(t)$ *of* (4.239) *is either oscillatory or* $x(t) \to 0$ *as* $t \to \infty$.

Theorem 4.4.22 *Assume that* (4.240) *and* (4.241) *hold. Again, assume that*

$$\limsup_{t\to\infty} \frac{1}{t^n} \int_\sigma^t (t-s)^n \left[\rho(s) r_3(s) - \frac{r_1^\gamma(s)(\rho'(s))_+^{\gamma+1}}{(\gamma+1)^{\gamma+1}(\delta(s))^\gamma \rho^\gamma(s)} \right] ds = \infty, \quad n > 1, \tag{4.242}$$

where $(\rho'(s))_+ = \max\{0, \rho'(s)\}$. *Then every solution* $x(t)$ *of* (4.239) *is oscillatory or* $x(t) \to 0$ *as* $t \to \infty$.

The integral condition (4.242), given in Theorem 4.4.22 is known as Kamenev-type integral criteria. This type of integral criteria was sufficiently developed for second-order differential equations. It is recently that Kamenev-type integral criteria for third-order differential equations have been given a serious study.

Theorem 4.4.23 *Assume that* (4.240) *and* (4.241) *hold. Furthermore, assume that there exists a* $\rho \in C^1([\sigma,\infty),[0,\infty))$ *such that*

$$\limsup_{t\to\infty} \frac{1}{t^n}\int_{\sigma}^{t}\left[(t-s)^n\rho(s)r_3(s) - \frac{\rho(s)r_1^{\gamma}(s)}{(\delta(s))^{\gamma}(\gamma+1)^{\gamma+1}(t-s)^{n\gamma}}P^{\gamma+1}(t,s)\right]ds$$
$$=\infty,$$

where $(\rho'(s))_+ = \max\{0,\rho'(s)\}$ *and*

$$P(t,s) = (t-s)^n\frac{(\rho'(s))_+}{\rho(s)} - n(t-s)^{n-1}, \quad t\ge s\ge\sigma.$$

Then every solution $x(t)$ *of* (4.239) *is oscillatory or* $x(t)\to 0$ *as* $t\to\infty$.

Finally, we give some integral average condition of Philos-type for the oscillation or asymptotic properties of nonoscillatory solutions of Eq. (4.239). Defining D_0, D_1, $H(t,s)$ and $h(t,s)$, as defined before in Theorem 4.4.8, Bakulikova et al. [4] proved Theorem 4.4.24.

Theorem 4.4.24 *Assume that* (4.240) *and* (4.241) *hold. Furthermore, assume that there exists a* $\rho \in C^1([\sigma,\infty),[0,\infty))$ *and* $H\in\mathcal{P}$ *such that*

$$\limsup_{t\to\infty}\frac{1}{H(t,t_0)}\int_{t_0}^{t}\left(H(t,s)\rho(s)r_3(s) - \frac{\rho(s)r_1^{\gamma}(s)Q^{\gamma+1}(t,s)}{(\gamma+1)^{\gamma+1}(\delta(s))^{\gamma}H^{\gamma}(t,s)}\right)ds = \infty, \tag{4.243}$$

where

$$Q(t,s) = \frac{(\rho'(s))_+}{\rho(s)}H(t,s) - h(t,s)\sqrt{H(t,s)}.$$

Then every solution $x(t)$ *of* (4.239) *is oscillatory or* $x(t)\to 0$ *as* $t\to\infty$.

The following two results provide alternative oscillation criteria when (4.243) of Theorem 4.4.24 fails.

Theorem 4.4.25 *Let all the assumptions, except* (4.243), *of Theorem* 4.4.24 *hold. Further, let*

$$0 < \inf_{s\ge\sigma}\left[\liminf_{t\to\infty}\frac{H(t,s)}{H(t,\sigma)}\right] \le \infty$$

and

$$\limsup_{t\to\infty}\frac{1}{H(t,\sigma)}\int_{\sigma}^{t}H(t,s)\rho(s)r_3(s)\,ds < \infty.$$

Let for some $\psi \in C([\sigma,\infty),[0,\infty))$,

$$\limsup_{t\to\infty}\int_{\sigma}^{t}\frac{\psi^{\frac{\gamma+1}{\gamma}}(s)\delta(s)}{\rho^{\frac{1}{\gamma}}(s)r_1(s)}ds = \infty$$

and for all $t \geq \sigma$ large enough, and

$$\liminf_{t\to\infty} \frac{1}{H(t,T)} \int_T^t \left[H(t,s)\rho(s)r_3(s) - \frac{\rho(s)r_1^{\gamma}(s)Q^{\gamma+1}(t,s)}{(\gamma+1)^{\gamma+1}\delta^{\gamma}(s)H^{\gamma}(t,s)} \right] ds \geq \psi(T),$$

where $H(t,s)$ and $Q(t,s)$ are same as in Theorem 4.4.24. *Then every solution $x(t)$ of* (4.239) *is oscillatory or $x(t) \to 0$ as $t \to \infty$.*

For the choice $H(t,s) = (t-s)^n$ and $h(t,s) = n(t-s)^{\frac{n-2}{2}}$ the Philos-type condition reduces to the Kamenev-type condition. Other choices of H include

$$H(t,s) = \left(\ln \frac{t}{s}\right)^n, \qquad h(t,s) = \frac{n}{s}\left(\ln \frac{t}{s}\right)^{\frac{n}{2}-1}$$

and

$$H(t,s) = \left(e^t - e^s\right)^n, \qquad h(t,s) = ne^s\left(e^t - e^s\right)^{\frac{n-2}{2}},$$

or more generally,

$$H(t,s) = \left(\int_s^t \frac{du}{\theta(u)}\right)^n, \qquad h(t,s) = \frac{n}{\theta(s)}\left(\int_s^t \frac{du}{\theta(u)}\right)^{\frac{n}{2}-1},$$

where $n > 1$ is an integer, and $\theta \in C([\sigma,\infty),[0,\infty))$ and satisfies $\lim_{t\to\infty}\int_\sigma^t \frac{du}{\theta(u)} = \infty$.

4.5 Open Problems and Discussions

Most of the results of this chapter have the conclusion that any solution of the considered third-order equation which has a zero, is oscillatory. Hence under the same condition, it follows that the nonoscillatory solution has no zero in $[\sigma,\infty)$.

In Sect. 4.2 of this chapter, we were able to study the behaviour of solutions of (4.9) or (4.10) under the cases given in (H_5) and (H_6). Thus, it remains to study the behaviour of solutions of (4.9) or (4.10) under the following cases:

(i) $a(t) \leq 0, b(t) \leq 0, c(t) < 0$
(ii) $a(t) \geq 0, b(t) \leq 0, c(t) < 0$
(iii) $a(t) \geq 0, b(t) \geq 0, c(t) > 0$
(iv) $a(t) \geq 0, b(t) \geq 0, c(t) < 0$
(v) $a(t) \leq 0, b(t) \geq 0, c(t) < 0$
(vi) $a(t) \leq 0, b(t) \geq 0, c(t) > 0$.

The cases (i) and (ii) may be studied directly. However, it seems that the nonoscillation of the second-order linear differential equation (2.110) will be a useful tool to study the asymptotic behaviour of nonoscillatory solutions of (4.9) and (4.10). It would be interesting to extend the results of Chap. 2 to Eqs. (4.9) and (4.10).

Other methods of studying (4.9) or (4.10) are the canonical transforms due to Trench [38]. An idea of canonical transformation and some comparison results are given in Chap. 6.

It would be interesting to extend the results of Sect. 4.3 either to (4.1) or (4.2) or (4.9) or (4.10). Most importantly, it is believed the results of Sect. 4.3 obtained for Eqs. (4.116) and (4.117) can be extended to (4.234) or

$$\big(r_2(t)\big(r_1(t)x'\big)'\big)' + r_3(t)x^\gamma = 0, \quad t \ge \sigma \tag{4.244}$$

for the cases when the equations are canonical and noncanonical. It is quite easier to study the behaviour of solutions, when the equations are canonical. Thus, it would be important to find a single result on the oscillation or asymptotic behaviour of nonoscillatory solution of (4.234) and (4.244) when they are noncanonical in nature. Theorems 4.4.12–4.4.16 may provide some indications in this direction.

4.6 Notes

Theorems 4.1.3–4.1.9 and 4.1.16 are taken from Parhi and Das [29], whereas Theorems 4.1.5–4.1.10 are due to Tiryaki and Yaman [37]. Theorems 4.1.16–4.1.21 are due to Škerlik [35]. Theorems 4.1.35–4.1.37 are proved by Cecchi et al. in [9]. Theorems 4.1.38–4.1.43 can be found in [6]. Greguš and Greguš Jr [14] proved Theorems 4.1.30–4.1.31. Theorems 4.1.32–4.1.34 are proved by Greguš [13]. Theorems 4.1.23–4.1.29 are taken from Greguš and Venko [15]. Theorems 4.1.44–4.1.49 are due to Tiryaki and Celebi [36]. Section 4.2 is incorporated from Parhi and Das [30]. Theorem 4.3.1 and Theorem 4.3.2 are proved in [30]. Theorems 4.3.3–4.3.5 are special cases of some results proved by Naito and Yano [27]. Theorems 4.3.10–4.3.16 are due to Cecchi and Marini [7]. Theorems 4.3.17–4.3.20 are taken from Kura [23]. Theorems 4.4.1–4.4.6 are proved in [34]. Aktas, Tiryaki and Zafer [2] proved Theorems 4.4.7–4.4.10. Theorems 4.4.11–4.4.14 can be found in Mojsej and Ohriska [25], whereas Theorems 4.4.15 to 4.4.20 are taken from Mojsej and Tartalova [26]. Theorems 4.4.21–4.4.25 are proved in [4].

References

1. S. Ahmad and A. C. Lazer; *On the oscillatory behavior of a class of linear third order differential equations*, Journal of Mathematical Analysis and Applications, 28(3) (1970), 681–689.
2. M. F. Aktas, A. Tiryaki and A. Zafer; *Integral criteria for oscillation of third order nonlinear differential equations*, Nonlinear Analysis; Theory Methods and Applications, 71(12) (2009), e1496–e1502.
3. I. Astashova; *On Izobov's problem for a nonlinear third order differential equation*, Czech-Georgian Workshop on Boundary Value Problems, Institute of Mathematics, Academy of Sciences of the Czech Republic, 5–9 December, 2011, Brno.
4. B. Baculiková, E. M. Elabbasy, S. H. Saker and J. Džurina; *Oscillation criteria for third-order nonlinear differential equations*, Mathematica Slovaca, 58(2) (2008), 201–220.

5. J. H. Barrett; *Oscillation theory of ordinary linear differential equations*, Advances in Mathematics, 3(4) (1969), 415–509. (Reprinted in *Lectures in Ordinary Differential Equations*, Edited by R. McKelvev, Academic Press, New York, 1970).
6. M. Bartušek, M. Cecchi and M. Marini; *On Kneser solutions of nonlinear third order differential equations*, Journal of Mathematical Analysis and Applications, 261(1) (2001), 72–84.
7. M. Cecchi and M. Marini; *On the oscillatory behavior of a third order nonlinear differential equation*, Nonlinear Analysis; Theory Methods and Applications, 15(2) (1990), 141–153.
8. M. Cecchi and M. Marini; *Oscillation results for Emden-Fowler type differential Equations*, Journal of Mathematical Analysis and Applications, 205(2) (1997), 406–422.
9. M. Cecchi, Z. Doslá and M. Marini; *On third order differential equations with Property A and B*, Journal of Mathematical Analysis and Applications, 231(2) (1999), 509–525.
10. L. Erbe; *Oscillation nonoscillation and asymptotic behaviour for third nonlinear differential equations*, Annali di Matematica Pura ed Applicata, 110(1) (1976), 373–391.
11. L. Erbe and V. S. H. Rao; *Nonoscillation results for third-order nonlinear differential equations*, Journal of Mathematical Analysis and Applications, 125(2) (1987), 471–482.
12. M. Greguš; *Third Order Linear Differential Equations*, D. Reidel Publishing Company, Boston, 1987.
13. M. Greguš; *On the oscillatory behaviour of certain third order nonlinear differential equations*, Archivum Mathematicum, 28(3–4) (1992), 221–228.
14. M. Greguš and M. Greguš Jr; *Remark concerning oscillatory properties of solutions of a certain nonlinear equation of the third order*, Archivum Mathematicum, 28(1–2) (1992), 51–55.
15. M. Greguš and J. Vencko; *On oscillatory properties of solutions of a certain nonlinear third order differential equation*, Czechoslovak Mathematical Journal, 42(117) (1992), 675–684.
16. M. Hanan; *Oscillation criteria for third order linear differential equations*, Pacific Journal of Mathematics, 11 (1961), 919–944.
17. P. Hartman; *Ordinary Differential Equations*, Wiley, New York, 1964 and Birkhäuser, Boston, 1982.
18. P. Hartman and A. Wintner; *Linear differential and difference equations with monotone solutions*, American Journal of Mathematics, 75 (1953), 731–743.
19. J. W. Heidel; *Qualitative behaviour of solutions of a third order nonlinear differential equation*, Pacific Journal of Mathematics, 27 (1968), 507–526.
20. I. T. Kiguradze; *On asymptotic properties of solutions of third order linear differential equations with deviating arguments*, Archivum Mathematicum, 30(1) (1994), 59–72.
21. I. T. Kiguradze and T. A. Chanturia; *Asymptotic Properties of Solutions of Nonautonomous Ordinary Differential Equations*, Kluwer Academic Publishers, Dordrecht, 1993.
22. R. G. Koplatadze; *On oscillatory properties of solutions of functional differential equations*, Memoires on Differential Equations and Mathematical Physics, 3 (1994), 1–179.
23. T. Kura; *Nonoscillation criteria for nonlinear ordinary differential equations of the third order*, Nonlinear Analysis; Theory Methods and Applications, 8(4) (1984), 369–379.
24. A. C. Lazer; *The behaviour of solutions of the differential equation* $y''' + p(x)y' + q(x)y = 0$, Pacific Journal of Mathematics, 17 (1966), 435–466.
25. I. Mojsej and J. Ohriska; *Comparison theorems for noncanonical third order nonlinear differential equations*, Central European Journal of Mathematics, 5(1) (2007), 154–163.
26. I. Mojsej and A. Tartalova; *On nonoscillatory solutions tending to zero of third order nonlinear differential equations*, Tatra Mountains Mathematical Publications, 48 (2011), 135–143.
27. M. Naito and K.Yano; *Positive solutions of higher order ordinary differential equations with general nonlinearities*, Journal of Mathematical Analysis and Applications, 250(1) (2000), 27–48.
28. J. L. Nelson; *A stability theorem for a third order nonlinear differential equation*, Pacific Journal of Mathematics, 24 (1968), 341–344.
29. N. Parhi and P. Das; *Oscillation criteria for a class of nonlinear differential equations of third order*, Annales Polonici Mathematici, LVII3.3 (1992), 219–229.
30. N. Parhi and P. Das; *Oscillatory and asymptotic behaviour of a class of nonlinear differential equations of third order*, Acta Mathematica Sinica, 18 (1998), 95–106.

31. N. Parhi and S. Parhi; *Oscillation and nonoscillation theorems for nonhomogeneous third order differential equations*, Bulletin of the Institute of Mathematics Academia Sinica, 11 (1983), 125–139.
32. J. Rovder; *On monotone solution of the third-order differential equation*, Journal of Computational and Applied Mathematics, 66(1–2) (1996), 421–432.
33. Y. P. Singh; *Some oscillation theorems for third order non-linear differential equations*, Yokohama Mathematical Journal, 18 (1970), 77–86.
34. A. Škerlik; *Oscillation theorems for third order nonlinear differential equations*, Mathematica Slovaca, 42 (1992), 471–484.
35. A. Škerlik; *Criteria of Property A for third order superlinear differential equations*, Mathematica Slovaca, 43(2) (1993), 171–183.
36. A. Tiryaki and A. O. Çelebi; *Nonoscillation and asymptotic behaviour for third order nonlinear differential equations*, Czechoslovak Mathematical Journal, 48(123) (1998), 677–685.
37. A. Tiryaki and S. Yaman; *Oscillatory behaviour of a class of nonlinear differential equations of third order*, Acta Mathematica Scientica, 21(B) (2001), 182–188.
38. W. F. Trench; *Canonical forms and principal systems for general disconjugate equations*, Transactions of the American Mathematical Society, 189 (1974), 319–327.
39. P. Waltman; *Oscillation criteria for third order nonlinear differential equations*, Pacific Journal of Mathematics, 18(2) (1966), 385–389.

Chapter 5
Oscillation and Nonoscillation of Nonlinear Nonhomogeneous Differential Equations of Third Order

This chapter is concerned with the study of oscillatory and asymptotic behaviour of nonoscillatory solutions of nonhomogeneous third-order differential equations of the form

$$x''' + a(t)x'' + b(t)x' + c(t)x = f\big(t, x, x', x''\big), \tag{5.1}$$

where a, b and $c \in C([\sigma, \infty), R)$ and $f : [\sigma, \infty) \times R^3 \to R$, $\sigma \in R$. Equation (5.1) may be written as

$$\big(r(t)x''\big)' + q(t)x' + p(t)x = r(t)f\big(t, x, x', x''\big), \tag{5.2}$$

where $r(t) = \exp(\int_\sigma^t a(s)\,ds)$, $q(t) = r(t)b(t)$ and $p(t) = r(t)c(t)$. Since Eqs. (5.1) and (5.2) are equivalent, we freely refer to either of them according to our convenience.

Consider the nonlinear nonhomogeneous differential equation

$$\big[r_2(t)\big(\big(r_1(t)x'\big)' + q_1(t)x^\alpha\big)\big]' + q_2(t)\big(r_1(t)x'\big)^\beta = f_1(t), \tag{5.3}$$

where r_1, r_2, q_1 and $q_2 \in C([\sigma, \infty), R)$, $\sigma \in R$, $r_1(t) > 0$ and $r_2(t) > 0$, and each of $\alpha > 0$ and $\beta > 0$ is a quotient of odd integers. Suppose that $\int_\sigma^\infty \frac{dt}{r_1(t)} = \infty$. The Liouville transformation $s = R(t)$, $y(s) = x(t)$, where $R(t) = \int_\sigma^t \frac{du}{r_1(u)}$, transforms (5.3) into an equation of the type

$$\big(r(t)y'' + p(t)y^\alpha\big)' + q(t)\big(y'\big)^\beta = f(t). \tag{5.4}$$

Nonlinear nonhomogeneous differential equations of the type

$$\big(r(t)x''\big)' + q(t)\big(x'\big)^\beta + p(t)x^\alpha = f\big(t, x, x', x''\big) \tag{5.5}$$

is a particular case of (5.4). Indeed, we may write Eq. (5.5) as

S. Padhi, S. Pati, *Theory of Third-Order Differential Equations*,
DOI 10.1007/978-81-322-1614-8_5, © Springer India 2014

$$\left[r(t)x'' + \left(\int_\sigma^t p(s)\,ds\right)x^\alpha\right]' + q(t)\left(x'\right)^\beta$$

$$= f\left(t, x, x', x''\right) + \alpha\left(\int_\sigma^t p(s)\,ds\right)x^{\alpha-1}x'. \tag{5.6}$$

Thus, if we set $f(t, x, x', x'') = g(t) - \alpha(\int_\sigma^t p(s)\,ds)x^{\alpha-1}x'$, then (5.6) reduces to an equation of the form (5.4).

Equations of the form (5.4) have lot of importance in the real-world applications. For example, equations of the type

$$x''' + xx'' + \lambda\left[1 - \left(x'\right)^2\right] = 0 \tag{5.7}$$

arise in the boundary layer theory in fluid mechanics [3, p. 520]. The particular case of (5.7), $x''' + xx'' = 0$, is known as the *Blasius equation*. An important phenomenon of the Blasius equation is that it contains both positive and negative solutions. For example, the equation $x''' + xx'' = 0$ admits both positive and negative solutions. The nonhomogeneous equation $x''' + xx'' = \frac{8}{t^4}$, $t \geq 1$, admits the positive bounded solution $x(t) = \frac{4}{t}$ and the equation $x''' + xx'' = e^{-t} + e^{-2t}$, $t \geq 0$ has a bounded negative solution $x(t) = -e^{-t}$. Thus it is quite interesting to give some results on the nonoscillation of the nonhomogeneous Blasius equation

$$x''' + xx'' = f(t). \tag{5.8}$$

The asymptotic behaviour of solutions of (5.7) has been studied by Hartman [3]. Equation (5.7) with $\lambda = \frac{1}{2}$ is often called the *Homann differential equation*.

In this chapter, we shall give sufficient conditions for the oscillation and nonoscillation of the above considered equations.

5.1 Oscillatory and Nonoscillatory Behaviour of Solutions of $x''' + a(t)x'' + b(t)x' + c(t)x = f(t, x, x', x'')$

In this section, we shall obtain sufficient conditions for the oscillation of solutions of (5.1). Throughout the section, we assume that

$$f(t, x_1, x_2, x_3)x_1 < 0 \quad \text{for all } (t, x_1, x_2, x_3) \in [\sigma, \infty) \times R^3, \quad x_1 \neq 0. \tag{5.9}$$

Two different cases have been considered in this section. Denote

(H_1) $a(t) \geq 0$, $b(t) \leq 0$ and $c(t) > 0$

and

(H_2) $a(t) \geq 0$, $b(t) \geq 0$ and $c(t) > 0$.

Theorem 5.1.1 *Let (5.9) hold,* $a(t) \geq 0$ *and* $2c(t) - a(t)b(t) - b'(t) \geq 0$, *and let the linear differential equation (2.1) be oscillatory. Then any solution of (3.1) which has a zero is oscillatory.*

Theorem 5.1.2 *Let* $a(t) \geq 0$, $c(t) \geq 0$ *and* $c(t)-a(t)b(t)-b'(t) \geq 0$ *for* $t \geq \sigma$, *and suppose that there exist positive constants m and M such that* $0 < m \leq b(t) \leq M$ *for all* $t \geq \sigma$. *Suppose, furthermore, that there exists a function* $h \in C^1([\sigma, \infty), (0, \infty))$ *such that*

$$\lim_{t\to\infty} h(t) = 0, \qquad h'(t) \leq 0 \quad \text{on } [t_0, \infty) \subseteq [\sigma, \infty), \qquad \int_{t_0}^{\infty} h(t)\,dt = \infty \tag{5.10}$$

and

$$f(t, x_1, x_2, x_3)\operatorname{sgn} x_1 \leq -h(t)|x_1| \quad \text{for } (t, x_1, x_2, x_3) \in [\sigma, \infty) \times R^3.$$

Then every solution of (5.1) with one zero is oscillatory, and every solution of (5.1) defined on $[T, \infty) \subset [\sigma, \infty)$ *with no zeros in this interval satisfies*

$$x(t) \to 0, \qquad x'(t) \to 0 \quad \text{and} \quad x''(t) \to 0 \quad \text{as } t \to \infty, \tag{5.11}$$

that is, (5.1) has property A.

Theorem 5.1.3 *Let* $a(t) \geq 0$, $c(t) \geq 0$ *and* $c(t) - a(t)b(t) - b'(t) \geq 0$ *for* $t \geq \sigma$, *and let* $b(t) \geq m > 0$ *for* $t \geq \sigma$. *Assume that the function f satisfies*

$$r(t)f(t, x_1, x_2, x_3)\operatorname{sgn} x_1 \leq -h(t)b(t)|x_1| \quad \text{for } (t, x_1, x_2, x_3) \in [\sigma, \infty) \times R^3,$$

where the function h satisfies the property (5.10). Then every solution of (5.1) with one zero is oscillatory, and for every solution $x(t)$ *of (5.1) with no zero in* $[T, \infty) \subset [\sigma, \infty)$ *satisfies the property (5.11), that is, (5.1) has property A.*

Theorem 5.1.4 *Let* $a(t) \geq 0$, $b(t) \leq 0$ *and* $2c(t) - a(t)b(t) - b'(t) \geq 0$ *for all* $t \in [\sigma, \infty)$, *and let f satisfy*

$$f(t, x_1, x_2, x_3)\operatorname{sgn} x_1 \leq -h(t)|x_1|$$

for $(t, x_1, x_2, x_3) \in [\sigma, \infty) \times R^3$, *where the function h satisfies (5.10). If (2.1) is oscillatory, then every solution of (5.1) with one zero is oscillatory, and every nonoscillatory solution* $x(t)$ *of (5.1) with no zeros on* $[T, \infty)$ *for some* $T \in [\sigma, \infty)$, *satisfies*

$$x(t) \to 0 \quad \text{and} \quad x'(t) \to 0 \quad \text{as } t \to \infty,$$

and

$$x(t)x'(t) < 0 \quad \text{and} \quad x(t)x''(t) > 0$$

for all $t \geq T$.

Theorem 5.1.5 *Let $a(t) \geq 0$, $b(t) \leq 0$ for $t \geq \sigma$, and let $2c(t) - a(t)b(t) - b'(t) \geq h(t)$ for all $t \geq \sigma$, where h satisfies* (5.10), *and let f satisfies the property* (5.9). *If* (2.1) *is oscillatory on $[\sigma, \infty)$, then every solution of* (5.1) *with one zero is oscillatory, and every nonoscillatory solution $x(t)$ of* (5.1) *with no zeros on $[T, \infty)$, for some $T \in [\sigma, \infty)$, satisfies*

$$x(t) \to 0 \quad \text{and} \quad x'(t) \to 0 \quad \text{as } t \to \infty$$

and

$$x(t)x'(t) < 0 \quad \text{and} \quad x(t)x''(t) > 0$$

for all $t \geq T$.

Remark 5.1.1 If, in addition, to the hypothesis of Theorem 5.1.4 or Theorem 5.1.5, we ask that $c(t) \geq 0$ for $t \geq \sigma$, then any solution $x(t)$ of (5.1) with no zeros on $[T, \infty) \subset [\sigma, \infty)$ has the property that $x''(t) \to 0$ as $t \to \infty$ as well.

Lemma 5.1.1 *Let* (H_1) *and* (5.9) *hold. If $x(t)$ is a nonoscillatory solution of* (5.1), *then there exists a $T \geq \sigma$ such that* (i) $x(t)x'(t) < 0$ *or* (ii) $x(t)x'(t) > 0$ *holds for $t \geq T$. Further, if* (i) *holds, then*

$$\operatorname{sgn} x(t) = \operatorname{sgn} x''(t) \neq \operatorname{sgn} x'(t) = \operatorname{sgn} x'''(t) \quad \text{for } t \geq T_1 \geq T$$

and

$$\lim_{t\to\infty} x'(t) = \lim_{t\to\infty} x''(t) = 0 \quad \text{and} \quad \lim_{t\to\infty} x(t) = \lambda, \quad \lambda \neq \pm\infty. \tag{5.12}$$

Proof Given that $x(t)$ is a nonoscillatory solution of (5.1). Suppose that $x(t) > 0$ for $t \geq t_0 \geq \sigma$. Let t_1 and t_2 be two consecutive zeros of $x'(t)$ such that $x''(t_1) \leq 0$, $x''(t_2) \geq 0$ and $x'(t) < 0$ for $t \in (t_1, t_2)$. Integrating (5.2) from t_1 to t_2, we obtain

$$\begin{aligned} 0 &< r(t_2)x''(t_2) - r(t_1)x''(t_1) + \int_{t_1}^{t_2} q(t)x'(t)\,dt + \int_{t_1}^{t_2} p(t)x(t)\,dt \\ &= \int_{t_1}^{t_2} r(t) f\big(t, x(t), x'(t), x''(t)\big)\,dt \\ &< 0, \end{aligned}$$

a contradiction. Hence there exists a $T \geq t_0$ such that either $x'(t) > 0$ or $x'(t) < 0$ for $t \geq T$. Suppose that $x'(t) < 0$ for $t \geq T$. Then from (5.2), we obtain $(r(t)x''(t))' < 0$ for $t \geq T$ and hence $x''(t) > 0$ or < 0 for $t \geq T_1 \geq T$. Clearly, $x''(t) < 0$ for $t \geq T_1$ implies that $x(t) < 0$ for large t, a contradiction. Thus $x''(t) > 0$ for $t \geq T_1$. Then (5.1) yields $x'''(t) < 0$ for $t \geq T_1$. It is clear that (5.12) holds.

The case $x(t) < 0$ for $t \geq t_0 \geq \sigma$ may be treated similarly. The proof is complete. □

Lemma 5.1.2 *Let* (H_1) *and* (5.9) *hold. If $x(t)$ is any solution of* (5.1), *satisfying $x(t_0) \geq 0$, $x'(t_0) \leq 0$ and $x''(t_0) > 0$, $t_0 > \sigma$, then $x(t) > 0$, $x'(t) < 0$ and $x''(t) > 0$ for $t \in [\sigma, t_0)$.*

Proof From the continuity of $x''(t)$, it follows that there exists a $\delta > 0$ such that $x''(t) > 0$ for $t \in (t_0 - \delta, t_0]$. Hence $x'(t) < 0$ and $x(t) > 0$ for $t \in (t_0 - \delta, t_0)$. We claim that $x''(t) > 0$ for $t \in [\sigma, t_0]$. If not, then there exists a $t_1 < t_0$ such that $x''(t_1) = 0$ and $x''(t) > 0$ for $t \in (t_1, t_0]$. Consequently, $x'(t) < 0$ and $x(t) > 0$ for $t \in [t_1, t_0)$. Integrating (5.2) from t_1 to t_0, we obtain

$$
\begin{aligned}
0 &< r(t_0)x''(t_0) - r(t_1)x''(t_1) + \int_{t_1}^{t_0} q(t)x'(t)\,dt + \int_{t_1}^{t_0} p(t)x(t)\,dt \\
&= \int_{t_1}^{t_0} r(t) f\bigl(t, x(t), x'(t), x''(t)\bigr)\,dt \\
&< 0,
\end{aligned}
$$

a contradiction. Hence our claim holds, that is, $x''(t) > 0$ for $t \in [\sigma, t_0]$. Consequently, $x'(t) < 0$ and $x(t) > 0$ for $t \in [\sigma, t_0)$. This completes the proof of the lemma. □

Theorem 5.1.6 *Let* (H_1) *and* (5.9) *hold. If* (2.1) *is oscillatory, then* (5.1) *is oscillatory.*

Proof Since (2.1) is of type C_I and oscillatory, by Lemma 1.5.10, any solution of (2.1) which has a zero is oscillatory.

Let $\{u_1, u_2, u_3\}$ be a solution basis for (2.1) with

$$
\begin{aligned}
&u_1(\alpha) = 1, \qquad && u_1'(\alpha) = 0, \qquad && u_1''(\alpha) = 0, \\
&u_2(\alpha) = 0, && u_2'(\alpha) = 1, && u_2''(\alpha) = 0, \\
&u_3(\alpha) = 0, && u_3'(\alpha) = 0, && u_3''(\alpha) = 1.
\end{aligned}
$$

Hence

$$
W(t) = W(u_1, u_2, u_3)(t) = \begin{vmatrix} u_1(t) & u_2(t) & u_3(t) \\ u_1'(t) & u_2'(t) & u_3'(t) \\ u_1''(t) & u_2''(t) & u_3''(t) \end{vmatrix} \equiv 1
$$

and $u_3(t)$ is oscillatory. By the method of "variation of parameters"

$$
x(t) = \int_\alpha^t W_1(t, s) f\bigl(s, x(s), x'(s), x''(s)\bigr)\,ds \tag{5.13}
$$

is a solution of (5.1), where

$$
W_1(t, s) = \begin{vmatrix} u_1(t) & u_2(t) & u_3(t) \\ u_1(s) & u_2(s) & u_3(s) \\ u_1'(s) & u_2'(s) & u_3'(s) \end{vmatrix}, \qquad \alpha \leq s < t.
$$

Since, for a fixed s, $W_1(t,s)$ is a solution of (2.1) with $W_1(s,s)=0$, $W_1'(s,s)=0$ and $W_1''(s,s)\equiv W(s)\equiv 1$, we have $x(\alpha)=0$, $x'(\alpha)=0$, $x''(\alpha)=0$ and $x'''(\alpha)=f(\alpha,x(\alpha),x'(\alpha),x''(\alpha))$.

To complete the proof of the theorem, it is enough to show that $x(t)$ is oscillatory. Since $u_3''(\alpha)=1>0$, from the continuity of $u_3''(t)$, it follows that there exists a $\delta>0$ such that $u_3(t)>0$ for $t\in(\alpha,\alpha+\delta)$. Let t_1 be the next zero of $u_3(t)$ to the right of α. This is possible since $u_3(t)$ is oscillatory. Clearly, $t_1\geq\alpha+\delta$. Since $W_1(t,s)$, $\alpha\leq s<t_1$, is a solution of (2.1) with a double zero at s, from Lemma 1.5.9, it follows that $W_1(t,s)\neq 0$ for $s<t<t_1$. But $W_1''(s,s)=1>0$ implies that $W_1(t,s)>0$ for $s<t<t_1$. We claim that $x(t)$ has a zero in (α,t_1). If possible, suppose that $x(t)>0$ for $t\in(\alpha,t_1)$. Let $\alpha<t_2<t_1$. Then from (5.13), we obtain, by using (5.9),

$$0<x(t_2)=\int_\alpha^{t_2} W_1(t_2,s)f\bigl(s,x(s),x'(s),x''(s)\bigr)\,ds<0,$$

a contradiction. A similar contradiction is obtained, if we assume $x(t)<0$ for $t\in(\alpha,t_1)$. Hence our claim holds, that is, $x(t)$ has a zero in (α,t_1). Let t_3 be the next zero of $x(t)$ to the right of α in (α,t_1). Let t_4 be next zero of $u_3(t)$ to the right of t_1. Clearly, $u_3(t)<0$ for $t\in(t_1,t_4)$, because $u_3(t)>0$ for $t\in(t_1,t_4)$ implies that $u_3(t_4)=0=u_3'(t_4)$. Since (2.1) is of type C_I, then $u_3(\alpha)=0$ is not possible. Then, by Lemma 1.5.9, $W_1(t,s)\neq 0$ for $s<t<t_4$, because $W_1(t,s)$ has a double zero at $s\in[t_1,t_4)$. Since $W_1''(s,s)=1>0$, we have $W_1(t,s)>0$ for $s<t<t_4$. We claim that $x(t)$ has a zero in (t_1,t_4). If $x(t)>0$ for $t\in(t_1,t_4)$, then from (5.13) we obtain, by using (5.9),

$$0<x(t_5)=\int_\alpha^{t_5} W_1(t_5,s)f\bigl(s,x(s),x'(s),x''(s)\bigr)\,ds<0,$$

where $t_5\in(t_1,t_4)$, a contradiction. A similar contradiction is obtained for $x(t)<0$ for $t\in(t_1,t_4)$. Hence $x(t)$ has a zero in (t_1,t_4). That is, $x(t)$ has a zero to the right of t_3.

Thus, continuing as above, we see that $x(t)$ has infinitely large zeros. Consequently, $x(t)$ is oscillatory. This completes the proof of the theorem. □

Remark 5.1.2 Theorem 5.1.6 improves Theorem 5.1.1. In Theorem 5.1.1, the condition $2c(t)-a(t)b(t)-b'(t)\geq 0$ is assumed to show that Eq. (2.1) is C_I. In Theorem 5.1.6, it follows directly from the assumptions $a(t)\geq 0$, $b(t)\leq 0$ and $c(t)>0$ that Eq. (2.1) is of type C_I (see Lemma 2.1.1).

If $y(t)$ is a solution of (5.1) such that $y(t)\neq 0$ for $t\geq t_y\geq\sigma$, then $y(t)$ is a solution of the linear differential equation

$$x'''+a(t)x''+b(t)x'+\left(c(t)-\frac{f(t,y(t),y'(t),y''(t))}{y(t)}\right)x=0 \tag{5.14}$$

for $t\geq t_y$.

Theorem 5.1.7 *Let* (H_1) *and* (5.9) *hold and* $(2c(t)-a(t)b(t)-b'(t))r(t) \geq d > 0$. *If* (2.1) *has property* A, *then* (5.1) *has property A and has an oscillatory solution.*

Proof Suppose that (2.1) has property A. Then (2.1) is oscillatory by Lemma 1.5.3. Consequently, by Theorem 5.1.6, Eq. (5.1) is oscillatory. To complete the proof of the theorem, it remains to show that every nonoscillatory solution of (5.1), along with their first- and second-order derivatives, tends to zero as $t \to \infty$. Let $x(t)$ be a nonoscillatory solution of (5.1) such that $x(t) \neq 0$ for $t \geq t_0 \geq \sigma$. Then $x(t)$ is a solution of (5.14). Since (5.9) holds,

$$\begin{aligned} r(t)&\left[2\left(c(t) - \frac{f(t, x(t), x'(t), x''(t))}{x(t)}\right) - a(t)b(t) - b'(t)\right] \\ &\geq r(t)\big(2c(t) - a(t)b(t) - b'(t)\big) \\ &\geq d \\ &> 0 \end{aligned}$$

implies that all the conditions of Theorem 2.1.14 are satisfied. Hence $x(t) \to 0$, $x'(t) \to 0$ and $x''(t) \to 0$ as $t \to \infty$. Consequently, (5.1) has property A. Thus the theorem is proved. □

Theorem 5.1.8 *Let* (H_1) *hold,* $2c(t) - a(t)b(t) - b'(t) \geq 0$ *and* f *satisfy*

$$r(t) f(t, x_1, x_2, x_3) \operatorname{sgn} x_1 \leq -d|x_1|, \quad d > 0. \tag{5.15}$$

If (2.1) *has property* A, *then* (5.1) *has property A and has an oscillatory solution.*

Proof Since (5.15) holds, (5.9) holds and

$$\begin{aligned} r(t)&\left[2\left(c(t) - \frac{f(t, x_1, x_2, x_3)}{x_1}\right) - a(t)b(t) - b'(t)\right] \\ &\geq -2r(t)\frac{f(t, x_1, x_2, x_3)}{x_1} \\ &\geq 2d \\ &> d. \end{aligned}$$

Now, the rest of the proof is same as in the proof of Theorem 5.1.7. □

Theorem 5.1.9 *Let* (H_2) *and* (5.9) *hold, and* $2c(t) - a(t)b(t) - b'(t) \geq 0$. *If* (2.1) *is oscillatory, then* (5.1) *is oscillatory.*

The proof of Theorem 5.1.9 is same as the proof of Theorem 5.1.6. However, here the condition $2c(t) - a(t)b(t) - b'(t) \geq 0$ is assumed to show that (2.1) is C_I.

Theorem 5.1.10 *Let* (H_2) *and* (5.9) *hold,* $c(t) \geq d > 0$, $\int_\sigma^\infty a(t)\,dt = \infty$ *and* $c(t) - a(t)b(t) - b'(t) \geq 0$. *If* (2.1) *is oscillatory, then* (5.1) *is oscillatory and has property* A.

Proof Clearly $2c(t) - a(t)b(t) - b'(t) \geq 0$. By Theorem 5.1.9, (5.1) is oscillatory. Let $x(t)$ be a nonoscillatory solution of (5.1) such that $x(t) \neq 0$ for $t \geq t_0 \geq \sigma$. Then $x(t)$ is a solution of (5.14) for $t \geq t_0$. Since (5.9) holds,

$$c(t) - \frac{f(t, x(t), x'(t), x''(t))}{x(t)} - a(t)b(t) - b'(t) \geq -\frac{f(t, x(t), x'(t), x''(t))}{x(t)} > 0.$$

Hence all the conditions of Theorem 2.5.11 are satisfied. Thus, by Theorem 2.5.11, every nonoscillatory solution of (5.14), along with their first and second derivatives tends to zero, as $t \to \infty$. In particular, $x(t) \to 0$, $x'(t) \to 0$ and $x''(t) \to 0$, as $t \to \infty$. Consequently, (5.1) has property A. This completes the proof of the theorem. □

Consider the third-order differential equation

$$x''' + a(t)x'' + b(t)x' + c(t)f\left(x, x', x''\right) = 0, \tag{5.16}$$

where $a \in C^1([\sigma, \infty), [0, \infty))$, $b \in C([\sigma, \infty), [0, \infty))$, $r \in L_{\text{loc}}([\sigma, \infty), [0, \infty))$, $f \in C(R^3, R)$, $c(t) \geq 0$, $\sigma \geq 0$ and

$$f(x_1, x_2, x_3)x_1 \geq 0 \quad \text{on } R^3. \tag{5.17}$$

A solution x defined on $[\sigma, \infty)$ is called singular, if there exists a $T_x > \sigma$, such that $x \equiv 0$ on $[T_x, \infty)$ and x is not trivial in any left neighbourhood of T_x.

One of significant problems is to study the existence of (proper) oscillatory solutions. Bartušek [1] gave sufficient conditions, under which

$$x(t) \neq 0, \quad t \in [\sigma, T_x), \tag{5.18}$$

holds for a singular solution $x(t)$ of (5.16).

The following example shows that the singular solution may exist, for which (5.18) does not hold.

Example 5.1.1 Let

$$x(t) = \begin{cases} t\{-15(t-1)^3 + 11(t-1)^2 - 5(t-1) + 1\} & \text{for } t \in [0, 1], \\ (2-t)^4 & \text{for } t \in (1, 2], \\ 0 & \text{for } t > 2, \end{cases}$$

$$c(t) \equiv 24, \qquad b(t) \equiv \frac{9}{2}\left[3 + \left(\frac{15}{4}\right)^{1/4}\right] \quad \text{on } [\sigma, \infty)$$

and

$$a(t)=\begin{cases}(-x'''-b(t)x'-c(t)x^{1/4})(x'')^{-1} & \text{for } t\in[0,1],\ t\neq\frac{2}{3},\\ \frac{15}{4}+(\frac{4}{15})^{\frac{1}{4}} & \text{for } t=\frac{2}{3},\\ \frac{b(t)}{3}(2-t) & \text{for } t\in\ [1,2],\\ 0 & \text{for } t>2.\end{cases}$$

Then $x(0)=0$, $x(t)>0$ on $(0,2)$ and $x(t)$ is the singular solution of

$$x'''+a(t)x''+b(t)x'+c(t)|x|^{\lambda}\operatorname{sgn} x=0$$

with $\lambda=\frac{1}{4}$. Thus (5.18) is not valid for $t\in[0,2)$.

The following theorem is often used and gives the sufficient condition under which singular solutions do not exist at all.

Theorem 5.1.11 *Let $\epsilon>0$ be such that*

$$\left|f(x_1,x_2,x_3)\right|\le\sum_{i=1}^{3}|x_i| \quad \text{on } [-\epsilon,\epsilon]^3. \tag{5.19}$$

Then there exists no singular solution of (5.16).

It is evident that the assumption of Theorem 5.1.11 is not valid for Emden–Fowler equation:

$$x'''+a(t)x''+b(t)x'+c(t)|x|^{\alpha}\operatorname{sgn} x=0, \qquad \lambda\in(0,1). \tag{5.20}$$

Moreover, singular solutions may exist for $a(t)\equiv b(t)\equiv 0$.

Lemma 5.1.3 *Let $I\subset[0,\infty)$, and let there exist a positive solution $h(t)$ of*

$$h''+a(t)h'+b(t)h=0 \tag{5.21}$$

on I. Then (5.16) *can be expressed on I in an equivalent form*

$$x^{[3]}+f_1\big(t,x^{[0]},x^{[1]},x^{[2]}\big)=0, \tag{5.22}$$

where $R(t)=e^{\int_0^t a(s)\,ds}$,

$$\left.\begin{aligned} x^{[0]}&=x,\\ x^{[1]}&=\frac{x'}{h(t)},\\ x^{[2]}&=R(t)h^2(t)\big(x^{[1]}\big)'=R(t)\big(x''h(t)-x'h'(t)\big),\\ x^{[3]}&=\big(x^{[2]}\big)', \end{aligned}\right\} \tag{5.23}$$

and

$$f_1(t,x_1,x_2,x_3)=c(t)h(t)R(t)f\left(x_1,h(t)x_2,\frac{x_3}{R(t)h(t)}+h'(t)x_2\right).$$

Moreover, if $t_0\in I$ *and* $x(t)$ *is a solution of Eq.* (5.22) (*and thus of Eq.* (5.16) *too*) *such that*

$$x(t_0)x^{[1]}(t_0)<0,\qquad x(t_0)x^{[2]}(t_0)>0, \tag{5.24}$$

then

$$x(t)x^{[1]}(t)<0,\qquad x(t)x^{[2]}(t)>0\quad \textit{for } t\in I,\ t\le t_0. \tag{5.25}$$

Proof The transformation of (5.16) into (5.22) can be obtained by direct computation. Further (5.17), (5.22) and (5.23) yield

$$x^{[3]}(t)x(t)\le 0\quad \text{almost everywhere on } I \tag{5.26}$$

and

$$x^{[i]}(t)\ge 0\ (\le 0)\quad \text{on } I_1\subset I,\ i\in\{1,2,3\}, \tag{5.27}$$

implies $x^{[i-1]}$ is nondecreasing (non-increasing) on I_1. Let (5.24) be valid with $x(t_0)>0$. For simplicity, put $I_2=\{t:t\in I, t<t_0\}$. We shall prove that

$$\prod_{i=0}^{2}\left|x^{[i]}(t)\right|>0\quad \text{on } I_2. \tag{5.28}$$

Thus, suppose indirectly that $\tau\in I_2$ is the maximal number, such that $\prod_{i=0}^{2}|x^{[i]}(\tau)|=0$. From (5.24), (5.26) and (5.27),

$$x(t)>0,\qquad x^{[1]}(t)<0,\qquad x^{[2]}(t)>0,\qquad x^{[3]}(t)\le 0\quad \text{on } (\tau,t_0]$$

and, evidently, $x^{[2]}(\tau)=0$. Thus

$$x^{[2]}(t_0)=x^{[2]}(t_0)-x^{[2]}(\tau)=\int_\tau^{t_0}x^{[3]}(s)\,ds\le 0.$$

This contradicts (5.24), and $x(t_0)>0$ proves that (5.28) holds. The conclusion (5.25) follows from Eqs. (5.24), (5.26)–(5.28). □

Theorem 5.1.12 *Let* $x(t)$ *be a singular solution of* (5.16), *and let* T_x *be the number from its definition. Then there exists a left-open neighbourhood* I *of* T_x *such that*

$$x(t)x'(t)<0\quad \textit{and}\quad x(t)x''(t)>0\quad \textit{on } I. \tag{5.29}$$

Proof Let $h(t)$ be a solution of Eq. (5.21) such that $h(T_x)=1$, $h'(T_x)=0$, and let J be a left-open neighbourhood of T_x on which $h(t)>0$ holds. Then according to Lemma 5.1.3, (5.16) can be expressed in the equivalent form (5.22). Further

$$x^{(i)}(T_x)=0, \quad i=0,1,2, \tag{5.30}$$

and we prove that

$$x(t)x^{[1]}(t)<0, \qquad x(t)x^{[2]}(t)>0 \quad \text{on } J. \tag{5.31}$$

First suppose that $x(t)\not\equiv 0$, say $x(t)>0$, in a left neighbourhood J_1 of T_x, $J_1\subset J$. Then (5.26) yields $x^{[3]}(t)\le 0$ almost everywhere on J_1. Moreover, $x^{[3]}(t)\not\equiv 0$ almost everywhere in any left neighbourhood J_2 of T_x. Otherwise, (5.23) and (5.30) yield $x(t)\equiv 0$ in J_2, that contradicts $x(t)\not\equiv 0$. From

$$x^{[2]}(t)=-\int_t^{T_x} x^{[3]}(s)\,ds>0 \quad \text{for } t\in J_1$$

and, by using (5.23) and (5.30), $x^{[1]}(t)<0$, $x(t)>0$ on J_1. The validity of (5.31) follows from Lemma 5.1.3.

The last possible case consists in the existence of an increasing sequence $\{\tau_k\}_1^\infty$ such that

$$x(t)\not\equiv 0 \quad \text{in any left neighbourhood of } T_x, \tag{5.32}$$

and $\tau_k\in J$, $\lim_{k\to\infty}\tau_k=T_x$ and $x(\tau_k)=0$ for $k=1,2,\dots$.

Let $J_2=[\alpha,T_x]$, $\alpha\in J$ be such an interval that

$$W=\frac{3}{2}\min_{s\in\bar{J}} h(s)-\frac{1}{2}\max_{s\in\bar{J}}\left|\left(R(s)h^3(s)\right)'\right|A(T_x)>0, \tag{5.33}$$

where $A(t)=\int_\alpha^t \frac{d\tau}{R(\tau)h^2(\tau)}$ and $\bar{J}$ is the closure of J. If

$$F(t)=-A(t)x^{[2]}(t)x(t)+\frac{1}{2}A(t)R(t)h^3(t)\left(x^{[1]}(t)\right)^2+x(t)x^{[1]}(t), \qquad t\in J_2,$$

then by using (5.23), (5.26) and (5.33), we have

$$\begin{aligned} F'(t)&=-A(t)x^{[3]}(t)x(t)+\left[\frac{3}{2}h(t)+\frac{1}{2}A(t)\left(R(t)h^3(t)\right)'\right]\left(x^{[1]}(t)\right)^2\\ &\ge W\left(x^{[1]}(t)\right)^2\\ &\ge 0 \quad \text{on } J_2. \end{aligned} \tag{5.34}$$

Thus F is nondecreasing and $F(T_x)=0$ yields $F(t)\le 0$ for $t\in J_2$. On the other hand, if $\tau_k\in J_2$, then $F(\tau_k)\ge 0$ and thus $F(t)\equiv 0$ on $[\tau_k,T_x]$. From this, and (5.34), we have

$$0=\int_{\tau_k}^{T_x} F'(s)\,ds\ge W\int_{\tau_k}^{T_x}\left(x^{[1]}(s)\right)^2 ds\ge 0.$$

Thus, by using (5.33), we obtain $x'(t) \equiv 0$ on $[\tau_k, T_x]$, which contradicts (5.30) and (5.32). This contradiction proves that (5.31) is valid, and applying (5.23), we obtain

$$x(t)x'(t) < 0, \quad t \in J. \tag{5.35}$$

Further suppose that there exists a sequence $\{\tau_k\}_1^\infty$ such that $\tau_k < T_x$, $k = 1, 2, \ldots, \lim_{k\to\infty} \tau_k = T_x$, $x''(\tau_k) = 0$.

Let $z(t)$ be the solution of

$$z'' - \big(za(t)\big)' + b(t)z = 0, \qquad z(T_x) = 1, \qquad z'(T_x) < a(T_x),$$

and $v(t)$ be the solution of

$$v'' - \big(va(t)\big)' + b(t)v = 3z' - 2a(t)z, \qquad v(T_x) = 1, \qquad v'(T_x) < 1 + a(T_x).$$

Let $J_3 \subset J$ be a left-open neighbourhood of T_x such that

$$\left.\begin{aligned} &z(t) > 0, && z'(t) - a(t)z(t) < 0, \\ &v(t) > 0, && z(t) - v'(t) + a(t)v(t) > 0, \end{aligned}\right\} \tag{5.36}$$

for $t \in J_3$. Define F on $J_3 \cup \{T_x\}$ by

$$\begin{aligned} F(t) = &-z(t)x''(t)x(t) + \big(z'(t) - a(t)z(t)\big)x'(t)x(t) \\ &+ \frac{1}{2}\big(z(t) - v'(t) + a(t)v(t)\big)\big(x'(t)\big)^2 + v(t)x''(t)x'(t). \end{aligned}$$

Then, by using (5.35) and (5.17)

$$\begin{aligned} F'(t) &= -v(t)c(t)x'(t)f\big(x(t), x'(t), x''(t)\big) \\ &\quad + z(t)c(t)x(t)f\big(x(t), x'(t), x''(t)\big) + v(t)\big(x''(t)\big)^2 \\ &\geq 0 \quad \text{on } J_3. \end{aligned}$$

From this, F is nondecreasing and by using $F(T_x) = 0$, we have $F(t) \leq 0$, $t \in J_3$. Let $\tau \in J_3$ be a zero of $x''(t)$. Then

$$F(\tau) = \big(z'(t) - a(t)z(t)\big)x'(t)x(t) + \frac{1}{2}\big(z(t) - v'(t) + a(t)v(t)\big)\big(x'(t)\big)^2\bigg|_{t=\tau} \leq 0$$

contradicts (5.35) and (5.36). Thus $x''(t) \neq 0$ on J_3, and it follows from (5.36) and $x^{(i)}(T_x) = 0$, $i = 0, 1, 2$, that $x(t)x''(t) > 0$, $t \in J_3$, holds. Now, set $I = J_3$ to get the desired result. □

Remark 5.1.3 For a singular solution $x(t)$ of (5.16), (5.31) is valid in a left neighbourhood of T_x. Further, the singular solution $x(t)$ of (5.16), fulfilling (5.29) in a left-open neighbourhood of T_x, is called a Kneser singular solution. Thus every singular solution is a Kneser singular solution.

Theorem 5.1.13 *Let $b(t) \le 0$ for $t \ge \sigma$. Then for a singular solution $x(t)$ of* (5.16), (5.18) *is valid.*

Proof Let $x(t)$ be a singular solution of (5.16). Then, there exists a positive solution $h(t) > 0$ of Eq. (5.21) on $[\sigma, \infty)$, and the assumptions of Lemma 5.1.3 are fulfilled with $I = [\sigma, \infty)$. According to Remark 5.1.3, the inequalities in (5.24) are valid for t_0 lying in a left neighbourhood of T_x. Hence (5.25) yields (5.18). □

Theorem 5.1.14 *Let $C \in R$, $a \in C^2([\sigma, \infty), [0, \infty))$, $b \in C^1([\sigma, \infty), [0, \infty))$,*

$$\big(1 - C(t)\big)\big\{a''(t) + 3C(t)a(t)a'(t) + C^2(t)a^3(t)\big\} - b'(t) - C(t)a(t)b(t) > 0$$

and either of the following holds:

$$\left.\begin{array}{l} C(t) > \dfrac{2}{3} \quad \text{and} \quad a(t) \ge 0, \\ \text{or} \quad C(t) < \dfrac{2}{3} \quad \text{and} \quad a(t) \le 0, \\ \text{or} \quad C(t) = \dfrac{2}{3}. \end{array}\right\} \tag{5.37}$$

Then for a singular solution $x(t)$ of (5.16), *the inequality* (5.18) *holds.*

As a consequence, we obtain the following: Let $b \in C^1([\sigma, \infty), R)$ and let one of the following assumptions hold:

(i) $a \in C^2([\sigma, \infty), R)$, $a(t) \le 0$, $a''(t) - b'(t) > 0$ for $t \ge \sigma$;
(ii) $a(t) \ge 0$, $b'(t) + a(t)b(t) < 0$ for $t \ge \sigma$;
(iii) $a \in C^2([\sigma, \infty), R)$, $9a''(t) + 18a(t)a'(t) + 4a^3(t) - 27b'(t) - 18a(t)b(t) > 0$ for $t \ge \sigma$.

Then for a singular solution $x(t)$ of (5.16), the relation (5.18) holds.

Theorem 5.1.15 *Let $\alpha > 0$ exists such that*

$$\big|f(x_1, x_2, x_3)\big| \ge \alpha|x_1| \quad \text{on } R^3. \tag{5.38}$$

Further, let one of the following assumptions be valid on $[\sigma, \infty)$;

(a) $a(t) \ge 0$, $b(t) < \sqrt{\alpha c(t)a(t)}$;
(b) $a(t) \ge 0$, $b \in C^1([\sigma, \infty), [0, \infty))$, $2\alpha c(t) - b'(t) - a(t)b(t) > 0$;
(c) $a(t) \le 0$, $a \in C^2([\sigma, \infty), [0, \infty))$, $b \in C^1([\sigma, \infty), [0, \infty))$, $\alpha c(t) + a''(t) - b'(t) > 0$.

Then for a singular solution of (5.16), *the inequality* (5.18) *holds.*

Proof Let $x(t)$ be a singular solution, $T_x \ge \sigma$.

(a) Suppose, on the contrary, that $\tau \in [0, T_x)$ exists such that $x(\tau) = 0$. Theorem 5.1.12 yields, for simplicity,

$$x(t) > 0, \qquad x'(t) < 0, \qquad x''(t) > 0 \tag{5.39}$$

in a left neighbourhood of T_x, and thus $\tau_1 \in (\tau, T_x)$ must exist such that $x'''(\tau_1) = 0$, and (5.39) is valid on $[\tau_1, T_x)$. In both of the following cases:

- x''' has the maximal zero on $[\tau, T_x)$,
- there exists an increasing sequence of zeros of $x'''(t)$ tending to T_x,

it is possible to choose τ_1 such that

$$\max_{\tau_1 \le t \le T_x} x''(t) = x''(\tau_1). \tag{5.40}$$

From this, we obtain

$$b(t)\left|x'(t)\right| = a(t)x''(t) + c(t)f\left(x(t), x'(t), x''(t)\right) \ge a(t)x''(t) + c(t)\alpha x(t)|_{t=\tau_1}. \tag{5.41}$$

On the other hand, by using $x(T_x) = x'(T_x) = 0$, (5.39) and (5.40), we have

$$\left(x'(\tau_1)\right)^2 = 2\int_{\tau_1}^{T_x} \left|x'(s)\right| x''(s)\, ds \le 2x''(\tau_1)x(\tau_1).$$

Thus, together with (5.41),

$$b(t)\left(2x''(t)x(t)\right)^{1/2} \ge a(t)x''(t) + c(t)\alpha x(t)|_{t=\tau_1}$$

and

$$\left(\sqrt{a(t)x''(t)} - \sqrt{\alpha c(t)x(t)}\right)^2 + \left(\sqrt{2\alpha a(t)c(t)} - b(t)\right)\sqrt{2x(t)x''(t)} \le 0\Big|_{t=\tau_1}$$

contradict the assumption $b(t) < \alpha a(t)c(t)$ on $[\sigma, \infty)$.

(b), (c) The statement can be proved similarly as in Theorem 5.1.14 for $C = 1$ or $C = 0$, respectively. Instead of (5.37), a more precise estimation of F' must be used, by using (5.37) and (5.38):

$$\begin{aligned} F'(t) = e^{C(t)\int_0^t a(s)\,ds}\Big\{x^2(t)\big[-b'(t) - C(t)a(t)b(t) + 3a(t)a'(t)\left(C(t) - C^2(t)\right) \\ + a^3(t)\left(C^2(t) - C^3(t)\right) + a''(t)\left(1 - C(t)\right) + 2\alpha c(t)\big]\Big\} \\ + e^{C(t)\int_0^t a(s)\,ds}\Big\{\left(x'(t)\right)^2\left[3C(t) - 2\right]a(t)\Big\}. \end{aligned}$$

□

5.2 Nonoscillation of Solutions of $(r_2(t)(r_1(t)x')')' + q(t)(x')^\beta + p(t)x^\alpha = f(t)$

This section is concerned with the oscillation and nonoscillation of the third-order nonlinear nonhomogeneous differential equation

$$\left(r(t)x''\right)' + q(t)\left(x'\right)^\beta + p(t)x^\alpha = f(t), \tag{5.42}$$

where r, p, q and f are real-valued continuous functions, such that $r(t) > 0$, $p(t) \leq 0$, $q(t) \leq 0$, $f(t) \geq 0$, for $t \geq \sigma$, $\alpha > 0$ is the ratio of odd integers and $\beta > 0$. We restrict our considerations to those real solutions of (5.42) which exist on the half line $[T, \infty)$, where $T \geq \sigma$ depends on the particular solution and are nontrivial in any neighbourhood of infinity.

Theorem 5.2.1 *If $p(t) \equiv 0$, then all solutions of* (5.42) *are nonoscillatory.*

Proof Let $x(t)$ be a solution of (5.42) on $[T, \infty)$, $T \geq \sigma$. If possible, let $x(t)$ be nonnegative z-type. Let a and b $(T \leq a < b)$ be consecutive double zeros of $x(t)$. So, there exists a $c \in (a, b)$ such that $x'(c) = 0$ and $x'(t) > 0$ for $t \in (a, c)$. Multiplying (5.42) throughout by $x'(t)$, we get

$$\left[r(t)x''(t)x'(t)\right]' = -q(t)\left(x'(t)\right)^{\beta+1} + f(t)x'(t) + r(t)\left(x''(t)\right)^2. \tag{5.43}$$

Integration of (5.43) from a to c yields

$$0 = \int_a^c \left[-q(t)\left(x'(t)\right)^{\beta+1} + f(t)x'(t) + r(t)\left(x''(t)\right)^2\right] dt > 0,$$

a contradiction. Similarly, it can be shown that $x(t)$ cannot be a nonpositive z-type solution.

Suppose that $x(t)$ is oscillatory. Let b_1, a and b $(T \leq b_1 < a < b)$ be any three consecutive zeros of $x(t)$ such that $x'(b_1) \leq 0$, $x'(a) \geq 0$, $x'(b) \leq 0$, $x(t) < 0$ for $t \in (b_1, a)$ and $x(t) > 0$ for $t \in (a, b)$. So, there exist points $c_1 \in (b_1, a)$ and $c \in (a, b)$ such that $x'(c_1) = 0$, $x'(c) = 0$ and $x'(t) > 0$ for $t \in (c_1, c)$. We consider two cases, viz. (i) $x''(a) \geq 0$ and (ii) $x''(a) < 0$. Suppose that $x''(a) \geq 0$. Integrating (5.43) from a to c, we obtain

$$0 \geq -r(a)x''(a)x'(a) = \int_a^c \left[-q(t)\left(x'(t)\right)^{\beta+1} + f(t)x'(t) + r(t)\left(x''(t)\right)^2\right] dt > 0,$$

a contradiction. Let $x''(a) < 0$. Integrating (5.43) from c_1 to a, we get

$$0 \geq r(a)x'(a)x''(a) = \int_{c_1}^a \left[-q(t)\left(x'(t)\right)^{\beta+1} + f(t)x'(t) + r(t)\left(x''(t)\right)^2\right] dt > 0,$$

a contradiction. Hence all solutions of (5.42) are nonoscillatory. The theorem is proved. □

Theorem 5.2.2 *Let β in (5.42) be the ratio of odd integers. If $p'(t) \geq 0$, $f'(t) \geq 0$ and $p(t) + (\alpha + 1)f(t) \geq 0$ then all solutions $x(t)$ of (5.42) for which $|x(t)| \leq 1$ ultimately are nonoscillatory.*

Proof Let $x(t)$ be a solution of (5.42) on $[T, \infty)$, $T \geq \sigma$, such that $|x(t)| \leq 1$ for $t \geq T_1 > T$. Proceeding as in Theorem 5.2.1, it can be shown that $x(t)$ cannot be of nonnegative z-type.

Suppose that $x(t)$ is of nonpositive z-type. Let a and b ($T_1 \leq a < b$) be consecutive double zeros of $x(t)$. Multiplying (5.42) throughout by $x'(t)$, we get

$$\begin{aligned}&\left[r(t)x''(t)x'(t) + \frac{1}{\alpha+1}p(t)x^{\alpha+1}(t) - f(t)x(t)\right]' \\ &\quad = r(t)\big(x''(t)\big)^2 + \frac{1}{\alpha+1}p'(t)x^{\alpha+1}(t) - f'(t)x(t) - q(t)\big(x'(t)\big)^{\beta+1}. \qquad (5.44)\end{aligned}$$

Integrating (5.44) from a to b, we get

$$0 = \int_a^b \left[r(t)\big(x''(t)\big)^2 + \frac{1}{\alpha+1}p'(t)x^{\alpha+1}(t) - f'(t)x(t) - q(t)\big(x'(t)\big)^{\beta+1}\right]dt > 0,$$

a contradiction.

If possible, let $x(t)$ be oscillatory. Let b_1, a and b ($T_1 \leq b_1 < a < b$) be any three consecutive zeros of $x(t)$ such that $x'(b_1) \leq 0$, $x'(a) \geq 0$, $x'(b) \leq 0$, $x(t) < 0$ for $t \in (b_1, a)$ and $x(t) > 0$ for $t \in (a, b)$. So there exist points $c_1 \in (b_1, a)$ and $c \in (a, b)$ such that $x'(c_1) = 0$, $x'(c) = 0$ and $x'(t) > 0$ for $t \in (c_1, c)$. As in Theorem 5.2.1, we consider two cases: (i) $x''(a) \geq 0$ and (ii) $x''(a) < 0$. Suppose that $x''(a) \geq 0$. Multiplying (5.42) throughout by $x'(t)$, we get

$$\big[r(t)x''(t)x'(t)\big]' = -q(t)\big(x'(t)\big)^{\beta+1} - p(t)x^{\alpha}(t)x'(t) + f(t)x'(t) + r(t)\big(x''(t)\big)^2. \qquad (5.45)$$

Integrating (5.45) from a to c, we obtain

$$\begin{aligned}0 &\geq -r(a)x''(a)x'(a) \\ &= \int_a^c \big[-q(t)\big(x'(t)\big)^{\beta+1} - p(t)x^{\alpha}(t)x'(t) + f(t)x'(t) + r(t)\big(x''(t)\big)^2\big]dt \\ &> 0,\end{aligned}$$

a contradiction. Let $x''(a) < 0$. Integration of (5.45) from c_1 to a yields

$$\begin{aligned}r(a)x''(a)x'(a) &= \int_{c_1}^a \big[-q(t)\big(x'(t)\big)^{\beta+1} + r(t)\big(x''(t)\big)^2\big]dt \\ &\quad + \int_{c_1}^a f(t)x'(t)\,dt - \int_{c_1}^a p(t)x^{\alpha}(t)x'(t)\,dt. \qquad (5.46)\end{aligned}$$

But

$$\int_{c_1}^{a} p(t)x^\alpha(t)x'(t)\,dt = \frac{1}{\alpha+1}\big[p(t)x^{\alpha+1}(t)\big]_{c_1}^{a} - \frac{1}{\alpha+1}\int_{c_1}^{a} p'(t)x^{\alpha+1}(t)\,dt$$
$$\le -\frac{1}{\alpha+1}p(c_1)x^{\alpha+1}(c_1)$$

and

$$\int_{c_1}^{a} f(t)x'(t)\,dt = \big[f(t)x(t)\big]_{c_1}^{a} - \int_{c_1}^{a} f'(t)x(t)\,dt \ge -f(c_1)x(c_1).$$

So

$$\int_{c_1}^{a} f(t)x'(t)\,dt - \int_{c_1}^{a} p(t)x^\alpha(t)x'(t)\,dt \ge -f(c_1)x(c_1) + \frac{1}{\alpha+1}p(c_1)x^{\alpha+1}(c_1)$$
$$\ge \frac{1}{\alpha+1}\big[x(c_1)+x^{\alpha+1}(c_1)\big]p(c_1)$$
$$\ge 0,$$

since $|x(t)| \le 1$ for $t \ge T_1$. Hence (5.46) yields

$$0 \ge r(a)x''(a)x'(a) > 0,$$

a contradiction. So $x(t)$ is nonoscillatory. This completes the proof of the theorem. □

Example 5.2.1 Consider

$$(tx'')' - (1+t^8+4t^3)(x')^3 - (t^{-3}+1)x^3 = t^2 - t^{-3}, \quad t \ge 2.$$

Clearly, $x(t) = t^{-1}$ is a nonoscillatory solution.

If β in (5.42) is the ratio of odd integers, then Theorem 5.2.1 holds for $f(t) \le 0$. If $f \equiv 0$ in Theorem 5.2.2, then $p \equiv 0$, and hence the theorem is not applicable to the homogeneous equation

$$(r(t)x'')' + q(t)(x')^\beta + p(t)x^\alpha = 0. \tag{5.47}$$

Theorem 5.2.3 *Consider* (5.42) *with* $\alpha = \beta = 1$. *If* $p'(t) \ge 0$, $f'(t) \ge 0$ *and* $p(t) - q'(t) \ge 0$, *then all solutions of* (5.42) *are nonoscillatory.*

Proof Let $x(t)$ be a solution of (5.42) on $[T,\infty)$, $T \ge \sigma$. The proof that $x(t)$ cannot be of nonnegative z-type or nonpositive z-type is similar to that of Theorem 5.2.2.

If possible, let $x(t)$ be oscillatory. Let b_1, a and b ($T \le b_1 < a < b$) be any three consecutive zeros of $x(t)$ such that $x'(b_1) \le 0$, $x'(a) \ge 0$, $x'(b) \le 0$, $x(t) < 0$

for $t \in (b_1, a)$ and $x(t) > 0$ for $t \in (a, b)$. So there exists a $c \in (a, b)$ such that $x'(c) = 0$ and $x'(t) > 0$ for $t \in (a, c)$. As in above theorems, we consider two cases: (i) $x''(a) \geq 0$ and (ii) $x''(a) < 0$. The proof that $x''(a)$ cannot be ≥ 0 is similar to that of Theorem 5.2.2. Let $x''(a) < 0$. We claim that there exists a point $c_1 \in [b_1, a)$ such that $x''(c_1) \geq 0$. If not, $x''(t) < 0$ for $t \in [b_1, a]$. So $x'(t)$ is decreasing in $[b_1, a]$ and hence $x'(a) < x'(b_1) \leq 0$. This contradicts the fact that $x'(a) \geq 0$. Hence our claim holds. Now integrating (5.42) from c_1 to a, we get

$$\begin{aligned}
0 &> r(a)x''(a) - r(c_1)x''(c_1) \\
&= -\int_{c_1}^{a} q(t)x'(t)\,dt - \int_{c_1}^{a} p(t)x(t)\,dt + \int_{c_1}^{a} f(t)\,dt \\
&= q(c_1)x(c_1) + \int_{c_1}^{a} \big(q'(t) - p(t)\big)x(t)\,dt + \int_{c_1}^{a} f(t)\,dt \\
&> 0,
\end{aligned}$$

a contradiction. Hence the theorem is proved. □

Example 5.2.2 By Theorem 5.2.3, the equation

$$x''' - \big(1 + t^4 + 6t^{-2}\big)x' - \big(t^{-1} + t^{-2}\big)x = t^2 - t^{-3}, \quad t \geq 2,$$

is nonoscillatory. In particular, $x(t) = t^{-1}$ is a nonoscillatory solution of this example.

Theorem 5.2.4 *Consider* (5.42) *with* $\beta = 1$ *and* $\alpha > 1$. *If* $p'(t) \geq 0$, $f'(t) \geq 0$ *and* $p(t) - q'(t) \geq 0$, *then all solutions* $x(t)$ *of* (5.42) *for which* $|x(t)| \leq 1$ *ultimately are nonoscillatory.*

Now, consider

$$x''' + q(t)x' + p(t)x^{\alpha} = f(t), \tag{5.48}$$

where p, q, f and α are same as in (5.42). Let $x_1(t)$, $x_2(t)$ and $x_3(t)$ be solutions of (5.48) on $[t_0, \infty)$, $t_0 \geq \sigma$, respectively, with initial conditions $x_1(t_0) = 0$, $x_1'(t_0) = 1$, $x_1''(t_0) = 0$; $x_2(t_0) = 1$, $x_2'(t_0) = 1$, $x_2''(t_0) = 0$ and $x_3(t_0) = 0$, $x_3'(t_0) = 0$, $x_3''(t_0) = 1$. This choice of $x_1(t)$, $x_2(t)$ and $x_3(t)$ yields the following theorem:

Theorem 5.2.5 *Let* $q(t)$ *be once continuously differentiable such that* $q'(t) \geq 0$. *If* $q(t_0) = 0$, *then* $x_1(t)$ *cannot meet* $x_2(t)$ *in* $[t_0, \infty)$. *If* $q(t_0) \neq 0$, *then* $x_1(t)$ *and* $x_2(t)$ *cannot meet in the strip* $[t_0, t_0 + \sqrt{2/-q(t_0)})$.

Proof Integrating

$$x_1'''(t) + q(t)x_1'(t) + p(t)x_1^{\alpha}(t) = f(t)$$

from t_0 to t, we get

$$x_1''(t) = -q(t)x_1(t) + \int_{t_0}^t q'(s)x_1(s)\,ds - \int_{t_0}^t p(s)x_1^\alpha(s)\,ds + \int_{t_0}^t f(s)\,ds.$$

Further integration from t_0 to t yields

$$\begin{aligned} x_1(t) = (t-t_0) &- \int_{t_0}^t \left(\int_{t_0}^\theta q(s)x_1(s)\,ds\right) d\theta + \int_{t_0}^t \int_{t_0}^u \int_{t_0}^\theta q'(s)x_1(s)\,ds\,d\theta\,du \\ &- \int_{t_0}^t \int_{t_0}^u \int_{t_0}^\theta p(s)x_1^\alpha(s)\,ds\,d\theta\,du + \int_{t_0}^t \int_{t_0}^u \int_{t_0}^\theta f(s)\,ds\,d\theta\,du. \end{aligned}$$

Integrating

$$x_2'''(t) + q(t)x_2'(t) + p(t)x_2^\alpha(t) = f(t)$$

from t_0 to t, we get

$$\begin{aligned} x_2(t) = 1 + (t-t_0) &- \int_{t_0}^t \int_{t_0}^\theta q(s)x_2(s)\,ds\,d\theta + q(t_0)\frac{(t-t_0)^2}{2} \\ &+ \int_{t_0}^t \int_{t_0}^u \int_{t_0}^\theta q'(s)x_2(s)\,ds\,d\theta\,du - \int_{t_0}^t \int_{t_0}^u \int_{t_0}^\theta p(s)x_2^\alpha(s)\,ds\,d\theta\,du \\ &+ \int_{t_0}^t \int_{t_0}^u \int_{t_0}^\theta f(s)\,ds\,d\theta\,du. \end{aligned}$$

Suppose that $x_1(t)$ meets $x_2(t)$ first at $t = t_1 > t_0$. So, $x_1(t) < x_2(t)$ for $t \in [t_0, t_1)$ and $x_1(t_1) = x_2(t_1)$. Consequently, we have

$$x_2(t_1) \ge 1 + q(t_0)\frac{(t_1-t_0)^2}{2} + x_1(t_1).$$

If $q(t_0) = 0$, the above inequality leads to a contradiction $1 \le 0$. If $q(t_0) \ne 0$, then $t_1 \ge t_0 + \sqrt{2/-q(t_0)}$. The theorem is proved. □

Theorem 5.2.6 *Suppose that the conditions of Theorem* 5.2.5 *are satisfied. Then* $x_1(t)$ *and* $x_3(t)$ *cannot meet in the strip* (t_0, t_0+2), *and* $x_2(t)$ *and* $x_3(t)$ *cannot meet the strip* $[t_0, t_0 + \frac{1+\sqrt{3-2q(t_0)}}{1-q(t_0)})$.

Lemma 5.2.1 *Let* $p \equiv 0$ *in* (5.42). *If* $x(t)$ *is a solution of* (5.42) *on* $[T, \infty)$, $T \ge \sigma$, *then there exists a* $c \ge T$ *such that* $x'(t) > 0$ *or* ≤ 0 *for* $t \ge c$.

Proof If possible, let $x'(t)$ be oscillatory. Let a and b $(T \le a < b)$ be consecutive zeros of $x'(t)$ such that $x'(t) > 0$ for $t \in (a, b)$, $x''(a) \ge 0$ and $x''(b) \le 0$. Integrating (5.42) from a to b, we get

$$0 > r(b)x''(b) - r(a)x''(a) = \int_a^b \left[-q(t)\left(x'(t)\right)^\beta - f(t)\right]dt > 0.$$

This contradiction proves the theorem. □

Lemma 5.2.2 *If $x(t)$ is a solution of (5.42) on $[T, \infty)$, $T \geq \sigma$, such that it is ultimately positive, then there exists a $t_0 \geq T$ such that $x'(t) > 0$ or ≤ 0 for $t \geq t_0$.*

The following lemma is due to Heidel [4].

Lemma 5.2.3 *Let $g : [t_0, \infty) \to R$, $t_0 \geq \sigma$. Suppose that $g(t) > 0$ and that $g'(t)$, $g''(t)$ exist for $t \geq t_0$. Suppose also that if $g'(t) \geq 0$ ultimately, then*

$$\lim_{t\to\infty} g(t) = A < \infty.$$

Then

$$\liminf_{t\to\infty} \left| t^\delta g''(t) - \delta t^{\delta-1} g'(t) \right| = 0$$

for any δ such that $0 \leq \delta \leq 2$.

Theorem 5.2.7 *Consider (5.42) with $r(t) \equiv 1$ and $\beta = 1$. For some fixed δ with $0 \leq \delta \leq 2$, suppose that $-\infty < -M < t^\delta q(t)$, $\int_\sigma^\infty t^\delta p(t)\,dt = -\infty$ and $\int_\sigma^\infty t^\delta f(t)\,dt < \infty$. If $x(t)$ is a nonoscillatory solution of (5.42) such that $x(t)x'(t) \leq 0$ ultimately then*

$$\lim_{t\to\infty} x(t) = 0.$$

Proof If $x(t)$ is ultimately positive, then one may proceed as in Theorem 2.3 of Heidel [4] to obtain

$$\lim_{t\to\infty} x(t) = 0.$$

Suppose that $x(t)$ is ultimately negative. So, there exists a $t_0 > \sigma$ such that $x(t) < 0$ and $x'(t) \geq 0$ for $t \geq t_0$. Let

$$\lim_{t\to\infty} x(t) = -A, \quad A > 0.$$

Multiplying (5.42) throughout by t^δ, $t \geq t_0$, and integrating the resulting identity from t_0 to t, we get

$$\begin{aligned} &\left[s^\delta x''(s)\right]_{t_0}^t - \delta\left[s^{\delta-1}x'(s)\right]_{t_0}^t + \delta(\delta-1)\int_{t_0}^t s^{\delta-2}x'(s)\,ds + \int_{t_0}^t s^\delta q(s)x'(s)\,ds \\ &\quad + \int_{t_0}^t s^\delta p(s)x^\alpha(s)\,ds \\ &= \int_{t_0}^t s^\delta f(s)\,ds, \end{aligned}$$

that is,

$$\begin{aligned}t^\delta x''(t) - \delta t^{\delta-1}x'(t) \le t_0^\delta x''(t_0) - \delta t_0^{\delta-1}x'(t_0) + M\int_{t_0}^t x'(s)\,ds \\ - \delta(\delta-1)\int_{t_0}^t s^{\delta-2}x'(s)\,ds - \int_{t_0}^t s^\delta p(s)x^\alpha(s)\,ds \\ + \int_{t_0}^t s^\delta f(s)\,ds.\end{aligned} \tag{5.49}$$

Clearly,

$$\lim_{t\to\infty}\int_{t_0}^t x'(s)\,ds = -A - x(t_0),$$

$$0 \le \lim_{t\to\infty}\int_{t_0}^t s^{\delta-2}x'(s)\,ds \le t_0^{\delta-2}\big(-A - x(t_0)\big)$$

and

$$\lim_{t\to\infty}\int_{t_0}^t s^\delta p(s)x^\alpha(s)\,ds = \infty.$$

Hence, taking lim inf in (5.49) as $t \to \infty$, we get

$$\liminf_{t\to\infty}\big(t^\delta x''(t) - \delta t^{\delta-1}x'(t)\big) = -\infty.$$

But by Lemma 5.2.3,

$$\liminf_{t\to\infty}\big(t^\delta x''(t) - \delta t^{\delta-1}x'(t)\big) = 0.$$

This contradiction proves the theorem. □

Theorem 5.2.8 *Consider* (5.42) *with* $r(t) \equiv 1$, $f(t) \equiv 0$ *and* β *as the ratio of odd integers. Suppose that* $\int_\sigma^\infty p(t)\,dt = -\infty$. *If* $x(t)$ *is a bounded nonoscillatory solution of* (5.42), *then* $x(t)x'(t) \le 0$ *ultimately.*

Proof Without any loss of generality, we can assume $x(t)$ to be ultimately positive. By Lemma 5.2.2, there exists a $t_0 > \sigma$ such that $x(t) > 0$ and $x'(t) > 0$ or ≤ 0 for $t \ge t_0$. Suppose that $x'(t) > 0$ for $t \ge t_0$. Since $x'''(t) \ge 0$ and $x(t)$ is bounded, $x''(t)$ is monotonic increasing and $x''(t) \le 0$. Integrating (5.42) from t_0 to t, we get

$$x''(t) \ge x''(t_0) - \int_{t_0}^t p(s)x^\alpha(s)\,ds \ge x''(t_0) - x^\alpha(t_0)\int_{t_0}^t p(s)\,ds.$$

So

$$\lim_{t\to\infty} x''(t) = \infty,$$

which contradicts the fact that

$$\lim_{t\to\infty} x''(t) \le 0.$$

Hence $x(t)x'(t) \le 0$ for $t \ge t_0$. The theorem is proved. □

Corollary 5.2.1 *In* (5.42), *let* $r(t) \equiv 1$, $f(t) \equiv 0$ *and* $\beta = 1$. *Suppose that* $\int_\sigma^\infty p(t)\,dt = -\infty$ *and* $-\infty < -M < t^\delta q(t)$ *for some fixed* δ *with* $0 \le \delta \le 2$. *If* $x(t)$ *is a bounded nonoscillatory solution of* (5.42), *then*

$$\lim_{t\to\infty} x(t) = 0.$$

This follows from Theorems 5.2.7 and 5.2.8.

Theorem 5.2.9 *Consider* (5.42) *with* $r(t) \equiv 1$, $p(t) \equiv 0$ *and* $\beta = 1$. *Suppose that* $\int_\sigma^\infty f(t)\,dt = \infty$ *and* $\int_\sigma^\infty tq(t)\,dt > -\infty$. *If* $x(t)$ *is a bounded nonoscillatory solution of* (5.42), *then* $x(t)x'(t) \le 0$ *ultimately.*

Proof Let $x(t)$ be ultimately positive. By Lemma 5.2.1, there exists a $t_0 > \sigma$ such that $x(t) > 0$ and $x'(t) > 0$ or ≤ 0 for $t \ge t_0$. Proceeding as in Theorem 5.2.8, we prove that $x'(t) \le 0$ for $t \ge t_0$.

Let $x(t)$ be ultimately negative. So there exists a $t_0 > \sigma$ such that $x(t) < 0$ and $x'(t) > 0$ or ≤ 0 for $t \ge t_0$. If possible, let $x'(t) \le 0$ for $t \ge t_0$. Since $x(t)$ is bounded, $x''(t)$ cannot be ≤ 0 for large t. We consider three cases for $x''(t)$, viz., $x''(t)$ is nonnegative z-type or $x''(t) > 0$ or $x''(t)$ is oscillatory. Suppose that $x''(t) \ge 0$ for $t \ge t_1 > t_0$. Since $\int_\sigma^\infty tq(t)\,dt > -\infty$, there exists a $t_2 \ge t_1$ such that $\int_{t_2}^\infty tq(t)\,dt > -1$. Multiplying (5.42) throughout by $t(\ge t_2)$ and integrating the resulting identity from t_2 to t, we get

$$\begin{aligned} tx''(t) - x'(t) + x'(t)\int_{t_2}^t sq(s)\,ds - \int_{t_2}^t x''(s)\left(\int_{t_2}^s \theta q(\theta)\,d\theta\right) ds \\ = t_2x''(t_2) - x'(t_2) + \int_{t_2}^t sf(s)\,ds, \end{aligned}$$

that is, $\int_{t_2}^t sf(s)\,ds \le tx''(t) - x'(t)$. Hence $\liminf_{t\to\infty}(tx''(t) - x'(t)) = \infty$. This contradicts Lemma 5.2.3. Suppose that $x''(t)$ is oscillatory. We proceed as in Theorem 2.4 of Heidel [4] to obtain a contradiction. So $x'(t) > 0$ for $t \ge t_0$. Hence the theorem is proved. □

Theorem 5.2.10 *Consider* (5.42) *with* $r(t) \equiv 1$, $p(t) \equiv 0$ *and* $\beta = 1$. *For some fixed* δ *with* $0 \le \delta \le 2$, *suppose that* $-\infty < -M < t^\delta q(t)$ *and* $\int_\sigma^\infty f(t)\,dt = \infty$. *If* $x(t)$ *is a nonoscillatory solution* (5.42) *such that* $x(t)x'(t) \le 0$ *ultimately, then*

$$\lim_{t\to\infty} x(t) = 0.$$

Proof We claim that $x(t)$ cannot be ultimately negative. If not, there exists a $t_0 > 0$ such that $x(t) < 0$ and $x'(t) \geq 0$ for $t \geq t_0$. From (5.42) it follows that $x''(t)$ is monotonic increasing. Clearly, $x''(t)$ cannot be > 0 for large t, because otherwise $x(t)$ will be ultimately positive. Suppose that $x''(t) < 0$ for large t. Integrating (5.42) from t_0 to t, we get

$$x''(t) \geq x''(t_0) + \int_{t_0}^{t} f(s)\,ds.$$

So

$$\lim_{t\to\infty} x''(t) = \infty,$$

a contradiction. This proves our assertion. Hence there exists a $t_1 > \sigma$ such that $x(t) > 0$ and $x'(t) \leq 0$ for $t \geq t_1$. Let

$$\lim_{t\to\infty} x(t) = A > 0.$$

Multiplying (5.42) throughout by t^δ, $t \geq t_1$, and integrating the resulting identity from t_1 to t, we get

$$\begin{aligned}&\bigl[s^\delta x''(s)\bigr]_{t_1}^{t} - \delta\bigl[s^{\delta-1}x'(s)\bigr]_{t_1}^{t} + \delta(\delta-1)\int_{t_1}^{t} s^{\delta-2}x'(s)\,ds + \int_{t_1}^{t} s^\delta q(s)x'(s)\,ds\\ &\quad= \int_{t_1}^{t} s^\delta f(s)\,ds.\end{aligned}$$

Using given conditions, we obtain

$$\begin{aligned}t^\delta x''(t) - \delta t^{\delta-1}x'(t) &> t_1^\delta x''(t_1) - \delta t_1^{\delta-1}x'(t_1) - \delta(\delta-1)\int_{t_1}^{t} s^{\delta-2}x'(s)\,ds\\ &\quad + M\int_{t_1}^{t} x'(s)\,ds + \int_{t_1}^{t} s^\delta f(s)\,ds.\end{aligned} \tag{5.50}$$

It is easy to see that

$$\lim_{t\to\infty}\int_{t_1}^{t} x'(s)\,ds = A - x(t_1),$$

and

$$0 \leq \lim_{t\to\infty}\biggl[-\int_{t_1}^{t} s^{\delta-2}x'(s)\,ds\biggr] \leq -t_1^{\delta-2}\bigl(A - x(t_1)\bigr).$$

Hence, from (5.50), it follows that

$$\liminf_{t\to\infty}\bigl[t^\delta x''(t) - \delta t^{\delta-1}x'(t)\bigr] = \infty,$$

a contradiction to Lemma 5.2.3. This completes the proof of the theorem. □

Corollary 5.2.2 *In (5.42), let $r(t) \equiv 1$, $p(t) \equiv 0$ and $\beta = 1$. Suppose that $\int_\sigma^\infty f(t)\,dt = \infty$, $\int_\sigma^\infty tq(t)\,dt > -\infty$ and $-\infty < -M < t^\delta q(t)$, for some fixed δ with $0 \le \delta \le 2$. If $x(t)$ is a bounded nonoscillatory solution of (5.42), then*

$$\lim_{t\to\infty} x(t) = 0.$$

Theorem 5.2.11 *Let $p(t) \ge 0$ and $q(t) \le 0$. If $q(t) + t^\alpha p(t) \le 0$ for large t, then all solutions of Eq. (5.42) with $\alpha = \beta$ are nonoscillatory.*

Theorem 5.2.12 *Let $p(t) \ge 0$ and $q(t) \le 0$. Let $q(t)$ be once continuously differentiable. If $q'(t) \ge 0$ such that $q'(t) + p(t) > 0$ for large t and $\lim_{t\to\infty} \frac{q'(t)}{p(t)} = \infty$, then all bounded solutions of Eq. (5.42) with $\alpha \ge 1$ and $\beta = 1$ are nonoscillatory.*

Theorem 5.2.13 *Let $p(t) \ge 0$ and $q(t) \le 0$. If $f(t) \ge M^\alpha p(t)$ for large t, where $M > 0$, then all solutions $x(t)$ of Eq. (5.42) which are bounded above by M in any interval where $x(t) > 0$ are nonoscillatory.*

Theorem 5.2.14 *Let $p(t) \ge 0$ and $q(t) \le 0$. Let p and f be once continuously differentiable functions. If $p'(t) \ge 0$, $f'(t) \le 0$ and $(\alpha + 1) f(t) \ge p(t)$, then all solutions $x(t)$ of Eq. (5.42) for which $|x(t)| \le 1$ ultimately are nonoscillatory.*

Proof Let $x(t)$ be a solution of Eq. (5.42) on $[T, \infty)$, $T \ge \sigma$, such that $|x(t)| \le 1$ for $t \ge T_1 > T$. Suppose that $x(t)$ is of nonnegative z-type. Let a and b $(T_1 \le a < b)$ be consecutive double zeros $x(t)$. Multiplying Eq. (5.42) throughout by $x'(t)$, we get

$$\begin{aligned}&\left[r(t)x'(t)x''(t) + \frac{1}{\alpha+1}p(t)x^{\alpha+1}(t) - f(t)x(t)\right]' \\ &\quad = r(t)\big(x''(t)\big)^2 + \frac{1}{\alpha+1}p'(t)x^{\alpha+1}(t) - f'(t)x(t) - q(t)\big(x'(t)\big)^{\beta+1}. \quad (5.51)\end{aligned}$$

Integrating Eq. (5.51) from a to b, we obtain

$$0 = \int_a^b \left[r(t)\big(x''(t)\big)^2 + \frac{1}{\alpha+1}p'(t)x^{\alpha+1}(t) - f'(t)x(t) - q(t)\big(x'(t)\big)^{\beta+1}\right] dt > 0$$

a contradiction.

Next suppose that $x(t)$ is of nonpositive z-type with consecutive double zeros at a and b $(T_1 \le a < b)$. So there exists a $c \in (a, b)$ such that $x'(c) = 0$ and $x'(t) > 0$ for $t \in (c, b)$. Multiplying Eq. (5.42) throughout by $x'(t)$, we get

$$r(t)x'(t)x''(t) = r(t)\big(x''(t)\big)^2 - q(t)\big(x'(t)\big)^{\beta+1} - p(t)x'(t)x^\alpha(t) + f(t)x'(t). \quad (5.52)$$

Integration of Eq. (5.52) from c to b yields

$$0 = \int_c^b \left[r(t)\big(x''(t)\big)^2 - q(t)\big(x'(t)\big)^{\beta+1} - p(t)x'(t)x^\alpha(t) + f(t)x'(t)\right] dt > 0$$

a contradiction.

If possible, let $x(t)$ be oscillatory. Let a, b and a_1 $(T_1 \le a < b < a_1)$ be any three consecutive zeros of $x(t)$ such that $x'(a) \le 0$, $x'(b) \ge 0$, $x'(a_1) \le 0$, $x(t) < 0$ for $t \in (a,b)$ and $x(t) > 0$ for $t \in (b, a_1)$. So there exist points $c \in (a,b)$ and $c_1 \in (b, a_1)$ such that $x'(c) = 0$, $x'(c_1) = 0$ and $x'(t) > 0$ for $t \in (c, c_1)$. We consider two cases, viz., $x''(b) \le 0$ and $x''(b) > 0$. Suppose that $x''(b) \le 0$. Integrating Eq. (5.52) from c to b, we obtain

$$\begin{aligned} 0 \ge r(b)x'(b)x''(b) - \int_c^b \big[r(t)\big(x''(t)\big)^2 - q(t)\big(x'(t)\big)^{\beta+1} - p(t)x'(t)x^\alpha(t) \\ + f(t)x'(t)\big]\,dt > 0 \end{aligned}$$

a contradiction. Let $x''(b) > 0$. Integration of Eq. (5.52) from b to c_1, we get

$$\begin{aligned} -r(b)x'(b)x''(b) = \int_b^{c_1} \big[r(t)\big(x''(t)\big)^2 - q(t)\big(x'(t)\big)^{\beta+1} - p(t)x'(t)x^\alpha(t) \\ + f(t)x'(t)\big]\,dt. \end{aligned} \tag{5.53}$$

But

$$\begin{aligned} \int_b^{c_1} p(t)x^\alpha(t)x'(t)\,dt &= \frac{1}{\alpha+1}\left[p(t)x^{\alpha+1}(t)\big|_b^{c_1} - \int_b^{c_1} p'(t)x^{\alpha+1}(t)\,dt\right] \\ &\le \frac{1}{\alpha+1}p(c_1)x^{\alpha+1}(c_1) \end{aligned}$$

and

$$\int_b^{c_1} f(t)x'(t)\,dt = f(t)x(t)\big|_b^{c_1} - \int_b^{c_1} f'(t)x(t)\,dt \ge f(c_1)x(c_1)$$

holds. So

$$\begin{aligned} &\int_b^{c_1} f(t)x'(t)\,dt - \int_b^{c_1} p(t)x^\alpha(t)x'(t)\,dt \\ &\ge f(c_1)x(c_1) - \frac{1}{\alpha+1}p(c_1)x^{\alpha+1}(c_1) \\ &\ge \frac{p(c_1)}{\alpha+1}x(c_1) - \frac{1}{\alpha+1}p(c_1)x^{\alpha+1}(c_1) \\ &= \frac{1}{\alpha+1}p(c_1)\big[x(c_1) - x^{\alpha+1}(c_1)\big] \\ &\ge 0. \end{aligned}$$

The reason is that $|x(t)| \le 1$ for $t \ge T_1$. Hence Eq. (5.53) yields

$$0 \ge -r(b)x'(b)x''(b) > 0$$

a contradiction. So $x(t)$ is nonoscillatory. This completes the proof of the theorem. □

Remark 5.2.1 If $f(t)=0$ in Theorem 5.2.14, then $p(t)=0$ and hence the theorem is not applicable to Eq. (5.47).

Theorem 5.2.15 *If $q(t)$ does not change sign for large t, $f(t)>0$ and $\lim_{t\to\infty}\frac{f(t)}{|p(t)|}=\infty$, then all bounded solutions of Eq.* (5.42) *are nonoscillatory.*

Proof Let $q(t)\le 0$ for $t\ge t_0$. Let $x(t)$ be a bounded solution of Eq. (5.42) such that $|x(t)|\le M$. From the given condition, it follows that there exists a $t_1\ge t_0$ such that $f(t)\ge M^\alpha|p(t)|$ for $t\ge t_1$. If possible, let $x(t)$ be oscillatory or z-type for $t\in(t_2,t_3)$, $t_2\ge t_1$. Now integrating Eq. (5.42) from t_2 to t_3, we obtain

$$\begin{aligned}0>r(t)x''(t)\big|_{t_2}^{t_3}&=\int_{t_2}^{t_3}f(t)\,dt-\int_{t_2}^{t_3}q(t)\big(x'(t)\big)^\beta\,dt-\int_{t_2}^{t_3}p(t)x^\alpha(t)\,dt\\&\ge\int_{t_2}^{t_3}f(t)\,dt-\int_{t_2}^{t_3}\big|p(t)x^\alpha(t)\big|\,dt\\&\ge\int_{t_2}^{t_3}\big(f(t)-M^\alpha\big|p(t)\big|\big)\,dt\\&>0\end{aligned}$$

a contradiction. So $x(t)$ is nonoscillatory.

If $q(t)\ge 0$ for $t\ge t_0$, then we can choose t_2 and $t_3(t_1<t_2<t_3)$ such that $x'(t_2)=0=x'(t_3)$ and $x'(t)\le 0$ for $t\in(t_2,t_3)$ to get a contradiction. □

Next, we consider the equation

$$\frac{d}{ds}\left[\sigma_1(s)\frac{d}{ds}\left(r_1(s)\frac{dy}{ds}\right)\right]+q_1(s)\left(\frac{dy}{ds}\right)^\beta+p_1(s)y^\alpha=f_1(s),\tag{5.54}$$

where σ, u, v, w and f_1 are real-valued continuous functions on $[\sigma,\infty)$ such that $\sigma_1(s)>0$, $r_1(s)>0$, and $f_1(s)\ge 0$, such that both $\alpha>0$ and $\beta>0$ are ratios of odd integers.

We reduce Eq. (5.54) to an equation of the form Eq. (5.42). Let $\int_\sigma^\infty\frac{du}{r_1(u)}=\infty$. Set $R(s)=\int_\sigma^s\frac{du}{r_1(u)}$. So $R:[\sigma,\infty)\to[0,\infty)$ is an increasing function such that $s\to\infty$ implies $R(s)\to\infty$. Consequently R^{-1} exists. Put $t=R(s)$, that is, $s=R^{-1}(t)$. Setting $x(t)=y(R^{-1}(t))=y(s)$, we get

$$r_1(s)\frac{dy}{ds}=\frac{dx}{dt}\quad\text{and}\quad\sigma_1(s)\frac{d}{ds}\left(r_1(s)\frac{dy}{ds}\right)=\frac{\sigma_1(s)}{r_1(s)}\frac{d^2x}{dt^2}.$$

So $\frac{d}{ds}[\sigma_1(s)\frac{d}{ds}(r_1(s)\frac{dy}{ds})] = \frac{1}{r_1(s)}\frac{d}{dt}[\frac{\sigma_1(s)}{r_1(s)}\frac{d^2x}{dt^2}]$. Putting $r(t) = \frac{\sigma_1(R^{-1}(t))}{r_1(R^{-1}(t))}$, Eq. (5.54) is reduced to Eq. (5.42) with

$$q(t) = \frac{q_1(R^{-1}(t))}{[r_1(R^{-1}(t))]^{\beta-1}}, \qquad p(t) = r_1\big(R^{-1}(t)\big)p_1\big(R^{-1}(t)\big)$$

and

$$f(t) = r_1\big(R^{-1}(t)\big)f_1\big(R^{-1}(t)\big).$$

In view of Theorems 5.2.11–5.2.15, we have the following theorems for Eq. (5.54).

Theorem 5.2.16 *Let $p_1(s) \geq 0$, $q_1(s) \leq 0$ and $R(s) = \int_0^s \frac{du}{r_1(u)}$. If $q_1(s) + R^\alpha(s)r_1^\alpha(s)p_1(s) \leq 0$ for large s, then all solutions of Eq. (5.54) with $\alpha = \beta$ are nonoscillatory.*

Theorem 5.2.17 *Let $p_1(s) \geq 0$, $q_1(s) \leq 0$. Let $q_1(s)$ be once continuously differentiable. If $q_1'(s) \geq 0$ such that $q_1'(s) + p_1(s) > 0$ for large s and $\lim_{s\to\infty} \frac{q_1'(s)}{p_1(s)} = \infty$, then all bounded solutions of Eq. (5.54) with $\alpha \geq 1$ and $\beta = 1$ are nonoscillatory.*

Theorem 5.2.18 *Let $p_1(s) \geq 0$, $q_1(s) \leq 0$. If $f_1(s) \geq M^\alpha p_1(s)$ for large s, where $M > 0$, then all solutions $y(s)$ of Eq. (5.54) which are bounded above by M in any interval where $y(s) > 0$, are nonoscillatory.*

Theorem 5.2.19 *Let $p_1(s) \leq 0$ and $q_1(s) \leq 0$. Let $r_1 p_1$ and $r_1 f_1$ be once continuously differentiable functions. If $(r_1 p_1)'(s) \geq 0$, $(r_1 f_1)'(s) \geq 0$ and $(\alpha + 1) f_1(s) + p_1(s) \geq 0$, then all solutions $y(s)$ of Eq. (5.54) for which $|y(s)| \leq 1$ ultimately are nonoscillatory.*

Theorem 5.2.20 *Let $p_1(s) \geq 0$ and $q_1(s) \leq 0$. Let $r_1 p_1$ and $r_1 f_1$ be once continuously differentiable functions. If $(r_1 p_1)'(s) \geq 0$, $(r_1 f_1)'(s) \leq 0$ and $(\alpha + 1) f_1(s) \geq p_1(s)$, then all solutions $y(s)$ of Eq. (5.54) for which $|y(s)| \leq 1$ ultimately are nonoscillatory.*

Theorem 5.2.21 *If $q_1(s)$ does not change sign for large s, $f_1(s) > 0$ and*

$$\lim_{s\to\infty} \frac{f_1(s)}{|p_1(s)|} = \infty,$$

then all bounded solutions of Eq. (5.54) are nonoscillatory.

Next we consider (5.3). The Liouville transformation transforms (5.3) to an equation of the type (5.4).

Theorem 5.2.22 *Let $\alpha = \beta \equiv 1$ in (5.4). If $p(t) \leq 0$ and $q(t) \leq 0$ for large t, then (5.4) is nonoscillatory.*

Proof Let $x(t)$ be a solution of (5.4) on $[\sigma, \infty)$. Let $p(t) \leq 0$ and $q(t) \leq 0$ for $t \geq t_0 \geq \sigma$. Let $x(t)$ be of nonnegative z-type with consecutive double zeros at t_1 and t_2 $(t_0 < t_1 < t_2)$. So, there exists a $b \in (t_1, t_2)$ such that $x'(b) = 0, x''(b) \leq 0$ and $x'(t) > 0$ for $t \in (t_1, b)$. Integrating (5.4) from t_1 to b, we get

$$0 \geq r(b)x''(b) + p(b)x(b) - r(t_1)x''(t_1) = \int_{t_1}^{b} f(t)\,dt - \int_{t_1}^{b} q(t)x'(t)\,dt > 0,$$

a contradiction, because $x''(t_1) \geq 0$. Suppose that $x(t)$ is a nonpositive z-type with consecutive double zeros at t_1 and t_2 $(t_0 < t_1 < t_2)$. Then there exists a $b \in (t_1, t_2)$ such that $x'(b) = 0$ and $x'(t) > 0$ for $t \in (b, t_2)$. We note that $x''(b) \geq 0$ and $x''(t_2) \leq 0$. Now, integrating (5.4) from b to t_2 yield

$$0 \geq r(t_2)x''(t_2) - r(b)x''(b) - p(b)x(b) = \int_{b}^{t_2} f(t)\,dt - \int_{b}^{t_2} q(t)x'(t)\,dt > 0,$$

a contradiction. Hence $x(t)$ cannot be of z-type.

Suppose that $x(t)$ is an oscillatory solution of (5.4) with consecutive zeros at t_1, t_2 and t_3 $(t_0 < t_1 < t_2 < t_3)$ such that $y(t) < 0$ for $t \in (t_1, t_2)$ and $x(t) > 0$ for $t \in (t_2, t_3)$. So there exist $b \in (t_1, t_2)$ and $c \in (t_2, t_3)$ such that $x'(b) = 0, x'(c) = 0, x'(t) > 0$ for $t \in (b, t_2)$ and $x'(t) > 0$ for $t \in (t_2, c)$. If $x''(t_2) \geq 0$, then integrating (5.4) from t_2 to c, we obtain

$$0 \geq r(c)x''(c) + p(c)x(c) - r(t_2)x''(t_2) = \int_{t_2}^{c} f(t)\,dt - \int_{t_2}^{c} q(t)x'(t)\,dt > 0,$$

a contradiction because $x''(c) \leq 0$. Furthermore, if $x''(t_2) < 0$, then integrating (5.4) from b to t_2, we get

$$0 > r(t_2)x''(t_2) - r(b)x''(b) - p(b)x(b) = \int_{b}^{t_2} f(t)\,dt - \int_{b}^{t_2} q(t)x'(t)\,dt > 0,$$

a contradiction, because $x''(b) \geq 0$. Hence $x(t)$ cannot be oscillatory. This completes the proof of the theorem. □

Theorem 5.2.23 *If $p(t) \leq 0$ and $q(t) \leq 0$, then* (5.4) *is nonoscillatory.*

Theorem 5.2.24 *Let $f(t) \geq 0$. All solutions of Eq.* (5.8) *are nonoscillatory.*

Proof Equation (5.8) may be written as

$$\left[x'' + xx'\right]' = \left(x'\right)^2 + f(t). \tag{5.55}$$

Let $x(t)$ be a solution of (5.8) on $[\sigma, \infty)$. Proceeding exactly as in Theorem 5.2.22, one may show that $x(t)$ cannot be of z-type or oscillatory. Hence $x(t)$ is nonoscillatory. □

The following examples illustrate the theorem.

Example 5.2.3 (i) The equation $x''' + xx'' = 0$ admits both positive and negative solutions $x_1(t) = t$ and $x_2(t) = -t$,

(ii) The equation $x''' + xx'' = \frac{8}{t^4}$, $t \geq 1$, admits the positive bounded solution $x(t) = \frac{4}{t}$,

(iii) $x(t) = -e^{-t}$ is a bounded negative solution of $x''' + xx'' = e^{-t} + e^{-2t}$, $t \geq 0$.

The asymptotic behaviour of solutions of Eq. (5.7) has been studied by Hartman [3]. Equation (5.7) with $\lambda = \frac{1}{2}$ is often called the Homann differential equation. In the sequel we give a theorem concerning nonoscillatory behaviour of solutions of nonhomogeneous equation associated with Eq. (5.7), that is,

$$x''' + xx'' + \lambda\left[1 - (x')^2\right] = f(t), \tag{5.56}$$

where $f \in C([\sigma, \infty), R)$ is such that $f(t) \geq 0$.

Theorem 5.2.25 *If* $-1 \leq \lambda < 0$ *then all solutions of Eq.* (5.56) *are nonoscillatory. If* $\lambda > 0$ *and* $\lim_{t\to\infty} f(t) = \infty$, *then all solutions of* (5.56) *are nonoscillatory. If* $\lambda < -1$ *and* $\lim_{t\to\infty} f(t) = \infty$, *then all solutions of* (5.56) *whose first derivatives are bounded are nonoscillatory.*

Proof Equation (5.56) can be written as

$$(x'' + xx')' = (1 + \lambda)(x')^2 + f(t) - \lambda.$$

In each case, we see that the right-hand side of the above identity is positive for sufficiently large t. Then proceeding as in Theorem 5.2.22, we can prove that all solutions of (5.56) are nonoscillatory. The theorem is complete. □

Example 5.2.4 (i) All solutions of

$$x''' + xx'' - \left[1 - (x')^2\right] = 6t^2 - 1, \quad t \geq 1,$$

are nonoscillatory. In particular, $x(t) = t^2$ is a nonoscillatory solution of the equation.

(ii) The equation

$$x''' + xx'' + \left[1 - (x')^2\right] = 1 + \frac{7}{t^4}, \quad t \geq 1,$$

is nonoscillatory with a particular nonoscillatory solution $x(t) = -\frac{1}{t}$.

(iii) The equation

$$x''' + xx'' + \left[1 - (x')^2\right] = 1 + e^t, \quad t \geq 0,$$

is nonoscillatory. In particular, $x(t) = e^t$ is a nonoscillatory solution of the equation.

5.3 Open Problems and Discussions

It seems, from the proof of Theorem 5.1.6 that the C_I nature of the homogeneous equation (2.1) plays an important role in proving the existence of an oscillatory solution of (5.1). On the other hand, no such result is known on the oscillation of (5.1), when (2.1) is of type C_{II}. Thus, it would be interesting to obtain some comparison theorems to prove the existence of an oscillatory solution of (5.1), when the corresponding homogeneous equation (2.1) is of type C_{II}. We have incorporated results only for the cases (i) $a(t) \geq 0, b(t) \leq 0, c(t) > 0$ and (ii) $a(t) \geq 0, b(t) \geq 0, c(t) > 0$. Hence, if we obtain a single result on the oscillation of (5.1) on the other possible cases, then it will be of immense importance.

5.4 Notes

Theorems 5.1.1–5.1.5 are due to Greguš, Graef and Gera [2]. Theorems 5.1.6–5.1.10 are taken from Padhi [5]. Theorems 5.1.11–5.1.15 are due to Bartušek [1]. Theorems 5.2.1–5.2.10 are proved by Parhi [6]. Temtek and Tiryaki [8] proved Theorem 5.2.11–5.2.21. Parhi [7] proved Theorems 5.2.23–5.2.25.

References

1. M. Bartušek; *On singular solutions of third order differential equations*, Mathematica Slovaca, 51(2) (2001), 231–239.
2. M. Greguš, J. R. Graef and M. Gera; *Oscillating nonlinear third order differential equations*, Nonlinear Analysis; Theory Methods and Applications, 28(10) (1997), 1611–1622.
3. P. Hartman; *Ordinary Differential Equations*, Wiley, New, York, 1964, and Birkhäuser, Boston, 1982.
4. J. W. Heidel; *Qualitative behaviour of solutions of a third order nonlinear differential equation*, Pacific Journal of Mathematics, 27 (1968), 507–526.
5. S. Padhi; *Contributions to the Oscillation Theory of Ordinary and Delay Differential Equations of Third Order*, Ph.D. Thesis, Berhampur University, India, 1998.
6. N. Parhi; *Nonoscillatory behaviour of solutions of nonhomogeneous third order differential equations*, Applicable Analysis, 12 (1981), 273–285.
7. N. Parhi; *On non-homogeneous canonical third-order linear differential equations*, Journal of the Australian Mathematical Society (Series A), 57 (1994), 138–148.
8. P. Temtek and A. Tiryaki; *Nonoscillation results for a class of third order nonlinear differential equations*, Applied Mathematics and Mechanics, 23(10) (2002), 1170–1175.

Chapter 6
Oscillatory and Asymptotic Behaviour of Solutions of Third-Order Delay Differential Equations

A good deal of work has been done and many interesting results have been given in Chaps. 2–4 concerning the asymptotic behaviour of solutions of third-order linear and nonlinear, homogeneous and nonhomogeneous ordinary differential equations. Several sufficient conditions were given in these chapters for the oscillation of Eqs. (2.1), (3.1), (4.1) and (4.2).

On the other hand, the theory of delay differential equations is complicated than the theory of ordinary differential equations. For example, see Sect. 1.3 of Chap. 1.

Let us consider the third-order linear delay differential equation of the form

$$x'''(t) + a(t)x''(t) + b(t)x'(t) + c(t)x\big(g(t)\big) = 0, \tag{6.1}$$

where a, b, c and $g \in C([\sigma, \infty), R)$, $\sigma \in R$, $g(t) \le t$ and $g(t) \to \infty$ as $t \to \infty$. Equation (6.1) can be rewritten as

$$\big(r(t)x''(t)\big)' + q(t)x'(t) + p(t)x\big(g(t)\big) = 0, \tag{6.2}$$

where $r(t) = \exp(\int_\sigma^t a(s)\,ds)$, $q(t) = r(t)b(t)$ and $p(t) = r(t)c(t)$.

We have seen in Chap. 1 that if the second-order differential equation (2.110) is nonoscillatory, then we can write Eq. (6.1) in the disconjugate form

$$\frac{1}{r_3(t)}\left(\frac{1}{r_2(t)}\left(\frac{1}{r_1(t)}\left(\frac{1}{r_0(t)}x(t)\right)'\right)'\right)' + p(t)x\big(g(t)\big) = 0, \tag{6.3}$$

where $r_i : [\sigma, \infty) \to (0, \infty)$, $i = 0, 1, 2, 3$ and $\int_\sigma^\infty r_i(t)\,dt = \infty$, $i = 1, 2$ irrespective of the sign of $b(t)$, provided that the second-order linear differential equation (2.110) or (2.113) is nonoscillatory. Thus in the first three sections of this chapter, we provide results on the oscillation and asymptotic behaviour of solutions of (6.1), (6.3) and

$$x'''(t) + p(t)x\big(g(t)\big) = 0, \tag{6.4}$$

which is the a particular case of (6.1) and (6.2).

S. Padhi, S. Pati, *Theory of Third-Order Differential Equations*,
DOI 10.1007/978-81-322-1614-8_6,

In this chapter, we will present some oscillation and asymptotic properties of nonoscillatory solutions of third-order delay differential equations, mainly including some contributions of the authors and their colleagues.

Section 6.1 deals with the asymptotic behaviour of solutions of (6.1). In the process, we shall present some recent results in the successive Sects. 6.4–6.6 on the properties of nonoscillatory solutions of the nonlinear equations

$$x'''(t) + q(t)x'(t) + p(t)f\big(x\big(g(t)\big)\big) = 0,$$

$$\big(r(t)\big(x''(t)\big)^{\gamma}\big)' + p(t)f\big(x\big(g(t)\big)\big) = 0,$$

$$\big(r(t)\big(x''(t)\big)^{\gamma}\big)' + p(t)x^{\gamma}\big(g(t)\big) = 0$$

and

$$r_3(t)\big(r_2(t)\big(r_1(t)x'(t)\big)'\big)' + q(t)x'(t) + p(t)f\big(x\big(g(t)\big)\big) = 0,$$

where r and r_i, $i = 0, 1, 2, 3$ are defined as earlier, $f : R \to R$ and $\gamma > 0$ is a ratio of odd positive integers. Third-order delay differential equations with distributed deviating arguments is considered in Sect. 6.7. Section 6.8 deals with the nonlinear delay differential equation

$$\big(r_2(t)\big(r_1(t)x'(t)\big)'\big)' + p(t)x^{\gamma}\big(g(t)\big) = 0.$$

Some interesting results on the oscillation of solutions of the nonlinear nonhomogeneous third-order delay differential equations are given in Sect. 6.9.

By a solution of the above considered equations, we mean a function $x(t) \in C([T_x, \infty), R)$, $T_x \geq \sigma$, that satisfies the considered equation on $[T_x, \infty)$.

6.1 Asymptotic Behaviour of Solutions of Linear Delay Differential Equations of the Form $x'''(t) + a(t)x''(t) + b(t)x'(t) + c(t)x(g(t)) = 0$

This section is concerned with the asymptotic behaviour of nonoscillatory solutions of the third-order delay differential equation (6.1). For our convenience, let us denote by

(H_1) $a(t) \geq 0, b(t) \geq 0, c(t) > 0;$
(H_2) $a(t) \geq 0, b(t) \leq 0, c(t) > 0;$
(H_3) $a(t) \leq 0, b(t) \leq 0, c(t) < 0.$

From the proof of the results given in Sects. 2.1, 2.3 and 2.5 of Chap. 2, it is clear that such techniques cannot be applied to derive similar results for (6.1). This is due to the presence of the delay in (6.1). However, the study of the asymptotic behaviour of solutions of (6.1) is possible for the cases (H_2) and (H_3) by using the canonical transformation due to Trench [42] and some comparison results by Kusano and Naito [22]. In a similar way, it is possible to study (6.1) by using Riccati's transformation Technique in the case (H_1) as we shall see in the preceding theorems.

Lemma 6.1.1 *Assume* (H_1). *Suppose that the second-order differential equation* (2.110) *is nonoscillatory. If* $x(t)$ *is a nonoscillatory solution of* (6.1), *then there exists a* $t_0 \in [\sigma, \infty)$ *such that* $x(t)x'(t) > 0$ *or* $x(t)x'(t) < 0$ *for* $t \geq t_0$.

Proof If $x(t)$ is a nonoscillatory solution of (6.1), then there exists a $t_1 \geq \sigma$ such that $x(t) > 0$ or $x(t) < 0$ for $t \geq t_1$. Let $t_2 > t_1$ be such that $g(t) > t_1$ for $t \geq t_2$. Then $x(g(t)) > 0$ or < 0 for $t \geq t_2$. Clearly, $x'(t)$ is a solution of the second-order nonhomogeneous equation

$$\big(r(t)z'\big)' + q(t)z = -p(t)x\big(g(t)\big), \quad t \geq t_2,$$

which has only nonoscillatory solutions, because (2.110) is nonoscillatory. Hence, $x'(t)$ is nonoscillatory. Consequently, there exists a $t_0 \geq t_2$ such that $x(t)x'(t) > 0$ or $x(t)x'(t) < 0$ for $t \geq t_0$. The lemma is proved. □

Theorem 6.1.1 *Assume* (H_1). *Let* $c(t) \geq d > 0$, $c(t) - a(t)b(t) - b'(t) \geq 0$ *and* $b(t)$ *be bounded. Further, suppose that* (2.110) *is nonoscillatory and* $\int_\sigma^\infty a(t)\,dt < \infty$. *If for any* $\mu \in (0, \frac{1}{2})$, *the third-order ordinary differential equation*

$$u''' + a(t)u'' + b(t)u' + \mu\left(\frac{g(t)}{t}\right)^2 c(t)u = 0 \tag{6.5}$$

admits an oscillatory solution, then every nonoscillatory solution of (6.1) *along with its first and second derivatives tends to zero as* $t \to \infty$.

Proof Let $x(t)$ be a nonoscillatory solution of (6.1). Without any loss of generality, we may assume that $x(t) > 0$ for $t \geq t_0 \geq \sigma$. Thus there exists a $t_1 \geq t_0$ such that $x(g(t)) > 0$ for $t \geq t_1$. By Lemma 6.1.1, there exists a $t_2 \geq t_1$ such that $x'(t) > 0$ or $x'(t) < 0$ for $t \geq t_2$. Suppose that $x'(t) > 0$ for $t \geq t_2$. Then by (6.2), $(r(t)x''(t))' < 0$ for $t \geq t_2$ and hence $x''(t) > 0$ or < 0 for $t \geq t_3 \geq t_2$. Clearly, the condition $\int_\sigma^\infty a(t)\,dt < \infty$ implies that $\int_\sigma^\infty \frac{1}{r(t)}\,dt = \infty$. If $x''(t) < 0$ for $t \geq t_3$, then the repeated integration of $(r(t)x''(t))' < 0$ from t_3 to t yields $x'(t) < 0$ for large t, a contradiction. Hence $x''(t) > 0$ for $t \geq t_3$. This, in turn, implies that $x'''(t) < 0$ for $t \geq t_3$. Then a simple calculation shows that, for every $\mu \in (0, \frac{1}{2})$, there exists a $T_\mu > t_3$ such that

$$\frac{x(g(t))}{x(t)} \geq \mu\left(\frac{g(t)}{t}\right)^2 \tag{6.6}$$

for $t \geq T_\mu$. Setting $z(t) = \frac{x'(t)}{x(t)}$ for $t \geq T_\mu$, we get $z'(t) + z^2(t) = x''(t)/x(t)$. Further, taking the transformation $u(t) = e^{\int_{T_\mu}^t z(s)\,ds}$ and by using (6.6), we obtain

$$u''' + a(t)u'' + b(t)u' + \mu\left(\frac{g(t)}{t}\right)^2 c(t)u \leq 0$$

for $t \geq T_\mu$, which implies by Lemma 4 due to Greguš and Gera [18] that (6.5) is disconjugate on $[T_x, \infty)$, a contradiction to the hypothesis of the theorem. Hence $x'(t) < 0$ for $t \geq t_2$. Consider the function

$$F[x(t)] = r(t)\big(x'(t)\big)^2 - 2r(t)x(t)x''(t) - q(t)x^2(t). \tag{6.7}$$

By using (6.2) in (6.7), we get

$$F'[x(t)] \geq r'(t)\big(x'(t)\big)^2 + \big(2p(t) - q'(t)\big)x^2(t) > 0. \tag{6.8}$$

Hence $F[x(t)] > 0$ or < 0 for $t \geq t_4 \geq t_2$. We claim that $F[x(t)] < 0$ for $t \geq t_4$. If not, then $F[x(t)] > 0$ for $t \geq t_4$. Since $x'(t) < 0$ for $t \geq t_2$, three possibilities on $x''(t)$ do occur, i.e., there exists $t_5 \geq t_2$ such that $x''(t) > 0$ or < 0 or $x''(t)$ changes the sign for $t \geq t_5$. Let $t_6 = \max\{t_4, t_5\}$. Clearly, $x''(t) < 0$ for $t \geq t_6$ implies that $x(t) < 0$ for large t, a contradiction. If $x''(t) > 0$ for $t \geq t_6$, then $\lim_{t\to\infty} x'(t)$ exists and ≤ 0. From (6.8), we obtain for $t \geq t_6$ that

$$F[x(t)] \geq F[x(t_6)] + \int_{t_6}^{t} \big(2c(s) - a(s)b(s) - b'(s)\big)r(s)x^2(s)\,ds \tag{6.9}$$

and hence

$$r(t)\big(x'(t)\big)^2 \geq \int_{t_6}^{t} \big(2c(s) - a(s)b(s) - b'(s)\big)r(s)x^2(s)\,ds. \tag{6.10}$$

Taking limit as $t \to \infty$ in (6.10), we have

$$\lim_{t\to\infty} r(t)\big(x'(t)\big)^2 = \lambda > 0.$$

Hence $\lim_{t\to\infty} x'(t) = \lambda_1$, $\lambda_1 < 0$. This, in turn, implies that $x(t) < 0$ for large t, a contradiction. If $x''(t)$ changes sign for $t \geq t_6$, then $x'(t)$ has maxima for arbitrarily large t. We claim that $\limsup_{t\to\infty} x'(t) = 0$. If not, then $\limsup_{t\to\infty} x'(t) < 0$. Then for $0 < \epsilon < -k$, there exists $T \geq t_6$ such that $x'(t) < k + \epsilon$ for $t \geq T$, which again implies that $x(t) < 0$ for large t, a contradiction. Hence our claim holds, that is, $\limsup_{t\to\infty} x'(t) = 0$. Let $\{t_n\}$ be the sequence of maxima of $x'(t)$. So, $\limsup_{t\to\infty} x'(t_n) = 0$. Clearly $\{t_n\}$ contains a subsequence $\{s_n\}$, $s_n \to \infty$ as $n \to \infty$ and $\lim_{n\to\infty} x'(s_n) = 0$. We may note that $x''(s_n) = 0$, $n = 1, 2, 3, \ldots$. Hence from (6.10), we obtain

$$0 = \lim_{n\to\infty} r(s_n)\big(x'(s_n)\big)^2 \geq \lim_{n\to\infty} \int_{t_6}^{s_n} \big(2c(s) - a(s)b(s) - b'(s)\big)r(s)x^2(s)\,ds > 0,$$

a contradiction. Thus, our claim holds, that is, $F[x(t)] < 0$ for $t \geq t_4$. As $r(t) > 1$ and $F[x(t)] < 0$ for $t \geq t_4$, then

$$0 \leq \big(x'(t)\big)^2 \leq r(t)\big(x'(t)\big)^2 < x(t)\big(2r(t)x''(t) + q(t)x(t)\big). \tag{6.11}$$

Thus

$$2r(t)x''(t)+q(t)x(t)>0, \quad t \ge t_4. \tag{6.12}$$

Consequently, from (6.2) and (6.12), we obtain

$$\begin{aligned} 0 &\le r(t)x''(t)+q(t)x(t) \\ &= r(t_4)x''(t_4)+q(t_4)x(t_4)+\int_{t_4}^{t}\big(q'(s)-p(s)\big)x(s)\,ds \\ &\le k, \end{aligned} \tag{6.13}$$

where $k=r(t_4)x''(t_4)+q(t_4)x(t_4)$. Then

$$2r(t)x''(t)+q(t)x(t) \le 2k. \tag{6.14}$$

By using (6.11) in (6.14), we get

$$\big(x'(t)\big)^2 \le 2kx(t). \tag{6.15}$$

Since $x(t)$ is bounded, then from (6.15) it follows that $x'(t)$ is bounded. Further, one may obtain from (6.9) with t_6 replaced by t_4 that

$$\int_{t_4}^{t} x^2(s)\,ds \le \frac{F[x(t_4)]}{d} < \infty.$$

Hence

$$\int_{t_4}^{\infty} x^2(s)\,ds < \infty.$$

Then Lemma 1.5.15 implies that $x(t)\to 0$ as $t\to\infty$. Thus from (6.15), it follows that $x'(t)\to 0$ as $t\to\infty$. Since (6.1) and (6.2) are equivalent, from (6.2) we have

$$\begin{aligned} \big[r(t)x''(t)+q(t)x(t)\big]' &\le -\big(p(t)-q'(t)\big)x(t) \\ &\le -\big(c(t)-a(t)b(t)-b'(t)\big)r(t)x(t) \\ &\le 0 \end{aligned}$$

for large t. Hence $(r(t)x''(t)+q(t)x(t))>0$ or <0, for large t. Clearly $(r(t)x''(t)+q(t)x(t))<0$ for large t implies that $x(t)<0$ for large t, a contradiction. Hence $(r(t)x''(t)+q(t)x(t))>0$ for large t. Consequently,

$$\lim_{t\to\infty}\big[r(t)x''(t)+q(t)x(t)\big]=l, \quad 0\le l<\infty.$$

Since $q(t)$ is bounded, we have $\lim_{t\to\infty} r(t)x''(t)=l$. If $l>0$, then $x'(t)>0$ for large t. This contradiction proves that $l=0$. Consequently, $\lim_{t\to\infty} x''(t)=0$. This completes the proof of the theorem. □

Using Theorems 2.5.4 and 6.1.1, we can obtain the following corollary.

Corollary 6.1.1 *Assume* (H_1). *Let* $c(t) \geq d > 0$, $\int_\sigma^\infty a(t)\,dt < \infty$, $b(t)$ *be bounded and* $\int_\sigma^\infty (\frac{g(t)}{t})^2 c(t)\,dt = \infty$. *If for any* $\mu \in (0, \frac{1}{2})$, $\mu(\frac{g(t)}{t})^2 c(t) - a(t)b(t) - b'(t) \geq 0$, $a'(t) \leq 0$ *and* (2.110) *is nonoscillatory, then every nonoscillatory solution of* (6.1) *along with its first and second derivatives tends to zero as* $t \to \infty$.

Example 6.1.1 Consider

$$x'''(t) + \frac{1}{t^2}x''(t) + \frac{1}{(t-1)}\left(\frac{2}{t^2} - \frac{1}{t^3}\right)x'(t) + \frac{1}{e}\left(1 - \frac{1}{t^2} + \frac{(2t-1)}{(t-1)t^3}\right)x(t-1) = 0 \tag{6.16}$$

for $t \geq 2$. From Theorem 2.6.14, the second-order linear differential equation

$$z'' + \frac{1}{t^2}z' + \frac{1}{(t-1)}\left(\frac{2}{t^2} - \frac{1}{t^3}\right)z = 0$$

is nonoscillatory. One may verify that all the conditions of Corollary 6.1.1 are satisfied and hence nonoscillatory solutions of (6.16) along with their first and second derivatives tends to zero as $t \to \infty$. In particular, $x(t) = e^{-t}$ is a nonoscillatory solution of (6.16).

Now, we consider the second-order linear differential equation

$$\big(r(t)v'\big)' + q(t)v = 0 \tag{6.17}$$

with $q(t) > 0$ and $R(t) = \int_\sigma^t \frac{1}{r(s)}\,ds \to \infty$ as $t \to \infty$. If $v(t)$ is a positive solution of (6.17), then (6.2) can be rewritten as

$$\frac{1}{v(t)}\left(\frac{r(t)}{v^{-2}(t)}\left(\frac{1}{v(t)}x'(t)\right)'\right)' + p(t)x\big(g(t)\big) = 0. \tag{6.18}$$

Indeed, the following calculation proves (6.18).

$$\frac{1}{v(t)}\left(\frac{r(t)}{v^{-2}(t)}\left(\frac{1}{v(t)}x'(t)\right)'\right)' = \big(r(t)x''(t)\big)' + q(t)x'(t). \tag{6.19}$$

Note that (6.18) is in canonical form if

$$\int_\sigma^\infty v(t)\,dt = \infty \tag{6.20}$$

and

$$\int_\sigma^\infty \frac{1}{r(t)v^2(t)}\,dt = \infty \tag{6.21}$$

hold. In Lemma 1.4.1, we have seen that if $q(t) \le 0$, then (6.17) admits a positive increasing solution $v(t)$ such that

$$\int_\sigma^\infty v(t)\,dt = \infty \quad \text{and} \quad \int_\sigma^\infty \frac{dt}{v^2(t)r(t)} < \infty \tag{6.22}$$

hold. If (6.17) is nonoscillatory and $q(t) \ge 0$, then it is always possible to choose a solution $v(t)$ of (6.17) for which (6.21) holds. Indeed, if $v_1(t)$ is a positive solution of (6.17) for which (6.21) is violated, then choose

$$v_2(t) = v_1(t) \int_\sigma^\infty \frac{1}{r(s)v^2(s)}\,ds$$

as another positive solution of (6.17) and, for $v_2(t)$, (6.21) holds.

Suppose that $v(t)$ is a positive solution of (6.17). Then it is easy to check that $v'(t) > 0$, $(r(t)v'(t))' < 0$, the property (6.20) holds and there exists a constant $c_0 > 0$ such that $v(t) \le c_0 R(t)$. It is important to find sufficient conditions under which (6.17) possesses a positive solution. In consequence, Sturm's comparison theorem proves that if

$$R^2(t)r(t)q(t) \le \frac{1}{4}, \tag{6.23}$$

then (6.17) possesses a positive solution $v(t)$. Thus the following lemma is an immediate consequence of the above discussions.

Lemma 6.1.2 *Assume* (H_1). *Let* (6.23) *holds. Then* (6.2) *can be represented in its canonical form* (6.18).

Denote $\Gamma_0 x = x$, $\Gamma_1 x = \frac{1}{v(t)}(\Gamma_0 x)'$, $\Gamma_2 x = v^2(t)r(t)(\Gamma_1 x)'$ and $\Gamma_3 x = (\Gamma_2 x)'$. Then, (6.18) can be expressed as

$$\Gamma_3 x(t) + v(t)p(t)x\big(g(t)\big) = 0. \tag{6.24}$$

Assume that (6.20) and (6.21) are satisfied. Then (6.18) or (6.24) is in canonical form. If $x(t)$ is a positive nonoscillatory solution of (6.18) or (6.24), then either

$$\Gamma_0 x(t) > 0, \qquad \Gamma_1 x(t) < 0, \qquad \Gamma_2 x(t) > 0 \quad \text{and} \quad \Gamma_3 x(t) < 0 \tag{6.25}$$

or

$$\Gamma_0 x(t) > 0, \qquad \Gamma_1 x(t) > 0, \qquad \Gamma_2 x(t) > 0 \quad \text{and} \quad \Gamma_3 x(t) < 0 \tag{6.26}$$

holds for large t, say for $t \ge t_0 \ge \sigma$. A nonoscillatory solution $x(t)$ of (6.18) or (6.24) that satisfies (6.25) is called $x(t)$ is of *degree* 0 and a nonoscillatory solution $x(t)$ of (6.18) or (6.24) that satisfies (6.26) is called $x(t)$ is of *degree* 2.

Definition 6.1.1 We say that (6.18) or (6.24) has property $\overline{A}$ if every nonoscillatory solution $x(t)$ of the equation is of *degree* 0, that is, it satisfies (6.25).

Let $x(t)$ be a nonoscillatory solution of (6.2) that satisfies the property

$$x(t)x'(t) < 0 \tag{6.27}$$

for some $t \geq t_0 \geq \sigma$. Theorem 6.1.2 shows that this property (6.27) is equivalent to the property $\overline{A}$ of (6.18) or (6.24) in the sense that $x(t)$ satisfies (6.27), if and only if $x(t)$ satisfies (6.25).

Theorem 6.1.2 *Let* (H_1) *and* (6.23) *hold. Assume that* $v(t)$ *is a positive solution of* (6.17), *satisfying* (6.20) *and* (6.21). *Then* (6.18) *or* (6.24) *has property* $\overline{A}$, *if and only if every nonoscillatory solution of Eq.* (6.2) *satisfies the property* (6.27).

Proof First, suppose that (6.24) has property $\overline{A}$ and $x(t)$ is a positive solution of (6.2). We need to show that $x'(t) < 0$. Since $x(t)$ is also a solution of (6.24), it satisfies (6.25). Therefore, $\Gamma_1 x(t) = x'(t)/v(t) < 0$ holds, that is, $x'(t) < 0$.

Next, suppose that a solution $x(t)$ of (6.2) satisfies the property (6.27). Then $x(t)$ is also a solution of (6.24). We shall show that $x(t)$ satisfies (6.25). Since $x'(t) < 0$, we have $\Gamma_1 x(t) < 0$. Then it follows from (6.24) that $\Gamma_3 x(t) < 0$. Thus $\Gamma_2 x(t)$ is decreasing and $\Gamma_2 x(t) > 0$ or $\Gamma_2 x(t) < 0$ eventually. If $\Gamma_2 x(t) < 0$ eventually, say for $t \geq t_1 \geq \sigma$, then $\Gamma_1 x(t)$ is decreasing, and integrating the inequality $\Gamma_1 x(t) < \Gamma_1 x(t_1)$ from t_1 to t, we get

$$x(t) < x(t_1) + \Gamma_1 x(t_1) \int_{t_1}^{t} v(s)\,ds \to -\infty \quad \text{as } t \to \infty.$$

This contradiction proves the theorem. □

Theorem 6.1.3 *Let* (6.23) *holds,* $p(t) > 0$ *and* $q(t) \geq 0$. *Assume that* $v(t)$ *is a positive solution of* (6.17) *satisfying* (6.20) *and* (6.21). *Let* $g \in C^1([\sigma, \infty), R)$, $g'(t) > 0$. *If the third-order ordinary differential equation*

$$\left(\frac{r(t)}{v^{-2}(t)}\left(\frac{1}{v(t)}x'(t)\right)'\right)' + \frac{v(g^{-1}(t))p(g^{-1}(t))}{g'(g^{-1}(t))}x(t) = 0 \tag{6.28}$$

has property $\overline{A}$, *then* (6.24) *has property* $\overline{A}$.

Combining Theorems 6.1.2 and 6.1.3, we obtain the following corollary.

Corollary 6.1.2 *Let* (6.23) *holds,* $p(t) > 0$, $q(t) \geq 0$, $g \in C^1([\sigma, \infty), R)$ *and* $g'(t) > 0$. *Assume that* $v(t)$ *is a positive solution of* (6.17), *satisfying* (6.20) *and* (6.21). *If* (6.28) *has property* $\overline{A}$, *then every nonoscillatory solution of* (6.2) *satisfies the property* (6.27).

Thus employing any known or future result for property $\overline{A}$ of (6.28), then in view of Corollary 6.1.2, we immediately obtain: every nonoscillatory solution of (6.2) satisfies the property (6.27).

Example 6.1.2 Consider the equation

$$\left(\frac{1}{t}x''(t)\right)' + \frac{\alpha(2-\alpha)}{t^3}x'(t) + p(t)x\big(g(t)\big) = 0, \tag{6.29}$$

where $0 < \alpha < 1$ and $g(t)$ satisfies the property $g \in C^1([\sigma, \infty), R)$ and $g'(t) > 0$. The corresponding second-order differential equation (6.17) takes the form

$$\left(\frac{1}{t}v'(t)\right)' + \frac{\alpha(2-\alpha)}{t^3}v(t) = 0$$

which has a pair of solutions $v(t) = t^\alpha$ and $v(t) = t^{2-\alpha}$. Here, $v(t) = t^\alpha$ is our desirable solution, which allows us to rewrite (6.29) in its canonical form. Then, by Corollary 6.1.2, every positive nonoscillatory solution of Eq. (6.29) satisfies the property (6.27), if the canonical equation

$$\big(t^{2\alpha-1}\big(t^{-\alpha}x'(t)\big)'\big)' + \frac{(g^{-1}(t))^\alpha p(g^{-1}(t))}{g'(g^{-1}(t))}x(t) = 0 \tag{6.30}$$

has property $\overline{A}$.

Corollary 6.1.3 *Let* (6.23) *holds*, $p(t) > 0$, $q(t) \geq 0$, $g \in C^1([\sigma, \infty), R)$ *and* $g'(t) > 0$. *Assume that* $v(t)$ *is a positive solution of* (6.17), *satisfying* (6.20) *and* (6.21). *If* (6.28) *has property* $\overline{A}$ *and*

$$\int_\sigma^\infty v(s_3) \int_{s_3}^\infty \frac{1}{r(s_2)v^2(s_2)} \int_{s_2}^\infty v(s_1)p(s_1)\,ds_1\,ds_2\,ds_3 = \infty, \tag{6.31}$$

then every nonoscillatory solution $x(t)$ *of* (6.2) *satisfies the property*

$$\lim_{t\to\infty} x(t) = 0.$$

Proof Let $x(t)$ be a positive solution of (6.2). Then, it follows from Corollary 6.1.2 that $x'(t) < 0$ for some $t \geq t_0 \geq \sigma$. Thus $\lim_{t\to\infty} x(t) = l \geq 0$. Assume that $l > 0$. Since $x(t)$ is also a solution of (6.18), in view of Theorem 6.1.2, it has to be of *degree* 0, that is, (6.25) is satisfied. Integrating (6.18) from t ($\geq t_0$) to ∞, we get

$$\Gamma_2 x(t) \geq \int_t^\infty v(s)p(s)x\big(g(s)\big)\,ds \geq l\int_t^\infty v(s)p(s)\,ds.$$

Multiplying this inequality by $\frac{1}{r(t)v^2(t)}$ and then integrating from t to ∞, we have

$$-\Gamma_1 x(t) \geq l\int_t^\infty \frac{1}{r(s_2)v^2(s_2)} \int_{s_2}^\infty v(s_1)p(s_1)\,ds_1\,ds_2.$$

Multiplying this by $v(t)$ and integrating from t_0 to t, we obtain

$$x(t_1) \geq l \int_{t_0}^{t} v(s_3) \int_{s_3}^{\infty} \frac{1}{r(s_2)v^2(s_2)} \int_{s_2}^{\infty} v(s_1)p(s_1)\,ds_1\,ds_2\,ds_3 \to \infty \quad \text{as } t \to \infty.$$

This contradiction proves that $l = 0$ and the proof is complete. □

Example 6.1.3 Consider (6.29). It is easy to see that (6.31) takes the form

$$\int_{\sigma}^{\infty} s_3^{\alpha} \int_{S_3}^{\infty} s_2^{1-2\alpha} \int_{s_2}^{\infty} s_1^{\alpha} p(s_1)\,ds_1\,ds_2\,ds_3 = \infty.$$

Then, by Corollary 6.1.3, very nonoscillatory solution of (6.29) tends to zero as $t \to \infty$, provided that the above integration holds and (6.30) has property $\overline{A}$.

Kusano and Naito [22] proved the following comparison result, which plays an important role in giving property $\overline{A}$ of (6.24).

Lemma 6.1.3 (Theorem 2, [22]) *Let $\int^{\infty} r_1(t)\,dt = \infty$. The second-order differential equation*

$$\frac{1}{r_2(t)}\left(\frac{1}{r_1(t)}\left(\frac{u(t)}{r_0(t)}\right)'\right)' + p(t)u\big(g(t)\big) = 0$$

has a nonoscillatory solution $u(t)$, satisfying

$$u(t) > 0, \qquad u'(t) < 0 \quad \text{and} \quad \left(\frac{1}{r_1(t)}\left(\frac{u(t)}{r_0(t)}\right)'\right)' < 0,$$

if and only if the differential inequality

$$\left\{\frac{1}{r_2(t)}\left(\frac{1}{r_1(t)}\left(\frac{z(t)}{r_0(t)}\right)'\right)' + p(t)z\big(g(t)\big)\right\} \operatorname{sgn} z\big(g(t)\big) \leq 0$$

has a solution $z(t)$ satisfying the property

$$z(t) > 0, \qquad z'(t) < 0 \quad \text{and} \quad \left(\frac{1}{r_1(t)}\left(\frac{z(t)}{r_0(t)}\right)'\right)' < 0.$$

Theorem 6.1.4 *Let* (6.23) *holds, $p(t) > 0$, $q(t) \geq 0$, $g \in C^1([\sigma, \infty), R)$ and $g'(t) > 0$. Assume that $v(t)$ is a positive solution of* (6.17) *satisfying* (6.20) *and* (6.21). *Let*

$$\int^{\infty} v(t)p(t)\,dt < \infty. \tag{6.32}$$

If the second-order differential equation

$$\big(r(t)v^2(t)z'(t)\big)' + \left(v\big(g(t)\big)g'(t)\int_{t}^{\infty} v(s)p(s)\,ds\right)z\big(g(t)\big) = 0 \tag{6.33}$$

is oscillatory, then (6.18) *or* (6.24) *has property $\overline{A}$.*

Proof Assume that $x(t)$ be a positive solution of (6.24). Then $x(t)$ is either of *degree* 0 or of *degree* 2. Let $x(t)$ satisfies the property (6.26). An integration of (6.24) from t $(\geq t_1)$ to ∞ yields

$$\Gamma_2 x(t) \geq \int_t^\infty v(s)p(s)x\big(g(s)\big)\,ds.$$

On the other hand,

$$x(t) \geq \int_{t_1}^t v(s)\Gamma_1 x(s)\,ds$$

holds. Combining the last two inequalities, we get

$$\begin{aligned}
\Gamma_2 x(t) &\geq \int_t^\infty v(s)p(s)\int_{t_1}^{g(s)} v(\theta)\Gamma_1 x(\theta)\,d\theta\,ds \\
&\geq \int_t^\infty v(s)p(s)\int_{g(t)}^{g(s)} v(\theta)\Gamma_1 x(\theta)\,d\theta\,ds \\
&= \int_{g(t)}^\infty \Gamma_1 x(\theta)v(\theta)\int_{g^{-1}(\theta)}^\infty v(s)p(s)\,ds\,d\theta.
\end{aligned}$$

Integrating the above inequality from t_1 to t, we see that $\omega(t) \equiv \Gamma_1 x(t)$ satisfies

$$\omega(t) \geq \omega(t_1) + \int_{t_1}^t \frac{1}{r(s)v^2(s)}\int_{g(s)}^\infty \Gamma_1 x(\theta)v(\theta)\int_{g^{-1}(\theta)}^\infty v(u)p(u)\,du\,d\theta\,ds. \quad (6.34)$$

Set

$$z(t) = \omega(t_1) + \int_{t_1}^t \frac{1}{r(s)v^2(s)}\int_{g(s)}^\infty \Gamma_1 x(\theta)v(\theta)\int_{g^{-1}(\theta)}^\infty v(u)p(u)\,du\,d\theta\,ds.$$

Then it is easy to see that $z(t) > 0$ and satisfies

$$\begin{aligned}
0 &= \big(r(t)v^2(t)z'(t)\big)' + \left(v\big(g(t)\big)g'(t)\int_t^\infty v(s)p(s)\,ds\right)\omega\big(g(t)\big) \\
&\geq \big(r(t)v^2(t)z'(t)\big)' + \left(v\big(g(t)\big)g'(t)\int_t^\infty v(s)p(s)\,ds\right)z\big(g(t)\big).
\end{aligned}$$

Then Lemma 6.1.3 implies that (6.33) has a positive solution, contradicting the hypothesis of the theorem. Hence $x(t)$ satisfies (6.25). The theorem is proved. □

The following theorem gives another sufficient condition when (6.32) fails to hold.

Theorem 6.1.5 *Let* (6.23) *hold,* $p(t) > 0$, $q(t) \geq 0$, $g \in C^1([\sigma,\infty), R)$ *and* $g'(t) > 0$. *Assume that* $v(t)$ *is a positive solution of* (6.17) *satisfying* (6.20) *and* (6.21). *Let*

$$\int^{\infty} v(t)p(t)\,dt = \infty,$$

then (6.24) *has property* $\overline{A}$.

Proof If $x(t)$ is a positive solution of (6.24) such that it is of *degree* 2, then an integration of (6.24) yields

$$\begin{aligned} \Gamma_2 x(t_1) &\geq \int_{t_1}^{t} v(s)p(s)x\big(g(s)\big)\,ds \\ &\geq x\big(g(t_1)\big)\int_{t_1}^{t} v(s)p(s)\,ds \\ &\to \infty \quad \text{as } t \to \infty, \end{aligned}$$

which is a contradiction. Thus, $x(t)$ is of *degree* 0. The proof is complete. □

Corollary 6.1.4 *Let* (6.23) *and* (6.32) *hold,* $p(t) > 0$, $q(t) \geq 0$, $g \in C^1([\sigma,\infty), R)$ *and* $g'(t) > 0$. *Assume that* $v(t)$ *is a positive solution of* (6.17) *satisfying* (6.20) *and* (6.21). *If* (6.33) *is oscillatory, then every nonoscillatory solution of* (6.2) *satisfies the property* (6.27).

Corollary 6.1.5 *Let* (6.23) *and* (6.32) *hold,* $p(t) > 0$, $q(t) \geq 0$, $g \in C^1([\sigma,\infty), R)$ *and* $g'(t) > 0$. *Assume that* $v(t)$ *is a positive solution of* (6.17) *satisfying* (6.20) *and* (6.21). *If*

$$\liminf_{t\to\infty}\left(\int_{\sigma}^{g(t)} \frac{1}{r(s)v^2(s)}\,ds\right)\left(\int_{t}^{\infty} v\big(g(s)\big)g'(s)\int_{s}^{\infty} v(u)p(u)\,du\,ds\right) > \frac{1}{4}, \tag{6.35}$$

then every nonoscillatory solution of (6.2) *satisfies the property* (6.27).

Corollary 6.1.6 *Let* (6.23) *hold,* $p(t) > 0$, $q(t) \geq 0$, $g \in C^1([\sigma,\infty), R)$ *and* $g'(t) > 0$. *Assume that* $v(t)$ *is a positive solution of* (6.17) *satisfying* (6.20) *and* (6.21). *If* (6.31) *and* (6.35) *hold, then every nonoscillatory solution of* (6.2) *tends to zero as* $t \to \infty$.

Example 6.1.4 Consider (6.29) again. By Corollary 6.1.6, every nonoscillatory solution of (6.29) tends to zero as $t \to \infty$, provided that the integration given in Example 6.1.3 holds and

$$\liminf_{t\to\infty} g^{2-2\alpha}(t)\left(\int_{t}^{\infty} g^{\alpha}(s)g'(s)\int_{s}^{\infty} u^{\alpha}p(u)\,du\,ds\right) > \frac{2-2\alpha}{4}.$$

For a special case of (6.29), namely, for

$$\left(\frac{1}{t}x''(t)\right)' + \frac{\alpha(2-\alpha)}{t^3}x'(t) + \frac{a}{t^4}x(\lambda t) = 0, \tag{6.36}$$

with $0 < \alpha < 1$, $0 < \lambda < 1$ and $\alpha > 0$, we see that every nonoscillatory solution of this equation tends to zero as $t \to \infty$ provided that

$$\frac{a\lambda^{3-\alpha}}{(3-\alpha)(1-\alpha)^2} > 1. \tag{6.37}$$

If we set $a = \beta[(\beta+1)(\beta+3) + \alpha(2-\alpha)]\lambda^{\beta}$, where $\beta > 0$, then one such solution of (6.36) is $x(t) = t^{-\beta}$. On the other hand, if for some $\gamma \in (1+\alpha, 3-\alpha)$, we have $a = \gamma[(\gamma-1)(3-\gamma) + \alpha(\alpha-2)]\lambda^{-\gamma} > 0$, then (6.37) is violated, and (6.36) has a nonoscillatory solution $x(t) = t^{\gamma}$ which is of degree 2.

Theorem 6.1.6 *Let* (6.23) *hold,* $p(t) > 0$, $q(t) \geq 0$, $g \in C^1([\sigma,\infty), R)$ *and* $g'(t) > 0$. *Assume that* $v(t)$ *is a positive solution of* (6.17) *satisfying* (6.20) *and* (6.21). *If every nonoscillatory solution of*

$$\big(r(t)x''(t)\big)' + q(t)x'(t) + \frac{v(g^{-1}(t))p(g^{-1}(t))}{v(t)g'(g^{-1}(t))}x(t) = 0 \tag{6.38}$$

satisfies the property (6.27), *then so does Eq.* (6.2).

Property $\overline{A}$ of (6.18) or (6.24) is sometimes formulated in stronger form (let us denote it by $\overline{A^s}$) as that every nonoscillatory solution of *degree* zero of (6.18) satisfies $\lim_{t\to\infty} x(t) = 0$. Then Theorem 6.1.2 becomes trivially for $\overline{A^s}$ and since Theorem 6.1.3 works also for the property $\overline{A^s}$, the conclusion of Theorem 6.1.6 can be formulated in a stronger form as the following theorem shows.

Theorem 6.1.7 *Let* (6.23) *hold,* $p(t) > 0$, $q(t) \geq 0$, $g \in C^1([\sigma,\infty), R)$ *and* $g'(t) > 0$. *Assume that* $v(t)$ *is a positive solution of* (6.17) *satisfying* (6.20) *and* (6.21). *If every nonoscillatory solution* $x(t)$ *of* (6.38) *satisfies the property* $\lim_{t\to\infty} x(t) = 0$, *then so does Eq.* (6.2).

This comparison principles of Theorems 6.1.6 and 6.1.7 allow us to obtain required properties of nonoscillatory solution of the delay equations from that of the equations without delay so that we can immediately extend the known results for (6.38) to (6.2).

Suppose that $a'(t) \leq 0$. Then in view of Theorems 2.5.4 and 2.5.11, one can obtain the following corollary.

Corollary 6.1.7 *Suppose that* $a'(t) \leq 0, c(t) \geq d > 0, c(t) - a(t)b(t) - b'(t) \geq 0$ *and* $\int_{\sigma}^{\infty} a(t)\,dt = \infty$. *Then every nonoscillatory solution* $x(t)$ *of Eq.* (2.1) *satisfies the property*

$$\lim_{t\to\infty} x(t) = \lim_{t\to\infty} x'(t) = \lim_{t\to\infty} x''(t) = 0. \tag{6.39}$$

Then Theorem 6.1.7 yields the following theorem.

Theorem 6.1.8 *Let* (6.23) *hold,* $p(t) > 0$, $q(t) \geq 0$, $g \in C^1([\sigma, \infty), R)$ *and* $g'(t) > 0$. *Assume that* $v(t)$ *is a positive solution of* (6.17) *satisfying* (6.20) *and* (6.21). *If* $a'(t) \leq 0, c^*(t) \geq d > 0, c^*(t) - a(t)b(t) - b'(t) \geq 0$ *and* $\int_\sigma^\infty a(t)\,dt = \infty$, *then every nonoscillatory solution* $x(t)$ *of Eq.* (6.2) *satisfies the property*

$$\lim_{t\to\infty} x(t) = 0,$$

where

$$c^*(t) = \frac{p(g^{-1}(t))v(g^{-1}(t))}{r(t)v(t)g'(g^{-1}(t))}. \tag{6.40}$$

In view of Corollary 2.5.2, we can obtain the following theorem.

Theorem 6.1.9 *Let* (6.23) *hold,* $p(t) > 0$, $q(t) \geq 0$, $g \in C^1([\sigma, \infty), R)$ *and* $g'(t) > 0$. *Assume that* $v(t)$ *is a positive solution of* (6.17) *satisfying* (6.20) *and* (6.21). *If* $b(t) \geq l > 0, c^*(t) \geq d > 0, c^*(t) - a(t)b(t) - b'(t) \geq 0$ *and* $\int_\sigma^\infty a(t)\,dt = \infty$, *then every nonoscillatory solution* $x(t)$ *of Eq.* (6.2) *satisfies the property*

$$\lim_{t\to\infty} x(t) = 0,$$

where $c^*(t)$ *is given in* (6.40).

In view of Theorems 2.5.4 and 2.5.12, one can obtain the following corollary.

Corollary 6.1.8 *Let* $c(t) \geq d > 0, c(t) - a(t)b(t) - b'(t) \geq 0$ *and there exist constants* l *and* L *be such that* $0 < l \leq b(t) \leq L$, *then every nonoscillatory solution of* (2.1) *satisfies the property* (6.39).

Thus, the following theorem is an immediate consequences of Corollary 6.1.7.

Theorem 6.1.10 *Let* (6.23) *hold,* $p(t) > 0$, $q(t) \geq 0$, $g \in C^1([\sigma, \infty), R)$ *and* $g'(t) > 0$. *Assume that* $v(t)$ *is a positive solution of* (6.17) *satisfying* (6.20) *and* (6.21). *If* $0 < l \leq b(t) \leq L, c^*(t) \geq d > 0$ *and* $c^*(t) - a(t)b(t) - b'(t) \geq 0$, *then every nonoscillatory solution* $x(t)$ *of Eq.* (6.2) *satisfies the property*

$$\lim_{t\to\infty} x(t) = 0,$$

where $c^*(t)$ *is given in* (6.40).

Theorems 2.5.1 and 2.5.11 yield the following corollary.

Corollary 6.1.9 *Let* $c(t) \geq d > 0$, $c(t) - a(t)b(t) - b'(t) \geq 0$, $ta(t) \geq 3$, $2ta(t) - t^2b(t) + t^2a'(t) > 2$ *and* (2.17) *holds. Then every nonoscillatory solution of* (2.1) *satisfies the property* (6.39).

Applying Corollary 6.1.9 to (6.38), we obtain the following theorem.

Theorem 6.1.11 *Let* (6.23) *hold,* $p(t) > 0$, $q(t) \geq 0$, $g \in C^1([\sigma, \infty), R)$ *and* $g'(t) > 0$. *Assume that* $v(t)$ *is a positive solution of* (6.17) *satisfying* (6.20) *and* (6.21). *Further, assume that* $c^*(t) \geq d > 0$, $c^*(t) - a(t)b(t) - b'(t) \geq 0$, $ta(t) \geq 3$, $2ta(t) - t^2b(t) + t^2a'(t) > 2$ *and*

$$\int_\sigma^\infty \left[\frac{2t^2a^3(t)}{27} - \frac{t^2a(t)b(t)}{3} + t^2c^*(t) - \frac{2}{3}a(t) + tb(t) + \frac{t^2a(t)a'(t)}{3} - ta'(t) - \frac{2}{3\sqrt{3}t}\left(1 - t^2b(t) + \frac{1}{3}t^2a^2(t) + t^2a'(t)\right)^{\frac{3}{2}}\right] dt = \infty,$$

then every nonoscillatory solution $x(t)$ *of Eq.* (6.2) *satisfies the property*

$$\lim_{t\to\infty} x(t) = 0,$$

where $c^*(t)$ *is given in* (6.40).

Theorem 6.1.12 *Let* (6.23) *hold,* $a \in C^1([\sigma, \infty), R)$, $a'(t) \geq 0$, $p(t) > 0$, $q(t) \geq 0$, $g \in C^1([\sigma, \infty))$ *and* $g'(t) > 0$. *Assume that* $v(t)$ *is a positive solution of* (6.17) *satisfying* (6.20) *and* (6.21). *If*

$$\int_\sigma^\infty R^2(t)r(t)p(t)\,dt = \infty,$$

then every nonoscillatory solution $x(t)$ *of* (6.2) *satisfying* (6.27) *also satisfies*

$$\lim_{t\to\infty} x(t) = 0.$$

Proof Without any loss of generality, assume that a nonoscillatory solution $x(t)$ of (6.2) satisfies the property $x(t) > 0$ and $x'(t) < 0$ for some $t \geq t_0 \geq \sigma$. Then there exists a $t_1 \geq t_0$ such that $x(g(t)) > 0$ for $t \geq T_1$. Then $\lim_{t\to\infty} x(t) = l \geq 0$. Let $l > 0$. Then $x(g(t)) > l$ for $t \geq t_2 \geq t_1$. Observe that

$$\begin{aligned} &R^2(t)r(t)\big(r(t)x''(t)\big)' + \frac{1}{4}x'(t) + l^2R^2(t)r(t)p(t) \\ &\quad \leq R^2(t)r(t)\big(r(t)x''(t)\big)' + R^2(t)r(t)q(t)x'(t) + R^2(t)r(t)p(t)x\big(g(t)\big) = 0, \end{aligned}$$

that is,

$$R^2(t)r(t)\big(r(t)x''(t)\big)' + \frac{1}{4}x'(t) \leq -l^2R^2(t)r(t)p(t).$$

Integrating the above inequality from t_2 to t, we get

$$\int_{t_2}^{t} R^2(s)r(s)\big(r(s)x''(s)\big)'\,ds + \frac{1}{4}x(t) - \frac{1}{4}x(t_2) \le -l^2 \int_{t_2}^{t} R^2(s)r(s)p(s)\,ds,$$

or

$$R^2(t)r^2(t)x''(t) - k + \frac{1}{4}x(t) - \frac{1}{4}x(t_2) \le -l^2 \int_{t_2}^{t} R^2(s)r(s)p(s)\,ds,$$

which, in turn, implies that $x''(t) < 0$ for large t. Consequently, $x(t) < 0$ for large t, a contradiction. Thus $l = 0$. This completes the proof of the theorem. □

Theorem 6.1.13 *Let* (6.23) *hold,* $p(t) > 0$, $q(t) \ge 0$, $g \in C^1([\sigma,\infty), R)$ *and* $g'(t) > 0$. *Assume that* $v(t)$ *is a positive solution of* (6.17), *satisfying* (6.20) *and* (6.21). *If the first-order delay differential inequality*

$$z'(t) + v(t)p(t)\left(\int_{\sigma}^{g(t)} v(s) \int_{\sigma}^{s} \frac{1}{r(u)v^2(u)}\,du\,ds\right) z\big(g(t)\big) \le 0 \tag{6.41}$$

has no eventually positive solutions, then every nonoscillatory solution $x(t)$ *of* (6.2) *satisfies the property* (6.27).

Proof Assume that $x(t)$ is a nonoscillatory solution of (6.2). Then $x(t)$ is also a solution of (6.18) or (6.24). Without any loss of generality, we may assume that $x(t) > 0$ for $t \ge t_0 \ge \sigma$. Then there exists a $t_1 \ge t_0$ such that $x(g(t)) > 0$ for $t \ge t_1$. Then either (6.25) or (6.26) holds for some $t \ge t_2 \ge t_1$. We claim that (6.25) holds. If not, then $\Gamma_0 x(t) > 0$, $\Gamma_1 x(t) > 0$, $\Gamma_2 x(t) > 0$ and $\Gamma_3 x(t) < 0$ for $t \ge t_2$.

Clearly, the function defined by $z(t) = r(t)v^2(t)(\frac{1}{v(t)}x'(t))'$ is positive and decreasing, which leads to

$$\begin{aligned}\frac{1}{v(t)}x'(t) &\ge \int_{t_2}^{t}\left(\frac{1}{v(u)}x'(u)\right)'du = \int_{t_2}^{t}\frac{1}{r(u)v^2(u)}\left[r(u)v^2(u)\left(\frac{1}{v(u)}x'(u)\right)'\right]du\\ &\ge z(t)\int_{t_2}^{t}\frac{1}{r(u)v^2(u)}\,du.\end{aligned}$$

Integrating the above inequality from t_2 to t, we obtain

$$x(t) \ge \int_{t_2}^{t} z(s)v(s)\int_{t_2}^{s}\frac{1}{r(u)v^2(u)}\,du\,ds \ge z(t)\int_{t_2}^{t} v(s)\int_{t_2}^{s}\frac{1}{r(u)v^2(u)}\,du\,ds,$$

which implies that

$$x\big(g(t)\big) \ge z\big(g(t)\big)\int_{t_2}^{g(t)} v(s)\int_{t_2}^{s}\frac{1}{r(u)v^2(u)}\,du\,ds.$$

The above inequality together with (6.18) shows that $z(t)$ is a solution of the first-order delay differential inequality

$$-z'(t) = v(t)p(t)x\big(g(t)\big) \geq \left[v(t)p(t)\int_{t_2}^{g(t)} v(s)\int_{t_2}^{s} \frac{1}{r(u)v^2(u)}\,du\,ds\right]z\big(g(t)\big),$$

that is, $z(t)$ is a positive solution of

$$z'(t) + \left[v(t)p(t)\int_{t_2}^{g(t)} v(s)\int_{t_2}^{s} \frac{1}{r(u)v^2(u)}\,du\,ds\right]z\big(g(t)\big) \leq 0,$$

a contradiction to the hypothesis of the theorem. Hence $x(t)$ satisfies the property (6.25), which in turn implies that $x(t)$ satisfies the property (6.27). The proof is complete. □

In view of Lemma 1.5.20, Theorem 6.1.13 yields the following corollary.

Corollary 6.1.10 *Let* (6.23) *hold,* $p(t) > 0$, $q(t) \geq 0$, $g \in C^1([\sigma,\infty), R)$ *and* $g'(t) > 0$. *Assume that* $v(t)$ *is a positive solution of* (6.17), *satisfying* (6.20) *and* (6.21). *Assume that*

$$\liminf_{t\to\infty}\int_{g(t)}^{t} v(u)p(u)\int_{\sigma}^{g(t)} v(s)\int_{\sigma}^{s}\frac{1}{r(\theta)v^2(\theta)}\,d\theta\,ds\,du > \frac{1}{e}, \tag{6.42}$$

then every nonoscillatory solution $x(t)$ *of* (6.2) *satisfies the property* (6.27).

Theorem 6.1.14 *Let* (6.23) *hold,* $p(t) > 0$, $q(t) \geq 0$, $g \in C^1([\sigma,\infty), R)$ *and* $g'(t) > 0$. *Assume that* $v(t)$ *is a positive solution of* (6.17), *satisfying* (6.20) *and* (6.21). *Further, assume that there exists a function* $\tau(t) \in C^1([\sigma,\infty))$ *such that* $\tau'(t) > 0$, $\tau(t) > t$ *and* $\phi(t) = g(\tau(\tau(t))) < t$. *If the first-order delay differential inequalities* (6.41) *and*

$$z'(t) + \left[v(t)\int_{t}^{\tau(t)}\frac{1}{r(s)v^2(s)}\int_{s}^{\tau(s)} v(u)p(u)\,du\,ds\right]z\big(\phi(t)\big) \leq 0 \tag{6.43}$$

have no positive solutions, then all solutions of (6.2) *are oscillatory.*

Proof If $x(t)$ is a nonoscillatory solution of (6.2), then $x(t)$ is also a solution of (6.18) or (6.24). Without any loss of generality, we may assume that $x(t) > 0$ for $t \geq t_0 \geq \sigma$. Then there exists a $t_1 \geq t_0$ such that $x(g(t)) > 0$ for $t \geq t_1$. Then either (6.25) or (6.26) holds for some $t \geq t_2 \geq t_1$. Theorem 6.1.13 implies that $x(t)$ satisfies (6.25). Consequently, $x'(t) < 0$ for $t \geq t_2$. Integration of (6.18) from t ($\geq t_2$) to $\tau(t)$ yields

$$r(t)v^2(t)\left(\frac{1}{v(t)}x'(t)\right)' \geq \int_{t}^{\tau(t)} v(s)p(s)x\big(g(s)\big)\,ds$$

$$\geq x\big(g\big(\tau(t)\big)\big)\int_t^{\tau(t)} v(s)p(s)\,ds,$$

that is,

$$\left(\frac{1}{v(t)}x'(t)\right)' \geq \frac{x(g(\tau(t)))}{r(t)v^2(t)}\int_t^{\tau(t)} v(s)p(s)\,ds,$$

which, upon integration from t to $\tau(t)$, gives

$$-\frac{1}{v(t)}x'(t) \geq x\big(\phi(t)\big)\int_t^{\tau(t)} \frac{1}{r(s)v^2(s)}\int_s^{\tau(s)} v(\theta)p(\theta)\,d\theta\,ds.$$

A final integration of the above inequality from t to ∞ yields

$$x(t) \geq \int_t^{\infty} x\big(\phi(u)\big)v(u)\int_u^{\tau(u)} \frac{1}{r(s)v^2(s)}\int_s^{\tau(s)} v(\theta)p(\theta)\,d\theta\,ds\,du.$$

Denoting

$$z(t) = \int_t^{\infty} x\big(\phi(u)\big)v(u)\int_u^{\tau(u)} \frac{1}{r(s)v^2(s)}\int_s^{\tau(s)} v(\theta)p(\theta)\,d\theta\,ds\,du,$$

we see that $x(t) \geq z(t) > 0$, and one may easily verify that $z(t)$ is a positive solution of the first-order delay differential inequality (6.43), a contradiction. Hence (6.2) has no nonoscillatory solutions. Consequently, all solutions of (6.2) are oscillatory. The proof is complete. □

Corollary 6.1.11 *Let* (6.23) *holds,* $p(t) > 0$, $q(t) \geq 0$, $g \in C^1([\sigma,\infty), R)$ *and* $g'(t) > 0$. *Assume that* $v(t)$ *is a positive solution of* (6.17) *satisfying* (6.20) *and* (6.21). *Assume that* (6.42) *and*

$$\liminf_{t\to\infty}\int_{\phi(t)}^{t} v(u)\int_u^{\tau(u)} \frac{1}{r(s)v^2(s)}\int_s^{\tau(s)} v(\theta)p(\theta)\,d\theta\,ds\,du > \frac{1}{e},$$

then all solutions of (6.2) *are oscillatory.*

Now, we consider the case (H_2), where $a \in C^1([\sigma,\infty), R), b, c \in C([\sigma,\infty), R)$ with $a(t) \geq 0, b(t) \leq 0$ and $c(t) > 0$. Here, we provide some sufficient conditions in terms of the coefficient functions $a(t), b(t)$ and $c(t)$ so that every nonoscillatory solution of (6.1) satisfies the property

$$\operatorname{sgn} x(t) = \operatorname{sgn} x''(t) \neq \operatorname{sgn} x'(t) = \operatorname{sgn} x'''(t) \quad \text{for large } t,$$

$$\lim_{t\to\infty} x(t) = 0, \qquad \lim_{t\to\infty} x'(t) = 0 \quad \text{and} \quad \lim_{t\to\infty} x''(t) = 0.$$

It follows from Lemma 1.5.4 and Theorem 2.1.2 that if the conditions of Theorem 2.1.2 are satisfied, then every nonoscillatory solution of Eq. (2.1) has the properties

$$\left.\begin{aligned} &x'''(t)x''(t)x'(t)x(t)\neq 0, \quad t\in[\sigma,\infty),\\ &\operatorname{sgn} x(t)=\operatorname{sgn} x''(t)\neq \operatorname{sgn} x'(t)=\operatorname{sgn} x'''(t),\\ &\lim_{t\to\infty} x(t)=\lambda\neq\pm\infty, \qquad \lim_{t\to\infty} x'(t)=\lim_{t\to\infty} x''(t)=0, \end{aligned}\right\} \tag{6.44}$$

By Corollary 2.1.1, it follows that if the conditions of Theorem 2.1.2 are satisfied and either $\lim_{t\to\infty} t^2 b(t)\neq 0$ or $\int_\sigma^\infty c(t)\,dt=\infty$, then every nonoscillatory solution of (2.1) tends to zero as $t\to\infty$.

Here, we shall give results similar to the above observations for the delay differential equation (6.1) for the case (H_2). A Riccati-type transformation was used in Theorem 6.1.1, while considering the case (H_1). Here, in particular, for the case (H_2), we shall use a canonical transformation due to Trench [42] and Lemmas 1.4.1 and 1.4.2 to prove the above observations. Note that the methods used in Theorem 2.1.2 and Corollary 2.1.1 for Eq. (2.1) cannot be applied to (6.1).

Setting

$$Lx=\big(r(t)x''\big)'+q(t)x', \tag{6.45}$$

one may write (6.2) as

$$Lx(t)+p(t)x\big(g(t)\big)=0. \tag{6.46}$$

The operator (6.45) may be written in the form

$$Lx\equiv\frac{1}{v(t)}\left(\frac{r(t)}{v^{-2}(t)}\left(\frac{x'}{v(t)}\right)'\right)', \tag{6.47}$$

where $v(t)$ is a positive solution of the second-order linear differential equation (2.110) or (2.113). Since $b(t)\le 0$, we have $q(t)\le 0$ and hence Lemma 1.4.1 implies that (2.113) admits a positive increasing solution $v(t)$ satisfying the property (6.22). Then proceeding as in Theorem 1.4.1, it can be shown that (6.46) can be represented essentially in the unique way in a canonical form

$$Lx(t)+p(t)x\big(g(t)\big)=0, \tag{6.46c}$$

where

$$Lx=\frac{1}{r_3(t)}\left(\frac{1}{r_2(t)}\left(\frac{1}{r_1(t)}\left(\frac{x}{r_0(t)}\right)'\right)'\right)', \tag{6.48}$$

$r_i\in C([\sigma,\infty),R)$ such that $r_i(t)>0$, $0\le i\le 3$, and

$$\int_\sigma^\infty r_i(t)\,dt=\infty, \quad i=1 \text{ and } 2.$$

Here, $r_i(t)$, $i = 0, 1, 2, 3$ are given by

$$\left.\begin{aligned} r_0(t) &= \tilde{r}_0(t) \int_t^{\infty} \tilde{r}_1(s)\, ds, \\ r_1(t) &= \tilde{r}_1(t) \left(\int_t^{\infty} \tilde{r}_1(s)\, ds \right)^{-2}, \\ r_2(t) &= \tilde{r}_2(t) \int_t^{\infty} \tilde{r}_1(s)\, ds \\ r_3(t) &= \tilde{r}_3(t) \end{aligned}\right\} \tag{6.49}$$

if

$$\int_{\sigma}^{\infty} \tilde{r}_1(t)\, dt < \infty,$$

where

$$\left.\begin{aligned} \tilde{r}_0(t) &= 1, \\ \tilde{r}_1(t) &= v(t) \int_t^{\infty} \frac{ds}{r(s) v^2(s)}, \\ \tilde{r}_2(t) &= \frac{1}{r(t) v^2(t)} \left(\int_t^{\infty} \frac{ds}{r(s) v^2(s)} \right)^{-2}, \\ \tilde{r}_3(t) &= v(t) \int_t^{\infty} \frac{ds}{r(s) v^2(s)} \end{aligned}\right\} \tag{6.50}$$

and $v(t)$ is a positive increasing solution of (2.113), satisfying (6.22). If

$$\int_{\sigma}^{\infty} \tilde{r}_1(t)\, dt = \infty,$$

then $r_i(t) = \tilde{r}_i(t)$, $i = 0, 1, 2, 3$.

Lemma 6.1.4 *Consider* (H$_2$). *If $x(t)$ is a nonoscillatory solution of* (6.2), *then there exists a $t_1 \geq \sigma$ such that either* (I) $x(t)x'(t) < 0$ *or* (II) $x(t)x'(t) > 0$ *holds for $t \geq t_1$. Further, if* (I) *holds, then*

$$(-1)^i x(t) x^{(i)}(t) > 0, \quad 0 \leq i \leq 3 \tag{6.51}$$

for $t \geq t_2 \geq t_1$.

Proof Without any loss of generality, we may assume that $x(t) > 0$ and $x(g(t)) > 0$ for $t \geq t_0 \geq \sigma$. If possible, let α and β ($t_0 \leq \alpha < \beta$) be consecutive zeros of $x'(t)$ such that $x''(\alpha) \leq 0$, $x''(\beta) \geq 0$ and $x'(t) < 0$ for $t \in (\alpha, \beta)$. Integrating (6.2) from

α to β, we obtain

$$0 \le r(\beta)x''(\beta) - r(\alpha)x''(\alpha) = -\int_{\alpha}^{\beta} q(t)x'(t)\,dt - \int_{\alpha}^{\beta} p(t)x\big(g(t)\big)\,dt < 0,$$

a contradiction. Hence there exists a $t_1 \ge t_0$ such that $x'(t) > 0$ or $x'(t) < 0$ for $t \ge t_1$. Thus $x(t)x'(t) < 0$ or $x(t)x'(t) > 0$ for $t \ge t_1$. Suppose that $x'(t) < 0$ for $t \ge t_1$. Then (6.2) implies that $(r(t)x''(t))' \le 0$ for $t \ge t_1$. Hence $x''(t)$ is nonoscillatory. If $x''(t) < 0$ for large t, then $x(t) < 0$ for large t, a contradiction. Thus $x''(t) > 0$ for $t \ge t_2 \ge t_1$. Since $r'(t) > 0$, we have $x'''(t) < 0$ for $t \ge t_2$. Hence (6.51) holds. This completes the proof of the lemma. □

Definition 6.1.2 Equation (6.46c) is said to have property A, if every nonoscillatory solution $x(t)$ of the equation satisfies

$$(-1)^i x(t)L_i x(t) > 0, \quad 0 \le i \le 3 \tag{6.52}$$

for large t and

$$\lim_{t\to\infty} L_i x(t) = 0, \quad i = 0, 1, 2. \tag{6.53}$$

We recall that (6.1) or (6.2) is said to have property A if every nonoscillatory solution $x(t)$ of the equation satisfies

$$(-1)^i x(t)x^{(i)}(t) > 0, \quad 0 \le i \le 3 \tag{6.54}$$

for large t, and

$$\lim_{t\to\infty} x^{(i)}(t) = 0, \quad i = 0, 1, 2. \tag{6.55}$$

This property A demands that Eq. (6.1) or (6.2) or (6.46c) should admit a nonoscillatory solution (in both cases, viz., $g(t) = t$ and $g(t) < t$). Theorem 2.1.8 gives the existence of a nonoscillatory solution of (6.1) with $g(t) = t$, and one may refer to Sect. 2.7 for the existence of a nonoscillatory solution of (6.46c) with $g(t) = t$. In fact, Lemma 2.7.3 shows that (6.46c) with $g(t) = t$ is of type C_I and hence it admits a nonoscillatory solution. Hence it is interesting to obtain sufficient conditions for the existence of nonoscillatory solutions of (6.1) or (6.46c) with $g(t) < t$. If one proceeds as in Lemma 1.4.2, it can be proved that if $x(t)$ is a nonoscillatory solution of (6.46c), then either

$$\operatorname{sgn} L_0x(t) = \operatorname{sgn} L_1x(t) = \operatorname{sgn} L_2x(t) \ne \operatorname{sgn} L_3x(t) \tag{6.56}$$

or

$$\operatorname{sgn} L_0x(t) = \operatorname{sgn} L_2x(t) \ne \operatorname{sgn} L_1x(t) = \operatorname{sgn} L_3x(t) \tag{6.57}$$

holds for large t. This property remains true when $\int_{\sigma}^{\infty} \tilde{r}_1(t)\,dt = \infty$, that is, $r_i(t) = \widetilde{r}_i(t)$, $0 \le i \le 3$. Note that $x(t)$ is a solution of (6.2), if and only if $x(t)$ is a solution of (6.46c). We can use the Definition 6.1.1 to Eq. (6.46c), where the operator Lx is given in (6.48). We define them here again. *Equation (6.1) (either $g(t) = t$ or $g(t) < t$) is said to have property $\overline{A}$, if every nonoscillatory solution $x(t)$ of the equation satisfies* (6.51) *for large t. Similarly, Eq. (6.46c) (either $g(t) = t$ or $g(t) < t$) is said to have property $\overline{A}$, if every nonoscillatory solution $x(t)$ of the equation satisfies* (6.57).

Theorem 6.1.15 *Let* (H_2) *be satisfied. If* (6.46c) *has the property $\overline{A}$, then* (6.2) *has the property $\overline{A}$.*

Proof Let $x(t)$ be a nonoscillatory solution of (6.2) such that $x(t) > 0$ and $x(g(t)) > 0$ for $t \geq t_0 \geq \sigma$. Since $x(t)$ is a solution of (6.46c) and it has property $\overline{A}$, we have $L_0x(t) > 0$, $L_1x(t) < 0$, $L_2x(t) > 0$ and $L_3x(t) < 0$ for $t \geq t_1 \geq t_0$. As

$$L_1x(t) = \frac{1}{r_1(t)}\big(L_0x(t)\big)' = \frac{1}{r_1(t)}\left(\frac{x'(t)}{r_0(t)} + \frac{\tilde{r}_1(t)x(t)}{r_0^2(t)}\right),$$

then $x'(t) < 0$ for $t \geq t_1$. From Lemma 6.1.4, it follows that $x(t)$ satisfies (6.51). Hence (6.2) has property $\overline{A}$. This completes the proof. □

Theorem 6.1.16 *Let* (H_2) *be satisfied. Let $g \in C^1([\sigma, \infty), R)$ be such that $g'(t) > 0$. If the ordinary differential equation*

$$Lx + \frac{p(g^{-1}(t))r_3(g^{-1}(t))}{g'(g^{-1}(t))r_3(t)}x = 0 \tag{6.58}$$

has the property $\overline{A}$, then so does the delay differential equation (6.46c), *where Lx is given in* (6.48).

Proof If possible, suppose that (6.46c) does not have property $\overline{A}$. Hence it admits a nonoscillatory solution $x(t)$ which fails to satisfy (6.57). Without any loss of generality, we may assume that $x(t) > 0, x(g(t)) > 0$ for $t \geq t_0 \geq \sigma$. Thus $L_0x(t) > 0$, $L_1x(t) > 0$, $L_2x(t) > 0$ and $L_3x(t) < 0$ for $t \geq t_1 \geq t_0$. An integration of (6.46c) from $t (\geq t_1)$ to ∞ gives

$$\big(L_1x(t)\big)' \geq r_2(t)\int_t^\infty r_3(s)p(s)x\big(g(s)\big)\,ds.$$

Further integration of the above inequality twice from t_1 to t yields

$$L_0x(t) \geq \lambda + \int_{t_1}^t r_1(s_1)\left(\int_{t_1}^{s_1} r_2(s_2)\left(\int_{s_2}^\infty r_3(s_3)p(s_3)x\big(g(s_3)\big)\,ds_3\right)ds_2\right)ds_1,$$

where $\lambda = L_0x(t_1) > 0$. The above integral inequality gives

$$
\begin{aligned}
L_0x(t) &\geq \lambda + \int_{t_1}^{t} r_1(s_1)\left(\int_{t_1}^{s_1} r_2(s_2)\right.\\
&\quad \times \left.\left(\int_{s_2}^{\infty} r_3(s_3)p(s_3)r_0\big(g(s_3)\big)L_0x\big(g(s_3)\big)\,ds_3\right)ds_2\right)ds_1\\
&\geq \lambda + \int_{t_1}^{t} r_1(s_1)\left(\int_{t_1}^{s_1} r_2(s_2)\right.\\
&\quad \times \left.\left(\int_{g(s_2)}^{\infty} \frac{r_3(g^{-1}(\theta))p(g^{-1}(\theta))r_0(\theta)L_0x(\theta)}{g'(g^{-1}(\theta))}\,d\theta\right)ds_2\right)ds_1\\
&\geq \lambda + \int_{t_1}^{t} r_1(s_1)\left(\int_{t_1}^{s_1} r_2(s_2)\right.\\
&\quad \times \left.\left(\int_{s_2}^{\infty} \frac{r_3(g^{-1}(\theta))p(g^{-1}(\theta))r_0(\theta)L_0x(\theta)}{g'(g^{-1}(\theta))}\,d\theta\right)ds_2\right)ds_1\\
&\geq \lambda + \int_{t_1}^{t} r_1(s_1)\left(\int_{t_1}^{s_1} r_2(s_2)\right.\\
&\quad \times \left.\left(\int_{s_2}^{s_1} \frac{r_3(g^{-1}(\theta))p(g^{-1}(\theta))r_0(\theta)L_0x(\theta)}{g'(g^{-1}(\theta))}\,d\theta\right)ds_2\right)ds_1\\
&\quad + \int_{t_1}^{t} r_1(s_1)\left(\int_{t_1}^{s_1} r_2(s_2)\right.\\
&\quad \times \left.\left(\int_{s_1}^{\infty} \frac{r_3(g^{-1}(\theta))p(g^{-1}(\theta))r_0(\theta)L_0x(\theta)}{g'(g^{-1}(\theta))}\,d\theta\right)ds_2\right)ds_1
\end{aligned}
$$

since $g(s_2) \leq s_2 \leq s_1$. Clearly

$$
\begin{aligned}
&\int_{t_1}^{\infty} \frac{r_3(g^{-1}(\theta))p(g^{-1}(\theta))r_0(\theta)L_0x(\theta)}{g'(g^{-1}(\theta))}\,d\theta\\
&\quad = \int_{g^{-1}(t_1)}^{\infty} r_3(s)p(s)r_0\big(g(s)\big)L_0x\big(g(s)\big)\,ds\\
&\quad = \lim_{t\to\infty} \int_{g^{-1}(t_1)}^{t} r_3(s)p(s)x\big(g(s)\big)\,ds\\
&\quad < L_2x\big(g^{-1}(t_1)\big)\\
&\quad < \infty
\end{aligned}
$$

because $L_2x(t) > 0$ and $L_3x(t) < 0$ for $t \geq t_1$. Hence, from Lemma 1.5.16, it follows that (we may note that Lemma 1.5.16 holds for $g(t) = t$) the integral equation

$$v(t) = \lambda + \int_{t_1}^{t} r_1(s_1)\Bigg(\int_{t_1}^{s_1} r_2(s_2) \times \left(\int_{s_2}^{s_1} \frac{r_3(g^{-1}(\theta))p(g^{-1}(\theta))r_0(\theta)v(\theta)}{g'(g^{-1}(\theta))}\,d\theta\right)ds_2\Bigg)ds_1 + \int_{t_1}^{t} r_1(s_1)\Bigg(\int_{t_1}^{s_1} r_2(s_2) \times \left(\int_{s_1}^{\infty} \frac{r_3(g^{-1}(\theta))p(g^{-1}(\theta))r_0(\theta)v(\theta)}{g'(g^{-1}(\theta))}\,d\theta\right)ds_2\Bigg)ds_1,$$

that is,

$$v(t) = \lambda + \int_{t_1}^{t} r_1(s_1)\Bigg(\int_{t_1}^{s_1} r_2(s_2) \times \left(\int_{s_2}^{\infty} \frac{r_3(g^{-1}(\theta))p(g^{-1}(\theta))r_0(\theta)v(\theta)}{g'(g^{-1}(\theta))}\,d\theta\right)ds_2\Bigg)ds_1$$

has a solution $v(t)$, satisfying

$$\begin{aligned} L_0x(t) &\geq v(t) \\ &\geq \lambda + \int_{t_1}^{t} r_1(s_1)\Bigg(\int_{t_1}^{s_1} r_2(s_2) \times \left(\int_{s_2}^{s_1} \frac{r_3(g^{-1}(\theta))p(g^{-1}(\theta))r_0(\theta)v(\theta)}{g'(g^{-1}(\theta))}\,d\theta\right)ds_2\Bigg)ds_1, \\ &> \lambda \end{aligned}$$

for $t \geq t_1$. Setting $z(t) = r_0(t)v(t)$, we notice that $L_0z(t) = v(t) > 0$, $L_1z(t) > 0$, $L_2z(t) > 0$ and $L_3z(t) < 0$ for $t \geq t_1$ and

$$Lz + \frac{p(g^{-1}(t))r_3(g^{-1}(t))}{g'(g^{-1}(t))r_3(t)}z = 0,$$

a contradiction, because $z(t)$ is an ultimately positive solution of (6.58) which has property $\overline{A}$. Hence (6.46c) has property $\overline{A}$. The theorem is proved. □

Suppose that $\int_{\sigma}^{\infty} \tilde{r}_1(t)\,dt = \infty$. Then Eq. (6.46c) has the operator Lx defined by

$$Lx = \frac{1}{\tilde{r}_3(t)}\left(\frac{1}{\tilde{r}_2(t)}\left(\frac{1}{\tilde{r}_1(t)}\left(\frac{x}{\tilde{r}_0(t)}\right)'\right)'\right)', \tag{6.59}$$

where $\tilde{r}_i(t)$, $0 \leq i \leq 3$ are defined in Eq. (6.50).

Theorem 6.1.17 *Let* (H$_2$) *hold. Let* $g \in C^1([\sigma, \infty), R)$ *be such that* $g'(t) > 0$. *Suppose that* $\int_\sigma^\infty \tilde{r}_1(t)\,dt = \infty$ *and* $v(t)$ *is a positive increasing solution of* (2.113) *satisfying* (6.22). *If*

$$\big(r(t)x''\big)' + q(t)x' + \frac{p(g^{-1}(t))\tilde{r}_3(g^{-1}(t))}{g'(g^{-1}(t))\tilde{r}_3(t)}x = 0 \tag{6.60}$$

has property $\overline{A}$, *then*

$$Lx + \frac{p(g^{-1}(t))\tilde{r}_3(g^{-1}(t))}{g'(g^{-1}(t))\tilde{r}_3(t)}x = 0 \tag{6.61}$$

has property $\overline{A}$, *where* Lx *is given in* (6.59).

Proof Let $x(t)$ be a nonoscillatory solution of (6.61) on $[\sigma, \infty)$. We may assume that $x(t) > 0$ for $t \ge t_0 \ge \sigma$. Clearly, $x(t)$ is a solution of (6.60). Since (6.60) has property $\overline{A}$, we have $x'(t) < 0$, $x''(t) > 0$ and $x'''(t) < 0$ for $t \ge t_1 \ge t_0$. Clearly Lx takes the form (6.59). From (6.61), it follows that $L_0x(t) > 0$ and $L_3x(t) < 0$. Further,

$$L_1x(t) = \frac{1}{\tilde{r}_1(t)}\big(L_0x(t)\big)' = \frac{1}{\tilde{r}_1(t)}\left(\frac{x(t)}{\tilde{r}_0(t)}\right)' = \frac{x'(t)}{\tilde{r}_1(t)} < 0$$

for large t. Consequently, (6.57) holds. Thus, (6.61) has property $\overline{A}$. This completes the proof of the theorem. □

If $\int_\sigma^\infty \tilde{r}_1(t)\,dt < \infty$ in Theorem 6.1.17, then the operator Lx in (6.61) is given by (6.48) with $r_i(t), 0 \le i \le 3$ defined in (6.49). Here

$$\begin{aligned}L_1x(t) &= \frac{1}{r_1(t)}\big(L_0x(t)\big)' = \frac{1}{r_1(t)}\left(\frac{x(t)}{r_0(t)}\right)' \\ &= \frac{1}{r_1(t)}\left[\frac{r_0(t)x'(t) - r_0'(t)x(t)}{r_0^2(t)}\right].\end{aligned}$$

Since $r_0'(t) = -\tilde{r}_1(t) < 0$, we cannot conclude that $L_1x(t) < 0$ for large t if $x'(t) < 0$ and $x(t) > 0$ ultimately. This forces us to consider $\int_\sigma^\infty \tilde{r}_1(t)\,dt = \infty$ in Theorem 6.1.17. In view of this observation, $Lx(t)$ in (6.46c) or (6.58) is of the form (6.59) and, $r_3(t)$ in (6.58) is replaced by $\tilde{r}_3(t)$.

Theorem 6.1.18 *Suppose that the conditions of Theorem* 6.1.17 *are satisfied. If* (6.60) *has property* $\overline{A}$, *then* (6.2) *has property* $\overline{A}$.

Proof Since (6.60) has property $\overline{A}$, from Theorem 6.1.17, it follows that (6.61) has property $\overline{A}$. This, in turn, implies in view of Theorem 6.1.16 that (6.46c) has property $\overline{A}$. Thus, by Theorem 6.1.15, (6.2) has property $\overline{A}$. Hence the theorem is proved. □

Corollary 6.1.12 *Suppose that the conditions of Theorem* 6.1.17 *are satisfied. If*

$$x''' + a(t)x'' + b(t)x' + c^*(t)x = 0 \tag{6.62}$$

has property $\overline{A}$, *then* (6.1) *has property* $\overline{A}$, *where* $c^*(t)$ *is given by*

$$c^*(t) = \frac{p(g^{-1}(t))\tilde{r}_3(g^{-1}(t))}{g'(g^{-1}(t))\tilde{r}_3(t)r(t)} > 0. \tag{6.63}$$

This follows from Theorem 6.1.18.

In view of Theorem 2.1.2, we have the following:

Theorem 6.1.19 *Suppose that the conditions of Theorem* 6.1.17 *are satisfied. If* $a'(t) \le 0$ *and*

$$\int_\sigma^\infty \left[\frac{2a^3(t)}{27} - \frac{a(t)b(t)}{3} + c^*(t) - \frac{2}{3\sqrt{3}}\left(\frac{a^2(t)}{3} - b(t)\right)^{3/2}\right] dt = \infty,$$

then (6.1) *has property* $\overline{A}$, *where* $c^*(t)$ *is given by* (6.63).

From Theorem 6.1.19, it follows that

$$\lim_{t\to\infty} x'(t) = \lim_{t\to\infty} x''(t) = 0 \quad \text{and} \quad \lim_{t\to\infty} x(t) = \lambda \ne \pm\infty$$

for every nonoscillatory solution $x(t)$ of (6.1). Now, we obtain sufficient conditions under which all nonoscillatory solutions of (6.1) tend to zero as $t \to \infty$.

Theorem 6.1.20 *Let* (H_2) *holds. Let* $g \in C^1([\sigma,\infty), R)$ *be such that* $g'(t) > 0$. *Suppose that* $\int_\sigma^\infty \tilde{r}_1(t)\,dt = \infty$. *If* (6.60) *has property* A, *then* (6.46c) *has property* A, *where* Lx *is given by* (6.59).

Proof From Theorem 6.1.16, it follows that (6.46c) has property $\overline{A}$. Let $x(t)$ be a nonoscillatory solution of (6.46c). We may assume that $x(t) > 0$ and $x(g(t)) > 0$ for $t \ge t_0 \ge \sigma$. Thus $L_0x(t) > 0$, $L_1x(t) < 0$, $L_2x(t) > 0$ and $L_3x(t) < 0$ for $t \ge t_1 \ge t_0$, where $L_0x(t) = \frac{x(t)}{\tilde{r}_0(t)} = x(t)$ and $L_ix(t) = \frac{(L_{i-1}x(t))'}{\tilde{r}_i(t)}$, $1 \le i \le 3$. Since $\int_\sigma^\infty \tilde{r}_i(t)\,dt = \infty$, $i = 1, 2$, we have $\lim_{t\to\infty} L_1x(t) = 0$, $\lim_{t\to\infty} L_2x(t) = 0$ and $\lim_{t\to\infty} L_0x(t) = \lambda$, $0 \le \lambda < \infty$. Suppose that $\lambda > 0$. Integrating (6.46c) from t ($\ge t_1$) to ∞, we get

$$L_2x(t) = \int_t^\infty \tilde{r}_3(s)p(s)x\big(g(s)\big)\,ds.$$

Further integration of this identity twice yields

$L_0x(t)$

$$
\begin{aligned}
&= \lambda + \int_t^\infty \tilde{r}_1(s_1)\left(\int_{s_1}^\infty \tilde{r}_2(s_2)\left(\int_{s_2}^\infty \tilde{r}_3(s_3)p(s_3)x\big(g(s_3)\big)\,ds_3\right)ds_2\right)ds_1\\
&= \lambda + \int_t^\infty \tilde{r}_1(s_1)\left(\int_{s_1}^\infty \tilde{r}_2(s_2)\left(\int_{g(s_2)}^\infty \frac{\tilde{r}_3(g^{-1}(\theta))p(g^{-1}(\theta))x(\theta)}{g'(g^{-1}(\theta))}\,d\theta\right)ds_2\right)ds_1\\
&= \lambda + \int_t^\infty \tilde{r}_1(s_1)\left(\int_{s_1}^\infty \tilde{r}_2(s_2)\left(\int_{s_2}^\infty \frac{\tilde{r}_3(g^{-1}(\theta))p(g^{-1}(\theta))L_0x(\theta)}{g'(g^{-1}(\theta))}\,d\theta\right)ds_2\right)ds_1.
\end{aligned}
$$

We may observe that

$$
\begin{aligned}
&\int_{t_1}^\infty \tilde{r}_1(s_1)\left(\int_{s_1}^\infty \tilde{r}_2(s_2)\left(\int_{s_2}^\infty \frac{\tilde{r}_3(g^{-1}(\theta))p(g^{-1}(\theta))L_0x(\theta)}{g'(g^{-1}(\theta))}\,d\theta\right)ds_2\right)ds_1\\
&\quad = \lim_{t\to\infty}\int_{t_1}^t \tilde{r}_1(s_1)\left(\int_{s_1}^\infty \tilde{r}_2(s_2)\left(\int_{g^{-1}(s_2)}^\infty \tilde{r}_3(s_3)p(s_3)L_0x\big(g(s_3)\big)\,ds_3\right)ds_2\right)ds_1\\
&\quad \le \lim_{t\to\infty}\int_{t_1}^t \tilde{r}_1(s_1)\left(\int_{s_1}^\infty \tilde{r}_2(s_2)\left(\int_{s_2}^\infty \tilde{r}_3(s_3)p(s_3)L_0x\big(g(s_3)\big)\,ds_3\right)ds_2\right)ds_1\\
&\quad \le x(t_1) - \lambda < \infty,
\end{aligned}
$$

since $g^{-1}(s_2) \ge s_2$ and (6.46c) is satisfied. Hence from Lemma 1.5.17, it follows that (this lemma holds for $g(t) = t$ too) the integral equation

$$
v(t) = \lambda + \int_t^\infty \tilde{r}_1(s_1)\left(\int_{s_1}^\infty \tilde{r}_2(s_2)\left(\int_{s_2}^\infty \frac{\tilde{r}_3(g^{-1}(\theta))p(g^{-1}(\theta))v(\theta)}{g'(g^{-1}(\theta))}\,d\theta\right)ds_2\right)ds_1,
$$

for $t \ge t_1$, admits a solution $v(t)$ on $[t_1, \infty)$ satisfying

$$
\lambda \le v(t) \le L_0x(t) = x(t).
$$

Thus $\lim_{t\to\infty} v(t) = \lambda > 0$, $L_0v(t) = v(t) > 0$, $L_1v(t) < 0$, $L_2v(t) > 0$, $L_3v(t) < 0$ for $t \ge t_1$ and $v(t)$ is a solution of (6.60). This leads us to a contradiction, because by assumption $\lim_{t\to\infty} v(t) = 0$. Hence $\lambda = 0$, that is, $\lim_{t\to\infty} x(t) = 0$. The proof is complete. □

Theorem 6.1.21 *Suppose that the conditions of Theorem* 6.1.17 *are satisfied. If* (6.60) *has property* A, *then* (6.2) *has property* A.

Proof From Theorem 6.1.17, it follows that (6.61) has property $\overline{A}$. Since every solution of (6.61) is a solution of (6.61), every nonoscillatory solution of (6.61) tends to zero as $t \to \infty$. Thus from Theorem 6.1.20, every nonoscillatory solution of (6.46c) with Lx given by (6.59) tends to zero as $t \to \infty$. Hence every nonoscillatory solution of (6.2) tends to zero as $t \to \infty$. Consequently, (6.2) has property A. Hence the theorem is proved. □

Corollary 6.1.13 *Suppose that the conditions of Theorem* 6.1.17 *are satisfied. If* (6.62) *has property* A, *then* (6.1) *has property* A.

Theorem 6.1.22 *Suppose that the conditions of Theorem* 6.1.19 *are satisfied. If either* $\lim_{t\to\infty} t^2 b(t) \neq 0$ *or* $\int_\sigma^\infty c^*(t)\,dt = \infty$, *then every nonoscillatory solution of* (6.1) *tends to zero as* $t \to \infty$.

Example 6.1.5 Consider

$$x'''(t) + \frac{1}{t}x''(t) - \frac{1}{t^2}x'(t) + e^t x(t-1) = 0, \quad t \geq 1. \tag{6.64}$$

This equation may be written as

$$\big(tx''(t)\big)' - \frac{1}{t}x'(t) + te^t x(t-1) = 0, \quad t \geq 1. \tag{6.65}$$

The second-order equation associated with (6.65) is given by

$$\big(tv'\big)' - \frac{1}{t}v = 0,$$

which has a solution $v(t) = t$, satisfying (6.22). Thus $\tilde{r}_0(t) = 1, \tilde{r}_1(t) = \tilde{r}_3(t) = \frac{1}{2t}$ and $\tilde{r}_2(t) = 4t$. Clearly $\int_1^\infty \tilde{r}_1(t)\,dt = \infty$ and $\int_1^\infty \tilde{r}_2(t)\,dt = \infty$. It is easy to see that

$$c^*(t) = \frac{p(g^{-1}(t))\tilde{r}_3(g^{-1}(t))}{g'(g^{-1}(t))\tilde{r}_3(t)r(t)} = e^{t+1}.$$

Hence

$$\begin{aligned}
&\int_\sigma^\infty \left[\frac{2a^3(t)}{27} - \frac{a(t)b(t)}{3} + c^*(t) - \frac{2}{3\sqrt{3}}\left(\frac{a^2(t)}{3} - b(t)\right)^{3/2}\right]dt \\
&= \int_1^\infty \left[\frac{2}{27t^3} + \frac{1}{3t^3} + e^{t+1} - \frac{16}{27}\cdot\frac{1}{t^3}\right]dt \\
&= \infty.
\end{aligned}$$

Thus from Theorem 6.1.19, it follows that (6.64) has property A. Since $\int_1^\infty c^*(t)\,dt = \infty$, every nonoscillatory solution of Eq. (6.64) tends to zero as $t \to \infty$, from Theorem 6.1.22. Note that $\lim_{t\to\infty} t^2 b(t) = -1 \neq 0$ also holds here.

If $\int_\sigma^\infty \tilde{r}_1(t)\,dt < \infty$, then we cannot proceed as above to obtain Theorems 6.1.19 and 6.1.22. However, we can obtain (6.46c), where $\int_\sigma^\infty r_i(t)\,dt = \infty$, $i = 1, 2$.

We shall conclude this section by considering the case (H$_3$), where $a(t) \in C^2([\sigma,\infty), R)$, $b \in C^1([\sigma,\infty), R)$ and $c \in C([\sigma,\infty), R)$ with $a(t) \leq 0, b(t) \leq 0$ and $c(t) < 0$. Since $b(t) \leq 0$, we have $q(t) \leq 0$ and hence Lemma 1.4.1 implies that (2.113) admits a positive increasing solution $v(t)$ satisfying the property (6.22).

Then proceeding as in Theorem 1.4.1, it can be shown that (6.46) can be represented essentially in a unique way in a canonical form (6.46c).

Let (H_3) holds. Then from Lemma 1.5.19 and Theorem 2.3.2, it follows that if the conditions of Theorem 2.3.2 are satisfied, then every nonoscillatory solution of (2.1) satisfies the property

$$\left.\begin{aligned} & x'''(t)x''(t)x'(t)x(t) \neq 0, \quad t \geq t_0 \geq \sigma, \\ & \operatorname{sgn} x(t) = \operatorname{sgn} x'(t) = \operatorname{sgn} x''(t) = \operatorname{sgn} x'''(t), \quad t \geq t_0 \geq \sigma. \\ & \text{and} \quad \lim_{t\to\infty} |x(t)| = \lim_{t\to\infty} |x'(t)| = \infty. \end{aligned}\right\} \tag{6.66}$$

Further, if $\lim_{t\to\infty} |c(t)| \neq 0$ holds, then

$$\lim_{t\to\infty} |x''(t)| = \lim_{t\to\infty} |x'''(t)| = \infty.$$

Here, in the case (H_3), we give results similar to the above observation for the delay differential equation (6.1).

We say that Eq. (6.1) *(with $g(t) = t$ or $g(t) < t$) has property B, if every nonoscillatory solution $x(t)$ of the equation satisfies*

$$x(t)x^{(i)}(t) > 0, \quad 0 \leq i \leq 3 \tag{6.67}$$

for $t \geq t_0 \geq \sigma$. Similarly, (6.46c) *(with $g(t) = t$ or $g(t) < t$) is said to have property B, if every nonoscillatory solution $x(t)$ of the equation satisfies*

$$x(t)L_i x(t) > 0, \quad 0 \leq i \leq 3 \tag{6.68}$$

for $t \geq t_0 \geq \sigma$.

Property B of (6.1) or (6.46c) demands that (6.1) or (6.46c) should admit a nonoscillatory solution (in both cases $g(t) = t$ and $g(t) < t$). The existence of nonoscillatory solution of (6.1) with $g(t) = t$, that is, of (2.1) with $a(t) \leq 0$, $b(t) \leq 0$ and $c(t) < 0$, follows from Lemma 2.3.1, and the existence of nonoscillatory solution of (6.46c) with $g(t) = t$ can be found in Sect. 2.7 of Chap. 2. However, it is not clear that this nonoscillatory solution satisfies the property (6.67). Now we show that (6.46c) admits a nonoscillatory solution, satisfying the property (6.67).

Consider

$$L_3 x(t) + p(t)x(t) = 0, \quad p(t) < 0.$$

This admits a solution $x(t)$ on $[\sigma, \infty)$ with $L_0 x(\sigma) = 0$, $L_1 x(\sigma) = 0$ and $L_2 x(\sigma) > 0$. Since $L_2 x(t)$ is continuous, there exists a $\delta > 0$ such that $L_2 x(t) > 0$ for $t \in [\sigma, \sigma + \delta)$. We claim that $L_2 x(t) > 0$ for $t \geq \sigma$. If not, then there exists a $t_1 > \sigma$ such that $L_2 x(t_1) = 0$ and $L_2 x(t) > 0$ for $t \in [\sigma, t_1)$. Hence $L_1 x(t) > 0$ and $L_0 x(t) > 0$ for $t \in (\sigma, t_1]$. Integrating

$$\big(L_2 x(t)\big)' = -r_3(t)p(t)r_0(t)L_0 x(t)$$

from σ to t_1 we get

$$0 > L_2x(t_1) - L_2x(\sigma) = -\int_\sigma^{t_1} r_3(t)p(t)r_0(t)L_0x(t)\,dt > 0,$$

a contradiction. Hence our claim holds, that is, $L_2x(t) > 0$ for $t \geq \sigma$. Consequently, $L_0x(t) > 0$, $L_1x(t) > 0$ and $L_3x(t) > 0$ for $t > t_0$. In particular, $x(t) > 0$ for $t > \sigma$. This $x(t)$ is the required solution. Next, we show that (6.46c) with $g(t) < t$ admits a nonoscillatory solution, satisfying the property (6.68). Just now we have shown that

$$L_3x(t) + p_1(t)x(t) = 0, \quad p_1(t) < 0$$

admits a nonoscillatory solution $x(t)$ satisfying the property (6.68), where $p_1(t) = \frac{p(t)r_0(g(t))}{r_0(t)}$. We may assume that $x(t) > 0$ and $x(g(t)) > 0$ for $t \geq t_0 > \sigma$. Hence there exists a $t_1 \geq t_0$ such that $L_0x(t) > 0$, $L_1x(t) > 0$, $L_2x(t) > 0$ and $L_3x(t) > 0$ for $t \geq t_1$. Integrating

$$L_3x(t) = -p_1(t)x(t) = \frac{|p(t)|r_0(g(t))x(t)}{r_0(t)}$$

from t_1 to t, we get

$$L_2x(t) \geq \int_{t_1}^t r_3(s)\frac{|p(s)|r_0(g(s))x(s)}{r_0(s)}\,ds.$$

Further integration of the above inequality twice yields

$$L_0x(t) \geq K + \int_{t_1}^t r_1(s_1)\int_{t_1}^{s_1} r_2(s_2)\int_{t_1}^{s_2} r_3(s_3)\big|p(s_3)\big|r_0\big(g(s_3)\big)L_0x(s_3)\,ds_3\,ds_2\,ds_1,$$

where $K = L_0x(t_1) > 0$. Hence

$$\begin{aligned}&L_0x(t)\\&\quad\geq K + \int_{t_1}^t r_1(s_1)\int_{t_1}^{s_1} r_2(s_2)\int_{t_1}^{s_2} r_3(s_3)\big|p(s_3)\big|r_0\big(g(s_3)\big)L_0x\big(g(s_3)\big)\,ds_3\,ds_2\,ds_1.\end{aligned}$$

Then by Lemma 1.5.18, the integral equation

$$u(t) = K + \int_{t_1}^t r_1(s_1)\int_{t_1}^{s_1} r_2(s_2)\int_{t_1}^{s_2} r_3(s_3)\big|p(s_3)\big|r_0\big(g(s_3)\big)u\big(g(s_3)\big)\,ds_3\,ds_2\,ds_1$$

has a solution $u(t)$, satisfying $0 < K \leq u(t) \leq L_0x(t)$. Setting $z(t) = r_0(t)u(t)$, we see that $z(t) > 0$ is a solution of (6.46c), satisfying the property (6.68).

Thus, it remains to obtain sufficient conditions under which (6.1) with $g(t) < t$ admits a nonoscillatory solution satisfying the property (6.67).

One may proceed as in Lemma 1.4.3 to prove that if $x(t)$ is a nonoscillatory solution of (6.46c), then either

$$\operatorname{sgn} L_0x(t) = \operatorname{sgn} L_1x(t) = \operatorname{sgn} L_2x(t) = \operatorname{sgn} L_3x(t) \tag{6.69}$$

or

$$\operatorname{sgn} L_0x(t) = \operatorname{sgn} L_1x(t) = \operatorname{sgn} L_3x(t) \neq \operatorname{sgn} L_2x(t) \tag{6.70}$$

for $t \geq t_0 \geq \sigma$.

Theorem 6.1.23 *Let $p(t) < 0$. Let $g \in C^1([\sigma, \infty), R)$ be such that $g'(t) > 0$ for $t \geq \sigma$. If the ordinary differential equation* (6.58) *has the property B, then the delay differential equation* (6.46c) *has property B.*

Proof Let $x(t)$ be a nonoscillatory solution of (6.46c). Without any loss of generality, we may assume that $x(t) > 0$ and $x(g(t)) > 0$ for $t \geq t_0 > \sigma$. Hence $L_0x(t) > 0$, $L_1x(t) > 0$ and $L_3x(t) > 0$ for large t. Thus, $L_2x(t) > 0$ or $L_2x(t) < 0$ for large t. To complete the proof of the theorem, it is enough to prove that $L_2x(t) > 0$ for large t. If possible, suppose that $L_2x(t) < 0$ for $t \geq t_1 > t_0$. Integrating (6.46c), we obtain

$$-L_2x(t) > \int_t^\infty r_3(s_3)\big|p(s_3)\big|x\big(g(s_3)\big)\,ds_3.$$

Further integration yields

$$L_1x(t) > \int_t^\infty r_2(s_2)\bigg(\int_{s_2}^\infty r_3(s_3)\big|p(s_3)\big|x\big(g(s_3)\big)\,ds_3\bigg)\,ds_2.$$

Integrating the above inequality from t_1 to t $(t > t_1)$, we get

$$\begin{aligned}
L_0x(t) &> K + \int_{t_1}^t r_1(s_1)\bigg(\int_{s_1}^\infty r_2(s_2)\bigg(\int_{s_2}^\infty r_3(s_3)\big|p(s_3)\big|x\big(g(s_3)\big)\,ds_3\bigg)\,ds_2\bigg)\,ds_1 \\
&> K + \int_{t_1}^t r_1(s_1)\bigg(\int_{s_1}^\infty r_2(s_2) \\
&\quad \times \bigg(\int_{g(s_2)}^\infty \frac{r_3(g^{-1}(\theta))|p(g^{-1}(\theta))|x(\theta)}{g'(g^{-1}(\theta))}\,d\theta\bigg)\,ds_2\bigg)\,ds_1 \\
&> K + \int_{t_1}^t r_1(s_1)\bigg(\int_{s_1}^\infty r_2(s_2) \\
&\quad \times \bigg(\int_{s_2}^\infty \frac{r_3(g^{-1}(\theta))|p(g^{-1}(\theta))|x(\theta)}{g'(g^{-1}(\theta))}\,d\theta\bigg)\,ds_2\bigg)\,ds_1 \\
&> K + \int_{t_1}^t r_1(s_1)\bigg(\int_{s_1}^\infty r_2(s_2) \\
&\quad \times \bigg(\int_{s_2}^\infty \frac{r_3(g^{-1}(\theta))|p(g^{-1}(\theta))|r_0(\theta)L_0x(\theta)}{g'(g^{-1}(\theta))}\,d\theta\bigg)\,ds_2\bigg)\,ds_1,
\end{aligned}$$

where $K = L_0 x(t_1) > 0$. Thus, from Lemma 1.5.18, it follows that the integral equation

$$u(t) = K + \int_{t_1}^{t} r_1(s_1)\left(\int_{s_1}^{\infty} r_2(s_2) \times \left(\int_{s_2}^{\infty} \frac{r_3(g^{-1}(\theta))|p(g^{-1}(\theta))|r_0(\theta)u(\theta)}{g'(g^{-1}(\theta))}\,d\theta\right)ds_2\right)ds_1$$

admits a solution $u \in C([t_1,\infty),(0,\infty))$, satisfying

$$K \le u(t) \le L_0 x(t), \quad t \ge t_1.$$

Setting $z(t) = r_0(t)u(t)$, $t \ge t_1$, we notice that $z(t) > 0$ for $t \ge t_1$ and it satisfies the equation

$$L_0 z(t) = K + \int_{t_1}^{t} r_1(s_1)\left(\int_{s_1}^{\infty} r_2(s_2) \times \left(\int_{s_2}^{\infty} \frac{r_3(g^{-1}(\theta))|p(g^{-1}(\theta))|z(\theta)}{g'(g^{-1}(\theta))}\,d\theta\right)ds_2\right)ds_1 > 0.$$

Hence

$$L_1 z(t) = \int_{t}^{\infty} r_2(s_2)\left(\int_{s_2}^{\infty} \frac{r_3(g^{-1}(\theta))|p(g^{-1}(\theta))|z(\theta)}{g'(g^{-1}(\theta))}\,d\theta\right)ds_2 > 0,$$

and

$$L_2 z(t) = -\int_{t}^{\infty} \frac{r_3(g^{-1}(\theta))|p(g^{-1}(\theta))|z(\theta)}{g'(g^{-1}(\theta))}\,d\theta < 0$$

and thus $z(t)$ is a solution to the equation

$$L_3 z + \frac{r_3(g^{-1}(t))p(g^{-1}(t))}{g'(g^{-1}(t))r_3(t)} z = 0.$$

Since $L_2 z(t) < 0$, we get a contradiction to the assumption that Eq. (6.58) has the property B. Hence the theorem is proved. □

We recall that $L_3 x$ is given by (6.59) if $\int_\sigma^\infty \tilde{r}_1(t)\,dt = \infty$, and it is given by (6.48) if $\int_\sigma^\infty \tilde{r}_1(t)\,dt < \infty$.

Theorem 6.1.24 *Suppose that* $\int_\sigma^\infty \tilde{r}_1(t)\,dt = \infty$ *and* (6.17) *admits a nonoscillatory solution* $v(t)$ *satisfying* (6.22) *and*

$$v'(t) \le \big(v(t)r(t)\big)^{-1}\left(\int_t^\infty \frac{ds}{r(s)v^2(s)}\right)^{-1}. \tag{6.71}$$

Let $g \in C^1([\sigma,\infty),R)$ be such that $g'(t)>0$ for $t\geq\sigma$. If (6.60) *has property B, then* (6.61) *has property B.*

Proof Let $x(t)$ be a nonoscillatory solution of (6.61). Without any loss of generality, we may assume that $x(t)>0$ for $t\geq t_0>\sigma$. Since $x(t)$ is a solution of (6.60), we have $x'(t)>0$, $x''(t)>0$ and $x'''(t)>0$ for $t\geq t_1\geq t_0$. It is clear that $L_0x(t)>0$ and $L_3x(t)>0$ for $t\geq t_1$. Further,

$$L_1x(t)=\frac{1}{\tilde{r}_1(t)}\big(L_0x(t)\big)'=\frac{1}{\tilde{r}_1(t)}\left(\frac{x(t)}{\tilde{r}_0(t)}\right)'>0$$

for $t\geq t_1$. From the assumption (6.71), it follows that $\tilde{r}_1'(t)\leq 0$ for $t\geq t_1$. Hence

$$\begin{aligned}L_2x(t)&=\frac{1}{\tilde{r}_2(t)}\big(L_1x(t)\big)'=\frac{1}{\tilde{r}_2(t)}\left(\frac{x'(t)}{\tilde{r}_1(t)}\right)'\\&=\frac{1}{\tilde{r}_2(t)}\left[\frac{\tilde{r}_1(t)x''(t)-\tilde{r}_1'(t)x'(t)}{(\tilde{r}_1(t))^2}\right]>0\end{aligned}$$

for $t\geq t_1$. Thus (6.61) has property B. This completes the proof of the theorem. □

Theorem 6.1.25 *Suppose that $g\in C^1([\sigma,\infty),R)$ such that $g'(t)>0$ for $t\geq\sigma$, $a'(t)\geq 0$, $b(t)-2a'(t)\leq 0$ and $c^*(t)-b'(t)+a''(t)<0$. If*

$$\int_\sigma^\infty\Bigg[-\frac{2}{27}a^3(t)+\frac{1}{3}a(t)b(t)-c^*(t)-\frac{2}{3}a(t)a'(t)+b'(t)-a''(t)\\-\frac{2}{3\sqrt{3}}\left(\frac{a^2(t)}{3}-b(t)+2a'(t)\right)^{3/2}\Bigg]dt=\infty,$$

then (6.60) *has property B, where*

$$c^*(t)=\frac{\tilde{r}_3(g^{-1}(t))p(g^{-1}(t))}{g'(g^{-1}(t))\tilde{r}_3(t)r(t)}.$$

Theorem 6.1.26 *Suppose that $\int_\sigma^\infty\tilde{r}_1(t)\,dt=\infty$ and $0\leq\lim_{t\to\infty}\tilde{r}_1(t)<\infty$. Let $g\in C^1([\sigma,\infty),R)$ be such that $g'(t)>0$ for $t\geq\sigma$. If* (6.60) *has property B, then* (6.61) *has property B, where L_3x is given by* (6.59).

Proof Let $x(t)$ be a nonoscillatory solution of (6.61). Let $x(t)>0$ for $t\geq t_0>\sigma$. Since $x(t)$ is also a solution of (6.60), from the given condition, it follows that $x'(t)>0$, $x''(t)>0$ and $x'''(t)>0$ for $t\geq t_1>t_0$. Thus $\lim_{t\to\infty}x'(t)=\infty$. On the other hand, it is clear that $L_0x(t)>0$ and $L_3x(t)>0$ for $t\geq t_1$. Since $\tilde{r}_0(t)=1$, we have $L_1x(t)>0$ for $t\geq t_1$. Hence $L_2x(t)>0$ or $L_2x(t)<0$ for some $t\geq t_2\geq t_1$. To complete the proof of the theorem, it is enough to show that $L_2x(t)>0$ for $t\geq t_2$

holds. If not, then $L_2x(t) < 0$ for $t \geq t_2$. Then $L_1x(t)$ is a positive decreasing function of t, and hence $\lim_{t\to\infty} L_1x(t)$ exists and $0 \leq \lim_{t\to\infty} L_1x(t) < \infty$. Consequently, $\lim_{t\to\infty} x'(t) = \lim_{t\to\infty} \tilde{r}_1(t)L_1x(t) < \infty$, a contradiction. This completes the proof of the theorem. □

Theorem 6.1.27 *Suppose that the conditions of Theorem* 6.1.25 *are satisfied. Let* $\int_\sigma^\infty \tilde{r}_1(t)\,dt = \infty$ *hold. If either* $0 \leq \lim_{t\to\infty} \tilde{r}_1(t) < \infty$ *or* (6.17) *admits a solution satisfying* (6.22) *and* (6.71), *then* (6.46c) *has property* B, *where* L_3x *is given by* (6.59).

Example 6.1.6 Consider

$$x'''(t) - \frac{1}{t}x''(t) - \frac{3}{t^2}x'(t) - e\left(1 - \frac{1}{t} - \frac{3}{t^2}\right)x(t-1) = 0, \quad t \geq 3.$$

The associated second-order differential equation

$$\left(\frac{3}{t}v'\right)' - \frac{9}{t^3}v = 0$$

admits a solution $v(t) = t^3$ which satisfies (6.22) and (6.71). Clearly, $\tilde{r}_1(t) = \frac{1}{12t} = \tilde{r}_3(t), \tilde{r}_2(t) = 48t^3$ and $\tilde{r}_0(t) = 1$. Thus $\int_\sigma^\infty \tilde{r}_1(t)\,dt = \infty$, and the given equation may be written in the canonical form as

$$12t\left(\frac{1}{48t^3}\left(12tx'(t)\right)'\right)' - 3e\left(1 - \frac{1}{t} - \frac{3}{t^2}\right)x(t-1) = 0, \quad t \geq 3.$$

As $g(t) = t-1, g^{-1}(t) = t+1$ and $p(t) = -3e(\frac{1}{t} - \frac{1}{t^2} - \frac{3}{t^3})$, then $c^*(t) = -et^2(\frac{1}{(t+1)^2} - \frac{1}{(t+1)^3} - \frac{3}{(t+1)^4})$. It is easy to see that all the conditions of Theorem 6.1.25 are satisfied. Thus, from Theorem 6.1.27, it follows that the above canonical equation has property B. In particular, $x(t) = e^t$ is a positive solution of the equation such that $L_0x(t) = e^t > 0$, $L_1x(t) = 12te^t > 0$, $L_2x(t) = \frac{1}{48t^3}(12e^t + 12te^t) > 0$ and $L_3x(t) = e^t(1 - \frac{1}{t} - \frac{3}{t^2}) > 0$.

Theorem 6.1.28 *Let* $\int_\sigma^\infty \tilde{r}_1(t)\,dt < \infty$ *and* $g \in C^1([\sigma,\infty), R)$ *be such that* $g'(t) > 0$ *for* $t \geq \sigma$. *If*

$$\left(r(t)x''\right)' + q(t)x' + \frac{p(g^{-1}(t))r_3(g^{-1}(t))}{g'(g^{-1}(t))r_3(t)}x = 0 \tag{6.72}$$

has property B, *then* (6.58) *has property* B, *where* L_3x *is given by* (6.48).

Proof Let $x(t)$ be a nonoscillatory solution of (6.58). We may assume that $x(t) > 0$ for $t \geq t_0 > \sigma$. Since $x(t)$ is also a solution of (6.72) which has property B, we have $x(t) > 0, x'(t) > 0, x''(t) > 0$ and $x'''(t) > 0$ for $t \geq t_1 \geq t_0$. Thus

$\lim_{t\to\infty} x(t) = \infty$. Clearly, $L_0x(t) > 0$ and $L_3x(t) > 0$ for $t \ge t_1$. Hence, in view of the properties (6.69) and (6.70), it follows that $L_1x(t) > 0$ and $L_2x(t) > 0$ or < 0 for $t \ge t_1$. To complete the proof of the theorem, it is enough to prove that $L_2x(t) < 0$ for $t \ge t_1$ does not hold. If possible, let $L_2x(t) < 0$ for $t \ge t_1$. Hence $0 \le \lim_{t\to\infty} L_1x(t) < \infty$. Further, $(L_0x(t))' = r_1(t)L_1x(t)$ implies that

$$\begin{aligned} r_1(t)L_1x(t) &= \frac{x'(t)}{r_0(t)} - \frac{x(t)r_0'(t)}{r_0^2(t)} \\ &= \frac{x'(t)}{r_0(t)} + \frac{\tilde{r}_1(t)x(t)}{r_0^2(t)} \\ &> \frac{\tilde{r}_1(t)x(t)}{r_0^2(t)} = r_1(t)x(t). \end{aligned}$$

Thus, $L_1x(t) > x(t)$. Taking limit as $t \to \infty$ both sides, we get $\infty = \lim_{t\to\infty} x(t) \le \lim_{t\to\infty} L_1x(t) < \infty$, a contradiction. Hence $L_2x(t) > 0$ for $t \ge t_1$ and this completes the proof. □

Theorem 6.1.29 *Let $g \in C^1([\sigma,\infty), R)$ be such that $g'(t) > 0$ for $t \ge \sigma$, $a'(t) \ge 0$, $b(t) - 2a'(t) \le 0$ and $c^{**}(t) - b'(t) + a''(t) < 0$. If*

$$\int_\sigma^\infty \Bigg[-\frac{2}{27}a^3(t) + \frac{1}{3}a(t)b(t) - c^{**}(t) - \frac{2}{3}a(t)a'(t) + b'(t) - a''(t) \\ - \frac{2}{3\sqrt{3}}\left(\frac{a^2(t)}{3} - b(t) + 2a'(t)\right)^{3/2}\Bigg] dt = \infty,$$

then (6.72) *has property B, where*

$$c^{**}(t) = \frac{r_3(g^{-1}(t))p(g^{-1}(t))}{g'(g^{-1}(t))r_3(t)r(t)}.$$

Theorem 6.1.30 *Let the conditions of Theorem* 6.1.29 *hold. If $\int_\sigma^\infty \tilde{r}_1(t)\,dt < \infty$, then* (6.46c) *has property B, where $L_3(x)$ is given by* (6.48).

Theorem 6.1.31 *Let $\int_\sigma^\infty \tilde{r}_1(t)\,dt < \infty$ and $\int_\sigma^\infty p(t)\,dt = -\infty$, where $p(t) = c(t)\exp(\int_\sigma^t a(s)\,ds)$. If* (6.46c) *has property B with L_3x as in* (6.48), *then* (6.1) *has property B.*

Proof Let $x(t)$ be a nonoscillatory solution of (6.1). So, it is a solution of (6.2). Without any loss of generality, we may assume that $x(t) > 0$ and $x(g(t)) > 0$ for $t \ge t_0 > \sigma$. Since $x(t)$ is a solution of (6.46c) which has property B, we have $L_0x(t) > 0$, $L_1x(t) > 0$, $L_2x(t) > 0$ and $L_3x(t) > 0$ for $t \ge t_1 \ge t_0$. Hence $\lim_{t\to\infty} L_1x(t) = \infty$. if $\beta > \alpha > t_1$ be such that $x'(\alpha) = 0$, $x''(\alpha) \ge 0$, $x'(\beta) = 0$, $x''(\beta) \le 0$ and

$x'(t) > 0$ for $t \in (\alpha, \beta)$, then integrating (6.2) from α to β, we obtain

$$0 \geq r(\beta)x''(\beta) - r(\alpha)x''(\alpha) = -\int_{\alpha}^{\beta} q(t)x'(t)\,dt - \int_{\alpha}^{\beta} p(t)x\big(g(t)\big)\,dt > 0,$$

a contradiction. Hence $x'(t) > 0$ or ≤ 0 for $t \geq t_2 > t_1$. If $x'(t) \leq 0$ for $t \geq t_2$, then $0 \leq \lim_{t\to\infty} x(t) < \infty$. On the other hand, $(L_0x(t))' = r_1(t)L_1x(t)$ implies that

$$\begin{aligned} r_1(t)L_1x(t) &= \frac{x'(t)}{r_0(t)} - \frac{x(t)r_0'(t)}{r_0^2(t)} \\ &= \frac{x'(t)}{r_0(t)} + \frac{\tilde{r}_1(t)x(t)}{r_0^2(t)} \\ &< \frac{\tilde{r}_1(t)x(t)}{r_0^2(t)} = r_1(t)x(t). \end{aligned}$$

Thus $x(t) > L_1x(t)$ and hence $\lim_{t\to\infty} x(t) = \infty$, a contradiction. Hence $x'(t) > 0$ for $t \geq t_2 < t_1$. Then from (6.2), we have $(r(t)x''(t))' > 0$ for $t \geq t_2$. Hence $x''(t) >$ 0 or < 0 for $t \geq t_3 \geq t_2$. If $x''(t) < 0$ for $t \geq t_3$, then integrating (6.2) from t_3 to t, we get

$$\begin{aligned} r(t)x''(t) &= r(t_3)x''(t_3) - \int_{t_3}^{t} q(s)x'(s)\,ds - \int_{t_3}^{t} p(s)x\big(g(s)\big)\,ds \\ &> r(t_3)x''(t_3) - x\big(g(t_3)\big)\int_{t_3}^{t} p(s)\,ds. \end{aligned}$$

Hence $x''(t) > 0$ for large t, a contradiction. Thus, $x''(t) > 0$ for $t \geq T_3$. Then from (6.1), it follows that $x'''(t) > 0$ for $t \geq t_3$. Consequently, (6.1) has property B. The theorem is proved. □

Corollary 6.1.14 *Let the conditions of Theorem* 6.1.29 *hold. If* $\int_{\sigma}^{\infty} \tilde{r}_1(t)\,dt < \infty$ *and* $\int_{\sigma}^{\infty} p(t)\,dt = -\infty$, *then* (6.1) *has property* B, *where* $p(t) = c(t)\exp(\int_{\sigma}^{t} a(s)\,ds)$. *Further,*

$$\lim_{t\to\infty} \left|x(t)\right| = \lim_{t\to\infty} \left|x'(t)\right| = \infty$$

and

$$\lim_{t\to\infty} \left|x''(t)\right| = \lim_{t\to\infty} \left|x'''(t)\right| = \infty$$

if

$$\lim_{t\to\infty} c(t) \neq 0.$$

Example 6.1.7 Consider

$$x'''(t) - \frac{1}{t}x''(t) - \frac{8}{t^2}x'(t) - e\left(1 - \frac{1}{t} - \frac{8}{t^2}\right)x(t-1) = 0, \quad t \geq 4. \tag{6.73}$$

This equation may be written as

$$\left(\frac{4}{t}x''(t)\right)' - \frac{32}{t^3}x'(t) - \frac{4e}{t}\left(1 - \frac{1}{t} - \frac{8}{t^2}\right)x(t-1) = 0.$$

The associated second-order equation

$$\left(\frac{4}{t}v'(t)\right)' - \frac{32}{t^3}v(t) = 0$$

admits a solution $v(t) = t^4$ satisfying (6.22). It is easy to verify that $r_0(t) = \frac{1}{24t}$, $r_1(t) = 24$, $r_2(t) = 6t^4$, and $\tilde{r}_0(t) = 1$, $\tilde{r}_1(t) = \frac{1}{24t^2}$, $\tilde{r}_2(t) = 144t^5$ and $\tilde{r}_3(t) = \frac{1}{24t^2}$. Then $\int_4^\infty \tilde{r}_1(t)\,dt < \infty$. Observe that $r_3(t) = \frac{1}{24t^2}$. Thus (6.73) may be written in the canonical form as

$$24t^2\left(\frac{1}{6t^4}\left(\frac{1}{24}(24tx(t))'\right)'\right)' - 4e\left(\frac{1}{t} - \frac{1}{t^2} - \frac{8}{t^3}\right)x(t-1) = 0, \quad t \geq 4.$$

Clearly,

$$\int_4^\infty p(t)\,dt = -4e\int_4^\infty \left(\frac{1}{t} - \frac{1}{t^2} - \frac{8}{t^3}\right)dt = -\infty,$$

$$c^{**}(t) = -et^3\left(\frac{1}{(t+1)^3} - \frac{1}{(t+1)^4} - \frac{8}{(t+1)^5}\right),$$

$a'(t) \geq 0$, $c^{**}(t) - b'(t) + a''(t) < 0$ and

$$\begin{aligned}
&\int_\sigma^\infty \Bigg[-\frac{2}{27}a^3(t) + \frac{1}{3}a(t)b(t) - c^{**}(t) - \frac{2}{3}a(t)a'(t) + b'(t) - a''(t) \\
&\quad - \frac{2}{3\sqrt{3}}\left(\frac{a^2(t)}{3} - b(t) + 2a'(t)\right)^{3/2}\Bigg]dt \\
&= \int_4^\infty \Bigg[\frac{2}{27t^3} + \frac{8}{3t^3} + et^3\left(\frac{1}{(t+1)^3} - \frac{1}{(t+1)^4} - \frac{8}{(t+1)^5}\right) \\
&\quad + \frac{2}{3t^3} + \frac{16}{t^3} + \frac{2}{t^3} - \frac{2}{3\sqrt{3}}\left(\frac{1}{3t^2} + \frac{8}{t^2} + \frac{2}{t^2}\right)^{\frac{3}{2}}\Bigg]dt \\
&= \infty.
\end{aligned}$$

From Theorem 6.1.30, it follows that the above canonical equation has property B. In particular, $x(t) = e^t$ is a positive solution with $L_0x(t) > 0$, $L_1x(t) > 0$, $L_2x(t) >$

0 and $L_3x(t) > 0$. Further, by Corollary 6.1.14, (6.73) has property B. In particular, $x(t) = e^t$ is a positive solution of (6.73) with $x'(t) > 0$, $x''(t) > 0$ and $x'''(t) > 0$.

Theorem 6.1.32 *Suppose that* $\int_\sigma^\infty \tilde{r}_1(t)\,dt = \infty$ *holds and* $0 < \lim_{t\to\infty} \tilde{r}_1(t)$. *If* (6.46c) *has property B with* L_3x *as in* (6.59), *then* (6.1) *has property B.*

Proof Let $x(t)$ be a nonoscillatory solution of (6.1). So, it is a solution of (6.2). Without any loss of generality, we may assume that $x(t) > 0$ and $x(g(t)) > 0$ for $t \ge t_0 > \sigma$. Since $x(t)$ is a solution of (6.46c), we have $L_0x(t) > 0$, $L_1x(t) > 0$, $L_2x(t) > 0$ and $L_3x(t) > 0$ for $t \ge t_1 > t_0$. Thus $\lim_{t\to\infty} L_1x(t) = \infty$. Further, $0 < L_1x(t) = \frac{x'(t)}{\tilde{r}_1(t)}$ implies that $x'(t) > 0$ for $t \ge t_1$. From (6.2), it follows that $x''(t) > 0$ or < 0 for large t. If $x''(t) < 0$ for large t, then $0 \le \lim_{t\to\infty} x'(t) < \infty$. On the other hand, $x'(t) = \tilde{r}_1(t)L_1x(t)$ implies that $\lim_{t\to\infty} x'(t) = \infty$, a contradiction. Hence $x''(t) > 0$ for $t \ge t_2 \ge t_1$. Consequently, from (6.1), it follows that $x'''(t) > 0$ for $t \ge t_2$. Thus, (6.1) has property B. The theorem is proved. □

Corollary 6.1.15 *Let the conditions of Theorem* 6.1.29 *hold. If* $\int_\sigma^\infty \tilde{r}_1(t)\,dt = \infty$ *and* $0 < \lim_{t\to\infty} \tilde{r}_1(t) < \infty$, *then* (6.1) *has property B. Further,*

$$\lim_{t\to\infty} |x(t)| = \lim_{t\to\infty} |x'(t)| = \infty$$

and

$$\lim_{t\to\infty} |x''(t)| = \lim_{t\to\infty} |x'''(t)| = \infty$$

if

$$\lim_{t\to\infty} c(t) \ne 0.$$

Theorem 6.1.33 *If g is monotonic increasing,* $\int_\sigma^\infty \tilde{r}_1(t)\,dt = \infty$ *and*

$$\int_\sigma^\infty c(t)\exp\left(\int_\sigma^t a(s)\,ds\right)dt = -\infty,$$

then (6.1) *has property B.*

Proof Let $x(t)$ be a nonoscillatory solution of (6.1). Without any loss of generality, we may assume that $x(t) > 0$ and $x(g(t)) > 0$ for $t \ge t_0 > \sigma$. Clearly, $x(t)$ satisfies (6.2) and (6.46c), where L_3x is given by (6.59). Since $L_0x(t) = \frac{x(t)}{\tilde{r}_0(t)} > 0$, $t \ge t_0$, it follows from (6.69) and (6.70) that $L_1x(t) > 0$, that is, $x'(t) > 0$ for $t \ge t_1 > t_0$. Integrating (6.2) from t_1 to t, we obtain

$$\begin{aligned} r(t)x''(t) &= r(t_1)x''(t_1) - \int_{t_1}^t q(s)x'(s)\,ds - \int_{t_1}^t p(s)x\big(g(s)\big)\,ds \\ &> r(t_1)x''(t_1) - x\big(g(t_1)\big)\int_{t_1}^t p(s)\,ds. \end{aligned}$$

Taking limit as $t \to \infty$, we see that $x''(t) > 0$ for large t. Hence from (6.1), it follows that $x'''(t) > 0$ for large t. Consequently, (6.1) has property B. This completes the proof of the theorem. ❑

Example 6.1.8 Consider the equation

$$x'''(t) - e^{t/2} x\left(\frac{t}{2}\right) = 0, \quad t \ge 1.$$

The associated second-order differential equation $v'' = 0$ has a solution $v(t) = t$, satisfying (6.22). In this case, $\tilde{r}_0(t) = \tilde{r}_1(t) = \tilde{r}_2(t) = \tilde{r}_3(t) = 1$. Hence $\int_\sigma^\infty \tilde{r}_1(t)\,dt = \infty$. Since $c^*(t) = -2e^t$, all the conditions of Corollary 6.1.15 are satisfied. Thus the equation has property B. In particular, $x(t) = e^t$ is a positive solution of the equation with $x'(t) > 0$, $x''(t) > 0$ and $x'''(t) > 0$.

Further, since $\int_1^\infty p(t)\,dt = \infty$, the conditions of Theorem 6.1.33 are satisfied.

Theorem 6.1.34 *Suppose that* $\int_\sigma^\infty \tilde{r}_1(t)\,dt < \infty$ *holds and* $g \in C^1([\sigma, \infty), R)$, *such that* $g'(t) > 0$ *for* $t \ge \sigma$. *If*

$$z'(t) + F(t) z\big(g(t)\big) \ge 0 \tag{6.74}$$

has no eventually negative solutions, then (6.46c) *has property* B, *where*

$$\begin{aligned} F(t) = {} & -r_3(t) p(t) r_0\big(g(t)\big)\big[R_1\big(g\big(g(t)\big)\big) - R_1\big(g\big(g\big(g(t)\big)\big)\big)\big] \\ & \times \big[R_2\big(g(t)\big) - R_2\big(g\big(g(t)\big)\big)\big] \end{aligned}$$

and $R_i(t) = \int_\sigma^t r_i(s)\,ds$, $i = 1, 2$.

Proof Let $x(t)$ be a nonoscillatory solution of (6.46c), where $L_3 x$ is given by (6.48). Without any loss of generality, we may assume that $x(t) > 0$ and $x(g(t)) > 0$ for $t \ge t_0 > \sigma$. From (6.69) and (6.70), it follows that $L_0 x(t) > 0$, $L_1 x(t) > 0$ and $L_3 x(t) > 0$ for $t \ge t_1 > t_0$. We claim that $L_2 x(t) > 0$ for $t \ge t_1$. If not, then $L_2 x(t) < 0$ for $t \ge t_1$. From (6.50c) we get, for $t \ge t_1$,

$$\big(L_2 x(t)\big)' = r_3(t)\big|p(t)\big| r_0\big(g(t)\big) L_0 x\big(g(t)\big).$$

Integrating $r_1(t) L_1 x(t) = (L_0 x(t))'$ from $g(g(t))$ to $g(t)$, we get

$$\int_{g(g(t))}^{g(t)} r_1(s) L_1 x(s)\,ds = L_0 x\big(g(t)\big) - L_0 x\big(g\big(g(t)\big)\big) < L_0 x\big(g(t)\big) < L_0 x(t),$$

for $t \ge t_2 > t_1$. Since $L_1 x(t)$ is a decreasing function of t, the above inequality yields

$$L_0 x(t) > L_1 x\big(g(t)\big) \int_{g(g(t))}^{g(t)} r_1(s)\,ds > L_1 x\big(g(t)\big)\big[R_1\big(g(t)\big) - R_1\big(g\big(g(t)\big)\big)\big]$$

for $t \geq t_2$. Thus, for $t \geq t_3 > t_2$, we have

$$\big(L_2x(t)\big)' > r_3(t)\big|p(t)\big|r_0\big(g(t)\big)L_1x\big(g\big(g(t)\big)\big)\big[R_1\big(g\big(g(t)\big)\big) - R_1\big(g\big(g\big(g(t)\big)\big)\big)\big]. \tag{6.75}$$

On the other hand, integration of $(L_1x(t))' = r_2(t)L_2x(t)$ from t to $g^{-1}(t)$ gives

$$\begin{aligned}-L_1x(t) < L_1x\big(g^{-1}(t)\big) - L_1x(t) &= \int_t^{g^{-1}(t)} r_2(s)L_2x(s)\,ds \\ &< L_2x\big(g^{-1}(t)\big)\int_t^{g^{-1}(t)} r_2(s)\,ds \\ &< L_2x\big(g^{-1}(t)\big)\big[R_2\big(g^{-1}(t)\big) - R_2(t)\big]\end{aligned}$$

for $t \geq t_1$, that is,

$$L_1x\big(g(t)\big) > -L_2x(t)\big[R_2(t) - R_2\big(g(t)\big)\big]$$

for $t \geq t_4 > t_1$. Thus, for $t \geq t_5 > t_4$, we have

$$L_1x\big(g\big(g(t)\big)\big) > -L_2x\big(g(t)\big)\big[R_2\big(g(t)\big) - R_2\big(g\big(g(t)\big)\big)\big]. \tag{6.76}$$

Hence (6.75) and (6.76) yield

$$\begin{aligned}\big(L_2x(t)\big)' > {}& -r_3(t)|p(t)|r_0\big(g(t)\big)L_2x\big(g(t)\big) \\ &\times \big[R_1\big(g\big(g(t)\big)\big) - R_1\big(g\big(g\big(g(t)\big)\big)\big)\big] \\ &\times \big[R_2\big(g(t)\big) - R_2\big(g\big(g(t)\big)\big)\big].\end{aligned}$$

Setting $z(t) = L_2x(t)$, we see from the above inequality that $z(t) < 0$ is a solution of (6.74), a contradiction. Hence our claim holds, that is, $L_2x(t) > 0$ for $t \geq t_1$. Consequently, (6.46c) has property B. The theorem is proved. □

Corollary 6.1.16 *Let the conditions of Theorem* 6.1.34 *hold. If*

$$\int_\sigma^\infty c(t)\exp\left(\int_\sigma^t a(s)\,ds\right)dt = -\infty,$$

then (6.1) *has property* B.

Lemma 6.1.5 *If* $g(t) < t$ *and*

$$\liminf_{t\to\infty}\int_{g(t)}^t F(s)\,ds > \frac{1}{e}, \tag{6.77}$$

then (6.74) *has no eventually negative solutions, where* $F(s)$ *is given in Theorem* 6.1.34.

In fact, the existence of a negative solution of (6.74) is equivalent to the existence of a positive solution of

$$y'(t) + F(t)y\big(g(t)\big) \le 0,$$

and the rest follows from Lemma 1.5.20.

Corollary 6.1.17 *If $\int_\sigma^\infty c(t)\exp(\int_\sigma^t a(s)\,ds)\,dt = -\infty$, $g(t) < t$ and (6.77) holds, then (6.1) has property B.*

Theorem 6.1.35 *Suppose that $\int_\sigma^\infty \tilde{r}_1(t)\,dt = \infty$ and $g \in C^1((\sigma,\infty), R)$ such that $g'(t) > 0$ for $t \ge \sigma$. If*

$$z'(t) + \tilde{F}(t)z\big(g(t)\big) \ge 0$$

does not admit eventually negative solutions, then (6.46c) has property B, where L_3x is given by (6.59) and

$$\tilde{F}(t) = -\tilde{r}_3(t)\tilde{r}_0\big(g(t)\big)\big[\tilde{R}_1\big(g\big(g(t)\big)\big) - \tilde{R}_1\big(g\big(g\big(g(t)\big)\big)\big)\big] \times \big[\tilde{R}_2\big(g(t)\big) - \tilde{R}_2\big(g\big(g(t)\big)\big)\big]$$

and $\tilde{R}_i(t) = \int_\sigma^t \tilde{r}_i(s)\,ds$, $i = 1, 2$.

Corollary 6.1.18 *Let $\int_\sigma^\infty \tilde{r}_1(t)\,dt = \infty$ and $g \in C^1((\sigma,\infty), R)$ be such that $g'(t) > 0$ for $t \ge \sigma$, and $0 < \lim_{t\to\infty} \tilde{r}_1(t)$ holds. If*

$$\liminf_{t\to\infty} \int_{g(t)}^t \tilde{F}(s)\,ds > \frac{1}{e},$$

then (6.1) has property B, where $\tilde{F}$ is same as in Theorem 6.1.35.

Theorem 6.1.36 *Let $\tau \in C([\sigma,\infty), R)$ be such that $\tau(t) > t$ and $\omega(t) = g(\tau(t)) < t$. If either*

$$\liminf_{t\to\infty} \int_{\omega(t)}^t Q(s)\,ds > \frac{1}{e}$$

or

$$\limsup_{t\to\infty} \int_{\omega(t)}^t Q(s)\,ds > 1,$$

then (6.46c) has property B, where

$$Q(t) = -r_2(t)\int_t^{\tau(t)} r_3(s)p(s)r_0\big(g(s)\big)\big[R_1\big(g(s)\big) - R_1(t_1)\big]\,ds$$

for sufficiently large t with $g(t) > t_1$.

Theorem 6.1.37 *Let $\tau \in C([\sigma,\infty),R)$ be such that $\tau(t) > t$ and $\omega(t) = g(\tau(t)) < t$. If either*

$$\liminf_{t\to\infty} \int_{\omega(t)}^{t} \tilde{Q}(s)\,ds > \frac{1}{e}$$

or

$$\limsup_{t\to\infty} \int_{\omega(t)}^{t} \tilde{Q}(s)\,ds > 1$$

then (6.46c) *has property B, where*

$$\tilde{Q}(t) = -r_2(t) \int_{t}^{\tau(t)} r_3(s)p(s)r_0\big(g(s)\big)R_1\big(g(s)\big)\,ds.$$

Example 6.1.9 Consider

$$x'''(t) - \frac{128}{t^3}x\left(\frac{t}{2}\right) = 0, \quad t > 1. \tag{6.78}$$

Here the associated second-order differential equation is given by $v''(t) = 0$ and $\tilde{r}_0(t) = \tilde{r}_1(t) = \tilde{r}_2(t) = \tilde{r}_3(t) = 1$. Further, $g(t) = \frac{t}{2}$ implies that $g(g(t)) = \frac{t}{4}$ and $g(g(g(t))) = \frac{t}{8}$. Thus $\tilde{R}_1(t) = \tilde{R}_2(t) = t-1$, $\tilde{F}(t) = \frac{4}{t}$ and hence

$$\liminf_{t\to\infty} \int_{t/2}^{t} \tilde{F}(s)\,ds = 4\log 2 > \frac{1}{e}.$$

From Corollary 6.1.18, it follows that (6.78) has property B. However, neither Theorem 6.1.36 nor Theorem 6.1.37 cannot be applied to (6.78).

Finally, we consider the equation

$$\big(r(t)x'(t)\big)'' + q(t)x'(t) + p(t)x\big(g(t)\big) = 0 \tag{6.79}$$

and the corresponding second-order differential equation

$$v''(t) + \frac{q(t)}{r(t)}v(t) = 0, \tag{6.80}$$

where r, q and $g \in C([\sigma,\infty),R)$ such that $q(t) \le 0, r(t) > 0, p(t) > 0$ and $g \in C^1([\sigma,\infty),R)$ such that $g(t) \le t$, $g'(t) > 0$ and $g(t) \to \infty$ as $t \to \infty$.

If $v(t)$ is a positive solution of (6.80), then (6.79) can be rewritten as

$$\left(v^2(t)\left(\frac{r(t)}{v(t)}x'(t)\right)'\right)' + p(t)v(t)x\big(g(t)\big) = 0. \tag{6.81}$$

Set $\Theta_0 x = x$, $\Theta_1 x = \frac{r(t)}{v(t)}(\Theta_0 x)'$, $\Theta_2(x) = v^2(t)(\Theta_1(x))'$ and $\Theta_3 x = (\Theta_2 x)'$. We shall provide some structure of nonoscillatory solutions of (6.81) provided that

(6.81) is in canonical form. We say that (6.81) is in canonical form if

$$\int^{\infty}\frac{1}{v^2(t)}\,dt=\infty \quad\text{and}\quad \int^{\infty}\frac{v(t)}{r(t)}\,dt=\infty.$$

Definition 6.1.3 Let (6.81) be in canonical form. We say that (6.81) has property $\overline{A}$, if every positive solution of (6.81) satisfies the property

$$\Theta_0 x(t)>0,\qquad \Theta_1 x(t)<0,\qquad \Theta_2 x(t)>0\quad\text{and}\quad \Theta_3 x(t)<0 \tag{6.82}$$

for large t. We say that (6.79) is said to have property $\overline{A}$, if every nonoscillatory solution $x(t)$ of (6.79) satisfies the property

$$x(t)>0,\qquad x'(t)<0,\qquad \big(r(t)x'(t)\big)'>0\quad\text{and}\quad \big(r(t)x'(t)\big)''<0$$

for large t.

Let $v(t)$ be a positive solution of (6.80) with $v'(t)\le 0$ and $v''(t)\ge 0$ for large t such that

$$\int^{\infty}\frac{v(t)}{r(t)}\,dt=\infty. \tag{6.83}$$

Then (6.83) implies, in view of the decreasing nature of $v(t)$ that $\int^{\infty}\frac{1}{r(t)}\,dt=\infty$. It is clear that $x(t)$ is a solution of (6.79), if and only if $x(t)$ is a solution of (6.81). Let $x(t)$ be a positive solution of (6.81) which has property $\overline{A}$. Then $x(t)>0$ and $x'(t)<0$ eventually. From (6.79), it follows that $(r(t)x'(t))''<0$. So, $(r(t)x'(t))'$ is of constant sign. If $(r(t)x'(t))'<0$ eventually, then $r(t)x'(t)$ decreasing and so $r(t)x'(t)\le r(t_1)x'(t_1)=c_1<0$. Integrating this inequality from t_1 to t and by using the property that $\int^{\infty}\frac{1}{r(t)}\,dt=\infty$, we obtain $x(t)<0$ for large t, a contradiction. Hence $(r(t)x'(t))'>0$ for large t. Consequently, (6.79) has property $\overline{A}$.

On the other hand, $x(t)>0$ and $x'(t)<0$ imply that $\Theta_0 x(t)>0$ and $\Theta_1 x(t)<0$. Then from (6.81), it follows that $\Theta_3 x(t)<0$. Then $\Theta_2 x(t)$ is decreasing. If we assume $\Theta_2 x(t)<0$ for $t\ge t_1$, then $\Theta_1 x(t)\le -l\le 0$ and by integrating this identity from t_1 to t and taking limit $t\to\infty$, we may get $x(t)<0$ for large t, a contradiction. Hence $\Theta_2 x(t)>0$ for large t. Consequently, (6.81) has property $\overline{A}$.

The above observation can be formulated in the following theorem.

Theorem 6.1.38 *Let $v(t)$ be a positive solution of* (6.80) *with $v'(t)\le 0$ and $v''(t)\ge 0$ for large t such that* (6.83) *holds. Then* (6.79) *has property $\overline{A}$, if and only if* (6.81) *has property $\overline{A}$.*

Theorem 6.1.39 *Let $v(t)$ be a positive solution of* (6.80) *with $v'(t)\le 0$ and $v''(t)\ge 0$ for large t such that* (6.83) *holds. If the differential inequality*

$$\big(v^2(t)z'(t)\big)'+\left[\frac{v(g(t))}{r(g(t))}g'(t)\int_t^{\infty}v(s)p(s)\,ds\right]z\big(g(t)\big)\le 0$$

has no positive solution, then (6.79) *has property $\overline{A}$.*

Corollary 6.1.19 *Let $v(t)$ be a positive solution of* (6.80) *with $v'(t) \le 0$ and $v''(t) \ge 0$ for large t such that* (6.83) *holds. If the second-order differential equation*

$$\left(v^2(t)z'(t)\right)' + \left[\frac{v(g(t))}{r(g(t))} g'(t) \int_t^\infty v(s)p(s)\,ds\right] z\left(g(t)\right) = 0$$

is oscillatory, then (6.79) *has property $\overline{A}$.*

Theorem 6.1.40 *Let $v(t)$ be a positive solution of* (6.80) *with $v'(t) \le 0$ and $v''(t) \ge 0$ for large t such that* (6.83) *holds. If*

$$\liminf_{t\to\infty} \left(\int_{t_1}^{g(t)} \frac{1}{v^2(s)}\,ds\right) \times \left(\int_t^\infty \frac{v(g(\theta))}{r(g(\theta))} g'(\theta) \int_\theta^\infty v(s)p(s)\,ds\,d\theta\right) > \frac{1}{4}$$

then (6.79) *has property $\overline{A}$.*

Theorem 6.1.41 *Let all the conditions of Theorem* 6.1.40 *hold. If, moreover,*

$$\int_\sigma^\infty \frac{v(s_3)}{r(s_3)} \int_{s_3}^\infty \frac{1}{v^2(s_2)} \int_{s_2}^\infty v(s_1)p(s_1)\,ds_1\,ds_2\,ds_3 = \infty,$$

then every nonoscillatory solution of (6.79) *tends to zero as $t \to \infty$.*

Theorem 6.1.42 *Assume that $\int^\infty \frac{s}{r(s)} q(s)\,ds > -\infty$ and $\int^\infty \frac{1}{r(t)}\,dt = \infty$. If*

$$\liminf_{t\to\infty} g(t) \left(\int_t^\infty \frac{1}{r(g(s))} g'(s) \int_s^\infty p(\theta)\,d\theta\,ds\right) > \frac{1}{4},$$

then (6.79) *has property $\overline{A}$. If, moreover,*

$$\int_\sigma^\infty \frac{1}{r(s_3)} \int_{s_3}^\infty \int_{s_2}^\infty p(s_1)\,ds_1\,ds_2\,ds_3 = \infty,$$

then every nonoscillatory solution of (6.79) *tends to zero as $t \to \infty$.*

6.2 Oscillation and Asymptotic Properties of Solutions of the Linear Delay Differential Equation $\frac{1}{r_3(t)}(\frac{1}{r_2(t)}(\frac{1}{r_1(t)}(\frac{x(t)}{r_0(t)})')')' + p(t)x(g(t)) = 0$

Consider the equation

$$\frac{1}{r_3(t)}\left(\frac{1}{r_2(t)}\left(\frac{1}{r_1(t)}\left(\frac{x(t)}{r_0(t)}\right)'\right)'\right)' + p(t)x\left(g(t)\right) = 0, \tag{6.84}$$

where $p, r_i \in C([\sigma, \infty), R)$ such that $r_i(t) > 0$, $0 \le i \le 3$, and $g \in C([\sigma, \infty), R)$ is increasing, such that $g(t) < t$, $g(t) \to \infty$ as $t \to \infty$ and $p(t) \ge$ for $t \ge \sigma$.

Setting $L_0x(t) = \frac{x(t)}{r_0(t)}$, $L_ix(t) = \frac{1}{r_i(t)}(L_{i-1}x(t))'$, $1 \le i \le 3$, one may rewrite (6.84) in the form (6.46c).

The motivation of this section has come from the works incorporated in Sect. 6.1 of this chapter. We shall provide two methods, one with $g(t)$ and the other with higher delay $w(t)$, and incorporating them.

Assume in the sequel that

$$\int_\sigma^\infty r_1(t)\,dt = \infty \quad \text{and} \quad \int_\sigma^\infty r_2(t)\,dt = \infty. \tag{6.85}$$

Let $p(t) \ge 0$. Notice that if $x(t)$ is a nonoscillatory solution of (6.84) on $[T, \infty)$, $T \ge \sigma$, then there is an integer $l \in \{0, 2\}$ and $t_0 > T$ such that

$$\left.\begin{aligned} &x(t)L_ix(t) > 0, \quad 0 \le i \le l,\\ &(-1)^{i-l}x(t)L_ix(t) > 0, \quad l \le i \le 3 \end{aligned}\right\} \tag{6.86}$$

hold for all $t \ge t_0$. If N denotes the set of all nonoscillatory solutions of (6.84) and N_l, $l \in \{0, 2\}$ denotes the set of nonoscillatory solutions of (6.84) satisfying (6.86), then $N = N_0 \cup N_2$. In this case, we say that (6.84) is said to have property $\overline{A}$ if $N = N_0$.

Next, suppose that $p(t) \le 0$. If $x(t)$ is a nonoscillatory solution of (6.84) on $[T, \infty)$, $T \ge \sigma$, then there is an integer $l \in \{1, 3\}$ and $t_0 > T$ such that (6.86) is satisfied. If $\overline{N}$ denotes the set of all nonoscillatory solutions of (6.84) and $\overline{N_l}$, $l \in \{1, 3\}$, denotes the set of all nonoscillatory solutions of (6.84) such that (6.86) is satisfied, then $\overline{N} = \overline{N_1} \cup \overline{N_3}$. In this case, we say that (6.84) is said to have property B if $\overline{N} = \overline{N_3}$.

Define $I_0 = 1$,

$$I_1(t, s; u) = \int_s^t u(\theta)\,d\theta \tag{6.87}$$

and

$$I_2(t, s; u, v) = \int_s^t u(\theta_1) \int_s^{\theta_1} v(\theta_2)\,d\theta_2\,d\theta_1,$$

$t, s \in [\sigma, \infty)$. It is obvious that $I_1(t, s; u) = -I_1(s, t; u)$. Further, the following holds:

$$\left.\begin{aligned} &I_2(t, s; u, v) = I_2(s, t; v, u),\\ &I_2(t, s; u, v) = \int_s^t v(\theta_1) \int_{\theta_1}^t u(\theta_1)\,d\theta_1\,d\theta_2. \end{aligned}\right\} \tag{6.88}$$

Suppose that either $p(t) \geq 0$ or ≤ 0 holds. If $x(t)$ is a nonoscillatory solution of (6.84) on $[T, \infty)$, $T \geq \sigma$, then

$$L_i x(t) = \sum_{j=1}^{k} (-1)^{j-i} L_j x(s) I_{j-i}(s, t; r_j, \dots, r_{i+1}) \\ + (-1)^{k-i+1} \int_t^s I_{k-i}(\theta, t; r_k, \dots, r_{i+1}) r_{k+1}(\theta) L_{k+1} x(\theta)\, d\theta, \quad (6.89)$$

for $0 \leq i \leq k \leq 2$ and $t, s \in [T, \infty)$. This is a generalisation of Taylor's formula with remainders, and with $n = 3$.

Theorem 6.2.1 *Let $p(t) > 0$. Suppose that the first-order delay differential inequality*

$$x'(t) + r_3(t) Q(t, T) x\big(g(t)\big) \leq 0 \quad (6.90)$$

does not admit eventually positive solutions for all $T > \sigma$, where

$$\left.\begin{aligned} Q(t, T) &= p(t) r_0\big(g(t)\big) \int_T^{g(t)} r_1(s)\big[R_2(s) - R_2(T)\big]\, ds \\ R_2(t) &= \int_\sigma^t r_2(s)\, ds, \end{aligned}\right\} \quad (6.91)$$

and $g(t) > T$ for $t \geq T_0 \geq T$. Then (6.84) *has property $\overline{A}$.*

Proof Let $x(t)$ be a nonoscillatory solution of (6.84) such that $x(t) > 0$ and $x(g(t)) > 0$ for $t \geq t_0 > T$. Then either (6.56) or (6.57) holds. If possible, suppose that (6.56) holds. Thus $L_0 x(t) > 0$, $L_1 x(t) > 0$, $L_2 x(t) > 0$ and $L_3 x(t) < 0$ for $t \geq t_1 > t_0$. Integrating $(L_0 x(t))' = r_1(t) L_1 x(t)$ from t_1 to t, we obtain

$$L_0 x(t) > \int_{t_1}^t r_1(s) L_1 x(s)\, ds.$$

Further integration of $(L_1 x(t))' = r_2(t) L_2 x(t)$ yields

$$L_1 x(t) > \int_{t_1}^t r_2(s) L_2 x(s)\, ds.$$

Thus, for $t \geq t_2 > t_1$, we have $g(t) > t_1$ and

$$\begin{aligned} L_0 x\big(g(t)\big) &> \int_{t_1}^{g(t)} r_1(s) L_1 x(s)\, ds \\ &> \int_{t_1}^{g(t)} r_1(s) \left(\int_{t_1}^s r_2(\theta) L_2 x(\theta)\, d\theta \right) ds. \end{aligned}$$

From (6.46c), we get, for $t \ge t_2$,

$$
\begin{aligned}
-L_3x(t) &= p(t)r_0\big(g(t)\big)L_0x\big(g(t)\big) \\
&> p(t)r_0\big(g(t)\big)\int_{t_1}^{g(t)} r_1(s)\left(\int_{t_1}^{s} r_2(\theta)L_2x(\theta)\,d\theta\right)ds \\
&> p(t)r_0\big(g(t)\big)\int_{t_1}^{g(t)} r_1(s)L_2x(s)\left(\int_{t_1}^{s} r_2(\theta)\,d\theta\right)ds \\
&> p(t)r_0\big(g(t)\big)L_2x\big(g(t)\big)\int_{t_1}^{g(t)} r_1(s)\left(\int_{t_1}^{s} r_2(\theta)\,d\theta\right)ds \\
&> Q(t,t_1)L_2x\big(g(t)\big).
\end{aligned}
$$

Thus, $L_2x(t)$ is a positive solution of

$$x'(t) + r_3(t)Q(t,t_1)x\big(g(t)\big) < 0$$

for $t \ge t_2$, a contradiction. This completes the proof of the theorem. □

Using Lemma 1.5.20, we can prove the following theorem.

Theorem 6.2.2 *Let $g(t) < t$ and $\lim_{t\to\infty} g(t) = \infty$. If*

$$\liminf_{t\to\infty} \int_{g(t)}^{t} r_3(s)Q(s,T)\,ds > \frac{1}{e}$$

or

$$\limsup_{t\to\infty} \int_{g(t)}^{t} r_3(s)Q(s,T)\,ds > 1$$

for every $T > \sigma$, then (6.1) *has property $\overline{A}$.*

Proof From Lemma 1.5.20, it follows that (6.90) does not admit eventually positive solutions for all $T > \sigma$. Hence (6.84) has property $\overline{A}$ by Theorem 6.2.1. Then the theorem follows from Theorem 6.1.15. □

The following example illustrates the above theorem.

Example 6.2.1 Consider

$$x'''(t) + \frac{3}{t}x''(t) - \frac{1}{t^2}x'(t) + \frac{2^{4+\sqrt{2}}}{t^3}x\left(\frac{t}{2}\right) = 0, \quad t \ge 1. \tag{6.92}$$

This equation may be written in the form

$$\big(t^3x''(t)\big)' - tx'(t) + 2^{4+\sqrt{2}}x\left(\frac{t}{2}\right) = 0, \quad t \ge 1. \tag{6.93}$$

The associated second-order equation

$$\left(t^3 v'\right)' - tv = 0$$

has a solution $v(t) = t^{\sqrt{2}-1}$, satisfying the conditions (6.22). Equation (6.93) may be written in the following canonical form

$$2\sqrt{2}t^{\sqrt{2}+1}\left(\frac{1}{2t^{\sqrt{2}-1}}\left(\frac{1}{4\sqrt{2}t^{\sqrt{2}-1}}\left(4t^{\sqrt{2}}x(t)\right)'\right)'\right)' + 2^{4+\sqrt{2}}x\left(\frac{t}{2}\right) = 0.$$

In this case, $\int_1^\infty \tilde{r}_1(t)\,dt < \infty$, $\int_1^\infty \tilde{r}_2(t)\,dt = \infty$, $\int_1^\infty r_1(t)\,dt = \infty$ and $\int_1^\infty r_2(t)\,dt = \infty$. Clearly, for $T > 1$,

$$\begin{aligned}\int_{t/2}^{t} r_3(s)Q(s,T)\,ds = 4\log 2 &+ \frac{2^{3+\sqrt{2}}T^{\sqrt{2}}}{\sqrt{2}}\frac{1}{t^{\sqrt{2}}} - \frac{2^{3+\sqrt{2}}}{\sqrt{2}T^{\sqrt{2}}}\frac{1}{t^{\sqrt{2}}} \\ &- \frac{2^{2+\sqrt{2}}T^{2\sqrt{2}}}{\sqrt{2}}\frac{1}{t^{2\sqrt{2}}} + \frac{2^{2+\sqrt{2}}T^{2\sqrt{2}}}{\sqrt{2}}\frac{1}{t^{2\sqrt{2}}}.\end{aligned}$$

Thus

$$\liminf_{t\to\infty} \int_{t/2}^{t} r_3(s)Q(s,T)\,ds = 4\log 2 > \frac{1}{e}$$

for every $T \geq 1$. Hence (6.92) has property $\overline{A}$.

Theorem 6.2.3 *If the delay differential inequality*

$$z'(t) + Q(t,T)z\big(w(t)\big) \leq 0$$

does not admit any eventually positive solutions for large $T > \sigma$, then (6.84) *has property $\overline{A}$, where $Q(t,T)$ is given in* (6.91), *w is a real-valued continuous function on $[\sigma,\infty)$ such that $\tau(t) > t$ and $w(t) = g(\tau(t)) < t$.*

Theorem 6.2.4 *Let $g(t) < t$ and $\lim_{t\to\infty} g(t) = \infty$. If*

$$\liminf_{t\to\infty} \int_{w(t)}^{t} Q(s,T)\,ds > \frac{1}{e}$$

or

$$\limsup_{t\to\infty} \int_{w(t)}^{t} Q(s,T)\,ds > 1$$

for every large $T > \sigma$, then (6.84) *has property $\overline{A}$.*

Theorem 6.2.5 *A necessary and sufficient condition for* (6.84) *to have a maximal solution $x(t)$ satisfying*

(i) $L_0x(t) > 0$, $L_1x(t) > 0$, $L_2x(t) > 0$ *and* $L_3x(t) < 0$ *for* $t \geq t_0 \geq \sigma$ *is that*

$$\int_\sigma^\infty p(s_1)r_3(s_1)r_0\big(g(s_1)\big)\int_\sigma^{g(s_1)} r_1(s_2)\int_\sigma^{s_2} r_2(s_3)\,ds_3\,ds_2\,ds_1 < \infty,$$

(ii) $L_0x(t) > 0$, $L_1x(t) < 0$, $L_2x(t) > 0$ *and* $L_3x(t) < 0$ *for* $t \geq t_0 \geq \sigma$ *is that*

$$\int_\sigma^\infty p(t)r_0\big(g(t)\big)r_3(t)\int_\sigma^t r_2(s)\int_\sigma^s r_1(\theta)\,d\theta\,ds\,dt < \infty.$$

Džurina in [10] proved a result similar to Theorem 6.2.1, where $Q(t,T)$ in (6.91) is replaced by an equivalent function

$$Q_1(t,T) = p(t)r_0\big(g(t)\big)r_3(t)\int_T^{g(t)} r_2(s_2)\int_T^{s_2} r_1(s_1)\,ds_1\,ds_2.$$

Theorem 6.2.6 *Suppose that either*

$$\liminf_{t\to\infty}\int_{g(t)}^t Q_1(s,T)\,ds > \frac{1}{e}$$

or

$$\limsup_{t\to\infty}\int_{g(t)}^t Q_1(s,T)\,ds > 1,$$

then (6.84) *has property* $\overline{A}$.

Theorem 6.2.7 *Suppose that either*

$$\liminf_{t\to\infty}\int_{g(t)}^t p(s)r_0\big(g(s)\big)r_3(s)\int_T^{g(s)} r_2(\theta)\big[R_1\big(g(s)\big) - R_1(\theta)\big]\,d\theta\,ds > \frac{1}{e}$$

or

$$\limsup_{t\to\infty}\int_{g(t)}^t p(s)r_0\big(g(s)\big)r_3(s)\int_T^{g(s)} r_2(\theta)\big[R_1\big(g(s)\big) - R_1(\theta)\big]\,d\theta\,ds > 1,$$

then (6.84) *has property* $\overline{A}$.

Example 6.2.2 Let $p(t) > 0$, $0 < g(t) \leq t$ and $\lim_{t\to\infty} g(t) = \infty$. If either

$$\liminf_{t\to\infty}\int_{g(t)}^t g^2(s)p(s)\,ds > \frac{2}{e}$$

or

$$\limsup_{t\to\infty}\int_{g(t)}^t g^2(s)p(s)\,ds > 2,$$

then Eq. (6.4) has property $\overline{A}$.

According to the generalisation of the result due to Hille [7, Theorem 11], Eq. (6.4) has property $\overline{A}$, if

$$\liminf_{t\to\infty} g^2(t)\int_t^\infty p(s)\,ds > \frac{1}{2\sqrt{3}}. \tag{6.94}$$

On the other hand, by the criterion due to Kusano and Naito [22], Eq. (6.4) has property $\overline{A}$, if

$$\liminf_{t\to\infty} g(t)\int_t^\infty \big(g(s)-g(t)\big)p(s)\,ds > \frac{1}{4}. \tag{6.95}$$

The following example shows that Theorem 6.2.7 is not included in the above-mentioned results.

Example 6.2.3 Consider the equation

$$x'''(t)+\frac{a}{t^3}x(\lambda t)=0, \quad t\ge 1,\ a>0,\ 0<\lambda<1. \tag{6.96}$$

By Theorem 6.2.7, Eq. (6.96) has property $\overline{A}$ if

$$a\lambda^2\ln\left(\frac{1}{\lambda}\right) > \frac{2}{e}.$$

On the other hand, the condition (6.94) for (6.96) takes the form

$$a\lambda^2 > \frac{2}{3\sqrt{3}}.$$

Hence Theorem 6.2.7 provides a better sufficient condition than (6.94), if

$$0<\lambda<e^{-3\sqrt{3}/e}.$$

On the other hand, the condition (6.95) takes the form

$$a\lambda^2 > \frac{1}{2}$$

and hence Theorem 6.2.7 provide a better sufficient condition than (6.95), if

$$0<\lambda<e^{-4/e}.$$

Theorem 6.2.8 *Equation* (6.84) *has a solution* $x(t)$, *satisfying the property* (6.52) *if and only if the differential inequality*

$$\big\{L_3x(t)+p(t)x\big(g(t)\big)\big\}\operatorname{sgn}\big\{x\big(g(t)\big)\big\}\le 0$$

has a solution satisfying the property (6.52).

Theorem 6.2.9 *Suppose that $g(t) \equiv t$. If*

$$\int_{\sigma}^{\infty} p(t) r_0(t) r_1(t)\, dt = \infty, \tag{6.97}$$

then (6.84) *has property A.*

Theorem 6.2.10 *Let $g(t) \equiv t$. Suppose that the integrals in* (6.97) *converge. Assume that the second-order delay equation*

$$\left(\frac{z'(t)}{r_2(t)}\right)' + q(t) z\bigl(\tau(t)\bigr) = 0, \quad i = 0, 2$$

or the ordinary equation

$$\left(\frac{z'(t)}{r_2(t)}\right)' + \tilde{q}(t) z(t) = 0$$

is oscillatory. Then (6.84) *has property $\overline{A}$, where*

$$q(t) = r_3(t) p(t) r_0(t) \int_{\tau(t)}^{t} r_1(s)\, ds,$$

$$\tilde{q}(t) = r_1(t) \int_{t}^{\infty} p(s) r_0(s) r_3(s)\, ds,$$

and $\tau : [\sigma, \infty) \to R$ is continuous function satisfying $\tau(t) \to \infty$ as $t \to \infty$ and $\tau(t) < t$.

Theorem 6.2.11 *Assume that* (6.84) *has property $\overline{A}$. Then every nonoscillatory solution of* (6.84) *satisfies*

$$\lim_{t \to \infty} \frac{x(t)}{r_0(t)} = 0,$$

if and only if

$$\int_{\sigma}^{\infty} r_0\bigl(g(t)\bigr) p(t) r_3(t) \int_{\sigma}^{t} r_2(s_2) \int_{\sigma}^{s_2} r_1(s_1)\, ds_1\, ds_2\, dt = \infty.$$

For the special case of Eq. (6.84), namely, for the equation

$$\left(\frac{1}{r_2(t)}\left(\frac{1}{r_1(t)} x'(t)\right)'\right)' + p(t) x\bigl(g(t)\bigr) = 0, \tag{6.98}$$

we have the following theorem:

Theorem 6.2.12 *Let $g \in C^1([\sigma, \infty), R)$, $g'(t) > 0$ and $g(t) \le t$. Further, suppose that at least one of the following conditions holds*:

(i) $\int_\sigma^\infty (\int_\sigma^t r_1(s)\,ds)\frac{p(g^{-1}(t))}{g'(g^{-1}(t))}\,dt = \infty$;

(ii) *the equation*

$$\left(\frac{z'(t)}{r_2(t)}\right)' + \left(\frac{p(g^{-1}(t))}{g'(g^{-1}(t))}\int_{\tau(t)}^t r_1(s)\,ds\right)z\big(\tau(t)\big) = 0$$

is oscillatory, where $\tau : [\sigma, \infty) \to R$ *is continuous,* $\tau(t) \to \infty$ *as* $t \to \infty$ *and* $\tau(t) \le t$;

(iii) *the equation*

$$\left(\frac{z'(t)}{r_2(t)}\right)' + \left(r_1(t)\int_t^\infty \frac{p(g^{-1}(s))}{g'(g^{-1}(s))}\,ds\right)z(t) = 0$$

is oscillatory.

Then (6.98) *has property* $\overline{A}$.

Theorem 6.2.12 can be extended to Eq. (6.84). Moreover, if $g \in C^1([\sigma, \infty), R)$, $g'(t) > 0$ and $g(t) \le t$ and at least one of the following conditions holds:

(i) $\int_\sigma^\infty (\int_\sigma^t r_1(s)\,ds)\frac{p(g^{-1}(t))}{g'(g^{-1}(t))} r_3(g^{-1}(t))\,dt = \infty$;

(ii) the equation

$$\left(\frac{z'(t)}{r_2(t)}\right)' + \left(\frac{p(g^{-1}(t))}{g'(g^{-1}(t))} r_3\big(g^{-1}(t)\big)\int_{\tau(t)}^t r_1(s)\,ds\right)z\big(\tau(t)\big) = 0$$

is oscillatory, where $\tau : [\sigma, \infty) \to R$ is continuous, $\tau(t) \to \infty$ as $t \to \infty$ and $\tau(t) \le t$;

(iii) the equation

$$\left(\frac{z'(t)}{r_2(t)}\right)' + \left(r_1(t)\int_t^\infty \frac{p(g^{-1}(s))}{g'(g^{-1}(s))} r_3\big(g^{-1}(s)\big)\,ds\right)z(t) = 0$$

is oscillatory,

then (6.84) has property $\overline{A}$.

Example 6.2.4 Let us consider the third-order Euler equation

$$\big(t^{1/2}x''\big)' + \frac{a}{t^{5/2}}x = 0, \quad t > 1,\ a \in R. \tag{6.99}$$

Set $\tau(t) = \frac{t}{3}$. Then by Theorem 6.2.12, (6.99) has property $\overline{A}$ if the second-order delay equation

$$\big(t^{1/2}v'(t)\big)' + \frac{2a}{3t^{3/2}}v(t/3) = 0$$

is oscillatory. By a generalisation of the well-known criterion of Hille [20], it comes if

$$a > \frac{1}{8\sqrt{3}},$$

and moreover, by Theorem 6.2.12, if $a > \frac{1}{8\sqrt{3}}$, then every nonoscillatory solution $x(t)$ of (6.99) satisfies

$$\lim_{t\to\infty} x(t) = 0.$$

If $r_1(t) = r_2(t) = r(t)$, then (6.98) takes the form

$$\left(\frac{1}{r(t)}\left(\frac{1}{r(t)}x'(t)\right)'\right)' + p(t)x\big(g(t)\big) = 0. \tag{6.100}$$

Theorem 6.2.13 *Let*

$$R(t) = \int_\sigma^t r(s)\,ds \to \infty \quad \text{as } t \to \infty. \tag{6.101}$$

Assume that $g \in C^1([\sigma, \infty), R)$, $0 < g(t) \le t$ *and* $g'(t) > 0$. *Then* (6.100) *has property* $\overline{A}$ *if*

$$\liminf_{t\to\infty} R^2\big(g(t)\big) \int_t^\infty p(s)\,ds > \frac{1}{3\sqrt{3}}.$$

Theorem 6.2.14 *Let* $g(t) \le t$, *and*

$$R_i(t) = \int_\sigma^t r_i(s)\,ds \to \infty \quad \text{as } t \to \infty,\ i = 1, 2. \tag{6.102}$$

Suppose that there exists a real $\lambda > 1$ *such that*

$$r_1(t)R_2(t) \ge \lambda r(t)R(t),$$

and $\frac{r(t)}{r_1(t)}$ *is a nonincreasing function. Then Eq.* (6.98) *has property* $\overline{A}$, *if so does Eq.* (6.100).

Theorem 6.2.15 *Let* $g \in C^1([\sigma, \infty), R)$, $g(t) \le t$ *and* $g'(t) > 0$. *Suppose that*

$$\liminf_{t\to\infty}\left(\int_\sigma^{g(t)} r_1(s)R_2(s)\,ds\right)\left(\int_t^\infty p(s)\,ds\right) > \frac{1}{6\sqrt{3}}, \tag{6.103}$$

and

$$R_2(t)\left(\int_\sigma^t r_1(s)R_2(s)\,ds\right)^{-1/2} \text{ is nondecreasing.} \tag{6.104}$$

Then (6.98) *has property* $\overline{A}$, *where*

$$R_2(t) = \int_\sigma^t r_2(s)\,ds.$$

Proof Choose $\lambda > 1$ such that

$$\liminf_{t\to\infty}\left(\int_\sigma^{g(t)} r_1(s)R_2(s)\,ds\right)\left(\int_t^\infty p(s)\,ds\right) > \frac{\lambda}{6\sqrt{3}}. \tag{6.105}$$

Consider Eq. (6.100) with $r(t)$ defined by

$$r(t)R(t) = \frac{1}{\lambda} r_1(t)R_2(t),$$

where $R(t) = \int_\sigma^t r(s)\,ds$. Integrating this relation and extracting the square root of the resulting identity, we arrive at

$$R(t) = \frac{\sqrt{2}}{\lambda}\left(\int_\sigma^t r_1(s)R_2(s)\,ds\right)^{1/2} \to \infty \tag{6.106}$$

as $t \to \infty$. It is easy to see that

$$r(t) = R'(t) = \frac{1}{\sqrt{2}\lambda} r_1(t)R_2(t)\left(\int_\sigma^t r_1(s)R_2(s)\,ds\right)^{-1/2}.$$

Hence the function $\frac{r(t)}{r_1(t)}$ is nonincreasing, if and only if (6.104) holds. From (6.106), we conclude that (6.105) is equivalent to the condition

$$\liminf_{t\to\infty} R^2\big(g(t)\big)\int_t^\infty p(s)\,ds > \frac{1}{3\sqrt{3}},$$

which is, as we see from Theorem 6.2.13, a sufficient condition for (6.100) to have property $\overline{A}$. Hence our assertion follows from Theorem 6.2.14. The proof is complete. □

Corollary 6.2.1 *Assume that the hypotheses of Theorem* 6.1.15 *hold, except the relation* (6.104) *is replaced by one of the following conditions*:

$$2r(t)\int_\sigma^t r_1(s)R_2(s)\,ds \le r_1(t)\big(R_2(t)\big)^2, \quad t \ge 0$$

or $\frac{r_1(t)}{r_2(t)}$ *is nondecreasing. Then* (6.98) *has property* $\overline{A}$.

Example 6.2.5 Let us consider the equation

$$\big(t^{1/2}\big(t^{1/2}x'(t)\big)'\big)' + \frac{a}{t^{13/6}}x(bt) = 0, \quad t \ge 1$$

and $b \in (0, 1]$. By Corollary 6.2.1, this equation has property $\overline{A}$ if $a > 49/(432\sqrt{36^{7/6}})$.

One can apply Theorem 6.2.14 to find some new sufficient conditions for the property $\overline{A}$ of the equation

$$x''' + p(t)x = 0. \tag{6.107}$$

One may find several oscillation criteria for (6.107) in Chap. 2. Kiguradze [21] proved that if

$$\liminf_{t\to\infty} t\int_t^\infty sp(s)\,ds > \frac{2}{3\sqrt{3}}$$

or

$$\limsup_{t\to\infty} t\int_t^\infty sp(s)\,ds > 2,$$

then (6.107) has property $\overline{A}$. The following theorem may be treated as an extension of the above inequalities to Eq. (6.98).

Theorem 6.2.16 *Assume that* $g(t) \equiv t$ *and* (6.104) *holds. Also, let* (6.85) *and* (6.104) *hold. Further, suppose that*

$$\liminf_{t\to\infty}\left(\int_\sigma^t r_2(s)R_1(s)\,ds\right)^{1/2}\int_t^\infty p(s)\left(\int_\sigma^s r_2(\theta)R_1(\theta)\,d\theta\right)^{1/2} ds > \frac{1}{3\sqrt{3}}$$

or

$$\limsup_{t\to\infty}\left(\int_\sigma^t r_2(s)R_1(s)\,ds\right)^{1/2}\int_t^\infty p(s)\left(\int_\sigma^s r_2(\theta)R_1(\theta)\,d\theta\right)^{1/2} ds > 1.$$

Then (6.98) *has property* $\overline{A}$.

Theorem 6.2.16 yields the following interesting sufficient condition.

Theorem 6.2.17 *Assume that* (6.85) *holds,* $g \in C^1([\sigma,\infty), R)$, $0 < g(t) \le t$ *and* $g'(t) > 0$ *and* (6.104) *holds. Further, suppose that*

$$\liminf_{t\to\infty}\left(\int_\sigma^{g(t)} r_2(s)R_1(s)\,ds\right)^{1/2}\int_t^\infty p(s)\left(\int_\sigma^{g(s)} r_2(\theta)R_1(\theta)\,d\theta\right)^{1/2} ds > \frac{1}{3\sqrt{3}}$$

or

$$\limsup_{t\to\infty}\left(\int_\sigma^{g(t)} r_2(s)R_1(s)\,ds\right)^{1/2}\int_t^\infty p(s)\left(\int_\sigma^{g(s)} r_2(\theta)R_1(\theta)\,d\theta\right)^{1/2} ds > 1.$$

Then (6.98) *has property* $\overline{A}$.

Now, we give some sufficient conditions for the oscillation or asymptotic behaviour of nonoscillatory solutions of the equation

$$\big(r(t)\big(r(t)x'(t)\big)'\big)' + p(t)x\big(g(t)\big) = 0, \quad t \geq \sigma. \tag{6.108}$$

We assume that $\int^{\infty} \frac{1}{r(t)}\, dt = \infty$ holds, $g \in C^1([\sigma,\infty), R)$ such that $0 < g(t) \leq t$ and $g(t) \to \infty$ as $t \to \infty$. Let us use the notation

$$P(t) = \int_t^{\infty} p(s)\, ds, \quad t \geq \sigma$$

when $\int^{\infty} p(t)\, dt$ converges.

It is known (cf. [35]) that under the condition either

$$\int^{\infty} \frac{P(t)}{r(t)}\, dt = \infty \quad \text{or} \quad \int^{\infty} \frac{1}{r(t)} \int_t^{\infty} \frac{1}{r(s)} P(s)\, ds\, dt = \infty, \tag{6.109}$$

every bounded solution $x(t)$ of (6.108) is oscillatory or such that

$$\lim_{t\to\infty} x(t) = \lim_{t\to\infty} r(t)x'(t) = \lim_{t\to\infty} r(t)\big(r(t)x'(t)\big)' = 0$$

monotonically.

Moreover (cf. [35]), if either

$$\int^{\infty} R\big(g(t)\big)p(t)\, dt = \infty \quad \text{or} \quad \int^{\infty} \frac{1}{r(t)} \int_t^{\infty} R\big(g(s)\big)P(s)\, ds\, dt = \infty, \tag{6.110}$$

then for every unbounded nonoscillatory solution $x(t)$ of (6.108), we have

$$\lim_{t\to\infty} \frac{x(t)}{R(t)} = \pm\infty.$$

Here, we give conditions, under which every solution of (6.108) is oscillatory or tending to zero at ∞.

Lemma 6.2.1 *Let $u(t)$ be a nonnegative function on an interval $[T,\infty)$, $T \geq \sigma$, such that $(r(t)(r(t)u')')'$ exists on $[T,\infty)$. Suppose that $u' \geq 0$ and $(r(t)(r(t) \times u')')' \leq 0$ on $[T,\infty)$. Then the function*

$$U(t) = u(t)\left(\int_T^t \frac{ds}{r(s)}\right)^{-2}, \quad t > T$$

is decreasing.

Theorem 6.2.18 *Under the condition*

$$\min\left\{\liminf_{t\to\infty} P(t)R^2\big(g(t)\big), \frac{1}{2}\liminf_{t\to\infty} r(t)p(t)R(t)R^2\big(g(t)\big)\right\} > \frac{1}{3\sqrt{3}} \tag{6.111}$$

every solution $x(t)$ of (6.108) is oscillatory or such that

$$\lim_{t\to\infty} x(t) = \lim_{t\to\infty} r(t)x'(t) = \lim_{t\to\infty} r(t)\big(r(t)x'(t)\big)' = 0$$

monotonically.

Proof Condition (6.111) implies (6.109), and so it is enough to prove the nonexistence of unbounded nonoscillatory solutions of (6.108). Moreover, for the study of the nonoscillatory solutions of (6.108), it suffices to deal only with positive ones.

Let $x(t)$ be a positive unbounded solution on an interval $[T_0, \infty)$, $T_0 \geq \sigma$ of (6.108) and let $T \geq T_0$ be chosen so that $g(t) > T_0$ for every $t \geq T$. Then the function $(r(t)(r(t)x'(t))')' \leq 0$ for $t \geq T_1 \geq T$. Thus, by $\int^\infty \frac{1}{r(t)}\,dt = \infty$, we observe that $(r(t)x'(t))' > 0$ for $t \geq T$. Moreover, $x'(t)$ is eventually positive. We suppose, without any loss of generality, that $x'(t) > 0$ for $t \geq T$. Equation (6.108) gives

$$r(t)\big(r(t)x'(t)\big)' \geq \int_t^\infty p(s)x\big(g(s)\big)\,ds, \quad t \geq T.$$

Put $v(t) = r(t)R(t)x'(t) - x(t)$, $t \geq T$. Then $v'(t) = R(t)(r(t)x'(t))' > 0$ for $t \geq T$ and consequently, $x(t)$ is either > 0 or < 0 eventually. The case $v(t) < 0$ for $t \geq T$ is impossible. Indeed, in this case

$$\left(\frac{x(t)}{R(t)}\right)' = \frac{V(t)}{r(t)R^2(t)} < 0 \quad \text{for all } t \geq T$$

and hence $\lim_{t\to\infty} \frac{x(t)}{R(t)} < \infty$, which is a contradiction, since (6.111) implies (6.110). Thus, $v(t)$ is eventually positive, that is,

$$r(t)R(t)x'(t) - x(t) > 0 \quad \text{for all large } t.$$

Now, by (6.111), we consider two constants c_1, c_2 with $\frac{2}{3\sqrt{3}} < c_1 < c < 2$ and

$$\min\Big\{2\liminf_{t\to\infty} P(t)R^2\big(g(t)\big), \ \liminf_{t\to\infty} r(t)p(t)R(t)R^2\big(g(t)\big)\Big\} > c_2.$$

Furthermore, we choose a $T_2 > T$ so that for every $t \geq T_2$

$$\begin{aligned}
&g(t) > T,\\
&\left(\int_T^{g(t)} \frac{1}{r(s)}\,ds\right)^2 R^{-2}\big(g(t)\big) \geq \frac{c_1}{c},\\
&\min\big\{2P(t)R^2\big(g(t)\big), r(t)p(t)R(t)R^2\big(g(t)\big)\big\} \geq c,\\
&r(T_2)R(T_2)x'(T_2) - x(T_2) > 0.
\end{aligned}$$

Let k be a number such that

$$1 < k < \frac{1}{2} + \frac{c}{4} + \frac{1}{2}\sqrt{\left(1 + \frac{c^2}{4} + 3c\right)}$$

and

$$r(T_2)R(T_2)x'(T_2) - kx(T_2) > 0.$$

We shall prove that the function $X_k = \frac{x}{R^k}$ is increasing on $[T_2, \infty)$. To do this, we first remark that

$$\lim_{t\to\infty} x\big(g(t)\big)P(t) = 0,$$

since for every $t \geq T_2$,

$$x\big(g(t)\big)P(t) = x\big(g(t)\big)\int_t^\infty p(s)\,ds \leq \int_t^\infty p(s)x\big(g(s)\big)\,ds.$$

So, for every $t \geq T_2$, we obtain

$$\begin{aligned}
&r\big(g(t)\big)\big(rX_k'\big)'\big(g(t)\big)R^k\big(g(t)\big) + 2kr\big(g(t)\big)X_k'\big(g(t)\big)R^{k-1}\big(g(t)\big)\\
&\quad + k(k-1)X_k\big(g(t)\big)R^{k-2}\big(g(t)\big)\\
&\quad = r\big(g(t)\big)\big(rx'\big)'\big(g(t)\big)\\
&\quad \geq r(t)\big(r(t)x'(t)\big)'\\
&\quad \geq \int_t^\infty p(s)x\big(g(s)\big)\,ds\\
&\quad = -\int_t^\infty x\big(g(s)\big)\,dP(s) - x\big(g(s)\big)P(s)\big|_t^\infty + \int_t^\infty P(s)x'\big(g(s)\big)g'(s)\,ds\\
&\quad = -\lim_{s\to\infty} x\big(g(s)\big)P(s) + P(t)x\big(g(t)\big) + \int_t^\infty P(s)x'\big(g(s)\big)g'(s)\,ds\\
&\quad = P(t)x\big(g(t)\big) + \int_t^\infty P(s)x'\big(g(s)\big)g'(s)\,ds\\
&\quad \geq P(t)x\big(g(t)\big) + \frac{1}{r(g(t))}x'\big(g(t)\big)\int_t^\infty r\big(g(s)\big)P(s)g'(s)\,ds\\
&\quad \geq \frac{c}{2}R^{-2}\big(g(t)\big)x\big(g(t)\big) + \frac{c}{2}r\big(g(t)\big)x'\big(g(t)\big)\int_t^\infty \frac{R^{-2}(g(s))}{r(g(s))}g'(s)\,ds\\
&\quad = \frac{c}{2}r\big(g(t)\big)X_k'\big(g(t)\big)R^{k-1}\big(g(t)\big) + \frac{c}{2}(k+1)X_k\big(g(t)\big)R^{k-2}\big(g(t)\big).
\end{aligned}$$

Thus, for $t \geq T_2$, we have

$$r(t)\big(r(t)X_k'(t)\big)' + \left(2k - \frac{c}{2}\right)\frac{r(t)X_k'(t)}{R(t)} + \left[k^2 - \left(1 + \frac{c}{2}\right)k - \frac{c}{2}\right]X_k(t)R^{-2}(t) \geq 0.$$

It is easy to see that $k^2 + (1 + \frac{c}{2})k - \frac{1}{2}c \leq 0$ and hence

$$\big(r(t)X_k'(t)\big)' + \left(2k - \frac{c}{2}\right)\frac{X_k'(t)}{R(t)} \geq 0 \quad \text{for all } t \geq T_2. \tag{6.112}$$

If $\bar{t} \geq T_2$ is such that the function $r(t)X_k'(t)$ takes a local minimum at $\bar{t}$, then $(r(t)X_k'(t))|_{t=\bar{t}} = 0$ and so, since $2k - \frac{c}{2} > 0$, the above inequality gives $X_k'(\bar{t}) = 0$. Moreover,

$$X_k'(T_2) = \frac{[R(T_2)r(T_2)x'(T_2) - kx(T_2)]}{r(T_2)R^{k+1}(T_2)} > 0.$$

Thus, $X_k'(t)$ is either ≥ 0 on $[T_2, \infty)$ or such that for some $\overline{T} > T$,

$$X_k'(\overline{T}) = 0 \quad \text{and} \quad X_k'(t) < 0 \quad \text{for } t > \overline{T}.$$

But the latter case is impossible, since then (6.112) gives

$$\big(r(t)X_k'(t)\big)' > 0 \quad \text{for all } t \geq \overline{T},$$

which is a contradiction. We have thus proved that $X_k'(t) \geq 0$ on $[T_2, \infty)$, which means that $X_k(t)$ is increasing on $[T_2, \infty)$.

Next, we consider the set K of all numbers k, $1 < k < 2$, for which the function $X_k = \frac{x}{R^k}$ is increasing on $[T_k, \infty)$ for some $T_k \geq T_2$. We observe that the set K is nonempty, and we put $k_0 = \sup K$.

Let k be an arbitrary number in K and $T_k \geq T_2$ be such that $X_k'(t) \geq 0$ on $[T_k, \infty)$. We shall prove that the function $r(t)X_k'(t)$ is decreasing on $[T_k', \infty)$ for some $T_k' \geq T_k$. To this end, by using Lemma 6.2.1, for every $t \geq T_k$, we get

$$\begin{aligned}
&\big[r(t)\big(r(t)X_k'(t)\big)'\big]'R^k(t) + 3k\big[r(t)X_k'(t)\big]'R^{k-1}(t) + 3k(k-1)X_k'(t)R^{k-2}(t)\\
&\quad + k(k-1)(k-2)\frac{X_k(t)}{r(t)}R^{k-3}(t)\\
&= \big[r(t)\big(r(t)x'(t)\big)'\big]'\\
&= -p(t)x\big(g(t)\big)\\
&\leq -c\frac{1}{r(t)R(t)}x\big(g(t)\big)R^{-2}\big(g(t)\big)\\
&= -c\frac{1}{r(t)R(t)}\left\{\left[\int_T^{g(t)}\frac{ds}{r(s)}\right]^2 R^{-2}\big(g(t)\big)\right\}\left\{x\big(g(t)\big)\left[\int_T^{g(t)}\frac{ds}{r(s)}\right]^{-2}\right\}
\end{aligned}$$

$$\le -c_1 \frac{1}{r(t)R(t)} x(t) \left[\int_T^t \frac{ds}{r(s)} \right]^{-2}$$

$$\le -c_1 \frac{X_k(t)}{r(t)} R^{k-3}(t),$$

that is, for all $t \ge T_k$, we have

$$\left[r(t)\left(r(t)X_k'(t)\right)'\right]' R^3(t) + 3k\left[r(t)X_k'(t)\right]' R^2(t) + 3k(k-1)X_k'(t)R(t)$$
$$+ \left[k(k-1)(k-2) + c_1\right] \frac{X_k(t)}{r(t)} \le 0.$$

Note that the maximum of the function $\rho(\theta) = -\theta(\theta-1)(\theta-2)$, $1 \le \theta \le 2$ is $\frac{2}{3\sqrt{3}}$, which is less than c_1. Hence

$$k(k-1)(k-2) + c_1 \ge 0.$$

Moreover, we observe that $k(k-1) > 0$. Thus

$$\left[r(t)\left(r(t)X_k'(t)\right)'\right]' R(t) + 3k\left[r(t)X_k'(t)\right]' \le 0, \quad t \ge T_k. \tag{6.113}$$

This inequality implies that $(r(t)X_k'(t))' \le 0$ eventually.

To prove this assertion, we first assume that $(r(t)X_k'(t))' > 0$ on some interval $[\overline{T}_k, \infty)$, $\overline{T}_k \ge T_k$. Then

$$r(t)X_k'(t) > r(\overline{T}_k)X_k'(T_k) \quad \text{for } t \ge \overline{T}_k$$

and consequently, there exist a $T_k^* > \overline{T}_k$ and a positive constant α such that

$$r(t)X_k'(t) \ge \alpha \quad \text{for all } t \ge T_k^*.$$

So,

$$X_k(t) - X\left(T_k^*\right) \ge \alpha \int_{T_k^*}^t \frac{ds}{r(s)}, \quad t \ge T_k^*,$$

which gives

$$\liminf_{t\to\infty} \frac{x(t)}{R^{k+1}(t)} \equiv \liminf_{t\to\infty} \frac{X_k(t)}{R(t)} \ge \alpha > 0.$$

But, by Lemma 6.2.1 and the fact that $k > 1$, we have

$$\lim_{t\to\infty} \frac{x(t)}{R^{k+1}(t)} = 0,$$

i.e., a contradiction. Next, if the function $r(t)(r(t)X_k'(t))'$ takes a local maximum at $\tilde{t} \ge T_k$, then $(r(rX_k')')'(\tilde{t}) = 0$, and hence (6.113) gives $(rX_k')'(\tilde{t}) \le 0$. We have thus proved that the only possible case is that $(rX_k')'(\tilde{t}) \le 0$ on $[T_k', \infty)$ for some $T_k' \ge T_k$.

Next, we choose $\tilde{T}_k \ge T_k'$ so that

$$g(t) \ge T_k' \quad \text{for all } t \ge \tilde{T}_k.$$

Then for every $t \ge \tilde{T}_k$, we obtain

$$\begin{aligned}
& r\big(g(t)\big)\big(rX_k'\big)\big(g(t)\big)R^k\big(g(t)\big) + 2kr\big(g(t)\big)X_k'\big(g(t)\big)R^{k-1}\big(g(t)\big) \\
&\quad + k(k-1)X_k\big(g(t)\big)R^{k-2}\big(g(t)\big) \\
&\quad = r\big(g(t)\big)\big(rx'\big)'\big(g(t)\big) \ge r(t)\big(r(t)x'(t)\big)' \\
&\quad \ge \int_t^\infty p(s)x\big(g(s)\big)\,ds \\
&\quad = P(t)x\big(g(t)\big) + \int_t^\infty P(s)x'\big(g(s)\big)g'(s)\,ds \\
&\quad = P(t)X_k\big(g(t)\big)R^k\big(g(t)\big) + \int_t^\infty P(s)X_k'\big(g(s)\big)R^k\big(g(s)\big)g'(s)\,ds \\
&\qquad + k\int_t^\infty P(s)\frac{X_k(g(s))}{r(g(s))}R^{k-1}\big(g(s)\big)g'(s)\,ds \\
&\quad \ge \frac{c}{2}X_k\big(g(t)\big)R^{k-2}\big(g(t)\big) + \frac{kc}{2}X_k\big(g(t)\big)\int_t^\infty \frac{R^{k-3}(g(s))}{r(g(s))}g'(s)\,ds \\
&\quad = \frac{c}{2}\left(1 - \frac{k}{k-2}\right)X_k\big(g(t)\big)R^{k-2}\big(g(t)\big).
\end{aligned}$$

Therefore,

$$r(t)X_k'(t) \ge \frac{k^3 - 3k^2 + 2k + c}{2k(2-k)}\frac{X_k(t)}{R(t)} \quad \text{for } t \ge \tilde{T}_k.$$

Since the minimum of the function

$$\rho_1(\theta) = \theta^3 - 3\theta^2 + 2\theta + c, \quad 1 \le \theta \le 2$$

is $c - \frac{2}{3\sqrt{3}} > 0$, we get

$$r(t)X_k'(t) \ge m\frac{X_k(t)}{R(t)} \quad \text{for every } t \ge \tilde{T}_k,$$

where

$$m = \frac{1}{2}\left(c - \frac{2}{3\sqrt{3}}\right).$$

By the last inequality, it is easy to see that the function $X_{k+m} = \frac{X_k}{R^m}$ is increasing on $[\tilde{T}_k, \infty)$. This, as k can be chosen arbitrarily close to k_0, is a contradiction. The theorem is proved. □

Consider the third-order delay differential equation

$$\frac{1}{r_3(t)}\left(\frac{1}{r_2(t)}\left(\frac{1}{r_1(t)}\left(\frac{x(t)}{r_0(t)}\right)'\right)'\right)' - p(t)x\big(g(t)\big) = 0, \tag{6.114}$$

where $g : [\sigma, \infty) \to R$, $0 < g(t) \le t$, $g(t) \to \infty$ as $t \to \infty$ and $p \in C([\sigma, \infty), R)$ such that $p(t) > 0$, and (6.85) is satisfied.

We assume the operators L_i to be as usual, that is,

$$\begin{aligned}
L_0x(t) &= \frac{x(t)}{x_0(t)},\\
L_1x(t) &= \frac{1}{r_1(t)}\big(L_0x(t)\big)' = \frac{1}{r_1(t)}\left(\frac{1}{r_0(t)}x'(t)\right)',\\
L_2x(t) &= \frac{1}{r_2(t)}\big(L_1x(t)\big)' = \frac{1}{r_2(t)}\left(\frac{1}{r_1(t)}\left(\frac{x(t)}{r_0(t)}\right)'\right)'\\
L_3x(t) &= \frac{1}{r_3(t)}\big(L_2x(t)\big)' = \frac{1}{r_3(t)}\left(\frac{1}{r_2(t)}\left(\frac{1}{r_1(t)}\left(\frac{x(t)}{r_0(t)}\right)'\right)'\right)'.
\end{aligned}$$

Then (6.114) can be rewritten as

$$L_3x(t) - p(t)x\big(g(t)\big) = 0.$$

Theorem 6.2.19 *Let $g'(t) \ge 0$. Assume that the linear differential inequality*

$$z'(t) + q(t)z\big(g(t)\big) \le 0$$

has no eventually positive solutions. Then (6.114) *has property B, where*

$$q(t) = r_2(t)\left(\int_t^{\infty} r_3(s)p(s)r_0\big(g(s)\big)\,ds\right)\left(\int_T^{g(t)} r_1(s)\,ds\right), \quad T \ge \sigma. \tag{6.115}$$

Theorem 6.2.20 *Let $g \in C^1([\sigma, \infty), R)$ be such that $0 < g(t) \le t$ and $g'(t) > 0$. Define a function $f = f(\lambda)$ for $0 \le \lambda \le \frac{1}{e}$ by*

$$fe^{-\lambda f} = 1, \quad 1 \le f \le e.$$

Assume that either

$$\liminf_{t\to\infty} \int_{g(t)}^{t} q(s)\,ds > \frac{1}{e}$$

or

$$\limsup_{t\to\infty}\int_{g(t)}^{t} q(s)\,ds > 1,$$

or when $0 < d \le \frac{1}{e}$ *and* $c \le 1$, *the following condition holds*

$$f(d)(1-\sqrt{1-c})^2 > 1.$$

Then (6.114) *has property* B, *where* $q(t)$ *is same as in* (6.115).

Example 6.2.6 Consider the third-order equation

$$x'''(t) - \frac{a}{t^2\sqrt[3]{t}}x\big(\sqrt[3]{t}\big) = 0, \quad a > 0,\ t \ge 1. \tag{6.116}$$

Here $g(t) = \sqrt[3]{t}$, $p(t) = \frac{a}{t^2\sqrt[3]{t}}$ and $q(t) = (g(t) - T)\int_t^\infty p(s)\,ds$. By Theorem 6.2.20, Eq. (6.116) has property B. On the other hand, by the comparison result, i.e., Theorem 6.1.23, Eq. (6.116) has property B, if the ordinary differential equation

$$x'''(t) - \frac{3a}{t^5}x = 0 \tag{6.117}$$

has property B. But, as (6.117) has no property B, Theorem 6.1.23 fails for (6.116).

Consider the function $\tau(t)$ and $w(t)$ satisfying

$$\tau(t) \in C\big([\sigma,\infty),[0,\infty)\big), \quad \tau(t) > t, \quad w(t) = g\big(\tau(t)\big) < t. \tag{6.118}$$

Theorem 6.2.21 *Let* (6.118) *hold. Assume that the linear differential inequality*

$$z'(t) + q_1(t)z\big(w(t)\big) \le 0$$

has no eventually positive solutions. Then (6.114) *has property* B, *where*

$$q_1(t) = r_2(t)\int_t^{\tau(t)} p(s)r_3(s)r_0\big(g(s)\big)\big[R_1\big(g(s)\big) - R_1(T)\big]\,ds. \tag{6.119}$$

Theorem 6.2.22 *Let* (6.118) *hold. Assume that either*

$$\liminf_{t\to\infty}\int_{w(t)}^{t} q_1(s)\,ds > \frac{1}{e}$$

or

$$\limsup_{t\to\infty}\int_{w(t)}^{t} q_1(s)\,ds > 1.$$

Then (6.114) *has property* B, *where* $q_1(t)$ *is same as in* (6.119).

Theorem 6.2.23 *Let* (6.118) *hold. Define*

$$\tilde{q}_1(t) = r_2(t) \int_t^{\tau(t)} p(s) r_0\big(g(s)\big) r_3(s) R_1\big(g(s)\big)\, ds.$$

Assume that either

$$\liminf_{t\to\infty} \int_{w(t)}^{t} \tilde{q}_1(s)\, ds > \frac{1}{e}$$

or

$$\limsup_{t\to\infty} \int_{w(t)}^{t} \tilde{q}_1(s)\, ds > 1.$$

Then (6.114) *has property B.*

The following example shows how to choose the function $\tau(t)$ satisfying (6.118).

Example 6.2.7 Consider the equation

$$x'''(t) - \frac{a}{t^2\sqrt{t}} x(\sqrt{t}) = 0, \quad a > 0,\ t \geq 1.$$

Here $g(t) = \sqrt{t}$. Let us put $\tau(t) = 2t$. Then $w(t) = \sqrt{2t} < t$, that is, (6.118) is satisfied. Clearly

$$\tilde{q}_1(t) = \int_t^{2t} \frac{a}{s^2\sqrt{s}} \sqrt{s}\, ds = \frac{a}{2t}.$$

By Theorem 6.2.23, this equation has property B if

$$\liminf_{t\to\infty} \int_{\sqrt{2t}}^{t} \tilde{q}_1(s)\, ds > \frac{1}{e},$$

which is true for any $a > 0$.

Theorem 6.2.24 *Suppose that*

$$\limsup_{t\to\infty} \frac{g(t)}{t} < c < 1.$$

Assume that either

$$\liminf_{t\to\infty} \int_{ct}^{t} r_2(s) \int_s^{g^{-1}(cs)} p(\theta) r_3(\theta) r_0\big(g(\theta)\big) R_1\big(g(\theta)\big)\, d\theta\, ds > \frac{1}{e}$$

or

$$\limsup_{t\to\infty} \int_{ct}^{t} r_2(s) \int_s^{g^{-1}(cs)} p(\theta) r_3(\theta) r_0\big(g(\theta)\big) R_1\big(g(\theta)\big)\, d\theta\, ds > 1$$

holds. Then (6.114) *has property B.*

One may see Sect. 6.1 on the property B of (6.46c) with $p(t) < 0$.

Property B of (6.114) demands the existence of a nonoscillatory solution of the equation. In the following, we provide a result guarantying the existence of a nonoscillatory solution of (6.114) satisfying the property $x(t)L_i x(t) > 0, 0 \le i \le 3$.

Theorem 6.2.25 *Equation* (6.114) *admits a nonoscillatory solution satisfying*

$$x(t)L_i x(t) > 0, \quad 0 \le i \le 3.$$

Proof From a result due to Kusano et al. [23], it follows that the equation

$$L_3 y - p^*(t) y = 0, \tag{6.120}$$

where $p^*(t) = (r_0(t))^{-1} p(t) r_0(g(t)) r_3(t)$, admits a nonoscillatory solution $y(t)$ satisfying

$$y(t)L_i y(t) > 0, \quad 0 \le i \le 3$$

for large t. We may assume, without any loss of generality, that $y(t) > 0$ for $t \ge t_0 > \sigma$. Hence $L_i y(t) > 0$ for $t \ge t_0$ and $y(g(t)) > 0$ for $t \ge t_1 > t_0$ and $0 \le i \le 3$. Successive integration of (6.120) from t_1 to t yields

$$\begin{aligned} L_0 y(t) &\ge K + \int_{t_1}^{t} r_1(s_1) \int_{t_1}^{s_1} r_2(s_2) \int_{t_1}^{s_2} r_3(s_3) p^*(s_3) y(s_3)\, ds_3\, ds_2\, ds_1 \\ &\ge K + \int_{t_1}^{t} r_1(s_1) \int_{t_1}^{s_1} r_2(s_2) \int_{t_1}^{s_2} r_3(s_3) p^*(s_3) r_0(s_3) L_0 y(s_3)\, ds_3\, ds_2\, ds_1, \end{aligned}$$

where $K = L_0 y(t_1) > 0$. Since $L_1 y(t) > 0$ for $t \ge t_1$, we have

$$L_0 x(t) \ge K + \int_{t_1}^{t} r_1(s_1) \int_{t_1}^{s_1} r_2(s_2) \int_{t_1}^{s_2} r_3(s_3) p^*(s_3) r_0(s_3) L_0 y\big(g(s_3)\big)\, ds_3\, ds_2\, ds_1.$$

By Lemma 1.5.18, it follows that the integral equation

$$V(t) = K + \int_{t_1}^{t} r_1(s_1) \int_{t_1}^{s_1} r_2(s_2) \int_{t_1}^{s_2} r_3(s_3) p^*(s_3) r_0(s_3) V\big(g(s_3)\big)\, ds_3\, ds_2\, ds_1$$

admits a solution $V(t)$, $t \ge t_1$, satisfying

$$K \le V(t) \le L_0 y(t), \quad t \ge t_1.$$

Hence $V(t) > 0$ for $t \ge t_1$. Setting $x(t) = r_0(t)V(t)$, we obtain $x(t) > 0$ for $t \ge t_1$ and

$$L_0 x(t) = K + \int_{t_1}^{t} r_1(s_1) \int_{t_1}^{s_1} r_2(s_2) \int_{t_1}^{s_2} r_3(s_3) p(s_3) x\big(g(s_3)\big)\, ds_3\, ds_2\, ds_1.$$

Successive differentiation shows that $x(t)$ is a positive solution of (6.114) satisfying $L_i x(t) > 0$ for $t \ge t_0$, $0 \le i \le 3$. Hence the theorem is proved. □

6.3 Oscillation and Asymptotic Behaviour of Solutions of the Linear Delay Differential Equation $x'''(t) + p(t)x(g(t)) = 0$

In this section, we shall provide some basic and interesting results on the oscillation and asymptotic behaviour of nonoscillatory solutions of the third-order linear delay differential equation of the form (6.4), under certain conditions on the coefficient function $p \in C((\sigma, \infty), [0, \infty))$, $\sigma \in R$ and $g \in C([\sigma, \infty), R)$ such that $g(t) \leq t$ and $g(t) \to \infty$ as $t \to \infty$.

We provide some sufficient conditions for oscillation of all solutions of (6.4) while dealing with property $\overline{A}$ of the equation. Moreover, these sufficient conditions are consistent with the situation when $p(t)$ is a constant.

A function $x : [g(\sigma), \infty) \to R$ is said to be a *proper solution* of (6.4) if it is absolutely continuous on (t_0, ∞), $t_0 \geq \sigma$ along with its derivatives up to the second order and satisfies (6.4) almost everywhere on (t_0, ∞) and $\sup\{|x(s)| : s \geq t\} > 0$ for $t \geq t_0$. A proper solution of (6.4) is called *oscillatory*, if it has a sequence of zeros tending to infinity. Otherwise, it is called *nonoscillatory*. Equation (6.4) with $g(t) = t$ is said to be disconjugate on $[\sigma, \infty)$, if no nontrivial solution of the equation has more than two zeros, counting multiplicities.

Let $x(t)$ be a positive solution of (6.4) for $t \geq t_0 \geq \sigma$. Then there exists a $t_1 > t_0$ such that $x(g(t)) > 0$ for $t \geq t_1$. Since $x'''(t) \leq 0$ for $t \geq t_1$, there exists an $l = 0$ or 2 such that

$$\left.\begin{aligned} &x(t)x^{(i)}(t) > 0, \quad 0 \leq i \leq l, \\ &(-1)^{i+l}x(t)x^{(i)}(t) > 0, \quad l+1 \leq i \leq 3 \end{aligned}\right\} \tag{6.121}$$

is satisfied for large t. Again, for this $l = 2$, the following inequality holds for large t, say for $t \geq t_2$.

$$|x(t)| \geq \frac{(t-t_2)^2}{2.1}|x''(t)|, \quad t \geq t_2. \tag{6.122}$$

The following conjecture is given in [21], which we state as a problem. (In fact, this problem was given by Kiguradze for the nth-order differential equation, and we define the problem here for third-order equations.)

Problem 6.3.1 *Let $M_{3^*} = \max(\lambda(\lambda-1)(\lambda-2))$. If*

$$\int^{\infty} t^2\left[p(t) - \frac{M_{3^*}}{t^3}\right]dt = \infty,$$

then (6.4) *with $g(t) = t$ has property $\overline{A}$.*

Lemma 6.3.1 *For any nonoscillatory solution $x(t)$ of* (6.4), *let $x(t) > 0, x'(t) > 0, x''(t) > 0$ and $x'''(t) < 0$ hold for large t. Then*

$$\int_{t_1}^{\infty} |x'''(s)|\,ds < \infty, \tag{6.123}$$

and

$$x^{(i)}(t) \geq x^{(i)}(t_1) + \frac{1}{(1-i)!}\int_{t_1}^{t} (t-s)^{1-i} x^{(i)}(s)\,ds \tag{6.124}$$

for $t \geq t_1$, $i = 0, 1$ *and*

$$x''(t) \geq \frac{1}{(1-i)!}\int_{t}^{\infty} \left|x'''(s)\right| ds \tag{6.125}$$

for $t \geq t_1$. *If in addition*

$$\int_{t_1}^{\infty} s\left|x'''(s)\right| ds = \infty, \tag{6.126}$$

then there exists $t_2 \geq t_1$ *such that*

$$x'(t) \geq t\int_{t}^{\infty} \left|x'''(s)\right| ds \tag{6.127}$$

for $t \geq t_2$ *and*

$$ix'(t) \geq tx^{(3-i)}(t) \geq (i-1)x^{(2-i)}(t) \tag{6.128}$$

for $t \geq t_2$, $i \in \{1, 2\}$.

Theorem 6.3.1 *If*

$$\int_{\sigma}^{\infty} g(t)p(t)\,dt = \infty,$$

then (6.4) *has property* $\overline{\mathrm{A}}$. *Moreover, if* $x(t)$ *is a solution of* (6.4), *then*

$$\lim_{t\to\infty} x(t) = \lim_{t\to\infty} x'(t) = \lim_{t\to\infty} x''(t) = 0.$$

Proof Let $x(t)$ be a nonoscillatory solution of (6.4). Without any loss of generality, we may assume that $x(t) > 0$ and $x(g(t)) > 0$ for $t \geq t_0 > \sigma$. If possible, let $x(t) > 0$, $x'(t) > 0$, $x''(t) > 0$ and $x'''(t) < 0$ for $t \geq t_1 > t_0$. Clearly, $x'(t) \geq x'(t_1)$ for $t \geq t_1$ implies that $x(t) > (t - t_1)x'(t_1)$. There exists a $t_2 > t_1$ such that $g(t) > t_1$ for $t \geq t_2$. Hence $x(g(t)) > (g(t) - t_1)x'(t_1)$ for $t \geq t_2$. Integrating (6.4) from t_2 to t, we obtain

$$x''(t_2) > \int_{t_2}^{t} p(s)x\big(g(s)\big)\,ds > x'(t_1)\int_{t_2}^{t} p(s)\big(g(s) - t_1\big)\,ds. \tag{6.129}$$

Hence

$$\int_{t_2}^{\infty} p(t)\big(g(t) - t_1\big)\,dt < \infty.$$

However, (6.129) yields

$$x''(t_2) > x'(t_1)\big(g(t_2) - t_1\big) \int_{t_2}^{t} p(s)\,ds,$$

which implies that

$$\int_{t_2}^{\infty} p(t)\,dt < \infty.$$

Thus

$$\int_{t_2}^{\infty} p(t)g(t)\,dt < \infty,$$

a contradiction to the hypothesis of the theorem. Hence (6.4) has property $\overline{A}$. Moreover,

$$\lim_{t\to\infty} x'(t) = \lim_{t\to\infty} x''(t) = 0$$

holds and $\lim_{t\to\infty} x(t) = \lambda$, $0 \le \lambda < \infty$. If possible, let $\lambda > 0$. From Lemma 1.2 due to Kiguradze and Chanturia [21], it follows that

$$\int_{T}^{\infty} t^2 \big|x'''(t)\big|\,dt < \infty.$$

There exists a $T_1 > \max\{T, t_1\}$ such that $x(g(t)) > \alpha$ for $t \ge T_1$. Hence

$$\begin{aligned} \infty &> \int_{T_1}^{\infty} t^2 x'''(t)\,dt \\ &= \int_{T_1}^{\infty} t^2 p(t)x\big(g(t)\big)\,dt \\ &\ge T_1\alpha \int_{T_1}^{\infty} tp(t)\,dt \\ &\ge T_1\alpha \int_{T_1}^{\infty} g(t)p(t)\,dt \end{aligned}$$

a contradiction. Hence $\lim_{t\to\infty} x(t) = 0$. This completes the proof of the theorem. □

Theorem 6.3.2 *Let* $g(t) < t$ *and*

$$\limsup_{t\to\infty} \big(t - g(t)\big)^2 \int_{g^{-1}(t)}^{\infty} p(s)\,ds > 2! \tag{6.130}$$

hold. Then (6.4) *has property* $\overline{A}$.

Proof Let $x(t)$ be a nonoscillatory solution of (6.4). Without any loss of generality, we may assume that $x(t) > 0$ for $t \ge t_0 > \sigma$. Thus there exists a $T_1 \ge t_0$ such that $x(g(t)) > 0$ for $t \geq T_1$. Consequently, from (6.4), it follows that $x'''(t) \leq 0$ for $t \geq T_1$. Then, there exists a $l \in \{0, 2\}$ such that (6.121) holds for some $t \geq t_1 > T_1$. We claim that $l = 0$. If not, then $l = 2$. Putting $i = 0$ in (6.124), we get

$$x(t) \geq \int_{t_1}^{t} (t - s)x''(s)\,ds, \quad t \geq t_1. \tag{6.131}$$

We can find a $t_2 \geq t_1$ such that $g(t) > t_1$ for $t \geq t_2$. Hence, for $t \geq t_2$

$$x(t) \geq x''(t) \int_{g(t)}^{t} (t - s)\,ds \geq x''(t) \cdot \frac{(t - g(t))^2}{2};$$

that is,

$$x(t) \geq \frac{(t - g(t))^2}{2!} x''(t). \tag{6.132}$$

Using (6.125) in (6.132), we obtain

$$\begin{aligned} x(t) &\geq \frac{(t - g(t))^2}{2!} \int_{t}^{\infty} \left|x'''(s)\right| ds \\ &\geq \frac{(t - g(t))^2}{2!} \int_{g^{-1}(t)}^{\infty} \left|x'''(s)\right| ds \\ &\geq \frac{(t - g(t))^2}{2!} \int_{g^{-1}(t)}^{\infty} p(s)x\big(g(s)\big)\,ds \\ &\geq \frac{(t - g(t))^2}{2!} x(t) \int_{g^{-1}(t)}^{\infty} p(s)\,ds \end{aligned}$$

for $t \geq t_2$, which is a contradiction to the hypothesis of the theorem. Hence (6.4) has property $\overline{A}$. This completes the proof of the theorem. □

Theorem 6.3.3 *Suppose that*

$$\limsup_{t\to\infty} t^2 \int_{g^{-1}(t)}^{\infty} p(s)\,ds > 8, \tag{6.133}$$

holds. Then (6.4) *has property* $\overline{A}$.

Proof Let $x(t)$ be a nonoscillatory solution of (6.4). Without any loss of generality, we may assume that $x(t) > 0$ for $t \geq t_0 > \sigma$. Then there exists a $t_1 \geq t_0$ such that $x(g(t)) > 0$ for $t \geq t_1$. Consequently, it follows from (6.4) that $x'''(t) \leq 0$ for $t \geq t_1$, and (6.121) holds. If possible, suppose that (6.4) has no property $\overline{A}$. Then $l = 2$. Clearly (6.122) holds for some $t \geq t_2 \geq t_1$. Since $x'(t) > 0$, for $t \geq t_2$, we have

$$x(t) \geq \frac{1}{8} t^2 x''(t). \tag{6.134}$$

On the other hand, integrating (6.4) from t $(\geq t_2)$ to ∞,we have

$$\begin{aligned} x''(t) &> \int_t^\infty p(s)x\big(g(s)\big)\,ds \\ &> \int_{g^{-1}(t)}^\infty p(s)x\big(g(s)\big)\,ds \\ &> x(t)\int_{g^{-1}(t)}^\infty p(s)\,ds. \end{aligned}$$

Then (6.134) gives

$$1 \geq \frac{1}{8}t^2\int_{g^{-1}(t)}^\infty p(s)\,ds$$

for $t \geq t_2$, which contradicts (6.133). Hence (6.4) has property $\overline{A}$. The theorem is proved. □

Theorem 6.3.4 *Suppose that $g(t) < t$, and the following inequality*

$$\limsup_{t\to\infty}\int_{g(t)}^t (t-s)\int_{g^{-1}(g^{-1}(s))}^\infty p(u)\,du\,ds > 1 \tag{6.135}$$

holds. Then (6.4) *has property* $\overline{A}$.

Proof Let $x(t)$ be a nonoscillatory solution of (6.4). Without any loss of generality, we may assume that $x(t) > 0$ and $x(g(t)) > 0$ for $t \geq t_0 > \sigma$. Thus (6.121) holds for some $t \geq t_1 > t_0$. Suppose that $l = 2$. Putting $i = 0$ in (6.124), we get

$$x(t) \geq \int_{t_1}^t (t-s)x''(s)\,ds. \tag{6.136}$$

From (6.124), we obtain

$$x''(t) \geq \int_t^\infty p(s)x\big(g(s)\big)\,ds. \tag{6.137}$$

Then from (6.136) and (6.137), we obtain

$$x(t) \geq \int_{t_1}^t (t-s)\int_s^\infty p(u)x\big(g(u)\big)\,du\,ds. \tag{6.138}$$

We can find a $t_2 \geq t_1$ such that $g(t) > t_1$ for $t \geq t_2$. Thus, for $t \geq t_2$

$$x(t) \geq \int_{g(t)}^t (t-s)\int_{g^{-1}(g^{-1}(s))}^\infty p(u)x\big(g(u)\big)\,du\,ds$$

which, in turn, yields

$$1 \geq \int_{g(t)}^{t} (t-s) \int_{g^{-1}(g^{-1}(s))}^{\infty} p(u)\, du\, ds.$$

Taking limit sup., we obtain a contradiction. Consequently, (6.4) has property $\overline{A}$. Hence the theorem is proved. □

Theorem 6.3.5 *Let $g(t) < t$ and*

$$\limsup_{t\to\infty} \int_{g(t)}^{t} \big(s-g(t)\big)^2 p(s)\, ds > 2. \tag{6.139}$$

Then (6.4) *has no solution satisfying the property $(-1)^i x^{(i)}(t) > 0$ for large t.*

Proof If possible, suppose that (6.4) has a nonoscillatory solution $x(t)$ satisfying the property $(-1)^i x^{(i)}(t) > 0$ for large t. Then $l = 0$ in (6.121). Suppose that $x(g(t)) > 0$ and $x(t) > 0$ for some $t \geq t_1 > \sigma$. From Lemma 1.2 due to Kiguradze and Chanturia [21], it follows for $i = 0$ that

$$\begin{aligned} x(t) &\geq \frac{1}{2}\int_t^{\infty} (s-t)^2 p(s) x\big(g(s)\big)\, ds \\ &\geq \frac{1}{2}\int_t^{g^{-1}(t)} (s-t)^2 p(s) x\big(g(s)\big)\, ds \\ &\geq \frac{x(t)}{2}\int_t^{g^{-1}(t)} (s-t)^2 p(s)\, ds, \end{aligned}$$

that is,

$$2 \geq \int_t^{g^{-1}(t)} (s-t)^2 p(s)\, ds,$$

for some $t \geq t_2 \geq t_1$. Then there exists a $t_3 \geq t_2$ such that $g(t) > t_2$ for $t \geq t_3$. Hence for $t \geq t_3$, we have

$$2 \geq \int_{g(t)}^{t} \big(s-g(t)\big)^2 p(s)\, ds.$$

Taking limit sup., we obtain a contradiction. Hence $l \neq 0$. The theorem is proved. □

Corollary 6.3.1 *Suppose that $g(t) < t$,* (6.139) *holds and either* (6.130) *or* (6.133) *or* (6.135) *is satisfied. Then every solution of* (6.4) *oscillates.*

Example 6.3.1 Consider

$$x'''(t) + \frac{30}{t^3} x\big(t/2^{1/3}\big) = 0, \quad t \geq 2. \tag{6.140}$$

By Theorem 6.3.3, (6.140) has property $\overline{A}$. In particular, $x(t) = 1/t^3$ is a nonoscillatory solution of (6.140).

Example 6.3.2 Consider

$$x'''(t) + \frac{82}{t^3}x(t/3) = 0, \quad t \geq 1. \tag{6.141}$$

Theorem 6.3.2 can be applied to this example, whereas Theorem 6.3.4 fails to hold. On the other hand, (6.139) is satisfied. Hence by Corollary 6.3.1, all solutions of (6.141) are oscillatory.

Example 6.3.3 Inequality (6.135) to the equation

$$x'''(t) + \frac{63}{t^3}x(t/2) = 0, \quad t \geq 1 \tag{6.142}$$

is satisfied, whereas (6.130) fails to hold. Hence Theorem 6.3.4 can be applied to (6.142). Further, since (6.139) is satisfied, all solutions of (6.142) are oscillatory, by Corollary 6.3.1.

Remark 6.3.1 Let $p(t) = p > 0$ be a constant and $g(t) = t - \tau$, $\tau > 0$ be a constant. Then (6.4) becomes

$$x'''(t) + px(t - \tau) = 0. \tag{6.143}$$

Clearly, the conditions (6.130), (6.133) and (6.135) are consistent with $p(t) = p$ and $g(t) = t - \tau$. Hence form Corollary 6.3.1, it follows that, if

$$p\tau^3 > 6, \tag{6.144}$$

then (6.143) is oscillatory.

The characteristic equation associated with (6.143) is given by

$$\lambda^3 + pe^{-\tau\lambda} = 0. \tag{6.145}$$

Setting $F(\lambda) = \lambda^3 + pe^{-\tau\lambda}$, we see that $F(\lambda) > 0$ for $\lambda \geq 0$. Suppose that $\lambda < 0$. We claim that $F(\lambda) > 0$ for $\lambda < 0$. If possible suppose that $F(\lambda) \leq 0$ for $\lambda < 0$. Then $\lambda^3 \leq -pe^{-\tau\lambda}$. Then $\lambda^3\tau^3 \leq -6e^{-\tau\lambda}$. Let $\lambda = -\gamma$, $\gamma > 0$. Then $\gamma^3\tau^3 \geq 6e^{\tau\gamma}$. Setting $\tau\gamma = \beta$, we see that $\beta^3 \geq 6e^{\beta}$, a contradiction. Hence our claim holds, that is, $F(\lambda) > 0$ for $\lambda < 0$. Thus (6.144) implies that all solutions of (6.143) are oscillatory.

Remark 6.3.2 Although the conditions in Theorems 6.3.2 and 6.3.4 are legitimate, these are not efficient. When $g(t)$ is close to t, the conditions (6.130) and (6.135) fail to hold. This is evident from the following examples:

If we replace $g(t) = \frac{t}{3}$ in (6.141) by $g(t) = \frac{3t}{4}$, then the equation becomes

$$x'''(t) + \frac{82}{t^3} x\left(\frac{3t}{4}\right) = 0, \quad t \geq 1. \tag{6.146}$$

Condition (6.130) fails to hold, and hence Theorem 6.3.2 cannot be applied to (6.146). Similarly, consider the equation

$$x'''(t) + \frac{46}{t^3} x\left(\frac{t}{2}\right) = 0, \quad t \geq 1. \tag{6.147}$$

Theorem 6.3.4 can be applied to this example. On the other hand, if $g(t) = \frac{t}{2}$ in (6.147) is replaced by $g(t) = \frac{10t}{11}$, then (6.147) becomes

$$x'''(t) + \frac{46}{t^3} x\left(\frac{10t}{11}\right) = 0, \quad t \geq 1, \tag{6.148}$$

then (6.135) fails, and hence Theorem 6.3.4 cannot be applied. The following theorems provide sufficient conditions for (6.4) to have property $\overline{A}$ when $g(t)$ is close to t.

Theorem 6.3.6 *Assume that $g(t) < t$ and $t - g(t) \to \infty$ as $t \to \infty$. If*

$$\limsup_{t\to\infty} \big(g(t)\big)^2 \int_{g^{-1}(t)}^{\infty} p(s)\,ds > 2 \tag{6.149}$$

holds, then (6.4) *has property* $\overline{A}$.

Proof We can find a $t_2 > t_1$ such that $t - g(t) > t_1$ for $t \geq t_2$. Hence for $t \geq t_2$, (6.131) gives

$$x(t) \geq x''(t) \int_{t-g(t)}^{t} (t - s)\,ds \geq \frac{g^2(t)}{2} x''(t).$$

Using (6.125) in the above inequality, we obtain a contradiction. The proof is complete. □

Corollary 6.3.2 *Suppose that the conditions of Theorems* 6.3.5 *and* 6.3.6 *are satisfied. then all solutions of* (6.4) *oscillates.*

Example 6.3.4 By Theorem 6.3.6, (6.146) has property $\overline{A}$.

Theorem 6.3.7 *Let $g(t) < t$ and $t - g(t) \to \infty$ as $t \to \infty$. If*

$$\limsup_{t\to\infty} \int_{t-g(t)}^{t} (t - s) \int_{g^{-1}(g^{-1}(s))}^{\infty} p(u)\,du\,ds > 1 \tag{6.150}$$

holds, then (6.4) *has property* $\overline{A}$.

Proof Proceeding as in the proof of Theorem 6.3.4, we arrive at (6.138) for $t \geq t_1$. Then we can find a $t_2 \geq t_1$ such that $t - g(t) > t_1$ for $t \geq t_2$. Hence from (6.138), we obtain

$$x(t) \geq \int_{t-g(t)}^{t} (t-s) \int_{g^{-1}(g^{-1}(s))}^{\infty} p(u)x\big(g(u)\big)\,du\,ds,$$

which further yields

$$1 \geq \int_{t-g(t)}^{t} (t-s) \int_{g^{-1}(g^{-1}(s))}^{\infty} p(u)\,du\,ds.$$

Taking limit sup. both sides in the above inequality, we obtain a contradiction. This completes the proof of the theorem. □

Corollary 6.3.3 *Suppose that the conditions of Theorems* 6.3.5 *and* 6.3.7 *are satisfied. Then all solutions of* (6.4) *are oscillatory.*

Example 6.3.5 By Theorem 6.3.7, (6.148) has property $\overline{A}$.

Let $x(t)$ be a nonoscillatory solution of (6.4) such that (6.131) holds for $t \geq t_1$. Then for $t > t_2 \geq 2t_1$, (6.131) gives

$$x(t) \geq \int_{t/2}^{t} (t-s)x''(s)\,ds, \quad t \geq t_1.$$

Using (6.125) and the above inequality, we obtain the following theorem.

Theorem 6.3.8 *Let* $g(t) \leq t$ *and*

$$\limsup_{t\to\infty} \int_{t/2}^{t} (t-s) \int_{g^{-1}(g^{-1}(s))}^{\infty} p(u)\,du\,ds > 1 \tag{6.151}$$

holds, then (6.4) *has property* $\overline{A}$.

Proof Proceeding as in the proof of Theorem 6.3.4, we obtain (6.138). Then for $t \geq t_2 > 2t_1$, (6.138) yields a contradiction. This completes the proof of the theorem. □

We note that when $g(t) = t/2$, then Theorems 6.3.4, 6.3.7 and 6.3.8 give same sufficient conditions to have property $\overline{A}$ of (6.4).

Corollary 6.3.4 *Suppose that the conditions of Theorem* 6.3.5 *are satisfied. If the conditions of Theorem* 6.3.8 *hold, then all solutions of* (6.4) *are oscillatory.*

Example 6.3.6 Consider

$$x'''(t) + \frac{44}{t^3} x\left(\frac{3t}{5}\right) = 0, \quad t \geq 1.$$

Theorems 6.3.2 and 6.3.6 can be applied to this example, whereas Theorem 6.3.3 cannot be applied to this example.

Example 6.3.7 Consider

$$x'''(t) + \frac{160}{t^3} x\left(\frac{t}{3}\right) = 0, \quad t \geq 1.$$

By Theorem 6.3.3, this equation has property $\overline{A}$, whereas Theorem 6.3.6 fails.

Theorem 6.3.9 *Let $g'(t) > 0$. If*

$$\int^{\infty} H(t)\, dt = \infty, \tag{6.152}$$

then every solution of (6.4) *is either oscillates or tend to zero as $t \to \infty$, in particular,* (6.4) *has property $\overline{A}$, where*

$$H(t) = t^2 p(t) - \frac{2}{g'(t) g(t)}. \tag{6.153}$$

Remark 6.3.3 Let $g(t) = t$ and $n = 2$. From Theorem 6.3.9, it follows that, if

$$\int^{\infty} \left[t p(t) - \frac{1}{4t} \right] dt = \infty, \tag{6.154}$$

then

$$x'' + p(t)x = 0 \tag{6.155}$$

is oscillatory. This gives a partial answer to Problem 6.3.1. Further, Theorem 6.3.9 improves the results due to Kneser [38, p. 45] and Hille and Kneser [38, Theorem 2.41]. We note that Theorem 6.3.9 holds for (6.4) with $g(t) = t$ for $n = 2$ and $n = 3$.

Now, suppose that $n = 3$ and $g(t) = t$. then Theorem 6.3.9 yields the following: if

$$\int^{\infty} \left[t^2 p(t) - \frac{2}{t} \right] dt = \infty,$$

then

$$x''' + p(t)x = 0 \tag{6.156}$$

has property $\overline{A}$. On the other hand, from Hanan [19, Theorem 5.7], and Kiguradze and Chanturia [21, Theorem 1.1], it follows that (6.156) has property $\overline{A}$ if

$$\int^{\infty} \left[t^2 p(t) - \frac{2}{3\sqrt{3}t} \right] dt = \infty. \tag{6.157}$$

Hence Theorem 6.3.9 is yet to be improved.

Proof of Theorem 6.3.9 If possible, suppose that (6.4) dose not have property $\overline{A}$. Then (6.4) admits a nonoscillatory solution $x(t)$ such that $x(t) > 0$, $x'(t) > 0$, $x''(t) > 0$ and $x'''(t) < 0$ for large t, that is, (6.121) holds with $l = 2$. We may assume, without any loss of generality, that $x(t) > 0$ and $x(g(t)) > 0$ for $t \geq t_1 > \sigma$.

Set $z(t) = \frac{t^2 x''(t)}{x(g(t))}$. Then

$$z'(t) = -t^2 p(t) + \frac{2}{t} z(t) - g'(t) \frac{x'(g(t))}{x(g(t))} z(t). \tag{6.158}$$

Putting $i = 1$ in (6.124), we obtain, for $t \geq t_1$

$$x'(t) \geq (t - t_1) x''(t).$$

Hence for $t \geq 2t_1$, we get

$$x'(t) \geq \frac{t}{2} x''(t).$$

Thus, for $t \geq t_2 > 2t_1$,

$$x'\big(g(t)\big) \geq \frac{g(t)}{2} x''(t).$$

Using the above inequality, (6.158) yields

$$z'(t) \leq -F(t), \tag{6.159}$$

where

$$F(t) = t^2 p(t) - \frac{2}{t} z(t) + \frac{g'(t) g(t)}{2t^2} z^2(t),$$

which as a function of z, attains the minimum $H(t)$ given in (6.153). Now, the integration of (6.159) from t_2 to t yields $z(t) < 0$ for large t, a contradiction. This completes the proof of the theorem. □

Example 6.3.8 Consider

$$x'''(t) + \frac{24(t-1)^2}{t^5} x(t-1) = 0, \quad t \geq 2. \tag{6.160}$$

All the conditions of Theorem 6.3.9 are satisfied. Hence (6.160) has property $\overline{A}$. In particular, $x(t) = 1/t^2$ is a nonoscillatory solution of (6.160).

Corollary 6.3.5 *Suppose that the conditions of Theorems* 6.3.5 *and* 6.3.9 *are satisfied. Then all solutions of* (6.4) *are oscillatory.*

Now, we consider the following ordinary differential equations associated with the delay differential equations (6.140)–(6.142), and (6.160):

$$x''' + \frac{30}{t^3} x = 0, \quad t \geq 2, \tag{6.161}$$

$$x''' + \frac{82}{t^3}x = 0, \quad t \geq 1, \tag{6.162}$$

$$x''' + \frac{63}{t^3}x = 0, \quad t \geq 1, \tag{6.163}$$

$$x''' + \frac{24(t-1)^2}{t^3}x = 0, \quad t \geq 2. \tag{6.164}$$

From Hanan [19, Theorem 5.7], it follows that (6.161)–(6.164) are oscillatory. We note that a third-order ordinary differential equation is said to be *oscillatory*, if it has an oscillatory solution; otherwise, it is called *nonoscillatory*. However, all solutions of (6.161)–(6.164) are not oscillatory. This is because (6.161)–(6.164) are of Class I or C_I and hence admits a nonoscillatory solution. We may note that Eq. (6.156) is said to be of Class I or C_I if any of its solutions $x(t)$ for which $x(t_0) = x'(t_0) = 0$ and $x''(t_0) > 0$, $(\sigma < t_0 < \infty)$ satisfies $x(t) > 0$ for $t \in [\sigma, t_0)$. It seems that the presence of delay in (6.141) and (6.142) is responsible for the change in the qualitative behaviour of solutions of the equations. It is easy to construct an example of a third-order delay differential equation, all solutions of which are oscillatory, but it is difficult to construct such an example of a third-order ordinary differential equation. It is evident from the following examples due to Dolan [6] and Example 2.6.5 in Chap. 2, respectively.

Example 6.3.9 (Dolan [6]) All solutions of

$$\left\{\left[z' - \frac{r'(t)}{r(t)}z\right] + r(t)z\right\}' = 0$$

are oscillatory, where $r(t) = [1 + \sqrt{2}\epsilon \sin(t + \frac{\pi}{4})]^{-1} > 0$, $t \geq 0$, $0 < \epsilon < \frac{1}{\sqrt{2}}$.

Theorem 6.3.10 *Suppose that for any $\mu \in (0, 1/2)$, the third-order ordinary differential equation*

$$u''' + G(t)u = 0, \tag{6.165}$$

admits an oscillatory solution, where

$$G(t) = \mu\left(\frac{g(t)}{t}\right)^2 p(t). \tag{6.166}$$

Then (6.4) *has property* $\overline{A}$.

Proof If (6.4) has no property $\overline{A}$, then it admits a nonoscillatory solution $x(t)$ such that (6.121) is satisfied with $l = 2$. We may assume, without any loss of generality, that $x(t) > 0$ and $x(g(t)) > 0$ for some $t \geq t_1 > t_0 > \sigma$. Then $x'(t) > 0$, $x''(t) > 0$ and $x'''(t) < 0$ for $t \geq t_2 \geq t_1$. For any $\mu \in (0, 1/2)$, there exists a $T_\mu \geq t_2$ such that

$$\frac{x(g(t))}{x(t)} \geq \mu\left(\frac{g(t)}{t}\right)^2 \tag{6.167}$$

for $t \geq T_\mu$ (see Theorem 2.2 in [16]). Setting $z(t) = \frac{x'(t)}{x(t)}$ for $t \geq T_\mu$, we get

$$z'(t) = \frac{x''(t)}{x(t)} - z^2(t). \tag{6.168}$$

Further, assuming $u(t) = \exp(\int_{T_\mu}^t z(s)\,ds)$ and by using (6.167) and (6.168), we obtain

$$u'''(t) + \mu\left(\frac{g(t)}{t}\right)^2 p(t)u(t) \leq 0$$

for $t \geq T_\mu$. From Lemma 4 in [18], it follows that (6.165) is disconjugate on $[T_\mu, \infty)$, a contradiction. Hence $l = 0$, that is, (6.4) has property $\overline{A}$. The theorem is proved. □

Corollary 6.3.6 *Suppose that $g(t) < t$. If all the conditions of Theorems* 6.3.5 *and* 6.3.10 *are satisfied, then all solutions of* (6.4) *are oscillatory.*

Example 6.3.10 Consider

$$x'''(t) + e^{-1}x(t-1) = 0, \quad t \geq 2. \tag{6.169}$$

As $\liminf_{t\to\infty} \mu e^{-1}t(t-1)^2 > \frac{2}{3\sqrt{3}}$, then, for every $\mu \in (0, 1/2)$, the equation

$$u''' + \mu e^{-1}\left(\frac{t-1}{t}\right)^2 u = 0, \quad t \geq 2$$

admits an oscillatory solution by Theorem 5.7 of [19]. From Theorem 6.3.10, it follows that (6.169) has property $\overline{A}$. In particular, $x(t) = e^{-t}$ is a solution of (6.169) for $t \geq 2$.

Remark 6.3.4 Consider Eqs. (6.141) and (6.142). For $0 < \mu < \frac{\sqrt{3}}{82}$, it follows that $\lim_{t\to\infty} t^3 \mu \frac{82}{9t^3} < \frac{2}{\sqrt{3}}$ and hence

$$u''' + \mu\frac{82}{9t^3}u = 0$$

is nonoscillatory [19, by Theorem 5.7]. Similarly, for $0 < \mu < \frac{4}{189\sqrt{3}}$, the equation $u''' + \mu\frac{63}{4t^3}u = 0$ is nonoscillatory. Hence Corollary 6.3.6 cannot be applied to (6.141) and (6.142).

Theorem 6.3.11 *Let $x(t)$ be a solution of* (6.4). *If*

$$\int_\sigma^\infty \left[g^2(t)p(t) - \frac{2}{3\sqrt{3}t}\right]dt = \infty,$$

then either $x(t)$ is oscillatory or $\lim_{t\to\infty} x(t) = 0$.

Corollary 6.3.7 *Let $x(t)$ be a solution of* (6.4). *If*

$$\liminf_{t\to\infty} g^2(t)tp(t) > \frac{2}{3\sqrt{3}},$$

then either $x(t)$ is oscillatory or $\lim_{t\to\infty} x(t) = 0$.

Example 6.3.11 Consider the third-order linear delay differential equation

$$x'''(t) + \frac{2}{3\sqrt{3}g^3(t)} x\big(g(t)\big) = 0, \quad t \geq 1,$$

where $g(t) = \frac{t}{2}$. By Theorem 6.3.11, if $x(t)$ is a solution of this example, then either $x(t)$ is oscillatory or $\lim_{t\to\infty} x(t) = 0$. In fact, the corresponding characteristic equation, which is given by

$$2^{m-4} m(m-1)(m-2) + \frac{1}{3\sqrt{3}} = 0$$

has only a negative real root given by -1.0232. This implies that the corresponding solution tends to zero as $t \to \infty$.

Now, we shall provide some sufficient conditions on the property B of

$$x'''(t) - p(t)x\big(g(t)\big) = 0 \tag{6.170}$$

under certain conditions on the coefficient function $p \in C((\sigma, \infty), [0, \infty))$, $\sigma \in R$ and $g \in C([\sigma, \infty), R)$, such that $g(t) \leq t$ and $g(t) \to \infty$ as $t \to \infty$. One may proceed as above to study the property B of Eq. (6.170).

Theorem 6.3.12 *If*

$$\limsup_{t\to\infty} (t-T)^2 \int_{2g^{-1}(t)}^{\infty} p(s)\,ds > 2$$

for every $T \geq \sigma$, then (6.170) *has property B.*

Theorem 6.3.13 *If*

$$\int_{\sigma}^{\infty} p(t)\,dt < \infty \quad \textit{and} \quad \int_{\sigma}^{\infty} tp(t)\,dt = \infty,$$

then (6.170) *has property B.*

6.4 Asymptotic Behaviour of Solutions of Nonlinear Delay-Differential Equation of the Form $x'''(t) + q(t)x'(t) + p(t)f(x(g(t))) = 0$

This section is concerned with the behaviour of nonoscillatory solutions of the third-order nonlinear delay differential equations of the form

$$x'''(t) + q(t)x'(t) + p(t)f\bigl(x\bigl(g(t)\bigr)\bigr) = 0, \tag{6.171}$$

where $p, q \in C([\sigma, \infty), R), \sigma \geq 0, g \in C([\sigma, \infty), R)$ such that $g(t) \leq t, g(t) \to \infty$ as $t \to \infty$, $f \in C(R, R)$ such that either $\frac{f(u)}{u} \geq \beta > 0$ or $uf(u) > 0$ for $u \neq 0$, whichever is required.

We assume that under the initial conditions $x(t) = \psi(t)$, $t \leq t_0$ and $x^{(i)}(t_0) = x_i^0, i = 1, 2$, where $t_0 \in [\sigma, \infty)$, Eq. (6.171) admits solutions which can be continued to $[t_0, \infty)$.

Theorem 6.4.1 *Let* $\frac{f(u)}{u} \geq \beta > 0$ *for* $u \neq 0$, $q(t) \leq 0$ *and* $p(t) > 0$. *If*

$$\int_\sigma^\infty \left[\beta t^2 p(t) + tq(t) - \frac{2}{3\sqrt{3}t}\bigl(1 - t^2 q(t)\bigr)^{3/2}\right] dt = \infty, \tag{6.172}$$

then every nonoscillatory solution of (6.171) *tends either to zero or* $\pm\infty$ *as* $t \to \infty$.

Proof Let $x(t)$ be a nonoscillatory solution of (6.171) on $[t_0, \infty)$, $t_0 \geq \sigma$. Suppose that $x(t)$ is eventually positive. The case when $x(t)$ is eventually negative may be treated similarly. So there exists a $t_1 \geq t_0$ such that $x(t) > 0$ and $x(g(t)) > 0$ for $t \geq t_1$. Since $q(t) \leq 0$, $x'(t)$ is nonoscillatory.

Suppose that $x'(t) > 0$ for $t \geq t_2 \geq t_1$. We claim that $\lim_{t\to\infty} x(t) = \infty$. If not, then $\lim_{t\to\infty} x(t) = K$, where $0 < K < \infty$ is a real constant. From the continuity of f, we have

$$\lim_{t\to\infty} \frac{f(x(g(t)))}{x(t)} = \frac{f(K)}{K}. \tag{6.173}$$

Hence, there exists a $T > t_2$ such that, for $t > T$, $\frac{f(x(g(t)))}{x(t)} > \beta$. We define $u(t) = \frac{t^2 x'(t)}{x(t)}$ for $t \geq T$. Clearly, $u(t)$ is a solution of

$$\begin{aligned}
&\left(u' + \frac{3}{2}t^{-2}u^2 - 4t^{-1}u\right)' \\
&\quad = -\left[t^{-4}u^3 - 3t^{-3}u^2 + \bigl(2t^{-2} + q(t)\bigr)u + t^2 p(t)\frac{f(x(g(t)))}{x(t)}\right] \\
&\quad \leq \left[t^{-4}u^3 - 3t^{-3}u^2 + \bigl(2t^{-2} + q(t)\bigr)u + \beta t^2 p(t)\right] \\
&\quad \leq \left[\beta t^2 p(t) + tq(t) - \frac{2}{3\sqrt{3}t}\bigl(1 - t^2 q(t)\bigr)^{3/2}\right],
\end{aligned}$$

because

$$\beta t^2 p(t) + t q(t) - \frac{2}{3\sqrt{3t}}\big(1 - t^2 q(t)\big)^{3/2}$$

is the minimum of

$$t^{-4}u^3 - 3t^{-3}u^2 + \big(2t^{-2} + q(t)\big)u + \beta t^2 p(t)$$

for $u(t) > 0$ at

$$u(t) = t\big[1 + 3t^{-1/2}\big(1 - t^2 q(t)\big)^{1/2}\big].$$

Integrating the above inequality from T to t, we get

$$\begin{aligned} & u'(t) + \frac{3}{2}t^{-2}u^2(t) - 4t^{-1}u(t) \\ & \quad \le K_0 - \int_T^t \left[\beta s^2 p(s) + s q(s) - \frac{2}{3\sqrt{3s}}\big(1 - s^2 q(s)\big)^{3/2}\right] ds, \end{aligned}$$

where K_0 is a constant. Since

$$\frac{3}{2}t^{-2}u^2(t) - 4t^{-1}u(t) \ge -\frac{8}{3},$$

integration of the above inequality from T to t $(t \ge T)$, yields

$$u(t) \le K_1 + K_2 t - \int_T^t \int_T^s \left[\beta\theta^2 p(\theta) + \theta q(\theta) - \frac{2}{3\sqrt{3\theta}}\big(1 - \theta^2 q(\theta)\big)^{3/2}\right] d\theta\, ds, \tag{6.174}$$

where $K_1 = u(T) + \frac{8}{3}T - K_0 T$, $K_2 = K_0 + \frac{8}{3}$. So it follows from (6.172) and (6.174) that $u(t) < 0$ for sufficiently large t, a contradiction. Hence $\lim_{t\to\infty} x(t) = \infty$.

Next, suppose that $x'(t) < 0$ for $t \ge t_2$. So $\lim_{t\to\infty} x(t)$ exists. If possible, let $\lim_{t\to\infty} x(t) = \lambda$, where $0 < \lambda < \infty$. Clearly, $A^2(t) - B^2(t) \le 0$ for $t \ge \sigma$, where

$$A(t) = t q(t) \quad \text{and} \quad B(t) = \frac{2}{3\sqrt{3t}}\big(1 - t^2 q(t)\big)^{3/2}.$$

So $A(t) - B(t) \le 0$ for $t \ge \sigma$. Consequently, from (6.172), it follows that

$$\int_\sigma^\infty \beta t^2 p(t)\, dt = \infty.$$

Multiplying Eq. (6.171) by t^2 and integrating the resulting identity from t_2 to t $(t \ge t_2)$, we obtain

$$t^2 x''(t) \le K - \frac{\lambda}{2}\beta \int_{t_2}^t s^2 p(s)\, ds.$$

This, in turn, implies that $x''(t) < 0$ for large t. Consequently, $x(t) < 0$ for large t, a contradiction. Hence $\lim_{t\to\infty} x(t) = 0$. The theorem is proved. □

Example 6.4.1 The equation

$$x'''(t) - \frac{t+2}{t}x'(t) + 2x(t - \ln t) = 0, \quad t \geq 2$$

illustrates Theorem 6.4.1. Clearly, $x(t) = e^t$ is a nonoscillatory solution of the equation which tends to infinity as $t \to \infty$.

Example 6.4.2 The equation

$$x'''(t) - \frac{1}{t^2}x'(t) + \frac{5(t-1)}{t^4}x(t-1) = 0, \quad t \geq 0$$

admits a nonoscillatory solution $x(t) = \frac{1}{t}$ which tends to zero as $t \to \infty$.

From Theorem 6.4.1, we have the result for the equation

$$x'''(t) + q(t)x'(t) + p(t)x^\gamma\big(g(t)\big) = 0, \tag{6.175}$$

where $\gamma \geq 1$ is a quotient of odd integers.

Corollary 6.4.1 *If*

$$\int_\sigma^\infty \left[\mu t^2 p(t) + tq(t) - \frac{2}{3\sqrt{3}t}\big(1 - t^2 q(t)\big)^{3/2}\right] dt = \infty \tag{6.176}$$

for all $\mu > 0$, *then every nonoscillatory solution of* (6.175) *tends either to zero or to* $\pm\infty$ *as* $t \to \infty$.

Lemma 6.4.1 *Let* $q(t) \geq 0$ *and* $p(t) > 0$. *Suppose that the* $q \in C^1([\sigma,\infty), R)$ *and* $2\beta p(t) - q'(t) \geq 0$ *for* $t \in [\sigma,\infty)$, *and not identically zero on any subinterval* $[\sigma,\infty)$. *Let* $x(t)$ *be a nonoscillatory solution of* (6.171) *which is eventually nonnegative with*

$$G\big(x(t_0)\big) = 2x(t)x''(t) - \big(x'(t)\big)^2 + q(t)x^2(t)\big|_{t=t_0} < 0, \quad t_0 \in [\sigma,\infty).$$

Then there exists a number $d \geq t_0$ *such that* $x(t) > 0, x'(t) > 0, x''(t) > 0$ *and* $x'''(t) \leq 0$ *for* $t \geq d$.

Using the above lemma, one can prove the following theorem with small modifications.

Theorem 6.4.2 *Let* $q(t) \geq 0$, $p(t) > 0$ *and the hypotheses of Lemma* 6.4.1 *hold, and* $t^2 q(t) \leq \frac{1}{4}$ *for all* $t > 0$. *If* (6.172) *holds, then every nonoscillatory solution of* (6.171) *tends to* $\pm\infty$ *as* $t \to \infty$.

Corollary 6.4.2 *Let $q(t) \geq 0$ and $p(t) > 0$. Suppose that the hypotheses of Lemma 6.4.1 hold and $t^2 q(t) \leq \frac{1}{4}$ for all $t > 0$. Suppose that (6.176) holds. Then any nonoscillatory solution $x(t)$ of (6.175) is unbounded.*

Lemma 6.4.2 *Let $q(t) \geq 0$ and $p(t) > 0$. Suppose that the second-order differential equation*

$$z'' + q(t)z = 0 \tag{6.177}$$

is nonoscillatory. If $x(t)$ is a nonoscillatory solution of (6.171), then there exists a $t_0 \in [\sigma, \infty)$ such that $x(t)x'(t) < 0$ or $x(t)x'(t) > 0$ for $t \geq t_0$.

Theorem 6.4.3 *Let $q(t) \geq 0$ and $p(t) > 0$. Further, suppose that the hypotheses of Lemma 6.4.2 hold and $t^2 q(t) \leq \frac{1}{4}$ for all $t > 0$. If (6.172) holds, then every nonoscillatory solution of (6.171) tends either to zero or to $\pm\infty$ as $t \to \infty$.*

Corollary 6.4.3 *Let $q(t) \geq 0$, $p(t) > 0$ and the hypotheses of Lemma 6.4.2 hold and $t^2 q(t) \leq \frac{1}{4}$ for all $t > 0$. If (6.176) holds, for all $\mu > 0$, then every nonoscillatory solution of (6.175) tends either to zero or to $\pm\infty$ as $t \to \infty$.*

Thus, if the second-order differential equation (6.177) is nonoscillatory, then the above theorem, lemma and corollary provide some results on the asymptotic behaviour of nonoscillatory solutions of (6.171).

Now, we assume that the second-order equation (6.177) is oscillatory. We mention here some results on the asymptotic behaviour of nonoscillatory solutions of (6.171).

Theorem 6.4.4 *Suppose that the second-order differential equation (6.177) is oscillatory. Then any nonoscillatory solution $x(t)$ of (6.171) either satisfies*

$$x(t)x'(t) \leq 0, \qquad x(t) \neq 0 \quad \text{for large } t$$

or

$$x(t) \neq 0 \quad \text{for large } t \text{ and } x'(t) \text{ changes sign.}$$

Lemma 6.4.3 *Let the second-order differential equation (6.177) be oscillatory. Then any solution $x(t)$ of (6.171) satisfies*

$$\liminf_{t\to\infty} x'(t) = 0 \quad \text{and} \quad \liminf_{t\to\infty} x''(t) = 0.$$

Lemma 6.4.4 *Let the second-order differential equation (6.177) be oscillatory. Let $q(t) \geq 0$ for large t. If $x(t)$ is a solution of (6.171) such that $x(t) > 0$ for large t and the function*

$$\omega(t) = x''(t) + q(t)x(t)$$

is nonincreasing for large t, then

$$\lim_{t\to\infty} \omega(t) = \omega_0 \geq 0.$$

Theorem 6.4.5 *Let (6.177) be oscillatory. Further, assume that $q(t)$ is continuously differentiable and is bounded away from zero, that is, $q(t) \geq q_0 > 0$. If*

$$\int_0^\infty \left|q'(t)\right| dt < \infty$$

and

$$\int_0^\infty p(t)\, dt < \infty,$$

then for any $c \in R\backslash\{0\}$ there exists a solution $x(t)$ of (6.171) satisfying

$$\lim_{t\to\infty} x(t) = c, \qquad \lim_{t\to\infty} x^{(i)}(t) = 0, \quad i = 1, 2.$$

Theorem 6.4.5 gives a necessary condition for the oscillation of (6.171), the following theorem provide a sufficient condition for the oscillation of (6.171).

Theorem 6.4.6 *Let (6.177) be oscillatory. Assume that*

$$\liminf_{x\to\infty} \left|f(x)\right| > 0 \tag{6.178}$$

and q is continuously differentiable for large t, satisfying $q(t) > 0$ and $q'(t) \leq 0$. If

$$\int_\sigma^\infty p(t)\, dt = \infty, \tag{6.179}$$

then any (nonoscillatory) solution of (6.171) satisfies

$$\liminf_{t\to\infty} \left|x(t)\right| = 0.$$

Moreover, any nonoscillatory solution with $x(t)x'(t) \leq 0$ for large t satisfies

$$\lim_{t\to\infty} x^{(i)}(t) = 0, \quad i = 0, 1, 2.$$

Corollary 6.4.4 *Let (6.177) be oscillatory. Assume that (6.178) holds and q is continuously differentiable for large t satisfying $q(t) > 0$ and $q'(t) \leq 0$. Then the condition (6.179) is a necessary and sufficient condition in order to every solution of (6.171) is either oscillatory or satisfies the property*

$$\liminf_{t\to\infty} \left|x(t)\right| = 0.$$

Corollary 6.4.5 *Let* (6.177) *be oscillatory. Assume that q is continuously differentiable for large t, satisfying $q(t) > 0$ and $q'(t) \leq 0$. If* (6.178) *and* (6.179) *are satisfied, then for any nonoscillatory solution $x(t)$ of* (6.171), *we may have*

$$F(t) > 0, \qquad F'(t) < 0$$

for large t, where

$$F(t) = F\big[x(t)\big] = 2x''(t)x(t) - \big(x'(t)\big)^2 + q(t)x^2(t). \tag{6.180}$$

Theorem 6.4.7 *Let $g(t) \geq t$ and* (6.177) *be oscillatory. Assume that q is continuously differentiable for large t satisfying $q(t) > 0$ and $q'(t) \leq 0$. If* (6.178) *and* (6.179) *are satisfied, then for any nonoscillatory solution $x(t)$ of* (6.171) *defined for $t \geq t_x \geq \sigma$, we have $x(t) \neq 0$ for $t \geq t_x$. In particular, any continuable solution of* (6.171) *with one zero is oscillatory.*

Theorem 6.4.8 *Let* (6.177) *be oscillatory. Assume that q is continuously differentiable satisfying $q(t) \geq q_0 > 0$ and $q'(t) \leq 0$ for large t. Then any nonoscillatory solution $x(t)$ of* (6.171) *satisfies*

$$\limsup_{t\to\infty} \big|x^{(i)}(t)\big| < \infty, \qquad i = 0, 1, 2.$$

Theorem 6.4.9 *Let* (6.177) *be oscillatory. Assume that q is continuously differentiable satisfying $q(t) \geq q_0 > 0$ and $q'(t) \leq 0$ for large t. Further assume that* (6.178) *and* (6.179) *hold. If*

$$\liminf_{x\to 0} \frac{f(x)}{x} > 0 \tag{6.181}$$

and

$$\int_\sigma^\infty p(t)\big|t - g(t)\big|\,dt < \infty, \tag{6.182}$$

then any nonoscillatory solution $x(t)$ of (6.171) *defined on $[t_x, \infty)$ satisfies*

$$\int_{t_x}^\infty p(t)x^2(t)\,dt < \infty.$$

If, moreover,

$$\frac{q(t)}{p(t)} \leq C < \infty$$

for large t, then $x(t)$ satisfies

$$\int_{t_x}^\infty \big(x'(t)\big)^2\,dt < \infty.$$

Theorem 6.4.10 *Let* (6.177) *be oscillatory and* (6.178) *hold. Further assume that* (6.181), (6.183) *and* q *are continuously differentiable satisfying* $q(t) \geq q_0 > 0$ *and* $q'(t) \leq 0$ *for large* t. *If there exists a real constant* $M_0 > 0$ *such that* $p(t) > M_0$ *for large* t, *then* (6.171) *has property* A, *that is, any nonoscillatory solution* $x(t)$ *of* (6.171) *satisfies*

$$\lim_{t\to\infty} x^{(i)}(t) = 0, \quad i = 0, 1, 2.$$

Corollary 6.4.6 *Consider the assumptions of Theorem* 6.4.10, $g(t) = t$, *and for some* $K > 0$

$$\left|f(x)\right| \geq K|x| \quad \text{for } x \in R,$$

then any nonoscillatory solution $x(t)$ *of* (6.171) *satisfies* $x(t)x'(t) < 0$ *for large* t.

Lemma 6.4.5 *Let* (6.177) *be oscillatory and* q *is continuously differentiable satisfying* $q(t) > 0$ *and* $q'(t) \geq 0$ *for large* t, *and* $p(t) < 0$. *Let* $x(t)$ *be a solution of* (6.171) *such that* $x(t) > 0$ *and* $x(g(t)) > 0$ *for* $t \geq t_0 > \sigma$. *Then the function* F *defined by* (6.180) *is increasing on* $[t_0, \infty)$.

Lemma 6.4.6 *Let* (6.177) *be oscillatory and* q *is continuously differentiable satisfying* $q(t) > 0$, $q(t) < q_0$ *and* $q'(t) \geq 0$ *for large* t, *and* $p(t) < 0$. *Assume that*

$$\int_\sigma^\infty p(t)\,dt = -\infty \tag{6.183}$$

holds. Then any bounded nonoscillatory solution $x(t)$ *of* (6.171) *satisfies*

$$\liminf_{t\to\infty} x(t) = 0.$$

One can obtain the following theorem by using Lemma 6.4.6.

Theorem 6.4.11 *Let* (6.177) *be oscillatory and* q *is continuously differentiable satisfying* $q(t) > 0$, $q(t) < q_0$ *and* $q'(t) \geq 0$ *for large* t, $p(t) < 0$ *and*

$$\liminf_{t\to\infty} \left|f(x)\right| > 0.$$

Then any nonoscillatory solution of (6.171) *is bounded, if and only if* (6.183) *holds.*

Theorem 6.4.12 *Let* (6.177) *be oscillatory and* q *is continuously differentiable satisfying* $q(t) > 0$ *and* $q(t) < q_0$ *for large* t, $p(t) < 0$. *Assume that*

$$\lim_{x\to\infty} \left|f(x)\right| = \infty, \tag{6.184}$$

and $p(t) \leq -p_0 < 0$. *Then any nonoscillatory solution of* (6.171) *is unbounded together with its first derivative.*

Theorem 6.4.13 *Let* (6.177) *be oscillatory and* q *is continuously differentiable satisfying* $q(t) > 0$ *and* $q(t) < q_0$ *for large* t, $p(t) < 0$. *Assume that* (6.184) *holds. Further assume that*

$$p(t) \leq -tp_0, \quad p_0 > 0.$$

Then every nonoscillatory solution of (6.171) *satisfies the property*

$$\lim_{t\to\infty} x^{(i)}(t) = \infty, \quad i = 0, 1, 2,$$

that is, (6.171) *has property* B.

6.5 Oscillation of Solutions of the Nonlinear Delay Differential Equation $(r(t)(x''(t))^\gamma)' + p(t)x^\gamma(g(t)) = 0$ and $(r(t)(x''(t))^\gamma)' + p(t)f(x(g(t))) = 0$

The objective of this section is to present up-to-date results systematically on the oscillation and asymptotic behaviour of nonoscillatory solutions of the third-order nonlinear delay differential equations of the form

$$\big(r(t)\big(x''(t)\big)^\gamma\big)' + p(t)x^\gamma\big(g(t)\big) = 0 \tag{6.185}$$

and

$$\big(r(t)\big(x''(t)\big)^\gamma\big)' + p(t)f\big(x\big(g(t)\big)\big) = 0, \tag{6.186}$$

where r, p and $g \in C([\sigma, \infty), R)$ such that $r(t) > 0$, $p(t) > 0$, $0 < g(t) \leq t$, $\lim_{t\to\infty} g(t) = \infty$ and $f \in C(R, R)$. The following notations are used in the sequel.

(J_1) $\int_\sigma^\infty \frac{ds}{r^{1/\gamma}(s)} = \infty$

(J_2) $\int_\sigma^\infty \frac{ds}{r^{1/\gamma}(s)} < \infty$

(J_3) γ is a quotient of odd positive integers

(J_4) $xf(x) > 0$, $f'(x) \geq 0$ for $x \neq 0$ and $-f(-xy) \geq f(xy) \geq f(x)f(y)$ for $xy \neq 0$.

Let $T_0 = \min\{g(t); t \geq 0\}$ and $g^{-1}(t) = \sup\{s \geq 0;\ g(s) \leq t\}$ for $t \geq T_0$. Clearly, $g^{-1}(t) \geq t$ for $t \geq T_0$, $g^{-1}(t)$ is nondecreasing and coincides with the inverse of $g(t)$, when the latter exists. By a solution of the above-mentioned equations, we mean a nontrivial real-valued function $x(t)$ which has the properties $x'(t) \in C^1([g^{-1}(\sigma), \infty))$ and $r(t)(x''(T))^\gamma \in C^1([g^{-1}(\sigma), \infty))$. We restrict our attention to the solutions of the above equations, which exist on some half line $[t_x, \infty)$, $t_x \geq \sigma$ and satisfy $\sup\{|x(t)|; t \geq t_1\} > 0$ for any $t_1 \geq t_x$. Further, we make a standing hypothesis that a solutions of the above equations exist.

Lemma 6.5.1 *Let* (J_1) *and* (J_3) *hold. If* $x(t)$ *is a nonoscillatory solution of* (6.185), *then there exists a* $t_1 \geq \sigma$ *such that either*

$$x(t) > 0, \qquad x'(t) > 0, \qquad x''(t) > 0 \tag{6.187}$$

or

$$x(t) > 0, \qquad x'(t) < 0, \qquad x''(t) > 0 \tag{6.188}$$

holds for $t \geq t_1$.

Lemma 6.5.2 *Let* (J_1) *and* (J_3) *hold. If* $x(t)$ *is a nonoscillatory solution of* (6.185) *satisfying the property* (6.188). *If*

$$\int_\sigma^\infty \int_t^\infty \left[\frac{1}{r(s)} \int_s^\infty p(\theta)\, d\theta\right]^{1/\gamma} ds\, dt = \infty, \tag{6.189}$$

then $\lim_{t\to\infty} x(t) = 0$.

Lemma 6.5.3 *Assume that* $z(t) > 0, z'(t) \geq 0$ *and* $z''(t) \leq 0$ *for* $t > \sigma$. *Then for each* $l \in (0, 1)$, *there exists a* $T_1 \geq \sigma$ *such that*

$$\frac{z(g(t))}{g(t)} \geq l \frac{z(t)}{t} \quad \textit{for } t \geq T_1.$$

Lemma 6.5.4 *Assume that* $x(t) > 0, x'(t) > 0, x''(t) > 0$ *and* $x'''(t) \leq 0$ *for* $t > T_1 > \sigma$. *Then*

$$\frac{x(t)}{x'(t)} \geq \frac{t - T_1}{2} \quad \textit{for } t \geq T_1.$$

Lemma 6.5.5 *Assume that* $x'(t) > 0, x''(t) > 0$ *and* $x'''(t) \leq 0$ *for* $t > T_1 > \sigma$. *Then*

$$(t - T_1)\frac{x''(t)}{x'(t)} \leq 1 \quad \textit{for } t \geq T_1.$$

Let us introduce the following notations:

$$p_* = \liminf_{t\to\infty} \frac{t^\gamma}{r(t)} \int_t^\infty P_l(s)\, ds, \qquad q_* = \limsup_{t\to\infty} \frac{1}{t} \int_\sigma^t \frac{s^{\gamma+1}}{r(s)} P_l(s)\, ds, \tag{6.190}$$

where

$$P_l(t) = l^\gamma p(s)\left(\frac{g(s)}{s}\right)^\gamma \left(\frac{1}{2}(g(s) - T_1)\right)^\gamma$$

with $l \in (0, 1)$ arbitrarily chosen and T_1 is large enough. Moreover, if $x(t)$ satisfies (6.187), then we define a function

$$\omega(t) = r(t)\left(\frac{x''(t)}{x'(t)}\right)^\gamma, \tag{6.191}$$

and

$$r = \liminf_{t\to\infty} \frac{t^\gamma \omega(t)}{r(t)} \quad \text{and} \quad R = \limsup_{t\to\infty} \frac{t^\gamma \omega(t)}{r(t)}. \tag{6.192}$$

Theorem 6.5.1 *Let* (J_1) *and* (J_3) *hold. Assume that* $r'(t) \geq 0$. *Let* $x(t)$ *be a positive solution of* (6.185). *The following hypotheses hold:*

(I) *Let* $p_* < \infty$ *and* $q_* < \infty$. *Suppose that* $x(t)$ *satisfies* (6.187). *Then*

$$p_* \leq r - r^{1+1/\gamma} \quad \text{and} \quad p_* + q_* \leq 1. \tag{6.193}$$

(II) *If* $p_* = \infty$ *or* $q_* = \infty$, *then* $x(t)$ *cannot satisfy* (6.187).

Proof Part (I): Assume that $x(t)$ satisfies (6.187). Clearly, $r'(t) \geq 0$ together with $(r(t)(x''(t))^\gamma)' \leq 0$ implies that $x'''(t) \leq 0$. So there exists a $T \geq \sigma$ such that $x(t)$ satisfies

$$x\big(g(t)\big) > 0, \qquad x'(t) > 0, \qquad x''(t) > 0, \qquad x'''(t) \leq 0, \quad \text{for } t \geq T.$$

From the definition of $\omega(t)$, we see that $\omega(t)$ is positive and satisfies

$$\begin{aligned}\omega'(t) &= \frac{(x'(t))^\gamma (r(t)(x''(t))^\gamma)' - (r(t)(x''(t))^\gamma)\gamma(x'(t))^{\gamma-1}x''(t)}{(x'(t))^{2\gamma}} \\ &= \frac{(r(t)(x''(t))^\gamma)'}{(x(g(t)))^\gamma}\frac{(x(g(t)))^\gamma}{(x'(t))^\gamma} - \gamma\frac{(r(t)(x''(t))^\gamma)}{(x'(t))^\gamma}\frac{x''(t)}{x'(t)} \\ &= -p(t)\frac{(x(g(t)))^\gamma}{(x'(t))^\gamma} - \frac{\gamma}{r^{1/\gamma}(t)}\omega^{1+1/\gamma}(t).\end{aligned} \tag{6.194}$$

From Lemma 6.5.3 with $z(t) = x'(t)$, we have, for some l as in P_l,

$$\frac{1}{x'(t)} \geq l\frac{g(t)}{t}\frac{1}{x'(g(t))}, \qquad t \geq T_1,$$

which together with (6.194) gives

$$\omega'(t) \leq -l^\gamma p(t)\left(\frac{g(t)}{t}\right)^\gamma \frac{(x(g(t)))^\gamma}{(x'(g(t)))^\gamma} - \frac{\gamma}{r^{1/\gamma}(t)}\omega^{1+1/\gamma}(t).$$

Using the fact from Lemma 6.5.4 that $x(t) \geq \frac{1}{2}(t - T_1)x'(t)$, we have

$$\omega'(t) + P_l(t) + \frac{\gamma}{r^{1/\gamma}(t)}\omega^{1+1/\gamma}(t) \leq 0. \tag{6.195}$$

Since $P_l(t) > 0$ and $\omega(t) > 0$ for $t \geq T_1$, we have from (6.195) that $\omega'(t) \leq 0$, and

$$-\left(\frac{\omega'(t)}{\gamma\omega^{\frac{(\gamma+1)}{\gamma}}(t)}\right) > \frac{1}{r^{1/\gamma}(t)} \quad \text{for } t \geq T_1. \tag{6.196}$$

This implies that

$$\left(\frac{1}{\omega^{1/\gamma}(t)}\right)' > \frac{1}{r^{1/\gamma}(t)}. \tag{6.197}$$

Integrating the last inequality from T_1 to t, we obtain

$$\omega(t) < \frac{1}{(\int_{T_1}^{t} ds/r^{1/\gamma}(s))^{\gamma}}, \tag{6.198}$$

which in view of (J_1) and (J_3) implies that $\lim_{t\to\infty}\omega(t) = 0$. On the other hand, from the definition of $\omega(t)$ and Lemma 6.5.5, we see that

$$0 \le r \le R < k < \infty. \tag{6.199}$$

Now, we prove that the first inequality in (6.193) holds. Let $\epsilon > 0$ be given. By the definition of p_* and r, we can pick $t_2 \in [T_2, \infty)$ sufficiently large so that

$$\frac{t^{\gamma}}{r(t)}\int_t^{\infty} P_l(s)\,ds \ge p_* - \epsilon, \quad \text{and} \quad \frac{t^{\gamma}\omega(t)}{r(t)} \ge r - \epsilon \quad \text{for } t \in [t_2, \infty).$$

Integrating (6.195) from t to ∞ and by using $\lim_{t\to\infty}\omega(t) = 0$, we have

$$\omega(t) \ge \int_t^{\infty} P_l(s)\,ds + \gamma\int_t^{\infty} \frac{\omega^{1+1/\gamma}(s)}{r^{1/\gamma}(s)}\,ds, \quad \text{for } t \in [t_2, \infty). \tag{6.200}$$

By virtue of the fact that $r'(t) \ge 0$, it follows from (6.200) that

$$\frac{t^{\gamma}}{r(t)}\omega(t) \ge (p_* - \epsilon) + (r - \epsilon)^{1+1/\gamma} t^{\gamma}\int_t^{\infty} \frac{\gamma}{s^{\gamma+1}}\,ds. \tag{6.201}$$

From (6.201), we have

$$\frac{t^{\gamma}\omega(t)}{r(t)} \ge (p_* - \epsilon) + (r - \epsilon)^{1+1/\gamma}.$$

Taking lim inf both sides as $t \to \infty$, we get

$$r \ge p_* - \epsilon + (r - \epsilon)^{1+1/\gamma}.$$

Since $\epsilon > 0$ is arbitrary, we get the desired result

$$p_* \le r - r^{1+1/\gamma}.$$

To complete the part (I), it remains to prove the second inequality in (6.193). Multiplying (6.195) by $\frac{t^{\gamma+1}}{r(t)}$ and integrating from t_2 to t, we get

$$\int_{t_2}^{t} \frac{s^{\gamma+1}}{r(s)}\omega'(s)\,ds \le -\int_{t_2}^{t} \frac{s^{\gamma+1}}{r(s)}P_l(s)\,ds - \gamma\int_{t_2}^{t}\left(\frac{s^{\gamma}\omega(s)}{r(s)}\right)^{1+1/\gamma} ds. \tag{6.202}$$

The property $r'(t) \geq 0$ implies that

$$\left(\frac{s^{\gamma+1}}{r(s)}\right)' \leq \frac{(\gamma+1)s^{\gamma}}{r(s)}.$$

Then using integration by parts in (6.202) and using the above inequality, we obtain

$$\frac{t^{\gamma+1}}{r(t)}\omega(t) \leq \frac{t_2^{\gamma+1}\omega(t_2)}{r(t_2)} - \int_{t_2}^{t} \frac{s^{\gamma+1}}{r(s)} P_l(s)\,ds + \int_{t_2}^{t}\left[(\gamma+1)\frac{s^{\gamma}\omega(s)}{r(s)} - \gamma\left(\frac{s^{\gamma}\omega(s)}{r(s)}\right)^{(\gamma+1)/\gamma}\right]ds.$$

Using the inequality

$$Bu - Au^{(\gamma+1)/\gamma} \leq \frac{\gamma^{\gamma}}{(\gamma+1)^{\gamma+1}}\frac{B^{\gamma+1}}{A^{\gamma}}$$

with $u(s) = s^{\gamma}\omega'(s)/r(s)$, and positive constants $A = \gamma$, $B = \gamma + 1$, we get

$$\frac{t^{\gamma+1}}{r(t)}\omega(t) \leq \frac{t_2^{\gamma+1}\omega(t_2)}{r(t_2)} - \int_{t_2}^{t} \frac{s^{\gamma+1}}{r(s)} P_l(s)\,ds + (t - t_2).$$

It follows that

$$\frac{t^{\gamma}}{r(t)}\omega(t) \leq \frac{1}{t}\frac{t_2^{\gamma+1}\omega(t_2)}{r(t_2)} - \frac{1}{t}\int_{t_2}^{t} \frac{s^{\gamma+1}}{r(s)} P_l(s)\,ds + \frac{t-t_2}{t}. \tag{6.203}$$

Taking the lim sup of both sides as $t \to \infty$, we obtain

$$R \leq -q_* + 1.$$

Combining this with the first inequality in (6.199), we get

$$p_* \leq r - r^{1+1/\gamma} \leq r \leq R \leq -q_* + 1,$$

which gives the desired second inequality in (6.193). The proof of part (I) is complete.

One can prove the Part (II) with the help of (6.203). □

Theorem 6.5.2 *Let* (J_1) *and* (J_3) *hold and* $r'(t) \geq 0$. *Assume that* (6.189) *holds. Let* $x(t)$ *be a solution of* (6.185). *If*

$$p_* = \liminf_{t\to\infty} \frac{t^{\gamma}}{r(t)} \int_t^{\infty} P_l(s)\,ds > \frac{\gamma^{\gamma}}{(\gamma+1)^{\gamma+1}},$$

then $x(t)$ *is oscillatory or satisfies* $\lim_{t\to\infty} x(t) = 0$.

Theorem 6.5.3 *Let* (J$_1$) *and* (J$_3$) *hold and* $r'(t) \geq 0$. *Assume that* (6.189) *holds. Let* $x(t)$ *be a solution of* (6.185). *If*

$$p_* + q_* > 1,$$

then $x(t)$ *is oscillatory or satisfies* $\lim_{t\to\infty} x(t) = 0$.

Corollary 6.5.1 *Let* (J$_1$) *and* (J$_3$) *hold and* $r'(t) \geq 0$. *Assume that* (6.189) *holds. Let* $x(t)$ *be a solution of* (6.185). *If*

$$q_* = \liminf_{t\to\infty} \frac{1}{t} \int_\sigma^t \frac{s^{\gamma+1}}{r(s)} P_l(s)\, ds > 1,$$

then $x(t)$ *is oscillatory or satisfies* $\lim_{t\to\infty} x(t) = 0$.

Theorem 6.5.4 *Let* (J$_1$) *and* (J$_3$) *hold and* $r'(t) \geq 0$. *Assume that* (6.189) *holds. Let* $x(t)$ *be a solution of* (6.185). *If*

$$\liminf_{t\to\infty} \int_{g(t)}^t \frac{p(s)}{r(g(s))} \left(\frac{(g(s)-T)^2}{2} \right)^\gamma ds > \frac{1}{e}$$

or

$$\limsup_{t\to\infty} \int_{g(t)}^t \frac{p(s)}{r(g(s))} \left(\frac{(g(s)-T)^2}{2} \right)^\gamma ds > 1,$$

then $x(t)$ *is oscillatory or satisfies* $\lim_{t\to\infty} x(t) = 0$.

Theorem 6.5.5 *Let* (J$_2$) *and* (J$_3$) *hold and* $r'(t) \geq 0$. *Assume that* (6.189) *holds. Further, assume that*

$$\limsup_{t\to\infty} \int_{t_2}^t \left[l^\gamma s p(s) \left(\frac{g(s)-T_l}{2} \frac{g(s)}{s} \right)^\gamma - \frac{1}{(\gamma+1)^{\gamma+1}} \frac{r(s)}{s^\gamma} \right] ds = \infty$$

for some $l \in (0,1)$ *and for sufficiently large* $T_l \geq \sigma$, $t_2 \geq T_l$. *Furthermore, assume that*

$$\limsup_{t\to\infty} \int_{t_3}^t \left[k^\gamma p(s) g^\gamma(s) \delta^\gamma(s) - \left(\frac{\gamma}{\gamma+1} \right)^{\gamma+1} \frac{1}{\delta(s) r^{1/\gamma}(s)} \right] ds = \infty$$

holds for some $k \in (0,1)$ *and for sufficiently large* $t_3 \geq \sigma$, *where*

$$\delta(t) = \int_t^\infty \frac{1}{r^{1/\gamma}(s)} ds.$$

Then (6.185) *is oscillatory.*

Now we consider Eq. (6.186). The following theorems give some sufficient conditions for the oscillation of (6.186).

Theorem 6.5.6 *Let* (J_2) *and* (J_4) *holds. Further, assume that*

$$\int_\sigma^\infty \left(\frac{1}{r(u)}\int_\sigma^u p(s)f\big(g(s)\big)f\left[\int_{g(s)}^\infty r^{-1/\gamma}(v)\,dv\right]ds\right)^{1/\gamma} du = \infty. \tag{6.204}$$

If the first-order delay equation

$$y'(t) + p(t)f\left[\int_\sigma^{g(t)} \big(g(t)-u\big)r^{-1/\gamma}(u)\,du\right]f\big(y^{1/\gamma}\big(g(t)\big)\big) = 0 \tag{6.205}$$

is oscillatory, then every solution of (6.186) *is either oscillatory or strongly decreasing.*

Theorem 6.5.7 *Let* (J_2) *and* (J_4) *holds. Assume that* $x(t)$ *is a strongly decreasing solution of* (6.186). *If*

$$\int_\sigma^\infty \int_t^\infty \left[\frac{1}{r^{1/\gamma}(s)}\int_s^\infty p(\theta)\,d\theta\right]^{1/\gamma} ds\,dt = \infty, \tag{6.206}$$

then $x(t)$ *tends to zero as* $t \to \infty$.

Theorem 6.5.8 *Let* (J_2) *and* (J_4), (6.204) *and* (6.206) *hold. If Eq.* (6.205) *is oscillatory, then every solution of* (6.186) *is oscillatory or tends to zero as* $t \to \infty$.

Corollary 6.5.2 *Let* (J_2) *and* (J_4) *and* (6.206) *hold and*

$$\int_\sigma^\infty \left(\frac{1}{r(u)}\int_\sigma^u p(s)g^\beta(s)\left(\int_{g(s)}^\infty r^{-1/\gamma}(v)\,dv\right)^\beta ds\right)^{1/\gamma} du = \infty, \tag{6.207}$$

where β *is a quotient of odd positive integers. If the delay equation*

$$y'(t) + p(t)f\left[\int_\sigma^{g(t)} \big(g(t)-u\big)r^{-1/\gamma}(u)\,du\right]y^{\beta/\gamma}\big(g(t)\big) = 0 \tag{6.208}$$

is oscillatory, then every solution of the equation

$$\big(r(t)\big(x''(t)\big)^\gamma\big)' + p(t)x^\beta\big(g(t)\big) = 0 \tag{6.209}$$

is oscillatory or tends to zero as $t \to \infty$.

Corollary 6.5.3 *Let* (J_2) *and* (J_4) *and* (6.206) *hold and*

$$\int_\sigma^\infty \left(\frac{1}{r(u)}\int_\sigma^u p(s)g^\gamma(s)\left(\int_{g(s)}^\infty r^{-1/\gamma}(v)\,dv\right)^\gamma ds\right)^{1/\gamma} du = \infty. \tag{6.210}$$

If

$$\liminf_{t\to\infty} \int_{g(t)}^t p(u)f\left[\int_\sigma^{g(u)} \big(g(u)-s\big)r^{-1/\gamma}(s)\,ds\right]^{1/\gamma} du > \frac{1}{e}, \tag{6.211}$$

then every solution of (6.185) *is oscillatory or tends to zero as* $t \to \infty$.

Theorem 6.5.9 *Let* (J_2) *and* (J_4) *and* (6.204) *hold and* $g'(t) > 0$. *Assume that there exists a function* $\xi \in C^1([\sigma, \infty), R)$ *such that* $\xi'(t) \geq 0$, $t < \xi(t)$ *and* $\eta(t) = g(\xi(\xi(t))) < t$. *If both the first-order equations* (6.205) *and*

$$z'(t) + \left[\int_t^{\xi(t)} \frac{1}{r^{1/\gamma}(s_2)} \left(\int_{s_2}^{\xi(s_2)} p(s_1)\,ds_1\right)^{1/\gamma} ds_2\right] f^{1/\gamma}\big(z(\eta(t))\big) = 0 \qquad (6.212)$$

are oscillatory, then (6.186) *is oscillatory.*

Corollary 6.5.4 *Let* (J_2) *and* (J_4) *and* (6.207) *hold and* $g'(t) > 0$. *Let* β *be a quotient of odd positive integers. Assume that there exist a function* $\xi(t)$ *such that* $\xi \in C^1([\sigma, \infty), R)$ *such that* $\xi'(t) \geq 0$, $t < \xi(t)$ *and* $\eta(t) = g(\xi(\xi(t))) < t$. *If both* (6.208) *and*

$$z'(t) + \left[\int_t^{\xi(t)} \frac{1}{r^{1/\gamma}(s_2)} \left(\int_{s_2}^{\xi(s_2)} p(s_1)\,ds_1\right)^{1/\gamma} ds_2\right] z^{\beta/\gamma}\big(\eta(t)\big) = 0$$

are oscillatory, then (6.209) *is oscillatory.*

6.6 Oscillation and Asymptotic Behaviour of Solutions of $(r_2(t)(r_1(t)x'(t))')' + q(t)x'(t) + p(t)f(x(g(t))) = 0$

In this section, we shall be concerned with the oscillation and asymptotic behaviour of nonoscillatory solutions of the third-order nonlinear delay equation

$$\big(r_2(t)\big(r_1(t)x'(t)\big)'\big)' + q(t)x'(t) + p(t)f\big(x\big(g(t)\big)\big) = 0, \qquad (6.213)$$

where r_1, r_2, p and $q \in C([\sigma, \infty), R)$, $\sigma \geq 0$, $\sigma \in R$, $r_i(t) > 0, i = 1, 2$, $p(t) \geq 0$, $q(t) \geq 0$ and $p(t) \not\equiv 0$ in the neighbourhood of infinity, $g \in C^1([\sigma, \infty), R)$, $0 < g(t) \leq t$, $g'(t) \geq 0$ and $g(t) \to \infty$ as $t \to \infty$, and $f \in C(R, R)$ satisfies $uf(u) > 0$ for $u \neq 0$. We note that (6.213) is more general than (6.2). Indeed, (6.2) is obtained from (6.213) with $r_2(t) = r(t)$, $r_1(t) = 1$ and $f(x) = x$.

Define $D_0x(t) = x(t)$, $D_ix(t) = r_i(t)(D_{i-1}x(t))'$, $i = 1, 2$ and $D_3x(t) = (D_2x(t))'$ for $t \geq \sigma$. Then (6.213) can be rewritten as

$$D_3x(t) + q(t)x'(t) + p(t)f\big(x\big(g(t)\big)\big) = 0.$$

If $x(t)$ is a solution of (6.213), then $y = -x$ is a solution of the equation

$$D_3y(t) + q(t)y'(t) + p(t)f^*\big(y\big(g(t)\big)\big) = 0,$$

where $f^*(y) = -f(-y)$ and $zf^*z > 0$ for $z \neq 0$. Thus, concerning the nonoscillatory solutions of (6.213), we can restrict our attention to the positive ones.

Define the functions

$$R_1(t,T)=\int_T^t \frac{ds}{r_1(s)}, \qquad R_2(t,T)=\int_T^t \frac{ds}{r_2(s)}, \qquad R_{12}(t,T)=\int_T^t \frac{R_2(s,T)}{r_1(s)}\,ds$$

for $\sigma \le T \le t < \infty$, with the assumptions that

$$R_1(t,\sigma)\to\infty, \quad \text{as } t\to\infty \tag{6.214}$$

and

$$R_2(t,\sigma)\to\infty, \quad \text{as } t\to\infty. \tag{6.215}$$

Definition 6.6.1 Let $x(t)$ be a solution of (6.213). We say that the solution x has property V_2 (one may see Sect. 4.4 of Chap. 4) on $[T,\infty)$, $T\ge\sigma$, if and only if

$$D_0x(t)D_kx(t)>0, \quad k=0,1,2; \quad \text{and} \quad D_0x(t)D_3x(t)\le 0$$

for every $t\in[T,\infty)$.

Theorem 6.6.1 *Suppose that the second-order differential equation*

$$\left(r_2(t)z'\right)'+\frac{q(t)}{r_1(t)}z=0 \tag{6.216}$$

is nonoscillatory. Let f satisfy the condition

$$\frac{f(u)}{u}\ge K>0 \quad \text{for } u\ne 0. \tag{6.217}$$

Assume that (6.214) *and* (6.215) *hold. Further, suppose that either*

$$\left.\begin{aligned} &\limsup_{t\to\infty}\int_T^t \frac{1}{r_1(s)}\left(\int_s^t \frac{1}{r_2(u)}\left(\int_u^\infty \left(Kp(\theta)-q'(\theta)\right)d\theta\right)du\right)ds=\infty\\ &\text{and}\quad \limsup_{t\to\infty}\int_T^t \frac{1}{r_1(s)}\left(\int_s^t \frac{q(u)}{r_2(u)}\,du\right)ds<\infty \end{aligned}\right\} \tag{6.218}$$

or

$$\limsup_{t\to\infty}\int_T^t \left(Kp(s)-q'(s)\right)ds=\infty, \tag{6.219}$$

where $Kp(t)-q'(t)\ge 0$ for $t\ge\sigma$ and not identically zero in any subinterval of $[\sigma,\infty)$. If there exists a differentiable positive function ρ such that

$$\limsup_{t\to\infty}\int_T^t \left[K\rho(s)p(s)-\frac{r_1(g(s))(\rho'(s)r_1(s)-\rho(s)q(s)R_2(g(s),T))^2}{4\rho(s)R_2(g(s),T)g'(s)r_1^2(s)}\right]ds=\infty \tag{6.220}$$

for every T, then any solution $x(t)$ of (6.213) *is either oscillatory or satisfies $x(t)\to 0$ as $t\to\infty$.*

From Theorem 6.6.1, we have the following result for the equation

$$D_3x(t) + q(t)x'(t) + p(t)x^{\gamma}\big(g(t)\big) = 0, \tag{6.221}$$

where $\gamma > 0$ is a quotient of odd integers.

Corollary 6.6.1 *Assume that* $\gamma \geq 1$, (6.214), (6.215) *hold. Equation* (6.216) *is nonoscillatory, and either*

$$\left.\begin{aligned} &\limsup_{t\to\infty} \int_T^t \frac{1}{r_1(s)}\left(\int_s^t \frac{1}{r_2(u)}\left(\int_u^\infty \big(K_1 p(\theta) - q'(\theta)\big)\,d\theta\right)du\right)ds = \infty \\ &\textit{and} \quad \limsup_{t\to\infty} \int_T^t \frac{1}{r_1(s)}\left(\int_s^t \frac{q(u)}{r_2(u)}\,du\right)ds < \infty \end{aligned}\right\} \tag{6.222}$$

or

$$\limsup_{t\to\infty} \int_T^t \big(K_1 p(s) - q'(s)\big)\,ds = \infty, \tag{6.223}$$

where $K_1 p(t) - q'(t) \geq 0$ *for* $t \geq \sigma$ *and not identically zero in any subinterval of* $[\sigma, \infty)$. *If there exists a differentiable positive function* ρ *such that*

$$\limsup_{t\to\infty} \int_T^t \left[K_1\rho(s)p(s) - \frac{r_1(g(s))(\rho'(s)r_1(s) - \rho(s)q(s)R_2(g(s),T))^2}{4\rho(s)R_2(g(s),T)g'(s)r_1^2(s)}\right]ds = \infty \tag{6.224}$$

for every T *and* $K_1 > 0$, *then any solution* $x(t)$ *of* (6.221) *is either oscillatory or satisfies* $x(t) \to 0$ *as* $t \to \infty$.

Corollary 6.6.2 *Assume that* $0 < \gamma < 1$, (6.214), (6.215) *hold. Equation* (6.216) *is nonoscillatory, and either* (6.222) *or* (6.223) *holds. If there exists a differentiable function* ρ *such that* (6.224) *holds, then any bounded solution of* $x(t)$ *of* (6.221) *is oscillatory or tends to zero as* $t \to \infty$.

Remark 6.6.1 When $\rho'(t) \leq 0$, we can take

$$\limsup_{t\to\infty} \int_T^t \frac{1}{r_1(s)}\left(\int_s^t \frac{1}{r_2(u)}\left(\int_u^\infty p(\theta)\,d\theta\right)du\right)ds = \infty$$

and

$$\limsup_{t\to\infty} \int_T^t \frac{1}{r_1(s)}\left(\int_s^t \frac{q(u)}{r_2(u)}\,du\right)ds < \infty$$

or

$$\limsup_{t\to\infty} \int_T^t p(s)\,ds = \infty,$$

to replace (6.222) or (6.223), respectively, in Corollaries 6.6.1 and 6.6.2.

Theorem 6.6.2 *Assume* (6.214) *and* (6.215). *Let* $f'(x) \geq 0$ *for* $x \neq 0$ *and there exists a constant* $k_0 > 0$ *such that* $f'(x) \geq k_0$ *for* $|x| \geq D$ *for some* $D > 0$. *Set* ρ *be a sufficiently smooth positive function defined on* $[\sigma, \infty)$, *and let*

$$\phi(t) = \big(r_2(t)\rho'(t)\big)' r_1(t) + \rho(t)q(t).$$

Suppose that there exists a $T_1 \geq \sigma$ *such that* $\rho'(t) \geq 0$, $\phi(t) \geq 0$, $\phi'(t) \leq 0$ *and*

$$\int_{T_1}^{\infty} \rho(s)p(s)\,ds = \infty \tag{6.225}$$

and (6.216) *is nonoscillatory. Assume that there exists a continuous function* $g_*(t)$ *defined on* $[\sigma, \infty)$ *such that* $g_*(t) \leq g(t)$ *for all* $t \geq \sigma$ *and such that* $\lim_{t\to\infty} g_*(t) = \infty$. *If there exists an eventually positive function* $\rho_1 \in C^1([\sigma, \infty)$ *such that*

$$\limsup_{t\to\infty} \int_T^t \left[\rho_1(s)p(s) - \frac{B^2(s)}{4A(s)}\right] ds = \infty \tag{6.226}$$

for some $T > \sigma$, *where*

$$A(t) = \frac{k_0 R_2(g(t), g_*(t))g'(t)}{r_1(g(t))\rho_1(t)} \quad \text{and} \quad B(t) = \frac{\rho_1'(t)}{\rho_1(t)} - q(t)\frac{R_2(g(t), g_*(t))}{r_1(t)},$$

then every solution $x(t)$ *of* (6.213) *either oscillates or converges to zero as* $t \to \infty$.

Lemma 6.6.1 *Assume that* (6.216) *is nonoscillatory and* f *is nondecreasing. If* $x(t)$ *is a nonoscillatory solution of* (6.213) *defined on* $[T, \infty)$, $T \geq \sigma$, *then there exists a* $t_1 \in [T, \infty)$ *such that either* $x(t)D_1x(t) > 0$ *or* $x(t)D_1x(t) < 0$ *for* $t \geq t_1$.

Lemma 6.6.2 *Let* ρ *be a sufficiently smooth positive function defined on* $[\sigma, \infty)$ *and set*

$$\phi(t) = \big(r_2(t)\rho'(t)\big)' r_1(t) + \rho(t)q(t)$$

and f *satisfies* (6.217). *Further, suppose that there exists a* $t_1 \geq T \geq \sigma$ *such that* $\phi(t) \geq 0$, $\phi'(t) > 0$, $\rho'(t) \geq 0$ *and*

$$\int_{t_1}^{\infty} \big(K\rho(s)p(s) - \phi'(s)\big)\,ds = \infty, \tag{6.227}$$

where $K\rho(t)p(t) - \phi'(t) \geq 0$ *for all* $t \in [t_1, \infty)$ *and not identically zero in any subinterval of* $[t_1, \infty)$. *If* (6.214) *holds and* $x(t)$ *be a nonoscillatory solution of* (6.213) *which satisfies* $x(t)D_1x(t) \leq 0$ *for all* $t \geq t_1$, *then*

$$\lim_{t\to\infty} x(t) = 0.$$

If

$$\phi'(t) \leq 0 \tag{6.228}$$

in Lemma 6.6.2, we can take

$$\int^{\infty} \rho(s)p(s)\,ds = \infty \tag{6.229}$$

to replace (6.227). Hence the condition (6.217) fails.

Lemma 6.6.3 *Let* (6.215) *holds. If $x(t)$ is a nonoscillatory solution of* (6.213) *which satisfies $x(t)D_1x(t) \geq 0$ for all large t, then there exists a $t_1 \geq \sigma$ such that $x(t)$ has the property V_2 for all $t \geq t_1$.*

Theorem 6.6.3 *Let the hypotheses of Lemmas* 6.6.1–6.6.3,

$$-f(-uv) \geq f(uv) \geq f(u)f(v) \quad \textit{for } uv > 0 \tag{6.230}$$

and (6.228) *hold. If the first-order delay equation*

$$y'(t) + \frac{q(t)}{r_1(t)}R_2\big(g(t),T\big)y\big(g(t)\big) + p(t)f\big(R_{12}\big(g(t),T\big)\big)f\big(y\big(g(t)\big)\big) = 0$$

is oscillatory for every $T \geq \sigma$, then every solution $x(t)$ of (6.213) *is either oscillatory or satisfies* $\lim_{t\to\infty} x(t) = 0$.

Corollary 6.6.3 *Let the hypotheses of Lemmas* 6.6.1–6.6.3 *hold. If the first-order delay equation*

$$y'(t) + \left(Kp(t)R_{12}\big(g(t),T\big) + \frac{q(t)}{r_1(t)}R_2\big(g(t),T\big)\right)y\big(g(t)\big) = 0$$

for some $K > 0$ and every $T \geq \sigma$ is oscillatory, then every solution $x(t)$ of (6.213) *is either oscillatory or satisfies* $\lim_{t\to\infty} x(t) = 0$.

Theorem 6.6.4 *Let the hypotheses of Lemmas* 6.6.1–6.6.3 *hold. If*

$$\limsup_{t\to\infty} \int_{g(t)}^{t} \left(Kp(s)R_{12}\big(g(s),T\big) + \frac{q(s)}{r_1(s)}R_2\big(g(s),T\big)\right)ds > 1$$

for some $K > 0$ and every $T \geq \sigma$, then every solution $x(t)$ of (6.213) *is either oscillatory or satisfies* $\lim_{t\to\infty} x(t) = 0$.

Theorem 6.6.5 *Let the hypotheses of Lemmas* 6.6.1–6.6.3 *hold. If*

$$\liminf_{t\to\infty} \int_{g(t)}^{t} \left(Kp(s)R_{12}\big(g(s),T\big) + \frac{q(s)}{r_1(s)}R_2\big(g(s),T\big)\right)ds > \frac{1}{e}$$

for some $K > 0$ and every $T \geq \sigma$, then every solution $x(t)$ of (6.213) *is either oscillatory or satisfies* $\lim_{t\to\infty} x(t) = 0$.

Theorem 6.6.6 *Let the hypotheses of Lemmas* 6.6.1–6.6.3 *and* (6.228), (6.230) *hold, and*

$$\frac{u}{f(u)} \to 0$$

as $u \to 0$. *If*

$$\limsup_{t\to\infty} Q(t) \int_{g(t)}^{t} p(s) f\big(R_{12}\big(g(s), T\big)\big)\, ds > 0 \tag{6.231}$$

where $Q(t) = 1/(1 - \int_{g(t)}^{t} \frac{q(s)}{r_1(s)} R_2(g(s), T)\, ds) \geq 0$ *for every* $t \geq T \geq \sigma$ *and not identically zero in any subinterval of* $[T, \infty)$, *then every solution* $x(t)$ *of* (6.213) *is either oscillatory or satisfies* $\lim_{t\to\infty} x(t) = 0$.

When Theorem 6.6.6 doesn't have the condition (6.228), we can take either

$$\limsup_{t\to\infty} \int_{g(t)}^{t} \bigg(Kp(s) f\big(R_{12}\big(g(s), T\big)\big) + \frac{q(s)}{r_1(s)} R_2\big(g(s), T\big)\bigg)\, ds > 1$$

or

$$\limsup_{t\to\infty} \int_{g(t)}^{t} \bigg(K^2 p(s) R_{12}\big(g(s), T\big) + \frac{q(s)}{r_1(s)} R_2\big(g(s), T\big)\bigg)\, ds > 1$$

or

$$\limsup_{t\to\infty} K^2 Q(t) \int_{g(t)}^{t} p(s) f\big(R_{12}\big(g(s), T\big)\big)\, ds > 1$$

to replace (6.231).

Theorem 6.6.7 *Let the hypotheses of Lemmas* 6.6.1–6.6.3 *and* $g(t) \leq t$, (6.228), (6.230) *hold. If the second-order equation*

$$\big(r_2(t) y'(t)\big)' + \frac{q(t)}{r_1(t)} y\big(g(t)\big) + p(t) f\bigg(\frac{R_{12}(g(t), T)}{R_2(g(t), T)}\bigg) f\big(y\big(g(t)\big)\big) = 0$$

for every $T \geq \sigma$ *is oscillatory, then every solution* $x(t)$ *of* (6.213) *is either oscillatory or satisfies* $\lim_{t\to\infty} x(t) = 0$.

Corollary 6.6.4 *Let the hypotheses of Lemmas* 6.6.1–6.6.3 *and* $g(t) \leq t$. *If the second-order equation*

$$\big(r_2(t) y'(t)\big)' + \bigg(Kp(t)\frac{R_{12}(g(t), T)}{R_2(g(t), T)} + \frac{q(t)}{r_1(t)}\bigg) y\big(g(t)\big) = 0$$

for some $K > 0$ *and every* $T \geq \sigma$, *then every solution* $x(t)$ *of* (6.213) *is either oscillatory or satisfies* $\lim_{t\to\infty} x(t) = 0$.

Theorem 6.6.8 *Let the hypotheses of Lemmas* 6.6.1–6.6.3 *and* $g(t) \le t$, (6.228), (6.230) *hold, and*

$$\int_{\pm\epsilon}^{\pm\infty} \frac{du}{f(u)} < \infty \tag{6.232}$$

for every $\epsilon > 0$. *If*

$$\int_T^\infty p(s) R_2\big(g(s), T\big) f\left(\frac{R_{12}(g(s), T)}{R_2(g(s), T)}\right) ds = \infty$$

holds for some $T \ge \sigma$, *then every solution* $x(t)$ *of* (6.213) *is either oscillatory or satisfies* $\lim_{t\to\infty} x(t) = 0$.

Theorem 6.6.9 *Let the hypotheses of Lemmas* 6.6.1–6.6.3 *and* $g(t) \le t$, (6.228), (6.230) *and* (6.232) *hold. Let there exists a nondecreasing function* $G \in C(R, R)$ *such that* $f(x) = |x| G(x)$ *for* $x \in R$. *Then, if*

$$\int_T^\infty p(s) R_2^2\big(g(s), T\big) f\left(\frac{R_{12}(g(s), T)}{R_2(g(s), T)}\right) \left(\int_{g(s)}^\infty p(u) f\left(\frac{R_{12}(g(u), T)}{R_2(g(u), T)}\right) du\right) ds$$
$$= \infty,$$

for $T \ge \sigma$, *and*

$$\int_{\pm\epsilon}^{\pm\infty} \frac{dx}{G(x)} < \infty,$$

for every $\epsilon > 0$, *then every solution* $x(t)$ *of* (6.213) *is either oscillatory or satisfies* $\lim_{t\to\infty} x(t) = 0$.

Theorem 6.6.10 *Let the hypotheses of Lemmas* 6.6.1–6.6.3 *and* $g(t) \le t$, (6.228), (6.230) *hold, and*

$$\int_0^{\pm\epsilon} \frac{du}{f(u)} < \infty \tag{6.233}$$

for every $\epsilon > 0$. *If*

$$\int^\infty p(s) f\big(R_{12}\big(g(s), T\big)\big)\, ds = \infty$$

for some $T \ge \sigma$, *then every solution* $x(t)$ *of* (6.213) *is either oscillatory or satisfies* $\lim_{t\to\infty} x(t) = 0$.

Theorem 6.6.11 *Let* $g(t) \le t$, *and the function* f *satisfies the condition*

$$\liminf_{|u|\to\infty} \big|f(u)\big| > 0.$$

If

$$\int^{\infty} p(t)\,dt = \infty,$$

then every solution $x(t)$ *of* (6.213) *is either oscillatory or satisfies* $\lim_{t\to\infty} x(t) = 0$.

6.7 Oscillation of Solutions of Third-Order Differential Equations with Distributed Deviating Arguments

Consider the third-order delay differential equation

$$x'''(t) + p(t)x''(t) + \int_a^b q(t,\xi) f\big(x(t), x\big(t - g(t,\xi)\big)\big)\,d\sigma(\xi) = 0, \tag{6.234}$$

where f, p, q, g are given continuous functions, and for the Stieltjes integral σ is nondecreasing. We consider only nonconstant solutions which we assume exist on some interval $[t_0, \infty)$. Thus we can define the following terms:

A solution is said to be *oscillatory*, if it has zeros of arbitrarily large value; otherwise the solution is called nonoscillatory. A function is called *eventually positive*, it there exists a value t_0 such that $x(t) > 0$ for all $t \geq t_0$. Similarly, we define *eventually negative, eventually nonnegative, eventually nonpositive, eventually nonnegative*, and eventually zero.

In this article we assume the following:

(A1) The integral kernel q is nonnegative and not identically zero on any set $[t_0, +\infty) \times [a, b]$. The coefficient p is continuous, nonnegative, and satisfies the condition (6.235) below.
(A2) The delay g is a continuous function such that $g_0(t) \leq g(t,\xi) \leq g_1(t)$ for all ξ in $[a, b]$;
(A3) The (nonlinear) function f is continuous in both variables having $f(0,0) = 0$, and a positive real number f_1 such that

$$f(x,y) \geq f_1 y \quad \text{when } x \text{ and } y \text{ are positive,}$$
$$f(x,y) \leq f_1 y \quad \text{when } x \text{ and } y \text{ are negative.}$$

For short notation, we define the function

$$Q(t) = \int_a^b q(t,\xi)\,d\sigma(\xi).$$

Lemma 6.7.1 *If* x *is a solution of* (6.234) *which is positive on some interval* $[t_0, +\infty)$ *and*

$$\int_{t_0}^{\infty} \exp\left(-\int_{t_0}^{s} p(s)\,ds\right) ds = \infty, \tag{6.235}$$

then $x''(t)$ is eventually positive. Similarly, if $x(t)$ is negative, $x''(t)$ is eventually negative.

Proof To complete the proof, we show only the case when $x(t) > 0$; the other case is similar. Select a large t_0 so that $x(t) > 0$, $x(t - g_1) > 0$, $Q(t) > 0$ for all $t \ge t_0$. Then multiply (6.234) by the integrating factor $r(t) = \exp(\int_{t_0}^t p(s)\,ds)$, which is positive. Using (A3), so that the integral in (6.234) is positive, we have

$$\big(r(t)x''(t)\big)' < 0.$$

Since $r(t)x''(t)$ is decreasing, there are only two possibilities: When $r(t)x''(t)$ is positive for all large t, $x''(t)$ is eventually positive and the proof is complete. When $r(t)x''(t)$ is negative for all large t, we argue as follows. Since $r(t)x''(t)$ is decreasing and $r(t)$ is positive,

$$x''(t) \le r(t_0)x''(t_0)\frac{1}{r(t)}.$$

Then integrating,

$$x'(t) \le x'(t_0) + r(t_0)x''(t_0)\int_{t_0}^t \frac{1}{r(s)}\,ds.$$

Due to the condition (6.235), the integral approaches ∞, and $x'(t)$ is eventually negative. Once $x''(t) < 0$ and $x'(t) < 0$ then $x(t) < 0$ for large t, which is a contradiction, and the proof is complete. □

Lemma 6.7.2 *Let $x(t)$ be an eventually positive function and $x'''(t)$ nonpositive and not eventually zero. Then each derivative of $x(t)$ is eventually positive or eventually negative. Also, there exist a time t_1 and an integer ℓ, $\ell \in \{0, 2\}$, and for all $t \ge t_1$, (6.121) holds.*

As a consequence of the above lemma, the set N of nonoscillatory solutions to (6.234) can be decomposed as the union of disjoint sets N_ℓ consisting of nonoscillatory solutions satisfying (6.235). That is, $N = N_0 \cup N_2$.

The following lemma is part of [21, Lemma 1.3], and its results are obtained by using repeated integration by parts.

Lemma 6.7.3 *Under the conditions of Lemma 6.7.2, for $t \ge t_1$, the following inequalities hold:*

$$\int_{t_0}^\infty s^{2-\ell}\big|x'''(s)\big|\,ds < \infty, \tag{6.236}$$

$$x(t) \ge x(t_1) + \frac{1}{(\ell-1)!}\int_{t_1}^t (t-s)^{\ell-1}x^{(\ell)}(s)\,ds, \tag{6.237}$$

$$x^{(\ell)}(t) \ge \frac{1}{(2-\ell)!}\int_t^\infty (s-t)^{2-\ell}\big(-x'''(s)\big)\,ds. \tag{6.238}$$

We remark that inequalities similar the ones above can be stated for eventually negative functions.

Lemma 6.7.4 *Assume* (A1)–(A3), *and that there exists a constant* $0 < \beta \leq 1$ *such that* $\beta \leq (t - g_1(t))'$. *Also assume that for each* l *in* $\{0, 2\}$,

$$\limsup_{t\to\infty}(t-t_1)^l \int_{t+t/\beta}^{\infty} (s-t)^{2-l} Q(s)\, ds > \frac{l!(2-l)!}{f_1}. \tag{6.239}$$

If there exists a nonoscillatory solution of (6.234), *then it belongs to* N_0; *i.e.* $\ell = 0$ *in* (6.121).

Proof Let $x(t)$ be a nonoscillatory solution of (6.234). This proof is done only for x eventually positive, since the negative case is similar. If necessary, increase the value of t_1 in Lemma 6.7.2 so that $x(t) > 0$, $x''(t) > 0$, $t - g_1(t) > 0$, $x(t - g_1) > 0$, for all $t \geq t_1$. Now, we claim that $\ell = 0$ in (6.121). On the contrary, assume that $\ell = 2$. Since $x'''(t)$ is negative, $x''(t)$ is decreasing and by (6.237),

$$x(t) \geq x''(t)\frac{1}{2!}\int_{t_1}^{t} (t-s)\, ds = \frac{x''(t)(t-t_1)^2}{2!}. \tag{6.240}$$

Note that this inequality also follows from the Taylor expansion of x about t_1 with remainder of order 2. Using (6.238),

$$x(t) \geq \frac{(t-t_1)^2}{2!}\int_{t}^{\infty} \big(-x'''(s)\big)\, ds. \tag{6.241}$$

From (6.234), (A2), (A3), and that x is increasing, we obtain

$$\begin{aligned} -x'''(s) &\geq \int_a^b q(s,\xi) f\big(x(s), x\big(s - g(s,\xi)\big)\big)\, d\xi \\ &\geq \int_a^b q(s,\xi) f_1 x\big(s - g(s,\xi)\big)\, d\xi \\ &\geq f_1 x\big(s - g_1(s)\big) Q(s). \end{aligned} \tag{6.242}$$

Recall that $(s - g_1(s))' \geq \beta > 0$ and that $t - g_1(t) > 0$. For $s \geq t + t/\beta$, we have

$$\begin{aligned} t \leq (s-t)\beta &= \int_t^s \beta\, dw \\ &\leq \int_t^s \big(w - \tau_1(w)\big)'\, dw \\ &\leq t - g_1(t) + \int_t^s \big(w - g_1(w)\big)'\, dw = s - g_1(s). \end{aligned}$$

Since x is increasing, $x(t) \leq x(s - g_1(s))$. Then the inequality $\int_t \leq \int_{t+t/\beta}$, (6.241) and (6.242) imply

$$x(t) \geq \frac{(t - t_1)^l f_1 x(t)}{2!} \int_{t+t/\beta}^{\infty} Q(s)\, ds,$$

which contradicts (6.239) and completes the proof. □

Remark 6.7.1 The conditions of Lemma 6.7.4 are satisfied, if for example,

$$Q(s) \geq (s - t)^{-3} \frac{2!2^2}{f_1 \beta^2}$$

for $s \geq t + t/\beta$. That is, $Q(s)$ does not decay faster than s^{-3}. Note that for a fixed t and for the function $G(s) \geq (s - t)^{-3}$, we have

$$\int_{t+t/\beta}^{\infty} G(s)\, ds \geq \int_{t+t/\beta}^{\infty} (s - t)^{-3}\, ds = \frac{\beta^2}{2t^2}.$$

Multiplying $G(s)$ by $2!\,2^2/(f_1 \beta^l)$ and using $\lim_{t\to\infty}(t - t_1)/t = 1$, we obtain an inequality that implies the condition (6.239).

Lemma 6.7.5 *Assume* (A2)–(A3), *and that* $\tau_0(t)$ *in* (A2) *is nondecreasing and positive for t large. Also, assume that there exists a positive constant b such that*

$$\limsup_{t\to\infty} \int_t^{t+\tau_0(t)+b} Q(s)\, ds > \frac{2!\,3}{2 f_1}. \tag{6.243}$$

Then nonoscillatory solutions of (6.234) *cannot belong to* N_0; *i.e.,* $\ell > 0$ *in* (6.121).

Proof Let $x(t)$ be a nonoscillatory solution of (6.234). Without any loss of generality, we assume that y is eventually positive; the negative case is proved similarly. If necessary, increase the value of t_1 in Lemma 6.7.2 so that $x(t) > 0$, $x'''(t) < 0$, $g_0(t) > 0$, and g_0 is increasing, for all $t \geq t_1$. Now, we claim that $\ell = 2$ in (6.121). On the contrary assume that $\ell = 0$. Then $x'(t) < 0$ and $x(t)$ is decreasing. From (6.234), (A2), (A3) and τ_0 being nondecreasing, we have

$$\begin{aligned} -x'''(s) &\geq \int_a^b q(s,\xi) f\big(x(s), x\big(s - g(s,\xi)\big)\big)\, d\xi \\ &\geq \int_a^b q(s,\xi) f_1 x\big(s - g(s,\xi)\big)\, d\xi \\ &\geq f_1 x\big(s - g_0(s)\big) Q(s). \end{aligned} \tag{6.244}$$

From (6.238) and the inequality $\int_t^{\infty} \geq \int_t^{t+\tau_0+b}$,

$$\begin{aligned} x(t) &\geq \frac{1}{2!}\int_t^{\infty}(s-t)^2\big(-x'''(s)\big)\,ds \\ &\geq \frac{f_1}{2!}\int_t^{t+g_0(t)+b}(s-t)^2 x\big(s-g_0(s)\big)Q(s)\,ds \\ &\geq \frac{f_1}{2!}x(t+b)\int_t^{t+g_0(t)+b}(s-t)^2 Q(s)\,ds. \end{aligned}$$

Therefore,

$$\frac{x(t)}{x(t+b)} \geq \frac{f_1}{2!}\int_t^{t+g_0(t)+b}(s-t)^2 Q(s)\,ds.$$

Since $x(t)$ is positive and decreasing, the $\lim_{t\to\infty} x(t)$ exists. If this limit is nonzero, then $\lim_{t\to\infty} x(t)/x(t+b) = 1$. On the other hand, if this limit is zero, we apply L'Hopital's Rule: that $x(t)$ is decreasing while $x'(t)$ is increasing, to obtain

$$1 \leq \lim_{t\to\infty}\frac{x(t)}{x(t+b)} = \lim_{t\to\infty}\frac{x'(t)}{x'(t+b)} \leq 1.$$

Thus there exists a t_2 such that $\frac{1}{2} < \frac{x(t)}{x(t+b)} < \frac{3}{2}$ and

$$\frac{3f_1}{2!2} \geq \int_t^{t+g_0(t)+b}(s-t)^2 Q(s)\,ds$$

for $t \geq t_2$. This contradicts (6.243) and completes the proof. □

We remark that if there exists a positive constant b such that

$$Q(s) \geq 3\frac{3!}{g_0(t+b)2f_1} \quad \text{for all } s \geq t,$$

then condition (6.243) is satisfied. That is, Q does not decay faster than $1/g_0$. Certainly this condition is stronger than the condition in Remark 6.7.1.

Theorem 6.7.1 *Under the assumptions of Lemmas* 6.7.3 *and* 6.7.4, *all solutions of* (6.234) *are oscillatory.*

Theorem 6.7.2 *Assume* (A1)–(A3), *and that* $g_0(t)$ *in* (A2) *is nondecreasing and positive for large* t. *Also assume that for some value* t_0,

$$\int_{t_0}^{\infty}(s-t_0)^2 Q(s)\,ds = \infty. \tag{6.245}$$

If x *is a nonoscillatory solution of* (6.234), *belonging to* N_0, *then* $\lim_{t\to\infty} x(t) = 0$.

Proof Certainly the limit exists because $x(t)$ is positive and decreasing. Now if $\lim_{t\to\infty} x(t) = x_1 \neq 0$, then there exists a value t_2 such that $x_1/2 < x(t) < 3x_1/2$ for all $t \geq t_2$. From (6.238), (6.244) and this inequality, we have

$$\frac{3}{2}x_1 \geq x(t) \geq \frac{f_1}{2!}\int_t^\infty (s-t)^2 x\big(s-g_0(s)\big)Q(s) \geq \frac{f_1 x_1}{2!2}\int_t^\infty (s-t)^2 Q(s).$$

Note that the integral in (6.245) is an increasing function of t_0 (its derivative is negative). Then the above inequality contradicts (6.245), and hence $\lim_{t\to\infty} x(t) = 0$. □

Lemma 6.7.6 *Under the conditions of Lemma 6.7.2, we have*

$$x^{(l)}(t) \geq \frac{1}{(2-\ell)!}\prod_{i=1}^{2-\ell}\big(2^i-1\big)t^{2-\ell}x''\big(2^{2-\ell}t\big), \quad l = 0, 2, \tag{6.246}$$

where the product with no factors, $\prod_{i=1}^0$, *is understood as* 1.

Proof If $\ell = 2$, the inequality (6.246) becomes equality and is true. Integrate $x''(t)$, which is positive and decreasing, to obtain

$$\begin{aligned} -x'(t) &= -x'(2t) + \int_t^{2t} x''(s)\,ds \\ &\geq \int_t^{2t} x''(s)\,ds \\ &\geq \big(2^2-1\big)tx''(2t). \end{aligned}$$

Integrating $x'(t)$ and using the fact that $x''(t)$ is decreasing, the above inequality yields

$$\begin{aligned} x(t) &= x(2t) - \int_t^{2t} x'(s)\,ds \\ &\geq \int_t^{2t} sx''(2s)\,ds \\ &\geq x''\big(2^2t\big)\int_t^{2t} s\,ds \\ &= \frac{1}{2!}\big(2^2-1\big)t^2x''\big(2^2t\big). \end{aligned}$$

The proof is now complete. □

Lemma 6.7.7 *Assume* (A1)–(A3) *and that there exists a constant* $0 < \beta \le 1$ *such that* $\beta \le (t - g_1(t))'$. *Also assume that*

$$\limsup_{t\to\infty} \int_{t_1}^{t} (t-s)^1 \int_{(s+s/\beta)}^{\infty} Q(u)\,du\,dt > \frac{1}{f_1}. \tag{6.247}$$

If there exists a nonoscillatory solution of (6.234), *then it belongs to* N_0; *i.e.* $\ell = 0$.

Proof Let $x(t)$ be a nonoscillatory solution of (6.234). Without any loss of generality, we assume that $x(t)$ is eventually positive. The negative case is proved similarly. If necessary, increase the value of t_1 in Lemma 6.7.2 so that $x(t) > 0$, $x''(t) > 0$, $t - g_1(t) > 0$, $x(t - g_1) > 0$, for all $t \ge t_1$. Now, we claim that $\ell = 0$ in (6.121). On the contrary, assume that $\ell = 2$. Integrating $x''(t)$ from t to ∞ and using $x''(t) > 0$,

$$x''(t) \ge \int_t^{\infty} \bigl(-x'''(s)\bigr)\,ds.$$

By (6.242), we have

$$x''(t) \ge x(t) \int_{(t+t/\beta)}^{\infty} Q(s)\,ds.$$

Using (6.237),

$$1 \ge f_1 \int_{t_1}^{t} (t-s) \int_{(s+s/\beta)}^{\infty} Q(u)\,du\,dt,$$

which contradicts (6.247) and completes the proof. □

Lemma 6.7.8 *Let r be a continuous nonnegative function and w be a positive, twice continuous differentiable function that satisfies*

$$w''(t) + r(t)w(t) \le 0 \quad \textit{for } t \ge t_1. \tag{6.248}$$

Then the differential equation

$$x''(t) + r(t)x(t) = 0 \tag{6.249}$$

does not have oscillatory solutions.

Theorem 6.7.3 *Under the assumptions of Lemmas* 6.7.5 *and Lemma* 6.7.7, *all solutions of* (6.234) *are oscillatory.*

Now, we consider the equation

$$\bigl(r_2(t)\bigl(r_1(t)x'(t)\bigr)'\bigr)' + \int_c^d q(t,\xi) f\bigl(x\bigl(\sigma(t,\xi)\bigr)\bigr)\,d\xi = 0, \tag{6.250}$$

where $a, b \in C([t_0,\infty),\mathbb{R})$, $f \in C(R,R)$. Suppose that the following conditions hold unless stated otherwise

(a) $r_1(t) > 0$, $r_2(t) > 0$, $r_2'(t) \geq 0$, $\int^\infty \frac{dt}{r_1(t)} = \infty$, $\int^\infty \frac{dt}{r_2(t)} = \infty$,
(b) $q(t,\xi) \in C([t_0,\infty) \times [c,d], \mathbb{R})$, $q(t,\xi) > 0$,
(c) $\frac{f(x)}{x} \geq \epsilon > 0$, for $x \neq 0$, ϵ is a constant,
(d) $\sigma(t,\xi) \in C([t_0,\infty) \times [c,d], \mathbb{R})$, $\sigma(t,\xi) < t$, $\xi \in [c,d]$, $\sigma(t,\xi)$ is nondecreasing with respect to t and ξ and

$$\lim_{t\to\infty} \min_{\xi\in[c,d]} \sigma(t,\xi) = \infty.$$

Theorem 6.7.4 *If*

$$\int_{t_1}^{\infty} \int_c^d q(s,\xi)\, d\xi\, ds = \infty \tag{6.251}$$

and

$$\epsilon \int_{g(t)}^{t} \left[\int_{g(t)}^{r} \frac{1}{r_1(u)} du \left(\int_u^r \frac{1}{r_2(v)} dv \right) \right] \int_c^d q(r,\xi)\, d\xi\, dr > 1, \tag{6.252}$$

where $g(t) = \sigma(t,d)$. *Then Eq.* (6.250) *is oscillatory.*

Proof Suppose that $x(t)$ is nonoscillatory solution of (6.250). Without loss of generality, we may assume that $x(t)$ is eventually positive. From (6.250), we have

$$\big(r_2(t)\big(r_1(t)x'(t)\big)'\big)' = -\int_c^d q(t,\xi) f\big(x\big(\sigma(t,\xi)\big)\big)\, d\xi. \tag{6.253}$$

It is easy to obtain

$$\big(r_2(t)\big(r_1(t)x'(t)\big)'\big)' \leq 0, \qquad \big(r_1(t)x'(t)\big)' > 0, \qquad \big(r_1(t)x'(t)\big)'' \leq 0$$

for large enough t. Thus, $x(t)$, $x'(t)$ and $(r_1(t)x'(t))'$ are monotone and eventually one-signed. From the condition (c),

$$f\big(x\big(\sigma(t,\xi)\big)\big) \geq \epsilon x\big(\sigma(t,\xi)\big) > 0.$$

Therefore,

$$0 \geq \big(r_2(t)\big(r_1(t)x'(t)\big)'\big)' + \epsilon \int_c^d q(t,\xi) x\big(\sigma(t,\xi)\big)\, d\xi. \tag{6.254}$$

Now, consider again two cases.

Suppose that $x'(t)$ is eventually positive, say $x'(t) > 0$ for $t > t_2$. Now, we can choose a constant $k > 0$ such that $x(k) > 0$. By the condition (d), there exists a sufficiently large T such that $\sigma(t,\xi) > k$ for $t > T$, $\xi \in [c,d]$. Therefore,

$$x\big(\sigma(t,\xi)\big) \geq x(k).$$

Thus,

$$\big(r_2(t)\big(r_1(t)x'(t)\big)'\big)' + \epsilon x(k)\int_c^d q(t,\xi)\,d\xi \le 0. \tag{6.255}$$

Integrating Eq. (6.255) from t_1 to t, we get

$$r_2(t)\big(r_1(t)x'(t)\big)' \le r_2(t_1)\big(r_1(t_1)x'(t_1)\big)' - \epsilon x(k)\int_{t_1}^t\int_c^d q(s,\xi)\,d\xi\,ds. \tag{6.256}$$

Taking the limit both sides as $t \to \infty$ and using (6.251), the inequality (6.256) leads to a contradiction to $(r_1(t)x'(t))' > 0$. Now suppose $x'(t)$ is eventually negative. Integrating Eq. (6.250) three times, we get

$$\int_t^\infty\bigg[\int_t^r \frac{1}{r_1(u)}\,du\bigg(\int_u^r \frac{1}{r_2(v)}\,dv\bigg)\bigg]\int_c^d q(r,\xi)f\big(x\big(\sigma(r,\xi)\big)\big)\,d\xi\,dr \le x(t). \tag{6.257}$$

Using the condition (c) in (6.257), we obtain

$$\int_t^\infty\bigg[\int_t^r \frac{1}{r_1(u)}\,du\bigg(\int_u^r \frac{1}{r_2(v)}\,dv\bigg)\bigg]\epsilon\int_c^d q(r,\xi)x\big(\sigma(r,\xi)\big)\,d\xi\,dr \le x(t). \tag{6.258}$$

Replace t by $g(t)$ in (6.258), where $g(t) = \sigma(t,d)$, then we have

$$\epsilon\int_{g(t)}^t\bigg[\int_{g(t)}^r \frac{1}{r_1(u)}\,du\bigg(\int_u^r \frac{1}{r_2(v)}\,dv\bigg)\bigg]\int_c^d q(r,\xi)x\big(\sigma(r,\xi)\big)\,d\xi\,dr \le x\big(g(t)\big). \tag{6.259}$$

Since $x(t)$ is decreasing and positive,

$$\epsilon\int_{g(t)}^t\bigg[\int_{g(t)}^r \frac{1}{r_1(u)}\,du\bigg(\int_u^r \frac{1}{r_2(v)}\,dv\bigg)\bigg]\int_c^d q(r,\xi)\,d\xi\,dr \le 1.$$

This is a contradiction to (6.252). Therefore, the proof is complete. □

Example 6.7.1 Consider the following functional differential equation

$$x''' + \int_{2/7\pi}^{1/2\pi} \frac{2e^{-1/\xi}}{\xi^2}x\bigg(t - \frac{1}{\xi}\bigg)\,d\xi = 0$$

so that $r_1(t) = 1$, $r_2(t) = 1$, $f(x) = x$, $q(t,\xi) = \frac{2e^{-1/\xi}}{\xi^2}$, $\sigma(t,\xi) = t - \frac{1}{\xi}$. We can easily see that the conditions of Theorem 6.7.4 are satisfied. It is easy to verify that $x(t) = e^{-t}\sin t$ is a solution of this problem.

Theorem 6.7.5 *Suppose* (6.252) *holds. In addition to that suppose there exists a* $p \in C([t_0,\infty), R)$, $p(t) > 0$ *such that*

$$\int^\infty\bigg[\Gamma(t)p(t) - \frac{r_1(\sigma(t,c))r_2(t)(p'(t))^2}{(\sigma(t,c) - T)\sigma'(t,c)4p(t)}\bigg]dt = \infty,$$

where $\Gamma(t) = \epsilon\int_c^d q(t,\xi)\,d\xi$. *Then Eq.* (6.250) *is oscillatory.*

Proof Suppose that $x(t)$ is nonoscillatory solution of (6.250). We can assume that $x(t)$ is eventually positive. The case of $x(t)$ is eventually negative can be proved by the same arguments. Proceeding as in the proof of Theorem 6.7.4, we have

$$\big(r_2(t)\big(r_1(t)x'(t)\big)'\big)' \le 0,$$
$$\big(r_1(t)x'(t)\big)' > 0 \quad \text{and} \quad \big(r_1(t)x'(t)\big)'' \le 0.$$

Thus, $x(t)$, $x'(t)$ and $(r_1(t)x'(t))'$ are monotone and eventually one-signed. From the condition (c),

$$\begin{aligned} &f\big(x\big(\sigma(t,\xi)\big)\big) \ge \epsilon x\big(\sigma(t,\xi)\big) > 0, \\ &\big(r_2(t)\big(r_1(t)x'(t)\big)'\big)' + \epsilon \int_c^d q(t,\xi)x\big(\sigma(t,\xi)\big)\,d\xi \le 0. \end{aligned} \tag{6.260}$$

If $x'(t)$ is eventually positive, then we can define

$$z(t) = \frac{r_2(t)(r_1(t)x'(t))'}{x(\sigma(t,c))}p(t).$$

It is obvious that $z(t) > 0$ for $t \ge t_2$ and $z'(t)$ is

$$z'(t) = \frac{(r_2(t)(r_1(t)x'(t))')'}{x(\sigma(t,c))}p(t) + \frac{p'(t)}{p(t)}z(t) - \frac{x'(\sigma(t,c))\sigma'(t,c)}{x(\sigma(t,c))}z(t). \tag{6.261}$$

Clearly, we have

$$a(t)x'(t) \ge (t-T)\big(a(t)x'(t)\big)'.$$

Since $(r_1(t)x(t))'$ is nonincreasing, we have

$$r_1\big(\sigma(t,c)\big)x'\big(\sigma(t,c)\big) \ge \big(\sigma(t,c)-T\big)\big(r_1(t)x'(t)\big)',$$

then

$$x'\big(\sigma(t,c)\big) \ge \frac{(\sigma(t,c)-T)(r_1(t)x'(t))'}{r_1(\sigma(t,c))}.$$

Plug the above inequality in (6.261), then we obtain

$$z'(t) = \frac{(r_2(t)(r_1(t)x'(t))')'}{x(\sigma(t,c))}p(t) + \frac{p'(t)}{p(t)}z(t) - \frac{(\sigma(t,c)-T)\sigma'(t,c)}{p(t)r_2(t)r_1(\sigma(t,c)))}z^2(t).$$

Completing the square leads to

$$z'(t) \le -\Gamma(t)p(t) + \frac{r_2(t)r_1(\sigma(t,c))(p'(t))^2}{(\sigma(t,c)-T)\sigma'(t,c))4p(t)}. \tag{6.262}$$

Integrating (6.262) from T to t and letting $t \to \infty$, we see that $\lim_{t\to\infty} z(t) = -\infty$. This contradicts that $z(t)$ is eventually positive.

If $x'(t)$ is eventually negative, the proof is exactly the same as in the second part of the proof of Theorem 6.7.4. □

6.8 Oscillation and Asymptotic Behaviour of Nonoscillatory Solutions of $(r_2(t)(r_1(t)x'(t))')' + p(t)f(x(g(t))) = 0$

Consider the third-order nonlinear delay differential equations

$$\big(r_2(t)\big(r_1(t)x'(t)\big)'\big)' + p(t)x^\gamma\big(g(t)\big) = 0, \tag{6.263}$$

$$\big(r_2(t)\big(r_1(t)x'(t)\big)'\big)' + p(t)f\big(x\big(g(t)\big)\big) = 0, \tag{6.264}$$

$$\big(r_1(t)\big(r_2(t)x'(t)\big)'\big)' - p(t)f\big(x\big(g(t)\big)\big) = 0, \tag{6.265}$$

where r_1, r_2, p and $g \in C([\sigma,\infty), R)$, $r_1(t) > 0, r_2(t) > 0, p(t) > 0, 0 < g(t) < t$, $g(t)$ is nondecreasing and $\lim_{t\to\infty} g(t) = \infty$, $\gamma > 0$ is a ratio of odd positive integers, $f \in C(R,R)$ is such that $uf(u) > 0$ for $u \neq 0$.

Theorem 6.8.1 *Let*

$$\int_\sigma^\infty \frac{dt}{r_2(t)} < \infty, \qquad \int_\sigma^\infty \frac{dt}{r_1(t)} = \infty. \tag{6.266}$$

Assume that there exist numbers $\alpha \le \gamma$, $\beta \ge \gamma$, and two functions $\xi, \rho \in C([\sigma,\infty), R)$ such that α, β are the ratios of odd positive integers, $\xi(t) > t$; $\xi(t)$ is nondecreasing, $g(\xi(\xi(t))) < t, \rho(t)$ is nondecreasing, and $\rho(t) > t$. If for all sufficiently large $t_1 \ge \sigma$ and for $t_2 > t_1$, the first-order delay differential equation

$$\omega'(t) + c_1^{\gamma-\alpha} p(t)\left(\int_{t_2}^{g(t)} \frac{\int_{t_1}^{s} \frac{du}{r_2(u)}}{r_1(s)}\,ds\right)^\alpha \omega^\alpha\big(g(t)\big) = 0$$

is oscillatory for all constants $c_1 > 0$, the first-order delay differential equation

$$v'(t) + \left(\frac{1}{r_1(t)}\int_t^{\xi(t)} \frac{1}{r_2(s_2)}\int_{s_2}^{\xi(s_2)} p(s_1)\,ds_1\,ds_2\right)v^\gamma\big(g\big(\xi\big(\xi(t)\big)\big)\big) = 0$$

is oscillatory and the first-order advanced differential equation

$$z'(t) - c_2^{\gamma-\beta} p(t)\left(\int_{\rho(t)}^\infty \frac{ds}{r_2(s)}\right)^\beta \left(\int_{t_1}^{g(t)} \frac{ds}{r_1(s)}\right)^\gamma z^\beta\big(\rho(t)\big) = 0$$

is oscillatory for all constants $c_2 > 0$, then (6.263) is oscillatory.

Corollary 6.8.1 *Let* (6.266) *hold and* $\gamma = 1$. *Assume that there exist two functions* $\xi, \rho \in C([\sigma, \infty), R)$ *such that* $\xi(t) > t$, $\xi(t)$ *is nondecreasing*, $g(\xi(\xi(t))) < t$, $\rho(t)$ *is nondecreasing, and* $\rho(t) > t$. *If for all sufficiently large* $t_1 \geq \sigma$ *and* $t_2 > t_1$,

$$\liminf_{t\to\infty} \int_{g(t)}^{t} p(s) \int_{t_2}^{g(s)} \frac{\int_{t_1}^{u} \frac{d\theta}{r_2(\theta)}}{r_1(u)}\, du\, ds > \frac{1}{e},$$

$$\liminf_{t\to\infty} \int_{g(\xi(\xi(t)))}^{t} \frac{1}{r_1(s)} \int_{s}^{\xi(s)} \frac{1}{r_2(s_2)} \int_{s_2}^{\xi(s_2)} p(s_1)\, ds_1\, ds_2\, ds > \frac{1}{e},$$

and

$$\liminf_{t\to\infty} \int_{t}^{\rho(t)} p(s) \int_{\rho(s)}^{\infty} \frac{du}{r_2(u)} \int_{t_1}^{g(s)} \frac{du}{r_1(u)}\, ds > \frac{1}{e},$$

then (6.263) *is oscillatory.*

Theorem 6.8.2 *Let the conditions of Theorem* 6.8.1 *hold with* (6.266) *is replaced by*

$$\int_{\sigma}^{\infty} \frac{du}{r_2(u)} < \infty, \qquad \int_{\sigma}^{\infty} \frac{du}{r_1(u)} < \infty. \tag{6.267}$$

If

$$\int_{\sigma}^{\infty} \frac{1}{r_1(v)} \int_{\sigma}^{v} \frac{1}{r_2(u)} \int_{\sigma}^{u} p(s) B^{\gamma}\big(g(s)\big)\, ds\, du\, dv = \infty,$$

then (6.263) *is oscillatory, where*

$$B(t) = \int_{t}^{\infty} \frac{du}{r_1(u)}.$$

Now, we give some sufficient conditions for the oscillation of (6.264) with $g(t) = t - \tau$, $\tau > 0$ is a real number.

Theorem 6.8.3 *Assume that* $g(t) = t - \tau$, *where* $\tau > 0$ *is a real number, and*

$$\int_{\sigma}^{\infty} \frac{ds}{r_1(s)} = \int_{\sigma}^{\infty} \frac{ds}{r_2(s)} = \infty,$$

and the function f *satisfies the condition* (6.217). *Let*

$$\limsup_{t\to\infty} \int_{\sigma}^{t} \frac{1}{r_1(s)} \int_{s}^{\infty} \frac{1}{r_2(u)} \int_{u}^{\infty} p(\theta)\, d\theta\, du\, ds = \infty.$$

Furthermore, assume that there exists a positive function $\rho \in C^1([\sigma, \infty), [0, \infty))$ *such that*

$$\limsup_{t\to\infty} \int_{\sigma}^{t} \left(K\rho(s)p(s) - \frac{(\rho'(s))^2 r_1(s-\tau)}{4\rho(s)\delta(s-\tau)} \right) ds = \infty, \tag{6.268}$$

where $\delta(t) = \int_{\sigma}^{t} \frac{ds}{r_2(s)}$. *Then every solution of* (6.264) *oscillates.*

Theorem 6.8.4 *Let all the assumptions of Theorem* 6.8.3 *hold, except the condition* (6.268) *is replaced by*

$$\limsup_{t\to\infty}\frac{1}{t^n}\int_\sigma^t (t-s)^n\left[K\rho(s)p(s)-\frac{(\rho'(s))^2 r_1(s-\tau)}{4\rho(s)\delta(s-\tau)}\right]ds=\infty.$$

Then every solution $x(t)$ of (6.264) *is oscillatory, where $\delta(t)$ is given in Theorem* 6.8.3.

Theorem 6.8.5 *Assume that*

$$\liminf_{|u|\to\infty}\frac{f(u)}{u}>0 \tag{6.269}$$

and f is nondecreasing in R, $|u|\geq|f(u)|$ for all $u\in R$ and $g(t)\leq t$. Then the following hypotheses hold.

(a) *Let $\int_\sigma^\infty p(t)\int_\sigma^{g(t)}\frac{1}{r_1(s)}ds\,dt=\infty$. If Eq.* (6.265) *has property B, then* (6.264) *has property A.*
(b) *Let $\int_\sigma^\infty p(t)\int_\sigma^t\frac{1}{r_1(s)}ds\,dt=\infty$, $\int_\sigma^\infty p(t)\int_{g(t)}^t\frac{1}{r_2(s)}\int_\sigma^s\frac{1}{r_1(u)}du\,ds\,dt<\infty$. If Eq.* (6.264) *has property A, then* (6.265) *has property B.*

Theorem 6.8.6 *Assume that* (6.269) *holds and f is nondecreasing in R, $|u|\geq|f(u)|$ for all $u\in R$ and $g(t)\leq t$. Further, suppose that*

$$\int_\sigma^\infty p(t)\int_\sigma^{g(t)}\frac{1}{r_1(s)}ds\,dt=\infty,\qquad \int_\sigma^\infty p(t)\int_{g(t)}^t\frac{1}{r_2(s)}\int_\sigma^s\frac{1}{r_1(u)}du\,ds\,dt<\infty.$$

Then Eq. (6.264) *has property A, if and only if* (6.265) *has property B.*

Theorem 6.8.7 *Assume that* (6.269) *holds and f is nondecreasing in R, $|u|\geq|f(u)|$ for all $u\in R$ and $g(t)\leq t$. Further, let*

$$\int_\sigma^\infty p(t)\int_\sigma^{g(t)}\frac{1}{r_1(s)}ds\,dt=\infty,\qquad \int_\sigma^\infty p(t)\int_\sigma^{g(t)}\frac{1}{r_2(s)}ds\,dt=\infty.$$

Then Eq. (6.264) *has property A, and Eq.* (6.265) *has property B.*

In the sequel, we consider a function $\tau(t)$ such that

$$\tau\in C\big([\sigma,\infty),R\big),\quad \tau(t)>t,\quad\text{and}\quad g\big(\tau(t)\big)\leq t, \tag{6.270}$$

$$C_\sigma(t)=\int_t^{\tau(t)}p(s)\int_\sigma^{g(s)}\frac{1}{r_1(u)}\int_\sigma^u\frac{1}{r_2(\theta)}d\theta\,du\,ds, \tag{6.271}$$

and

$$S_f=\max\left\{\limsup_{u\to\infty}\frac{u}{f(u)},\limsup_{u\to-\infty}\frac{u}{f(u)}\right\}<\infty. \tag{6.272}$$

Theorem 6.8.8 *Let* (6.270) *and* (6.272) *hold. Assume that*

$$\limsup_{t\to\infty} C_\sigma(t) > S_f. \tag{6.273}$$

If the linear equation

$$\big(r_2(t)\big(r_1(t)x'(t)\big)'\big)' + p(t)x\big(g(t)\big) = 0, \tag{6.274}$$

has property A, then Eq. (6.264) *has property A.*

Corollary 6.8.2 *Assume that* (6.272) *holds, and $g(t)$ is an increasing function. Let $g^{-1}(t)$ be the function inverse to $g(t)$ and*

$$\limsup_{t\to\infty} \int_t^{g^{-1}(t)} p(s) \int_\sigma^{g(s)} \frac{1}{r_1(u)} \int_\sigma^u \frac{1}{r_2(\theta)}\, ds\, du\, ds > S_f.$$

If (6.274) *has property A, then* (6.264) *has property A.*

Theorem 6.8.8 provides a nice result for the superlinear case of Eq. (6.264), namely, for the equation

$$\big(r_2(t)\big(r_1(t)x'(t)\big)'\big)' + p(t)\big|x\big(g(t)\big)\big|^\gamma \operatorname{sgn} x\big(g(t)\big) = 0, \quad t \ge \sigma,\ \gamma > 1. \tag{6.275}$$

Corollary 6.8.3 *Let* (6.270) *hold. Assume that*

$$\limsup_{t\to\infty} C_\sigma(t) > 0.$$

If (6.274) *has property A, then* (6.275) *has property A.*

6.9 Nonoscillation of Solutions of Nonlinear Nonhomogeneous Delay Differential Equations of Third Order

Consider the equations

$$\big(r(t)x''(t)\big)' + q(t)k\big(x'(t)\big) + p(t)x^\gamma\big(g(t)\big) = f(t), \quad t \ge \sigma, \tag{6.276}$$

and

$$\big(r(t)x''(t)\big)' + q(t)k\big(x'(t)\big) + p(t)h\big(x\big(g(t)\big)\big) = f(t), \quad t \ge \sigma, \tag{6.277}$$

where $\sigma \ge 0$ is a fixed real number, f, p, q, r and $g \in C([0,\infty), R)$ such that $r(t) > 0$ and $f(t) \ge 0$ for all $t \ge 0$, $k, h \in C(R, R)$ are nondecreasing such that $h(x)x > 0$, $h(x')x' > 0$ for all $x \ne 0$, $\gamma > 0$ is a quotient of odd integers, $0 < g(t) \le t$ and $\lim_{t\to\infty} g(t) = \infty$.

Theorem 6.9.1 *Let $q(t) \le 0$. If $\lim_{t\to\infty} \frac{f(t)}{|p(t)|} = \infty$, then all bounded solutions of* (6.276) *are nonoscillatory.*

Proof Let $x(t)$ be a bounded solution of (6.276) on $[T_x, \infty)$, $T_x \ge 0$, such that $|x(t)| \le m$ for $t \ge T_x$. Since $\lim_{t\to\infty} g(t) = \infty$, there exists a $t_1 > T_x$ such that $g(t) > T_x$ for $t \ge t_1$. In view of the assumption $\lim_{t\to\infty} \frac{f(t)}{|p(t)|} = \infty$, it follows that there exists a $t_2 \ge t_1$ such that $f(t) > M^\gamma |p(t)|$ for $t \ge t_2$.

If possible, let $x(t)$ be nonnegative z-type with consecutive double zeros at a and b ($t_2 < a < b$) such that $x(t) > 0$ for $t \in (a, b)$. So, there exists $c \in (a, b)$ such that $x'(c) = 0$ and $x'(t) > 0$ for $t \in (a, c)$. Multiplying (6.276) by $x'(t)$, we get

$$\begin{aligned}\big(r(t)x'(t)x''(t)\big)' &= r(t)\big(x''(t)\big)^2 - q(t)k\big(x'(t)\big)x'(t) - p(t)x^\gamma\big(g(t)\big)x'(t)\\ &\quad + f(t)x'(t). \end{aligned} \tag{6.278}$$

Integrating (6.278) from a to c, we obtain

$$\begin{aligned}0 &= \int_a^c \big[r(t)\big(x''(t)\big)^2 - q(t)k\big(x'(t)\big)x'(t) - p(t)x^\gamma\big(g(t)\big)x'(t) + f(t)x'(t)\big]\,dt\\ &\ge \int_a^c \big[f(t) - p(t)x^\gamma\big(g(t)\big)\big]x'(t)\,dt\\ &\ge \int_a^c \big[f(t) - M^\gamma\big|p(t)\big|\big]x'(t)\,dt > 0,\end{aligned}$$

a contradiction. One may obtain a similar contradiction, when $x(t)$ is of nonpositive z-type.

Finally, suppose that $x(t)$ is oscillatory with consecutive zeros at a, b and a_1 ($t_2 < a < b < a_1$) such that $x'(a) \le 0, x'(b) \ge 0, x'(a_1) \le 0$ and $x(t) < 0$ for $t \in (a, b)$ and $x(t) > 0$ for $t \in (b, a_1)$. Therefore, there exist points $c \in (a, b)$ and $c_1 \in (b, a_1)$ such that $x'(c) = 0$, $x'(c_1) = 0$, $x'(t) > 0$ for $t \in (c, b)$ and $x'(t) > 0$ for $t \in (b, c_1)$. Now integrating (6.278) from c to c_1, we get

$$\begin{aligned}0 &= \int_c^{c_1} \big[r(t)\big(x''(t)\big)^2 - q(t)k\big(x'(t)\big)x'(t) - p(t)x^\gamma\big(g(t)\big)x'(t) + f(t)x'(t)\big]\,dt\\ &\ge \int_c^b \big[f(t) - p(t)x^\gamma\big(g(t)\big)\big]x'(t)\,dt + \int_b^{c_1} \big[f(t) - p(t)x^\gamma\big(g(t)\big)\big]x'(t)\,dt\\ &\ge \int_b^c \big[f(t) - \big|p(t)\big|\big|x^\gamma\big(g(t)\big)\big|\big]x'(t)\,dt + \int_b^{c_1} \big[f(t) - \big|p(t)\big|\big|x^\gamma\big(g(t)\big)\big|\big]x'(t)\,dt\\ &\ge \int_c^b \big[f(t) - M^\gamma\big|p(t)\big|\big]x'(t)\,dt + \int_b^{c_1} \big[f(t) - M^\gamma\big|p(t)\big|\big]x'(t)\,dt > 0,\end{aligned}$$

a contradiction. Hence $x(t)$ must be oscillatory and this completes the proof. □

Theorem 6.9.2 *let $0 \leq p(t) < f(t)$ and $q(t) \leq 0$. If $x(t)$ is a solution of* (6.276) *which satisfies the inequality*

$$1 - z^{\gamma}\big(g(t)\big) \geq 0$$

on any interval where $x'(t) > 0$, then $x(t)$ is nonoscillatory.

One may find several criteria for nonoscillation of (6.276) for the particular case $k(x') = (x')^{\beta}$ and $x^{\gamma}(g(t)) = x'(t)$. One may refer to Chap. 5 of this monograph for some important results.

Theorem 6.9.3 *Suppose that $q(t) \leq 0$ and $0 \leq p(t) < f(t)$. If $x(t)$ is a solution of* (6.277) *such that it satisfies the inequality*

$$1 - h\big(z\big(g(t)\big)\big) > 0$$

on any interval where $x'(t) > 0$, then $x(t)$ is nonoscillatory.

Theorem 6.9.4 *Suppose that $0 \leq q(t) \leq p(t) \leq f(t)$ and $q \neq 0$ on any subinterval $[T_x, \infty)$, $T_x \geq 0$. If $x(t)$ is a solution of* (6.277) *satisfying the inequality*

$$1 - k\big(z'\big) - h(z) > 0$$

on any subinterval of $[T_x, \infty)$, $T_x \geq 0$, where $x'(t) > 0$, then $x(t)$ is nonoscillatory.

Theorem 6.9.5 *Let $p(t) \geq 0$, $q(t) \leq 0$ and $h(x) \leq x$ for all $x > 0$. If $p(t)$ and $f(t)$ are once continuously differentiable functions such that $p'(t) \geq 0$, $f'(t) \leq 0$ and $2f(t) - p(t) \geq 0$, then all solutions $x(t)$ of* (6.277) *for which $|x(t)| < 1$ ultimately are nonoscillatory.*

6.10 Open Problems and Discussions

- One may find some sufficient conditions for the existence of nonoscillatory solutions of (2.1) satisfying the property (6.51) and existence of nonoscillatory solution of (6.46c) with $g(t) \equiv t$ satisfying the property (6.57) in Chap. 2 of this book. However, no such result exists, that is, the existence of nonoscillatory solutions of (6.1) or (6.46c) can be found in the literature. Thus, it would be interesting to find some sufficient conditions of the existence of nonoscillatory solutions of (6.1) or (6.46c) with $g(t) < t$. It seems that one has to apply some fixed-point theorems to find the results by transforming the third-order equations into a system of first-order equations, or writing the equation into an equivalent integral equation.
- The asymptotic behaviour of solutions of (6.1) with $a(t) \leq 0$, $b(t) \leq 0$ and $c(t) < 0$ has not been studied directly. It has been studied by using some canonical transformations due to Trench, as discussed in Sect. 6.1 of this chapter. It

would be interesting to study the asymptotic behaviour of the solutions of (6.1) directly. Further, when $a(t) \leq 0, b(t) \leq 0$ and $c(t) < 0$, then no result exists as regards finding sufficient conditions under which (6.1) with $g(t) < t$ admits a nonoscillatory solution satisfying the property (6.67). This still remains an open problem.

- Obtain oscillation criteria for (6.263), without requiring conditions that g and ξ are nondecreasing, $\xi(t) > t$ and $g(\xi(\xi(t))) < t$.
- Obtain results similar to Theorems 6.5.1–6.5.4, the lemmas and corollaries given between them to cover the case $\int_\sigma^\infty dt/r^{1/\gamma}(t) < \infty$.

6.11 Notes

Theorem 6.1.1 is due to Padhi [28]. Theorems 6.1.2–6.1.5 are taken from Džurina and Komarikova [13]. Theorems 6.1.6 and 6.1.7 are due to Džurina and Kotorová [15]. Theorems 6.1.8–6.1.14 are the consequences of some comparison results due to Padhi [27]. A particular case, that is, (6.2) with $r(t) \equiv 1$ in Theorems 6.1.8–6.1.14, can be found in [4] and [15]. Theorems 6.1.15–6.1.22 are taken from [27, 31]. Theorems 6.1.23–6.1.35 are due to Parhi and Padhi [32]. Theorem 6.1.36 and 6.1.37 are taken from Džurina [11]. The proof of Theorems 6.1.39–6.1.42 can be found in [14]. Theorems 6.2.1–6.2.4 are due to Parhi and Padhi [31, 32], and their generalisations can be found in [33]. Theorem 6.2.5 is a particular case of a result due to Fink and Kusano [17]. Theorem 6.2.6 is proved by Džurina [10]. Theorems 6.2.8–6.2.12 are proved by Džurina [12]. Theorems 6.2.13–6.2.17 are taken from Džurina [8]. Theorem 6.2.18 is due to Philos and Sficas [35]. Theorems 6.2.19 and 6.2.20 can be found in [9]. Theorems 6.2.21–6.2.24 are taken from [11]. Theorem 6.2.25 is due to Padhi [27, 33]. Results of Sect. 6.3 are taken from Padhi [29] and Parhi and Padhi [34]. Theorems 6.4.1–6.4.3 are due to Tiryaki and Yamen [41]. Theorems 6.5.1–6.5.4 are taken from Saker and Džurina [37]. Theorem 6.5.5 is due to Li et al. [24]. Theorems 6.5.6–6.5.9 are proved in [3]. Theorem 6.6.1 can be found in [39]. Theorem 6.6.2 is due to Aktas, Tiryaki and Zafer [2]. Theorems 6.6.3–6.6.11 are proved by Agarwal et al. [1]. Lemmas 6.7.1–6.7.7 and Theorems 6.7.1–6.7.3 are taken from Padhi and Dix [30], whereas Theorem 6.7.4–6.7.5 are due to Candan and Dahiya [5]. Theorems 6.8.3 and 6.8.4 are proved in [36]. Theorem 6.8.5 to Theorem 6.8.7 can be found in [25]. Theorem 6.8.8 to Corollary 6.8.3 are proved by Mojsej [26]. Section 6.9 is taken from Tunc [43].

References

1. R. P. Agarwal, M. F. Aktas and A. Tiryaki; *On oscillation criteria for third order nonlinear delay differential equations*, Archivum Mathematicum, 45(1) (2009), 1–18.
2. M. F. Aktas, A. Tiryaki and A. Zafer; *Oscillation criteria for third-order nonlinear functional differential equations*, Applied Mathematics Letters, 23(7) (2010), 756–762.

3. B. Baculiková and J. Džurina; *Oscillation of third order functional differential equations*, Electronic Journal of Qualitative Theory of Differential Equations, 43 (2010), 1–10.
4. B. Baculiková and J. Džurina; *Comparison theorems for the third-order delay trinomial differential equations*, Advances in Difference Equations, Vol. 2010, Article ID 160761, 12 pages.
5. T. Candan and R. S. Dahiya; *Oscillation of third order functional differential equations with delay*, Fifth Mississippi Conference on Differential Equations and Computational Simulations, Electronic Journal of Differential Equations, 10 (2003), 79–88.
6. J. M. Dolan; *On the relationship between the oscillatory behaviour of a linear third-order differential equation and its adjoint*, Journal of Differential Equations, 7 (1970), 367–388.
7. J. Džurina; *Comparison theorems for nonlinear ODEs*, Mathematica Slovaca, 42 (1992), 299–315.
8. J. Džurina; *Asymptotic properties of third-order differential equations with deviating argument*, Czechoslovak Mathematical Journal, 44 (1994), 163–172.
9. J. Džurina; *Property (B) of differential equations with deviating argument*, Archivum Mathematicum, (BRNO) 30(3) (1994), 207–213.
10. J. Džurina; *A comparison theorem for linear delay differential equations*, Archivum Mathematicum, (BRNO) 31(2) (1995), 113–120.
11. J. Džurina; *Asymptotic properties of third order delay differential equations*, Czechoslovak Mathematical Journal, 45 (1995), 443–448.
12. J. Džurina; *Comparison theorems for differential equations with deviating argument*, Mathematica Slovaca, 45(3) (1995), 79–89.
13. J. Džurina and R. Komariková; *Asymptotic properties of third-order delay trinomial differential equations*, Abstract and Applied Analysis, Vol. 2011 (2011), 10 pages.
14. J. Džurina and R. Kotorová; *Asymptotic properties of trinomial delay differential equations*, Archivum Mathematicum, 44(2) (2008), 149–158.
15. J. Džurina and R. Kotorová; *Properties of the third order trinomial differential equations with delay argument*, Nonlinear Analysis; Theory Methods and Applications, 71(5–6) (2009), 1995–2002.
16. L. Erbe, Q. Kong and B. G. Zhang; *Oscillation Theory for Functional Differential Equations*, Marcel Dekker Inc., New York, 1995.
17. A. M. Fink and T. Kusano; *Nonoscillation theorems for differential equations with general deviating arguments*, Lecture Notes in Mathematics #1032, 224–239, Springer, Berlin, 1983.
18. M. Greguš and M. Gera; *Some results in the theory of a third order linear differential equation*, Annales Polonici Mathematici, 42 (1983), 93–101.
19. M. Hanan; *Oscillation criteria for third order linear differential equations*, Pacific Journal of Mathematics, 11 (1961), 919–944.
20. E. Hille; *Non-oscillation theorems*, Transactions of the American Mathematical Society, 64 (1948), 234–252.
21. I. T. Kiguradze and T. A. Chanturia; *Asymptotic Properties of Solutions of Nonautonomous Ordinary Differential Equations*, Kluwer Academic Publishers, Dordrecht, 1993.
22. T. Kusano and M. Naito; *Comparison theorems for functional differential equations with deviating arguments*, Journal of the Mathematical Society of Japan, 3 (1981), 509–532.
23. T. Kusano, M. Naito and K. Tanaka; *Oscillatory and asymptotic behaviour of solutions of a class of linear ordinary differential equations*, Proceedings of the Royal Society of Edinburgh, 90A (1981), 25–40.
24. T. Li, C. Zhang, B. Baculiková and J. Džurina; *On the oscillation of third-order quasi-linear delay differential equations*, Tatra Mountains Mathematical Publications, 48 (2011), 117–123.
25. I. Mojsej; *Comparison theorems for third order nonlinear differential equations with deviating argument*, Tatra Mountains Mathematical Publications, 38 (2007), 201–213.
26. I. Mojsej; *Asymptotic properties of solutions of third order nonlinear differential equations with deviating arguments*, Nonlinear Analysis; Theory Methods and Applications, 68 (2008), 3581–3591.
27. S. Padhi; *Contributions to the Oscillation Theory of Ordinary and Delay Differential Equations of Third Order*, Ph.D. Thesis, Berhampur University, India, 1998.

28. S. Padhi; *On the asymptotic behaviour of solutions of third order delay differential equations*, Georgian Mathematical Journal, 12 (2005), 369–376.
29. S. Padhi; *Asymptotic behaviour of solutions to n-th order functional differential equations*, Electronic Journal of Differential Equations, 2005(65) (2005), 1–14.
30. S. Padhi and J.G. Dix; *Oscillation of functional differential equations of n-th order with distributing deviating arguments*, Journal of Contemporary Mathematics, 1 (2008), 11–24.
31. N. Parhi and S. Padhi; *On asymptotic behaviour of delay-differential equations of third order*, Nonlinear Analysis; Theory Methods and Applications, 34 (1998), 391–403.
32. N. Parhi and S. Padhi; *Asymptotic behaviour of a class of third order delay differential equations*, Mathematica Slovaca, 50 (2000), 315–333.
33. N. Parhi and S. Padhi; *Asymptotic behaviour of solutions of delay differential equations of n-th order*, Archivum Mathematicum, 37(2) (2001), 81–101.
34. N. Parhi and S. Padhi; *Asymptotic behaviour of solutions of third order delay differential equations*, Indian Journal of Pure and Applied Mathematics, 33 (2002), 1609–1620.
35. Ch. G. Philos and Y. G. Sficas; *Oscillatory and asymptotic behaviour of second and third order retarded differential equations*, Czechoslovak Mathematical Journal, 32 (1982), 169–182.
36. S.H. Saker; *Oscillation criteria of third-order nonlinear delay differential equations*, Mathematica Slovaca, 56(4) (2006), 433–450.
37. S.H. Saker and J. Džurina; *On the oscillation of certain class of third-order nonlinear delay differential equations*, Mathematica Bohemica, 135(3) (2010), 225–237.
38. C. A. Swanson; *Comparison and Oscillation Theory of Linear Differential equations*, Academic Press, New York, 1968.
39. A. Tiryaki and M. F. Aktas; *Oscillation criteria of a certain class of third order nonlinear delay differential equations with damping*, Journal of Mathematical Analysis and Applications, 325(1) (2007), 54–68.
40. A. Tiryaki and S. Yaman; *Oscillatory behaviour of a class of nonlinear differential equations of third order*, Acta Mathematica Scientia, 21(B) (2001), 182–188.
41. A. Tiryaki and S. Yaman; *Asymptotic behaviour of a class of nonlinear functional differential equations of third order*, Applied Mathematics Letters, 14(3) (2001), 327–332.
42. W. F. Trench; *Canonical forms and principal systems for general disconjugate equations*, Transactions of the American Mathematical Society, 189 (1974), 319–327.
43. E. Tunc; *New results on the nonoscillation of solutions of some nonlinear differential equations of third order*, Journal of Inequalities and Applications, Vol. 2009, Article ID 896934, 11 pages.

Chapter 7
Stability of Third-Order Differential Equations

"Stability theory" is a very important problem in the theory and applications of differential equations. The most effective method to study the stability of nonlinear differential equation is "Lyapunov's direct method". Many stability results have been established on the second-order nonlinear differential equations. However, the stability of third-order linear and nonlinear differential equations are relatively scarce in the literature. This is perhaps due to the construction of a suitable Lyapunov function.

In this chapter, we shall give some results on the asymptotic behaviour of solutions of the unperturbed and the perturbed third-order nonlinear differential equations

$$x''' + \psi(x, x')x'' + f(x, x') = 0 \tag{7.1}$$

and

$$x''' + \psi(x, x')x'' + f(x, x') = p(t), \tag{7.2}$$

where $\psi, f, \psi_x, f_x \in C(R \times R, R)$ and $p \in C([0, \infty), R)$.

This chapter has been divided into four sections. Section 7.1 deals with the stability of Eq. (7.1), whereas the stability of (7.2) is dealt with in Sect. 7.2. Stability of equations of the form

$$x''' + \psi(x, x', x'')x'' + f(x, x') = p(t, x, x', x'') \tag{7.3}$$

is considered in Sect. 7.3, where

$$\begin{aligned} &\psi \in C(R \times R \times R, R), \qquad f \in C(R \times R, R) \quad \text{and} \\ &p \in C\big([0, \infty) \times R \times R \times R, R\big). \end{aligned} \tag{7.4}$$

On the way, we shall give some new results on the stability of zero solutions of the autonomous equation

$$x'''(t) = p_1 x''(t) + p_2 x''(t - \tau) + q_1 x'(t) + q_2 x'(t - \tau) + r_1 x(t) + r_2 x(t - \tau), \quad t \geq 0 \tag{7.5}$$

S. Padhi, S. Pati, *Theory of Third-Order Differential Equations*,
DOI 10.1007/978-81-322-1614-8_7, © Springer India 2014

with the initial condition

$$x(t)=\phi(t), \quad t\in[-\tau,0], \tag{7.6}$$

where p_1, p_2, q_1, q_2, r_1 and r_2 are real constants, $\tau>0$ is a real number and $\phi\in C([-\tau,0),R)$ is an initial function. Results related to Eq. (7.5) is incorporated in Sect. 7.4.

7.1 Stability of Solutions of $x'''+\psi(x,x')x''+f(x,x')=0$

We shall give some results on the stability of solutions of (7.1) in this section. We start with the following theorem.

Theorem 7.1.1 *Assume that*

(i) $f(x,0)x>0$ *for* $x\neq 0$,
(ii) $\int_0^y f(0,v)\,dv>0$ *for* $y\neq 0$,
(iii) $y\psi_x(x,y)\le 0$ *for* $y\neq 0$

and there is a positive number B such that

(iv) $\psi(x,y)\ge B$,
(v) $B[f(x,y)-f(x,0)]y>y\int_0^y f_x(x,v)\,dv$ *for* $y\neq 0$,
(vi) $4B\int_0^x f(u,0)\,du\int_0^y[f(x,v)-f(x,0)]\,dv-y^2f^2(x,0)>0$ *for* $xy\neq 0$,
(vii) $\lim_{|x|\to\infty}\{B\int_0^x f(u,0)\,du+\int_0^y f(x,v)\,dv\}=\infty$, *for all fixed y*.

Then the trivial solution of Eq. (7.1) *is globally asymptotically stable.*

Theorem 7.1.2 *Assume that*

(1) $xf(x,0)>0$ *for* $x\neq 0$,
(2) $\int_0^y f(0,v)\,dv>0$ *for* $y\neq 0$,
(3) $\lim_{|x|\to\infty}\int_0^x f(u,0)\,du=\infty$

and there is a positive number B such that

(4) $\psi(x,y)\ge B$,
(5) $B[f(x,y)-f(x,0)-\int_0^y\psi_x(x,v)v\,dv]y\ge y\int_0^y f_x(x,v)\,dv$,
(6) $B[f(x,y)-f(x,0)-\int_0^y\psi_x(x,v)v\,dv]y+\psi(x,y)>\int_0^y f_x(x,v)\,dv+B$ *for* $y\neq 0$,
(7) $4B\int_0^x f(u,0)\,du\{\int_0^y[f(x,v)-f(x,0)]\,dv+B\int_0^y[\psi(x,v)-B]v\,dv\}>y^2f^2(x,0)>0$, *for* $xy\neq 0$.

Then the trivial solution of Eq. (7.1) *is globally asymptotically stable.*

Proof Clearly, Eq. (7.1) is equivalent to the system

$$x'=y, \qquad y'=z, \qquad z'=-f(x,y)-\psi(x,y)z \tag{7.7}$$

and (0, 0, 0) is a solution of (7.7). Now, consider the Lyapunov function

$$V(x, y, z) = BF(x) + yf(x, 0) + \phi(x, y) + \frac{(z + By)^2}{2} + B\int_0^y \big(\psi(x, v) - B\big)v\,dv, \tag{7.8}$$

where

$$F(x) = \int_0^x f(u, 0)\,du \quad \text{and} \quad \phi(x, y) = \int_0^y \big[f(x, v) - f(x, 0)\big]\,dv.$$

Observe that

$$V'_{2.1}(x, y, z) = y\left\{B\left[f(x, 0) - f(x, y) + \int_0^y \psi_x(x, v)v\,dv\right] + \int_0^y f_x(x, v)\,dv\right\} + \big[B - \psi(x, y)\big]z^2. \tag{7.9}$$

From the hypotheses, we see that $V'_{2.1}(x, y, z) \le 0$.

In the following, we are going to show that V is a positive definite function, every positive semi-trajectory of (7.7) is bounded, and the set $\{(x, y, z) : V'_{2.1}(x, y, z) = 0\}$ does not contain the whole trajectory of any solution of (7.1) except the trivial solution (0, 0, 0).

First, we show that V is a positive definite function. Observe that when $x = 0$,

$$V(0, y, z) = \phi(0, y) + \frac{(z + By)^2}{2} + B\int_0^y \big(\psi(0, v) - B\big)v\,dv. \tag{7.10}$$

Hence, in view of (2) and (4), it is easy to see that $V(0, y, z)$ is a positive definite function of y and z. Next, assume that $x \ne 0$. Then $F(x) > 0$ and so $V(x, y, z)$ can be written in the form

$$V(x, y, z) = \left[\sqrt{BF(x)} + \frac{yf(x, 0)}{2\sqrt{BF(x)}}\right]^2 + \frac{(z + By)^2}{2} + \left[\phi(x, y) + B\int_0^y \big(\psi(x, v) - B\big)v\,dv - \frac{y^2 f^2(x, 0)}{4BF(x)}\right]. \tag{7.11}$$

If $y = 0$, then

$$V(x, 0, z) = BF(x) + \frac{z^2}{2} \ge BF(x) > 0.$$

If $y \ne 0$, then

$$V(x, y, z) \ge \phi(x, y) + B\int_0^y \big(\psi(x, v) - B\big)v\,dv - \frac{y^2 f^2(x, 0)}{4BF(x)},$$

which, in view of (7), implies that $V(x, y, z) > 0$. Hence, V is a positive definite function.

Next, we show that the positive semi-trajectory of every solution of (7.7) is bounded. To this end, consider the set

$$D(r, M, N) = \left\{(x, y, z) | V(x, y, z) < r^2,\ x^2 < M^2,\ y^2 < N^2\right\},$$

where r is any fixed positive number and $M = M(r)$ and $N = N(r)$ are positive constants which satisfy

$$BF(\pm M) > r^2 \tag{7.12}$$

and

$$N > \frac{\sqrt{2}r}{B}, \tag{7.13}$$

respectively. The existence of M is guaranteed by (3). From the expression $V(x, y, z)$, we see that every point $(x, y, z) \in D$ satisfies the condition

$$(z + By)^2 \le 2r^2, \tag{7.14}$$

which implies that z is bounded since y is bounded. Hence, D is a bounded set. Now, we show that any positive semi-trajectory Γ^+ of (7.7) setting off from a point in D cannot go out of D. By noting the fact that $V'_{(1,2)}(x, y, z) \le 0$, we see that if Γ^+ goes out of D, it must go through either σ_1, or σ_2, or σ_3, or σ_4 where $\sigma_1, \sigma_2, \sigma_3$ and σ_4 are the segments of the planes $x = M, x = -M$, $y = N$ and $y = -N$ on the boundary δD and D, respectively. We claim that this is impossible, too. First, we show that Γ^+ cannot go through σ_1, or σ_2. Observe that when $(x, y, z) \in \sigma_1 \cup \sigma_2$, $x = M$ or $x = -M$ and satisfies

$$\left|\sqrt{BF(x)} + \frac{yf(x,0)}{2\sqrt{BF(x)}}\right| \le r. \tag{7.15}$$

Hence, by noting (7.12), we see that $yf(x, 0)|_{(x,y,z)\in\sigma_1\cup\sigma_2} < 0$. Then it follows that

$$xy\frac{f(x,0)}{x}\bigg|_{(x,y,z)\in\sigma_1\cup\sigma_2} < 0,$$

which, in view of (1), implies that $xy|_{(x,y,z)\in\sigma_1\cup\sigma_2} < 0$. Therefore,

$$\frac{dx^2}{dt}\bigg|_{(x,y,z)\in\sigma_1\cup\sigma_2} = 2xy < 0.$$

From this fact, we see that the direction field of (7.7) enters D on σ_1 and σ_2 and so Γ^+ cannot go through σ_1 or σ_2.

Next, we show that Γ^+ cannot go through either σ_3 or σ_4. In fact, from (7.13) and (7.14), we see that if there is a moment T such that $|y(t)| = N$, then $z(T)y(T) < 0$. By noting $y'(T) = z(T)$, it follows that $y'(T)y(T) < 0$ which implies that the direction field of (7.7) enters D on σ_3 and σ_4. Hence, Γ^+ cannot go through σ_3 or σ_4.

Therefore, any positive semi-trajectory of (7.7) setting off from a point in D cannot go out of D and so is bounded. Since, for any bounded region Ω in the space, we can choose a sufficient large r such that $\Omega \subset D$, it follows that every positive semi-trajectory is bounded.

Finally, we show that the set

$$\Lambda = \left\{(x, y, z) : V'_{(2,1)}(x, y, z) = 0\right\}$$

does not contain the whole trajectory of any solution of (7.7), except the trivial solution $(0, 0, 0)$. Assume that $(x(t), y(t), z(t))$ is a solution of (7.7) which satisfies

$$V'_{(2,1)}\big(x(t), y(t), z(t)\big) = 0 \quad \text{for all } t. \tag{7.16}$$

We claim that $y(t) \equiv 0$. Otherwise, by the continuity of $y(t)$, there are T_1 and T_2 with $T_1 < T_2$ such that $y(t) \neq 0$ for $t \in (T_1, T_2)$. Then in view of (4), (5) and (7.9), it follows from (7.16) that

$$B\left[f\big(x(t), 0\big) - f\big(x(t), y(t)\big) + \int_0^{y(t)} \psi_x(x, v)|_{x=x(t)}\, v\, dv\right] + \int_0^{y(t)} f_x(x, v)|_{x=x(t)}\, dv \equiv 0 \quad \text{for } t \in (T_1, T_2).$$

Hence, by noting (6), we find that

$$\psi\big(x(t), y(t)\big) - B > 0 \quad \text{for } t \in (T_1, T_2)$$

and so (7.16) yields $z(t) \equiv 0$ for $t \in (T_1, T_2)$. Then it follows from (7.7) that

$$y(t) \equiv c \quad \text{and} \quad x(t) = ct + d \quad \text{for } t \in (T_1, T_2),$$

where c is nonzero constant and d is a constant. By using the continuation theorem of solutions, it is easy to see that

$$x(t) = ct + d, \qquad y(t) = c, \qquad z(t) = 0 \quad \text{for } t \in (-\infty, \infty).$$

However, it has been shown the positive trajectory of every solution of (7.7) is bounded. Hence $c = 0$, which is a contradiction and so $y(t) \equiv 0$ for $t \in (-\infty, \infty)$. Then, it follows from (7.7) that

$$x(t) \equiv d, \qquad z(t) \equiv 0 \quad \text{for } t \in (-\infty, \infty) \quad \text{and} \quad f(d, 0) = 0,$$

where d is a constant which, in view of (1), implies that $d = 0$ and so $x(t) \equiv 0$ for $t \in (-\infty, \infty)$. Therefore, the set Λ does not contain the whole trajectory of any solution of (7.7), except the trivial solution $(0, 0, 0)$.

Hence, by the Barbashin–Krasovskii theorem, the trivial solution of (7.7) is globally asymptotically stable. The proof is complete. □

Now, we apply Theorem 7.1.2 to some special cases of (7.1). First, we consider the third-order differential equation

$$x''' + \psi\big(x, x'\big)x'' + \phi\big(x'\big) + g(x) = 0, \tag{7.17}$$

where

$$\psi, \psi_x \in C[R \times R, R] \quad \text{and} \quad \phi, g' \in C[R, R].$$

Theorem 7.1.3 *Assume that*

(i) $g(x)/x \ge c_0 > 0$ *for* $x \ne 0$, $g'(x) \le c$;
(ii) $\phi(y)/y \ge b > 0$ *for* $y \ne 0$;
(iii) $\psi(x, y) \ge a > c/b$;
(iv) $y\psi_x(x, y) \le 0$ *for* $y \ne 0$,

where a, b, c *and* c_0 *are positive constants. Then the trivial solution of* (7.17) *is globally asymptotically stable.*

Theorem 7.1.4 *Assume that*

(i) $xg(x) > 0$ *for* $x \ne 0$;
(ii) $[a\phi(y) - g'(x)y]y > 0$ *for* $y \ne 0$;
(iii) $\psi(x, y) \ge a > 0$, $y\psi_x(x, y) \le 0$;
(iv) $U(x, y) = a\int_0^x g(u)\,du + yg(x) + \int_0^y \phi(v)\,dv$ *positive definite and* $\lim_{x^2+y^2\to\infty} U(x, y) = \infty$,

where a *is a positive constant. Then the trivial solution of Eq.* (7.17) *is globally asymptotically stable.*

However, by noting that Eq. (7.17) is a special case of (7.1) with $f(x, y) = g(x) + \phi(y)$ and by applying Theorem 7.1.3, we have the following corollary.

Corollary 7.1.1 *Assume that*

(1) $xg(x) > 0$ *for* $x \ne 0$,
(2) $\int_0^y \phi(v)\,dv > 0$ *for* $y \ne 0$,
(3) $\lim_{|x|\to\infty} \int_0^x g(u)\,du = \infty$

and there is a positive number B *such that*

(4) $\psi(x, y) \ge B$,
(5) $B[\phi(y) - \int_0^y \psi_x(x, v)v\,dv]y \ge g'(x)y^2$,
(6) $\psi(x, y) + B[\phi(y) - \int_0^y \psi_x(x, v)v\,dv]y > B + g'(x)y^2$ *for* $y \ne 0$,
(7) $4B\int_0^x g(u)\,du\{\int_0^y \phi(v)\,dv + B\int_0^y [\psi(x, v) - B]v\,dv\} > g^2(x)y^2$ *for* $xy \ne 0$.

Then the trivial solution of Eq. (7.17) *is globally asymptotically stable.*

We claim that both Theorems 7.1.3 and 7.1.4 are direct consequences of Corollary 7.1.1. Let us consider Theorem 7.1.3 first. Suppose all the hypotheses of Theorem 7.1.3 hold. It suffices to show that all the hypotheses in Corollary 7.1.1 hold, too. Clearly, (i) and (ii) imply (1)–(3). Take $B = c/b$. Then (iii) implies (4). By noting (ii) and (iv), we see that $(\int_0^y \psi_x(x, v)v\,dv)y \leq 0$ and

$$B\left[\phi(y) - \int_0^y \psi_x(x, v)v\,dv\right]y \geq B\phi(y)y \geq Bby^2 = cy^2 \geq g'(x)y^2.$$

So (5) holds. Then, (6) follows from (5) and the fact $\psi(x, y) > B$. From (i), we see that

$$g'(x)g(x) \geq cg(x) \quad \text{for } x > 0 \quad \text{and} \quad g'(x)g(x) \leq cg(x) \quad \text{for } x < 0$$

and so it follows that

$$\int_0^x g(u)\,du \geq \frac{1}{2}g^2(x) > 0 \quad \text{for } x \neq 0.$$

In addition, observe that (ii) implies that

$$\int_0^y \phi(v)\,dv \geq \frac{1}{2}by^2.$$

Hence

$$\begin{aligned} &4B\int_0^x g(u)\,du\left[\int_0^y \phi(v)\,dv + B\int_0^y \big[\psi(x, v) - B\big]v\,dv\right] \\ &> 4B\int_0^x g(u)\,du\int_0^y \phi(v)\,dv \\ &\geq g^2(x)y^2 \quad \text{for } xy \neq 0 \end{aligned}$$

and so (7) holds. Therefore, all the hypotheses of Corollary 7.1.1 hold.

Next, assume that all the hypotheses of Theorem 7.1.4 hold. We show that all the hypotheses in Corollary 7.1.1 hold, too. Clearly, (1)–(3) follow from (i) and (iv) immediately. Now take $B = a$, and so (iii) implies (4). Then by noting (ii) and (iii), we see that

$$B\left[\phi(y) - \int_0^y \psi_x(x, v)v\,dv\right]y \geq B\phi(y)y > g'(x)y^2 \quad \text{for } y \neq 0,$$

which implies (5) and (6). Clearly (ii) yields

$$a\int_0^y \phi(v)\,dv > \frac{1}{2}g'(x)y^2 \quad \text{for } y \neq 0.$$

Hence, by noting (i), also

$$ag(x)\int_0^y \phi(v)\,dv > \frac{1}{2}g(x)g'(x)y^2 \quad \text{for } x>0 \text{ and } y\neq 0$$

and

$$ag(x)\int_0^y \phi(v)\,dv < \frac{1}{2}g(x)g'(x)y^2 \quad \text{for } x<0 \text{ and } y\neq 0.$$

Then it follows that

$$a\int_0^x g(u)\,du\int_0^y \phi(v)\,dv > \frac{1}{4}g^2(x)y^2 \quad \text{for } xy\neq 0,$$

which implies that (7) holds. Hence, all the hypotheses in Corollary 7.1.1 hold.

Example 7.1.1 Consider the equation

$$x''' + \left[(\sin x)x' + \left(x'\right)^2 + 2\right]x'' + \left(x'\right)^3 + x' + \frac{x}{1+x^2} = 0. \tag{7.18}$$

Equation (7.18) is in the form (7.17) with

$$\psi(x,y) = (\sin x)y + y^2 + 2, \qquad g(x) = \frac{x}{1+x^2}, \qquad \phi(y) = y^3 + y.$$

Take $B = 1$. Observe that

$$\begin{aligned}\left[\phi(y) - \int_0^y \psi_x(x,v)v\,dv\right]y &= \left[y^3 + y - \frac{1}{3}(\cos x)y^3\right]y \\ &> y^2\frac{1-x^2}{(1+x^2)^2} = y^2 g'(x) \quad \text{for } y\neq 0\end{aligned}$$

and

$$\begin{aligned}4\int_0^x g(u)\,du\int_0^y \phi(v)\,dv &= 2\ln\left(1+x^2\right)\left(\frac{1}{4}y^4 + \frac{1}{2}y^2\right) \\ &> \frac{x^2}{(1+x^2)^2}y^2 \\ &= g^2(x)y^2 \quad \text{for } xy\neq 0.\end{aligned}$$

Then it is easy to check that all the hypotheses of Corollary 7.1.1 are satisfied and so the trivial solution of (7.18) is globally asymptotically stable. However, both Theorems 7.1.3 and 7.1.4 cannot be applied to (7.18). Indeed,

$$y\psi_x(x,y) = (\cos x)y^2$$

does not satisfy the condition $y\psi_x(x,y)\leq 0$.

When $g(x) = cx$ and $\phi(y) = by$, where b and c are positive constants, then Eq. (7.17) reduces to

$$x''' + \psi(x, x')x'' + bx' + cx = 0. \tag{7.19}$$

Applying Theorem 7.1.2 to (7.19), we have the following corollary:

Corollary 7.1.2 *Assume that*

(1) $b\psi(x, y) \geq c$,
(2) $y \int_0^y \psi_x(x, v)v\,dv \leq 0$,
(3) $\psi(x, y) - y \int_0^y \psi_x(x, v)v\,dv > c/b$ *for* $y \neq 0$,
(4) $\int_0^y [\psi(x, v) - \frac{c}{b}]v\,dv > 0$ *for* $y \neq 0$.

Then the trivial solution of (7.19) *is globally asymptotically stable.*

Next, consider the third-order differential equation

$$x''' + h(x')x'' + \phi(x') + g(x) = 0, \tag{7.20}$$

where

$$h,\ \phi,\ g \text{ and } g' \in C[R, R].$$

Equation (7.20) is a special case of (7.17) with $\psi(x, y) = h(y)$, and so by using Corollary 7.1.1 we can establish the following result.

Corollary 7.1.3 *Assume that*

(1) $xg(x) > 0$ *for* $x \neq 0$,
(2) $\lim_{|x|\to\infty} \int_0^x g(x)\,dx = \infty$,

and there is a positive number B such that

(3) $h(y) \geq B$,
(4) $B\phi(y)y \geq g'(x)y^2$,
(5) $h(y) + B\phi(y)y > B + g'(x)y^2$ *for* $y \neq 0$.

Then the trivial solution of (7.20) *is globally asymptotically stable.*

Proof Clearly, it suffices to show that (2) and (7) in Corollary 7.1.1 hold. From (1), we see that there is an $x_0 \neq 0$ such that $g'(x_0) > 0$. Hence, it follows from (4) that

$$B\phi(y)y \geq g'(x_0)y^2 > 0 \quad \text{for } y \neq 0$$

which implies that (2) in Corollary 7.1.1 holds. In view of (3)–(5), we see that

$$B\phi(y) + B^2\big(h(y) - B\big)y > g'y \quad \text{for } y > 0$$

and

$$B\phi(y) + B^2\big(h(y) - B\big)y < g'(x)y \quad \text{for } y < 0.$$

Then it follows that

$$B\left[\int_0^y \phi(v)\,dv + B\int_0^y \big[h(y) - B\big]y\,dy\right] > \frac{1}{2}g'(x)y^2 \quad \text{for } y \neq 0.$$

By (1), we see that

$$Bg(x)\left[\int_0^y \phi(v)\,dv + B\int_0^y \big[h(y) - B\big]y\,dy\right] > \frac{1}{2}g'(x)g(x)y^2$$
$$\text{for } x > 0 \text{ and } y \neq 0,$$

and

$$Bg(x)\left[\int_0^y \phi(v)\,dv + B\int_0^y \big[h(y) - B\big]y\,dy\right] < \frac{1}{2}g'(x)g(x)y^2$$

for $x < 0$ and $y \neq 0$. Hence

$$B\int_0^x g(u)\,du\left[\int_0^y \phi(v)\,dv + B\int_0^y \big[h(y) - B\big]y\,dy\right] > \frac{1}{4}g^2(x)y^2 \quad \text{for } xy \neq 0,$$

which implies that (7) in Corollary 7.1.1 holds. This completes the proof. □

Another particular case of Eq. (7.1) is the third-order equation

$$x''' + h\big(x'\big)x'' + \mu\big(x'\big)x' + k(x)x = 0, \tag{7.21}$$

where h, μ, k and $k_x \in C[R, R]$. Applying Corollary 7.1.3 to (7.21), we have the following result.

Corollary 7.1.4 *Assume that*

(1) $k(x) > 0$ *for* $x \neq 0$, $\lim_{|x|\to\infty} \int_0^x k(u)u\,du = \infty$,

and there is a positive number B *such that*

(2) $h(y) \geq B$,
(3) $B\mu(y) \geq (xk(x))'$,
(4) $h(y) + B\mu(y) > B + (xk(x))'$.

Then the trivial solution of (7.21) *is globally asymptotically stable.*

Example 7.1.2 Consider the equation

$$x''' + \left(1 + \frac{1}{1 + (x')^2}\right)x'' + x' + x = 0. \tag{7.22}$$

Equation (7.22) is in the form of (7.21) with $h(y) = 1 + \frac{1}{1+y^2}$, $\mu(y) \equiv 1$ and $k(x) \equiv 1$. By taking $B = 1$, it is easy to check that all the hypotheses of Corollary 7.1.4 are satisfied, and so the trivial solution of (7.22) is globally asymptotically stable.

7.2 Stability of Solutions of $x''' + \psi(x,x')x'' + f(x,x') = p(t)$

We shall provide some earlier interesting results on the global asymptotic stability of Eq. (7.2) in this section. Most of the results of this section are taken from Qian [3].

Theorem 7.2.1 *Assume that*

(1) $\int_0^x f(u,0)\,du > 0$ *for* $x \neq 0$,
(2) $\lim_{|x|\to\infty} \sup \int_0^x f(u,0)\,du = \infty$,
(3) $\int_0^y f(0,v)\,dv \geq 0$,
(4) $\int_0^\infty |p(t)|\,dt < \infty$,

and there is a positive number B *such that*

(5) $\psi(x,y) \geq B$,
(6) $B[f(x,y) - f(x,0) - \int_0^y \psi_x(x,v)v\,dv]y \geq y\int_0^y f_x(x,v)\,dv$,
(7) $4B\int_0^x f(u,0)\,du[\int_0^y [f(x,v) - f(x,0)]\,dv + B\int_0^y [\psi(x,v) - B]v\,dv] \geq y^2 f^2(x,0)$.

Then any solution $x(t)$ *of* (7.2), *there are constants* c_1, c_2 *and* c_3 *such that*

$$\left|x(t)\right| < c_1, \qquad \left|x'(t)\right| < c_2 \quad \text{and} \quad \left|x''(t)\right| < c_3 \quad \text{for } t \geq 0.$$

Proof Clearly, Eq. (7.2) is equivalent to the system

$$x' = y, \qquad y' = z, \qquad z' = -f(x,y) - \psi(x,y)z + p(t), \tag{7.23}$$

and it suffices to show that every solution of (7.23) is bounded. To this end, consider the function

$$V(t,x,y,z) = e^{-P(t)}\Bigg[BF(x) + yf(x,0)\phi(x,y) \\ + B\int_0^y \big(\psi(x,v) - B\big)v\,dv + \frac{(z+By)^2}{2} + 2\Bigg], \tag{7.24}$$

where

$$P(t) = \int_0^t \left|p(s)\right| ds, \qquad F(x) = \int_0^x f(u,0)\,du$$

and

$$\phi(x,y) = \int_0^y \big[f(x,v) - f(x,0)\big]\,dv.$$

We claim that V is a positive function. For the sake of convenience, let

$$G(x,y) = BF(x) + yf(x,0) + \phi(x,y) + B\int_0^y \big(\psi(x,v) - B\big)v\,dv, \tag{7.25}$$

and then it follows that

$$V(t,x,y,z) = e^{-P(t)}\left[G(x,y) + \frac{(z+By)^2}{2} + 2\right]. \tag{7.26}$$

Hence, to show that V is a positive function, it suffices to show that G is nonnegative. In fact, if $x=0$, then by noting (3), (5) and $f(0,0)=0$ (since (1) holds and f is continuous), we see that

$$G(0,y) = \phi(0,y) + B\int_0^y \big(\psi(0,v) - B\big)v\,dv \ge 0;$$

if $x \neq 0$, then by rewriting $G(x,y)$ in the form

$$\begin{aligned} G(x,y) = {} & \left[\sqrt{BF(x)} + \frac{yf(x,0)}{2\sqrt{BF(x)}}\right]^2 + \left[\phi(x,y) + B\int_0^y \big(\psi(x,v) - B\big)v\,dv \right. \\ & \left. - \frac{y^2 f^2(x,0)}{4BF(x)}\right], \end{aligned} \tag{7.27}$$

and by noting (7), we find that $G(x,y) \ge 0$ also. Hence, V is a positive function.

Next, let $(\frac{dV}{dt})|_{(7.23)}$ denote the derivative of V along the solutions of (7.23), and observe that

$$\begin{aligned} \left.\frac{dV}{dt}\right|_{(7.23)} = {} & e^{-P(t)}\left[-|p(t)|\left(G(x,y) + \frac{(z+By)^2}{2} + 2\right)\right. \\ & + y\left[B\left(f(x,0) - f(x,y) + \int_0^y \psi_x(x,v)v\,dv\right) + \int_0^y f_x(x,v)\,dv\right] \\ & \left. + \big[B - \psi(x,y)\big]z^2 + (z+By)p(t)\right]. \end{aligned} \tag{7.28}$$

We claim that

$$\left.\frac{dV}{dt}\right|_{(7.23)} \le 0. \tag{7.29}$$

Clearly, if $|z+By| < 2$, then $(z+By)p(t) \le 2|p(t)|$. If $|z+By| \ge 2$, then $(z+By)p(t) \le ((z+By)^2/2)|p(t)|$. Hence, for any t, x and y,

$$(z+By)p(t) \le \left(2 + \frac{(z+By)^2}{2}\right)|p(t)|$$

and so

$$\left.\frac{dV}{dt}\right|_{(7.23)} \le e^{-P(t)}\Bigg[-|p(t)|G(x,y) \\ + y\bigg[B\bigg(f(x,0) - f(x,y) + \int_0^y \psi_x(x,v)v\,dv\bigg) + \int_0^y f_x(x,v)\,dv\bigg] \\ + [B - \psi(x,y)]z^2\Bigg]. \tag{7.30}$$

Then by (5), (6) and $G(x,y) \ge 0$, we see that (7.29) holds.

Now, we are ready to show that every solution of (7.23) is bounded. Suppose that $(x(t), y(t), z(t))$ is a solution if (7.23) with the initial condition $(x(0), y(0), z(0)) = (x_0, y_0, z_0)$. First, we show that there are two positive constants L_1 and L_2 such that $-L_2 < x_0 < L_1$,

$$F(L_1) > \frac{e^{P(\infty)}}{B} V(0, x_0, y_0, z_0), \qquad f(L_1, 0) > 0 \tag{7.31}$$

and

$$F(-L_2) > \frac{e^{P(\infty)}}{B} V(0, x_0, y_0, z_0), \qquad f(-L_2, 0) < 0. \tag{7.32}$$

In fact, by (2), we see that there is a positive constant K such that

$$K > x_0 \quad \text{and} \quad F(K) > \frac{e^{-P(\infty)}}{B} V(0, x_0, y_0, z_0).$$

If $f(K,0) > 0$, then choose $L_1 = K$, and we are done. If $f(K,0) \le 0$, then by (2) again there must be a number $L_1 > K$ such that $f(L_1, 0) > 0$ and $f(L_1) \ge F(K)$. Hence, (7.31) holds. Similarly, we can show that there is a positive number $L_2 > 0$ such that $-L_2 < x_0$ and (7.32) holds.

Next, choose a positive number M such that

$$|y_0| < M \quad \text{and} \quad M > \frac{1}{B}\sqrt{2e^{P(\infty)} V(0, x_0, y_0, z_0)} \tag{7.33}$$

and then choose a positive number N such that

$$|z_0| < N \quad \text{and} \quad N \ge 2BM. \tag{7.34}$$

Now, consider the box

$$\Sigma = \big\{(x,y,z)| -L_2 \le x \le L_1,\ -M \le y \le M,\ -N \le z \le N\big\}.$$

In what follows, we show that

$$\big(x(t), y(t), z(t)\big) \in \Sigma \quad \text{for all } t \ge 0.$$

Let us use $\sigma_1, \sigma_2, \ldots, \sigma_6$ to represent the surfaces $x = L_1, x = -L_2$, $y = M$, $y = -M$, $z = N$ and $z = -N$ on Σ, respectively.

First, we show that $(x(t), y(t), z(t))$ cannot leave Σ through σ_1. Suppose that there exists a moment t_1 such that $x(t_1) = L_1$. Then it follows from (7.27) and (7) that

$$G(x, y) \geq \left[\sqrt{BF(L_1)} + \frac{y(t_1)f(L_1, 0)}{2\sqrt{BF(L_1)}}\right]^2.$$

Hence, (7.26) yields

$$V\big(t_1, x(t_1), y(t_1), z(t_1)\big) \geq e^{-P(\infty)}\left[\sqrt{BF(L_1)} + \frac{y(t_1)f(L_1, 0)}{2\sqrt{BF(L_1)}}\right]^2.$$

By (7.29) and (7.31), we have

$$y(t_1)f(L_1, 0) < 0$$

and so $y(t_1) < 0$. Then it follows from (7.23) that $\frac{dx(t_1)}{dt} = y(t_1) < 0$, which clearly implies that $(x(t), y(t), z(t))$ cannot exit Σ through σ_1. By using a similar argument, we can show that $(x(t), y(t), z(t))$ cannot leave Σ through σ_2 either.

Next, we show that $(x(t), y(t), z(t))$ cannot go out of Σ through σ_3 or σ_4. Suppose that there exists a moment t_2 such that $|y(t_2)| = M$. Observe that

$$V\big(t_2, x(t_2), y(t_2), z(t_2)\big) \geq e^{-P(\infty)}\frac{(z(t_2) + By(t_2))^2}{2}.$$

Then by (7.29) and (7.33), we see that $y(t_2)z(t_2) < 0$. Since $y'(t_2) = z(t_2)$, then from (7.23), it follows that $y'(t_2)y(t_2) < 0$, which clearly implies that $(x(t), y(t), z(t))$ cannot go out of Σ through σ_3 or σ_4.

Finally, we show that $(x(t), y(t), z(t))$ cannot leave Σ through σ_5 or σ_6. Suppose that there is a moment t_3 such that $|z(t_3)| = N$. Observe that

$$\big|z(t_3) + By(t_3)\big| \geq \big|z(t_3)\big| - \big|By(t_3)\big|$$

and since $|y(t_3)| \leq M$, (7.34) gives

$$\big|z(t_3) + By(t_3)\big| \geq N - BM \geq BM > \sqrt{2e^{P(\infty)}V(0, x_0, y_0, z_0)}.$$

Hence

$$V\big(t_3, x(t_3), y(t_3), z(t_3)\big) \geq e^{-P(\infty)}\frac{(z(t_3) + By(t_3))^2}{2} > V(0, x_0, y_0, z_0),$$

which contradicts (7.29), and so $(x(t), y(t), z(t))$ cannot leave Σ through σ_5 or σ_6. Therefore, $(x(t), y(t), z(t))$ will stay in Σ forever, and so it is bounded. The proof is now complete. □

Example 7.2.1 Consider Eq. (7.2) with $\psi(x, y) = \ln(1 + x^2) + 1$, $f(x, y) = \frac{x}{1+x^2}(1 + y^2) + y + \frac{1}{3}y^3$ and $p(t) = \frac{\sin t}{1+t^2}$. It is easy to check that the hypotheses

(1)–(4) of Theorem 7.2.1 are satisfied. Take $B = 1$. Since $\psi(x,y) \geq 1$ and

$$\begin{aligned}
&\left[f(x,y) - f(x,0) - \int_0^y \psi_x(x,v)v\,dv\right]y \\
&= \left(\frac{x}{1+x^2}y^2 + y + \frac{1}{3}y^3 - \frac{x}{1+x^2}y^2\right)y \\
&= y^2 + \frac{1}{3}y^4 \\
&\geq \frac{1-x^2}{(1+x^2)^2}\left(y^2 + \frac{1}{3}y^4\right) \\
&= y\int_0^y f_x(x,v)\,dv,
\end{aligned}$$

we see that (5) and (6) of Theorem 7.2.1 also hold. Now, observe that

$$\begin{aligned}
&4\int_0^x f(u,0)\,du\left[\int_0^y \left[f(x,v) - f(x,0)\right]dv + \int_0^y \left[\psi(x,v) - 1\right]v\,dv\right] \\
&= 2\ln\left(1+x^2\right)\left[\frac{x}{3(1+x^2)}y^3 + \frac{1}{2}y^2 + \frac{1}{12}y^4 + \frac{1}{2}\ln\left(1+x^2\right)y^2\right] \\
&= y^2\ln\left(1+x^2\right)\left[1 + h(x,y)\right],
\end{aligned} \tag{7.35}$$

where

$$h(x,y) = \ln\left(1+x^2\right) + \frac{2x}{3(1+x^2)}y + \frac{1}{6}y^2.$$

We claim that $h(x,y) \geq 0$. In fact, (0, 0) is the critical point of h. It is easy to see that h has the minimum value at (0, 0). Hence $h(x,y) \geq h(0,0) = 0$. Then it follows from (7.35) that

$$\begin{aligned}
&4\int_0^x f(u,0)\,du\left[\int_0^y \left[f(x,v) - f(x,0)\right]dv + \int_0^y \left[\psi(x,v) - 1\right]v\,dv\right] \\
&\geq y^2\ln\left(1+x^2\right).
\end{aligned}$$

In addition, it is easy to check that

$$\ln\left(1+x^2\right) \geq \frac{x^2}{(1+x^2)^2}.$$

Then by the last two inequalities and

$$y^2 f^2(x,0) = y^2\frac{x^2}{(1+x^2)^2},$$

we have

$$4\int_0^x f(u,0)\,du\left[\int_0^y \left[f(x,v) - f(x,0)\right]dv + \int_0^y \left[\psi(x,v) - 1\right]v\,dv\right] \geq y^2 f^2(x,0),$$

that is, (7) of Theorem 7.2.1 holds. By Theorem 7.2.1, for every solution $x(t)$ of (7.2) there are constants c_1, c_2 and c_3 such that

$$|x(t)| < c_1, \qquad |x'(t)| < c_2, \quad \text{and} \quad |x''(t)| < c_3 \quad \text{for } t \geq 0.$$

Lemma 7.2.1 *Let Q be an open set in R^n and $I = [0, \infty)$. Consider the differential system*

$$\frac{d\boldsymbol{x}}{dt} = H(\boldsymbol{x}) + G(t, \boldsymbol{x}), \tag{7.36}$$

where H is continuous in Q, G is continuous in $I \times Q$ and for any continuous and bounded function $\boldsymbol{x}(t)$ on $t_0 \leq t < \infty$,

$$\int_{t_0}^{\infty} \left\| G\big(s, \boldsymbol{x}(s)\big) \right\| ds < \infty.$$

Assume that all the solutions of (7.36) are bounded, and that there exists a non-negative continuous function $V(t, \boldsymbol{x})$ which satisfies locally a Lipschitz condition with respect to $\boldsymbol{x}$ in Q such that $V'(t, \boldsymbol{x}) \leq -W(\boldsymbol{x})$, where $W(\boldsymbol{x})$ is positive definite with respect to a closed set Ω. Then all the solutions of (7.36) approach the largest semi-invariant set contained in Ω of the equation

$$\frac{d\boldsymbol{x}}{dt} = H(\boldsymbol{x}) \tag{7.37}$$

on Ω.

Theorem 7.2.2 *Assume that*

(1) $xf(x, 0) > 0$ *for* $x \neq 0$,
(2) $\lim_{|x| \to \infty} \int_0^x f(u, 0)\, du = \infty$,
(3) $\int_0^y f(0, v)\, dv \geq 0$,
(4) $\int_0^\infty |p(t)|\, dt < \infty$,

and there is a positive number B such that (5)–(7) in Theorem 7.2.1 hold and

(8) $B[f(x, y) - f(x, 0) - \int_0^y \psi_x(x, v) v\, dv]y + \psi(x, y) > y \int_0^y f_x(x, v)\, dv + B$ *for* $y \neq 0$.

Then every solution $x(t)$ of Eq. (7.2) satisfies

$$\lim_{t \to \infty} x(t) = 0, \qquad \lim_{t \to \infty} x'(t) = 0, \quad \text{and} \quad \lim_{t \to \infty} x''(t) = 0. \tag{7.38}$$

Proof Consider the system (7.23), and let V be defined by (7.24). Then by noting $G(x, y) \geq 0$, $G(x, y)$ is defined in (7.24), it follows from (7.30) that

$$\begin{aligned} \left.\frac{dV}{dt}\right|_{(7.23)} \leq -e^{P(\infty)} \Bigg[y \Bigg[B\bigg(f(x, y) - f(x, 0) - \int_0^y \psi_x(x, v) v\, dv \bigg) \\ - \int_0^y f_x(x, v)\, dv \Bigg] + \big[\psi(x, y) - B\big] z^2 \Bigg]. \end{aligned} \tag{7.39}$$

Set

$$W(x, y, z) = -e^{P(\infty)}\left[y\left[B\left(f(x, y) - f(x, 0) - \int_0^y \psi_x(x, v)v\,dv\right)\right.\right.$$
$$\left.\left. - \int_0^y f_x(x, v)\,dv\right] + \left[\psi(x, y) - B\right]z^2\right].$$

Conditions (5) and (7) yield $W(x, y, z) \geq 0$. Now, consider the set

$$\Omega = \{(x, y, z);\ W(x, y, z) = 0\}.$$

Since the function W is continuous, the set Ω is closed and W is positive definite with respect to Ω. Next, consider the system (7.7) whose asymptotic behaviour is studied in Theorem 7.1.2. With the same hypothesis, we find here, as it has been shown in Theorem 7.1.2, that $(0, 0, 0)$ is the largest semi-invariant set of (7.7) contained in Ω. In addition, since all the hypotheses of Theorem 7.2.1 are satisfied, every solution of (7.23) is bounded. Now let

$$\mathbf{x} = (x, y, z)^T,$$
$$H(\mathbf{x}) = \left(y, z, -f(x, y) - \psi(x, y)z\right)^T,$$
$$G(t, \mathbf{x}) = \left(0, 0, p(t)\right)^T.$$

We see that (7.23) is in the form (7.36), and from the above discussion, all the hypotheses in Lemma 7.2.1 are satisfied. Hence, by Lemma 7.2.1, every solution of (7.23) tends to the largest semi-invariant set contained in Ω of (7.7), that is, $(0, 0, 0)$. The proof is complete. □

Example 7.2.2 Consider (7.2) with $\psi(x, y) = 2\arctan x + y^2 + 2\pi$, $f(x, y) = \frac{x}{1+x^2}(1 + y^2) + 2y + y^3$ and $p(t) = \frac{\sin t}{1+t^2}$. It is easy to check that the hypotheses (1)–(4) of Theorem 7.2.2 are satisfied. Now, take $B = 2$ and observe that $\psi(x, y) > 2$,

$$\begin{aligned}
&2\left[f(x, y) - f(x, 0) - \int_0^y \psi_x(x, v)v\,dv\right]y \\
&= 2\left(\frac{x}{1+x^2}y^2 + 2y + y^3 - \frac{1}{1+x^2}y^2\right)y \\
&= \left(4 + 2\frac{x-1}{1+x^2}y + 2y^2\right)y^2 \\
&\geq \left[4 - 3|y| + 2y^2\right]y^2 \\
&\geq \left[1 + y^2 + \left(\frac{3}{2} - |y|\right)^2\right]y^2 \\
&\geq \left[1 + y^2\right]y^2
\end{aligned}$$

$$
\begin{aligned}
&\geq \frac{1-x^2}{(1+x^2)^2}\left(1+\frac{1}{3}y^2\right)y^2 \\
&= y\int_0^y f_x(x,v)\,dv \qquad (7.40)
\end{aligned}
$$

and

$$
\begin{aligned}
&4.2\int_0^x f(u,0)\,du\left[\int_0^y\left[f(x,v)-f(x,0)\right]dv+2\int_0^y\left[\psi(x,v)-2\right]v\,dv\right] \\
&\geq 8\int_0^x f(u,0)\,du\left[\int_0^y\left[f(x,v)-f(x,0)\right]dv\right] \\
&= 4\ln\left(1+x^2\right)\left[\frac{x}{3(1+x^2)}y^3+y^2+\frac{1}{4}y^4\right] \\
&= y^2\ln\left(1+x^2\right)\left[\frac{4x}{3(1+x^2)}y+4+y^2\right] \\
&\geq y^2\ln\left(1+x^2\right)\left\{y^2-2|y|+4\right\} \\
&= y^2\ln\left(1+x^2\right)\left\{3+\left(|y|-1\right)^2\right\} \\
&\geq y^2\ln\left(1+x^2\right) \\
&\geq y^2\frac{x^2}{(1+x^2)^2} \\
&= y^2 f^2(x,0).
\end{aligned}
$$

We see that the hypotheses (5)–(7) of Theorem 7.2.2 are satisfied. In addition, $\psi(x,y)>2$ and (7.40) clearly imply that (8) in Theorem 7.2.2 holds. Hence, all the hypotheses of Theorem 7.2.2 are satisfied, and so every solution $x(t)$ of (7.2) satisfies the property (7.38).

A special case of (7.2) is the nonlinear equation

$$x'''+\psi\left(x,x'\right)x''+\phi\left(x'\right)+g(x)=p(t), \qquad (7.41)$$

where

$$\psi,\psi_x\in C[R\times R,R],\qquad \phi,g'\in C[R,R],\qquad p\in C\big[[0,\infty),R\big].$$

Applying Theorem 7.2.2 to (7.41), we obtain the following corollary:

Corollary 7.2.1 *Assume that*

(1) $xg(x)>0$ *for* $x\neq 0$,
(2) $\lim_{|x|\to\infty}\int_0^x g(u)\,du=\infty$,

(3) $\int_0^y \phi(v)\,dv \geq 0$,
(4) $\int_0^\infty |p(t)|\,dt < \infty$,

and there is a positive number B such that

(5) $\psi(x, y) \geq B$,
(6) $B[\phi(y) - \int_0^y \psi_x(x, v)v\,dv]y \geq g'(x)y^2$,
(7) $4B\int_0^x g(u)\,du[\int_0^y \phi(v)\,dv + B\int_0^y [\psi(x, v) - B]v\,dv] \geq g^2(x)y^2$,
(8) $\psi(x, y) + B[\phi(y) - \int_0^y \psi_x(x, v)v\,dv]y \geq B + g'(x)y^2$ *for* $y \neq 0$.

Then every solution $x(t)$ *of* (7.41) *satisfies the property* (7.38).

When $g(x) = cx$ and $\phi(y) = by$, where b and c are positive constants, then (7.41) reduces to

$$x''' + \psi(x, x')x'' + bx' + cx = p(t). \tag{7.42}$$

Then by choosing $B = c/b$, we have the following result from Corollary 7.2.1.

Corollary 7.2.2 *Assume that*

(1) $\psi(x, y) \geq c/b$,
(2) $y\int_0^y \psi_x(x, v)v\,dv \leq 0$,
(3) $2\int_0^y \psi(x, v)v\,dv \geq (c/b)y^2$,
(4) $\psi(x, y) - y\int_0^y \psi_x(x, v)v\,dv > c/b$ *for* $y \neq 0$,
(5) $\int_0^\infty |p(t)|\,dt < \infty$.

Then every solution $x(t)$ *of Eq.* (7.42) *satisfies* (7.38).

Next, consider the equation

$$x''' + h(x')x'' + \phi(x') + g(x) = p(t), \tag{7.43}$$

which is a special case of (7.41) with $\psi(x, y) = h(y)$, where $h, \phi, g' \in C[R, R]$ and $p \in C[[0, \infty), R]$. Applying Corollary 7.2.1 to (7.43), we obtain

Corollary 7.2.3 *Assume that*

(1) $xg(x) > 0$ *for* $x \neq 0$,
(2) $\lim_{|x|\to\infty} \int_0^x g(u)\,du = \infty$,
(3) $\int_0^\infty |p(t)|\,dt < \infty$,

and there is a positive number B such that

(4) $h(y) \geq B$,
(5) $B\phi(y)y \geq g'(x)y^2$,
(6) $h(y) + B\phi(y)y > B + g'(x)y^2$ *for* $y \neq 0$.

Then every solution $x(t)$ *of* (7.43) *satisfies* (7.38).

In particular, for the equation

$$x''' + h(x')x'' + \mu(x')x' + k(x)x = p(t), \tag{7.44}$$

where $h, \mu, k' \in C[R, R]$ and $p \in C[[0, \infty), R]$, Corollary 7.2.3 yields the following Corollary 7.2.4.

Corollary 7.2.4 *Assume that*

(1) $k(x) > 0$ *for* $x \neq 0$,
(2) $\lim_{|x|\to\infty} \int_0^x k(u)\,du = \infty$,
(3) $\int_0^\infty |p(t)|\,dt < \infty$,

and there is a positive number B *such that*

(4) $h(y) \geq B$,
(5) $B\mu(y) \geq (xk(x))'$,
(6) $h(y) + B\phi(y) > B + (xk(x))'$ *for* $y \neq 0$.

Then every solution $x(t)$ *of* (7.44) *satisfies* (7.38).

Finally, when $h(y) = a$, $\mu(y) = b$ and $k(x) = c$ are all constants, then (7.44) reduces to the linear equation

$$x''' + ax''' + bx' + cx = p(t). \tag{7.45}$$

Then, by choosing $B = c/b$, the following result follows from Corollary 7.2.4.

Corollary 7.2.5 *Assume that*

(1) $a > 0$, $b > 0$, $c > 0$,
(2) $ab > c$,
(3) $\int_0^\infty |p(t)|\,dt < \infty$.

Then every solution $x(t)$ *of* (7.45) *satisfies* (7.38).

Clearly, (1) and (2) in Corollary 7.2.5 are well-known Routh–Hurwitz conditions for the asymptotic stability of the third-order linear differential equation with constant coefficients of the form (1.5).

7.3 Stability of Solutions of $x''' + \psi(x, x', x'')x'' + f(x, x') = p(t, x, x', x'')$

Consider Eq. (7.3) together with the condition (7.4). Clearly, (7.3) is equivalent to the system

$$x' = y, \qquad y' = z, \qquad z' = -\psi(x, y, z)z - f(x, y) + p(t, x, y, z). \tag{7.46}$$

Through this section, we shall require that $f(0,0) = 0$, the derivative $\frac{\partial \psi(x,y,z)}{\partial x} \equiv \psi_x(x, y, z)$, $\frac{\partial \psi(x,y,z)}{\partial z} \equiv \psi_z(x, y, z)$, $\frac{\partial f(x,y)}{\partial x} \equiv f_x(x, y)$ and $\frac{\partial f(x,y)}{\partial y} \equiv f_y(x, y)$ exist and are continuous, and the uniqueness of the solutions of (7.3) will be assumed.

Theorem 7.3.1 *Assume that*

(1) $\int_0^x f(u,0)\,du > 0$ *for* $u \neq 0$,
(2) $\limsup_{|x|\to\infty} \int_0^x f(u,0)\,du = \infty$,
(3) $\int_0^y f(0,v)\,dv \geq 0$,
(4) *The function* p *satisfies* $|p(t,x,y,z)| \leq |e(t)|$ *uniformly in* t, *where* $e(t)$ *is a continuous function of* t *such that* $\int_0^\infty |e(t)|\,dt < \infty$,
(5) *There is a positive constant* B *such that* $\psi(x,y,z) \geq B$,
(6) $B[f(x,y) - f(x,0) - \int_0^y \psi_x(x,v,0)v\,dv]y \geq y\int_0^y f_x(x,v)\,dv$,
(7) $4B\int_0^x f(u,0)\,du[\int_0^y [f(x,v) - f(x,0)]\,dv + B\int_0^y [\psi(x,v,0) - B]v\,dv] \geq y^2 f^2(x,0)$ *for all* $xy \neq 0$,
(8) $y\psi_z(x,y,z) \geq 0$.

Then for any solution $(x(t), y(t), z(t))$ *of the system* (7.46), *there are positive constants* c_1, c_2 *and* c_3 *such that*

$$|x(t)| \leq c_1, \qquad |y(t)| \leq c_2, \qquad |z(t)| \leq c_3, \quad t \geq 0.$$

Theorem 7.3.2 *Suppose the following hypotheses hold.*

(1) *There is a positive constant* B *such that the assumptions* (4)–(8) *in Theorem* 7.3.1 *hold.*
(2) $xf(x,0) > 0$ *for* $x \neq 0$,
(3) $\limsup_{|x|\to\infty} \int_0^x f(u,0)\,du = \infty$,
(4) $\int_0^y f(0,v)\,dv \geq 0$,
(5) $B[f(x,y) - f(x,0) - \int_0^y \psi_x(x,v,0)v\,dv]y + \psi(x,y,z) \geq y\int_0^y f_x(x,v)\,dv + B$ *for* $y \neq 0$.

Then every solution $(x(t), y(t), z(t))$ *of the system* (7.46) *satisfies*

$$\lim_{t\to\infty} x(t) = 0, \qquad \lim_{t\to\infty} y(t) = 0, \qquad \lim_{t\to\infty} z(t) = 0. \tag{7.47}$$

Theorem 7.3.3 *Let* $\delta_0 > 0, a > 0, b > 0, c > 0$ *be constants such that* $ab > c$. *Assume that*

(1) $\frac{f(x,0)}{x} \geq \delta_0 > 0$, $x \geq 0$,
(2) $f'(x,0) \leq c$,
(3) $f_y(x,\theta y) \geq b$ *for* $0 \leq \theta \leq 1$,
(4) $\psi(x,y,z) > a$,
(5) $y\psi_z(x,y,\theta z) \geq 0$ *for* $0 \leq \theta \leq 1$,
(6) $a[f(x,y) - f(x,0) - \int_0^y \psi_x(x,y,0)v\,dv]y \geq y\int_0^y f_x(x,v)\,dv$,
(7) *the function* p *satisfies* $|p(t,x,y,z)| \leq |e(t)|$ *uniformly in* t, *where* $e(t)$ *is a continuous function of* t *such that* $\int_0^\infty |e(t)|\,dt < \infty$.

Then for any solution $(x(t), y(t), z(t))$ *of* (7.46), *there are positive constants* c_1, c_2 *and* c_3 *such that*

$$|x(t)| \le c_1, \qquad |y(t)| \le c_2, \qquad |z(t)| \le c_3, \quad t \ge 0. \tag{7.48}$$

Proof Clearly (7.3) is equivalent to (7.46), and it suffices to show that every solution of (7.46) is bounded. To this end, consider the function

$$V(t, x, y, z) = e^{-P(t)} U(x, y, z), \tag{7.49}$$

where

$$U(x, y, z) = \int_0^x f(u, 0)\, du + \int_0^y \psi(x, v, 0) v\, dv + a^{-1} \int_0^y f(x, v)\, dv + \frac{1}{2} a^{-1} z^2 + yz + 2a^{-1}$$

and

$$P(t) = \int_0^t |e(s)|\, ds.$$

We claim that V is a positive function. To show this, it suffices to show that U is positive. We can rewrite U as

$$\begin{aligned} U(x, y, z) = {} & \frac{a^{-1}}{2}(ay + z)^2 + \frac{a^{-1}}{2b}\big(f(x, 0) + by\big)^2 + \int_0^y \big[\psi(x, v, 0) - a\big] v\, dv \\ & + a^{-1} \int_0^y \big[f_v(x, \theta v) - b\big] v\, dv + \int_0^x \left[1 - \frac{a^{-1}}{b} f'(u, 0)\right] f(u, 0)\, du \\ & + 2a^{-1}, \end{aligned}$$

where $f_v(x, \theta v) = v^{-1}\{f(x, v) - f(x, 0)\}$, $v \ne 0$.

Using the hypotheses (1)–(5) of this theorem, we get

$$U(x, y, z) \ge \frac{a^{-1}}{2}(ay + z)^2 + \frac{a^{-1}}{2b}\big(f(x, 0) + by\big)^2 + \frac{\delta_1}{2} x^2. \tag{7.50}$$

Combining (7.49) and (7.50), we obtain

$$V(t, x, y, z) \ge \frac{1}{2} e^{-P(t)} \left[a^{-1}(ay + z)^2 + \frac{a^{-1}}{b}\big(f(x, 0) + by\big)^2 + \delta_1 x^2\right].$$

Thus, there exists a constant $K > 0$ small enough such that

$$V(t, x, y, z) \ge K\big(x^2 + y^2 + z^2\big). \tag{7.51}$$

Hence $V(t, x, y, z)$ is a positive function.

Next, we show that the derivative of $V(t, x, y, z)$ with respect to t along the solution path of (7.46) satisfies

$$V'_{(7.46)} = \frac{dV}{dt}\bigg|_{(7.46)} = \frac{d}{dt}V\big(t, x(t), y(t), z(t)\big)\bigg|_{(7.46)} \le -D_1 \tag{7.52}$$

provided that $x^2 + y^2 + z^2 \ge D_2$, D_1, D_2 are some positive constants. Now

$$V'_{(7.46)} = e^{-P(t)}\big\{-\big|e(t)\big|U + U'_{(7.46)}\big\},$$

where $U = U(x(t), y(t), z(t))$ and $\frac{dU}{dt}|_{(7.46)} = \frac{d}{dt}U(x(t), y(t), z(t))|_{(7.46)}$. Then

$$\begin{aligned} V'_{(7.46)} = e^{-P(t)}\bigg[&-\big|e(t)\big|\bigg(\frac{a^{-1}}{2}(ay+z)^2 + \frac{a^{-1}}{2b}\big(f(x,0)+by\big)^2 \\ &+ \int_0^y \big(\psi(x,v,0)-a\big)v\,dv + a^{-1}\int_0^y \big(f_v(x,\theta v)-b\big)v\,dv \\ &+ \int_0^x \bigg(1-\frac{a^{-1}}{b}f'(u,0)\bigg)f(u,0)\,du + 2a^{-1}\bigg) \\ &+ a^{-1}y\int_0^y f_x(x,v)\,dv - \psi_z(x,y,\theta z)yz^2 - \big(a^{-1}\psi(x,y,z)-1\big)z^2 \\ &- \bigg(f(x,y) - f(x,0) - \int_0^y \psi_x(x,v,0)v\,dv\bigg)y \\ &+ a^{-1}(z+ay)p(t,x,y,z)\bigg], \end{aligned}$$

where

$$\psi_z(x, y, \theta z) = \frac{\psi(x,y,z) - \psi(x,y,0)}{z}, \qquad 0 \le \theta \le 1.$$

Clearly,

- if $|z + ay| < 2$, then $(z + ay)p(t, x, y, z) \le 2|p(t, x, y, z)| \le 2|e(t)|$;
- if $|z + ay| \ge 2$, then $(z + ay)p(t, x, y, z) \le \frac{1}{2}(z + ay)^2|p(t, x, y, z)| \le \frac{1}{2}(z + ay)^2|e(t)|$.

Hence, for any t, x and y

$$(z+ay)p(t,x,y,z) \le \bigg(2+\frac{1}{2}(z+ay)^2\bigg)\big|p(t,x,y,z)\big| \le \bigg(2+\frac{1}{2}(z+ay)^2\bigg)\big|e(t)\big|,$$

and so

$$V'_{(7.46)} \le e^{-P(t)}\bigg[-|e(t)|\bigg(\frac{a^{-1}}{2b}\big(f(x,0)+by\big)^2+\int_0^y\big(\psi(x,v,0)-a\big)v\,dv$$
$$+a^{-1}\int_0^y\big(f_v(x,\theta v)-b\big)v\,dv+\int_0^x\Big(1-\frac{a^{-1}}{b}f'(u,0)\Big)f(u,0)\,du\bigg)$$
$$+a^{-1}y\int_0^y f_x(x,v)\,dv-\psi_z(x,y,\theta z)yz^2-\big(a^{-1}\psi(x,y,z)-1\big)z^2$$
$$-\Big(f(x,y)-f(x,0)-\int_0^y\psi_x(x,v,0)v\,dv\Big)y\bigg]. \tag{7.53}$$

Then, by noting (1)–(6), we can find an $\eta > 0$ small enough so that

$$V'_{(7.46)} \le -e^{-P(t)}\bigg[\eta z^2+\frac{a^{-1}}{2b}\big(f(x,0)+by\big)^2+\frac{1}{2}\delta_1 x^2\bigg]|e(t)|,$$

where $\delta_1 = \delta_0(1-\frac{c}{ab})$. Thus, there exists a constant $D_3 > 0$ small enough such that

$$V'_{(7.46)} \le -D_3\big(x^2+y^2+z^2\big).$$

Hence $V'_{(7.46)} \le -D_4$ provided that $(x^2+y^2+z^2) \ge D_4D_3^{-1}$, and this completes the verification of (7.52).

Finally, we show that all solutions of (7.46) are bounded. For this, assume that $(x(t), y(t), z(t))$ is a solution of (7.46). Then, there is evidently a $t_0 \ge 0$ such that $x^2(t_0)+y^2(t_0)+z^2(t_0) < D_2$, where D_2 is a constant defined earlier. Otherwise, if $x^2(t)+y^2(t)+z^2(t) \ge D_2$, $t \ge 0$, then by (7.52), $V'_{(7.46)} \le -D_1 < 0$, $t \ge 0$. This, in turn, implies that $V(t) \to -\infty$ as $t \to \infty$, which contradicts (7.51). Hence to prove (7.48), it suffices to show that, if

$$x^2(t)+y^2(t)+z^2(t) < D_5 \quad \text{for } t = T, \tag{7.54}$$

where $D_5 \ge D_2$ is a finite constant, then there exists a constant $D_6 > 0$ depending on a, b, c, δ_0 and D_5 such that

$$x^2(t)+y^2(t)+z^2(t) \le D_6 \quad \text{for } t \ge T. \tag{7.55}$$

For any given constant $d > 0$, let $S(d)$ denote the surface: $x^2+y^2+z^2 = d$. Because V is continuous in t, x, y, z and tends to ∞ as $x^2+y^2+z^2 \to \infty$, there is evidently a constant $D_7 > 0$ depending on D_5 as well as a, b, c, δ_0 such that

$$\min_{(x,y,z)\in S(D_7)} V(t,x,y,z) > \max_{(x,y,z)\in S(D_5)} V(t,x,y,z). \tag{7.56}$$

It is easy to see from (7.54) and (7.56) that

$$x^2(t)+y^2(t)+z^2(t) < D_7, \quad t \ge T. \tag{7.57}$$

On the contrary, there is a $t > T$ such that

$$x^2(t) + y^2(t) + z^2(t) \geq D_7.$$

Then, by (7.54) and by the continuity of the quantities $x(t)$, $y(t)$ and $z(t)$, there exist t_1, t_2, $T < t_1 < t_2$ such that

$$x^2(t_1) + y^2(t_1) + z^2(t_1) = D_5, \tag{7.58}$$

$$x^2(t_2) + y^2(t_2) + z^2(t_2) = D_7, \tag{7.59}$$

and

$$D_5 \leq x^2(t) + y^2(t) + z^2(t) \leq D_7, \quad t_1 \leq t \leq t_2. \tag{7.60}$$

By writing $V(t) \equiv V(t, x(t), y(t), z(t))$ and the fact $D_5 \geq D_2$, (7.60) obviously implies (in view of (7.53)) that $V(t_2) < V(t_1)$, and this contradicts the conclusion (from (7.56), (7.58) and (7.59)) that $V(t_2) > V(t_1)$. Hence (7.57) holds. This completes the proof of (7.48). The theorem now follows. □

Example 7.3.1 Consider (7.3) with $\psi(x, y, z) = \ln(1 + x^2) + e^{yz} + 2$, $f(x, y) = x + \frac{x}{1+x^2}(1 + y^2) + y + \frac{1}{3}y^3$ and $p(t) = \frac{\sin t}{1+t^2}$. The coefficients satisfy all the hypotheses (1)–(5) of Theorem 7.3.3. Since $\psi(x, y, z) \geq 2$ and

$$\begin{aligned}
&2\left[f(x, y) - f(x, 0) - \int_0^y \psi_x(x, v, 0)v\,dv\right]y \\
&= 2\left[\frac{x}{1+x^2}y^2 + y + \frac{1}{3}y^3 - \frac{x}{1+x^2}y^2\right]y \\
&= 2\left(y^2 + \frac{1}{3}y^4\right) \\
&\geq y^2 + \frac{1-x^2}{(1+x^2)^2}\left(y^2 + \frac{1}{3}y^4\right) \\
&= y\int_0^y f_x(x, v)\,dv,
\end{aligned}$$

we see that (6) of Theorem 7.3.3 also hold. By Theorem 7.3.3, there is a constant $D > 0$ such that $|x(t)| < D$, $|x'(t)| < D$ and $|x''(t)| < D$ for $t \geq 0$.

Theorem 7.3.4 *Let $\delta_0 > 0$, $a > 0$, $b > 0$ and $c > 0$ be constants such that $ab > c$. Assume that*

(1) $\frac{f(x,0)}{x} \geq \delta_0$, $x \neq 0$,
(2) $f'(x, 0) \leq c$,
(3) $f_y(x, \theta y) \geq b$ *for* $0 \leq \theta \leq 1$,
(4) $\psi(x, y, z) > a$,
(5) $y\psi_z(x, y, \theta z) \geq 0$ *for* $0 \leq \theta \leq 1$,

(6) $a[f(x,y)-f(x,0)-\int_0^y \psi_x(x,v,0)v\,dv]y \geq y\int_0^y f_x(x,v)\,dv$,
(7) *the function* p *satisfies* $|p(t,x,y,z)| \leq |e(t)|$ *uniformly in* t, *where* $e(t)$ *is a continuous function of* t *such that* $\int_0^\infty |e(t)|\,dt < \infty$.

Then every solution $(x(t),y(t),z(t))$ *of the system* (7.46) *satisfies the property* (7.47).

Proof Consider the system (7.46) and let V be defined by (7.49). Then by noting

$$\frac{a^{-1}}{2}(ay+z)^2+\frac{a^{-1}}{2b}\big(f(x,0)+by\big)^2+\int_0^y \big[\psi(x,v,0)-a\big]v\,dv$$
$$+a^{-1}\int_0^y \big[f_v(x,\theta v)-b\big]v\,dv+\int_0^x \left[1-\frac{a^{-1}}{b}f'(x,0)\right]f(u,0)\,du \geq 0,$$

it follows from (7.53) that

$$\frac{dV}{dt}\bigg|_{(7.46)} \leq -e^{-P(\infty)}\bigg[y\bigg(f(x,y)-f(x,0)$$
$$-\int_0^y \psi_x(x,v,0)v\,dv-a^{-1}\int_0^y f_x(x,v)\,dv\bigg)$$
$$+\big[a^{-1}\psi(x,y,z)-1\big]z^2\bigg].$$

Set

$$W(x,y,z)=e^{-P(\infty)}\bigg[y\bigg(f(x,y)-f(x,0)$$
$$-\int_0^y \psi_x(x,v,0)v\,dv-a^{-1}\int_0^y f_x(x,v)\,dv\bigg)$$
$$+\big[a^{-1}\psi(x,y,z)-1\big]z^2\bigg].$$

Observe that $W(x,y,z)\geq 0$. Now, we consider the set

$$\Omega=\big\{(x,y,z):W(x,y,z)=0\big\}.$$

Since the function W is continuous, the set Ω is closed and W is positive definite with respect to Ω. Now, we consider the system

$$x'=y, \qquad y'=z, \qquad z'=-\psi(x,y,z)z-f(x,y). \tag{7.61}$$

Proceeding as earlier in previous sections, we can prove that $(0,0,0)$ is the largest semi-invariant set of (7.61) contained in Ω. In addition, all the hypothesis of Theorem 7.3.3 are satisfied, and hence every solution of the system (7.46) is bounded.

Now, let $\mathbf{x} = (x, y, z)^T$, $H(\mathbf{x}) = (y, z, -f(x, y) - \psi(x, y, z)z)^T$ and $G(t, \mathbf{x}) = (0, 0, p(t, x, y, z))^T$. Obviously, the system (7.46) is in the form (7.36). Hence by Lemma 7.2.1, every solution of (7.46) tends to the largest semi-invariant set contained in Ω of (7.61) on Q, that is, $(0, 0, 0)$. This completes the proof of the theorem. □

Example 7.3.2 Consider Eq. (7.3) with $\psi(x, y, z) = y \sin x + y^2 + e^{2yz} + 2$, $f(x, y) = y^3 + y + x + \frac{x}{1+x^2}$ and $p(t) = \frac{\sin t}{1+t^2}$. One may verify that the hypotheses (1)–(5) in Theorem 7.3.4 are satisfied. Since $\psi(x, y, z) > 1$ and

$$\begin{aligned}
&\left[f(x, y) - f(x, 0) - \int_0^y \psi_x(x, v, 0) v\, dv\right] y \\
&= \left[y^3 + y - \frac{1}{3}(\cos x) y^3\right] y \\
&> y^2\left(1 + \frac{1 - x^2}{(1 + x^2)^2}\right) \\
&= y \int_0^y f_x(x, v)\, dv,
\end{aligned}$$

we see that (6) of Theorem 7.3.4 holds. Hence by Theorem 7.3.4, all solutions of the equation satisfy the property (7.47).

7.4 Stability of Solutions of $x'''(t) = p_1 x''(t) + p_2 x''(t - \tau) + q_1 x'(t) + q_2 x'(t - \tau) + r_1 x(t) + r_2 x(t - \tau)$

The substitution $x_1 = x$, $x_2 = x'$, and $x_3 = x''$ transforms the linear delay differential equation (7.5) into the following equivalent first-order linear delay differential system:

$$\left.\begin{aligned}
x_1'(t) &= x_2(t), \\
x_2'(t) &= x_3(t), \\
x_3'(t) &= r_1 x_1(t) + r_2 x_1(t - \tau) + q_1 x_2(t) + q_2 x_2(t - \tau) + p_1 x_3(t) + p_2 x_3(t - \tau).
\end{aligned}\right\} \tag{7.62}$$

This delay system is considered in conjunction with the initial condition

$$\left.\begin{aligned}
x_1'(t) &= \phi(t) \quad \text{for } -\tau \le t \le 0, \\
x_2'(t) &= \phi'(t) \quad \text{for } -\tau \le t \le 0, \\
x_3'(t) &= \phi''(t) \quad \text{for } -\tau \le t \le 0.
\end{aligned}\right\} \tag{7.63}$$

Hence, the initial-value problem (7.5) and (7.6) is transformed into the equivalent initial-value problems (7.62) and (7.63).

On the basis of the above transformation of the delay differential equation (7.5) into the equivalent delay differential system (7.62), one can formulate the definitions of the notions of stability, the instability, and the asymptotic stability of the trivial solutions of (7.5). One may refer to the book due to Driver [1] for the notions of stability, the instability, and the asymptotic stability of the trivial solutions of (7.5).

Let $C([-\tau,0],R)$ be the Banach space of all continuous real-valued functions on the interval $[-\tau,0]$, endowed with the usual sup-norm

$$\|\psi\| = \max_{-\tau\le t\le 0} \left|\psi(t)\right| \quad \text{for } \psi\in C\big([-\tau,0],R\big).$$

Moreover, let $C^2([-\tau,0],R)$ be the set of all twice continuously differentiable real-valued function on the interval $[-\tau,0]$. This set is a Banach space with the norm

$$\nparallel \omega \nparallel = \max\big\{\|\omega\|, \|\omega'\|, \|\omega''\|\big\} \quad \text{for } \omega\in C^2\big([-\tau,0],R\big).$$

As it concerns with the initial-value problems (7.5) and (7.6), the initial function ϕ belongs to $C^2([-\tau,0],R)$. Thus we shall use the following norm:

$$\begin{aligned}\nparallel \phi \nparallel &= \max\big\{\|\phi\|, \|\phi'\|, \|\phi''\|\big\}\\ &= \max\Big\{\max_{-\tau\le t\le 0}\left|\phi(t)\right|, \max_{-\tau\le t\le 0}\left|\phi'(t)\right|, \max_{-\tau\le t\le 0}\left|\phi''(t)\right|\Big\}.\end{aligned}$$

Definition 7.4.1 The trivial solution of the delay differential equation (7.5) is said to be stable (at 0) if, for each $\epsilon>0$, there exists a $\delta\equiv\delta(\epsilon)>0$ such that for any $\phi\in C^2([-\tau,0],R)$ with $\nparallel \phi \nparallel<\delta$, i.e., $\max\{\|\phi\|,\|\phi'\|,\|\phi''\|\}<\delta$, the solution of the initial-valued problems (7.5) and (7.6) satisfy

$$\max\big\{\left|x(t)\right|, \left|x'(t)\right|, \left|x''(t)\right|\big\}<\epsilon \quad \text{for all } t\ge -\tau;$$

otherwise, the trivial solution of (7.5) is called "unstable" (at 0). Moreover, the trivial solution of (7.5) is called "asymptotically stable" (at 0), if it is stable (at 0) in the above sense and, in addition, there exists a $\delta_0>0$ such that for any $\phi\in C^2([-\tau,0],R)$ with $\nparallel \phi \nparallel<\delta_0$, i.e., $\max\{\|\phi\|,\|\phi'\|,\|\phi''\|\}<\delta_0$, the solution x of the initial-valued problems (7.5) and (7.6) satisfy

$$\lim_{t\to\infty}\max\big\{\left|x(t)\right|, \left|x'(t)\right|, \left|x''(t)\right|\big\}=0,$$

that is,

$$\lim_{t\to\infty} x(t) = \lim_{t\to\infty} x'(t) = \lim_{t\to\infty} x''(t) = 0.$$

The characteristic equation associated with Eq. (7.5) is

$$\lambda^3 = p_1\lambda^2 + p_2\lambda^2 e^{-\lambda\tau} + q_1\lambda + q_2\lambda e^{-\lambda\tau} + r_1 + r_2 e^{-\lambda\tau}. \tag{7.64}$$

Let $\lambda_0 \neq 0$ be a real root of the characteristic equation (7.64). Let $x(t)$ be a solution of Eqs. (7.5)–(7.6). Set $y(t) = e^{-\lambda_0 t}x(t),\ t \geq -\tau$. Then for every $t \geq 0$, Eq. (7.5) is equivalent to

$$\begin{aligned}
&\Big[y'(t) + (3\lambda_0 - p_1)y(t) - p_2 e^{-\lambda_0\tau} y(t-\tau) - \big(2\lambda_0 p_1 + q_1 - 3\lambda_0^2\big)\int_0^t y(s)\,ds \\
&\qquad - (2\lambda_0 p_2 + q_2)e^{-\lambda_0\tau}\int_0^{t-\tau} y(s)\,ds\Big]'' \\
&\quad = \big(p_1\lambda_0^2 + q_1\lambda_0 + r_1 - \lambda_0^3\big)y(t) + \big(p_2\lambda_0^2 + q_2\lambda_0 + r_2\big)e^{-\lambda_0\tau}y(t-\tau). \qquad (7.65)
\end{aligned}$$

The initial condition (7.6) can equivalently be written as

$$y(t) = e^{-\lambda_0 t}\phi(t), \quad t \in [-\tau, 0]. \qquad (7.66)$$

Integrating (7.65) from 0 to t, we have, noting that λ_0 is a root of (7.64),

$$\begin{aligned}
&y''(t) + (3\lambda_0 - p_1)y'(t) - p_2 e^{-\lambda_0\tau}y'(t-\tau) - \big(2\lambda_0 p_1 + q_1 - 3\lambda_0^2\big)y(t) \\
&\quad - (2\lambda_0 p_2 + q_2)e^{-\lambda_0\tau}y(t-\tau) \\
&= y''(0) + (3\lambda_0 - p_1)y'(0) - p_2 e^{-\lambda_0\tau}y'(-\tau) - \big(2\lambda_0 p_1 + q_1 - 3\lambda_0^2\big)y(0) \\
&\quad - (2\lambda_0 p_2 + q_2)e^{-\lambda_0\tau}y(-\tau) + \big(p_1\lambda_0^2 + q_1\lambda_0 + r_1 - \lambda_0^3\big)\int_0^t y(s)\,ds \\
&\quad + \big(p_2\lambda_0^2 + q_2\lambda_0 + r_2\big)e^{-\lambda_0\tau}\int_0^t y(s-\tau)\,ds \\
&= y''(0) + (3\lambda_0 - p_1)y'(0) - p_2 e^{-\lambda_0\tau}y'(-\tau) - \big(2\lambda_0 p_1 + q_1 - 3\lambda_0^2\big)y(0) \\
&\quad - (2\lambda_0 p_2 + q_2)e^{-\lambda_0\tau}y(-\tau) + \big(p_1\lambda_0^2 + q_1\lambda_0 + r_1 - \lambda_0^3\big)\int_0^t y(s)\,ds \\
&\quad + \big(p_2\lambda_0^2 + q_2\lambda_0 + r_2\big)e^{-\lambda_0\tau}\int_{-\tau}^{t-\tau} y(s)\,ds \\
&= y''(0) + (3\lambda_0 - p_1)y'(0) - p_2 e^{-\lambda_0\tau}y'(-\tau) - \big(2\lambda_0 p_1 + q_1 - 3\lambda_0^2\big)y(0) \\
&\quad - (2\lambda_0 p_2 + q_2)e^{-\lambda_0\tau}y(-\tau) - \big(p_2\lambda_0^2 + q_2\lambda_0 + r_2\big)e^{-\lambda_0\tau}\int_0^t y(s)\,ds \\
&\quad + \big(p_2\lambda_0^2 + q_2\lambda_0 + r_2\big)e^{-\lambda_0\tau}\int_0^{t-\tau} y(s)\,ds \\
&\quad + \big(p_2\lambda_0^2 + q_2\lambda_0 + r_2\big)e^{-\lambda_0\tau}\int_{-\tau}^{0} y(s)\,ds \\
&= L_0(\lambda_0;\phi) - \big(p_2\lambda_0^2 + q_2\lambda_0 + r_2\big)e^{-\lambda_0\tau}\int_{t-\tau}^{t} y(s)\,ds,
\end{aligned}$$

that is,

$$\begin{aligned} &y''(t) + (3\lambda_0 - p_1)y'(t) - p_2 e^{-\lambda_0\tau} y'(t-\tau) - \big(2\lambda_0 p_1 + q_1 - 3\lambda_0^2\big) y(t) \\ &\quad - (2\lambda_0 p_2 + q_2)e^{-\lambda_0\tau} y(t-\tau) \\ &\quad = L_0(\lambda_0;\phi) - \big(p_2\lambda_0^2 + q_2\lambda_0 + r_2\big)e^{-\lambda_0\tau} \int_{t-\tau}^{t} y(s)\,ds, \end{aligned} \tag{7.67}$$

where

$$\begin{aligned} L_0(\lambda_0;\phi) = {} & \phi''(0) + (\lambda_0 - p_1)\phi'(0) + \big(\lambda_0^2 - \lambda_0 p_1 - q_1\big)\phi_0 \\ & - p_2\phi'(-\tau) - (p_2\lambda_0 + q_2)\phi(-\tau) \\ & + \big(p_2\lambda_0^2 + q_2\lambda_0 + r_2\big)e^{-\lambda_0\tau} \int_{-\tau}^{0} e^{-\lambda_0 s}\phi(s)\,ds. \end{aligned} \tag{7.68}$$

Define

$$z(t) = y(t) - \frac{L_0(\lambda_0;\phi)}{\beta_{\lambda_0}}, \tag{7.69}$$

where

$$\beta_{\lambda_0} \equiv e^{-\lambda_0\tau}\big(\tau p_2\lambda_0^2 + \tau q_2\lambda_0 + \tau r_2 - 2\lambda_0 p_2 - q_2\big) - 2\lambda_0 p_1 - q_1 + 3\lambda_0^2 \neq 0. \tag{7.70}$$

Then from Eq. (7.67), we obtain

$$\begin{aligned} &z''(t) + (3\lambda_0 - p_1)z'(t) - p_2 e^{-\lambda_0\tau} z'(t-\tau) - \big(2\lambda_0 p_1 + q_1 - 3\lambda_0^2\big) z(t) \\ &\quad - (2\lambda_0 p_2 + q_2)e^{-\lambda_0\tau} z(t-\tau) + \big(p_2\lambda_0^2 + q_2\lambda_0 + r_2\big)e^{-\lambda_0\tau} \int_{t-\tau}^{t} z(s)\,ds = 0. \end{aligned} \tag{7.71}$$

The initial condition (7.66) becomes

$$z(t) = \phi(t)e^{-\lambda_0 t} - \frac{L_0(\lambda_0;\phi)}{\beta_{\lambda_0}}, \quad t \in [-\tau, 0]. \tag{7.72}$$

If we look for a solution of (7.71) of the form $e^{\delta t}$, $\delta \in R$, we see that δ is a root of the characteristic equation

$$\begin{aligned} &\delta^2 + (3\lambda_0 - p_1)\delta - p_2 e^{-\lambda_0\tau} e^{-\delta\tau} - \big(2\lambda_0 p_1 + q_1 - 3\lambda_0^2\big) \\ &\quad - (2\lambda_0 p_2 + q_2)e^{-\lambda_0\tau} e^{-\delta\tau} + \big(p_2\lambda_0^2 + q_2\lambda_0 + r_2\big)e^{-\lambda_0\tau}\delta^{-1}\big(1 - e^{-\delta\tau}\big) = 0. \end{aligned} \tag{7.73}$$

Let

$$F(\delta) = \delta^2 + (3\lambda_0 - p_1)\delta - p_2 e^{-\lambda_0\tau} e^{-\delta\tau} - \left(2\lambda_0 p_1 + q_1 - 3\lambda_0^2\right)$$
$$- (2\lambda_0 p_2 + q_2) e^{-\lambda_0\tau} e^{-\delta\tau} + \left(p_2\lambda_0^2 + q_2\lambda_0 + r_2\right) e^{-\lambda_0\tau} \delta^{-1}\left(1 - e^{-\delta\tau}\right).$$

Clearly, $\delta = 0$ is a removable singularity of $F(\delta)$, and hence we can regard $F(\delta)$ is an entire function with

$$F(0) = \beta_{\lambda_0} \neq 0.$$

Let $\delta_0 \neq 0$ is a real root of (7.73). Define for $\delta_0 \neq 0$,

$$v(t) = e^{-\delta_0 t} z(t), \qquad \text{for } t \geq -\tau.$$

Then from (7.71), we have

$$v''(t) + (3\lambda_0 - p_1 + 2\delta_0)v'(t) - p_2 e^{-(\lambda_0+\delta_0)\tau} v'(t-\tau) + \left[\delta_0^2 + \delta_0(3\lambda_0 - p_1)\right.$$
$$\left. - \left(2\lambda_0 p_1 + q_1 - 3\lambda_0^2\right)\right] v(t) - (2\lambda_0 p_2 + q_2 + \delta_0 p_2) e^{-(\lambda_0+\delta_0)\tau} v(t-\tau)$$
$$+ \left(p_2\lambda_0^2 + q_2\lambda_0 + r_2\right) e^{-\lambda_0\tau} \int_{t-\tau}^{t} e^{\delta_0(s-t)} v(s)\, ds = 0,$$

or

$$v''(t) + (3\lambda_0 - p_1 + 2\delta_0)v'(t) - p_2 e^{-(\lambda_0+\delta_0)\tau} v'(t-\tau) + \left[\delta_0^2 + \delta_0(3\lambda_0 - p_1)\right.$$
$$\left. - \left(2\lambda_0 p_1 + q_1 - 3\lambda_0^2\right)\right] v(t) - (2\lambda_0 p_2 + q_2 + \delta_0 p_2) e^{-(\lambda_0+\delta_0)\tau} v(t-\tau)$$
$$+ \left(p_2\lambda_0^2 + q_2\lambda_0 + r_2\right) e^{-\lambda_0\tau} \int_0^{\tau} e^{-\delta_0 s} v(t-s)\, ds = 0. \tag{7.74}$$

The initial condition (7.72) becomes

$$v(t) = e^{-(\lambda_0+\delta_0)t}\phi(t) - e^{-\delta_0 t}\frac{L_0(\lambda_0;\phi)}{\beta_{\lambda_0}}, \qquad -\tau \leq t \leq 0. \tag{7.75}$$

One may write Eq. (7.74) as

$$\left[v'(t) + (3\lambda_0 - p_1 + 2\delta_0)v(t) - p_2 e^{-(\lambda_0+\delta_0)\tau} v(t-\tau)\right]'$$
$$= \left[-\delta_0^2 - \delta_0(3\lambda_0 - p_1) + \left(2\lambda_0 p_1 + q_1 - 3\lambda_0^2\right)\right] v(t)$$
$$+ (2\lambda_0 p_2 + q_2 + \delta_0 p_2) e^{-(\lambda_0+\delta_0)\tau} v(t-\tau)$$
$$- \left(p_2\lambda_0^2 + q_2\lambda_0 + r_2\right) e^{-\lambda_0\tau} \int_0^{\tau} e^{-\delta_0 s} v(t-s)\, ds. \tag{7.76}$$

Integrating (7.76) from 0 to t, we have

$$v'(t) + (3\lambda_0 - p_1 + 2\delta_0)v(t) - p_2 e^{-(\lambda_0+\delta_0)\tau} v(t-\tau)$$
$$= v'(0) + (3\lambda_0 - p_1 + 2\delta_0)v(0) - p_2 e^{-(\lambda_0+\delta_0)\tau} v(-\tau)$$

$$
\begin{aligned}
&+ \left[-\delta_0^2 - \delta_0(3\lambda_0 - p_1) + \left(2\lambda_0 p_1 + q_1 - 3\lambda_0^2\right)\right] \int_0^t v(s)\,ds \\
&+ (2\lambda_0 p_2 + q_2 + \delta_0 p_2)e^{-(\lambda_0+\delta_0)\tau} \int_0^t v(s-\tau)\,ds \\
&- \left(p_2\lambda_0^2 + q_2\lambda_0 + r_2\right)e^{-\lambda_0\tau} \int_0^\tau e^{-\delta_0 s} \int_0^t v(u-s)\,du\,ds \\
= {}& v'(0) + (3\lambda_0 - p_1 + 2\delta_0)v(0) - p_2 e^{-(\lambda_0+\delta_0)\tau} v(-\tau) \\
&+ \left[-\delta_0^2 - \delta_0(3\lambda_0 - p_1) + \left(2\lambda_0 p_1 + q_1 - 3\lambda_0^2\right)\right] \int_0^t v(s)\,ds \\
&+ (2\lambda_0 p_2 + q_2 + \delta_0 p_2)e^{-(\lambda_0+\delta_0)\tau} \int_{-\tau}^{t-\tau} v(s)\,ds \\
&- \left(p_2\lambda_0^2 + q_2\lambda_0 + r_2\right)e^{-\lambda_0\tau} \int_0^\tau e^{-\delta_0 s} \int_{-s}^{t-s} v(u)\,du\,ds \\
= {}& v'(0) + (3\lambda_0 - p_1 + 2\delta_0)v(0) - p_2 e^{-(\lambda_0+\delta_0)\tau} v(-\tau) \\
&+ \left[-\delta_0^2 - \delta_0(3\lambda_0 - p_1) + \left(2\lambda_0 p_1 + q_1 - 3\lambda_0^2\right)\right] \int_0^t v(s)\,ds \\
&+ (2\lambda_0 p_2 + q_2 + \delta_0 p_2)e^{-(\lambda_0+\delta_0)\tau} \int_0^{t-\tau} v(s)\,ds \\
&+ (2\lambda_0 p_2 + q_2 + \delta_0 p_2)e^{-(\lambda_0+\delta_0)\tau} \int_{-\tau}^0 v(s)\,ds \\
&- \left(p_2\lambda_0^2 + q_2\lambda_0 + r_2\right)e^{-\lambda_0\tau} \int_0^\tau e^{-\delta_0 s} \int_0^{t-s} v(u)\,du\,ds \\
&- \left(p_2\lambda_0^2 + q_2\lambda_0 + r_2\right)e^{-\lambda_0\tau} \int_0^\tau e^{-\delta_0 s} \int_{-s}^0 v(u)\,du\,ds.
\end{aligned}
$$

Using (7.73), we have

$$
\begin{aligned}
&v'(t) + (3\lambda_0 - p_1 + 2\delta_0)v(t) - p_2 e^{-(\lambda_0+\delta_0)\tau} v(t-\tau) \\
&\quad = v'(0) + (3\lambda_0 - p_1 + 2\delta_0)v(0) - p_2 e^{-(\lambda_0+\delta_0)\tau} v(-\tau) \\
&\qquad - (2\lambda_0 p_2 + q_2 + \delta_0 p_2)e^{-(\lambda_0+\delta_0)\tau} \int_0^t v(s)\,ds \\
&\qquad + \left(p_2\lambda_0^2 + q_2\lambda_0 + r_2\right)e^{-\lambda_0\tau} \delta_0^{-1}\left(1 - e^{-\delta_0\tau}\right) \int_0^t v(s)\,ds \\
&\qquad + (2\lambda_0 p_2 + q_2 + \delta_0 p_2)e^{-(\lambda_0+\delta_0)\tau} \int_0^{t-\tau} v(s)\,ds
\end{aligned}
$$

$$
\begin{aligned}
&+ (2\lambda_0 p_2 + q_2 + \delta_0 p_2)e^{-(\lambda_0+\delta_0)\tau} \int_{-\tau}^{0} v(s)\,ds \\
&- \left(p_2\lambda_0^2 + q_2\lambda_0 + r_2\right)e^{-\lambda_0\tau} \int_0^{\tau} e^{-\delta_0 s} \int_0^{t-s} v(u)\,du\,ds \\
&- \left(p_2\lambda_0^2 + q_2\lambda_0 + r_2\right)e^{-\lambda_0\tau} \int_0^{\tau} e^{-\delta_0 s} \int_{-s}^{0} v(u)\,du\,ds \\
= {} & v'(0) + (3\lambda_0 - p_1 + 2\delta_0)v(0) - p_2 e^{-(\lambda_0+\delta_0)\tau} v(-\tau) \\
&- (2\lambda_0 p_2 + q_2 + \delta_0 p_2)e^{-(\lambda_0+\delta_0)\tau} \int_{t-\tau}^{t} v(s)\,ds \\
&+ \left(p_2\lambda_0^2 + q_2\lambda_0 + r_2\right)e^{-\lambda_0\tau} \delta_0^{-1}\left(1 - e^{\delta_0\tau}\right) \int_0^{t} v(s)\,ds \\
&+ (2\lambda_0 p_2 + q_2 + \delta_0 p_2)e^{-(\lambda_0+\delta_0)\tau} \int_{-\tau}^{0} v(s)\,ds \\
&- \left(p_2\lambda_0^2 + q_2\lambda_0 + r_2\right)e^{-\lambda_0\tau} \int_0^{\tau} e^{-\delta_0 s} \int_0^{t-s} v(u)\,du\,ds \\
&- \left(p_2\lambda_0^2 + q_2\lambda_0 + r_2\right)e^{-\lambda_0\tau} \int_0^{\tau} e^{-\delta_0 s} \int_{-s}^{0} v(u)\,du\,ds \\
= {} & v'(0) + (3\lambda_0 - p_1 + 2\delta_0)v(0) - p_2 e^{-(\lambda_0+\delta_0)\tau} v(-\tau) \\
&+ (2\lambda_0 p_2 + q_2 + \delta_0 p_2)e^{-(\lambda_0+\delta_0)\tau} \int_{-\tau}^{0} v(s)\,ds \\
&- \left(p_2\lambda_0^2 + q_2\lambda_0 + r_2\right)e^{-\lambda_0\tau} \int_0^{\tau} e^{-\delta_0 s} \int_{-s}^{0} v(u)\,du\,ds \\
&- (2\lambda_0 p_2 + q_2 + \delta_0 p_2)e^{-(\lambda_0+\delta_0)\tau} \int_{t-\tau}^{t} v(s)\,ds \\
&+ \left(p_2\lambda_0^2 + q_2\lambda_0 + r_2\right)e^{-\lambda_0\tau} \int_0^{\tau} e^{-\delta_0 s} \int_0^{t} v(u)\,du\,ds \\
&- \left(p_2\lambda_0^2 + q_2\lambda_0 + r_2\right)e^{-\lambda_0\tau} \int_0^{\tau} e^{-\delta_0 s} \int_0^{t-s} v(u)\,du\,ds \\
= {} & R_0(\lambda_0, \delta_0; \phi) - (2\lambda_0 p_2 + q_2 + \delta_0 p_2)e^{-(\lambda_0+\delta_0)\tau} \int_{t-\tau}^{t} v(s)\,ds \\
&+ \left(p_2\lambda_0^2 + q_2\lambda_0 + r_2\right)e^{-\lambda_0\tau} \int_0^{\tau} e^{-\delta_0 s} \int_{t-s}^{t} v(u)\,du\,ds,
\end{aligned}
\tag{7.77}
$$

where

$$R_0(\lambda_0,\delta_0;\phi)=v'(0)+(3\lambda_0-p_1+2\delta_0)v(0)-p_2e^{-(\lambda_0+\delta_0)\tau}v(-\tau)$$
$$+(2\lambda_0p_2+q_2+\delta_0p_2)e^{-(\lambda_0+\delta_0)\tau}\int_{-\tau}^{0}v(s)\,ds$$
$$-\big(p_2\lambda_0^2+q_2\lambda_0+r_2\big)e^{-\lambda_0\tau}\int_0^{\tau}e^{-\delta_0 s}\int_{-s}^{0}v(u)\,du\,ds.$$

Using (7.75), we have

$$\begin{aligned}R_0(\lambda_0,\delta_0;\phi)={}&\phi'(0)+(2\lambda_0+\delta_0-p_1)\phi(0)-p_2\phi(-\tau)\\&+(p_1-\delta_0-3\lambda_0)\frac{L_0(\lambda_0;\phi)}{\beta_{\lambda_0}}+p_2e^{-\lambda_0\tau}\frac{L_0(\lambda_0;\phi)}{\beta_{\lambda_0}}\\&+(2\lambda_0p_2+q_2+\delta_0p_2)e^{-(\lambda_0+\delta_0)\tau}\bigg[\int_{-\tau}^{0}e^{-(\lambda_0+\delta_0)s}\phi(s)\,ds\\&+\frac{L_0(\lambda_0;\phi)}{\beta_{\lambda_0}}\delta_0^{-1}\big(1-e^{-\delta_0\tau}\big)\bigg]\\&-\big(p_2\lambda_0^2+q_2\lambda_0+r_2\big)e^{-\lambda_0\tau}\bigg[\int_0^{\tau}e^{-\delta_0 s}\int_{-s}^{0}e^{-(\lambda_0+\delta_0)u}\phi(u)\,du\,ds\\&+\delta_0^{-1}\frac{L_0(\lambda_0;\phi)}{\beta_{\lambda_0}}\big(-\tau+\delta_0^{-1}\big(1-e^{-\delta_0\tau}\big)\big)\bigg].\end{aligned}\tag{7.78}$$

Next, we define

$$w(t)=v(t)-\frac{R_0(\lambda_0,\delta_0;\phi)}{\eta_{\lambda_0,\delta_0}}\quad\text{for } t\geq-\tau,\tag{7.79}$$

where

$$\begin{aligned}\eta_{\lambda_0,\delta_0}={}&3\lambda_0-p_1+2\delta_0+\big(\tau(2\lambda_0p_2+q_2+\delta_0p_2)-p_2\big)e^{-(\lambda_0+\delta_0)\tau}\\&-\big(p_2\lambda_0^2+q_2\lambda_0+r_2\big)e^{-\lambda_0\tau}\big[-\tau\delta_0^{-1}e^{-\delta_0\tau}+\delta_0^{-2}\big(1-e^{-\delta_0\tau}\big)\big]\\\neq{}&0.\end{aligned}\tag{7.80}$$

Then (7.77) is equivalent to

$$\begin{aligned}&w'(t)+(3\lambda_0-p_1+2\delta_0)w(t)-p_2e^{-(\lambda_0+\delta_0)\tau}w(t-\tau)\\&+(2\lambda_0p_2+q_2+\delta_0p_2)e^{-(\lambda_0+\delta_0)\tau}\int_{t-\tau}^{t}w(s)\,ds\\&-\big(p_2\lambda_0^2+q_2\lambda_0+r_2\big)e^{-\lambda_0\tau}\int_0^{\tau}e^{-\delta_0 s}\int_{t-s}^{t}w(u)\,du\,ds=0\end{aligned}\tag{7.81}$$

and the initial condition (7.75) is equivalent to

$$w(t)=\phi(t)e^{-(\lambda_0+\delta_0)t}-\frac{L_0(\lambda_0;\phi)}{\beta_{\lambda_0}}e^{-\delta_0 t}-\frac{R_0(\lambda_0,\delta_0;\phi)}{\eta_{\lambda_0,\delta_0}},\quad -\tau\le t\le 0. \tag{7.82}$$

If we look for a solution of (7.81) of the form $w(t)=e^{\mu t}$ for $t\in R$, we see that μ is a root of the characteristic equation

$$\begin{aligned}&\mu+(3\lambda_0-p_1+2\delta_0)-p_2e^{-(\lambda_0+\delta_0+\mu)\tau}\\&\quad+\frac{1}{\mu}(2\lambda_0p_2+q_2+\delta_0p_2)e^{-(\lambda_0+\delta_0)\tau}\big(1-e^{-\mu\tau}\big)\\&\quad-\frac{1}{\mu}\big(p_2\lambda_0^2+q_2\lambda_0+r_2\big)e^{-\lambda_0\tau}\big[\delta_0^{-1}\big(1-e^{-\delta_0\tau}\big)-(\delta_0+\mu)^{-1}\big(1-e^{-(\delta_0+\mu)\tau}\big)\big]\\&=0.\end{aligned}\tag{7.83}$$

Let

$$\begin{aligned}F_1(\mu)&=\mu+(3\lambda_0-p_1+2\delta_0)-p_2e^{-(\lambda_0+\delta_0+\mu)\tau}\\&\quad+\frac{1}{\mu}(2\lambda_0p_2+q_2+\delta_0p_2)e^{-(\lambda_0+\delta_0)\tau}\big(1-e^{-\mu\tau}\big)\\&\quad-\frac{1}{\mu}\big(p_2\lambda_0^2+q_2\lambda_0+r_2\big)e^{-\lambda_0\tau}\big[\delta_0^{-1}\big(1-e^{-\delta_0\tau}\big)\\&\quad-(\delta_0+\mu)^{-1}\big(1-e^{-(\delta_0+\mu)\tau}\big)\big].\end{aligned}$$

Then $F_1(\mu)$ can be expressed as

$$\begin{aligned}F_1(\mu)&=\mu+(3\lambda_0-p_1+2\delta_0)-p_2e^{-(\lambda_0+\delta_0+\mu)\tau}\\&\quad+\frac{1}{\mu}(2\lambda_0p_2+q_2+\delta_0p_2)e^{-(\lambda_0+\delta_0)\tau}\big(1-e^{-\mu\tau}\big)\\&\quad-\big(p_2\lambda_0^2+q_2\lambda_0+r_2\big)e^{-\lambda_0\tau}\int_0^\tau e^{-\delta_0 s}\mu^{-1}\big(1-e^{-\mu s}\big)\,ds.\end{aligned}$$

Since $\mu=0$ is a removable singularity of $F_1(\mu)$, then $F_1(\mu)$ may be regarded as an entire function with

$$\begin{aligned}F_1(0)&=3\lambda_0-p_1+2\delta_0-p_2e^{-(\lambda_0+\delta_0)\tau}+\tau(2\lambda_0p_2+q_2+\delta_0p_2)e^{-(\lambda_0+\delta_0)\tau}\\&\quad-\big(p_2\lambda_0^2+q_2\lambda_0+r_2\big)e^{-\lambda_0\tau}\big[-\tau\delta_0^{-1}e^{-\delta_0\tau}+\delta_0^{-2}\big(1-e^{-\delta_0\tau}\big)\big]\\&=\eta_{\lambda_0,\delta_0}\\&\ne 0.\end{aligned}$$

Hence $F_1(\mu)$ has a nonzero real root μ_0. Define

$$u(t)=e^{-\mu_0 t}w(t)\quad\text{for all } t\in[-\tau,\infty).$$

Then from (7.81), we obtain

$$\begin{aligned} &u'(t) + (3\lambda_0 - p_1 + 2\delta_0 + \mu_0)u(t) - p_2 e^{-(\lambda_0+\delta_0+\mu_0)\tau} u(t-\tau) \\ &\quad + (2\lambda_0 p_2 + q_2 + \delta_0 p_2) e^{-(\lambda_0+\delta_0)\tau} \int_0^\tau e^{-\mu_0 s} u(t-s)\,ds \\ &\quad - \big(p_2\lambda_0^2 + q_2\lambda_0 + r_2\big) e^{-\lambda_0\tau} \int_0^\tau e^{-\delta_0 s} \int_0^s e^{-\mu_0\theta} u(t-\theta)\,d\theta\,ds = 0. \end{aligned} \tag{7.84}$$

The initial condition (7.82) becomes

$$\begin{aligned} u(t) = \phi(t) e^{-(\lambda_0+\delta_0+\mu_0)t} &- \frac{L_0(\lambda_0;\phi)}{\beta_{\lambda_0}} e^{-(\delta_0+\mu_0)t} \\ &- \frac{R_0(\lambda_0,\delta_0;\phi)}{\eta_{\lambda_0,\delta_0}} e^{-\mu_0 t}, \quad -\tau \le t \le 0. \end{aligned} \tag{7.85}$$

Integrating (7.84) from 0 to t, we get

$$\begin{aligned} u(t) &= u(0) - (3\lambda_0 - p_1 + 2\delta_0 + \mu_0) \int_0^t u(s)\,ds + p_2 e^{-(\lambda_0+\delta_0+\mu_0)\tau} \int_0^t u(s-\tau)\,ds \\ &\quad - (2\lambda_0 p_2 + q_2 + \delta_0 p_2) e^{-(\lambda_0+\delta_0)\tau} \int_0^\tau e^{-\mu_0 s} \int_0^t u(\theta - s)\,d\theta\,ds \\ &\quad + \big(p_2\lambda_0^2 + q_2\lambda_0 + r_2\big) e^{-\lambda_0\tau} \int_0^\tau e^{-\delta_0 s} \int_0^s e^{-\mu_0\theta} \int_0^t u(\xi-\theta)\,d\xi\,d\theta\,ds \\ &= u(0) - (3\lambda_0 - p_1 + 2\delta_0 + \mu_0) \int_0^t u(s)\,ds + p_2 e^{-(\lambda_0+\delta_0+\mu_0)\tau} \int_{-\tau}^{t-\tau} u(s)\,ds \\ &\quad - (2\lambda_0 p_2 + q_2 + \delta_0 p_2) e^{-(\lambda_0+\delta_0)\tau} \int_0^\tau e^{-\mu_0 s} \int_{-s}^{t-s} u(\theta)\,d\theta\,ds \\ &\quad + \big(p_2\lambda_0^2 + q_2\lambda_0 + r_2\big) e^{-\lambda_0\tau} \int_0^\tau e^{-\delta_0 s} \int_0^s e^{-\mu_0\theta} \int_{-\theta}^{t-\theta} u(\xi)\,d\xi\,d\theta\,ds. \end{aligned}$$

That is,

$$\begin{aligned} u(t) &= -(3\lambda_0 - p_1 + 2\delta_0 + \mu_0) \int_0^t u(s)\,ds + p_2 e^{-(\lambda_0+\delta_0+\mu_0)\tau} \int_0^{t-\tau} u(s)\,ds \\ &\quad - (2\lambda_0 p_2 + q_2 + \delta_0 p_2) e^{-(\lambda_0+\delta_0)\tau} \int_0^\tau e^{-\mu_0 s} \int_0^{t-s} u(\theta)\,d\theta\,ds \\ &\quad + \big(p_2\lambda_0^2 + q_2\lambda_0 + r_2\big) e^{-\lambda_0\tau} \int_0^\tau e^{-\delta_0 s} \int_0^s e^{-\mu_0\theta} \int_0^{t-\theta} u(\xi)\,d\xi\,d\theta\,ds \\ &\quad + R_1(\lambda_0,\delta_0,\mu_0;\phi), \end{aligned} \tag{7.86}$$

where

$$\begin{aligned}
&R_1(\lambda_0, \delta_0, \mu_0; \phi) \\
&= u(0) + p_2 e^{-(\lambda_0+\delta_0+\mu_0)\tau} \int_{-\tau}^{0} u(s)\,ds \\
&\quad - (2\lambda_0 p_2 + q_2 + \delta_0 p_2) e^{-(\lambda_0+\delta_0)\tau} \int_0^{\tau} e^{-\mu_0 s} \int_{-s}^{0} u(\theta)\,d\theta\,ds \\
&\quad + \left(p_2\lambda_0^2 + q_2\lambda_0 + r_2\right) e^{-\lambda_0\tau} \int_0^{\tau} e^{-\delta_0 s} \int_0^{s} e^{-\mu_0\theta} \int_{-\theta}^{0} u(\xi)\,d\xi\,d\theta\,ds.
\end{aligned}$$

Using (7.85), $R_1(\lambda_0, \delta_0, \mu_0; \phi)$ can be expressed as

$$\begin{aligned}
&R_1(\lambda_0, \delta_0, \mu_0; \phi) \\
&= \phi(0) - \frac{L_0(\lambda_0; \phi)}{\beta_{\lambda_0}} - \frac{R_0(\lambda_0, \delta_0; \phi)}{\eta_{\lambda_0,\delta_0}} \\
&\quad + p_2 e^{-(\lambda_0+\delta_0+\mu_0)\tau} \biggl[\int_{-\tau}^{0} \phi(s) e^{-(\lambda_0+\delta_0+\mu_0)s}\,ds \\
&\quad + \frac{L_0(\lambda_0; \phi)}{(\delta_0+\mu_0)\beta_{\lambda_0}} \left(1 - e^{(\delta_0+\mu_0)\tau}\right) + \frac{R_0(\lambda_0, \delta_0; \phi)}{\mu_0 \eta_{\lambda_0,\delta_0}} \left(1 - e^{\mu_0\tau}\right) \biggr] \\
&\quad - (2\lambda_0 p_2 + q_2 + \delta_0 p_2) e^{-(\lambda_0+\delta_0)\tau} \biggl[\int_0^{\tau} e^{-\mu_0 s} \int_{-s}^{0} \phi(u) e^{-(\lambda_0+\delta_0+\mu_0)u}\,du\,ds \\
&\quad + \left\{ \frac{L_0(\lambda_0; \phi)}{(\delta_0+\mu_0)\beta_{\lambda_0}} + \frac{R_0(\lambda_0, \delta_0; \phi)}{\mu_0 \eta_{\lambda_0,\delta_0}} \right\} \mu_0^{-1} \left(1 - e^{-\mu_0\tau}\right) \\
&\quad - \delta_0^{-1} \frac{L_0(\lambda_0; \phi)}{(\delta_0+\mu_0)\beta_{\lambda_0}} \left(e^{\delta_0\tau} - 1\right) - \frac{R_0(\lambda_0, \delta_0; \phi)}{\mu_0 \eta_{\lambda_0,\delta_0}} \tau \biggr] \\
&\quad + \left(p_2\lambda_0^2 + q_2\lambda_0 + r_2\right) e^{-\lambda_0\tau} \\
&\quad \times \biggl[\int_0^{\tau} e^{-\delta_0 s} \int_0^{s} e^{-\mu_0\theta} \int_{-\theta}^{0} u(\xi) e^{-(\lambda_0+\delta_0+\mu_0)\xi}\,d\xi\,d\theta\,ds \\
&\quad + \left\{ \frac{L_0(\lambda_0; \phi)}{(\delta_0+\mu_0)\beta_{\lambda_0}} + \frac{R_0(\lambda_0, \delta_0; \phi)}{\mu_0 \eta_{\lambda_0,\delta_0}} \right\} \mu_0^{-1} \bigl(\delta_0^{-1} \left(1 - e^{-\delta_0\tau}\right) \\
&\quad - (\delta_0+\mu_0)^{-1} \left(1 - e^{-(\delta_0+\mu_0)\tau}\right) \bigr) \\
&\quad - \delta_0^{-1} \frac{L_0(\lambda_0; \phi)}{(\delta_0+\mu_0)\beta_{\lambda_0}} \left(\tau - \delta_0^{-1}\left(1 - e^{-\delta_0\tau}\right)\right) \\
&\quad - \frac{R_0(\lambda_0, \delta_0; \phi)}{\mu_0 \eta_{\lambda_0,\delta_0}} \left(-\tau\delta_0^{-1} e^{-\delta_0\tau} + \delta_0^{-2}\left(1 - e^{-\delta_0\tau}\right)\right) \biggr]. \qquad (7.87)
\end{aligned}$$

Using (7.83), (7.86) gives

$$\begin{aligned}u(t) &= R_1(\lambda_0,\delta_0,\mu_0;\phi) + p_2 e^{-(\lambda_0+\delta_0+\mu_0)\tau}\int_0^{t-\tau} u(s)\,ds\\&\quad + \Big[-p_2 e^{-(\lambda_0+\delta_0+\mu_0)\tau} + (2\lambda_0 p_2 + q_2 + \delta_0 p_2)e^{-(\lambda_0+\delta_0)\tau}\int_0^{\tau} e^{-\mu_0 s}\,ds\\&\quad - \big(p_2\lambda_0^2 + q_2\lambda_0 + r_2\big)e^{-\lambda_0\tau}\int_0^{\tau} e^{-\delta_0 s}\int_0^{s} e^{-\mu_0\theta}\,d\theta\,ds\Big]\int_0^{t} u(s)\,ds\\&\quad - (2\lambda_0 p_2 + q_2 + \delta_0 p_2)e^{-(\lambda_0+\delta_0)\tau}\int_0^{\tau} e^{-\mu_0 s}\int_0^{t-s} u(\theta)\,d\theta\,ds\\&\quad + \big(p_2\lambda_0^2 + q_2\lambda_0 + r_2\big)e^{-\lambda_0\tau}\int_0^{\tau} e^{-\delta_0 s}\int_0^{s} e^{-\mu_0\theta}\int_0^{t-\theta} u(\xi)\,d\xi\,d\theta\,ds,\end{aligned}$$

or

$$\begin{aligned}u(t) &= R_1(\lambda_0,\delta_0,\mu_0;\phi) - p_2 e^{-(\lambda_0+\delta_0+\mu_0)\tau}\int_{t-\tau}^{t} u(s)\,ds\\&\quad + (2\lambda_0 p_2 + q_2 + \delta_0 p_2)e^{-(\lambda_0+\delta_0)\tau}\int_0^{\tau} e^{-\mu_0 s}\int_{t-s}^{t} u(\theta)\,d\theta\,ds\\&\quad - \big(p_2\lambda_0^2 + q_2\lambda_0 + r_2\big)e^{-\lambda_0\tau}\int_0^{\tau} e^{-\delta_0 s}\int_0^{s} e^{-\mu_0\theta}\int_{t-\theta}^{t} u(\xi)\,d\xi\,d\theta\,ds. \qquad (7.88)\end{aligned}$$

Define

$$h(t) = u(t) - \frac{R_1(\lambda_0,\delta_0,\mu_0;\phi)}{\gamma_{\lambda_0,\delta_0,\mu_0}}, \qquad (7.89)$$

where

$$\begin{aligned}\gamma_{\lambda_0,\delta_0,\mu_0} &= 1 + \tau p_2 e^{-(\lambda_0+\delta_0+\mu_0)\tau}\\&\quad + (2\lambda_0 p_2 + q_2 + \delta_0 p_2)e^{-(\lambda_0+\delta_0)\tau}\left[\frac{-\tau e^{-\mu_0\tau}}{\mu_0} + \frac{(1-e^{\mu_0\tau})}{\mu_0^2}\right]\\&\quad + \big(p_2\lambda_0^2 + q_2\lambda_0 + r_2\big)e^{-\lambda_0\tau}\Big[\frac{\tau}{\mu_0(\delta_0+\mu_0)}e^{-(\delta_0+\mu_0)\tau}\\&\quad - \frac{(1-e^{-(\delta_0+\mu_0)\tau})}{\mu_0(\delta_0+\mu_0)^2} + \frac{(1-e^{-\delta_0\tau})}{\delta_0\mu_0^2} - \frac{(1-e^{-(\delta_0+\mu_0)\tau})}{\mu_0^2(\delta_0+\mu_0)}\Big] \neq 0. \qquad (7.90)\end{aligned}$$

Then from (7.88) we have

$$h(t) = -p_2 e^{-(\lambda_0+\delta_0+\mu_0)\tau} \int_{t-\tau}^{t} h(s)\,ds$$
$$+ (2\lambda_0 p_2 + q_2 + \delta_0 p_2)e^{-(\lambda_0+\delta_0)\tau} \int_0^{\tau} e^{-\mu_0 s} \int_{t-s}^{t} h(\theta)\,d\theta\,ds$$
$$- \big(p_2\lambda_0^2 + q_2\lambda_0 + r_2\big)e^{-\lambda_0\tau} \int_0^{\tau} e^{-\delta_0 s} \int_0^{s} e^{-\mu_0\theta} \int_{t-\theta}^{t} h(\xi)\,d\xi\,d\theta\,ds. \quad (7.91)$$

The initial condition (7.85) is equivalent to

$$h(t) = \phi(t)e^{-(\lambda_0+\delta_0+\mu_0)t} - \frac{L_0(\lambda_0;\phi)}{\beta_{\lambda_0}} e^{-(\delta_0+\mu_0)t} - \frac{R_0(\lambda_0,\delta_0;\phi)}{\eta_{\lambda_0,\delta_0}} e^{-\mu_0 t}$$
$$- \frac{R_1(\lambda_0,\delta_0,\mu_0;\phi)}{\gamma_{\lambda_0,\delta_0,\mu_0}}, \quad t \in [-\tau, 0]. \quad (7.92)$$

Now, we have the following theorem:

Theorem 7.4.1 *Let λ_0, δ_0 and μ_0 be real roots of the characteristic equations* (7.64), (7.73) *and* (7.83), *respectively, satisfying the property*

$$\alpha_{\lambda_0,\delta_0,\mu_0} = \tau|p_2|e^{-(\lambda_0+\delta_0+\mu_0)\tau} + |2\lambda_0 p_2 + q_2 + \delta_0 p_2|e^{-(\lambda_0+\delta_0)\tau}\left[\left|\frac{-\tau}{\mu_0}.e^{-\mu_0\tau}\right.\right.$$
$$\left.\left. + \frac{1}{\mu_0^2}\big(1 - e^{-\mu_0\tau}\big)\right|\right] + e^{-\lambda_0\tau}\big|p_2\lambda_0^2 + q_2\lambda_0 + r_2\big|\left[\left|\frac{\tau e^{-(\mu_0+\delta_0)\tau}}{\mu_0(\delta_0+\mu_0)}\right.\right.$$
$$\left.\left. - \frac{(1 - e^{-(\mu_0+\delta_0)\tau})}{\mu_0(\delta_0+\mu_0)^2} + \frac{(1 - e^{-\delta_0\tau})}{\delta_0\mu_0^2} - \frac{(1 - e^{-(\delta_0+\mu_0)\tau})}{\mu_0^2(\delta_0+\mu_0)}\right|\right] < 1, \quad (7.93)$$

$$\beta_{\lambda_0} \equiv e^{-\lambda_0\tau}\big(\tau p_2\lambda_0^2 + \tau q_2\lambda_0 + \tau r_2 - 2\lambda_0 p_2 - q_2\big) - 2\lambda_0 p_1 - q_1 + 3\lambda_0^2 \neq 0$$

and

$$\eta_{\lambda_0,\delta_0} = 3\lambda_0 - p_1 + 2\delta_0 + \big(\tau(2\lambda_0 p_2 + q_2 + \delta_0 p_2) - p_2\big)e^{-(\lambda_0+\delta_0)\tau}$$
$$- \big(p_2\lambda_0^2 + q_2\lambda_0 + r_2\big)e^{-\lambda_0\tau}\big[-\tau\delta_0^{-1}e^{-\delta_0\tau} + \delta_0^{-2}\big(1 - e^{-\delta_0\tau}\big)\big] \neq 0.$$

Then for any $\phi \in C([-\tau, 0], R)$, the solution $x(t)$ of (7.5)–(7.6) *satisfies the property*

$$\left|e^{-(\lambda_0+\delta_0+\mu_0)t}x(t) - \frac{L_0(\lambda_0;\phi)}{\beta_{\lambda_0}} e^{-(\delta_0+\mu_0)t} - \frac{R_0(\lambda_0,\delta_0;\phi)}{\eta_{\lambda_0,\delta_0}} e^{-\mu_0 t}\right.$$
$$\left. - \frac{R_1(\lambda_0,\delta_0,\mu_0;\phi)}{\gamma_{\lambda_0,\delta_0,\mu_0}}\right|$$
$$\leq M(\lambda_0,\delta_0,\mu_0;\phi)\alpha_{\lambda_0,\delta_0,\mu_0} \quad (7.94)$$

for all $t \geq 0$, *where*

$$M(\lambda_0, \delta_0, \mu_0; \phi) = \max_{-\tau \leq t \leq 0} \left| e^{-(\lambda_0+\delta_0+\mu_0)t}\phi(t) - \frac{L_0(\lambda_0; \phi)}{\beta_{\lambda_0}} e^{-(\delta_0+\mu_0)t} \right.$$
$$\left. - \frac{R_0(\lambda_0, \delta_0; \phi)}{\eta_{\lambda_0,\delta_0}} e^{-\mu_0 t} - \frac{R_1(\lambda_0, \delta_0, \mu_0; \phi)}{\gamma_{\lambda_0,\delta_0,\mu_0}} \right|. \tag{7.95}$$

Proof It is clear from (7.93) that $\gamma_{\lambda_0,\delta_0,\mu_0} > 0$. Applying the definitions of x, y, z, v, w, u and h, it is easy to see that (7.94) is equivalent to

$$|h(t)| \leq M(\lambda_0, \delta_0, \mu_0; \phi)\alpha_{\lambda_0,\delta_0,\mu_0}, \quad t \geq 0. \tag{7.96}$$

Hence, we will prove (7.96). From (7.92) and (7.95), we see that

$$|h(t)| \leq M(\lambda_0, \delta_0, \mu_0; \phi), \quad t \in [-\tau, 0].$$

Now, we prove that

$$|h(t)| \leq M(\lambda_0, \delta_0, \mu_0; \phi) \quad \text{for } t \geq -\tau. \tag{7.97}$$

To this end, let us consider an arbitrary $\epsilon > 0$. We claim that

$$|h(t)| \leq M(\lambda_0, \delta_0, \mu_0; \phi) + \epsilon \quad \text{for } t \in [-\tau, \infty]. \tag{7.98}$$

Otherwise, there exists a $t^* > 0$ such that

$$|h(t)| \leq M(\lambda_0, \delta_0, \mu_0; \phi) + \epsilon, \quad t < t^*$$

and

$$|h(t^*)| = M(\lambda_0, \delta_0, \mu_0; \phi) + \epsilon.$$

Then from (7.91), we obtain by using (7.93)

$$\begin{aligned}
&M(\lambda_0, \delta_0, \mu_0; \phi) + \epsilon \\
&\quad = |h(t^*)| \leq |p_2| e^{-(\lambda_0+\delta_0+\mu_0)\tau} \int_{t^*-\tau}^{t^*} |h(s)|\, ds \\
&\qquad + |2\lambda_0 p_2 + q_2 + \delta_0 p_2| e^{-(\lambda_0+\delta_0)\tau} \int_0^\tau e^{-\mu_0 s} \int_{t^*-s}^{t^*} |h(\theta)|\, d\theta\, ds \\
&\qquad + |p_2\lambda_0^2 + q_2\lambda_0 + r_2| e^{-\lambda_0\tau} \int_0^\tau e^{-\delta_0 s} \int_0^s e^{-\mu_0\theta} \int_{t^*-\theta}^{t^*} |h(\xi)|\, d\xi\, d\theta\, ds \\
&\quad \leq \left[\tau |p_2| e^{-(\lambda_0+\delta_0+\mu_0)\tau} + |2\lambda_0 p_2 + q_2 + \delta_0 p_2| e^{-(\lambda_0+\delta_0)\tau} \left\{ \left| \frac{-\tau}{\mu_0} e^{-\mu_0\tau} \right. \right. \right. \\
&\qquad \left. \left. + \frac{1}{\mu_0^2}\left(1 - e^{-\mu_0\tau}\right) \right| \right\} + e^{-\lambda_0\tau} |p_2\lambda_0^2 + q_2\lambda_0 + r_2| \left\{ \left| \frac{\tau e^{-(\mu_0+\delta_0)\tau}}{\mu_0(\delta_0+\mu_0)} \right. \right.
\end{aligned}$$

$$- \frac{(1-e^{-(\mu_0+\delta_0)\tau})}{\mu_0(\delta_0+\mu_0)^2} + \frac{(1-e^{-\delta_0\tau})}{\delta_0\mu_0^2}$$
$$- \frac{(1-e^{-(\delta_0+\mu_0)\tau})}{\mu_0^2(\delta_0+\mu_0)}\Bigg|\Bigg\}\Bigg]\big(M(\lambda_0,\delta_0,\mu_0;\phi)+\epsilon\big)$$
$$< M(\lambda_0,\delta_0,\mu_0;\phi)+\epsilon,$$

a contradiction. Hence (7.98) holds. Since $\epsilon > 0$ is arbitrary, then (7.97) holds. Now by using (7.91) and (7.97), we have

$$\begin{aligned} h(t) &\le |p_2|e^{-(\lambda_0+\delta_0+\mu_0)\tau}\int_{t-\tau}^{t}\big|h(s)\big|\,ds \\ &\quad + |2\lambda_0 p_2 + q_2 + \delta_0 p_2|e^{-(\lambda_0+\delta_0)\tau}\int_0^{\tau} e^{-\mu_0 s}\int_{t-s}^{t}\big|h(\theta)\big|\,d\theta\,ds \\ &\quad + \big|p_2\lambda_0^2 + q_2\lambda_0 + r_2\big|e^{-\lambda_0\tau}\int_0^{\tau} e^{-\delta_0 s}\int_0^{s} e^{-\mu_0\theta}\int_{t-\theta}^{t}\big|h(\xi)\big|\,d\xi\,d\theta\,ds \\ &\le M(\lambda_0,\delta_0,\mu_0;\phi)\alpha_{\lambda_0,\delta_0,\mu_0} \end{aligned}$$

for all $t \ge 0$. This completes the proof of the theorem. □

Theorem 7.4.2 *Let λ_0, δ_0 and μ_0 be real roots of the characteristic equations* (7.64), (7.73) *and* (7.83), *respectively. Further, assume that β_{λ_0} and $\eta_{\lambda_0,\delta_0}$ are as in Theorem* 7.4.1, *and* (7.93) *holds. Then for any $\phi \in C([-\tau,\infty], R)$, the solution $x(t)$ of* (7.5)–(7.6) *satisfy*

$$\begin{aligned} &\lim_{t\to\infty}\left\{e^{-(\lambda_0+\delta_0+\mu_0)t}x(t) - \frac{L_0(\lambda_0;\phi)}{\beta_{\lambda_0}}e^{-(\delta_0+\mu_0)t} - \frac{R_0(\lambda_0,\delta_0;\phi)}{\eta_{\lambda_0,\delta_0}}e^{-\mu_0 t}\right\} \\ &\quad = \frac{R_1(\lambda_0,\delta_0,\mu_0;\phi)}{\gamma_{\lambda_0,\delta_0,\mu_0}}. \end{aligned}$$

Proof To complete the proof of the theorem, it is enough to show that $\lim_{t\to\infty} h(t) = 0$. By using (7.91) and from (7.96), one can show, by an easy induction, that $h(t)$ satisfies

$$\big|h(t)\big| \le \alpha^n_{\lambda_0,\delta_0,\mu_0} M(\lambda_0,\delta_0,\mu_0;\phi)$$

for all $t \ge n\tau - \tau, n = 0, 1, 2, \ldots$. Since (7.93) guarantees that $0 \le \alpha_{\lambda_0,\delta_0,\mu_0} < 1$, then $h(t) \to 0$ as $t \to \infty$. Thus the theorem is proved. □

Theorem 7.4.3 *Let λ_0, δ_0 and μ_0 be real roots of the characteristic equations* (7.64), (7.73) *and* (7.83), *respectively. Further assume that the conditions of Theorem* 7.4.1 *hold. Then for any $\phi \in C([-\tau,0], R)$, the solution $x(t)$ of* (7.5)–(7.6) *satisfies*

$$\begin{aligned}\left|x(t)\right| &\le \frac{k_{\lambda_0}}{|\beta_{\lambda_0}|}e^{\lambda_0 t}N(\lambda_0,\delta_0,\mu_0;\phi)+\frac{H_{\lambda_0,\delta_0}}{|\eta_{\lambda_0,\delta_0}|}e^{(\lambda_0+\delta_0)t}N(\lambda_0,\delta_0,\mu_0;\phi)\\ &+\left[\left\{e_{\delta_0+\mu_0}\left(1+\frac{k_{\lambda_0}}{|\beta_{\lambda_0}|}\right)+\frac{H_{\lambda_0,\delta_0}}{|\eta_{\lambda_0,\delta_0}|}e_{\mu_0}\right\}\alpha_{\lambda_0,\delta_0,\mu_0}\right.\\ &\left.+(1+\alpha_{\lambda_0,\delta_0,\mu_0})\frac{J_{\lambda_0,\delta_0,\mu_0}}{|\gamma_{\lambda_0,\delta_0,\mu_0}|}\right]e^{(\lambda_0+\delta_0+\mu_0)t}N(\lambda_0,\delta_0,\mu_0;\phi), \qquad (7.99)\end{aligned}$$

where $\gamma_{\lambda_0,\delta_0,\mu_0}$ *is given in* (7.90),

$$\begin{aligned}k_{\lambda_0} &= 1+|\lambda_0-p_1|+|p_2|+\left|\lambda_0^2-q_1-\lambda_0 p_1\right|+|\lambda_0 p_2+q_2|\\ &\quad+\left|p_2\lambda_0^2+q_2\lambda_0+r_2\right|\tau e^{-\lambda_0\tau}, \qquad (7.100)\end{aligned}$$

$$\begin{aligned}&H_{\lambda_0,\delta_0}\\ &=1+|2\lambda_0+\delta_0-p_1|+|p_2|\left(1+e^{-\lambda_0\tau}\frac{k_{\lambda_0}}{|\beta_{\lambda_0}|}\right)+|p_1-\delta_0-3\lambda_0|\frac{k_{\lambda_0}}{|\beta_{\lambda_0}|}\\ &+|2\lambda_0 p_2+q_2+\delta_0 p_2|e^{-(\lambda_0+\delta_0)\tau}\left\{\left|\frac{(1-e^{-(\lambda_0+\delta_0)\tau})}{\lambda_0+\delta_0}\right|+\frac{k_{\lambda_0}}{|\beta_{\lambda_0}|}\left|\frac{(1-e^{-\delta_0\tau})}{\delta_0}\right|\right\}\\ &+\left|p_2\lambda_0^2+q_2\lambda_0+r_2\right|e^{-\lambda_0\tau}\left\{\left|\frac{1}{\lambda_0\delta_0}-\frac{1}{(\lambda_0+\delta_0)}\left(\frac{e^{-\delta_0\tau}}{\delta_0}+\frac{e^{\lambda_0\tau}}{\lambda_0}\right)\right.\right.\\ &\left.\left.+\left|\frac{k_{\lambda_0}}{|\beta_{\lambda_0}|}\right|\frac{1}{\delta_0}\left(-\tau+\frac{(1-e^{-\delta_0\tau})}{\delta_0}\right)\right|\right\}, \qquad (7.101)\end{aligned}$$

$$\begin{aligned}&J_{\lambda_0,\delta_0,\mu_0}\\ &=1+\frac{k_{\lambda_0}}{|\beta_{\lambda_0}|}+\frac{H_{\lambda_0,\delta_0}}{|\eta_{\lambda_0,\delta_0}|}+|p_2|e^{-(\lambda_0+\delta_0+\mu_0)\tau}\left\{\left|\frac{(1-e^{(\lambda_0+\delta_0+\mu_0)\tau})}{\lambda_0+\delta_0+\mu_0}\right|\right.\\ &\left.+\frac{k_{\lambda_0}}{|\beta_{\lambda_0}|}\left|\frac{(1-e^{(\delta_0+\mu_0)\tau})}{\delta_0+\mu_0}\right|+\frac{H_{\lambda_0,\delta_0}}{|\eta_{\lambda_0,\delta_0}|}\left|\frac{(1-e^{\mu_0\tau})}{\mu_0}\right|\right\}\\ &+|2\lambda_0 p_2+q_2+\delta_0 p_2|\cdot\frac{e^{-(\lambda_0+\delta_0)\tau}}{|\lambda_0+\delta_0+\mu_0|}\left\{\left|\frac{(1-e^{-\mu_0\tau})}{\mu_0}+\frac{(1-e^{(\lambda_0+\delta_0)\tau})}{\lambda_0+\delta_0}\right|\right.\\ &+\left(\frac{k_{\lambda_0}}{|\delta_0+\mu_0||\beta_{\lambda_0}|}+\frac{H_{\lambda_0,\delta_0}}{|\mu_0||\eta_{\lambda_0,\delta_0}|}\right)\left|\frac{(1-e^{-\mu_0\tau})}{\mu_0}\right|+\frac{k_{\lambda_0}}{|\beta_{\lambda_0}|}\left|\frac{(e^{\delta_0\tau}-1)}{\delta_0(\delta_0+\mu_0)}\right|\\ &\left.+\frac{H_{\lambda_0,\delta_0}}{|\eta_{\lambda_0,\delta_0}|}\left|\frac{\tau}{\mu_0}\right|\right\}+\left|p_2\lambda_0^2+q_2\lambda_0+r_2\right|e^{-\lambda_0\tau}\left\{\frac{1}{|\lambda_0+\delta_0+\mu_0|}\right.\\ &\times\left|\left(\frac{1}{\delta_0\mu_0}+\frac{1}{\delta_0(\delta_0+\mu_0)}\right)(1-e^{-\delta_0\tau})-\frac{(1-e^{-(\delta_0+\mu_0)\tau})}{\mu_0(\delta_0+\mu_0)}-\frac{(e^{\lambda_0\tau}-1)}{\lambda_0(\delta_0+\mu_0)}\right|\end{aligned}$$

$$+\left(\frac{k_{\lambda_0}}{|\delta_0+\mu_0||\beta_{\lambda_0}|}+\frac{H_{\lambda_0,\delta_0}}{|\mu_0||\eta_{\lambda_0,\delta_0}|}\right)\left|\frac{1}{\mu_0}\left(\frac{(1-e^{-\delta_0\tau})}{\delta_0}-\frac{(1-e^{-(\delta_0+\mu_0)\tau})}{\delta_0+\mu_0}\right)\right|$$
$$+\frac{k_{\lambda_0}}{|\delta_0+\mu_0||\delta_0||\beta_{\lambda_0}|}\left|\tau-\frac{(1-e^{-\delta_0\tau})}{\delta_0}\right|$$
$$+\frac{H_{\lambda_0,\delta_0}}{|\mu_0||\eta_{\lambda_0,\delta_0}|}\left|\frac{-\tau e^{-\delta_0\tau}}{\delta_0}+\frac{(1-e^{-\delta_0\tau})}{\delta_0^2}\right|\Bigg\}, \tag{7.102}$$

$$e_{\mu_0}=\max_{-\tau\le t\le 0}\left\{e^{-\mu_0 t}\right\}, \qquad e_{\delta_0+\mu_0}=\max_{-\tau\le t\le 0}\left\{e^{-(\delta_0+\mu_0)t}\right\}, \tag{7.103}$$

and

$$N(\lambda_0,\delta_0,\mu_0;\phi)=\max\Big\{\max_{-\tau\le t\le 0}\left|\phi''(t)\right|,\ \max_{-\tau\le t\le 0}\left|\phi'(t)\right|,$$
$$\max_{-\tau\le t\le 0}\left|\phi(t)\right|,\ \max_{-\tau\le t\le 0}\left|e^{-\lambda_0\tau}\phi(t)\right|\Big\}. \tag{7.104}$$

Moreover, the following holds:

(i) *If* $\lambda_0\le 0$, $\lambda_0+\delta_0\le 0$ *and* $\lambda_0+\delta_0+\mu_0\le 0$, *then the trivial solution of* (7.5) *is stable* (*at* 0).
(ii) *If* $\lambda_0<0$, $\lambda_0+\delta_0<0$ *and* $\lambda_0+\delta_0+\mu_0<0$, *then the trivial solution of* (7.5) *is asymptotically stable* (*at* 0).
(iii) *If either* $\lambda_0+\delta_0+\mu_0>0$, $\delta_0+\mu_0>0$ *and* $\mu_0<0$ *or* $\lambda_0+\delta_0+\mu_0>0$, $\delta_0+\mu_0<0$ *and* $\mu_0>0$ *or* $\lambda_0+\delta_0+\mu_0>0$, $\delta_0>0$ *and* $\mu_0<0$ *holds, then the trivial solution of* (7.5) *is unstable at* 0.

Proof Since the conditions of Theorem 7.4.1 hold, then (7.94) holds. Hence

$$e^{-(\lambda_0+\delta_0+\mu_0)t}\left|x(t)\right|$$
$$\le\frac{|L_0(\lambda_0;\phi)|}{|\beta_{\lambda_0}|}e^{-(\delta_0+\mu_0)t}+\frac{|R_0(\lambda_0,\delta_0;\phi)|}{|\eta_{\lambda_0,\delta_0}|}e^{-\mu_0 t}$$
$$+\frac{|R_1(\lambda_0,\delta_0,\mu_0;\phi)|}{|\gamma_{\lambda_0,\delta_0,\mu_0}|}+M(\lambda_0,\delta_0,\mu_0;\phi)\alpha_{\lambda_0,\delta_0,\mu_0}. \tag{7.105}$$

Furthermore, by using (7.100)–(7.104), from (7.68), (7.78), (7.87) and (7.95), we obtain

$$\left|L_0(\lambda_0;\phi)\right|\le k_{\lambda_0}.N(\lambda_0,\delta_0,\mu_0;\phi), \tag{7.106}$$
$$\left|R_0(\lambda_0,\delta_0;\phi)\right|\le H_{\lambda_0,\delta_0}.N(\lambda_0,\delta_0,\mu_0;\phi), \tag{7.107}$$
$$\left|R_1(\lambda_0,\delta_0,\mu_0;\phi)\right|\le J_{\lambda_0,\delta_0,\mu_0}.N(\lambda_0,\delta_0,\mu_0;\phi), \tag{7.108}$$

and

$$\left|M(\lambda_0,\delta_0,\mu_0;\phi)\right| \le \left\{ e_{\delta_0+\mu_0}\left(1+\frac{k_{\lambda_0}}{|\beta_{\lambda_0}|}\right) + \frac{H_{\lambda_0,\delta_0}}{|\eta_{\lambda_0,\delta_0}|}e_{\mu_0} + \frac{J_{\lambda_0,\delta_0,\mu_0}}{|\gamma_{\lambda_0,\delta_0,\mu_0}|}\right\} N(\lambda_0,\delta_0,\mu_0;\phi). \tag{7.109}$$

Using the inequalities (7.106)–(7.109), we obtain, from (7.105)

$$\begin{aligned} &e^{-(\lambda_0+\delta_0+\mu_0)t}\left|x(t)\right| \\ &\le \frac{k_{\lambda_0}}{|\beta_{\lambda_0}|}e^{-(\delta_0+\mu_0)t}N(\lambda_0,\delta_0,\mu_0;\phi) + \frac{H_{\lambda_0,\delta_0}}{|\eta_{\lambda_0,\delta_0}|}e^{-\mu_0 t}N(\lambda_0,\delta_0,\mu_0;\phi) \\ &+ \frac{J_{\lambda_0,\delta_0,\mu_0}}{|\gamma_{\lambda_0,\delta_0,\mu_0}|}N(\lambda_0,\delta_0,\mu_0;\phi) + \left\{ e_{\delta_0+\mu_0}\left(1+\frac{k_{\lambda_0}}{|\beta_{\lambda_0}|}\right) + \frac{H_{\lambda_0,\delta_0}}{|\eta_{\lambda_0,\delta_0}|}e_{\mu_0} \right. \\ &\left. + \frac{J_{\lambda_0,\delta_0,\mu_0}}{|\gamma_{\lambda_0,\delta_0,\mu_0}|}\right\}\alpha_{\lambda_0,\delta_0,\mu_0}N(\lambda_0,\delta_0,\mu_0;\phi), \quad t \ge 0, \end{aligned}$$

or

$$\begin{aligned} \left|x(t)\right| \le{}& \frac{k_{\lambda_0}}{|\beta_{\lambda_0}|}e^{\lambda_0 t}N(\lambda_0,\delta_0,\mu_0;\phi) + \frac{H_{\lambda_0,\delta_0}}{|\eta_{\lambda_0,\delta_0}|}e^{(\lambda_0+\delta_0)t}N(\lambda_0,\delta_0,\mu_0;\phi) \\ &+ \left[\left\{ e_{\delta_0+\mu_0}\left(1+\frac{k_{\lambda_0}}{|\beta_{\lambda_0}|}\right) + \frac{H_{\lambda_0,\delta_0}}{|\eta_{\lambda_0,\delta_0}|}e_{\mu_0}\right\}\alpha_{\lambda_0,\delta_0,\mu_0} \right. \\ &\left. + (1+\alpha_{\lambda_0,\delta_0,\mu_0})\frac{J_{\lambda_0,\delta_0,\mu_0}}{|\gamma_{\lambda_0,\delta_0,\mu_0}|}\right]e^{(\lambda_0+\delta_0+\mu_0)t}N(\lambda_0,\delta_0,\mu_0;\phi), \quad t \ge 0. \end{aligned}$$

Hence (7.99) holds for all $t \ge 0$.

First, suppose that $\lambda_0 \le 0$, $\lambda_0+\delta_0 \le 0$ and $\lambda_0+\delta_0+\mu_0 \le 0$ hold. Define $\|\phi\| = \max_{-\tau\le t\le 0}|\phi(t)|$. Clearly,

$$\|\phi\| \le N(\lambda_0,\delta_0,\mu_0;\phi).$$

From (7.99), we have

$$\begin{aligned} \left|x(t)\right| \le{}& \left[\frac{k_{\lambda_0}}{|\beta_{\lambda_0}|} + \frac{H_{\lambda_0,\delta_0}}{|\eta_{\lambda_0,\delta_0}|} + \left\{\left\{ e_{\delta_0+\mu_0}\left(1+\frac{k_{\lambda_0}}{|\beta_{\lambda_0}|}\right) + \frac{H_{\lambda_0,\delta_0}}{|\eta_{\lambda_0,\delta_0}|}e_{\mu_0}\right\}\alpha_{\lambda_0,\delta_0,\mu_0} \right.\right. \\ &\left.\left. + (1+\alpha_{\lambda_0,\delta_0,\mu_0})\frac{J_{\lambda_0,\delta_0,\mu_0}}{|\gamma_{\lambda_0,\delta_0,\mu_0}|}\right\}\right]N(\lambda_0,\delta_0,\mu_0;\phi) \end{aligned}$$

for every $t \ge 0$. Since $\frac{k_{\lambda_0}}{\beta_{\lambda_0}} > 1$, by taking into account the fact that

$$\begin{aligned} &\frac{k_{\lambda_0}}{|\beta_{\lambda_0}|} + \frac{H_{\lambda_0,\delta_0}}{|\eta_{\lambda_0,\delta_0}|} + \left\{\left\{ e_{\delta_0+\mu_0}\left(1+\frac{k_{\lambda_0}}{|\beta_{\lambda_0}|}\right) + \frac{H_{\lambda_0,\delta_0}}{|\eta_{\lambda_0,\delta_0}|}e_{\mu_0}\right\}\alpha_{\lambda_0,\delta_0,\mu_0} \right. \\ &\left. + (1+\alpha_{\lambda_0,\delta_0,\mu_0})\frac{J_{\lambda_0,\delta_0,\mu_0}}{|\gamma_{\lambda_0,\delta_0,\mu_0}|}\right\} > 1, \end{aligned}$$

we have

$$\left|x(t)\right| \le \left[\frac{k_{\lambda_0}}{|\beta_{\lambda_0}|} + \frac{H_{\lambda_0,\delta_0}}{|\eta_{\lambda_0,\delta_0}|} + \left\{\left\{e_{\delta_0+\mu_0}\left(1+\frac{k_{\lambda_0}}{|\beta_{\lambda_0}|}\right) + \frac{H_{\lambda_0,\delta_0}}{|\eta_{\lambda_0,\delta_0}|}e_{\mu_0}\right\}\alpha_{\lambda_0,\delta_0,\mu_0} + (1+\alpha_{\lambda_0,\delta_0,\mu_0})\frac{J_{\lambda_0,\delta_0,\mu_0}}{|\gamma_{\lambda_0,\delta_0,\mu_0}|}\right\}\right]N(\lambda_0,\delta_0,\mu_0;\phi)$$

for every $t \in [-\tau, \infty)$, which means that the trivial solution is stable at zero.

Next, suppose that $\lambda_0 < 0$, $\lambda_0+\delta_0 < 0$ and $\lambda_0+\delta_0+\mu_0 < 0$. Then from (7.99), it follows that $x(t) \to 0$ as $t \to \infty$. Hence the trivial solution is asymptotically stable at zero.

Finally, suppose that either $\lambda_0+\delta_0+\mu_0 > 0$, $\delta_0+\mu_0 > 0$ and $\mu_0 < 0$ or $\lambda_0 + \delta_0+\mu_0 > 0$, $\delta_0+\mu_0 < 0$ and $\mu_0 > 0$ or $\lambda_0+\delta_0+\mu_0 > 0$, $\delta_0 > 0$ and $\mu_0 < 0$ hold. We shall prove that the trivial solution is unstable at zero. Otherwise, there exists a number $l \equiv l(1) > 0$ such that for any $\phi \in C([-\tau, 0], R)$ with $\|\phi\| < l$, the solution $x(t)$ of (7.5)–(7.6) satisfies

$$\left|x(t)\right| < 1 \quad \text{for all } t \ge -\tau. \tag{7.110}$$

Define

$$\phi_0(t) = e^{(\lambda_0+\delta_0)t} - e^{\lambda_0 t}, \quad t \in [-\tau, 0].$$

Then using the definitions of $L_0(\lambda_0;\phi)$, $R_0(\lambda_0,\delta_0;\phi)$ and $R_1(\lambda_0,\delta_0,\mu_0;\phi)$ and by using (7.64) and (7.73), we see that

$$L_0(\lambda_0;\phi) \equiv -\beta_{\lambda_0},$$
$$R_0(\lambda_0,\delta_0;\phi) \equiv \eta_{\lambda_0,\delta_0}$$

and

$$R_1(\lambda_0,\delta_0,\mu_0;\phi) \equiv 0$$

respectively. Let $\phi \in C([-\tau, 0], R)$ be defined by

$$\phi = \frac{l_1}{\|\phi_0\|}\cdot\phi_0,$$

where l_1 is a number $0 < l_1 < l$. Moreover, let $x(t)$ be a solution of (7.5)–(7.6). Then from Theorem 7.4.2, it follows that

$$\lim_{t\to\infty}\left[e^{-(\lambda_0+\delta_0+\mu_0)t}x(t) - \frac{R_1(\lambda_0,\delta_0,\mu_0;\phi)}{\gamma_{\lambda_0,\delta_0,\mu_0}}\right]$$
$$= \lim_{t\to\infty}\left[\frac{L_0(\lambda_0;\phi)}{\beta_{\lambda_0}}e^{-(\delta_0+\mu_0)t} + \frac{R_0(\lambda_0,\delta_0;\phi)}{\eta_{\lambda_0,\delta_0}}e^{-\mu_0 t}\right]$$

which further yields

$$\lim_{t\to\infty}\left[e^{-(\lambda_0+\delta_0+\mu_0)t}x(t)+\frac{l_1}{\|\phi_0\|}\frac{R_1(\lambda_0,\delta_0,\mu_0;\phi_0)}{\gamma_{\lambda_0,\delta_0,\mu_0}}\right]$$
$$=\lim_{t\to\infty}\left[-\frac{l_1}{\|\phi_0\|}e^{-(\delta_0+\mu_0)t}+\frac{l_1}{\|\phi_0\|}e^{-\mu_0 t}\right],$$

that is,

$$\lim_{t\to\infty}\left\{e^{-(\lambda_0+\delta_0+\mu_0)t}x(t)\right\}=\frac{l_1}{\|\phi_0\|}\lim_{t\to\infty}\left\{\left(e^{-\mu_0 t}-e^{-(\delta_0+\mu_0)t}\right)\right\}.$$

Since $|x(t)|<1$ for $t\geq-\tau$, then by taking limit as $t\to\infty$, we obtain a contradiction. Hence $x(t)$ is unstable at zero.

This completes the proof of the theorem. □

Example 7.4.1 Consider the equation

$$x'''(t)=x'(t-\tau)-x(t)+x(t-\tau),\quad t\geq 1, \tag{7.111}$$

where $\tau=\log 3$. Clearly, $\lambda=\lambda_0=-1$ is a root of

$$\lambda^3=\lambda e^{-\lambda\tau}-1+e^{-\lambda\tau},$$

$\delta=\delta_0=1$ is a root of

$$\delta^2-3\delta+3-e^{(1-\delta)\tau}=0$$

and $\mu=\mu_0=-1$ is a root of

$$\mu-1+\frac{1}{\mu}\left(1-e^{-\mu\tau}\right)=0.$$

By a simple calculation, it is easy to verify that the condition (7.93), given in Theorem 7.4.1, is satisfied. Thus the trivial solution of (7.111) is stable.

Remark 7.4.1 It would be interesting to obtain stability, asymptotic stability and instability of the solutions of (7.5) for the signs on λ_0,δ_0 and μ_0, which are not covered in Theorem 7.4.3.

Now, we consider the third-order ordinary differential equation

$$x'''=p_1x''+q_1x'+r_1x, \tag{7.112}$$

where p_1,q_1 and r_1 are reals. Applying Theorems 7.4.1–7.4.3 to Eq. (7.112), we have

Theorem 7.4.4 *Let λ_0 be a real root of*

$$\lambda^3 = p_1\lambda^2 + q_1\lambda + r_1,$$

δ_0 be a real root of

$$\delta^2 + (3\lambda_0 - p_1)\delta - \left(2\lambda_0 p_1 + q_1 - 3\lambda_0^2\right) = 0,$$

and μ_0 be a real root of

$$\mu + (3\lambda_0 - p_1 + 2\delta_0) = 0.$$

Then the following hold:

(i) *If $\lambda_0 \le 0$, $\lambda_0 + \delta_0 \le 0$ and $\lambda_0 + \delta_0 + \mu_0 \le 0$, then the trivial solution of* (7.112) *is stable* (*at* 0).
(ii) *If $\lambda_0 < 0$, $\lambda_0 + \delta_0 < 0$ and $\lambda_0 + \delta_0 + \mu_0 < 0$, then the trivial solution of* (7.112) *is asymptotically stable* (*at* 0).
(iii) *If either $\lambda_0 + \delta_0 + \mu_0 > 0$, $\delta_0 + \mu_0 > 0$ and $\mu_0 < 0$ or $\lambda_0 + \delta_0 + \mu_0 > 0$, $\delta_0 + \mu_0 < 0$ and $\mu_0 > 0$, then the trivial solution of* (7.112) *is unstable at* 0.

Example 7.4.2 Consider

$$x''' = -6x'' - 11x' - 6x, \quad t \ge 1. \tag{7.113}$$

Clearly $\lambda = \lambda_0 = -1$ is a negative real root of

$$\lambda^3 = -6\lambda^2 - 11\lambda - 6,$$

$\delta = \delta_0 = -1$ is a negative real root of the equation

$$\delta^2 + 3\delta + 2 = 0$$

and $\mu = \mu_0 = -1$ is a negative real root of the equation

$$\mu + 1 = 0.$$

Then by Theorem 7.4.4, the trivial solution of (7.113) is asymptotically stable. We may note that $x_1(t) = e^{-t}$, $x_2(t) = e^{-2t}$ and $x_3(t) = e^{-3t}$ are the solutions of (7.113) with $x_i(t) \to 0$ as $t \to \infty$, $i = 1, 2, 3$.

7.5 Open Problems and Discussions

One may observe that a suitable Lyapunov function has been used in the first three sections to study the global asymptotic stability of the nonlinear equations, and then the result are applied to linear equations with constant coefficients. Thus, it would be interesting to study the global asymptotic stability of the third-order linear differential equations of the form (2.1) with variable and constant coefficients. It follows from Example 7.4.2 and Proposition 1.2.5 that the result given in Proposition 1.2.5 requires more improvement. However, we have not succeeded yet. This is still an open problem. Further, from the Theorem 7.2.2 and Corollary 7.2.4, it follows that Theorem 3.4.1 requires more improvement. It seems that a suitable Green's function is not enough to improve Theorem 3.4.1. However, a suitable Lyapunov function may provide required sufficient conditions for the global asymptotic stability of Eq. (3.1). A number of results exist in the literature on the global asymptotic stability of third-order nonlinear differential equations. However, we have considered Eqs. (7.1)–(7.3) and (7.5), keeping our interests on the third-order linear equations.

7.6 Notes

Results of Sect. 7.1 are taken from Qian [3], whereas the results in Sect. 7.2 are due to Qian [4]. Results in Sect. 7.3 are due to Omeike [2]. Results in Sect. 7.4 are new.

References

1. R. D. Driver; *Ordinary and Delay Differential Equations*, Springer, New York, 1977.
2. M. O. Omeike; *New results on asymptotic behavior of solutions of certain third-order nonlinear differential equations*, Kragujevac Journal of Mathematics, 35(1) (2011), 99–109.
3. C. Qian; *On global stability of third-order nonlinear differential equations*, Nonlinear Analysis; Theory Methods and Applications, 42 (2000), 651–661.
4. C. Qian; *Asymptotic behavior of a third-order nonlinear differential equation*, Journal of Mathematical Analysis and Applications, 284(1) (2003), 191–205.

Index

S. Padhi, S. Pati, *Theory of Third-Order Differential Equations*,
DOI 10.1007/978-81-322-1614-8, © Springer India 2014

About the Author

Seshadev Padhi is Associate Professor of Mathematics at Birla Institute of Technology, Mesra, Ranchi, India. He received his PhD on the topic "Oscillation Theory of Third-order Differential Equations". He was awarded the Better Opportunities for Young Scientists in Chosen Areas of Science and Technology (BOYSCAST) fellowship by the Department of Science and Technology (DST), Government of India in 2004, to visit Mississippi State University, USA, where he subsequently did his postdoctoral research. In addition, Dr. Padhi has visited several institutes of international repute: Florida Institute of Technology, Melbourne, Florida, USA, in 2006 to work in collaboration with Prof. T. Gnanabhaskar; Texas State University, San Marcos, Texas, USA, in 2009 to work in collaboration with Prof. Julio G. Dix; University of Tennessee, Chattanooga, Tennessee, USA, in 2011, 2012 and 2013 to work in collaboration with Prof. John R. Graef; University of Szeged, Szeged, Hungary, in 2007 and 2011 to work in collaboration with Prof. Tibor Krisztin. Besides these, he has also visited Eidgenössische Technische Hochschule (ETH) Zürich, Switzerland, under Borel Set Theory Programme in 2005, and many other countries to deliver lectures in international conferences and workshops. Dr. Padhi has published more than 60 research papers in international journals of repute and has been working as referee for Mathematical Review and for more than 30 international journals since 2006.

Smita Pati is a postdoctoral fellow in the Department of Applied Mathematics, Birla Institute of Technology, Mesra, Ranchi, India, working under the postdoctoral grant from the National Board for Higher Mathematics (NBHM), Department of Atomic Energy (DAE), Government of India. Besides, Dr. Pati has been working as a referee for seven international mathematics journals and for Mathematical Review. She was awarded with Marqui's Whos Who in the year 2012 and has published 15 research papers in international journals of repute, in addition to publishing several research papers in collaboration with Prof. Julio G. Dix, Texas State University, USA, and Prof. John R. Graef, University of Tennessee, Chattanooga, Tennessee, USA. She has visited several institutes in Hungary and Italy to deliver lectures in different conferences on Differential Equations.

Printed by Printforce, the Netherlands

USEFUL

D.AS